Insecticide Biochemistry and Physiology

Insecticide Biochemistry and Physiology

Edited by
C. F. Wilkinson
Cornell University

PLENUM PRESS · NEW YORK AND LONDON

Library of Congress Cataloging in Publication Data
Main entry under title:

Insecticide biochemistry and physiology.

Includes bibliographies and index.

1. Insecticides—Physiological effect. 2. Insecticides—Toxicology. I. Wilkinson, Christopher Foster, 1938- [DNLM: 1. Insecticides—Pharmacodynamics. 2. Insecticides—Toxicity. 3. Insects—Drug effects. WA240 I59]
SB951.5.I57 632'.951 76-10596
ISBN 0-306-30872-X

A Division of Plenum Publishing Corporation
227 West 17th Street, New York, N.Y. 10011

Printed in the United States of America

Contributors

Gerald Thomas Brooks
A.R.C. Unit of Invertebrate Chemistry and Physiology
Brighton, England

W. C. Dauterman
North Carolina State University
Raleigh, North Carolina

John Doull
The University of Kansas Medical Center
Kansas City, Kansas

Amira Toppozada Eldefrawi
Cornell University
Ithaca, New York

Lawrence Fishbein
National Center for Toxicological Research
Jefferson, Arkansas

Jun-ichi Fukami
The Institute of Physical and Chemical Research
Wako, Saitama, Japan

T. R. Fukuto
University of California
Riverside, California

Ernest Hodgson
North Carolina State University
Raleigh, North Carolina

R. M. Hollingworth
Purdue University
West Lafayette, Indiana

Wendell W. Kilgore
University of California
Davis, California

Ming-yu Li
University of California
Davis, California

Michael A. Morelli
State University of New York
Syracuse, New York

Tsutomu Nakatsugawa
State University of New York
Syracuse, New York

Toshio Narahashi
Duke University Medical Center
Durham, North Carolina

R. D. O'Brien
Cornell University
Ithaca, New York

F. J. Oppenoorth
Laboratory for Research on Insecticides
Wageningen, The Netherlands

D. L. Shankland
Purdue University
West Lafayette, Indiana

Laurence G. Tate
North Carolina State University
Raleigh, North Carolina

W. Welling
Laboratory for Research on Insecticides
Wageningen, The Netherlands

C. F. Wilkinson
Cornell University
Ithaca, New York

Raymond S. H. Yang
Albany Medical College
Albany, New York

Preface

Only four short decades ago, the control of insect pests by means of chemicals was in its early infancy. The pioneers in the area consisted largely of a group of dedicated applied entomologists working to the best of their abilities with a very limited arsenal of chemicals that included inorganics (arsenicals, fluorides, etc.), some botanicals (nicotine), and a few synthetic organics (dinitro-*o*-cresol, organothiocyanates). Much of the early research was devoted to solving practical problems associated with the formulation and application of the few existing materials, and although the discovery of new types of insecticidal chemicals was undoubtedly a pipe dream in the minds of some, little or no basic research effort was expended in this direction.

The discovery of the insecticidal properties of DDT by Paul Müller in 1939 has to be viewed as the event which marked the birth of modern insecticide chemistry and which has served as the cornerstone for its subsequent developement. DDT clearly demonstrated for the first time the dramatic potential of synthetic organic chemicals for insect control and provided the initial stimulus which has caused insecticide chemistry to become a field not only of immense agricultural and public health importance but also one that has had remarkable and unforseeable repercussions in broad areas of the physical, biological, and social sciences.

Indeed, there can be few other synthetic chemicals which will be judged in history to have had such a broad and telling impact on mankind as has DDT. Initially, of course, its discovery signaled the beginning of the intensive search for new insecticides that continues unabated to this day and that has led in part to the successful development and use of the other chlorinated hydrocarbons, the organophosphorus compounds, the carbamates, and the additional materials currently in use. The widespread use of these compounds during the last 20 years has played a major role in the development of our own modern agricultural system and is an integral component of the "green revolution" which it is hoped will serve to partially alleviate the nutritional needs of the rapidly expanding populations in other areas of the world. In addition, insecticides have brought spectacular social and economic gains to large areas of the world as a result of their role in the eradication of malaria and other arthropod-borne diseases. But in spite of these benefits the widespread use of insecticides, like so many other modern technological advances, has created some serious problems that were not immediately obvious.

Early optimism regarding the practical potential of DDT and other materials has been tempered to some extent by the emergence of insect resistance to insecticides as a real threat to continued control, and more recently there has been increasing recognition of the potentially deleterious effects of persistent insecticides on wildlife and other nontarget species in the environment. Furthermore, concern continues to be voiced over the possible toxicological hazards to man of chronic exposure to trace residues of insecticides in the food supply. As a result of these and other problems, many insecticides such as DDT, which initially proved so successful, are no longer considered acceptable and have been or are being phased out. The search for new replacements has been intensified since, despite frequent statements to the contrary, it seems clear that chemicals will continue to represent our major means of insect control in the foreseeable future. It is equally clear that in order to discover and develop these insecticides of the future a much greater degree of sophistication will be required than has hitherto been the case.

In recognition of this need, basic research in all aspects of the field has expanded by leaps and bounds during the last decade and has assumed many new dimensions with respect to the disciplines that it now encompasses. Indeed, the entire area of insect control with chemicals has come a long way from its rudimentary beginnings as a branch of applied entomology and is extremely difficult to define. It now reaches into almost every corner of the biological, physical, and life sciences and includes such disciplines as physiology, biochemistry, pharmacology, toxicology, organic and physical chemistry, and environmental studies as well as economic entomology. The current overall level of sophistication in the field closely reflects that of the individual disciplines of which it is composed.

The decision to prepare this book was in itself not arrived at lightly. The enormous breadth and diversity of the field and the voluminous literature with which it is now associated should be sufficient to give serious pause to any would-be author. Furthermore, in the last few years we have been bombarded with a plethora of volumes (books, monographs, symposia proceedings, committee reports, etc.) covering most major areas of the pesticide field. Among these are books dealing with specific groups of insecticides such as the pyrethroids (Casida, 1973), naturally occurring insecticides (Jacobson and Crosby, 1971), chlorinated hydrocarbons (Brooks, 1974), organophosphorus compounds (Eto, 1974), and chemosterilants (LaBrecque and Smith, 1968).[1]

[1]Casida, J. E. (ed.), 1973, *Pyrethrum, the Natural Insecticide*, Academic Press, New York. Jacobson, M., and Crosby, D. J. (eds.), 1971, *Naturally Occurring Insecticides*, Dekker, New York. Brooks, G. T., 1974, *Chlorinated Insecticides*, Vols. I and II, CRC Press, Cleveland. Eto, M., 1974, *Organophosphorus Pesticides: Organic and Biological Chemistry*, CRC Press, Cleveland. LaBrecque, G. C., and Smith, C. N. (eds.), 1968, *Principles of Insect Chemosterilization*, Appleton-Century-Crofts, New York.

Other recently published volumes include those on the chemistry (Melnikov, 1971), metabolism (Menzie, 1969), mode of action (Corbett, 1974; Aldridge and Reiner, 1972), formulation (Van Valkenberg, 1973), and environmental aspects (White-Stevens, 1971) of pesticides, and there are books on the so-called third generation pesticides, the insect hormones (Menn and Beroza, 1971; Sláma *et al.*, 1974).[2] Add to this incomplete list the numerous symposia proceedings (O'Brien and Yamamoto, 1970; Gillett, 1970; Metcalf and McKelvey, 1975)[3] and pesticide-related chapters in other texts and it is clear that the field cannot by any stretch of the imagination be said to suffer from underexposure. Consequently it is perhaps necessary from the outset to provide some justification (or apology) for yet another volume.

Many of the volumes referred to offer excellent and comprehensive coverage of the subject areas with which they are concerned, but in keeping with the evolution of the field and following the common trend of contemporary scientific literature they are frequently of a narrow, highly specialized nature. As a result, they often appear isolated from and out of context with the broader aspects and more general principles of the field of which they are a part. For this reason, it was decided to attempt to reverse the trend toward specialization and prepare a volume which would provide a broad (though admittedly selective) coverage of the basic interactions of insecticides with living organisms.

Although several excellent general texts are available (Martin, 1964; Metcalf, 1955; O'Brien, 1967; Hassall, 1969),[4] most are to some extent outdated, and there seems a notable reluctance on the part of the authors to revise them. Indeed, with one recent exception (Matsumura, 1975)[5] the task of compiling a general text seems to be rapidly passing the point at which it can be adequately accomplished by a single person. Consequently, in ascertaining the need for this book it was decided to adopt the now common format of a

[2]Melnikov, N. N., 1971, *Chemistry of Pesticides*, Springer-Verlag, New York. Menzie, C. M., 1969, *Metabolism of Pesticides*, Bureau of Sport Fisheries and Wildlife, Special Scientific Report—Wildlife No. 27, Washington, D.C. Corbett, J. R., 1974, *The Biochemical Mode of Action of Pesticides*, Academic Press, New York. Aldridge, W. W., and Reiner, E., 1972, *Enzyme Inhibitors as Substrates*, North-Holland, Amsterdam. Van Valkenberg, W., 1973, *Pesticide Formulations*, Dekker, New York. White-Stevens, R. (ed.), 1971, *Pesticides in the Environment*, Vols. I and II, Dekker, New York. Menn, J. J., and Beroza, M. (eds.), 1971, *Insect Juvenile Hormones: Chemistry and Action*, Academic Press, New York. Sláma, K., Romaňuk, M., and Šorm, F., 1974, *Insect Hormones and Bioanalogues*, Springer-Verlag, New York.

[3]O'Brien, R. D., and Yamamoto, I. (eds.), 1970, *Biochemical Toxicology of Insecticides*, Academic Press, New York. Gillett, J. W. (ed.), 1970, *The Biological Impact of Pesticides in the Environment*, Oregon State University, Corvallis. Metcalf, R. L., and McKelvey, J. (eds.), 1975, *Insecticides for the Future: Needs and Prospects*, Wiley, New York.

[4]Martin, H., 1964, *Scientific Principles of Crop Protection*, Metcalf, R. L., 1955, *Organic Insecticides*, Interscience, New York. O'Brien, R. D., 1967, *Insecticides: Action and Metabolism*, Academic Press, New York. Hassall, K., 1969, *World Crop Protection*, Vol. 2, CRC Press, Cleveland.

[5]Matsumura, F., 1975, *Toxicology of Insecticides*, Plenum Press, New York.

multiauthor volume written by outstanding authorities in the various areas to be included.

This cannot truly be considered a general text since it is concerned primarily with the basic biochemical and physiological events that determine the biological activity of an insecticide once it has reached the outer surface of an organism. In contrast to most of the existing books in the area, which are divided according to the various groups of insecticides discussed, (chlorinated hydrocarbons, carbamates, etc.), the first ten chapters of this volume are organized on a functional basis that emphasizes in the approximate order in which they occur the physiological processes and biochemical events taking place from the time the insecticide is administered to an organism to the time of its arrival and interaction at the target site; wherever possible an attempt has been made to stress the comparative aspects of the area under consideration.

As a logical consequence of the organization employed, the first chapter is concerned with the *Penetration and Distribution of Insecticides*. This is followed by a section consisting of four chapters that discuss the major biochemical mechanisms and pathways by which organisms metabolize a large variety of insecticides and other foreign compounds and thus afford themselves some degree of protection from the potentially hazardous effects of such compounds. Included here are chapters on *Microsomal Oxidation and Insecticide Metabolism, Cytochrome P450 Interactions, Extramicrosomal Metabolism of Insecticides*, and *Enzymatic Conjugation and Insecticide Metabolism*.

The next section of the book deals with the actual interactions of insecticides with their targets. Since most known insecticides are neurotoxicants, it is hoped that inclusion of the chapter on *The Nervous System: Comparative Physiology and Pharmacology* will provide a useful background for the subsequent discussions on *Acetylcholinesterase and Its Inhibition, The Acetylcholine Receptor and Its Interactions with Insecticides*, and the *Effects of Insecticides on Nervous Conduction and Synaptic Transmission*. The only other major insecticide target is covered in the chapter on *Insecticides as Inhibitors of Respiration*. An additional chapter on the *Physicochemical Aspects of Insecticidal Action* is included to emphasize the importance of physicochemical parameters in studies of structure–activity relationships and in the rational design of insecticides.

No book of this type would be complete without inclusion of the extremely important areas of insect resistance to insecticides and selective toxicity. These large and complex areas are discussed at length in chapters entitled *The Biochemical and Physiological Basis of Selective Toxicity* and *The Biochemistry and Physiology of Resistance*, both of which contain a complete cross-section of the material appearing in previous chapters.

The last section of the book contains selected topics of toxicological interest and importance and emphasizes the effects of insecticides on mammals. The evaluation of potentially hazardous chronic effects which occupies a central role in modern toxicology is discussed in *Teratogenic, Mutagenic, and Carcinogenic Effects of Insecticides*, and the various mechanisms by which the toxic effects of insecticides may be modified through combination with other insecticides or drugs are outlined in the chapter on *Insecticide Interactions*. The following chapter on the *Treatment of Insecticide Poisoning* is one not often found in books of this type. Its inclusion reflects the growing concern over the occupational exposure of agricultural workers to insecticide residues in the environment, a subject which among others is given additional attention in the final chapter on *Environmental Toxicology*.

Several other topics could have usefully been included in the book and there may be some disappointment over the arbitrary limits which were established as a basis for selection. As previously discussed, however, the field is large, and it is clear that some focus was required. With one or two notable exceptions, all chapters which were initially planned have been included, and the book represents a comprehensive account of the current status of basic biochemical and physiological research in the insecticide field. We hope that it will prove useful as a fairly advanced text for students in the field and become a well-used reference source for many others whose interests or professions are in this important area.

Thanks are owed to numerous individuals who have played an active role in the preparation of the book. These include Rona Springer, for patient and valuable secretarial assistance, and the editorial staff at Plenum, particularly Seymour Weingarten, Senior Editor, and Evelyn Grossberg. Of course, the book could not have been prepared without the expert cooperation of the various contributing authors, and I wish to express to them my sincere gratitude for their enthusiastic participation in this venture.

C. F. Wilkinson

Ithaca

Contents

Chapter 3

Cytochrome P450 Interactions

Ernest Hodgson and Laurence G. Tate

Chapter 4

Extramicrosomal Metabolism of Insecticides

W. C. Dauterman

Chapter 5

Enzymatic Conjugation and Insecticide Metabolism

Raymond S. H. Yang

Part III
Target Site Interactions

Chapter 6

The Nervous System: Comparative Physiology and Pharmacology

D. L. Shankland

Chapter 7

Acetylcholinesterase and Its Inhibition

R. D. O'Brien

Chapter 8

The Acetylcholine Receptor and Its Interactions with Insecticides

Amira Toppozada Eldefrawi

Chapter 9

Effects of Insecticides on Nervous Conduction and Synaptic Transmission

Toshio Narahashi

Chapter 10

Insecticides as Inhibitors of Respiration

Jun-ichi Fukami

Chapter 11

Physicochemical Aspects of Insecticidal Action

T. R. Fukuto

Part IV
Selectivity and Resistance

Chapter 12

The Biochemical and Physiological Basis of Selective Toxicity

R. M. Hollingworth

Chapter 13

Biochemistry and Physiology of Resistance

F. J. Oppenoorth and W. Welling

Part V

Insecticide Toxicology

Chapter 14

Teratogenic, Mutagenic, and Carcinogenic Effects of Insecticides

Lawrence Fishbein

Chapter 15
Insecticide Interactions
C. F. Wilkinson

Chapter 16
The Treatment of Insecticide Poisoning
John Doull

Chapter 17

Environmental Toxicology

Wendell W. Kilgore and Ming-yu Li

I
Penetration and Distribution

1

Penetration and Distribution of Insecticides

Gerald Thomas Brooks

1. Introduction

In order to exert its characteristic effects seen *in vivo*, a drug or toxicant must arrive at the site(s) of action [target(s)] in sufficient quantity for efficient interaction. If a method is available for measuring the interaction of a toxicant (insecticide) with the target itself on a molecular basis, then the potentially attainable toxicity (intrinsic toxicity) can be assessed, and such measurements provide important information about structure–activity relationships at the level of the site of action (Fig. 1). The development of anticholinesterase insecticides of the organophosphate and carbamate types provides some good examples of this approach, since intrinsic toxicities of candidate compounds are indicated by the extent to which they inhibit isolated preparations of insect acetylcholinesterase.

Unfortunately, the intrinsic toxicity measured in this way may bear little relationship to the effect seen *in vivo*, since from the time of its first contact with an organism the toxicant is subject to the numerous processes of Fig. 1 which reduce the amount arriving and its rate of arrival at the target. Therefore, molecular design has to take into account the effect of structure on these intermediary processes, as well as on fit to the target(s).

Gerald Thomas Brooks • Agricultural Research Council, Unit of Invertebrate Chemistry and Physiology, University of Sussex, Brighton, England.

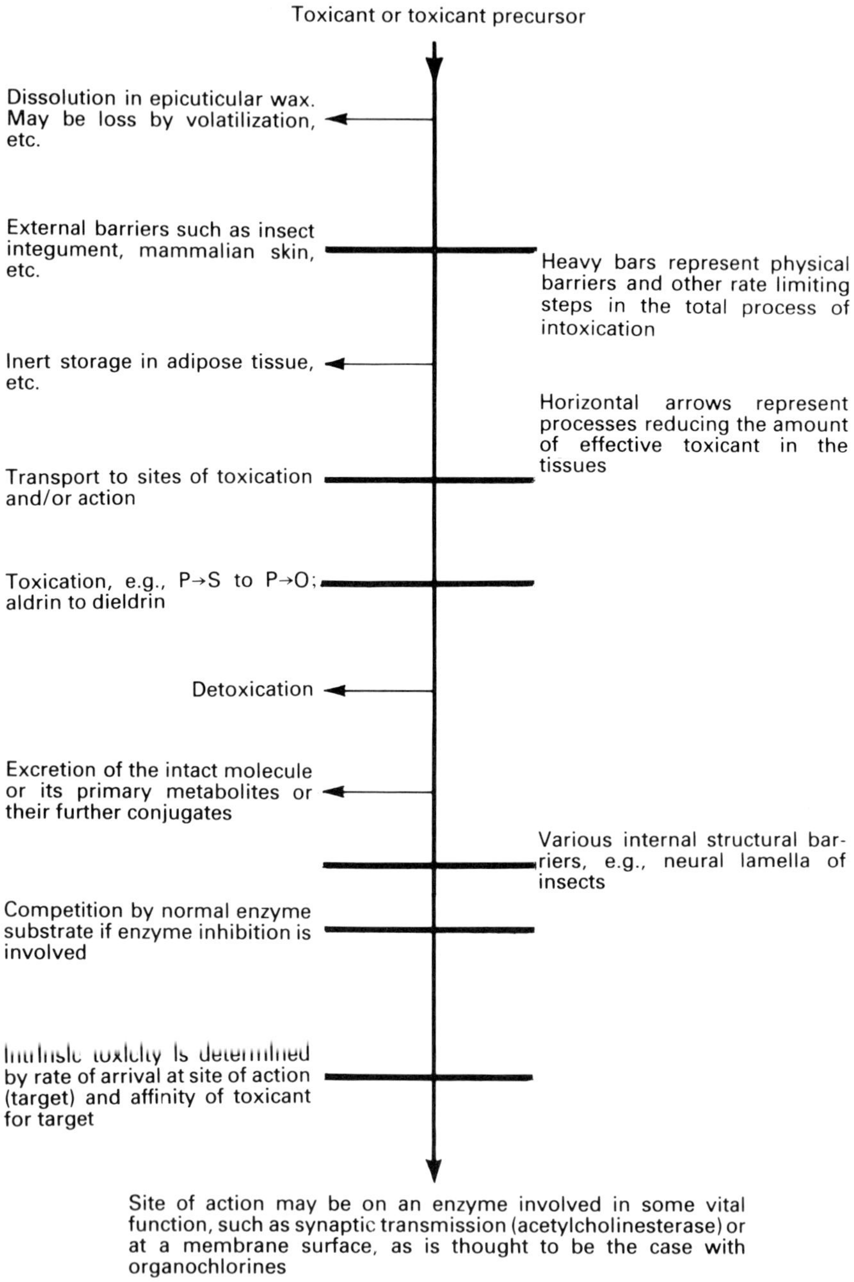

Fig. 1. Factors influencing insecticide dynamics in an animal. After Winteringham (1969).

According to Goodman and Gilman (1965), the study of absorption, distribution, biotransformation, and excretion of a drug as well as its mechanism of action at the target (i.e., all of the processes summarized in Fig. 1) is termed *pharmacodynamics.* In a somewhat different definition (Ariens, 1973), the first four properties comprise the *pharmacokinetic phase* of drug action, and this determines the amount of drug actually available for interaction with the target site (biological availability) in the ensuing *pharmacodynamic phase.* The present discussion adopts the second definition and will consider only the pharmacokinetic phase of insecticidal action, since interactions with the target(s) are considered in later chapters.

The most common means of contact between an insecticide and a living organism is through accidental or deliberate application to the integument (e.g., insect cuticle or vertebrate skin) or ingestion with the food followed by absorption through the walls of the alimentary tract. Also, inhalation may bring insecticides into contact with the pulmonary epithelium of higher animals, and the gills of fish take up toxicants very readily. Having passed these barriers, the insecticide enters the insect hemolymph or vertebrate bloodstream, in which it may be carried to all parts of the organism in solution or bound to proteins or dissolved in lipid particles, depending on its physical properties. During the penetration and distribution process, some of the toxicant will be taken up by inert tissues, such as adipose tissue, and so become biologically unavailable. At any time during both penetration and distribution there may be opportunities for both nonenzymatic chemical conversions and enzymatic biotransformations, leading to more toxic molecules (toxication or bioactivation) or less toxic ones (detoxication). Both the original insecticide and its conversion products may be excreted as such or following further conversion into water-soluble conjugates, so that enhanced excretion provides a means of protection for the organism against the toxicant. Finally, the distribution of the insecticide or its metabolic products will depend on their ability to cross semipermeable membranes separating the various biophases within the organism. It is obvious, therefore, that factors which affect the capacity for inert storage, membrane permeability, or enzymatic conversions may profoundly alter the pharmacokinetic behavior of an insecticide. Basic information on these topics is available in books on foreign compounds including insecticides (Parke, 1968), on insecticides in general (O'Brien, 1967*a*), and on chlorinated insecticides (Brooks, 1974). The specialist reports of The Chemical Society on foreign compound metabolism (including insecticides) also provide up-to-date comprehensive reviews of developments in this field (Hathway *et al.*, 1970, 1972).

2. Mode of Entry and Transfer Within the Organism

The processes of absorption, distribution, biotransformation, and excretion referred to above all involve insecticide transfer across biological barriers at some stage. Some barriers, such as insect cuticle, human skin (integumental barriers), and placenta, consist of several layers of cells. In contrast, the gastrointestinal epithelium and hepatic parenchyma consist of a single layer of cells, while other barriers may have the structure of the fundamental unit, the cell membrane. The cell or plasma membrane surrounds single cells, mitochondria, and cell nuclei and typically consists of a bimolecular lipoid sheet bounded on both sides by protein, with an overall thickness of about 100 Å. This structure contains minute, water-filled pores (usually about 4 Å in diameter) which allow the rapid penetration of small, water-soluble compounds, while the lipid membrane itself is readily penetrated by lipophilic substances. Thus compounds which are highly lipophobic or ionized do not readily pass such membranes unless they are small enough to move through the aqueous pores (molecular weight not greater than 100–200). Since the undissociated forms of ionizable toxicants are usually lipid soluble, the degree of ionization is important and their rate of transfer through membranes is therefore strongly dependent on pH. Most foreign compounds (e.g., drugs and insecticides; collectively called xenobiotics) move through the lipid membrane by *passive diffusion* down a concentration gradient; the diffusion of hydrophilic molecules and low molecular weight ions through aqueous pores also follows a concentration gradient and is called *filtration*. In two other processes, *facilitated diffusion* and *active transport*, compounds are presumed to be carried through the membranes in the form of complexes with endogenous substances. Facilitated diffusion occurs down a concentration gradient, whereas active transport occurs against a concentration gradient and requires metabolic energy. Certain drugs are thought to be transferred by these processes, although they generally show marked substrate specificity and apply largely to endogenous metabolites such as those involved in intermediary metabolism. Finally, invaginations of the cell membrane can engulf extracellular material, which therefore enters the cell in vacuoles or extracellular fluid. This is the process of *pinocytosis*.

2.1. Penetration Through the Integument

2.1.1. General Considerations

Penetration through the integument is of great importance because of its possible influence on the selective action of insecticides as between insects and mammals. At first sight, the morphological differences between insects and mammals are so great that selective toxicant action would be expected on these grounds alone. The obvious difference, for example, between the flexible mammalian skin and the rigid sclerotized chitin of the insect cuticle immediately suggests differences in their permeability properties. As far as insecticides are concerned, this seems to be borne out by the data of Table I. The difference between contact toxicity and toxicity measured by an injection

Table I. Acute Toxicities of Injected or Externally Applied Insecticides to Periplaneta americana and to Mammals[a]

Insecticide	Method of administration	Insect LD_{50} (ppm)	Mammal	Mammalian LD_{50} (ppm)
Acethion	Contact	700		
	Injection	375	Mouse	1280
Carbaryl	Contact	—	Rabbit	5000
	Injection	—	Rabbit	561 (oral)
DDT	Contact	10–30	Rat	3000
	Injection	5–18	Rat	40–50
Diazinon	Contact	2.0	Rabbit	4000
	Injection	0.75	Mouse	40
Dieldrin	Contact	1.3	Rat	>50
	Injection	1.1	Rat	<8
Dimethoate	Contact	2.0	Rat	<150–1150
	Injection	1.0	Mouse	140
Lindane	Contact	4–7	Rat	500
	Injection	3–7	Rat	<50
Malathion	Contact	23.6	Rat	>4000
	Injection	8.4	Rat	50
Parathion	Contact	1.2	Rat	11–21
	Injection	0.95	Rat	4–7
Pyrethrins	Contact	6.5	Rabbit	11,200 (for allethrin)
	Injection	6.0	Dog	<6–8
Rotenone	Contact	2000	Rat	—
	Injection	5–8	Rat	<6

[a] Adapted from Winteringham (1969).

bypassing the integument would appear to be a measure of the barrier effect of the latter. With the possible exception of parathion in rats, this effect appears to be smaller for the cockroach than for the mammal. Rotenone also is exceptional in being very much less toxic to the cockroach by contact action than by injection. However, comparative experiments on the penetration of insecticides through insect cuticle and mammalian skin reveal that these differences are not simply related to permeability. Thus DDT was once thought to pass readily through insect cuticle but not so readily through mammalian skin. However, there is evidence (Olson and O'Brien, 1963; O'Brien and Dannelley, 1965) that the half-times of its penetration through cockroach cuticle and rat skin are each about 26 hr. If, as indicated by injection experiments, the intrinsic toxicity of DDT is not greatly different to these two species, these data suggest that the slow penetration might afford opportunities for detoxication in the mammal that are not available to the insect.

An insecticide applied to insect cuticle comes first into contact with the outer layer of epicuticular wax which overlies more polar layers consisting largely of chitin and tanned protein. There is a general tendency for the best contact insecticides to be found among relatively nonpolar compounds, although Olson and O'Brien (1963) found an inverse relationship between lipophilicity (measured by oil–water partition coefficient) and rate of cuticular penetration (Table II). The explanation appears to be that the epicuticular wax becomes saturated with insecticide, which in order to penetrate must then partition into the inner, more polar layers of the cuticle. Such partitioning occurs more readily for those molecules having higher polarity (smaller oil–water partition coefficients) and applies when the toxicant is introduced directly into the epicuticular wax following topical application either in pure undiluted form or in a volatile solvent.

Thus phosphoric acid applied directly to cockroach cuticle penetrated very much more rapidly than DDT (Olson and O'Brien, 1963) but when applied in a droplet of water did not penetrate at all until the water evaporated. An insecticide applied to the cuticle in a nonvolatile solvent must partition from the solvent into the epicuticular wax before it can enter the more polar layers of the cuticle. Under these conditions, the partition coefficient of the insecticide between its carrier and the epicuticular wax determines the actual contact concentration, so that the nature of the carrier markedly affects the results obtained (see Section 3.3). This situation applies to animal integuments generally and may explain why there was a positive correlation between penetration rate and polarity when solutes were applied in a volatile solvent to meal beetles, two species of toad, and a chameleon (Buerger and O'Brien, 1965) but a negative correlation between these parameters for salicylate penetrating intact human skin (Brown and Scott, 1934) and for various compounds penetrating rabbit skin (Treherne, 1956) or excised locust integument (Treherne, 1957), all from aqueous solutions.

In these last experiments, partitioning of the solutes between water and the outermost nonpolar integumental layers is the rate-limiting step. However, O'Brien and Dannelley (1965) found no correlation between penetration and polarity for several insecticides topically applied in benzene to intact, shaved rat skin, the treated area of the integument being excised for analysis after the required time of exposure. For this mammal, different compounds showed

Table II. Relationship between Partition Coefficients and Absorption Rates of Insecticides Topically Applied in Acetone to the American Cockroach[a]

Insecticide	Partition coefficient (olive oil-water)	Half-time of penetration (min)
DDT	316	1584
Dieldrin	64	320
Paraoxon	4.06	55
Dimethoate	0.34	27

[a] Adapted from Olson and O'Brien (1963).

different penetration characteristics, with DDT and dieldrin (compounds with the higher oil–water partition coefficients) exhibiting, respectively, the longest (26 hr) and shortest (3.5 hr) penetration half-times. For such studies, use of the partition coefficient between water and the outer keratinized layer (stratum corneum) of the skin would be better (Poulsen, 1973), but this parameter would be difficult to measure. In a discussion of these phenomena in relation to insecticide selectivity, O'Brien (1967*a*) concludes that there is no evidence for any consistent difference between various animal classes in regard to the permeability/penetrant polarity relationship of their integuments, nor is there sound evidence that selective toxicity, between, for example, insects and mammals, is related to differences in integumental permeability.

One particularly important difference between insect and mammal is the large ratio of surface area to volume in the former and the greater accessibility of the insect central nervous system (CNS) to topically applied insecticides. Using labeled oils, Lewis (1962) showed that within 15 min of tarsal contact adults of *Phormia terraenovae* acquired a monolayer of material over their entire body surface. Since the oxygen supply to the insect CNS is by means of a tracheal system, the walls of which are continuous with the external cuticle, it is possible that oil-bound insecticides can gain rapid access to the CNS by this route, as well as by the obvious but less direct one involving penetration through the cuticle and transport through the hemolymph (blood). The slow diffusion of apolar insecticides from the epicuticular wax into the more polar inner strata of the cuticle presumably provides greater opportunity for their lateral diffusion in this waxy layer. If access to the CNS by the tracheal route is significant, the greater contact efficiency of apolar toxants might thus be explained. Burt (1970) points out that with this mode of entry into the CNS, passage of the toxicant through the outer sheath of the nervous system (neural lamella and perineurium), generally regarded as protective against insecticidal action, would be avoided. However, the toxicant must still traverse the glial cells that lie between the tracheoles and the nerve cells and must consequently be capable of membrane penetration. Another possibility which is perhaps more likely for apolar compounds is that insecticide may enter the hemolymph via the tracheal system.

The mechanism of penetration of insecticides through insect cuticle has become controversial following experiments by Gerolt (1969, 1970, 1972) which led him to conclude that the entry of dieldrin and some other compounds into the bodies of several insect species occurs entirely by the lateral spreading and tracheal route. These observations are partly supported by others showing that within a short time after the topical application of [^{14}C]lindane or [^{14}C]DDT to the prothorax of *Periplaneta americana* (Quraishi and Abdul Matin, 1962) or of [^{14}C]DDT to the mesonotum of houseflies (Quraishi and Poonawalla, 1969) the insecticides spread over the entire body surface (*cf.* Lewis, 1962, 1965). The housefly experiments, which included sectioning and autoradiographic studies, indicated that topical application of [^{14}C]DDT in benzene resulted in removal of the epicuticular wax at the point of application, so that the DDT became adsorbed in the underlying cuticle at this point and remained there without diffusing in the cuticle either laterally or inward. However, radiolabel was found in sutures, membranes, and other entry routes located in the cuticle at some distance from the point of application. Since the cuticle surrounding these areas contained no radiolabel, the insecticide could not have arrived at these points by lateral spreading from the inner layers of cuticle at the original point of application. It was surmised that DDT unadsorbed in the cuticle itself at the point of application entered the epicuticular wax layer at the edges of the "crater" left by the benzene and traversed the insect through this layer. Unfortunately, the fixing methods used removed the epicuticular wax and any radiolabel it contained, so that no evidence was obtained on this important point. Since insecticide did not appear to diffuse inward through the cuticle itself, but nevertheless was uniformly distributed in the entire alimentary canal 15 min after topical application to the mesonotum, one is led to conclude that (aside from possible entry into the tracheal system following vaporization) the epicuticular wax must have distributed the insecticide to the various portals of entry through the cuticle and into the tracheal system. It is possible that insecticide entering through one portal might be carried by the blood to others, but if this is so it is difficult to see why it is not also transported to some of the surrounding cuticle. At this point, the probable influence on entry routes of the differing physical properties of the various types of contact insecticides must be borne in mind, since the above observations relate to chlorinated insecticides (see Section 3.3).

Sun *et al.* (1967) found that when DDT, dieldrin, or the water-soluble organophosphate 3-hydroxy-*cis*-crotonamide dimethylphosphate (SD-11319) was injected (acetone–dimethylsulfoxide), topically applied (acetone), or infused (acetone–dimethylsulfoxide) into the housefly thorax, the LD_{50}s increased (toxicity decreased) in the same order. Infusion is equivalent to slow injection and for DDT or dieldrin, but not for SD-11319, the LD_{50} increased greatly with time of infusion. Thus the LD_{50} for dieldrin was increased a hundredfold by a 30-min infusion as compared with injection. This seems remarkable since only 50% of the LD_{50} for normal flies (0.02 μg) penetrated in 4 hr following topical application, and this route would therefore be expected to be the least effective. The results suggest that the infusion technique affords opportunities for the inert storage of chlorinated insecticides, but not for the organophosphate. Clearly, topical application is much more effective than infusion for the chlorinated insecticides and does not, as might be supposed, have the effect of a very slow infusion. In contrast, the percutaneous absorption of dieldrin by rats is well simulated by very slow intravenous infusion (Heath and Vandekar, 1964). For the water-soluble phosphate, partitioning into lipids is expected to be a much less significant factor and, in this case, the route of administration made little difference to the LD_{50} (see Section 3.3).

These results may relate to those of Gerolt (1969), who administered dieldrin and some other insecticides to houseflies in various nonphysiological ways. Since dieldrin undergoes very little metabolism or excretion, it is ideal for such studies. Large amounts of crystalline dieldrin placed in housefly abdomens produced toxic effects only slowly, as did 0.2 μg (10 × LD_{50}) adsorbed on filter paper and placed directly in the abdomen. Within 1 hr, 0.012 μg of the dieldrin from the filter paper appeared in the thorax without exerting any toxic effect, whereas 0.009 μg in the thorax (picked up from a similar surface deposit) produced 50% knockdown in this time. Similar experiments with chlorfenvinphos and methyl parathion also showed a four- to five-fold delay in knockdown time by the internal as compared with the external application. Thus it appears that these insecticides removed from filter paper by the hemolymph are less available to the target(s) in the thorax than when they enter via the external (physiological) route. The nonphysiological routes may offer better opportunities for retention of the insecticides in a bound, inactive form (which may be mobile), although this is not revealed by the experimental technique used. In injection experiments, the toxicity of dieldrin increased with the volume (nontoxic) of acetone used as a carrier, which suggests that this solvent may actually assist penetration of toxicant into the target, or may act as a dispersing agent to increase its bioavailability.

Various lines of evidence indicate that the speed of action of contact insecticides varies inversely with distance between the point of application and the target and is not consistently related to the degree of sclerotization. Gerolt (1969) argues that this is inconsistent with passage through the cuticle and transport in the rapidly circulating hemolymph. However, a point overlooked is that extra distance to travel in the blood may provide greater opportunities for inert storage and result in the reduced effect of a given dose. The apparent lack of penetration through isolated cuticles can be attributed to the use of nonphysiological bathing media (Olson, 1973), but certain other observations are less easily explained. Thus a wax barrier separating thorax from abdomen in houseflies protected them from the toxic effects of dieldrin applied to the abdomen. The presence of the barrier greatly reduced the amount of toxicant reaching the thorax and head and approximately doubled the recovery of dieldrin from the abdomen at various times after application. There is no reason to suppose that such treatment would reduce direct penetration through the abdominal cuticle into the hemolymph, but it should reduce lateral transfer, as suggested by Gerolt (1969). Similar results were obtained with German cockroaches and different insecticides (methylparathion, chlorfenvinphos, paraoxon, and Isolan). Olive oil (1 μl) injected into the thorax of houseflies is distributed throughout the body and might be expected to selectively absorb lipophilic insecticides circulating in the hemolymph. In Gerolt's experiments, this treatment had little effect on the speed of action of dieldrin, DDT, methyl parathion, or chlorfenvinphos picked up from glass surfaces and therefore entering the insect in a "physiological" manner. This observation argues against the blood as a transport medium. However, oil present in the aqueous blood in discrete droplets might not in fact compete adequately with transport mechanisms (such as binding with mobile proteins) that await insecticides which pass through the cuticle in a physiological fashion. In support of this view, Olson (1973) has shown that hemolymph (in contrast to water) should compete favorably with oil for dieldrin; also, an appreciable amount of topically applied endrin entered hemolymph through the isolated cuticles of susceptible tobacco budworm larvae (Polles and Vinson, 1972). Other work indicates that injected olive oil does protect houseflies against poisoning by DDT or dieldrin, although it has little effect on the toxicity of methyl parathion or dimethoate (Collins *et al.*, 1974).

Moriarty and French (1971) applied dieldrin separately to the pronotum and upper forewing of *P. americana* and found that 10% of the dose (25 μg per cockroach) entered the body in 76 hr. From careful analysis of wing and other tissues, they concluded that lateral spreading of the toxicant in the epicuticular lipids or in the procuticle is unimportant, and they favored transport in the blood as the distribution mechanism.

When equivalent doses (LD_{95}, 0.45 μg) of pyrethrin I were topically applied to the metathoracic sterna or introduced into the metathoracic spiracle of *P. americana*, there was little difference in the course of poisoning (Burt *et al.*, 1971). However, only excessive doses of this compound affected the spontaneous activity of the sixth abdominal ganglion when injected into the ventral trachea close to the ganglion. It was concluded from these experiments that the hemolymph plays a major part in the transport of insecticides to the nervous system, but participation of the tracheal system cannot be excluded.

In summary, this important question of the mode of penetration of insecticides into insects is still open, and for a given methodology the situation is undoubtedly influenced by interspecific differences in cuticle structure as well as by the properties of the toxicant.

2.1.2. Kinetics of Penetration

Toxicant penetration into intact insects is usually measured on the assumption that entry takes place *through* the cuticle into the hemolymph. Typically, groups of insects topically treated with a given amount of insecticide are rinsed at intervals with a solvent such as acetone, methanol, or hexane to determine the amount of chemical remaining on the "outside." These solvents tend to remove the epicuticular wax, so that what is actually measured is the loss of insecticide from this layer into the lower strata of the cuticle. A technique (O'Brien, 1967*a*) which actually measures loss into the body involves application of insecticide to the integument, which is later excised intact (allowing for lateral spread of the toxicant) from that part of the body and analyzed for residual insecticide. This method has been used with both insects and mammals. Less physiological methods (Treherne, 1956, 1957; Gerolt, 1969) involve measurement of the amount of insecticide passing into artificial media, such as water or saline, bathing the inner surface of isolated areas of integument fixed in a suitable cell.

The kinetics of penetration into insect cuticle or mammalian skin appear to be somewhat predictable from Fick's law of diffusion, which leads to a first-order form in which the rate of penetration at any time is proportional to the amount on the "outside" at that time. This is shown by equation (1), where C is the amount remaining "outside" at time t and k is the rate constant of penetration:

$$-dC/dt = kC \tag{1}$$

It follows from equation (1) that

$$C = C_0 e^{-kt} \tag{2}$$

where C_0 is the amount of insecticide applied. Thus a plot of the logarithm of the amount unpenetrated against the time of penetration should be linear. Some of the published results indicate that this simple form is followed throughout, as in the case of DDT penetrating into Madeira cockroaches or rats (Fig. 2), or for dieldrin, dimethoate, and paraoxon penetrating into American cockroaches (O'Brien, 1967*a*). In other cases, a biphasic or even more complicated form of penetration is observed (Fig. 3). Examples include the penetration of pyrethroid insecticides into mustard beetles (Elliott *et al.*, 1970), HCH isomers into American cockroaches (Kurihara, 1970), and malathion and dieldrin into rats (O'Brien, 1967*a*). The form of the penetration curve obtained will obviously depend on factors such as metabolism which influence the rate of removal of insecticide from the inner layers of the integument. Thus a fairly rapid initial transfer of insecticide from the outer to

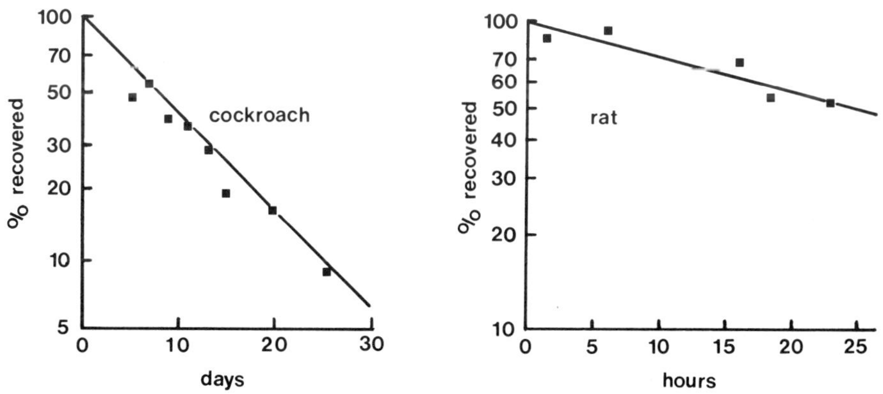

Fig. 2. Apparent monophasic penetration of DDT into the Madeira cockroach (measured by the solvent-rinsing method) and through the skin of the female rat (measured by the integument excision method). Adapted from O'Brien (1967*a*).

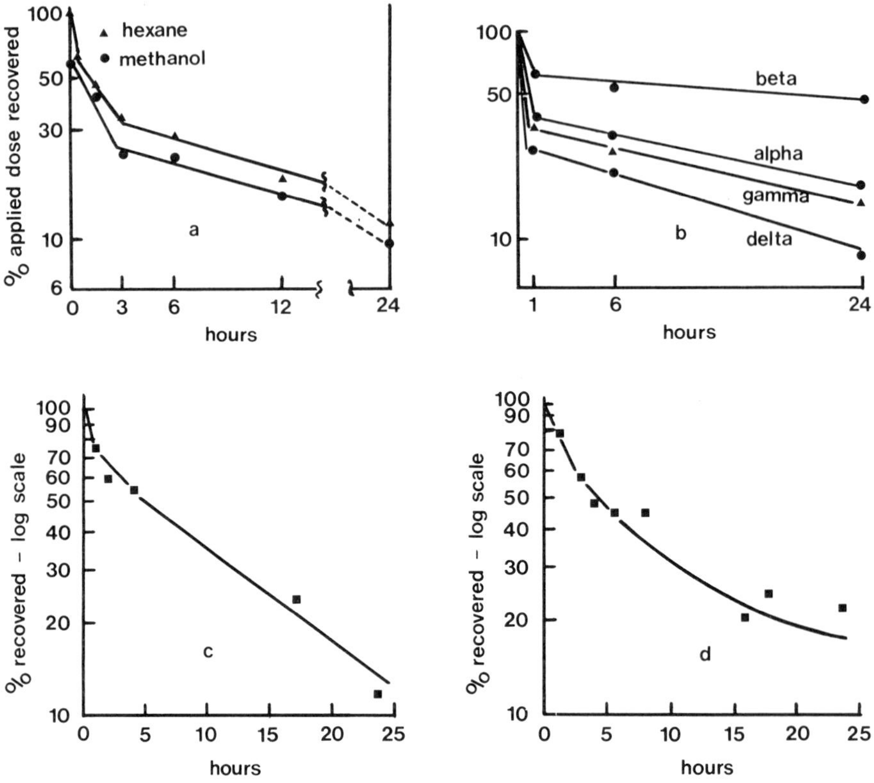

Fig. 3. Multiphasic penetration of (a) 2,3-dimethylbenzyl-(±)-*cis-trans*-chrysanthemate into mustard beetles (Elliott *et al.,* 1970), (b) HCH isomers into American cockroaches (Kurihara, 1970), (c) malathion and (d) dieldrin through the skin of the female rat (from O'Brien, 1967*a*; courtesy of Academic Press).

the inner layers of the integument would be expected, followed by a steady-state rate of inward diffusion as the system equilibrates. Whether the initial phase can be observed may depend on the animal investigated and the time scale of the observations. For the HCH isomers, the initial rapid phase of penetration into American cockroaches appears to be linear, and therefore first order.

In the case of pyrethroids entering mustard beetles, the initial phase itself seems to be biphasic and the linear part of the plot represents penetration governed by rate determining detoxication. When the external pyrethroid [2,3-dimethylbenzyl-(±)-*cis-trans*-chrysanthemate; 2,3-DMBC] was rinsed off with methanol in these experiments, 40% of the dose was apparently lost from the surface within seconds of its application (Fig. 3). With hexane, however, the "zero-time" recovery approached 100%, indicating that there is an initial rapid loss of pyrethroid into phases, presumably lipophilic, which are accessible to hexane but not to methanol. Apart from this initial difference, the rinsing solvent apparently has little effect on the course of penetration.

The negative correlation between penetration rate and partition coefficient (P) referred to in Section 2.1.1 is reflected in the negative slopes of the straight lines obtained when log (partition coefficient) is plotted against log (penetration rate constant); k becomes smaller as P increases. This relationship is evident for the penetration of HCH isomers (Kurihara, 1970) and other topically applied toxicants and solutes (Olson and O'Brien, 1963) into American cockroaches, as well as for the penetration of dimethoate analogues through the isolated guts of two other insect species (Shah *et al.*, 1972). The penetration data of Olson and O'Brien (1963) are approximated by the following equations (Penniston *et al.*, 1969), of which the second appears to be more adequate:

$$\log t_{0.5} = 0.543 \log P_{\text{olive oil}} + 1.639 \qquad (r = 0.978) \tag{3}$$

$$\log t_{0.5} = 0.124(\log P_{\text{olive oil}})^2 + 0.360 \log P_{\text{olive oil}} + 1.492 \qquad (r = 0.998) \tag{4}$$

where $t_{0.5}$ is the half-time of penetration and r is the correlation coefficient. The implication of the squared term in equation (4) is that the relationship between k and P is really parabolic, so that there is a value of P for which penetration rate is optimal. Thus there may be an apparent linear relationship between k and P at P values on either side of the optimum and the slope of the log–log plot may be either positive or negative.

The total amount of insecticide absorbed from an external dose within a given time by an insect does not increase indefinitely as the dose is increased but tends to approach an upper limit. Considering the penetration of DDT into DDT-resistant houseflies, Hewlett (1958) concluded that the relationship between the amount of insecticide absorbed (w) at a given time after topical application and the external dose (d) was best represented by the expression

$$w = a(1 - e^{-d/a}) \tag{5}$$

Thus the amount absorbed (w) approaches an upper limit (a) as d increases indefinitely, presumably due to tissue saturation effects. For DDT penetrating into houseflies during 24 hr a is about 10 when w and d are expressed in micrograms per housefly. These quantities are then represented by the type of curves shown in Fig. 4 (Winteringham, 1969). If, for simplicity, the amount absorbed (w) is regarded as the amount reaching the site of action, then for DDT-susceptible flies (curve A), intoxication occurs at some level below the upper limit (a) attainable by increasing the external dose. Curves B and C show the effect of changes in the various parameters influencing penetration such that first a half and then three-fourths of the dose which would normally arrive at the site of action is prevented from doing so. This is equivalent to the

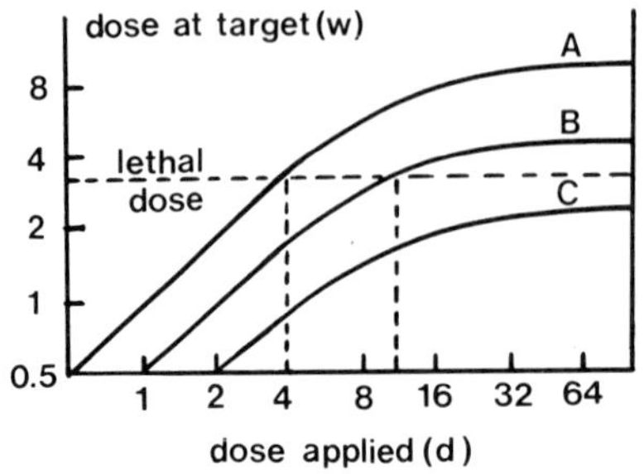

Fig. 4. Relation between dose of an insecticide topically applied and dose absorbed based on equation (5) (Hewlett, 1958) and the effect (curves B and C) of barriers or processes which restrict the amount arriving at the site of action. Adapted from Winteringham (1969).

successive addition of two barriers, each of which halves the amount of toxicant reaching the site of action. A resistant strain having the absorption characteristics of curve B will still be killed before the limiting internal concentration is reached, although a threefold increase in the external dose (d) is required. However, when both barriers are present (curve C), the lethal internal concentration can never be reached because the attainable concentration lies below it for all external doses. This shows how a combination of apparently minor changes, such as reduced cuticle permeability combined with a small alteration in the permeability characteristics of some internal phase, or with a seemingly minor detoxication mechanism, may produce total immunity to a formerly efficient toxicant. Such phenomena may also explain interspecific selectivity.

2.2. Transfer Through the Gut and Other Internal Membranes

2.2.1. Absorption Through the Gut

Since insecticides are part of the diverse class of foreign compounds referred to as drugs, they follow the general principles of membrane penetration outlined for drugs in general at the beginning of this section (Goodman and Gilman, 1965; Parke, 1968; Chasseaud, 1970). Following their oral ingestion by mammals, drugs may be absorbed from the mouth, stomach, or small intestine. The mucosal lining of the mouth behaves as a lipoid barrier through which nonionized compounds can diffuse rapidly into the bloodstream. This route avoids any destructive processes resulting from contact with the gastric juices or passage through the liver and can result in a higher concentration in the blood than when absorption occurs in the stomach or intestine.

The lining of the stomach also behaves as a typical lipoid barrier, so that nonionized substances can diffuse through it into the blood plasma. If the ionized portion of a weak electrolyte cannot pass through the aqueous pores, the distribution achieved will depend on the pK_a (negative logarithm of the dissociation constant, K_a) and on the pH gradient across the membrane. A typical situation is shown in Fig. 5 for a weak acid of p$K_a = 4.4$ entering the stomach. The degree of ionization of an organic electrolyte at any pH is given

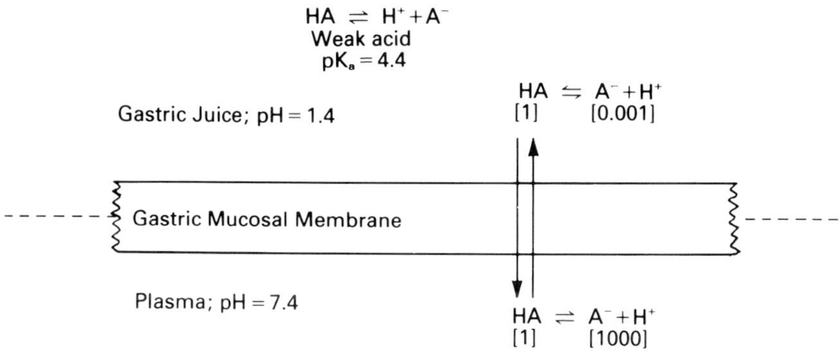

Fig. 5. Influence of pH on the distribution of a weak acid between plasma and gastric juice, separated by the lipoid gastric membrane.

by the Henderson–Hasselbach equation:

$$\log \frac{C_m}{C_i} = \mathrm{p}K_a - \mathrm{pH} \qquad \text{(for acids)} \tag{6}$$

$$\log \frac{C_i}{C_m} = \mathrm{p}K_a - \mathrm{pH} \qquad \text{(for bases)} \tag{7}$$

where C_m and C_i are concentrations of the molecular and ionized forms, respectively.

The relative total concentrations (ionized plus nonionized) of the electrolyte on either side of the membrane are given by the following equations:

$$\frac{C_g}{C_p} = \frac{1 + 10^{(\mathrm{pH}_g - \mathrm{p}K_a)}}{1 + 10^{(\mathrm{pH}_p - \mathrm{p}K_a)}} \qquad \text{(for a weak acid)} \tag{8}$$

$$\frac{C_g}{C_p} = \frac{1 + 10^{(\mathrm{p}K_a - \mathrm{pH}_g)}}{1 + 10^{(\mathrm{p}K_a - \mathrm{pH}_p)}} \qquad \text{(for a weak base)} \tag{9}$$

where C_g and C_p are the concentrations in gastric juice and plasma, respectively.

In Fig. 5, nonionized HA readily penetrates the membrane and achieves equal concentrations on either side at equilibrium. To maintain this equilibrium, HA must cross the membrane from the gastric juice to replace that which is ionized in the plasma. HA is largely unionized in gastric juice [equation (6)] and the ratio of total concentrations between the plasma and the gastric juice sides of the membrane is 1000 : 1 [equation (8)]. The situation is exactly reversed for a weak base of $\mathrm{p}K_a = 4.4$. Since gastric mucosal blood flow restricts the rate at which drugs can be delivered to or removed from the gastric juice, the concentration ratios found in practice are much smaller than this. Apart from the maintenance of the pH gradient, which requires the expenditure of energy, the process is a purely physical one depending on preferential permeability of the membrane to the nonionized form. The equations show that weak bases are largely ionized at the gastric juice pH and consequently are poorly absorbed. Clearly, the situation can be altered if the pH of the stomach contents is changed artificially. In the small intestine the pH is about 6.5, so that, in contrast to the situation in the stomach, weak bases pass the intestinal epithelium more readily than weak acids. The intestine is the major route of absorption for most foreign compounds. These considerations also apply to the passage of foreign compounds across other body membranes, including cell membranes, although there is usually only a small pH gradient between intracellular and extracellular fluids and the concentration gradient across cell membranes is small.

Although there is a great deal of published work on the distribution of insecticides and their conversion products in vertebrate and insect tissue, the number of investigations specifically designed to study transfer across internal barriers is more limited. Much of the work with vertebrates assumes that the usual patterns of drug transference will be followed and is designed to provide information about residue levels in tissues and overall rates of clearance from the animal body. Insofar as they usually relate to the mechanisms of action, selectivity, or resistance, investigations with insects have more fundamental aims, but have often been hampered by difficulties in experimental technique. Recent experiments specifically designed to measure penetration through the gut are those conducted with rats (Pekas, 1971*a,b*; Casper *et al.*, 1973) and with insects and mice (Shah and Guthrie, 1970, 1971; Shah *et al.*, 1972).

Two types of experiments with rats illustrate the methodology usually employed *in vivo* (Casper *et al.*, 1973). In the first type, [1-naphthyl-1-^{14}C] *N*-methylcarbamate (7.5 μmoles/kg body weight) was administered intragastrically to anesthetized, starved female rats with ligated pylori and intact portal circulation. At 22 or 67 min after dosing, the appropriately ligated gastrointestinal tracts were removed from the abdominal cavities so that the stomachs, small intestines, large intestines, and the remainder of the rats could be homogenized separately. Radioassay of the homogenates indicated a loss of $52.6 \pm 14.1\%$ and $81.7 \pm 15.7\%$ of the label from the stomachs at these respective intervals (total recovery $> 98\%$ in each case), presumably due to absorption, since the pylori were ligated. In the second type, the dose was passed into stomachs with esophagi and pylori ligated, arrangements being made for continuous collection of portal blood and its simultaneous replacement with rat donor blood via the vena cava. In this experiment, which departs from the physiological situation in that the absorbed label is progressively removed from the body, $69 \pm 25.9\%$ of the label was absorbed during 66 min of blood collection. The portal blood contained $97.2 \pm 1.9\%$ of the absorbed label, and this consisted largely (89.3%) of the unchanged carbamate. These observations are consistent with the nonionized nature and lipophilic character of carbaryl. Metabolism of the insecticide appears to have been of relatively minor importance in the last experiment, but may be expected to influence absorption under conditions in which the portal circulation is intact so that all of the absorbed chemical can be carried to the liver. Absorption will also be influenced by normal intakes of food and water.

Shah and Guthrie (1970, 1971) and Shah *et al.* (1972) studied the penetration of chlorinated, organophosphate, and carbamate insecticides through the isolated midgut of cockroaches (*Blaberus* sp.) and tobacco hornworm larvae (*Manduca sexta*), and through a segment of the small intestine of a mouse. In these experiments, the entire midgut of the insect or the intestinal segment of the mouse was tied to form a sac which could be suspended in an appropriate physiological buffer (serosal medium). The distribution of an insecticide introduced into the lumen of the gut was then measured at various time intervals by sampling serosal fluid, gut tissue and gut luminal fluid.

Thus the first measurement represents actual penetration through the gut and the second represents retention by the gut tissues after various times. The distribution of carbaryl in mouse gut 80 min after the introduction of 0.1 μg of the ^{14}C-labeled compound into the lumen resembled that observed in the second rat experiment of Casper *et al.* (1973). Of the 63% of the label which entered the serosal fluid, the majority (82%) was unchanged carbamate and the remainder mostly 1-naphthol (11%) plus water-soluble products. The gut tissue contained 12% of the original label and the proportions of carbaryl and 1-naphthol in this fraction and in the luminal fluid residue resembled that in the serosal fluid.

Only 25% of the original label entered the serosal fluid through cockroach gut and a further 13% was found in the gut tissues. The highest proportion of water-soluble metabolites was produced by *M. sexta* gut (all three fractions). In summary, more unchanged carbaryl penetrated (i.e., gut tissue plus serosal fluid content) mouse and hornworm gut (62%) than cockroach gut (31%). DDT and dieldrin penetrated the isolated guts of these animals only slowly and were strongly retained in their gut tissues. For each insecticide, the sum of gut tissue and serosal fluid content was greatest for *M. sexta* (Shah and Guthrie, 1971).

Similar studies have been conducted on the penetration of ^{14}C-labeled carbaryl, parathion, and dieldrin in isolated honey stomachs (foregut) of the honey bee (*Apis mellifera*) (Conner, 1974). The results indicated a rapid initial uptake of insecticide from the lumen of the foregut into the foregut tissue, the tissue level attained being proportional to the partition coefficient of the compound (dieldrin > parathion > carbaryl). Total penetration from the lumen through the foregut and into the serosal fluid, however, was not proportional to the lipophilic character of the insecticide, the relative order being parathion > carbaryl > dieldrin.

These results indicate that the physicochemical characteristics of parathion promote rapid penetration and suggest that the limiting factors with carbaryl and dieldrin may be their partitioning into and out of the foregut tissues, respectively. Following the initial rapid uptake by the foregut tissue, the rate of penetration of each insecticide into the serosal fluid followed first-order kinetics.

In general, lower polarity favors retention by the gut tissue, and penetration of straight chain dialkoxy analogues of dimethoate (except the dimethoxy analogue) through isolated guts was found to increase with polarity (Shah *et al.*, 1972), as found for insecticides penetrating insect cuticle (Olson and O'Brien, 1963). The penetration of the dimethoate analogues apparently follows first-order kinetics so that the rate constant of penetration (k) is related to the half-time of penetration ($t_{0.5}$) by equation (10), which is derived from equation (2):

$$k = 0.693/t_{0.5} \tag{10}$$

Although there appear to be only relatively small interspecific differences in the penetration of these analogues through isolated guts, such differences, if present *in vivo*, may nevertheless contribute significantly to overall selectivity (*cf.* Fig. 4).

In some cases, the penetration patterns are inevitably complicated by metabolism. With malathion, for example, the total potential toxicant entering the serosal fluid through the isolated gut is reduced by detoxication (Shah and Guthrie, 1971). The toxic effect depends on the amount of the active metabolite (malaoxon) that enters the serosal fluid and the efficiency with which it eventually reaches and interacts with acetylcholinesterase. Chlorinated insecticides differ in that they are only slowly released from the midgut into the serosal fluid but are rather stable toxicants which can accumulate progressively in target areas (Shah and Guthrie, 1971). Alternatively, slow release may allow the animal a greater opportunity to store the incoming toxicant in nontarget tissues. It is notable that chlorinated insecticides tend to show reduced toxicity to the later instars of lepidopterous larvae, which normally have large fat bodies and generally greater inert storage potential than the earlier instars (Sun *et al.*, 1967).

2.2.2. Uptake and Release by the Blood

While due allowance must be made for the nonphysiological nature of some of the *in vitro* experiments conducted on gut preparations, it would appear, for insects at least, that when compounds are not ionizable there is an optimal balance between polarity and lipophilicity for facile penetration *through* the gut wall. Compounds with lipophilic character in excess of the optimum tend to penetrate the gut tissues and accumulate there, passing only slowly into the serosal fluid beyond. On this basis, penetration *in vivo* might actually be retarded by modifications resulting in increased lipid solubility. What is not entirely clear is the extent to which the penetration of compounds such as the highly lipophilic chlorinated insecticides is influenced by their partitioning from gut tissues onto mobile blood proteins and cellular components *in vivo*. In some of the experiments described in Section 2.2.1, attempts were made to approximate more closely to physiological conditions by incorporating blood plasma in the serosal fluid for the mouse gut and hemolymph in that used with the tobacco hornworm gut. These incorporations made little difference to the penetration of dimethoate analogues (mouse) or carbaryl (hornworm).

It is well known, however, that foreign compounds entering the blood may become bound to plasma proteins and transported in this way (Goodman and Gilman, 1965; Parke, 1968). Albumin is the protein most commonly involved, but other plasma proteins may participate if the albumin becomes saturated. Various degrees of binding may occur, and if the foreign compound has a high affinity for such proteins the latter may become a significant storage site. Dieldrin and isobenzan, for example, are 4000 times more soluble in rabbit serum than in water (Hathway, 1965). In rabbit and rat blood, they are located primarily in the plasma and erythrocytes and not in the stroma, platelets, or leukocytes. The erythrocyte stroma appears to be a typical lipoid barrier which is rather permeable to these insecticides, since their distribution between blood plasma and erythrocytes *in vivo* remains approximately constant with time and is independent of overall concentration changes. Within the erythrocytes, the compounds are associated mainly with hemoglobin and an unidentified component (Moss and Hathway, 1964; Hathway, 1965); in the serum, albumin and globulins are the main carrier proteins. In order to enter the tissues, compounds carried in the blood must eventually dissociate from their carriers and pass through the walls of the blood capillaries. Except in the central nervous system, most substances cross blood capillary walls relatively more easily than other biological membranes and access to the tissues is limited by the rate of capillary blood flow rather than by permeability.

Insect blood circulates quite rapidly and insecticides which can enter it will undoubtedly become bound to both cellular components and soluble proteins, as in the case of vertebrates. For example, dieldrin in American cockroach blood binds to a protein with molecular weight of about 18,900 and to two groups of proteins with molecular weights of $\geq 160{,}000$. The hemocytes in whole cockroach blood contain 37% of the dieldrin which is in the blood (Olson, 1973).

2.2.3. Transfer Through Excretory Membranes

The processes of transfer through excretory membranes are most clearly understood for mammals, through investigations on the elimination of drugs. In mammals, blood containing foreign compounds absorbed from the gut is carried to the liver by the portal vein. Within the liver lobules, the blood enters the hepatic sinusoids, where it is brought into intimate contact with the hepatic parenchymal cells. The walls of the hepatic sinusoids behave as highly porous membranes which allow the rapid transfer of most molecules and ions smaller than proteins from blood plasma into the extracellular fluid of the hepatic cells. The pores of the lipoid membrane of hepatic cells are smaller than those of the blood sinusoids but large enough to permit the transfer of lipophobic substances that usually cannot penetrate other cells. Hepatic cells contain the enzymes responsible for the biotransformation of foreign compounds and are permeated by bile canaliculi with permeable walls. Thus the hepatic cells serve as a clearing house for foreign molecules, which may be modified and either returned to the sinusoidal blood for transfer to the kidney and excretion in the urine or passed into the bile for eventual excretion in the feces.

Effectively, the blood and bile are separated by a rather porous membrane, so that many substances enter the bile in concentrations resembling those in the plasma. However, bile salts and other highly polar compounds such as the conjugates of xenobiotics are transferred into the bile by an active transport process. Conjugation of foreign compounds is most frequently with glucuronic acid, sulfuric acid, or amino acids and the excretory route followed by these derivatives depends on their molecular weight and on the vertebrate species involved. For 16 different organic anions excreted by rat, rabbit, or guinea pig, excretion was mainly in bile for those with molecular weights greater than 500 and in urine for those with molecular weights of less than 300 (Hirom *et al.*, 1972). The estimated thresholds below which excretion is mainly in the urine appeared to be 325 ± 50, 400 ± 50, and 475 ± 50 for rat, guinea pig, and rabbit, respectively.

The excretion of insecticides and their metabolites is generally expected to follow the same principles. Thus the *trans*-diol formed by the hydrolytic epoxide ring opening of dieldrin (molecular weight 399) is excreted in the urine of rabbits but in the feces of rats. The pentachloroketone (known as Klein's ketone, mol wt 361) produced from dieldrin by concurrent molecular rearrangement and oxidative dechlorination in male rats is evidently in the marginal area for this animal; in fact, it appears in the urine. However, this metabolite occurs only in kidney and urine (Baldwin, 1971) and there is a possibility that it may be formed from dieldrin in the kidney. Rats treated with endrin oxidize the unchlorinated methylene bridge in this molecule to give a ketone (mol wt 395) which is stored in the tissues and excreted in the urine. This molecular weight exceeds the upper limit indicated for urinary excretion in rats, although Walker (1974) points out that the indicated thresholds were derived for molecules having relatively large size in relation to molecular weight; the cyclodiene insecticides combine high molecular weight with small volume and the thresholds for such molecules may well be higher.

At this point, it is appropriate to mention the phenomenon of enterohepatic cycling. An insecticide or drug may be metabolized and excreted in the bile as a conjugate. Once in the intestine, such conjugates may be hydrolyzed by various enzymatic mechanisms, including the action of gut microorganisms, and the parent compound or metabolite thus produced may be reabsorbed to return to the liver. Further metabolism may occur there, and some of the chemical may be passed to the urine, or the earlier cycle may be repeated, with gradual depletion as the products are removed in the feces and urine or lost to the tissues. Dieldrin administered to ruminants undergoes enterohepatic cycling (Cook, 1970). After administration, it enters the gastrointestinal tract via the saliva, bile, and pancreatic juice and probably through rumen clearance. A suitable adsorbent passed into the gut may trap this unchanged dieldrin, thereby preventing its reabsorption. Accordingly, granulated charcoal administered with the food increases the rate of excretion of dieldrin (Wilson and Cook, 1970).

In the mammalian kidney, glomerular filtration in the Bowman's capsule (glomerulus) at the proximate end of each nephron produces an ultrafiltrate of blood plasma. This ultrafiltrate, containing xenobiotics and their metabolites in about the same concentrations as in the plasma, then passes down the connecting renal tubule, where water, endogenous substances, and some inorganic ions are reabsorbed. The distal part of the tubule especially behaves as a typical lipoprotein membrane through which lipid-soluble nonionized xenobiotics pass readily in either direction, depending on the concentration gradient for them between urine and blood plasma. The principles described for absorption from the gut apply also to the penetration of ionizable molecules through the distal portion of the kidney tubule. Those compounds less ionized at urine pH than at the plasma pH tend to be reabsorbed into the blood through the tubule walls. Accordingly, the filtration of ionizable xenobiotics can be influenced by artificially manipulating the pH of the urine. Lipophobic ionized molecules such as conjugates of xenobiotics are more easily excreted than their precursors, since they are poorly reabsorbed from the gromerular filtrate during its passage down the tubule. Since a lipophilic xenobiotic entering the nephron unchanged is likely to be reabsorbed during its passage through the tubule, its further conversion into more polar metabolites, especially ones which can be conjugated, is an important factor in ensuring its efficient clearance from the body in urine. In addition to this process of passive tubular transfer, active transport mechanisms for the secretion of anions and cations from blood to urine are found in the proximal part of the tubule (Parke, 1968; Chasseaud, 1970).

Since much of the transfer of foreign compounds across membranes is due to passive diffusion, it follows that in appropriate circumstances these may pass from blood plasma into stomach or intestine, as indicated for dieldrin (Heath and Vandekar, 1964; Cook, 1970). In addition to this reversal of the major processes already described, excretion of xenobiotics can occur by diffusion across the epithelia of sweat glands, salivary glands, and milk glands, apparently in accordance with the same laws of transferral. Milk is more acidic than plasma, so that basic compounds may be relatively more concentrated in it than in plasma; nonionized compounds can enter the milk readily, regardless of pH gradient. The secretion of insecticide residues in cow's milk and human milk is a familiar problem (Tanabe, 1972). Lipophilic insecticides also enter the ovaries of birds quite readily, and for a species such as the domestic hen, which lays eggs continuously, this may be a significant route for the elimination of tissue residues. Finally, sufficiently volatile compounds may be eliminated with the expired air after diffusing through the pulmonary epithelium.

In insects, various organs participate in the excretory process insofar as they are able to metabolize foreign compounds through one or both stages of the process leading to water-soluble conjugates (Sections 3.2.2 and 3.3). The insect fat body has frequently been likened to the mammalian liver, since it stores fats, proteins, and carbohydrates (Kilby, 1963), is involved in protein and nucleic acid synthesis (Price, 1973), and is known to metabolize foreign compounds (Wilkinson and Brattsten, 1972). This organ is well situated to take up insecticides from the hemolymph since most of its surface is freely exposed to the hemolymph circulating in the body cavity. However, the mechanism by which this occurs has not been studied. Pericardial cells clustered around the insect heart and aorta can take up colloidal particles and high molecular weight dyes circulating in the blood, and a scavenging function is performed by phagocytic hemocytes and various other cells called nephrocytes which are distributed throughout the body (Maddrell, 1971).

The system comprising the Malpighian tubules and hindgut of insects apparently has an excretory function analogous to that of mammalian kidney. An ultrafiltrate of hemolymph is produced by the tubules and must be processed to remove water and useful solutes before it is finally excreted from the rectum. Little is known of the way in which this system deals with insecticides, but the indications are that endogenous substances and foreign compounds of small molecular weight may enter the vicinal tubules by passive diffusion. Also, certain dyestuffs are secreted into the lumen by an active transport process which is accelerated by increasing pH, suggesting the involvement of anionic species (Maddrell, 1971). The fluid from the Malpighian tubules is discharged into the hindgut, which typically consists of a narrow ileum leading into a thick-walled rectum at the posterior. In some insects the distal portion of the tubules may reabsorb water and useful substances, but this is mainly the function of the hindgut, which is lined with cuticle. Maddrell (1971) summarizes the evidence that the excretory fluid is concentrated during its passage through the hindgut toward the rectum, so that the latter must be exposed to quite high concentrations of unwanted substances.

The cuticular lining of the rectum should effectively prevent the return of large and ionized molecules (e.g., conjugates of insecticides) to the insect body; compounds with molecular weights greater than 300–500 penetrate very slowly, and studies on the cuticle of locust rectum (Phillips and Dockrill, 1968) have established that it is virtually impermeable to molecules with diameters greater than 1.0–1.2 nm. It is possible, however, that insecticide conjugates might be hydrolyzed to their precursors by enzymes passed backward from the midgut or by microbial activity, in which case some form of recycling of insecticides or their primary metabolites could occur.

Since insecticides taken up by tarsal contact appear to spread rapidly over the insect cuticle, it may be that they can also traverse the surface of the hindgut cuticle from outside. This would provide yet another means of entry into the insect body and might explain how topically applied DDT can appear mainly in the hindgut and midgut of houseflies without apparently penetrating the cuticle (Quraishi and Poonawalla, 1969; see Section 2.1.1).

2.2.4. Absorption Through the Placenta

Toxic chemicals may cross the placental membrane or membranes separating the maternal and fetal circulations and thus exert their toxic effects on the fetus. Transfer is mostly by simple diffusion, with penetration limited to compounds with molecular weights below 1000. Nonionized drugs with high fat solubility such as alcohol, chloral hydrate, many barbiturates, salicylates, and anesthetic gases can enter the fetal blood in this way. The appearance of characteristic withdrawal symptoms in the newborn infants of addicted mothers indicates the transplacental transfer of narcotics such as heroin and morphine, and the teratogenic effects of thalidomide are well known. As with other membranes, highly ionized compounds such as the quaternary ammonium muscle-relaxing drugs cross the placenta with difficulty. The placenta is an actively metabolizing tissue, so that xenobiotics may be metabolized during their passage through it. However, information to date indicates that the microsomal drug-metabolizing enzymes typical of mammalian liver are absent or occur at low levels in this membrane. There has naturally been much concern about the possible effects of organophosphorus and other insecticides on mammalian reproduction, and the subject has been discussed by Hathway and Amoroso (1972).

Hathway *et al.* (1967) studied the uptake of intravenously administered [^{14}C]dieldrin by rabbit blastocysts and fetuses at various stages after conception. The unimplanted blastocysts in 6-days-pregnant rabbits equilibrated with the maternal blood within 30–40 min after injection of 10 μCi of dieldrin per animal, when the amount of label in each tissue was about 30% of that present in the blood 5 min after injection. After 60 min, the concentration in blastocysts decreased rapidly, suggesting a return of [^{14}C]dieldrin to the mother. The blastocysts become implanted at about the ninth day of pregnancy They then take up dieldrin much more slowly and equilibration with maternal blood requires about 4 hr. The differentiation of blastocyst membranes may be partly responsible for the slow uptake of dieldrin at this stage. This insecticide is rapidly cleared from the blastocoelic cavity, and the authors comment that the accumulation of [^{14}C]dieldrin in blastocyst fluid is about a thousandfold lower than the corresponding accumulation of label following a similar dose of [^{14}C]thalidomide. By the sixteenth day of pregnancy, the placenta is established, organogenesis is complete, and the fetus is less sensitive to the effects of foreign compounds. At this stage, there is transplacental transfer of administered dieldrin, resulting in rapid equilibration between fetal tissue and maternal blood. There is two-way transplacental transfer of dieldrin between mother and 24-day fetuses since at this stage [^{14}C]dieldrin injected into a single fetus appeared both in maternal blood and in the blood of an untreated fetus in the same uterus. Following a maternal intravenous injection of [^{14}C]dieldrin at this stage, there is a steep rise in its concentration in fetal blood, after which the time course of concentration decay has the same shape as that for maternal blood.

It is interesting that in nonpregnant rabbits the secretion of dieldrin from the endometrium into the uterine lumen following intravenous injection was strongly influenced by the medium within the lumen; much more dieldrin entered a uterine loop containing plasma than when the luminal fluid was physiological saline. Furthermore, a Krebs–Ringer solution containing 3% albumin effected the transfer of rather more dieldrin than the physiological saline. This seems to be a rather clear indication of the role, mentioned previously, of proteinaceous macromolecules in the transport of insecticides of this type (Hathway *et al.*, 1967).

It is well known that following the ingestion of stable pesticides by the females, residues of these compounds may appear in birds' eggs. Studies of this phenomenon relate mainly to the amounts of such residues found and the consequences of their presence, rather than to the mechanisms of uptake, and have been reviewed (for organochlorines) by Robinson (1969*a*, 1970*a*). In the case of insects, sublethal doses of insecticides administered to females may also result in altered reproductive potential. Thus the appearance of DDE in the ovaries of resistant mosquitoes (*Culex pipiens*) following exposure to DDT at either the larval or adult stage shows that these compounds can cross the membranes of the reproductive tissue (Zaghloul and Brown, 1968). Again, the mechanisms of transference have not been examined. The literature on the consequences of sublethal exposure of insects to insecticides has been reviewed by Moriarty (1969).

2.2.5. Entry into the Nervous System

The nervous system of insects is the most obvious target for toxicants, and nerve poisons have the advantage that their primary effects, and also secondary ones arising from loss of functional coordination, usually result in rapid immobilization of the insect. Studies of the interactions between insect nerve tissue and insecticides stem from the need to understand mode of action (especially in relation to the design of new toxicants) and the reasons for insect resistance. There is also a need to understand the mode of action of nerve poisons in experimental mammals (in the hope that this information can be extrapolated to man) so that protective measures may be devised and toxicants designed that are selectively active against insects.

The concept of a vertebrate barrier which slows the transfer of many compounds, especially ions, from blood to brain (blood–brain barrier; BBB) originates from Ehrlich's observation in 1885 that of all the tissues of the body only the central nervous system (CNS) escaped staining by certain aniline dyes. In general, the entry of foreign compounds into the CNS and the cerebrospinal fluid (CSF) follows the principles applying to transfer across other membranes; passive diffusion occurs at rates related to lipid solubility, concentration in the plasma, and degree of ionization. The BBB is located between the plasma and the extracellular space of the brain cells, and not at their surface. Thus intercellular occlusions interposed between the plasma and the extraneuronal fluid appear to restrict access of certain types of molecule to this fluid. The barrier lies at the capillary endothelial cells or at the glial feet that surround the brain capillaries. The blood–CSF barrier is located at the choroid plexus, which allows the passage of small ions and nonelectrolytes such as urea, although at lower rates than for other membranes. Lipophilic substances pass through these barriers readily and, indeed, the highly lipophilic barbiturate thiopental enters the mammalian brain so rapidly following intravenous injection that penetration appears to be limited only by cerebral blood flow rates. Transfer through these barriers in the reverse direction is possible for lipophilic compounds, but solutes of all kinds can leave the CSF by nonspecific transfer through the arachnoid villi and there are active transportation processes for certain organic ions which are similar to those found in the renal tubules.

The idea that there is a blood–brain barrier separating insect hemolymph from the nervous tissue arose from the observations of Hoyle (1953) that locust nerve is remarkably insensitive to high external concentrations of potassium ions. This barrier has been extensively investigated since then, and the current status of the concept has been discussed in relation to insecticidal action (O'Brien, 1967*a*) and from a physiological standpoint (Treherne and Pichon, 1972). These last authors conclude that insects have a well-developed blood–brain barrier which is similar in some respects to that of vertebrates.

The presence of such a barrier in the insect nervous system is of great significance for insecticidal action. For example, the cholinesterase of insects is confined to the ganglia (O'Brien, 1967*a,b*) so that anticholinesterase insecticides must penetrate the ganglia to exert their effect. Assuming that the principles previously outlined for toxicant penetration of biological membranes apply in this case, the fraction of an ionizable toxicant that is nonionized at a given pH and pK_a can be derived from equations (6) and (7), and its magnitude should be positively correlated with insect toxicity. Investigating this idea on a comparative basis, O'Brien (1967*a*) found that the toxicities of 35 nerve poisons to the mouse and to five insect species were indicative of an ion-impermeable barrier protecting the nervous system of the insect but not that of the mammal. Nicotine, for example (more than 90% ionized at pH 7.0), behaved as though most of it (ionized plus nonionized) was effective against the mammal but only the nonionized fraction against the insects. Consequently, assuming that the target sites are similar in insects and mammals and that other factors such as target sensitivity and detoxication rates are equal, a progressive reduction in the degree of ionization of a series of nerve poisons at physiological pH should correspondingly increase their relative toxicities to insects.

One of the best demonstrations of this dependence of penetration on degree of ionization is afforded by examining the changes in penetration of radiolabeled ionizable compounds into intact nerve cords *in situ* in response to variations in the pH of the bathing medium (O'Brien, 1967*a*,*b*; Eldefrawi *et al.*, 1968). Penetration rates are measured by removing the nerve cords at appropriate intervals, rinsing to remove external medium containing the compound under study, digesting the nerve cord, and counting the radioactivity taken up. When anticholinesterases are examined, the inhibition of endogenous cholinesterase may also be used as a measure of the amount of toxicant penetrating. As an example, the organophosphate insecticide Amiton (Fig. 6) is a base of pK_a 8.5 which has high mammalian toxicity but generally low toxicity to insects. The low insect toxicity is presumed to be due to its high degree of ionization at pH 7.0, and for American cockroach nerve cord the shape of the penetration curve as the pH increases resembles that of the titration curve. The difference in penetration rates between the ionized and nonionized forms of Amiton is estimated to be about fifteenfold (O'Brien, 1967*a*). For the positively charged quaternary ammonium analogue of Amiton and for the nonionizable tetraethylpyrophosphate, penetration is independent of pH.

Following the observations of Hoyle (1953), various other investigations suggested that the insect nervous system is impermeable to cations. However, subsequent studies (Treherne, 1961*a*,*b*,*c*,*d*, 1962, 1965; Treherne and Smith,

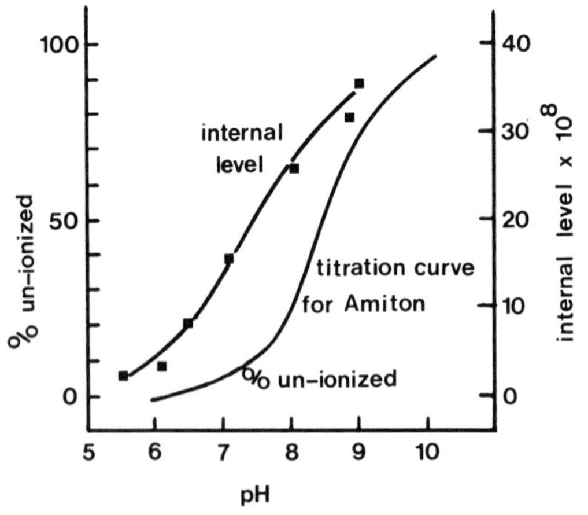

Fig. 6. Dependence on pH and ionization of the penetration of Amiton into American cockroach nerve cord. Adapted from O'Brien (1967*a*).

1965) have revealed that K^+, Na^+, Ca^{2+}, and acetylcholine penetrate readily into insect nerve, and it is now apparent that the barrier is by no means absolute. From an examination of the penetration into American cockroach nerve cord of the fatty acids from acetic to octanoic and of two series of quaternary ammonium ions, O'Brien (1967*a,b*) concluded that entry rate increases with apolarity. The penetration rates of homologous alcohols as far as butanol also increased with apolarity. With the exception of certain of the fatty acids, whose penetration is apparently enhanced by metabolism, the penetration rate constants among the nonionic, anionic, and cationic compounds examined varied only about sevenfold, showing that the influence of ionization is not large.

In a series of experiments (Eldefrawi *et al.*, 1968), the penetration of butanol and butyltrimethylammonium was found to be pH independent and that of butyric acid and butylamine to depend on pH in the expected manner (Fig. 7). Butyltrimethylammonium penetrates 6.5-fold more slowly than butylamine and the butyrate ion 3.5-fold more slowly than butyric acid; the general conclusion was that for molecules of comparable size, penetration is retarded five- to fifteenfold when a charge is present. Further, the polarity effect of the charge may be compensated by an appropriate increase in lipophilic character, as, for example, by increasing the number of methylene groups in an aliphatic chain.

When the cockroach nerve cord was desheathed in these experiments, there was a small increase in the penetration of cations such as butyltrimethylammonium, which suggests that the sheath itself retards cation penetration to some extent. However, there was little effect on the pH dependence of penetration for the other molecules examined, so the major barrier evidently does not lie in the sheath. Nevertheless, histological experiments (Eldefrawi *et al.*, 1968) show that the cationic dye methylene blue has little ability to

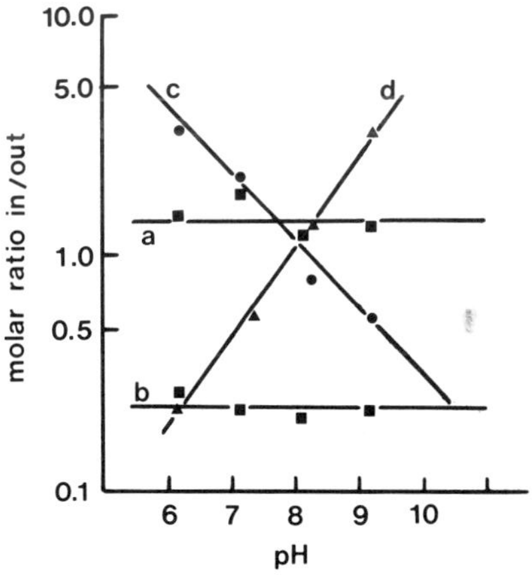

Fig. 7. Effect of pH on the influx of (a) butanol, (b) butyltrimethyl ammonium, (c) butyrate, and (d) butylamine into intact American cockroach nerve cord. Adapted from Eldefrawi *et al.* (1968).

penetrate beyond the sheath, the neuropile containing the nerve synapses remaining virtually unstained; similar results were reported earlier for the anionic dye, trypan blue (Wigglesworth, 1960). These results suggest that ionizable compounds undergo ion exchange only with the cortical layers external to the neuropile, but the location of the barrier is uncertain; it may be a function of the ganglion as a whole rather than of a particular layer of cells. Amiton certainly penetrates to the cholinesterase zone of the nerve cord in a pH-dependent manner, and it seems likely that other ionized compounds reach this zone up to 15 times more slowly than nonionized ones. A seemingly small barrier of this sort might have considerable effect in reducing toxicity if combined with one or more other barriers intervening between the toxicant and its target (see Section 2.1.2). Since histochemical studies indicate a correlation between anticholinesterase poisoning and enzyme inhibition in the peripheral rather than the neuropile region of the housefly thoracic ganglion (Booth and Metcalf, 1970; Brady, 1970), there is still some way to go in understanding these phenomena.

It appears, therefore, that the insect barrier behaves rather like the mammalian blood–brain barrier in that it delays the entry of large and polar molecules. The difference between insects and mammals is that in their peripheral nerves the latter have cholinesterase which is not protected by a membrane barrier against inhibition by ionic anticholinesterases. Accordingly, such inhibitors tend to be more toxic to mammals than to insects (O'Brien, 1967*a,b*).

Chlorinated insecticides pass the mammalian blood–brain barrier quite readily. Thus [^{36}Cl]dieldrin injected intravenously into mice (12 mg/kg) appeared in maximum concentration in the brain (16 μg/g) within 10 min, while the concentration in blood fell rapidly (Heath, 1962). The proportion of this compound entering rat brain depends on the rate of administration; it is particularly high after rapid intravenous infusion, whereas slow intravenous infusion results in preferential uptake by the fat (Heath and Vandekar, 1964). Accordingly, rats tolerated nearly as much dieldrin given by slow intravenous infusion as they could tolerate by oral administration. These results are reminiscent of those obtained when dieldrin is slowly infused into houseflies (Schaefer and Sun, 1967; Sun *et al.*, 1967); the toxicity of dieldrin is much reduced as compared with rapid injection (see Section 2.1.1). Hathway *et al* (1967) showed that when [^{14}C]dieldrin is given intravenously to rabbits in the final stage of pregnancy (24 days *post coitum*) the insecticide reaches maximum concentration in fetal brain, blood, and liver at about the same time (10–20 min after administration), indicating rapid transfer across the fetal blood–brain barrier.

Several studies of the binding of chlorinated insecticides to nervous tissue have been made by measuring the incorporation of the toxicants into subcellular fractions either directly *in vitro* or following initial application of the toxicants *in vivo* (Matsumura and Hayashi, 1966, 1969*a,b*; Hatanaka *et al.*, 1967; Brunnert and Matsumura, 1969; Telford and Matsumura, 1970). Some of these investigations are interesting in relation to the problem of insect resistance, but the present discussion is confined to interactions with intact nerve tissue.

Ray (1963) examined the uptake of dieldrin by the nerve cords of susceptible and resistant German cockroaches (*Blattella germanica*) following optical application of the toxicant and concluded that there was no difference between the strains in their rates of uptake. For the resistant strain, which tolerated much more dieldrin, it was shown that the amount of dieldrin entering the nerve cord increased with the amount applied externally, so that there was no evidence for a dieldrin-impermeable barrier surrounding the nerve cord of this strain. Matsumura and Hayashi (1969*a*) also compared a susceptible and a resistant strain of this insect and concluded that the nerve cord of the resistant one could tolerate much more internal dieldrin. Generally similar results were obtained for houseflies by Schaefer and Sun (1967), who found no significant differences between susceptible and resistant strains in the rate of uptake of [^{14}C]dieldrin by their nerve cords, either *in vivo* or *in vitro*. For the resistant strain a 1250-fold increase (to 25 μg) in the injected dose of dieldrin resulted after 24 hr in a 525-fold increase in the blood level and a 430-fold increase in the CNS content of dieldrin, compared to the corresponding levels when 0.02 μg was injected. These figures were obtained by the analysis of individuals (50% of the group) which survived the treatment, clearly showing that large quantities of dieldrin can enter the nerve cord and be tolerated by these insects.

Isolated American cockroach nerve cords exposed to 10^{-5} M [^{14}C]DDT for various periods absorbed it rapidly (30% of the total uptake) during the first minute, after which a slower phase of uptake followed first-order kinetics. The outward diffusion followed similar laws when nerve cords previously exposed to the same concentration of [^{14}C]DDT were transferred into fresh saline (Matsumura and O'Brien, 1966). As the concentration of [^{14}C]DDT in the external medium was increased, the plot of concentration vs. absorption by the cord exhibited plateaus corresponding to the saturation of different nerve cord phases at about 5×10^{-7} M and 2×10^{-5} M. Beyond about 6×10^{-5} M, the uptake became linear with increasing concentration. These and other observations are consistent with the formation of charge-transfer complexes between DDT and the nerve cord (O'Brien and Matsumura, 1964). Holan (1969) also advocates the involvement of charge-transfer complex formation in the mode of action of DDT and points out that the dissociation of such complexes with increasing temperature may explain the well-known negative temperature coefficient of insect intoxication by DDT.

Distribution studies involving tissue extraction or whole body autoradiography have shown that [^{14}C]α-HCH or [^{14}C]γ-HCH administered intraperitoneally to rats or mice (Koransky *et al.*, 1963; Koransky and Ullberg, 1964; Nakajima *et al.*, 1970) readily enters the CNS. In studies of the distribution of 4 μCi (22, 32, and 21 μg, respectively) of each of the [^{14}C]HCH isomers (α, β, and γ) at successive intervals for up to 72 hr after administration,

Nakajima *et al.* (1970) found a high accumulation of α-HCH in the white matter of the CNS within 5 min of dosing which persisted for more than 24 hr. The β- and γ-isomers were also taken up rapidly (but to a lesser extent than α-HCH) and their distribution was more uniform amongst brain tissues. They disappeared rapidly from this tissue.

Autoradiographic studies with whole American cockroaches (Kurihara *et al.*, 1970) showed that $[^{14}C]\beta$-HCH penetrated the cuticle slowly, whereas $[^{14}C]\gamma$-HCH had reached almost all parts of the CNS within 15 min of topical application. Surprisingly, $[^{14}C]\gamma$-HCH was found in the peripheral regions of the CNS and not inside it, whereas $[^{3}H]$nicotine penetrated rapidly and became distributed in both peripheral and internal regions. These statements imply that the radioisotope in the tissues is associated with the parent molecules. This may not necessarily be true, and it is important from a mode of action standpoint to know whether metabolites are present at the locations indicated by autoradiography. Chromatographic analysis of the *total* radioactive material recovered from these cockroaches revealed the presence of additional compounds in the case of γ-HCH and nicotine. Consequently, it cannot be concluded that all of the label at or in the CNS in these experiments was due to unchanged toxicant.

Similar comments apply to some other investigations of this type. Thus Telford and Matsumura (1971) used $[^{14}C]$dieldrin to investigate the distribution of this toxicant in the intact nerve tissues of susceptible and resistant German cockroaches. This combined electron microscopic and autoradiographic study provided some evidence for localization of the isotope on the axonic and glial cell membranes, and the authors concluded that the intensity was higher for the susceptible strain than for the resistant strain. However, the use of tritium-labeled insecticides is preferable for such purposes, since (provided that tritium exchange with endogenous compounds does not occur) the shorter path of the β-particle from this isotope provides better resolution in autoradiography and hence more specific localization of the labeled molecules. When investigating $[^{3}H]$dieldrin distribution in the thoracic ganglion of the resistant housefly following topical application to the thorax, Sellers and Guthrie (1971) found localization of the isotope in collagenlike fibers of the neural lamella and on axonic membranes in the neuropil region, but not in the motor neuron region between neural lamella and neuropil. The label clearly reached the synaptic region of the CNS without consequent harm to the resistant insect.

Since there is now evidence that dieldrin undergoes both hydroxylation and epoxide ring hydration in German and in American cockroaches (Nelson and Matsumura, 1973) and in the housefly (Nelson and Matsumura, 1973; Sellers and Guthrie, 1972), it becomes important to determine the precise nature of the labeled chemical present at these sites. Whether such hydroxylation and hydration products are detoxication or toxication products of dieldrin is open to question, since there is evidence that some of them, especially *cis*- and *trans*-6,7-dihydroxydihydroaldrin, have neuroactive effects on the nerve tissue of several species (Wang *et al.*, 1971; Matsumura, 1972; Akkermans *et al.*, 1973; Burt *et al.*, 1974). Furthermore, the American cockroach nerve cord and housefly thoracic ganglion contain hydrase enzymes which can open the epoxide ring of a dieldrin analogue *in vitro* (Brooks, unpublished results) and could conceivably attack dieldrin itself. This has not been demonstrated, but it would concur with the previously mentioned metabolism studies *in vivo* (Nelson and Matsumura, 1973; Sellers and Guthrie, 1972). Epoxide ring hydration apparently was not found during investigations of the fate of [^{14}C]dieldrin in housefly ganglia (Schaefer and Sun, 1967).

In another autoradiographic study, Coons and Guthrie (1972) examined the distribution of [^{3}H]DDT in the nervous system following its injection into female *P. americana* and sixth instar larvae of tobacco budworm (*Heliothis virescens*). The isotope was observed to be localized in the neural lamella of the ganglia of both species and in the collagenlike fibers beneath, but there was no apparent deposition in the nerve cell or glial cell bodies of either insect. There appeared to be discontinuous localization of the isotope on the membranes of axons and glial cell extensions in the neuropile of *H. virescens* ganglia (but not on those of *P. americana* ganglia), and interganglionic connectives of *H. virescens* showed the same deposition pattern as the ganglia. Thus the nervous system of the tobacco budworm is readily penetrated and the [^{3}H]DDT and/or its metabolites become localized on the nerve membranes. The situation with the cockroach is less clear, but few microtubules and neurotubules could be observed in these preparations, and there is a possibility that technical difficulties prevented observation of the true distribution in the CNS of this insect.

3. Distribution

Section 2 was mainly concerned with the permeability properties of various biological membranes and their effects on the transfer of insecticides. These effects, together with the processes of biotransformation (and excretion), result in a series of dynamic equilibria between the tissues and the circulatory system. In mammals, with their closed circulatory system and separate blood supplies to the different tissues and organs, the distribution patterns observed may have particularly complex origins and can be greatly influenced by the route of administration. In contrast, insects have an open circulation which bathes the tissues and organs with a common blood supply, and the situation might be expected to be somewhat simpler than in vertebrates. However, it will be evident from the previous section that the movement of insecticides in the insect body is by no means clearly understood. A further complication in distribution studies is that the aforementioned equilibria observed *in vivo* may be profoundly affected by factors such as the age, sex, and nutritional status of the animal and the presence in the tissues of xenobiotics other than the one administered which may cause induction or inhibition of the drug-metabolizing enzymes. The influence of these factors on pesticide toxicity has been reviewed by Durham (1967) and is discussed in Chapter 15.

3.1. Methodology

The methodology used in the determination of insecticide distribution *in vivo* varies considerably, depending on the experimental objective. Radiolabeled insecticides are almost indispensable for such studies (Casida, 1969) and there is a voluminous literature on their use. Summaries of a large number of these investigations have been provided by Menzie (1969) for pesticides in general, and the accounts by Chasseaud (1970, 1972) provide concise summaries of recent work on pesticides in mammals. The literature on cyclodiene insecticide metabolism has been reviewed by Brooks (1969, 1972*a*,*b*, 1974).

The primary objective in these studies is to obtain a complete and balanced account of the distribution of the insecticide in the body at various times after its administration to the animal *in vivo*. To this end, groups of animals are killed at appropriate times after administration of the toxicant and individual tissues are processed for analysis. Tissues may be subjected to various extraction procedures with organic solvents, or completely combusted to isotope-labeled gases, or dissolved directly in various media for direct radioassay in a scintillation spectrometer, or combinations of these techniques may be used. Urine and feces are collected and processed similarly. Further information may be gained by such methods as cannulation of the bile duct to demonstrate biliary excretion, cannulation of the thoracic duct to demonstrate transfer via the lymphatics, or perfusion of intact livers with insecticide-loaded blood to investigate hepatic biotransformations. In this way, the fate of a single dose of an insecticide may be investigated, but more valuable information can be obtained by long-term daily oral administration, which permits equilibrium to be established between intake, distribution, and excretion. With chlorinated insecticides, for example, the daily collection and analysis of urine and feces reveal a time after which the radioisotope excreted daily balances that ingested daily. This situation indicates the attainment of steady-state concentrations in the tissues, and the difference between the administered and excreted totals of isotope reveals the amount of insecticide (plus metabolites) retained in the body. If administration is now stopped, the decline of this steady-state internal level of insecticide can be monitored and the half-life of the insecticide in the body can be calculated from the derived plot of excretion vs. time.

The application of these techniques to insects is obviously limited by their size, and much of the earlier experimentation measured only whole body levels of toxicants. However, investigations with insects have become increasingly sophisticated in recent years, so that the measurement of toxicants and their metabolites in individual tissues and organs is quite common. Details of the methods used are to be found in the references relating to Section 2.

3.2. Distribution in Vertebrates

3.2.1. Compartmental Analysis

In spite of the complexity of the pharmacokinetic process, a number of experimental observations relating to drug behavior can be rationalized if the animal is treated as a number of connecting compartments which may hold and/or metabolize the drugs (Atkins, 1969). The compartments are considered to be related in such a manner that intercompartmental transfer or turnover through metabolism are first-order processes (i.e., transfer is by simple diffusion through membranes and metabolic rates are proportioned to the concentration of the drug in a compartment). In the catenary compartmental model, 1, 2, . . . , *n*, compartments are linearly connected, whereas in the mammillary model the individual compartments each connect with a central compartment (Fig. 8). For simple model systems with few compartments, when a steady state is reached

$$\frac{dP_n}{dt} = k_{(n-1)n}p_{n-1} - k_{n(n+1)}p_n \tag{11}$$

where P_n is the amount of pesticide in compartment n and $k_{(n-1)n}$ and $k_{n(n+1)}$ are the rate constants for transfer of P into and out of this compartment. The general solution of this equation for P_n at time t is of the form

$$P = A - \Sigma B_n e^{-\lambda_n t} \tag{12}$$

where A, B, and λ are constants.

Investigations of the distribution behavior of chlorinated insecticides by Robinson and his associates (Robinson, 1967, 1969*a*,*b*, 1970*a*,*b*; Robinson and Roberts, 1968; Robinson *et al.*, 1967, 1969) provide a good example of the application of compartmental analysis to pharmacokinetic problems. Their long-term feeding studies with organochlorine insecticides have examined (a) the relationship between the dietary intake of an insecticide and its concentration in tissues, (b) the relationships between the concentrations of the insecticide occurring in the different tissues, (c) the changes in the tissue concentra-

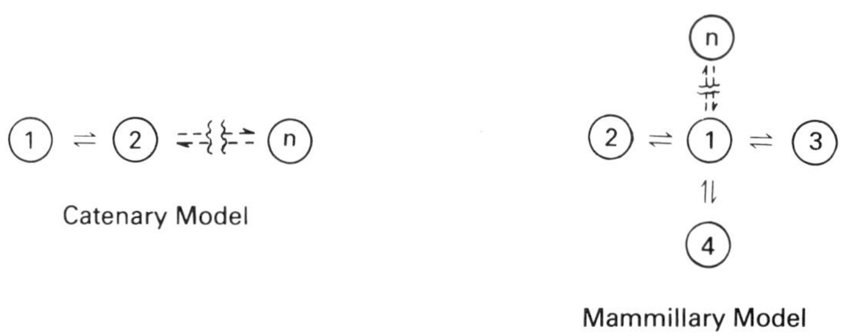

Fig. 8. Two types of mathematical models used in the study of drug pharmacokinetics.

tions as the time of dietary exposure increases, and (d) the changes in tissue concentrations when dietary intake ceases. From a consideration of the information of this kind available for several organochlorine insecticides (especially DDT and dieldrin) in vertebrates, Robinson (1967) concluded that (a) the concentration of an organochlorine insecticide in a particular tissue is a function of the daily intake, (b) there is a functional relationship between the concentrations in different tissues, (c) the concentrations in the tissues depend on the time of exposure, and (d) when exposure ceases the concentrations in the tissues decline at rates proportional to the concentrations in the tissues at any given time. Conclusion (b) is of importance, since when the nature of the relationship is known it permits the determination of concentrations in other tissues from, say, the concentration in blood. In (c), the relationship is curvilinear rather than rectilinear, indicating (for a specified daily intake) that tissue concentrations do not increase indefinitely but approach an upper limit with increasing time of exposure. This relationship, which is of great importance, arises from a combination of (a) with (d), the latter being the familiar first-order excretion process (though not always applicable in this simple form).

These four conclusions or postulates are consistent with the concept of the mammal as a two-compartment mammillary system in which a central compartment consisting of the blood and probably the liver (in which the toxicant can be metabolized) is in contact with a peripheral inert storage compartment (adipose tissue), in which no metabolism occurs. This is the simplest form of mammillary model. While it is essential that the circulating blood be considered as the central compartment, the compartments are otherwise conceptual and are not necessarily to be equated with particular tissues. Furthermore, this simple model is applicable only in the case of chronic exposure; the situation is likely to be more complex for acute exposures. In Fig. 9, C_1 and C_2 are the

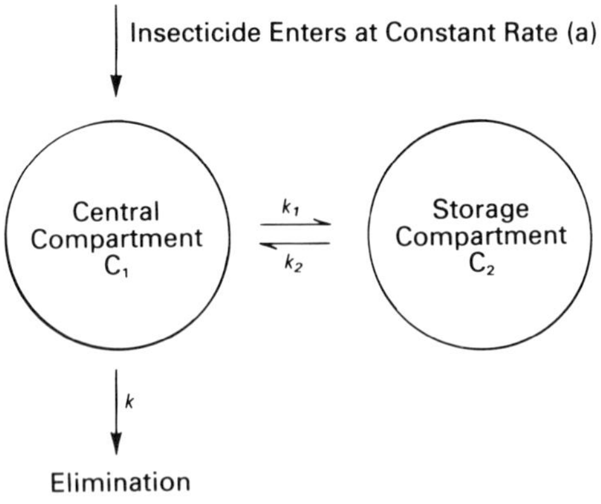

Fig. 9. Simple two-compartment mammillary model applied to the pharmacokinetics of an insecticide ingested and excreted by a mammal.

concentrations of insecticide in the respective compartments (assumed to be of constant volume during the experimental period) and k_1, k_2, and k are the rate constants for transfer between the compartments and for elimination, respectively. Then

$$\frac{dC_1}{dt} = a - k_1C_1 + k_2C_2 - kC_1 \tag{13}$$

and

$$\frac{dC_2}{dt} = k_1C_1 - k_2C_2 \tag{14}$$

Since there are two compartments, the general solutions of these equations for C_1 and C_2 each contain two exponential terms, in accordance with equation (12). Assuming the establishment of equilibrium in long-term feeding trials, equation (13) can be simplified to

$$\frac{dC_1}{dt} = a - kC_1, \qquad \text{since } k_1C_1 = k_2C_2 \tag{15}$$

Whence

$$C_1 = \frac{a}{k}(1 - e^{-kt}) + C_0\, e^{-kt} \tag{16}$$

where C_0 is the level of toxicant present at zero time. The important consequence of this equation is that C_1 approaches an upper limit of a/k as the time of exposure increases indefinitely. In some feeding trials of this type, the tissue levels have been observed to reach plateau levels and then decline. The reasons for this are unclear but may be a consequence of induction of the microsomal enzymes or of physiological changes with age that result in changes in compartmental size and/or the rate constants. The decline in toxicant level in the central compartment when exposure ceases may be represented by a similar equation:

$$C_1 = C_\infty + (C_0 - C_\infty)\, e^{-kt} \tag{17}$$

In this case, C_0 is the concentration present when exposure ceases (now regarded as zero time), C_1 that present at time t, and C_∞ the final low concentration reached at $t = \infty$. In some cases (rat adipose tissue is an example, Fig. 10), concentration decay appears to follow a simple first-order law, so that $dC_1/dt = -kC_1$ and a plot of log (concentration in the tissue) vs. time is linear.

In other cases, the situation is not so simple, and some investigators have used power function relationships (log–log plots) to express the results. Convincing arguments have been advanced (Robinson, 1967; Robinson and Roberts, 1968; Robinson *et al.*, 1969) against the use of power functions, and,

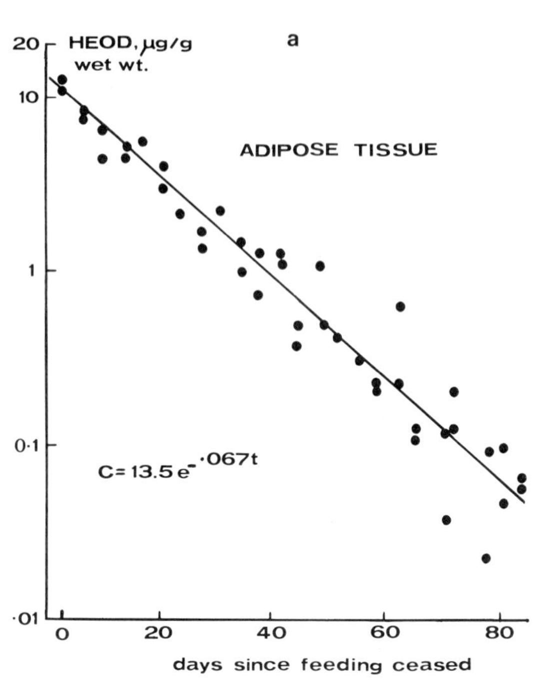

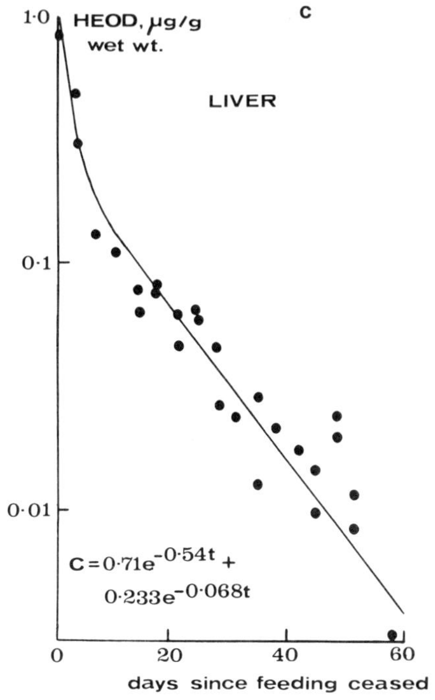

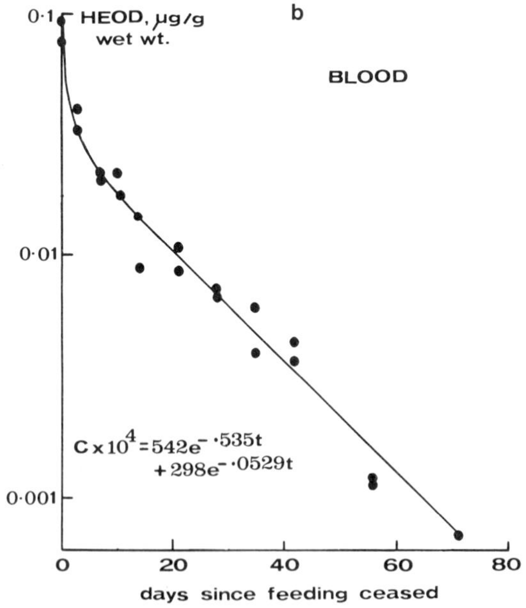

Fig. 10. Decline in HEOD (dieldrin) concentration in the tissues of rats fed 10 ppm of HEOD in the diet for 8 weeks followed by a basic diet for up to 84 days. Each point represents an individual rat. Adapted from Robinson *et al.* (1969), courtesy Pergamon Press.

as an example, the decline of dieldrin levels in the blood and liver of rats following prolonged exposure is best expressed by functions incorporating two exponential terms (Fig. 10), in accordance with the mammillary model. For both tissues, the concentration decay curves plotted on the log-linear basis are biphasic, with similar half-lives for the rapid (1.3 days) and slow (10.2 days) phases of decay. It is noteworthy that the decay constants in the first and second exponential terms for the blood decay relationship are similar to the corresponding terms in the relationship for liver. Moreover, the decline in concentration in adipose tissue (half-life 10.3 days) is adequately represented by a single exponential term (first-order decay) in which the rate constant is very similar to that for the slow phase of decline in blood and liver. The rate constants (k_1 and k_2) for transfer of dieldrin between blood and adipose tissue or between liver and adipose tissue (assumed to be the second compartment) were not significantly different (Robinson *et al.*, 1969), from which it appears that liver and blood are part of the same central compartment.

The implication of these observations is that there is always a rapid interchange of dieldrin between blood and liver. When exposure ceases, there is initially a rapid fall in concentration in this central compartment, after which the rate of elimination from the body is controlled by the rate of transfer from the storage compartment to the blood/liver compartment. It is conceivable that the brain is part of a third compartment, but the investigators (Robinson *et al.*, 1969) were unable to reach definite conclusions based on the small amount of data available.

The relationships discussed above appear to be generally applicable to the distribution of chlorinated insecticides in vertebrate tissues, including those of man (Hunter *et al.*, 1969). As a consequence of the functional relationship between daily intake of chlorinated insecticide and its concentration in the tissues, this intake can be assessed by analyzing samples of blood or adipose tissue. Thus, for dieldrin in man (Hunter *et al.*, 1969), the average total equivalent intake of HEOD (dieldrin) in μg/day is given by

mean (geometric) concentration of HEOD in blood (μg/ml) $\div$ 0.000086

or

mean (geometric) concentration of HEOD in adipose tissue (μg/g) $\div$ 0.0185

Furthermore, the functional relationships between the insecticide concentrations in the various tissues enable the concentration in a particular tissue to be calculated from that in blood or adipose tissue. Figure 11 shows some of these relationships for DDT-type compounds and for dieldrin in man (de Vlieger *et al.*, 1968).

3.2.2. Effects of Biotransformation

The principles of foreign compound biotransformation in vertebrates are now fairly well understood and have recently been discussed in various books

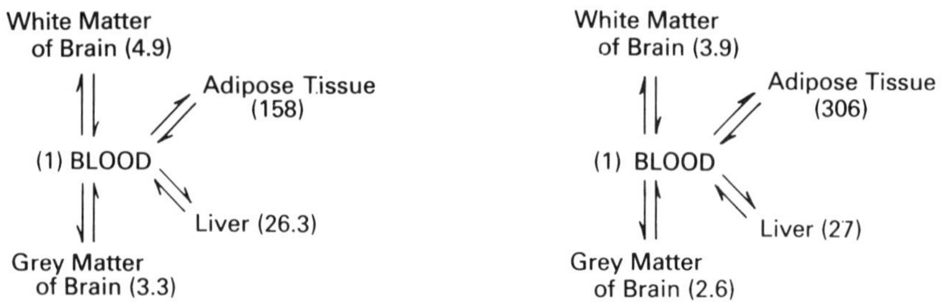

Dieldrin (HEOD) **DDT-Type Compounds**

Fig. 11. Distribution of HEOD (dieldrin) and DDT-type compounds between blood and other tissues of man. Tissue concentrations in parentheses are relative to those in blood. Data of de Vlieger *et al.* (1968).

and reviews for xenobiotics in general (Parke, 1968; Hathway *et al.*, 1970, 1972) and for pesticides in particular (Brooks, 1972*a*,*b*, 1974). Various aspects of this important topic are detailed in later chapters of this book. Therefore, it will be sufficient to state here that biotransformations consist of (1) "phase I" reactions, which are mainly oxidative, hydrolytic, or reductive, may be toxicative or detoxicative, and usually (but not invariably) make the parent molecules more polar, and (2) "phase II" reactions, in which polar groups such as hydroxyl introduced in phase I are combined with endogenous molecules such as sugars, glucuronic acid, and glutathione, to give water-soluble conjugates which can be excreted. In mammals, a major site for biotransformation is the endoplasmic reticulum of hepatic cells, but activity is also found in mitochondria, in the soluble fractions of cells and in blood plasma.

Some of the consequences of these transformations were mentioned in connection with transfer phenomena in Sections 2.1 and 2.2. Biotransformations may influence the overall pharmacokinetic process by producing concentration gradients across membranes (for example, if a metabolite is readily removed from the inner side of a membrane) and by bringing the distribution of stable metabolites (some of which may themselves be bioactive or toxic) into consideration. Referring again to the situation with chlorinated insecticides, it is evident from equations (16) and (17) that metabolic processes which lead to their eventual excretion are vitally important in preventing indefinite buildup in the tissues. It follows from the pharmacokinetic principles discussed in the previous section that animal populations exposed to constant concentrations of these insecticides in the diet will eventually achieve plateau levels in their tissues and these levels will rise or fall with fluctuations in the dietary exposure. This explains why there appears to have been little change in the residues of chlorinated insecticides found in human adipose tissue in the United States and the United Kingdom between about 1950 and 1965, and a tendency for these levels to fall in recent years in consequence of the restrictions placed on the use of these compounds in the early 1960s (Robinson, 1970*b*).

Insecticides present in the tissues can have important effects on biotransformation processes, by either inhibiting or stimulating the enzymes metabolizing other incoming insecticides. If the biotransformation of a particular insecticide results in toxication (for example, the phosphorothioate to phosphate conversion of organophosphorus insecticides), its inhibition may protect the organism, whereas its stimulation may result in increased toxicity for a given dose. Such interactions are considered elsewhere in this book.

Street (1969) found that when DDT and some other chlorinated insecticides and drugs were coadministered with dieldrin to rats there was reduced storage of dieldrin in adipose tissue and the excretion of hydrophilic metabolites was accelerated. Furthermore, dieldrin fed to rats at 200 ppm in the diet for 4 days induced a sixfold increase in the rate of *in vitro* liver microsomal metabolism of dieldrin, and in rats but not in mice pretreatment with small doses of dieldrin protected against subsequent larger doses (Wright *et al.*, 1972). In the compartmental analysis of dieldrin distribution in rats discussed in the previous section, the overall elimination constant (*k of* equation 15) for the elimination of dieldrin from the central compartment will include a rate constant for the elimination of dieldrin metabolites, as well as for unchanged dieldrin. Clearly, the metabolism rate constant may increase during long-term feeding experiments, with consequent effect on the analysis.

Street's observation (1968) that inhibitors of protein synthesis which are expected to block enzyme induction *in vivo* did not alter the action of DDT on dieldrin storage suggests the involvement of factors other than induction of the microsomal drug metabolizing enzymes. Thus it may be that the rate-limiting step in dieldrin metabolism is its release from storage sites (plasma proteins, liver proteins, adipose tissue, etc.) and consequent availability for metabolism, rather than the actual levels of the enzymes themselves. This view is supported by the pharmacokinetic analysis of dieldrin elimination by rats (Robinson *et al.*, 1969) discussed in Section 3.2.1, which indicates that dieldrin mobilization from the peripheral compartment (adipose tissue, etc.) into the central compartment is the rate-limiting step in its elimination. Accordingly, the relative affinities of the compounds competing for the various storage sites will be important, apart from any effects they may have on the levels of microsomal enzymes. The implication is that DDT has a higher affinity than dieldrin for mutually available storage sites and can displace it from these sites.

These findings have suggested several ways of reducing the levels of chlorinated insecticides in animal fat (Street, 1969). The rate of absorption of the orally ingested compounds from the gut is very dependent on the nature of the gut contents, and the use of charcoal to interrupt enterohepatic cycling of dieldrin (Wilson and Cook, 1970), thereby enhancing its excretion rate, was mentioned in Section 2.2.1. A second method would involve the administration of suitable drugs to induce increased microsomal enzyme activity and a third would involve a reduction in the size of fat depots by starvation or stimulation of the thyroid. The last method may be complicated by apparent differences between the mechanisms for the mobilization of fat and for the pesticide it contains, since Baron and Walton (1971) found that fat mobilization may actually concentrate the pesticide in fat depots before it is finally released.

3.3. Distribution in Insects

In contrast to the situation for mammals, our knowledge of the way in which insecticides are distributed in insects is rather rudimentary, although the same basic principles apply. Some of the problems attending studies of the mode of penetration of insecticides through insect membranes are apparent from the discussion in Sections 2.1 and 2.2. The balance among penetration, biotransformation, and excretion rates will determine the total level of an insecticide in the tissues at any time, and insect toxicologists are particularly concerned with the way in which this level is related to that at the site of action. Although many biochemical details remain to be elucidated, the mechanisms of biotransformation are evidently quite similar to those outlined for vertebrates in Section 3.2.2. Microsomal oxidase activity is particularly important (Wilkinson and Brattsten, 1972) and is found in several insect tissues, especially fat body, gut, and Malpighian tubules. Since an insecticide entering the hemolymph should be readily transported to all of these tissues, the opportunities for rapid metabolism should really be greater in insects than in vertebrates, for which routing through the hepatic circulation is of major importance. The remainder of this section will consider some representative investigations which have attempted to relate the distribution of insecticides in insect tissues to the toxic effects observed.

Burt and Lord (1968) found that 75% of the LD_{90} (2.6 μg) of diazoxon topically applied to American cockroaches penetrated the cuticle within 2 hr and that a maximum concentration of 1.4 μM was achieved in total body fluids after 1 hr (allowing for adsorption of 40% of the internal diazoxon on body solids). The uptake of diazoxon on cockroach solids was approximated from separate investigations of its partitioning between the solid and aqueous phases of whole-body homogenates in buffer at pH 7.0. Signs of poisoning appeared after about 1 hr, when between one-third and one-half of the dose had entered and the internal amount of diazoxon was maximal (one-fifth of the dose applied); the internal level declined thereafter, as the diazoxon was removed by detoxication. Analysis of the hemolymph indicated a median concentration of diazoxon of 1.8 μM when the internal diazoxon was maximal, suggesting that hemolymph and total body fluids are in approximate equilibrium with regard to diazoxon distribution. According to measurements made on the exposed nerve cord, the stage of poisoning reached in intact cockroaches after 1–2 hr was simulated by irrigating the nerve cord for this same time with 0.6–1.0μM diazoxon solutions. The similar values of these effective concentrations, as measured *in vivo* and *in vitro*, suggests that diazoxon reaches the nerve cord via the hemolymph and that the concentrations in these two tissues are in equilibrium.

In contrast to these results for diazoxon (which is a poor lipophil), only 30% of the LD_{95} (0.5 μg per insect) of the highly lipophilic pyrethrin I penetrated in 2 hr (Burt *et al.*, 1971). After 1 hr, when the insects were badly affected, 20% of the applied dose had penetrated and about half of this had been eliminated, leaving an internal level which then remained constant for a long time. Thus diazoxon penetrates the cuticle rapidly and is metabolized rather rapidly, whereas pyrethrin I penetrates slowly but is metabolized slowly, so that the tissues are exposed to it for a much longer period of time. Assuming uniform distribution in the insect tissues, the internal concentration of pyrethrin I 1 hr after treatment was about one-tenth (0.2 μM) that of diazoxon after the same interval. Pyrethrin I is very strongly, but reversibly, adsorbed by cockroach solids (solid–aqueous ratio $3 \times 10^4 : 1$) in contrast to diazoxon (3.7 : 1). This leads to estimated concentrations of about 1–5 μM in both insect solids and body fluids 1 hr after treatment with the LD_{95} for diazoxon. The concentration of pyrethrin I in the solids would be similar, but its concentration in the hemolymph would be very low (4×10^{-11} M). If the affinity of these toxicants for nerve cord tissue is similar to that for whole-body solids (as was shown to be approximately the case for pyrethrin I) and if adsorption by this tissue is a reflection of the amounts of the toxicants required at the target, then their intrinsic toxicities are about the same, although the LD_{95} for diazoxon is fivefold higher. This difference may be related to the rapid detoxication of diazoxon during the critical period.

The actual concentrations of pyrethrin I in hemolymph and nerve cord were below the respective detection limits of 0.02 μM and 0.5 μM during this period (calculated concentration in wet weight of cord about 10^{-7} M). Also, the calculated hemolymph concentration (4×10^{-11} M) of pyrethrin I during the critical phase of intoxication appeared too low to account for poisoning since 10–100 times this concentration in saline was required to produce effects on exposed nerve cords (Burt and Goodchild, 1971). However, adsorption on hemolymph solids may have considerable influence *in vivo* (Burt *et al.*, 1971), and since the presence of proteins also markedly increases the partitioning of lipophilic insecticides into the aqueous phase (Olson, 1973) the actual amount of pyrethrin I carried in hemolymph may be considerably higher. Much depends on the way in which the lipophilic insecticide is transferred from the hemolymph or its component solids to the solids of the nerve cord. Since the internal concentration of pyrethrin I reaches a steady state and the hemolymph is circulating rapidly, a low steady-state concentration in the hemolymph may actually correspond to a high turnover of toxicant between cuticle and nerve cord. The experiment with diazoxon clearly implicates the hemolymph in the transport of this toxicant. For pyrethrin I, the situation is less clear, but Burt *et al.* (1971) found no strong evidence for the involvement of the tracheae in the penetration of this compound.

Studies of the pharmacokinetics of pyrethroids in mustard beetles (*Phaedon cochleariae*) provide useful information about the relationship among the penetration, metabolism, and toxicity of these insecticides (Elliott *et al.*, 1970). A 24-hr comparison of the penetration of 4-allylbenzyl-(±)-*cis*-*trans*-chrysanthemate (ABC) and 2,6-dimethyl-4-allylbenzyl-(±)-)-*cis*-*trans*-chrysanthemate (DMABC) at five doses ranging from 0.1 to 25.6 μg per insect revealed that the less toxic compound ABC was more rapidly lost from the surface at all doses. Moreover, the plots of log (percent applied dose unpenetrated) vs. time in accordance with equation (2) were similar for all these doses of a particular compound, so that a single penetration curve for each represented the behavior at all doses. This implies strict proportionality between dose applied and penetration for the range of doses used for each compound, so that the limitations indicated by equation (5) had not been reached.

Similar results were obtained with 21 other methyl-substituted benzylchrysanthemates, so that for doses of 4 μg and 40 μg, generalized penetration curves illustrated their penetration behavior. The penetration curves at moderate doses consisted (Fig. 3) of an initial phase in which 40% of the dose left the surface within seconds of application, an intermediate phase in which the rate of loss became slower, and a third phase of loss according to first-order kinetics [equation (2)]. The rapid initial loss was seen when the cuticles were rinsed with methanol or acetone, but recovery was higher with hexane, indicating that the initial phase corresponds to a rapid transfer of insecticide into areas of the cuticle inaccessible to the more polar rinsing solvents. In the absence of metabolism, the rate of inward diffusion decreases as the external concentration decreases, so that equilibrium is eventually established between external toxicant, toxicant in the inner layers of the integument, and toxicant in the tissues. This results in an internal level of toxicant that is stationary with time, as seen when benzyl-(±)-*cis*-*trans*-chrysanthemate (BC) is applied to dead insects, in which degradation is virtually absent. Similar results were found with metabolically stable chlorinated insecticides such as dieldrin and the heptachlor epoxides in houseflies (Brooks, 1966). Insects can survive for extended periods in an atmosphere of nitrogen, and this treatment arrests insecticide metabolism due to microsomal oxidation (Brooks *et al.*, 1963). The effect of this treatment on the penetration of BC is seen in Fig. 12. An internal plateau level is achieved and the penetration is slower. However, the fact that penetration continues implies that some degree of metabolism still occurs, possibly by hydrolysis.

The second phase of penetration appears to correspond to the approach to the internal plateau level of toxicant. If a metabolic process is present, a steady state will first be achieved in which metabolic rate balances penetration rate. Then, as metabolism of the internal toxicant continues, the external toxicant remaining eventually becomes insufficient to maintain the steady-state internal

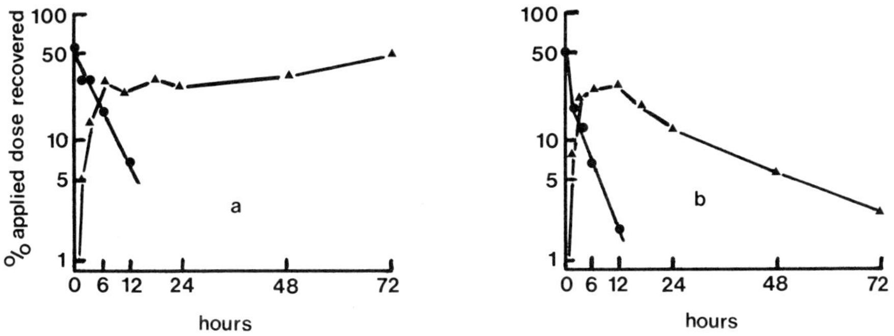

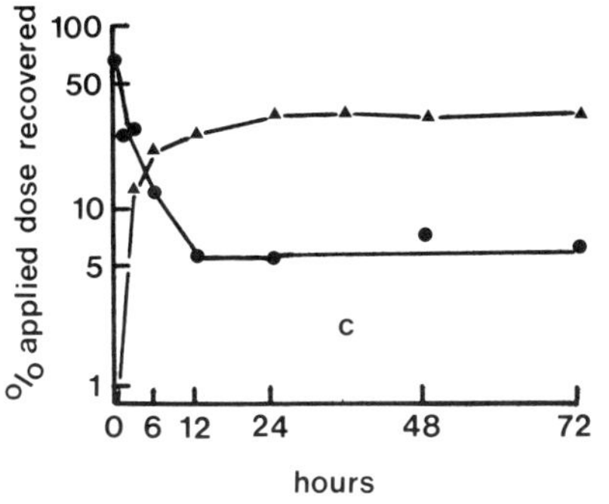

Fig. 12. Penetration of benzyl-(±)-*cis-trans*-chrysanthemate (BC) (4 μg/insect) into (a) mustard beetles in nitrogen, (b) beetles in air, (c) dead beetles. Symbols: ●, external rinse; ▲, internal extract. (Data of Elliott *et al.,* 1970.)

level, which therefore begins to decline from its maximum value. This is seen for BC in Fig. 12. However, even for BC, which is metabolized more rapidly than some others, the internal level remains constant for several hours after the application of 4 μg per insect. Recent investigations of the phenomenon of knockdown by pyrethroids indicate that the effect probably results from action on the CNS and is related to their rate of penetration (Burt and Goodchild, 1974; Ford and Pert, 1974).

For a number of compounds, the linear third phase of the penetration curve corresponds to the period of constant internal concentration, and the slope of this portion is therefore a measure of both penetration and metabolic rates. In general, it was found that steeper slopes (higher detoxication rates) were correlated with low toxicities. With the compound BC as a standard (slope 1.0), the relative slopes for ABC and DMABC were 0.48 and 0, respectively, DMABC being highly toxic to mustard beetles. However, the toxicities of ABC and DMABC were increased (23-fold for ABC, 2.3-fold for DMABC) to quite similar final levels by coapplication with a mixture of the synergists sesamex and DEF. This suggests that their intrinsic insecticidal activities are similar and that much of the observed difference in toxicities is due to detoxication. With the same combination, 2,3-dimethylbenzyl-(±)-*cis-trans*-chrysanthemate (2,3-DMBC) was synergized 24-fold, giving it a higher final toxicity than the other two compounds.

Since metabolism reduces the internal concentration of toxicant, it increases the concentration gradient between the outside and inside of the insect and accelerates the penetration of labile compounds such as ABC. When metabolism is relatively slow, as for pyrethrin I in cockroaches and DMABC in mustard beetles, the toxicant accumulates in the tissues and the metabolic rate may be rate limiting for penetration. The situation superficially resembles a series of consecutive first-order reactions (Glasstone, 1948) in which different compounds arise from a precursor [compare equations (11) and (12) in Section 3.2.1]:

$$\underset{C_1}{①} \xrightarrow{k_1} \underset{C_2}{②} \xrightarrow{k_2} \underset{C_3}{③} \tag{18}$$

In equation (18), C_1 and C_2 are the respective amounts of external and internal parent toxicant after time t, and C_3 is the amount of a metabolite, which may be either a toxication or a detoxication product.

Since $C_1 + C_2 + C_3 = a$ (the dose applied) and

$$\frac{dC_2}{dt} = k_1C_1 - k_2C_2 \tag{19}$$

it can be shown that

$$C_2 = a \cdot \frac{k_1}{k_2 - k_1}(e^{-k_1t} - e^{-k_2t}) \tag{20}$$

It follows from equation (19) that C_2 increases while the penetration rate k_1C_1 exceeds the conversion rate k_2C_2, reaches a maximum value when these two rates become equal ($dC_2/dt = 0$), and declines when k_2C_2 exceeds k_1C_1. The formation of the metabolite will approximate to a first-order reaction provided that C_2 is insufficient to saturate the enzyme. Clearly, modifications of the rate constants k_1 and k_2 can be critical for survival of the insect. Insecticide synergists, for example, increase C_2 by modifying k_2 and can affect penetration through the cuticle by the concentration feedback effect. According to Sun and Johnson (1972), they may also increase the penetration rate constant, k_1, for some insecticides by increasing the permeability of the cuticle. This effect, which was termed "quasisynergism" to distinguish it from synergistic effects due to metabolic inhibition, causes a rapid buildup (C_2) in the tissues (k_1C_1 large) so that intoxication occurs before the metabolic process (k_2C_2) can have any useful protective effect. Conversely, any factor which increases k_2 for a slowly penetrating toxicant will tend to protect the insect if this conversion is detoxicative.

Sun (1968) has used a graphical method to illustrate some of these phenomena. If a toxicant penetrates the cuticle at a rate P, is activated at a rate

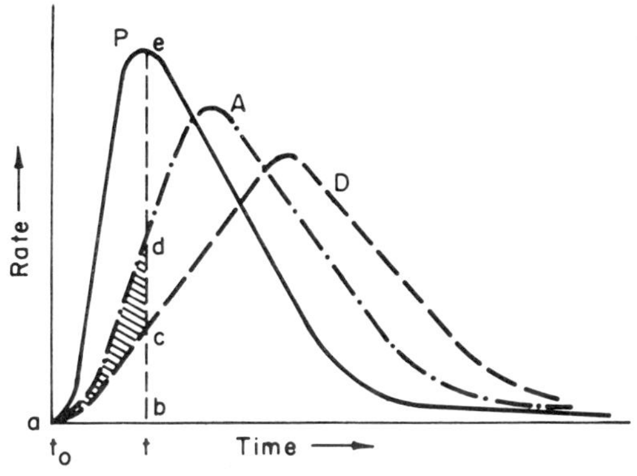

Fig. 13. Hypothetical curves for the penetration (*P*), activation (*A*), and detoxication (*D*) of an insecticide, each being considered as a first-order process. Adapted from Sun (1968).

A, and is detoxified at a rate *D*, all being first-order processes, then a series of maxima will be observed on a plot of rate vs. time, as the rate of each process passes through a maximum and then declines (Fig. 13). The amount of each process which has occurred at time *t* is the area under the appropriate curve at that time and so the net amount of toxication product in the tissues at time *t* (Fig. 13) is given by the area *acd*. An example is the oxidative conversion of a phosphorothioate to the corresponding phosphate (the active anticholinesterase agent) *in vivo*, which is balanced by metabolism of both the phosphate and its precursor. If the phosphate itself is applied instead (curve *P*), and curve *A* (instead of curve *D*) now represents detoxication, the tissue content of the phosphate at time *t* is represented by the larger area *ade*, which shows the advantage of applying the toxic phosphate rather than its nontoxic precursor.

Rates of penetration and detoxication for various compounds are derived from the more regular plots of percent penetrated or detoxified vs. time by measuring the slopes of these curves at appropriate time intervals. The derived data are then plotted as in Fig. 13, in terms of percent penetrating or detoxified per hour vs. time after dosing. Such plots reveal the maximum penetration rate and the time required to achieve it. The data from various sources analyzed by Sun (1968) showed that organophosphorus compounds such as metaphoxide, parathion, paraoxon, and dimethoate penetrated quite rapidly whereas chlorinated insecticides (DDT, dieldrin) and carbamates (carbaryl and 3-isopropylphenyl-*N*-methylcarbamate) penetrated more slowly.

In Fig. 14A, the comparison of DDT penetration and detoxication in susceptible and resistant housefly larvae shows that no DDT accumulates in resistant larvae because detoxication rate equals penetration rate throughout the period of observation. In contrast, the susceptible strain metabolizes DDT more slowly and its tissues contain the toxicant at all times. The situation is somewhat similar for 3-isopropylphenyl-*N*-methylcarbamate applied to resistant and susceptible houseflies (Fig. 14B). For dimethoate topically applied to houseflies (Fig. 14C), a high maximum penetration rate is quickly achieved,

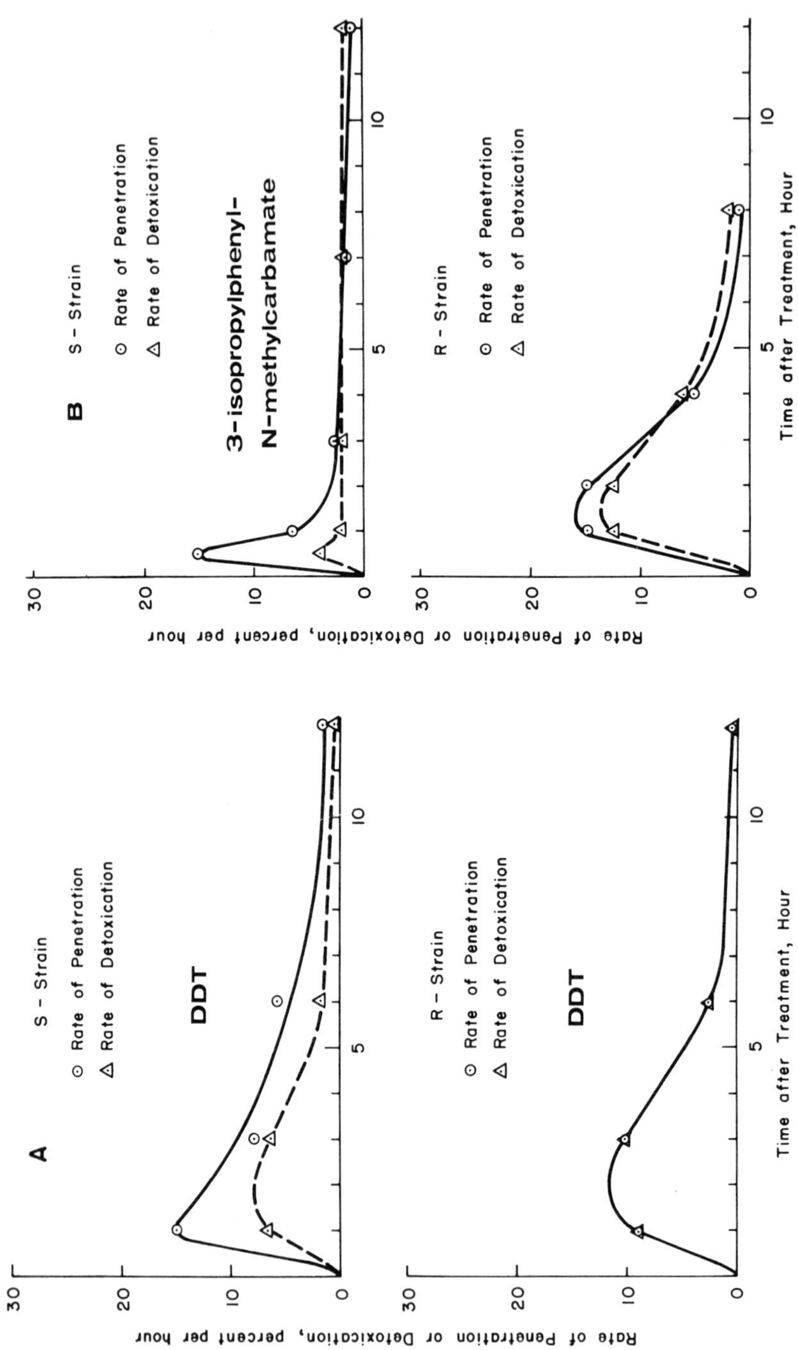

A
S - Strain
Rate of Penetration
Rate of Detoxication
DDT
R - Strain
Rate of Penetration
Rate of Detoxication
DDT
B
S - Strain
Rate of Penetration
Rate of Detoxication
3-isopropylphenyl-
N-methylcarbamate
R - Strain
Rate of Penetration
Rate of Detoxication
Rate of Penetration or Detoxication, percent per hour
Time after Treatment, Hour

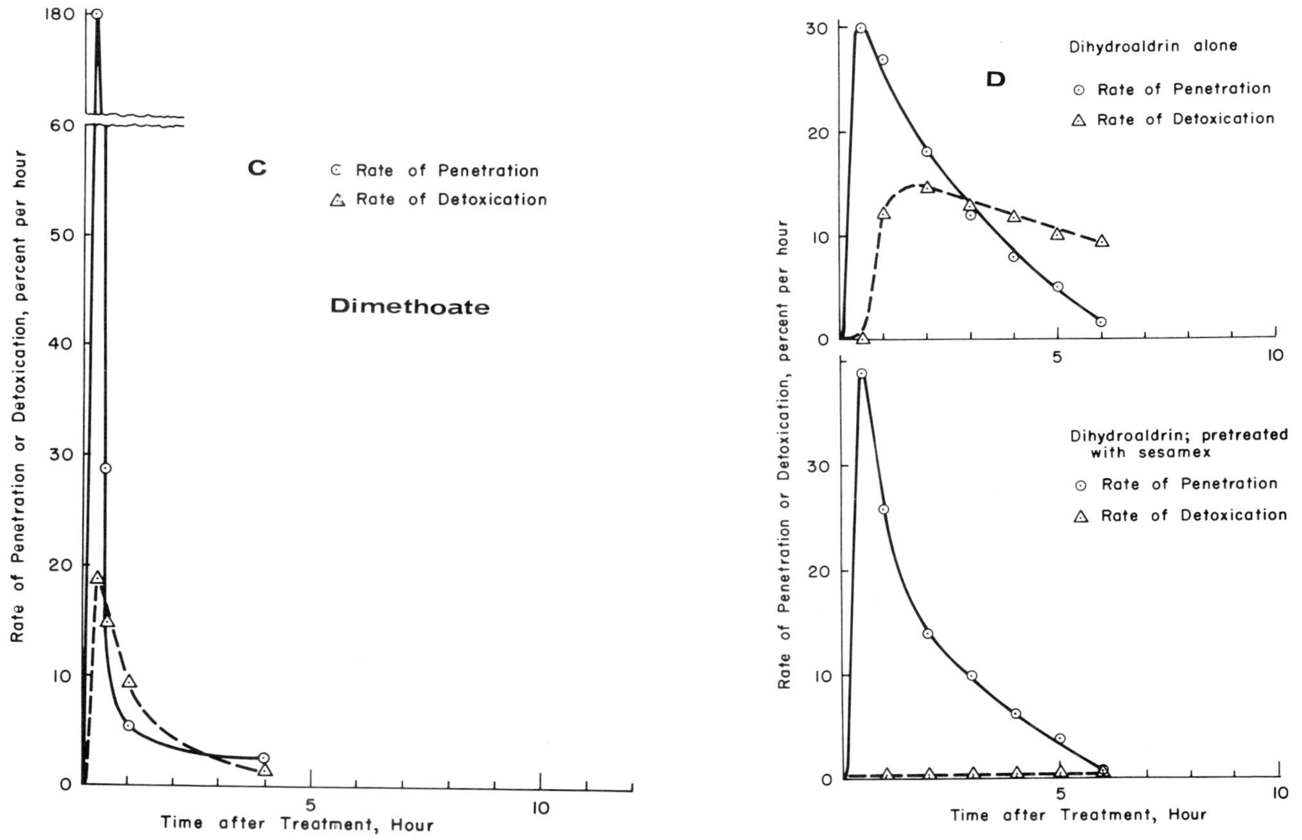

Fig. 14. Rates of penetration vs. detoxication for various insecticides in housefly larvae (A) and adults (B, C, D). A and B show the effect of enhanced detoxication in reducing internal levels of toxicant in resistant strains, C demonstrates the very rapid penetration of dimethoate, and D shows the marked action of an inhibitor of mixed function oxidases in suppressing detoxication. Adapted from Sun (1968) courtesy Entomological Society of America.

and since the detoxication rate is much lower a large quantity of toxicant appears in the tissues in a very short time. The high toxicity of dimethoate to houseflies and its rapid action are thereby explained. Figure 14D compares the uptake of 6,7-dihydroaldrin when applied to houseflies with or without the synergist sesamex, which apparently enhances its toxicity by suppressing enzymatic hydroxylation (Brooks and Harrison, 1966, 1969). Initially, sesamex increases the maximum penetration rate, but metabolism is virtually prevented and the penetration rate declines more rapidly in flies treated with the synergized combination than in flies treated with the insecticide alone.

Valuable information regarding the relationship between insecticide pharmacokinetics and toxicity is afforded by comparing housefly toxicities measured by different methods of administration and by observing the effects of synergists (Sun *et al.*, 1967; Schaefer and Sun, 1967; Sun, 1968; Sun and Johnson, 1969, 1971, 1972). Thus the water-soluble carbamate Isolan (1-isopropyl-3-methyl-5-pyrazolyl dimethylcarbamate) quickly knocks house-flies down when injected, but rapid recovery reflects the effect of efficient metabolism (detoxication) on the level of cholinesterase inhibition by this insecticide. Indeed, relatively large doses can be injected slowly (infused during 30 min) without producing either knockdown or mortality. Clearly, infusion fusion provides ample opportunity for protective detoxication, whereas rapid injection quickly loads the nervous system, producing intoxication before significant detoxication can occur. As expected, pretreatment of the flies with sesamex stabilizes quite small doses of Isolan, which then produce rapid and irreversible intoxication.

The effects of different methods of administration on the toxicities of DDT, dieldrin, and a water-soluble organophosphate (3-hydroxy-*cis*-crotonamide dimethylphosphate; SD-11319) were mentioned in Section 2.1.1. Besides the methods of topical application, injection, and infusion mentioned there, toxicities were measured following the spray application of kerosene solutions (Sun and Johnson, 1972). Dieldrin, SD-11319, and parathion had LD_{50}s for houseflies (measured by topical application in acetone) in the range 0.02–0.054 μg per insect; the LD_{50} for dieldrin was approximately doubled by injection or oil spray application, but the toxicities of the other compounds remained fairly constant. The relatively small change in toxicity associated with different methods of application appears to be quite general for organochlorine and organophosphorus insecticides (Sun and Johnson, 1972). It is attributed to the fact that dieldrin, for example, is stable and gives virtually the same end result by any method of administration; parathion penetrates rapidly and its "oxon" (paraoxon) is so toxic that detoxication effects are minimal; SD-11319 penetrates rapidly and also is relatively stable.

However, carbamates gave rather variable results when tested by the three different methods. Thus carbaryl was virtually inactive by topical application, but had an LD_{50} of about 1.0 μg per insect by each of the other methods, an increase in toxicity of more than 400-fold. Since injection bypasses the cuticle, the low toxicity by topical application is attributed to slow penetration, which affords ample opportunity for the very active detoxication mechanisms to destroy the incoming toxicant. This *natural* tolerance to topically applied carbaryl is therefore reminiscent of the situation found with 3-isopropylphenyl-*N*-methylcarbamate penetrating into *resistant* houseflies (Fig. 14B).

The results (Sun and Johnson, 1972) with oil sprays (solutions in refined kerosene) of carbaryl and other carbamates suggest that this method is akin to injection. Furthermore, 2,3,5-trimethylphenyl-*N*-methylcarbamate (SD-8786) and 2,4,5-trimethylphenyl-*N*-methylcarbamate (SD-9003) showed large increases in toxicity when 5% of kerosene was included in the acetone used for topical application of SD-8786, or when SD-9003 was formulated with five parts of the synergist thanite for topical application in acetone. The toxicities observed in these situations then approached those found when the carbamates are administered alone by either oil spray or injection. Thus kerosene appeared to act as a "synergist" and the effect of thanite was approximately equaled by that of an oil spray application of the carbamate alone. Thanite actually had an antagonistic effect on carbamate toxicity when incorporated in the oil sprays.

The position of substituent methyl groups is important, since 3,4,5-trimethylphenyl-*N*-methylcarbamate (SD-8530) showed similar toxicities by all three routes of administration and was only slightly synergized by topical application in acetone–kerosene (19 : 1). These authors concluded (although actual penetration rates were not reported) that kerosene and thanite are "quasisynergists" which increase cuticle permeability [k_1 of equation (19)] without necessarily altering the rate of metabolism (k_2). Assuming that any increased toxicity observed when synergist and toxicant are injected together should reflect only the inhibition of detoxication, while that observed on topical application should also include any effects on penetration, then

$$\begin{array}{l}\text{synergism due to}\\ \text{increased penetration}\end{array} = \frac{\text{synergism by topical application}}{\text{synergism by injection or oil spray}}$$

On this basis, sesamex was found to have relatively small effects on the penetration of several cyclodiene insecticides and analogues of Azodrin (3-hydroxy-*N*-methyl-*cis*-crotonamide dimethylphosphate) but quite large effects with the carbamates mentioned above, except SD-8530. Considering all three groups of insecticides, the synergistic factor assumed to be due to the inhibition of metabolism by sesamex ranged from less than 1 (slight antagonism) to nearly fortyfold. With carbaryl, the apparent "penetration factor" for piperonyl butoxide (23-fold) was similar to that for sesamex; that for thanite (>830-fold) was remarkably high due to its apparent antagonistic effect (synergistic factor 0.48) in the oil spray. These observations show that care is needed in interpreting the results of topical application. The choice of the vehicle for administration is clearly important and the use of acetone-kerosene solutions for topical treatments may have some advantages over acetone.

For a series of Azodrin analogues with similar intrinsic toxicities to houseflies (as indicated by their inhibition of fly brain cholinesterase), the ratio infusion toxicity/injection toxicity was positively correlated with the degree of synergism of each analogue measured for oil spray application with sesamex (Sun and Johnson, 1969). This is to be expected if the reduced toxicity of these water-soluble compounds by infusion is due to detoxication rather than to inert storage in lipoid material and if injection gives a measure of the intrinsic toxicity. In this series, unsynergized toxicities decrease markedly with the length of the *N*-alkyl substituent, which may reflect a progressive reduction in penetration rates with corresponding increases in opportunities for detoxication. For the unsubstituted parent of this series (SD-11319), which acts rapidly and is highly toxic, the dosage–mortality plots for injection and infusion are closely similar, in contrast to the situation for the lipophilic compounds dieldrin and DDT. For these, the reductions in toxicity by infusion result in flattened slopes similar to those found with resistant insects. This difference seems to be associated with the preferential adsorption of the water-insoluble chlorinated insecticides at inert sites when administration is by infusion.

4. Concluding Comments

The pharmacokinetics of insecticide behavior in both mammals and insects has been discussed in a semiquantitative way. Clearly, some method is needed by which the integrated effects (i.e., toxicity measured *in vivo*) of penetration, distribution (partitioning between the various biophases), metabolism, and final interaction with the target site can be described in terms of the molecular structure of toxicants. The general importance of toxicant partitioning between hydrophobic and aqueous phases of an insect is evident from Fig. 15, which assumes that the hemolymph is the major transporting

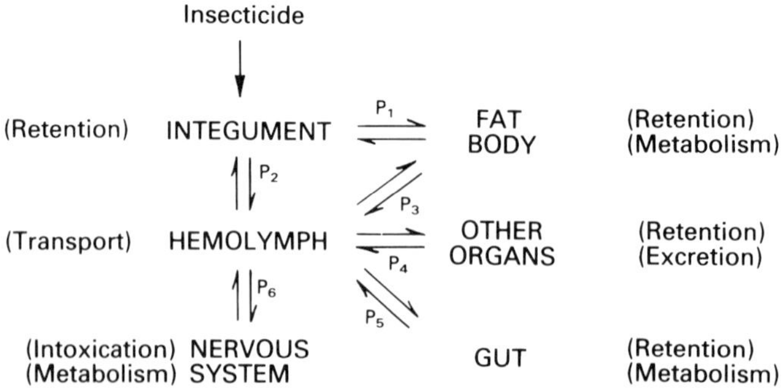

Fig. 15. Various equilibria which influence the fate of an insecticide. After Jones *et al.* (1969).

phase in all cases. For optimal toxicity, a compound must first partition readily from integument to hemolymph, and subsequently from hemolymph to nervous tissue (assuming this to be the target). Ideally, the partitioning between hemolymph and other tissues, especially those such as gut and fat body, which are known to be involved in detoxication, should be in favor of the hemolymph. Some of these requirements are mutually incompatible and the problem is well illustrated by the series of oxime carbamates discussed by Felton (1968). Efficient inhibition of cholinesterase requires a high affinity for the enzyme, which is correlated in part with lipid solubility. Unfortunately, excessive lipid solubility may retard penetration through the integument, favor distribution into nontarget tissues, and allow increased opportunities for metabolism, thus reducing the availability of the toxicant to the nervous system. Therefore, the ideal toxicant reflects a balance between these conflicting factors. The values of the various molecular parameters that will give the optimal balance for a particular class of insecticide are frequently predictable by modern mathematical techniques (Chapter 11) when a series of analogues is available for study.

5. References

Akkermans, L. M. A., van der Zalm, J. M., and van den Bercken, J., 1973, Is aldrin-transdiol the active form of the insecticide dieldrin? *Arch. Int. Pharmacodyn. Ther.* **206:**363.

Ariens, E. J., ed., 1973, *Drug Design*, Vol. IV, Academic Press, New York.

Atkins, G. L., 1969, *Multicompartment Models for Biological Systems*, Methuen, London.

Baldwin, M. K., 1971, The metabolism of the chlorinated insecticides aldrin, dieldrin, endrin and isodrin, Ph.D. thesis, University of Surrey.

Baron, R. L., and Walton, M. S., 1971, Dynamics of HEOD (dieldrin) in adipose tissue of the rat, *Toxicol. Appl. Pharmacol.* **18:**958.

Booth, G. M., and Metcalf, R. L., 1970, Histochemical evidence for localized inhibition of cholinesterase in the housefly, *Ann. Ent. Soc. Am.* **63:**197.

Brady, U. E., 1970, Localization of cholinesterase activity in housefly thoraces: inhibition of cholinesterase with organophosphate compounds, *Entomol. Exp. Appl.* **13:**423.

Brooks, G. T., 1966, Progress in metabolic studies of the cyclodiene insecticides and its relevance to structure–activity correlations, *World Rev. Pest Control* **5:**62.

Brooks, G. T., 1969, The metabolism of diene-organochlorine (cyclodiene) insecticides, *Residue Rev.* **27:**81.

Brooks, G. T., 1972*a*, The fate of chlorinated hydrocarbons in living organisms, *in: Pesticide Terminal Residues* (A. S. Tahori, ed.), pp. 111–136, Butterworths, London.

Brooks, G. T., 1972*b*, Pathways of enzymatic degradation of pesticides, *in: Environmental Quality and Safety*, Vol. 1 (F. Coulston and F. Korte, eds.), pp. 106–163, Georg Thieme, Stuttgart, Academic Press, New York.

Brooks, G. T., 1974, *Chlorinated Insecticides*, Vols. 1 and 2, Chemical Rubber Company, Cleveland.

Brooks, G. T., and Harrison, A., 1966, Metabolism of aldrin and dihydroaldrin by houseflies (*M. domestica* L.) *In vivo* and by houseflies and pig liver microsomes, *Life Sci.* **5:**2315.

Brooks, G. T., and Harrison, A., 1969, The oxidative metabolism of aldrin and dihydroaldrin by houseflies, housefly microsomes and pig liver microsomes and the effect of inhibitors, *Biochem. Pharmacol.* **18:**557.

Brooks, G. T., Harrison, A., and Cox, J. T., 1963, Significance of the epoxidation of the isomeric insecticides aldrin and isodrin by the adult housefly *in vivo*, *Nature* **197:**311.

Brown, E. W., and Scott, W. O., 1934, The comparative absorption of certain salicylate esters by the human skin, *J. Pharmacol. Exp. Ther.* **50:**373.

Brunnert, H., and Matsumura, F., 1969, Binding of 1,1,1-trichloro-2,2-di-*p*-chlorophenylethane (DDT) with subcellular fractions of rat brain, *Biochem. J.* **114:**135.

Buerger, A. A., 1966, A model for the penetration of integuments by non-electrolytes, *J. Theor. Biol.* **11:**131.

Buerger, A. A., 1967, A theory of integumental penetration, *J. Theor. Biol.* **12:**66.

Buerger, A. A., and O'Brien, R. D., 1965, Penetration of non-electrolytes through animal integuments, *J. Cell. Comp. Physiol.* **66:**227.

Burt, P. E., 1970, Biophysical aspects of nervous activity in relation to studies on the mode of action of insecticides, *Pestic. Sci.* **1:**88.

Burt, P. E., and Goodchild, R. E., 1971, The site of action of pyrethrin I in the nervous system of the cockroach *Periplaneta americana*, *Entomol. Exp. Appl.* **14:**179.

Burt, P. E., and Goodchild, R. E., 1974, Knockdown by pyrethroids: Its role in the intoxication process, *Pestic. Sci.* **5:**625.

Burt, P. E., and Lord, K. A., 1968, The influence of penetration, distribution, sorption and decomposition on the poisoning of the cockroach *Periplaneta americana* treated topically with diazinon, *Entomol. Exp. Appl.* **11:**55.

Burt, P. E., Lord, K. A., Forrest, J. M., and Goodchild, R. E., 1971, The spread of topically-applied pyrethrin I from the cuticle to the central nervous system of the cockroach *Periplaneta americana*, *Entomol. Exp. Appl.* **14:**255.

Burt, P. E., Brooks, G. T., and Goodchild, R. E., 1973, *in: Annual Report of the Rothamsted Experimental Station for 1973*, Agricultural Research Council, London, p. 171.

Casida, J. E., 1969, Radiotracer studies on metabolism, degradation, and mode of action, *Residue Rev.* **25:**149.

Casper, H. C., Pekas, J. C., and Dinusson, W. E., 1973, Gastric absorption of a pesticide (1-naphthyl *N*-methylcarbamate) in the fasted rat, *Pestic. Biochem. Physiol.* **2:**391.

Chasseaud, L. F., 1970, Processes of absorption, distribution and excretion, *in: Foreign Compound Metabolism in Mammals*, Vol. 1, pp. 1–33, Specialist Periodical Report, The Chemical Society, London.

Chasseaud, L. F., 1972, Transference of radioactively labelled foreign compounds, *in: Foreign Compound Metabolism in Mammals*, Vol. 2, pp. 62–162, Specialist Periodical Report, The Chemical Society, London.

Collins, W. J., Hornung, S. B., and Federle, P. F., 1974, The effect of topical and injected olive oil on knockdown by DDT, dieldrin, methyl parathion, and dimethoate in *Musca domestica*, *Pestic. Biochem. Physiol.* **4:**153.

Conner, W. E., 1974, Penetration of insecticides through the foregut of *Apis mellifera* L., M.S. dissertation, Cornell University, Ithaca, N.Y.

Cook, R. M., 1970, Dieldrin recycling from the blood to the gastrointestinal tract, *J. Agr. Food Chem.* **18:**434.

Coons, L. B., and Guthrie, F. E., 1972, High resolution radioautography of ^{3}H-DDT-treated insects, *J. Econ. Entomol.* **65:**1004.

de Vlieger, M., Robinson, J., Baldwin, M. K., Crabtree, A. N., and van Dijk, M. C., 1968, The organochlorine insecticide content of human tissues, *Arch. Environ. Health* **17:**759.

Durham, W. F., 1967, The interaction of pesticides with other factors, *Residue Rev.* **18:**21.

Eldefrawi, M. E., Toppozada, A., Salpeter, M. M., and O'Brien, R. D., 1968, The location of penetration barriers in the ganglia of the American cockroach, *Periplaneta americana* (L.), *J. Exp. Biol.* **48:**325.

Elliott, M., Ford, M. G., and Janes, N. F., 1970, Penetration of pyrethroid insecticides into mustard beetles (*Phaedon cochleariae*), *Pestic. Sci.* **1:**220.

Felton, J. C., 1968, Insecticidal activity of some oxime carbamates, *J. Sci. Fd. Agric. Suppl.*, pp. 32–38.

Ford, M. G., and Pert, D. R., 1974, Time–dose–response relationships of pyrethroid insecticides with special reference to knockdown, *Pestic. Sci.* **5:**635.

Gerolt, P., 1969, Mode of entry of contact insecticides, *J. Insect Physiol.* **15:**563.
Gerolt, P., 1970, The mode of entry of contact insecticides, *Pestic. Sci.* **1:**209.
Gerolt, P., 1972, Mode of entry of oxime carbamates into insects, *Pestic. Sci.* **3:**43.
Glasstone, S., 1948, *Textbook of Physical Chemistry*, 2nd ed., p. 1075, Macmillan, London.
Goodman, L. S., and Gilman, A., eds., 1965, *The Pharmacological Basis of Therapeutics*, 3rd ed., Macmillan, New York.
Hatanaka, A., Hilton, B. D., and O'Brien, R. D., 1967, The apparent binding of DDT to tissue components, *J. Agr. Food Chem.* **15:**854.
Hathway, D. E., 1965, The biochemistry of dieldrin and telodrin, Arch. Environ. Health **11:**380.
Hathway, D. E., and Amoroso, E. C., 1972, The effects of pesticides on mammalian reproduction, *in: Toxicology, Biodegradation and Efficacy of Livestock Pesticides* (M. A. Khan and W. O. Haufe, eds.), pp. 218–251, Swets and Zeitlinger, Amsterdam.
Hathway, D. E., Moss, J. A., Rose, J. A., and Williams, D. J. M., 1967, Transport of dieldrin from mother to blastocyst and from mother to foetus in pregnant rabbits, *Eur. J. Pharmacol.* **1:**167.
Hathway, D. E., Brown, S. S., Chasseaud, L. F., and Hutson, D. H., reporters, 1970, *Foreign Compound Metabolism in Mammals*, Vol. 1, Specialist Periodical Report, The Chemical Society, London.
Hathway, D. E., Brown, S. S., Chasseaud, L. F., Hutson, D. H., Moore, D. H., Sword, I. P., and Welling, P. G., reporters, 1972, *Foreign Compound Metabolism in Mammals*, Vol. 2, Specialist Periodical Report, The Chemical Society, London.
Heath, D. F., 1962, Cl^{36} dieldrin in mice, *in: Radioisotopes and Radiation in Entomology*, Proceedings of the Bombay Symposium, 1960, P. 83, International Atomic Energy Agency, Vienna.
Heath, D. F., and Vandekar, M., 1964, Toxicity and metabolism of dieldrin in rats, *Brit. J. Ind. Med.* **21:**269.
Hewlett, P. S., 1958, Interpretation of dosage–mortality data for DDT-resistant houseflies, *Ann. Appl. Biol.* **46:**37.
Hirom, P. C., Millburn, P., Smith, R. L., and Williams, R. T., 1972, Species variations in the threshold molecular-weight factor for the biliary excretion of organic ions, *Biochem. J.* **129:**1071.
Holan, G., 1969, New halocyclopropane insecticides and the mode of action of DDT, *Nature* **221:**1025.
Hoyle, G., 1953, Potassium ions and insect-nerve muscle, *J. Exp. Biol.* **30:**121.
Hunter, C. G., Robinson, J., and Roberts, M., 1969, Pharmacodynamics of dieldrin (HEOD); ingestion by human subjects for 18 to 24 months and post exposure for eight months, *Arch. Environ. Health* **18:**12.
Jones, R. L., Metcalf, R. L., and Fukuto, T. R., 1969, Use of the multiple regression equation in the prediction of the insecticidal activity of anticholinesterase insecticides, *J. Econ. Entomol.* **62:**801.
Kilby, B. A., 1963, The biochemistry of insect fat body, *in: Advances in Insect Physiology*, Vol. 1 (J. W. L. Beament, J. E. Treherne, and V. B. Wigglesworth, eds.), pp. 111–174, Academic Press, London.
Koransky, W., and Ullberg, S., 1964, Distribution in the brain of ^{14}C-benzenehexachloride: Autoradiographic study, *Biochem. Pharmacol.* **13:**1537.
Koransky, W., Portig, J., and Meunch, G. 1963, Absorption, distribution and elimination of α- and γ-benzene hexachloride, *Arch. Expl. Pathol. Pharmakol.* **244:**564.
Kurihara, N., 1970, BHC—Its toxicity and its penetration, translocation and metabolism in insects and mammals, *Botyu Kagaku* **35(11):**56.
Kurihara, N., Nakajima, E., and Shindo, H., 1970, Whole body autoradiographic studies on the distribution of BHC and nicotine in the American cockroach, *in: Biochemical Toxicology of Insecticides* (R. D. O'Brien and I. Yamamoto, eds.), pp. 41–50, Academic Press, New York.
Lewis, C. T., 1962, Diffusion of oil films over insects, *Nature* **193:**904.
Lewis, C. T., 1965, Influence of cuticle structure and hypodermal cells on DDT absorption by *Phormia terraenovae* R-D, *J. Insect Physiol.* **11:**683.

Maddrell, S. H. P., 1971, The mechanisms of insect excretory systems, *in: Advances in Insect Physiology*, Vol. 8 (J. W. L. Beament, J. E. Treherne, and V. B. Wigglesworth, eds.), pp. 200–331, Academic Press, New York.

Matsumura, F., 1972, Metabolism of insecticides in microorganisms and insects, *in: Environmental Quality and Safety*, Vol. 1 (F. Coulston and F. Korte, eds.), pp. 96–106, Georg Thieme, Stuttgart, Academic Press, New York.

Matsumura, F., and Hayashi, M., 1966, Dieldrin: Interaction with nerve components of cockroaches, *Science* **153:**757.

Matsumura, F., and Hayashi, M., 1969*a*, Dieldrin resistance—Biochemical mechanisms in the German cockroach, *J. Agr. Food Chem.* **17:**231.

Matsumura, F., and Hayashi, M., 1969*b*, Comparative mechanisms of insecticide binding with nerve components of insects and mammals, *Residue Rev.* **25:**265.

Matsumura, F., and O'Brien, R. D., 1966, Absorption and binding of DDT by the central nervous system of the American cockroach, *J. Agr. Food Chem.* **14:**36.

Menzie, C. M., 1969, *Metabolism of Pesticides*, Special Scientific Report, Wildlife, No. 127, 487 pp., Bureau of Sport Fisheries and Wildlife, U.S. Department of the Interior, Washington, D.C.

Moriarty, F., 1969, The sublethal effects of synthetic insecticides on insects, *Biol. Rev.* **44:**321.

Moriarty, F., and French, M. C., 1971, The uptake of dieldrin from the cuticular surface of *Periplaneta americana* L., *Pestic. Biochem. Physiol.* **1:**286.

Moss, J. A., and Hathway, D. E., 1964, Partition of dieldrin and telodrin between the cellular components and soluble proteins of blood, *Biochem. J.* **91:**384.

Nakajima, E., Shindo, H., and Kurihara, N., 1970, Whole body autoradiographic studies on the distribution of α-, β- and γ-BHC-^{14}C in mice, *Radioisotopes* **19:**60.

Nelson, J. O., and Matsumura, F., 1973, Dieldrin (HEOD) metabolism in cockroaches and houseflies, *Arch. Environ. Contam. Toxicol.* **1:**224.

O'Brien, R. D., 1967*a*, *Insecticides, Action and Metabolism*, Academic Press, New York.

O'Brien, R. D., 1967*b*, Barrier systems in insect ganglia and their implications for toxicology, *Fed. Proc.* **26:**1056.

O'Brien, R. D., and Dannelley, C. E., 1965, Penetration of insecticides through rat skin, *J. Agr. Food Chem.* **13:**245.

O'Brien, R. D., and Matsumura, F., 1964, DDT: A new hypothesis of its mode of action, *Science* **146:**657.

Olson, W. P., 1973, Dieldrin transport in the insect: An examination of Gerolt's hypothesis, *Pestic. Biochem. Physiol.* **3:**384.

Olson, W. P., and O'Brien, R. D., 1963, The relation between physical properties and the penetration of solutes into the cockroach cuticle, *J. Insect Physiol.* **9:**777.

Parke, D. V., 1968, *The Biochemistry of Foreign Compounds*, Pergamon Press, Oxford.

Pekas, J. C., 1971*a*, Uptake and transport of pesticidal carbamates by everted sacs of rat small intestine, *Can. J. Physiol. Pharmacol.* **49:**14.

Pekas, J. C., 1971*b*, Intestinal metabolism and transport of naphthyl *N*-methylcarbamate *in vitro* (rat), *Am. J. Physiol.* **220:**2008.

Penniston, J. T., Beckett, L., Bentley, D. L., and Hansch, C., 1969, Passive permeation of organic compounds through biological tissue: A non-steady state theory, *Mol. Pharmacol.* **5:**333.

Phillips, J. E., and Dockrill, A. A., 1968, Molecular sieving of hydrophilic molecules by the rectal intima of the desert locust (*Schistocerca gregaria*), *J. Exp. Biol.* **48**:521.

Polles, S. G., and Vinson, S. B., 1972, Penetration, distribution and metabolism of ^{14}C-endrin in resistant and susceptible tobacco budworm larvae, *J. Agr. Food Chem.* **20:**38.

Poulsen, B. J., 1973, Design of topical drug products: Biopharmaceutics, *in: Drug Design*, Vol. IV (E. J. Ariens, ed.), pp. 149–189, Academic Press, New York.

Price, G. M., 1973, Protein and nucleic acid metabolism in insect fat body, *Biol. Rev.* **48:**333.

Quraishi, M. S., and Abdul Matin, A. S. M., 1962, Mode of action of insecticides, CENTO Institute of Nuclear Science, Tehran, Iran. Unpublished report cited in Quraishi, M. S., and Poonawalla, Z. T., 1969, Radioautographic study of the diffusion of topically applied DDT-C^{14} into the house fly and its distribution in internal organs, *J. Econ. Entomol.* **62:**988.

Quraishi, M. S., and Poonawalla, Z. T., 1969, Radioautographic study of the diffusion of topically applied DDT-C^{14} into the house fly and its distribution in internal organs, *J. Econ. Entomol.* **62:**988.

Ray, J. W., 1963, Insecticide absorbed by the central nervous system of susceptible and resistant cockroaches exposed to dieldrin, *Nature* **197:**1226.

Robinson, J., 1967, Dynamics of organochlorine insecticides in vertebrates and ecosystems, *Nature* **215:**33.

Robinson, J., 1969*a*, Organochlorine insecticides and bird populations in Britain, *in: Chemical Fallout* (M. W. Miller and G. G. Berg, eds.), pp. 113–169, C. C. Thomas, Springfield, Ill.

Robinson, J., 1969*b*, The burden of chlorinated hydrocarbon pesticides in man, *Can. Med. Assoc. J.* **100:**180.

Robinson, J., 1970*a*, Birds and pest control chemicals, *Bird Study* **17:**195.

Robinson, J., 1970*b*, Persistent pesticides, *Ann. Rev. Pharmacol.* **10:**353.

Robinson, J., and Roberts, M., 1968, Accumulation, distribution and elimination of organochlorine insecticides by vertebrates, *in: Physicochemical and Biophysical Factors Affecting the Activity of Pesticides*, pp. 106–119, Society of Chemical Industry Monograph No. 29, London.

Robinson, J., Richardson, A., and Brown, V. K. H., 1967, Pharmacodynamics of dieldrin in pigeons, *Nature* **213:**734.

Robinson, J., Roberts, M., Baldwin, M., and Walker, A. I. T., 1969, The pharmacokinetics of HEOD (dieldrin) in the rat, *Food Cosmet. Toxicol.* **7:**317.

Schaefer, C. H., and Sun, Y.-P., 1967, A study of dieldrin in the house fly central nervous system in relation to dieldrin resistance, *J. Econ. Entomol.* **60:**1580.

Sellers, L. G., and Guthrie, F. E., 1971, Localization of dieldrin in housefly thoracic ganglion by electron microscopic autoradiography, *J. Econ. Entomol.* **64:**352.

Sellers, L. G., and Guthrie, F. E., 1972, Distribution and metabolism of ^{14}C-dieldrin in the resistant and susceptible house fly, *J. Econ. Entomol.* **65:**378.

Shah, A. H., and Guthrie, F. E., 1970, Penetration of insecticides through the isolated midgut of insects and mammals, *Comp. Gen. Pharmacol.* **1:**391.

Shah, A. H., and Guthrie, F. E., 1971, *In vitro metabolism of insecticides during midgut penetration, Pestic. Biochem. Physiol.* **1:**1.

Shah, P. V., Dauterman, W. C., and Guthrie, F. E., 1972, Penetration of a series of dialkoxy analogues of dimethoate through the isolated gut of insects and mammals, *Pestic. Biochem. Physiol.* **2:**324.

Street, J. C., 1968, Modification of animal responses to toxicants, *in: Enzymatic Oxidations of Toxicants* (E. Hodgson, ed.), pp. 197–226, North Carolina State University, Raleigh N.C.

Street, J. C., 1969, Methods of removal of pesticide residues, *Can. Med. Assoc. J.* **100:**154.

Sun, Y.-P., 1968, Dynamics of insect toxicology, a mathematical and graphical evaluation of the relationship between insect toxicity and rates of penetration and detoxication of insecticides, *J. Econ. Entomol.* **61:**949.

Sun, Y.-P., and Johnson, E. R., 1969, Relationship between structure of several Azodrin insecticide homologues and their toxicities to houseflies, tested by injection, infusion, topical application and spray methods with and without synergist, *J. Econ. Entomol.* **62:**1130.

Sun, Y.-P., and Johnson, E. R., 1971, A new technique for studying the toxicology of insecticides with houseflies by the infusion method, with comparable topical application and injection results, *J. Econ. Entomol.* **64:**75.

Sun, Y.-P., and Johnson, E. R., 1972, Quasi-synergism and penetration of insecticides, *J. Econ. Entomol.* **65:**349.

Sun, Y.-P., Schaefer, C. H., and Johnson, E. R., 1967, Effects of application methods on the toxicity and distribution of dieldrin in house flies, *J. Econ. Entomol.* **60:**1033.

Tanabe, H., 1972, Contamination of milk with chlorinated hydrocarbon pesticides, *in: Environmental Toxicology of Pesticides* (F. Matsumura, G. Mallory Bush, and T. Misato, eds.), pp. 239–256, Academic Press, New York.

Telford, J. N., and Matsumura, F., 1970, Dieldrin binding in subcellular nerve components of cockroaches: An electron microscopic and autoradiographic study, *J. Econ. Entomol.* **63**:795.

Telford, J. N., and Matsumura, F., 1971, Electron microscopic and autoradiographic studies on distribution of dieldrin in the intact nerve tissues of German cockroaches, *J. Econ. Entomol.* **64**:230.

Treherne, J. E., 1956, The permeability of skin to some nonelectrolytes, *J. physiol. (London)* **133**:171.

Treherene, J. E., 1957, The diffusion of non-electrolytes through the isolated cuticle of *Schistocerca gregaria, J. Insect Physiol.* **1**:178.

Treherne, J. E., 1961*a*, Sodium and potassium fluxes in the abdominal nerve cord of the cockroach, *Periplaneta americana, J. Exp. Biol.* **38**:315.

Treherne, J. E., 1961*b*, The movements of sodium ions in the isolated abdominal nerve cord of the cockroach, *Periplaneta americana, J. Exp. Biol.* **38**:629.

Treherne, J. E., 1961*c*, The efflux of sodium ions from the last abdominal ganglion of the cockroach, *Periplaneta americana, J. Exp. Biol.* **38**:729.

Treherne, J. E., 1961*d*, The kinetics of sodium transfer in the central nervous system of the cockroach, *Periplaneta americana, J. Exp. Biol.* **38**:737.

Treherne, J. E., 1962, The distribution and exchange of some ions and molecules in the central nervous system of *Periplaneta americana*, L., *J. Exp. Biol.* **39**:193.

Treherne, J. E., 1965, The distribution and exchange of inorganic ions in the central nervous system of the stick insect, *Carausius morosus, J. Exp. Biol.* **42**:7.

Treherne, J. E., and Pichon, Y., 1972, The insect blood–brain barrier, *in: Advances in Insect Physiology*, Vol. 9 (J. E. Treherne, M. J. Berridge, and V. B. Wigglesworth, eds.), pp. 257–312, Academic Press, London.

Treherne, J. E., and Smith, D. S., 1965, The penetration of acetylcholine into the central nervous tissues of an insect, *Periplaneta americana* L., *J. Exp. Biol.* **43**:13.

Walker, C. H., 1975, Variations in the intake and elimination of pollutants, *in: Organochlorine Insecticides: Persistent Pollutants* (F. Moriarty, ed.), Chap. 3, Academic Press, New York.

Wang, C. M., Narahashi, T., and Yamada, M., 1971, The neurotoxic action of dieldrin and its derivatives in the cockroach, *Pestic. Biochem. Physiol.* **1**:84.

Wigglesworth, V. B., 1960, The nutrition of the central nervous system in the cockroach *Periplaneta americana* L., *J. Exp. Biol.* **37**:500.

Wilkinson, C. F., and Brattsten, L. B., 1972, Microsomal drug metabolising enzymes in insects, *Drug. Metab. Rev.* **1**:153.

Wilson, K. A., and Cook, R. M., 1970, Use of activated carbon as an antidote for pesticide poisoning in ruminants, *J. Agr. Food Chem.* **18**:437.

Winteringham, F. P. W., 1969, Mechanisms of selective insecticidal action, *Ann. Rev. Entomol.* **14**:409.

Wright, A. S., Potter, D., Wooder, M. F., Donninger, C., and Greenland, R. D., 1972, The effects of dieldrin on the subcellular structure and function of mammalian liver cells, *Food Cosmet. Toxicol.* **10**:311.

Zaghloul, T. M. A., and Brown, A. W. A., 1968, Effects of sublethal doses of DDT on the reproduction and susceptibility of *Culex pipiens* L., *Bull. WHO* **38**:459.

II
Metabolism

2

Microsomal Oxidation and Insecticide Metabolism

Tsutomu Nakatsugawa and Michael A. Morelli

1. Introduction

Once in the animal body, organic insecticides are subject to metabolism by a variety of enzymes. Depending on their chemical structure, the compounds may be hydrolyzed, oxidized, conjugated with endogenous metabolites, or otherwise modified. Of particular importance are reactions mediated by the oxidative enzymes known as microsomal oxidases, so called because of their localization in the microsomal fraction of cell homogenates. Since these enzymes as a class have an extremely broad spectrum of substrates and catalyze a wide variety of biotransformations, they play a central role in the metabolism of insecticides. In fact, most of the biodegradability of current insecticides is dependent on their successful biotransformation by microsomal oxidases of various species. These enzymes are also intimately associated with the phenomena of synergism, enzyme induction, and insecticide resistance.

Despite early indications, the prevalence of biological oxidation in insecticide metabolism eluded recognition for some time because its importance was not obvious from the chemical structure and reactivity of insecticide molecules. Thus, although carbaryl is an ester and is extremely labile to alkaline hydrolysis, its primary biotransformation is initiated not by hydrolytic

---

Tsutomu Nakatsugawa and Michael A. Morelli • Department of Entomology, State University of New York, College of Environmental Science and Forestry, Syracuse, New York.

enzymes but by microsomal oxidases. Pyrethrins too were initially assumed to undergo hydrolytic degradation because of their ester structure, but more recent studies have established that their degradation is primarily oxidative. The involvement of microsomal oxidases is not always discernible from the nature of *in vivo* metabolites of an insecticide. For instance, urinary metabolites of parathion from the rat indicate extensive hydrolysis and yet careful enzymological studies show that a large portion of the "hydrolysis" is in fact the result of microsomal oxidation.

The first indication of the general importance of oxidation in insecticide metabolism was provided by Sun and Johnson (1960) using a simple bioassay technique. With great insight, they ascribed the action of methylenedioxyphenyl synergists to their inhibition of the oxidative metabolism of a number of insecticides, a conclusion which has been amply supported by subsequent enzymological studies. The primary role of the microsomal oxidases in insecticide metabolism has now been well confirmed, establishing a parallel with their previously recognized importance in drug metabolism.

Knowledge of the pathways and products of oxidative degradation is but one step toward an understanding of the role of microsomal oxidation as a major determinant of insecticide action. What are the mechanisms of these enzymatic reactions? How is the contribution of microsomal oxidation influenced by physiological and environmental factors? What are the physiological and toxicological consequences of the oxidative biotransformation of insecticides? These are some of the questions that must be considered before the significance of microsomal oxidation can be seen in proper perspective. The intent of this chapter is to provide a summary of the current concept of microsomal oxidation and review its role in insecticide metabolism.

2. Biochemistry of Microsomal Oxidases

The potential importance of microsomes as the locus of oxidative metabolism was first recognized following Axelrod's discovery of the deamination of amphetamine by rabbit liver microsomes (Axelrod, 1954). The classical review by Brodie *et al.* (1958) of the incipient studies of drug oxidations confirmed the universal *in vitro* requirement of these enzymes for O_2 and NADPH, and cited "hydroxylation" as the common feature of these reactions. The subsequent accumulation of detailed knowledge of these oxidases has made microsomes an increasingly useful tool for *in vitro* studies of drug and insecticide metabolism. Many accounts of this system have been published (Parke, 1968; Hodgson, 1968; Gillette *et al.*, 1969; LaDu *et al.*, 1971; Hathway, 1970, 1972; Estabrook *et al.*, 1973*a*; Cooper and Salhanick, 1973), including some specifically on insects (Hodgson and Plapp, 1970; Wilkinson and Brattsten, 1972; Agosin and Perry, 1974).

The modern concept of microsomes originated in the early work of Claude (1938) when it was first demonstrated that the deep staining of material of the cytoplasmic ground substance, hitherto designated cytoplasmic basophilia, was attributable to subcellular particles which could be isolated by differential centrifugation. "Microsomes" was proposed as a general term for these small particles (Claude, 1943). The advent of the electron microscope allowed better resolution of these particles and through the pioneering efforts of Claude, Porter, Siekevitz, and Palade resulted in a refinement of our knowledge of microsome morphology.

2.1. Morphological, Chemical, and Enzymatic Composition of Microsomes

Under the most rigorous preparative procedures, the microsomal fraction consists of the isolated form of the smooth and rough endoplasmic reticulum (Palade and Siekevitz, 1956) (Fig. 1). The transformation of the membranous endoplasmic reticulum into microsomes is a matter of conjecture, but a mechanism has been hypothesized wherein portions of the extended structure vesiculate by "pinching off" without loss of luminal contents in response to adverse conditions such as cell disruption during homogenization (Claude, 1969). If this is indeed the case, then microsomes are a reasonably accurate morphological and biochemical facsimile of the endoplasmic reticulum of the intact cell. Chemically they are composed predominantly of lipoprotein, the lipid of which accounts for approximately 40% of the pellet's weight, and consists of phospholipid, lipositol, plasmalogen, and fatty acids (Claude, 1969). They also contain 12% of the total cellular protein and approximately 50% of the ribonucleic acid. It is the high content of RNA associated with the ribosomes which imparts basophilic staining properties to the fraction. Although the smooth and rough endoplasmic reticulum are part of the same anastomosing system, they are morphologically and biochemically distinct. Both types perform microsomal oxidations, although smooth endoplasmic reticulum has higher activity (Gram and Fouts, 1968; Holtzman *et al.*, 1968). In both cases, activity is associated with the enzymatic components on the membrane itself since various solubilizers and organic solvents disrupt activity whereas ribonuclease has no effect (Gillette, 1966).

For most toxicological studies, microsomes are rarely obtained in a pure form and are usually defined simply as those vesicles precipitated by the high-speed centrifugation (e.g., 100,000g, 60 min) of a postmitochondrial supernatant. The protocol for differential centrifugation (choice of centrifugal force, length of time) may vary considerably for each different tissue and species (Wilkinson and Brattsten, 1972; Hodgson and Plapp, 1970; Siekevitz, 1965). Therefore, each procedure should be carefully determined on the basis of biochemical criteria and preferably also electron microscopic evidence. The homogenization medium is often important since its ionic strength, osmolarity,

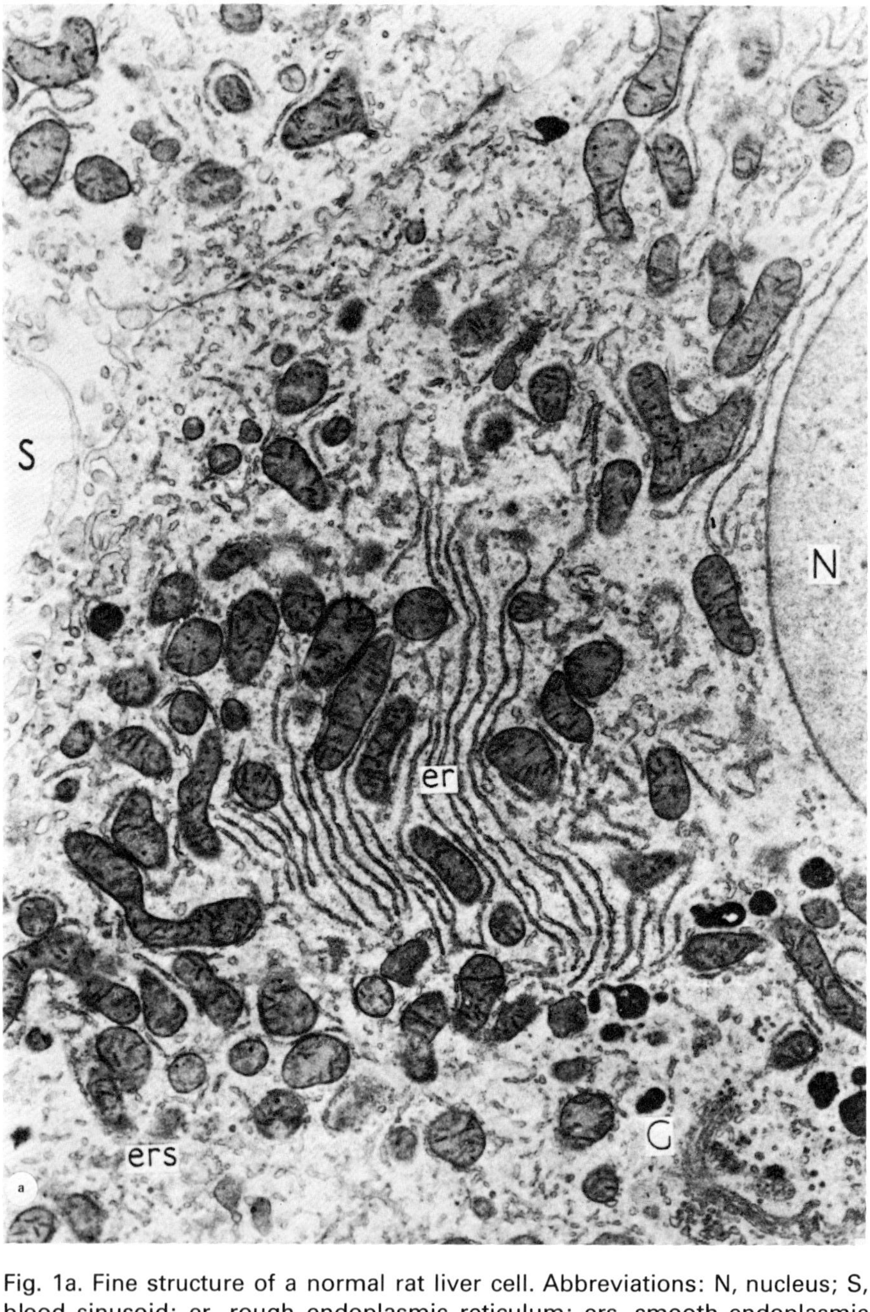

Fig. 1a. Fine structure of a normal rat liver cell. Abbreviations: N, nucleus; S, blood sinusoid; er, rough endoplasmic reticulum; ers, smooth endoplasmic reticulum; G, Golgi apparatus. From Porter (1961).

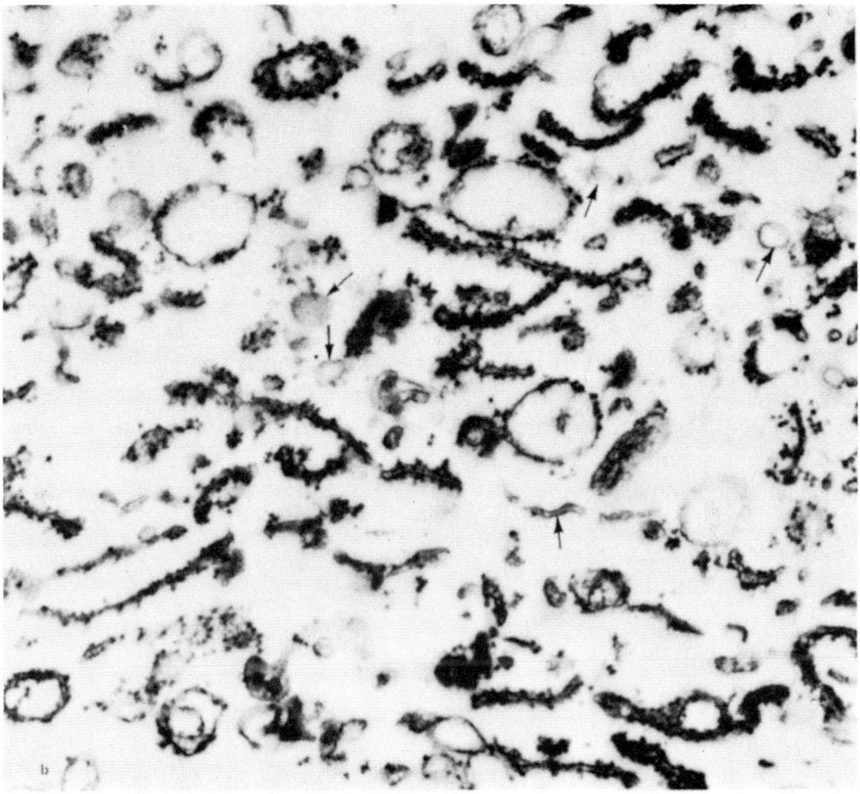

Fig. 1b. Micrograph of microsome pellet from rat liver. Vesicles derived from rough and smooth endoplasmic reticulum are visible. From Porter (1961).

and chemical composition may affect the structural integrity of the vesicles, their sedimentation rates, and enzymatic activities. Microsome preparations from some species may be further complicated by the presence of endogenous inhibitors of microsomal oxidases. Notable cases have been reported from insect studies (Wilkinson and Brattsten, 1972), where materials such as eye pigment (xanthommatin) (Schonbrod and Terriere, 1971; Wilson and Hodgson, 1972) and digestive proteases of the gut (Krieger and Wilkinson, 1970; Brattsten and Wilkinson, 1973) often present serious problems.

The procedure providing the highest recovery of microsomal oxidase activity often results in considerable contamination of the microsomal fraction with extramicrosomal components. Cleaner preparations can be obtained, but usually at the expense of leaving a considerable portion of microsomal oxidase activity in other subcellular fractions. In most toxicological studies, a higher recovery is favored and so the microsomal fraction is generally morphologically and biochemically heterogeneous. It may contain a multiplicity of membrane fragments from the plasma membrane, mitochondria, and Golgi apparatus in addition to various cellular inclusions such as lysosomes, peroxisomes, glycogen granules, and ribosomes (Estabrook and Cohen, 1969; Palade and Siekevitz, 1956). In addition to the known enzymes of the rough and smooth endoplasmic reticulum (e.g., NADPH-cytochrome *c* reductase, NADH-cytochrome b_5 reductase, glucose-6-phosphatase, cytochromes b_5 and P450, esterase, and nucleoside diphosphatase), the fraction is frequently contaminated with monoamine oxidase and cytochrome oxidase from the mitochondrial membrane and 5′-nucleotidase from the plasma membrane. Soluble enzymes like fumarase, aldolase, and glutamine synthetase also are commonly present because they become partially attached to the ribosomes (Amar-Costesec *et al.*, 1969). The diverse functions of the microsomal fraction, not all of which are associated with foreign compound metabolism, are manifestations of its inherent biochemical heterogeneity. The biosynthesis of steroids, phospholipids, and complex polysaccharides, protein synthesis, conjugation, and lipid peroxidation comprise part of the collection of metabolic processes characterizing these subcellular particles (Ernster and Orrenius, 1973; Parke, 1968; Nishibayashi *et al.*, 1967).

2.2 Components of Microsomal Oxidases

The characteristic requirement of the membrane-bound microsomal oxidases for a reducing agent under aerobic conditions was at first puzzling but suggested the involvement of electron carriers between NADPH and O_2. The participation of an oxygen-activating enzyme was indicated by $^{18}O_2$ studies (Hayaishi, 1962; Mason, 1957). In fact, the incorporation of one atom of atmospheric oxygen into the substrate and the presumed reduction of the other atom to H_2O is the basis for the general classification of microsomal oxidases as mixed-function oxidases (Mason, 1957) or monooxygenases (Hayaishi, 1962). These incipient studies led to the concept that microsomal oxidation involved an enzyme complex consisting of several components. The major components of the system, as we know them today, are a flavoprotein, NADPH-cytochrome *c* reductase, and a unique cytochrome, cytochrome P450, which plays the central role in oxidation (Fig. 3).

2.2.1. NADPH-Cytochrome c Reductase

NADPH-cytochrome *c* reductase is presently recognized as a mediator of electron flow from NADPH to the oxygen-activating enzyme. It is a stable flavoprotein, having a molecular weight of 70,000 and containing 2 moles of FAD per mole, and is commonly assayed by the reduction of artificially added electron acceptors such as cytochrome *c* or neotetrazolium in the presence of NADPH (Kamin and Masters, 1968). The major observations linking this already known enzyme to microsomal oxidation are that microsomal oxida-

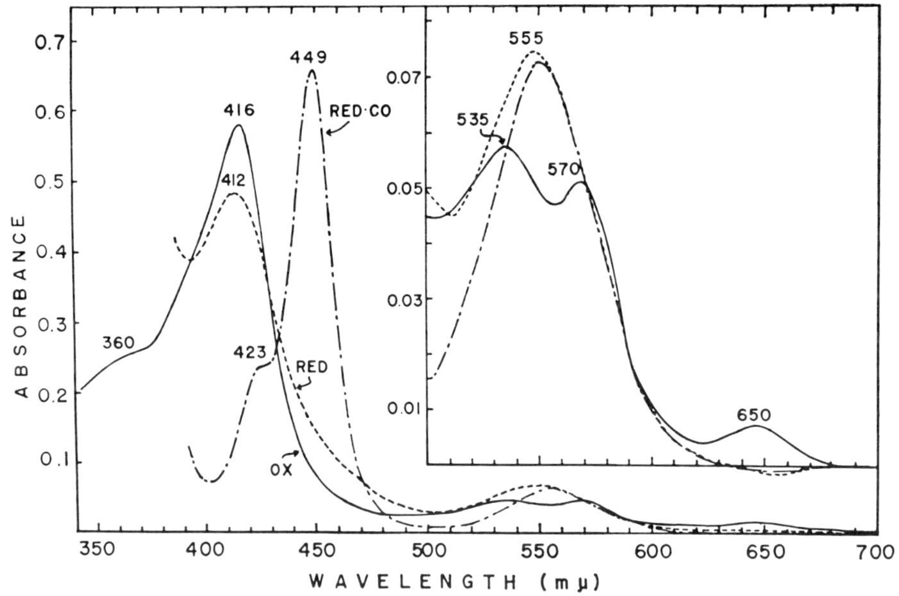

Fig. 2. Absolute spectra of P450 particles prepared by subjecting liver microsomes from phenobarbital-treated rabbits to a *Bacillus subtilis* protease to remove cytochrome b_5 and NADPH-specific flavoprotein. The spectra represent 5.64×10^{-6} M cytochrome P450 in phosphate buffer (pH 7.4) containing 30% glycerol. ——, Oxidized P450 particles; - - - -, particles reduced with a small amount of solid $Na_2S_2O_4$; – · – ·, particles reduced with $Na_2S_2O_4$ and bubbled with CO for about 30 sec. From Nishibayashi and Sato (1968).

tions are best effected by NADPH, that induction of microsomal oxidases results in an increased level of NADPH-cytochrome *c* reductase (Ernster and Orrenius, 1965), and that the antibody to NADPH-cytochrome *c* reductase inhibits oxidative metabolism (Masters *et al.*, 1973). NADPH-cytochrome *c* reductase has been found in the microsomal fraction of mammalian tissues such as liver, adrenal cortex, spleen, kidney, heart, and lung (Masters *et al.*, 1973) as well as in the tissues of insects (Hodgson and Plapp, 1970; Wilkinson and Brattsten, 1972).

2.2.2. Cytochrome P450

A breakthrough in the understanding of microsomal oxidation came when Estabrook *et al.* (1963) reported that a hemoprotein eventually named cytochrome P450, functioned as the terminal oxidase in steroid hydroxylation by bovine adrenal cortex microsomes. Utilizing a technique originally introduced by Warburg (1926) in his studies of cytochrome oxidase, they exploited the ability of monochromatic light to dissociate the hemoprotein/CO complex.

The resultant "photochemical action spectrum" for the reactivation of the CO-inhibited steroid hydroxylase had a peak at 450 nm which coincided with the absorption spectrum of the reduced cytochrome P450/CO complex. Subsequently, similar results were obtained for rat liver microsomal oxidases (Cooper *et al.*, 1965). Today, there is little doubt that cytochrome P450 is "the common oxygen-activating enzyme for the entire family of microsomal mixed-function oxidases" (Omura *et al.*, 1965). The localization of this hemoprotein, however, is not totally restricted to the microsomal fraction, nor, for that matter, to any specific tissue or group of animals. Microbes and insects, in addition to mammals, are known to contain cytochrome P450 (Sato *et al.*, 1973; Wilkinson and Brattsten, 1972). In mammals, it has been located in the microsomal fraction of extrahepatic organs such as the kidney, lung, and placenta (Sato *et al.*, 1973; Orrenius *et al.*, 1973; Burns and Gurtner, 1973) and has also been reported in the mitochondria of the adrenal cortex and corpus luteum (Sato *et al.*, 1973; McIntosh *et al.*, 1973). Cytochrome P450 is also found in the particulate fraction of yeast (Sato *et al.*, 1973).

The rigid association of cytochrome P450 with membranes has hindered detailed studies of its physicochemical properties. However, the discovery and subsequent purification of a soluble form of this pigment in camphor-grown *Pseudomonas putida* have greatly expanded our knowledge. Cytochrome $P450_{cam}$, as it is called, consists of a single polypeptide having a molecular weight of 45,000 daltons. It contains 1 mole of ferriprotoporphyrin IX, the iron of which is bound to four pyrrole nitrogens and two amino acid ligands, possibly cysteine and histidine (Lipscomb and Gunsalus, 1973). The oxidized form exhibits three absorption maxima at 417–418 nm, 535 nm, and 570–571 nm, the reduced form two maxima at 408–411 nm, and 540 nm, and the reduced cytochrome/CO complex two absorption peaks at 447 nm and 550 nm (Gillette *et al.*, 1972). Unlike cytochrome P450 from other sources, cytochrome $P450_{cam}$ hydroxylates only camphor and a few analogues (Gunsalus *et al.*, 1973).

Despite many technical difficulties, significant advances have also been made in recent years in investigations of the membrane-bound form of cytochrome P450 from adrenal cortex mitochondria and liver microsomes. The absolute absorption spectrum of membrane-bound cytochrome P450 is difficult to determine because the cytochrome is readily converted to an inactive product called cytochrome P420 and because of interference by another microsomal pigment, cytochrome b_5 (Sato *et al.*, 1973). Two approaches have been utilized to overcome this problem, each giving slightly different results. One method capitalizes on the different responses of the two hemoproteins to inducing agents. The treatment of rats with phenobarbital causes a two- to four-fold increase in the level of cytochrome P450, but only a 1.5-fold increase in cytochrome b_5 (Remmer *et al.*, 1968). Assuming there is no other pigment present, the absolute spectrum of cytochrome P450 can be obtained on a double-beam spectrophotometer by matching cytochrome b_5

concentration in the induced and control microsomes. Another method entails the removal of substantial amounts of the less tightly bound cytochrome b_5 from the membrane and results in an absolute spectrum of cytochrome P450 with only slight contamination (Sato *et al.*, 1969). Absolute spectra have also been recorded with purified cytochrome P450 preparations obtained by detergent solubilization followed by chromatographic separation (Sato *et al.*, 1973; Cooper *et al.*, 1973). The conversion of cytochrome P450 to 420, presumably due to the destruction of the hydrophobic environment of the heme prosthetic group, is minimized by utilizing polyols such as glycerol. Furthermore, purification must be conducted under a nitrogen atmosphere to avoid the alteration of the heme of cytochrome P450 by peroxidation of membrane lipids (Sato *et al.*, 1973).

A typical absolute spectrum of oxidized cytochrome P450 shows three major peaks at 570 nm, 535 nm, and 416 nm corresponding to the α, β, and γ (Soret) bands, respectively (Fig. 2). Upon reduction, the Soret band becomes broader, less intense, and shifted to a shorter wavelength, and there is only one peak in the visible region. Binding of reduced cytochrome P450 with CO results in a shift of the Soret band to a longer wavelength (Nishibayashi and Sato, 1968). Although the position of the Soret band of the reduced cytochrome/CO complex is responsible for the general name P450 (for "pigment-450") (Omura and Sato, 1964), significant variations are associated with different sources of this cytochrome (Fig. 2). There is now good evidence that some variations indicate the existence of multiple forms of hepatic cytochrome P450 (Mannering, 1971).

A different form of cytochrome P450 appears in the rat liver on induction with certain compounds such as 3-methylcholanthrene, and has been called P_1450, P448, and P446 (Nebert *et al.*, 1973). This cytochrome has different spectral properties and a different spectrum of substrate specificity from the P450 of control and/or phenobarbital-treated animals (Fujita and Mannering, 1971; Lu *et al.*, 1971; Lu and Levin, 1972). Both cytochromes P450 and P448 have been solubilized and characterized in partially purified form (Lu and Coon, 1968; Mitani *et al.*, 1971; Lu and Levin, 1972; Lu *et al.*, 1973). Studies of houseflies have also indicated differences in the cytochromes P450 (Philpot and Hodgson, 1971). A peak at 448 nm has been observed with a highly insecticide-resistant strain, compared with a peak at 452 nm of six less resistant strains (Perry *et al.*, 1971). There is also a report that a new species of P450 which has a maximum at 446 nm and different spectral characteristics is induced upon treatment of houseflies with naphthalene or phenobarbital (Capdevila *et al.*, 1973).

2.2.3. Additional Components

Although NADPH-cytochrome *c* reductase and cytochrome P450 are essential components of the microsomal oxidase system, they alone are not sufficient for effecting oxidative metabolism. The existence of additional components is evidenced by the lack of correlation between the levels of these components and the activity of the oxidase system. For example, although NADPH-cytochrome *c* reductase and cytochrome P450 of newborn rats reached adult levels by the end of the first week, the corresponding *N*-demethylase activity was only 5% of that observed in adult animals (Dallner *et al.*, 1966). One such component may be the phospholipid, phosphatidylcholine, which has been found essential in the reconstituted liver microsomal oxidase system, presumably because it facilitates the reaction of cytochrome P450 with the other electron transport components (Autor *et al.*, 1973).

An additional carrier between NADPH-cytochrome *c* reductase and cytochrome P450 has been invoked in an effort to explain the high ratio of cytochrome P450 to NADPH-cytochrome *c* reductase. The ratio, which is sometimes as high as 40, is considered contraindicative of the direct interaction of these two components. The participation of such a carrier has already been demonstrated in the oxidase system from adrenocortical mitochondria and in the bacteria cytochrome $P450_{cam}$ system. In both systems, the carrier is a nonheme, iron-sulfur protein called "adrenodoxin" and "putidaredoxin", respectively (Estabrook *et al.*, 1973*b*). To date, however, there is neither spectral nor immunochemical evidence for the existence of an iron-sulfur protein in liver microsomes (Masters *et al.*, 1973).

The transfer of reducing equivalents to cytochrome P450 via an NADPH-cytochrome P450 reductase system may be supplemented by a second microsomal electron transport system which transfers electrons from NADH to cytochrome P450 through NADH-cytochrome b_5 reductase and cytochrome b_5 (Estabrook and Cohen, 1969). Unlike NADPH, NADH cannot always support the microsomal oxidation by itself, but it can often synergize the effect of NADPH. Furthermore, the oxidation of reduced cytochrome b_5 is induced by the addition of substrate to the NADPH–microsome system (Gillette *et al.*, 1972).

2.3. Dynamics of Microsomes

Like other enzyme proteins, the microsomal oxidase system does not remain static during the lifetime of the cell but is in a state of continual flux of synthesis and degradation. Thus the half-life of total membrane protein and lipid in rat liver microsomes has been estimated at 108 and 97 hr, respectively (Omura *et al.*, 1967). The reported half-life times of microsomal oxidase components display wide variations and depend on whether the iron, heme, or protein moiety is radioactively labeled (Greim *et al.*, 1970); differences in animal species or strain may also be a contributing factor. In the case of NADPH-cytochrome *c* reductase, the half-life is 60–84 hr using radioactive amino acids (Kuriyama *et al.*, 1969). The half-lives of cytochrome b_5 of 96–120 hr (Kuriyama *et al.*, 1969) and 45 hr (Greim *et al.*, 1970) correspond to radioactive precursors of amino acid and heme, respectively, whereas a half-life of 22 hr for cytochrome P450 (Greim *et al.*, 1970; Remmer, 1971) has been reported based on heme labeling. Some experimental evidence suggests that intracellular degradation of membrane proteins may be preceded by their dissociation from the membrane structure (Taylor *et al.*, 1973).

Alterations in the delicate balance between synthesis and degradation occur in response to a variety of physiological and exogenous factors and are observed as changes in the levels of microsomal oxidase activity or the components of that system. Thus treatment of rats with phenobarbital results in an immediate increase in the levels of cytochrome P450 and NADPH-cytochrome *c* reductase with a concomitant increase in certain microsomal oxidases (Ernster and Orrenius, 1965). The enhancement of enzymatic activity by a variety of chemicals, a process termed "induction" (see Chapter 3), is thought to involve greater enzyme synthesis and reduced degradation rates (Ernster and Orrenius, 1973). Morphologically, induction is observed as a proliferation of the smooth endoplasmic reticulum which seems to arise from its rough counterpart (Claude, 1969).

The phenomenon of induction is not limited to the enzymes specifically associated with microsomal oxidation or even those of the endoplasmic reticulum. Within 24 hr of treatment with an inducer, increased levels of microsomal components such as cytochrome b_5, esterases, *N*-glucuronyltransferase, glucose-6-phosphatase, and nucleotidase are observed. Cytoplasmic enzymes such as UDP-glucose dehydrogenase, transaminase, and conjugases are also induced (Remmer *et al.*, 1968). Nor is induction characteristic of the liver of mammals since increased levels of the microsomal oxidase in kidney, intestine, lung, skin, and placenta have been reported following administration of various inducing agents (Alvares *et al.*, 1973; Lake *et al.*, 1973). Studies on induction in insects have been reviewed by Wilkinson and Brattsten (1972).

Microsomal oxidases of the rat hepatic system display a diurnal rhythm in which higher activity is generally associated with periods of darkness (Radzialowski and Bousquet, 1967; Nair and Casper, 1969; Jori *et al.*, 1971). Thus hexobarbital oxidase activity is 44% higher at 10 P.M than at 2 P.M. (Nair and Casper, 1969) and parallels a measurable *in vivo* variation in the duration of hexobarbital-induced sleeping time. Such fluctuations seem to be related to changes in the abundance of smooth endoplasmic reticulum (Chedid and Nair, 1972). It has been noted that diurnal rhythms can be eliminated by maintaining constant lighting conditions (either total darkness or light). When rats are reared with constant illumination, low enzymatic activity is established, whereas exposure to continuous darkness results in higher stable levels (Nair and Casper, 1969).

Dramatic changes in microsomal enzyme activity occur with age and stage of development in insects (Wilkinson and Brattsten, 1972). In general, eggs and pupae are devoid of measurable activity. In the larval or nymphal stages of several species, consistent rhythmic patterns of activity occur and appear to be closely linked with the molting cycle, activity always being low at the molt and high during the mid-instar period (Benke *et al.*, 1972). Studies on changes in microsomal enzyme activity and cytochrome P450 content occurring during the development of the housefly indicate their ephemeral existence during the larval stage (Perry and Buckner, 1970; Yu and Terriere, 1971).

2.4. Catalytic Events

The process of microsomal oxidation integrates the transfer of electrons from NADPH with the binding of substrate and oxygen at cytochrome P450 (Fig. 3). Two separate one-electron reductions are involved, the first occurring after the initial complexing of the substrate with oxidized cytochrome P450 and the second on formation of the reduced cytochrome P450/substrate/oxygen complex. Subsequent to catalysis, oxidized cytochrome P450 is regenerated by dissociation of the hydroxylated product and water. The scheme shown in Fig. 3 merely represents a working model compatible with the body of knowledge thus far accumulated from studies of the well-characterized NADH-dependent bacterial system and the NADPH-dependent mammalian system. Although the first three steps of the sequence are fairly well understood, the nature of oxygen attack on the substrate and mechanism by which the second reducing equalent is introduced remain speculative. In the following sections, the mechanism of microsomal oxidation is discussed in terms of three basic events: (1) substrate binding, (2) reduction, and (3) oxygen binding and activation.

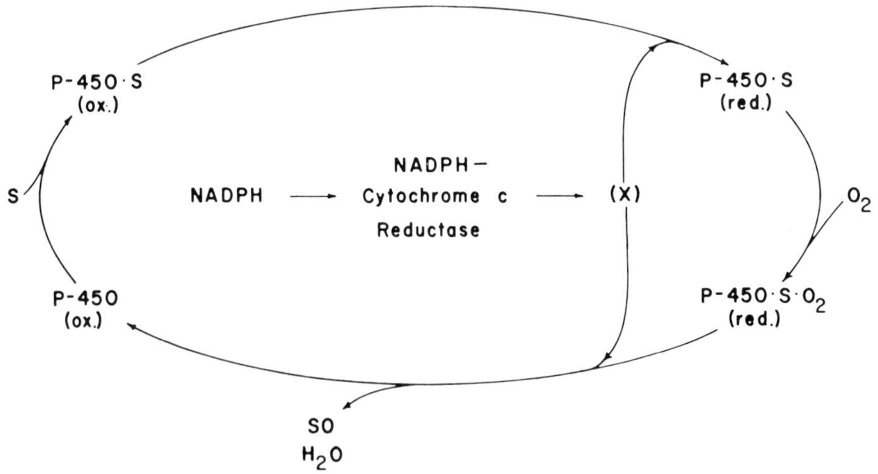

Fig. 3. Interaction of electrons, oxygen, and substrate in the microsomal oxidase system.

2.4.1. Substrate Binding

Binding as the initial reaction in the microsomal oxidase system is suggested by observations that the addition of NADPH to the system does not change the affinity of cytochrome P450 for a particular substrate. Complex formation of oxidized cytochrome P450 with any one of a number of structurally dissimilar lipophilic compounds results in a characteristic spectral alteration which can be measured by difference spectroscopy. Optical difference spectra obtained following the addition of various substrates to microsomal suspensions may be conveniently classified into several categories, the two major ones being types I and II. The majority of insecticides tested thus far are type I compounds characterized by a difference spectrum with a maximum at 390 nm and a minimum at 420 nm. In contrast, type II compounds such as nicotine, zinophos, and menazon display difference spectra with maxima at 430 nm and minima in the range 390–410 nm (Mailman and Hodgson, 1972). A more comprehensive discussion of difference spectra and their interpretation with respect to substrate–cytochrome P450 binding is provided in Chapter 3.

In some instances, the Michaelis constant (K_m) for the oxidation of a substrate can be equated with its spectral dissociation constant (K_s; defined as the concentration of substrate required for the half-maximal absorption change), and a correspondence has sometimes been noted between the magnitude of the spectral change and the metabolic rate (Remmer *et al.*, 1968). However, similar results have not been obtained with all compounds. Although spectral changes suggest substrate binding, they are not always indicative of metabolism, as in the case of perfluorohexane, a type I substrate, which undergoes no measurable hydroxylation *in vitro* by rat liver microsomes (Ullrich and Diehl, 1971). Furthermore, substrate binding to cytochrome P450 may not always produce a binding spectrum even though metabolism occurs, as is the case with benzene and barbital (Imai and Sato, 1967). Treatment of rat liver microsomes with phospholipase C abolishes the type I binding spectra of ethylmorphine and hexobarbital but reduces their hydroxylation by only 40% (Chaplin and Mannering, 1970).

2.4.2. Reduction

The overall process of microsomal oxidation requires the transfer of two electrons from NADPH through a series of redox components to cytochrome P450. The 1 : 1 ratio for substrate oxidation and NADPH oxidation expected from the general equation ($S + O_2 + XH_2 \rightarrow SO + H_2O + X$) has been demonstrated (Stripp *et al.*, 1972). Reduction, however, occurs in two separate one-electron steps. The first of these, involving the reduction of the cytochrome P450–substrate complex via NADPH-cytochrome *c* reductase, is monitored in the presence of carbon monoxide by the appearance of absorbance at 450 nm. The second electron is introduced at the level of the oxygenated cytochrome P450–substrate complex (oxycytochrome P450). Evidence for this step primarily stems from titration studies of bacterial cytochrome $P450_{cam}$ which show that, after the introduction of oxygen, excess reduced putidaredoxin causes the rapid decomposition of oxycytochrome P450 (Ishimura *et al.*, 1971).

The route of the second electron from a reduced pyridine nucleotide to oxycytochrome P450 reputedly involves cytochrome b_5, a component of the microsomal electron transport system usually associated with fatty acid desaturation. Some evidence supports the role of this pigment in microsomal oxidation. Cytochrome b_5, fully reduced under aerobic conditions when NADPH is present but substrate is not, is partially reoxidized on addition of substrate and the partial reoxidation is associated with increased oxygen uptake and product formation (Hildebrandt and Estabrook, 1971). Furthermore, the concentration of substrate (ethylmorphine) causing a half-maximal decrease in the level of reduced cytochrome b_5 approximates the K_m of the reaction. Nevertheless, the essentiality of cytochrome b_5 in NADPH-mediated microsomal oxidation is disputed by the fact that an antibody to cytochrome b_5 does not inhibit the reaction (specifically, the *N*-demethylation of ethylmorphine) (Gillette *et al.*, 1972; Mannering *et al.*, 1974).

Based on *in vitro* studies, NADPH is the preferred source of reducing equivalents for both reduction steps, although its *in vivo* role may be shared by other electron donors. One likely candidate, NADH, is capable of supporting the microsomal oxidation of certain substrates, but the reactions usually proceed at reduced rates compared with those observed in the presence of NADPH. This lower activity presumably reflects the unsuitability of NADH as an electron donor in the first reduction. However, NADH increases the microsomal oxidation of certain substrates in the presence of saturating concentrations of NADPH (Gillette *et al.*, 1972) and also decreases the rate of NADPH oxidation (Sasame *et al.*, 1973). The synergistic effect of NADH is apparently mediated by cytochrome b_5 since it can be abolished by an antibody to this cytochrome. Because NADH stimulates NADPH-supported reactions without altering the substrate affinity of the system and because it reduces the lifetime of oxycytochrome P450 in the presence of NADPH, it is thought by many to facilitate the second reduction step (Hildebrandt and Estabrook, 1971). On the other hand, donation of the second electron by NADH is inconsistent with the finding that the ratio of "CO-inhibitable" NADPH-oxidation to "CO-inhibitable" drug metabolism is not decreased by NADH (Gillette *et al.*, 1972).

2.4.3. Oxygen Binding and Activation

Binding of carbon monoxide to cytochrome P450 in competition with oxygen is an indication of the role of this cytochrome in oxygen activation. In fact, along with the requirements for NADPH and oxygen, inhibition by carbon monoxide is an important criterion for cytochrome P450-mediated microsomal oxidation. The formation of oxycytochrome P450 has been demonstrated in hepatic microsomes and adrenocortical mitochondria and has been characterized from a purified bacterial preparation (Baron *et al.*, 1973). Its formation in mammalian preparations was originally deduced from the observation of a peak at 440 nm in the differences spectra of type I substrates in the presence of NADPH and O_2 (Estabrook *et al.*, 1971). The magnitude of this peak was dependent on the type and concentration of the substrate employed. In the case of hexobarbital, a similarity was observed among the type I spectral dissociation constant (K_s), the Michaelis constant (K_m), and the substrate concentration required for half-maximal formation of the oxygenated complex (Baron *et al.*, 1973). However, similar results with other substrates have not yet been demonstrated.

Subsequent to its binding to cytochrome P450, the oxygen molecule is activated and split, one atom being inserted into the substrate and the other reduced to water. The mechanism by which cytochrome P450 effects the introduction of oxygen into a substrate could conceivably involve the generation of a free radical or the direct insertion of atomic or molecular oxygen into the substrate. Since many enzymatic hydroxylations are reminiscent of electrophilic displacement reactions (Daly, 1971), the active oxygen species may act as an electrophile. In line with this reasoning, an oxenoid mechanism involving an electron-deficient oxygen species (electrophilic or radical) has been proposed (Hamilton *et al.*, 1973). No direct evidence is available for or against this mechanism in the P450 system. Recent studies with the reconstituted P450 system implicate the participation of superoxide anion ($O_2^{\overline{\cdot}}$) (Coon *et al.*, 1973). Although benzphetamine hydroxylation was inhibited by superoxide dismutase and Tiron, superoxide quenchers, it has not been resolved whether the superoxide anion acts as an electron source or is related to the putative active oxygen.

In summary, the evidence to date has established cytochrome P450 as the enzyme which binds both substrate and oxygen. The cyclic sequence of reactions in which substrate binding and oxygen activation are intimately linked to two one-electron reductions has been gradually uncovered, although details of the precise mechanism of substrate oxidation, including the identity of activated oxygen, remain speculative.

3. Physiological Significance of Microsomal Oxidases

3.1. Substrate Spectrum

Since the evolution of animals could not have anticipated the introduction of insecticides and drugs, the enzymes responsible for the metabolism of these chemicals must play a role in the metabolism of other substances normally found in the animal. It is well known that bile pigments, bile acids, and various hormones are conjugated in the liver (Parke, 1968; see also Chaper 5), and fatty acids, bile acids and steroids are hydroxylated by hepatic microsomal oxidases (Conney and Kuntzman, 1971). Insect steroids may also undergo microsomal oxidation as indicated by the hydroxylation of α-ecdysone, a prohormone, to the active molting hormone, β-ecdysone (King, 1972). Although many normal body constituents are metabolized in the same manner as insecticides and drugs, it is not clear whether the same enzymes are involved. As Wilkinson and Brattsten (1972) have pointed out, the highly delicate physiological balance of steroid hormones seems incompatible with their regulation by nonspecific enzymes. In fact, steroid metabolism has been found in organs relatively deficient in drug metabolism (Cooper and Salhanick, 1973).

Regardless of their function in the metabolism of essential substances, the enzymes that metabolize nonessential, foreign compounds are an integral part of the homeostatic mechanism which contributes to the remarkable adaptability of animals. The levels of these enzymes may reflect the degree of exposure of animals to foreign compounds, or xenobiotics as they are often called, which continually find their way into animals via ingestion of food, by contact, or by inhalation, or may even originate in symbiotic microbes. Gordon (1961) suggested that the high tolerance of polyphagous insects to insecticides is probably the result of selection pressure from a variety of natural food plants. Krieger *et al.*, (1971) have provided experimental evidence for this hypothesis. In a survey of 35 species of lepidopterous larvae, they found that microsomal oxidase activity in the gut tissues tends to be higher in those species that feed on a greater variety of plant families.

Therefore, the normal substrates of drug- and insecticide-metabolizing enzymes are the lipophilic and often toxic xenobiotics of natural origin (Brodie, 1956). Insecticides and drugs, like other man-made chemicals including industrial pollutants, are thus only incidental substrates for the xenometabolic enzymes* which have developed through the evolution of each animal species.

The fundamental function of xenometabolic enzymes is to facilitate the elimination of lipophilic foreign compounds by converting them to more polar substances. In considering the elimination of various chemicals, it might be appropriate to liken the animal body to a chromatographic system in which the lipoprotein membranes constitute the stationary phase and the aqueous body fluids the mobile phase. Lipophilic substances tend to be retained by the system whereas polar materials are readily eliminated.

Nonoxidative xenometabolic enzymes, such as glutathione *S*-transferases, hydrolases, hydrases, and many conjugation enzymes, usually confer a high degree of polarity on certain xenobiotics. The effectiveness of these enzymes is limited, however, by the requirement for specific functional groups (OH, NH_2,

*The phrase "xenometabolic enzymes" was originally suggested by Williams (1967) in reference to the presumably foreign compound-specific "drug-metabolizing" enzymes of liver microsomes. In our discussions, however, "xenometabolism" is used simply for "foreign compound metabolism" without reference to substrate specificity or class of enzymes.

OCH_3, etc.) in their substrates to enable the characteristic nucleophilic displacement (SN_2) reactions to occur. Foreign compounds lacking such groups do not lend themselves to direct disposal by these enzymes. In contrast, microsomal oxidases can increase the polarity of most xenobiotics by virtue of their much wider substrate spectrum, often facilitating further introduction of hydrophilic groups by other enzymes. Sometimes the action of microsomal oxidases alone yields metabolites of sufficient polarity for direct excretion. The central role of microsomal oxidases in the elimination of xenobiotics is reflected in their generally low degree of development in aquatic organisms. For these animals, polarity should be of less importance since the high affinity of lipophilic substances for animal tissues is offset by a large volume of ambient water. Nevertheless, microsomal oxidases are vitally important in aquatic organisms in dealing with highly hydrophobic chemicals. Biomagnification of certain insecticides in aquatic organisms bears witness to the difficulty in eliminating such compounds without increasing their polarity (Nakatsugawa and Nelson, 1972).

Microsomal oxidases seem to be particularly adapted to the task of attacking lipophilic foreign compounds. The propensity of these enzymes for lipophilic compounds was first indicated by Gaudette and Brodie (1959), who showed that oxidative dealkylation of *N*-alkylamines by rabbit liver microsomes was limited to compounds that are lipid soluble at a physiological pH. Subsequent examination of a wider range of compounds, however, has demonstrated that even water-soluble compounds can be rapidly metabolized by these oxidases (Mazel and Henderson, 1965). The importance of the lipophilicity of a substrate seems to lie, not in influencing its reaction rate, but in governing its affinity for the enzymes. Thus Martin and Hansch (1971) found an excellent correlation ($r = 0.920$) between the lipophilicity of a series of 14 xenobiotics (in terms of octanol–water partition coefficients) and their affinity ($1/K_m$) for the microsomes determined from substrate-dependent NADPH oxidation. Such a property would allow the microsomal oxidases to metabolize more lipophilic xenobiotics preferentially when presented with alternative substrates and might explain the higher inhibition of microsomal oxidases by more lipophilic drugs (Kato *et al.*, 1969). Unfortunately much of the work cited was concerned only with *N*-alkyl compounds, the oxidation of which might also involve microsomal flavo protein *N*-oxidases (see Section 4). More work is needed in order to determine if the relationship found for *N*-alkyl compounds applies to the P450 oxidases in general.

The predilection of microsomal oxidases for diverse lipophilic xenobiotics might suggest the existence of a single nonspecific enzyme in each species. Lipophilicity, however, is not the only property that determines substrate specificity. The existence of a multiplicity of microsomal oxidases with different characteristics is suggested by the differential effects of inhibitors on various types of reactions catalyzed by a single enzyme preparation, by unequal degrees of enzyme induction for various oxidative pathways, and by demonstrations of different forms of cytochrome P450 in a single species. Since none of the putatively distinct enzymes has been isolated in pure form, the question of enzyme specificity remains unsettled. Meanwhile, the enzyme activity determined for a particular reaction may be considered as the sum of the activities of an unknown number of enzyme species with an overlapping substrate specificity. Such "an enzyme" is commonly defined according to the reaction it catalyzes, e.g., aldrin epoxidase and naphthalene hydroxylase.

3.2. Strategic Localization of Microsomal Oxidases

In general, foreign substances are inadvertently assimilated by the animal at one of three main portals of entry: the respiratory apparatus, the integument, or the digestive tract. The presence of microsomal oxidases at these strategic locations may therefore constitute the first line of defense against a variety of toxicants.

Low, but inducible, polycyclic hydrocarbon hydroxylase activity has been demonstrated in human skin, one of the largest organs of the body, using tissue culture techniques (Alvares *et al.*, 1973). Activity has also been detected by histological means in the sebaceous glands (Wattenberg and Leong, 1971). Microsomal oxidases in the lung may have significance in providing defense against airborne toxicants and gaseous materials. Both biochemical and histological studies indicate that *N*-demethylase and hydroxylase activities are localized in the alveolar epithelium (Bend *et al.*, 1973; Wattenberg and Leong, 1971). Moreover, these enzymes may have toxicological implications in the metabolism of toxicants already in the circulation. The activation of parathion in the lung, for example, may be of critical importance in determining its lethal action (Neal, 1972).

Since food is the major natural source of foreign compounds, the digestive tract must be the chief portal of entry. It is not surprising, therefore, that the digestive system is a common site of microsomal oxidases in animals. Substantial levels of polycyclic hydrocarbon hydroxylase activity are found throughout the entire gastrointestinal tract of the rat and even approach the level of liver enzymes in the proximal portion of the small intestine (Wattenberg and Leong, 1971). However, the digestive tract may be even more important in those species that lack a single liverlike organ of detoxication. Wilkinson and coworkers observed that microsomal oxidases are almost exclusively associated with the gut tissues in many species of lepidopterous larvae whose exposure to naturally occurring foreign compounds is mainly by way of the food. In contrast, insect species such as cockroaches, which tend to be more exposed to xenobiotics through the integument, have high microsomal oxidases in the fat body as well as the gut (Wilkinson and Brattsten, 1972; Benke *et al.*, 1972). The digestive tract is also the major tissue of microsomal oxidation in the common earthworm, *Lumbricus terrestris* L. (Nelson, 1974).

In vertebrates, by far the highest microsomal oxidase activity is located in the liver, and in the context of the foregoing discussion it is of interest to note that the liver is embryologically derived from the intestine. Having been absorbed with the food, toxicants proceed directly to the liver by way of the hepatic portal vein. The highly active microsomal oxidases of this organ are therefore ideally located as a second line of defense and attest to the dominance of this organ in detoxication. Enzymatic activity, however, may not be uniformly distributed among hepatic parenchymal cells. The majority of polycyclic hydrocarbon hydroxylase activity in uninduced animals is reported to reside in the central portion of the hepatic lobule (Wattenberg and Leong, 1971).

Although the liver is undoubtedly the central organ of detoxication in vertebrates, and tissues such as the lung, skin, and digestive tract also play a role, some activity probably occurs in many other tissues. For instance, benzyprene hydroxylase activity has been detected in the kidney and placenta (Wattenberg and Leong, 1971). Also noteworthy is the low level of parathion activation in the brain reported by Poore and Neal (1972), which may have special toxicological consequences.

4. *Microsomal Oxidation of Insecticides*

Most organic insecticides and synergists are subject to microsomal oxidation (Wilkinson, 1968; Dahm and Nakatsugawa, 1968; Casida, 1970; Brooks, 1972). Many of them possess multiple sites at which oxidation can occur, and consequently a combination of several transformations can take place with any particular compound (Fig. 4). Unfortunately, our knowledge is as yet insuffi-

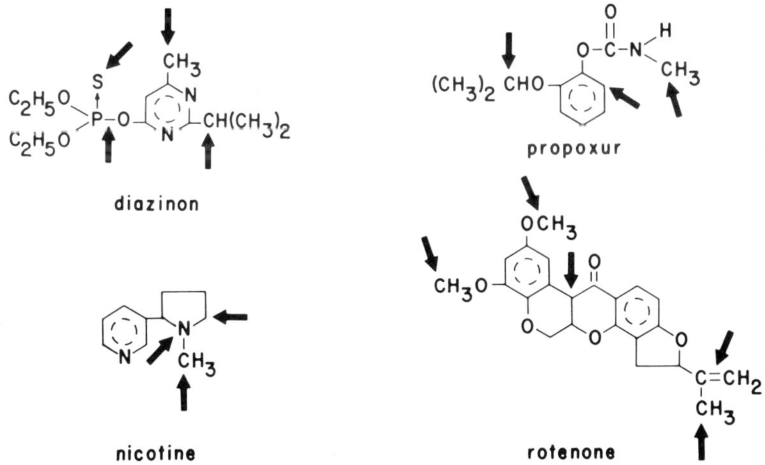

Fig. 4. Examples of multiple sites for microsomal oxidation.

cient to enable us to foretell which of the possible pathways will dominate in a given species *in vivo*. However, the recognition of characteristic transformations for various chemical structures has permitted some degree of prediction. In the following sections, we shall examine these reaction types and arrive at a generality that might prevail in the microsomal oxidation of insecticides.

4.1. Oxidation of Various Structures

The seemingly bewildering array of oxidative transformations of insecticides can be viewed as combinations of a few basic reaction types. For convenience, these reactions will be discussed in five categories: (1) *O*-, *S*-, and *N*-alkyl hydroxylation, (2) aliphatic hydroxylation and epoxidation, (3) aromatic hydroxylation, (4) ester oxidation of organophosphates, and (5) nitrogen and thioether oxidation.

4.1.1. O-, S-, and N-Alkyl Hydroxylation

An alkyl group adjacent to a hetero atom such as oxygen, sulfur, or nitrogen is a potential target for microsomal hydroxylation, but because of the electronegativity of the hetero atom the reaction often leads to dealkylation. Dealkylation of *O*-alkyl groups of the ester or ether structures of insecticides occurs readily, but does not take place by simple replacement of an alkoxy group with a hydroxy group (Renson *et al.*, 1965). Instead, an unstable α-hydroxy intermediate is produced which spontaneously releases an aldehyde in the case of a primary alkyl group and a ketone in the case of a secondary alkyl group (Fig. 5). Thus the isopropyl group of the carbamate, propoxur, is readily removed as acetone by the microsomal oxidase system of rats as well as several strains of houseflies (Oonnithan and Casida, 1968; Shrivastava *et al.*, 1969); deisopropylation is a dominant feature of propoxur metabolism. The demethylation of one or both methyl groups of methoxychlor appears to be the basis for its biodegradability. This pathway has been demonstrated in the mouse liver microsomal oxidase system and is suspected in houseflies because of the high degree of methoxychlor synergism by piperonyl butoxide (Kapoor *et al.*, 1970). Similar removal of the ethyl groups of ethoxychlor has been shown in a mouse liver homogenate system (Kapoor *et al.*, 1972).

Rotenone also contains two methoxy groups that are potential targets for microsomal dealkylation (Fig. 4). Recent studies with [3-methoxy-^{14}C] rotenone *in vivo* have shown that rats and mice release $^{14}CO_2$ extensively following oral or intraperitoneal administration (Unai *et al.*, 1973). This evidence strongly indicates that oxidative demethylation of rotenone and/or its metabolites is a major degradation pathway, a fact which was not obvious from the results of earlier *in vitro* studies with [6a-^{14}C]rotenone. While the studies of Unai *et al.* (1973) showed the fate of only the 3-methoxy group, similar

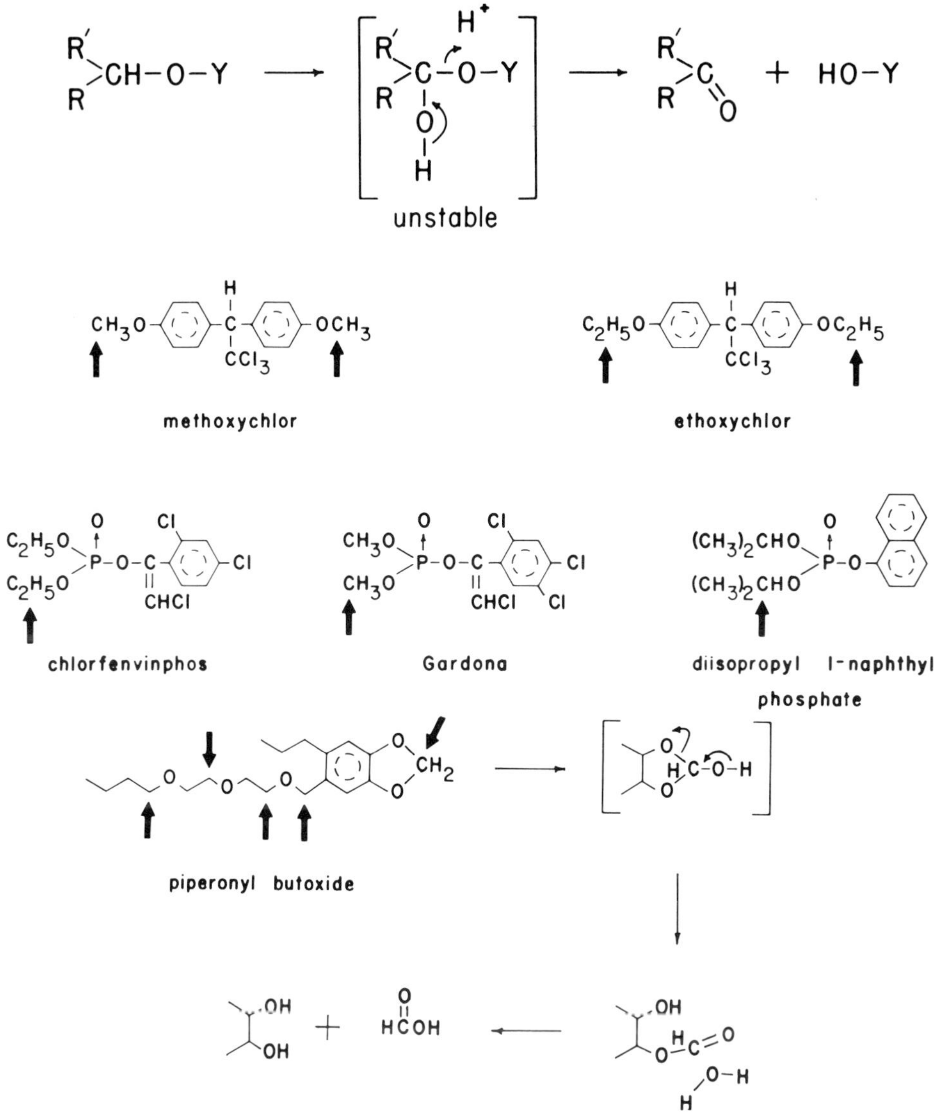

Fig. 5. *O*-Dealkylation.

demethylation of the 2-methoxy group seems possible in view of the dual dealkylation with methoxychlor and ethoxychlor.

Contrary to the earlier belief that the importance of dealkylation in organophosphate metabolism was limited to dimethyl esters, accumulating evidence indicates that the removal of alkyl groups is also a significant metabolic pathway for higher homologues (Appleton and Nakatsugawa,

1972). Microsomal oxidases seem to dealkylate a number of organophosphates, including certain dimethyl triesters (Fig. 5). Although the role of glutathione *S*-alkyltransferases in demethylation of organophosphates is well established (see Chapter 5), it is likely that the glutathione-linked and the oxidative demethylases have different substrate spectra and species variations. The deethylation of chlorfenvinphos by rabbit liver microsomes yielded acetaldehyde and the diester, and a dimethyl analogue, Gardona, was similarly demethylated to yield formaldehyde (Donninger *et al.*, 1972). The same system also dealkylated dimethyl-1-naphthylphosphate and its diethyl and diisopropyl homologues. The abdomen microsomes of the F_c strain of the housefly and rat liver microsomes deethylate paraoxon in a similar manner (Oppenoorth, 1971; Ku and Dahm, 1973). Although direct enzymological evidence is not available, microsomal oxidases may also be responsible for the formation of deisopropyl parathion and its phosphate analogue, which are observed in the urine of mice treated with isopropyl parathion (Camp *et al.*, 1969) (see also Section 4.2.4 for malaoxon deethylation).

An important variant of *O*-dealkylation is found in the microsomal demethylenation of methylenedioxyphenyl synergists (Fig. 5). Although an immediate hydroxylation product has not been isolated, it is likely to be an unstable hydroxy derivative which would yield the observed metabolite, formic acid. The other expected metabolite, a catechol, has been identified with several substrates (Casida *et al.*, 1966; Kamienski and Casida, 1970; Wilkinson and Hicks, 1969). All methylenedioxyphenyl compounds examined to date show characteristic demethylenation. For some compounds including piperonyl butoxide and sulfoxide diastereomers, this is the major pathway, whereas with compounds such as Tropital the polyether side chains on the aromatic ring are the preferred sites of oxidative attack. The latter pathway is probably a typical *O*-dealkylation.

Microsomal *S*-demethylation of several methylthio compounds has been reported (Gram, 1971), and its involvement in aldicarb metabolism has been inferred from the *in vivo* conversion of the methylthio carbon to carbon dioxide in the housefly (Metcalf *et al.*, 1967). However, no *S*-demethylation of aldicarb has been detected in the rat liver *in vitro* system (Andrawes *et al.*, 1967).

The hydroxylation of *N*-alkyl groups occurs in the metabolism of many organophosphates and carbamates. Unlike *O*-alkyl hydroxylation, this reaction often yields a fairly stable *N*-α-hydroxyalkyl derivative, probably because nitrogen is less electronegative than oxygen. The metabolite may then undergo nonoxidative cleavage to a dealkylated product and an aldehyde (Fig. 6). In the case of *N*-methyl hydroxylation, further oxidation to an *N*-formyl derivative has sometimes been noted, although the nature of the oxidase for this step has not been defined.

The activation of the phosphoramidate insecticide, schradan or OMPA (Fig. 6), to *N*-hydroxymethyl schradan is a classical example of *N*-alkyl hydroxylation (O'Brien, 1960). It is likely that the *N*-hydroxymethyl deriva-

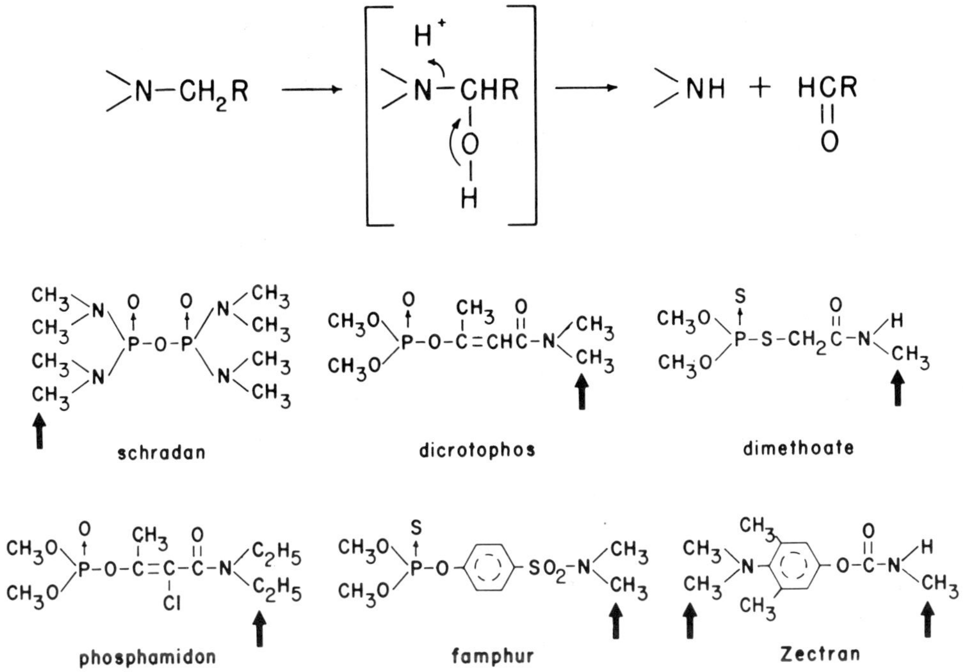

Fig. 6. *N*-Alkyl hydroxylation and dealkylation.

tive is an intermediate in stepwise *N*-demethylation reactions of this compound. More recent studies with the related chemical HMPA (hexamethylphosphoramide, a chemosterilant) have demonstrated a series of demethylations in the rat to yield *N,N′,N″*-trimethylphosphoramide (Jones and Jackson, 1968; Jones, 1970). Although small amounts of *N*-formyl derivatives were observed, indicating that the reaction is initiated by *N*-methyl hydroxylation, no *N*-hydroxymethyl intermediates were isolated. Similar dealkylations of the ethyl and *n*-propyl homologues of HMPA were also demonstrated.

The consecutive steps of *N*-methyl hydroxylation and demethylation of the *N,N*-dimethyl group of dicrotophos were initially indicated by *in vivo* studies (Bull and Lindquist, 1964). The sequence was later elucidated with an *in vitro* microsomal system using an elegant double-labeling (^{32}P and ^{14}C) technique (Menzer and Casida, 1965). Subsequent studies have shown an identical type of *N*-ethyl oxidation for a closely related insecticide, phosphamidon (Bull *et al.*, 1967; Clemons and Menzer, 1968; Lucier and Menzer, 1971). Dimethoate and famphur also undergo *N*-demethylation (Lucier and Menzer, 1970; O'Brien *et al.*, 1965).

An important, if not a major, metabolic pathway of the *N*-methyl- and *N,N*-dimethylcarbamates is the *N*-methyl hydroxylation of the carbamyl moiety (Fig. 6). *N*-Hydroxymethyl carbamates or their conjugates have been identified as metabolites of a number of carbamates *in vivo* and in microsomal oxidase systems of both mammals and insects (Hodgson and Casida, 1960; Oonnithan and Casida, 1968; Miyamoto *et al.*, 1969; Shrivastava *et al.*, 1969; Douch and Smith, 1971*a,b*; Strother, 1972). The corresponding *N*-demethylated metabolite is known in only a few cases (Douch and Smith, 1971*a*) since *N*-unsubstituted carbamates may be extremely labile to hydrolysis (Fahmy *et al.*, 1966).

The *N,N*-dimethyl group is a substituent on the aromatic ring of the carbamates Zectran and aminocarb (Figs. 6 and 17) and is subject to microsomal oxidative to mono- and bis-demethyl derivatives (Oonnithan and Casida, 1968). The methylformamido metabolite of Zectran has also been isolated *in vitro*. Related compounds, *N,N*-dialkylarylamines, are known to undergo microsomal dealkylation via *N*-oxides rather than via hydroxyalkyl derivatives (Willi and Bickel, 1973). However, the *N*-oxide is probably not an obligatory intermediate in dealkylation, so the general applicability of the oxide pathway has been discounted (Gram, 1971; Weisburger and Weisburger, 1971).

N-Demethylation appears to be an important primary step in chlordimeform metabolism (Fig. 17). The monodemethyl derivative has been detected *in vivo* in houseflies, rats, dogs, and goats (Knowles and Sen Gupta, 1970; Sen Gupta and Knowles, 1970; Knowles and Shrivastava, 1973) and was produced *in vitro* by a housefly abdomen microsomal system.

One of the primary metabolic reactions of nicotine is the hydroxylation at carbon 5′ (adjacent to *N*) of the pyrrolidine ring to yield 5′-hydroxynicotine (Fig. 7). Cotinine, a major metabolite of nicotine, probably arises via oxidation of 5′-hydroxynicotine by a cyanide-sensitive enzyme (Hucker *et al.*, 1960). Microsomal *N*-demethylation of nicotine and cotinine to nornicotine and demethylcotinine, respectively, has also been demonstrated (Papadopoulos and Kintzios, 1963).

4.1.2. Aliphatic Hydroxylation and Epoxidation

In addition to aliphatic C—H groups associated with hetero atoms, other aliphatic structures are readily attacked by microsomal oxidases. An aliphatic C—H group not adjacent to a hetero atom is often hydroxylated to a stable alcohol. An aliphatic carbon–carbon double bond may be converted to an epoxide or a dihydrodiol. These reactions constitute another characteristic set of oxidative transformations catalyzed by microsomes (Figs. 8–12). Alkyl groups such as methyl, ethyl, isopropyl, and tertiary-butyl are common substituents on aromatic nuclei of insecticides (Fig. 8). *In vivo* and *in vitro* studies indicate that microsomal oxidation of these groups is important in the degradation of carbamates such as Meobal (Miyamoto *et al.*, 1969). Butacarb (Douch and Smith, 1971*b*), and possibly Banol and UC 10854 (Oonnithan and Casida, 1968). The alcoholic metabolite thus formed may be

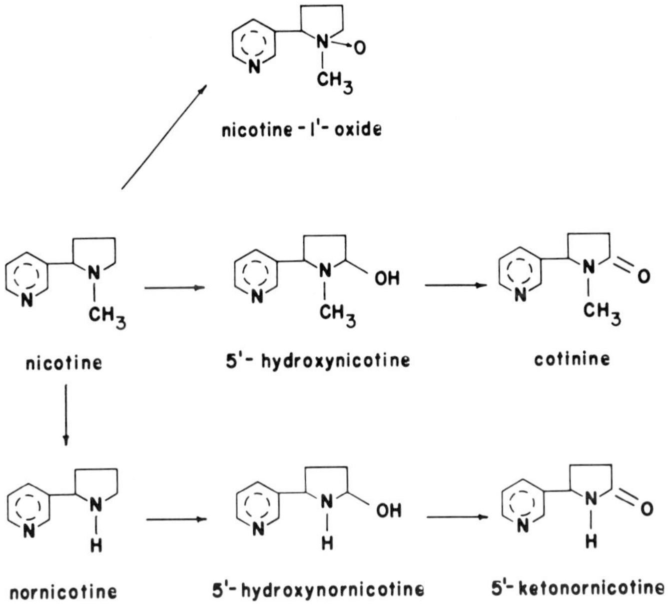

Fig. 7. Oxidative metabolism of nicotine.

further oxidized to a carboxylic acid by a soluble NAD-coupled oxidase or conjugated with glucuronic acid (Miyamoto *et al.*, 1969). A similar hydroxylation seems to be important in the metabolism of the organophosphate diazinon, since two isomers of hydroxydiazinon have been isolated from the urine and tissues of sheep after oral administration of diazinon (Machin *et al.*, 1971, 1972). Substantial amounts of alkyl-hydroxylated pyrimidinols which may be formed from hydroxydiazinon and/or hydroxydiazoxon have also been found in the urine of rats treated with diazinon (Mücke *et al.*, 1970). The microsomal hydroxylation of DDT to dicofol or a related metabolite (Tsukamoto, 1959; Agosin *et al.*, 1961; Morello, 1964, 1965; Dinamarca *et al.*, 1962) also belongs in this category of oxidation.

Microsomal epoxidation of the unchlorinated double bond (Fig. 9) is the most important metabolic reaction of certain cyclodiene insecticides such as heptachlor, aldrin, isodrin, and chlordene (Brooks, 1974). The epoxidation occurs rapidly both *in vivo* and *in vitro* (Davidow and Radomski, 1953; Brooks *et al.*, 1963; Wong and Terriere, 1965; Nakatsugawa *et al.*, 1965). Although these epoxides are often quite stable metabolically, they are subject to hydration to dihydrodiols by epoxide hydrases (Brooks and Harrison, 1969*a*; Brooks *et al.*, 1968) (Fig. 9). *Trans*-6,7-dihydroxydihydroaldrin (*trans*–aldrindiol) has been identified as a urinary metabolite of dieldrin (epoxide of aldrin) from rabbit and rat (Korte and Arent, 1965; Matthews *et al.*, 1971) and as an *in vitro* metabolite of several microsomal systems

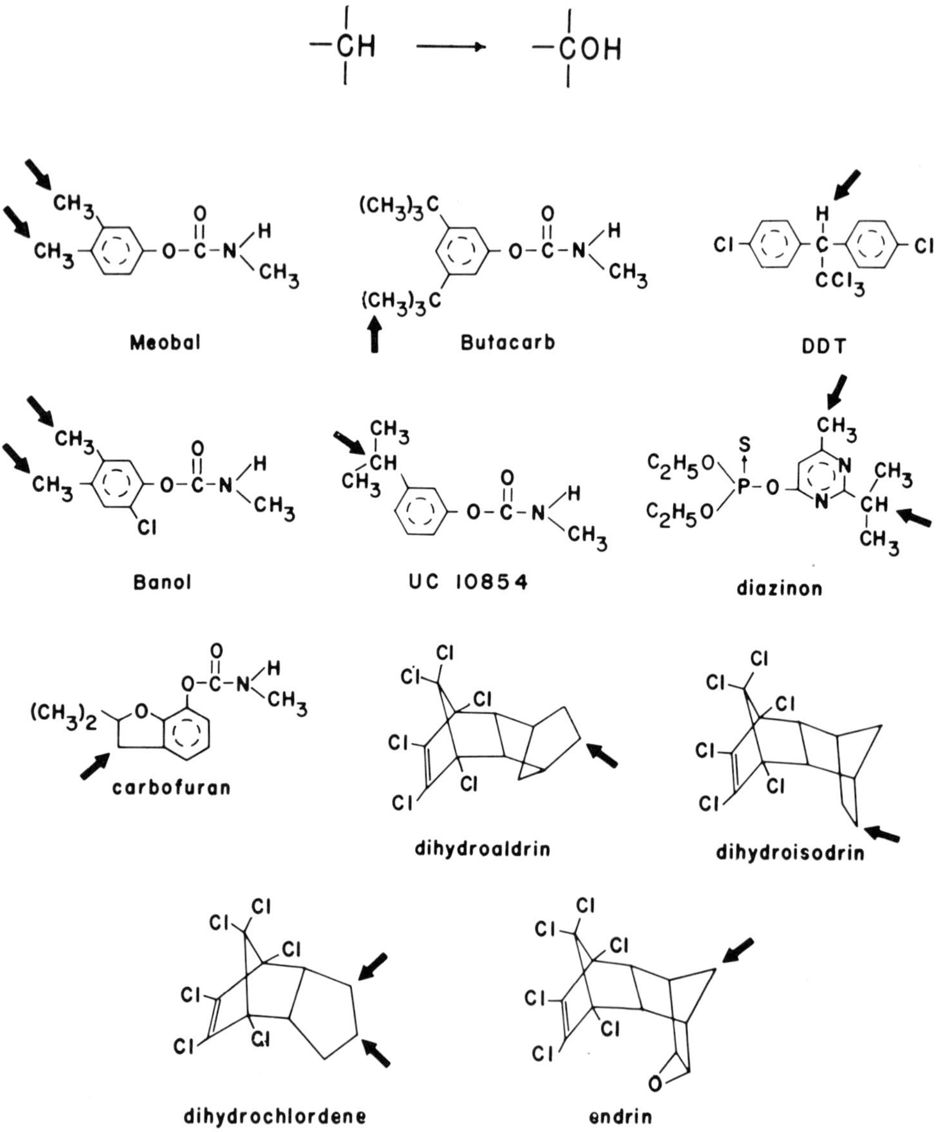

Fig. 8. Aliphatic hydroxylation.

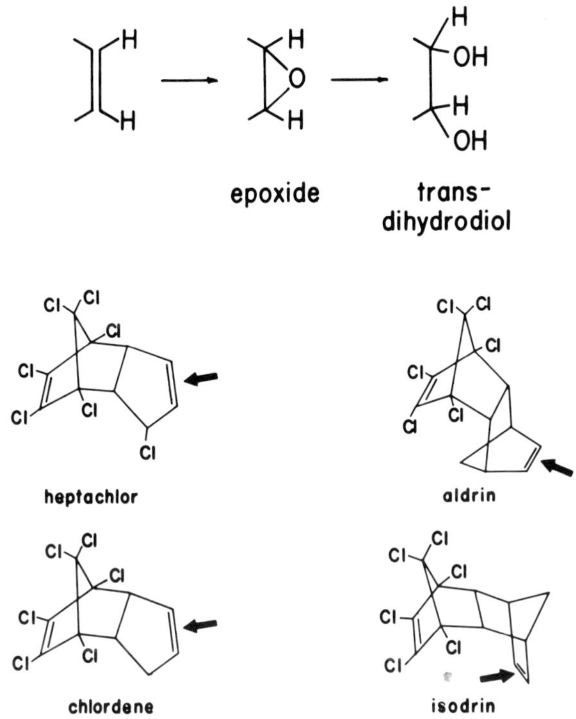

Fig. 9. Epoxidation and hydration of cyclodiene insecticides.

(Matthews and Matsumura, 1969; Brooks *et al.*, 1970). Whether the *trans* isomer is the direct metabolite of hydration is uncertain in view of the discovery in the rat liver of an epoxide epimerase which unidirectionally epimerizes the *cis* to the *trans* isomer (Matthews and McKinney, 1974).

Slow oxidative metabolism of epoxides also occurs, as indicated by the isolation of oxidative derivatives following oral administration of the epoxides dieldrin (Richardson *et al.*, 1968) and endrin (Baldwin *et al.*, 1970) to rats. Demonstration of such slow oxidative pathways of dieldrin metabolism *in vitro* was possible by a long (2 hr) incubation of a low concentration (2×10^{-6} M) of [^{14}C]-dieldrin with a rat liver microsome-plus-soluble fraction (Matthews and Matsumura, 1969) (Fig. 10). It appears that dieldrin undergoes hydration, hydroxylation at 4a, and more predominantly a complex hydroxylation to yield a derivative with a hydroxyl group at position 2. The 4a-hydroxyl derivative (F-1) is the major *in vivo* metabolite, mainly in the form of a conjugate (Richardson *et al.*, 1968; Matthews *et al.*, 1971). The 2-hydroxyl product (M-6) was either converted spontaneously to a keto derivative, known as "Klein's metabolite," or rapidly conjugated with glucuronic acid.

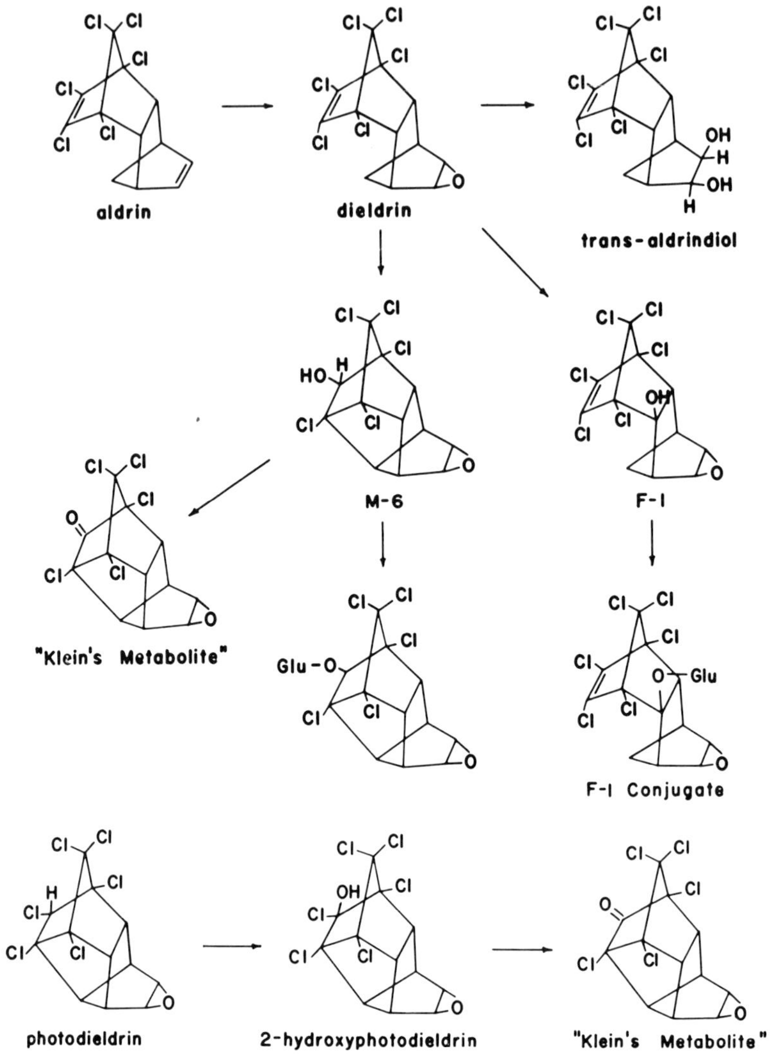

Fig. 10. Oxidation of dieldrin and photodieldrin.

The mechanism of oxidative metabolism leading to the keto derivative, detachment of a chlorine, and formation of the cage structure has not been clarified.

"Klein's metabolite" has also been isolated *in vivo* as a metabolite of photodieldrin in insects and rats (Baldwin and Robinson, 1969; Khan *et al.*, 1969; Klein *et al.*, 1970). It is likely that the metabolite is formed via 2-hydroxyphotodieldrin (Fig. 10). The same intermediate could conceivably arise from dieldrin through dieldrin epoxide.

Hydroxylation of the carbon skeleton of cyclic structures occurs more readily with the related compounds (Fig. 8). For example, 6,7-dihydroaldrin is hydroxylated by microsomal enzymes to both the *endo* and *exo* isomers of 6-hydroxy-6,7-dihydroaldrin (Brooks and Harrison, 1969*b*), whereas dihydroisodrin, the *endo,endo* isomer of dihydroaldrin, yields only the 6-*exo*-hydroxy derivative, possibly owing to the steric hindrance of the *endo,endo* structure (Krieger and Wilkinson, 1971). Several products have been found in the microsomal oxidation of dihydrochlordene; hydroxylation seems to occur at carbons 1 and 2 (Brooks and Harrison, 1967). An *in vivo* demonstration of 9-hydroxyendrin and 9-ketoendrin in a rat which had been fed endrin indicates that the endomethylene carbon is also a potential target of hydroxylation (Baldwin *et al.*, 1970). Hydroxylation of the cyclic skeleton is a major

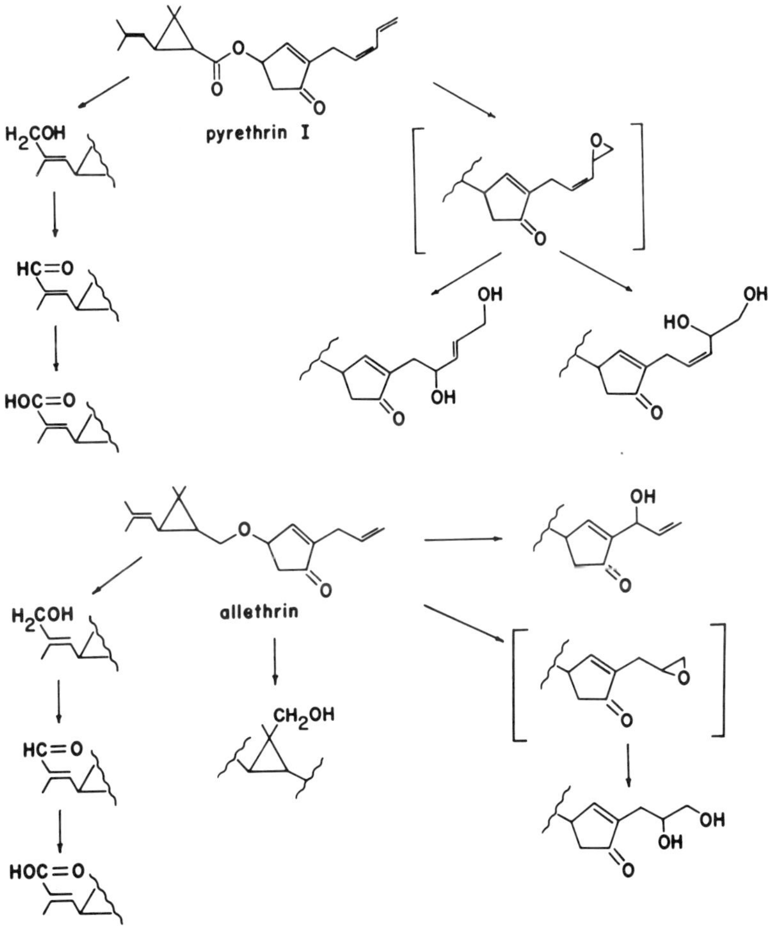

Fig. 11. Oxidative metabolism of pyrethrin I and allethrin.

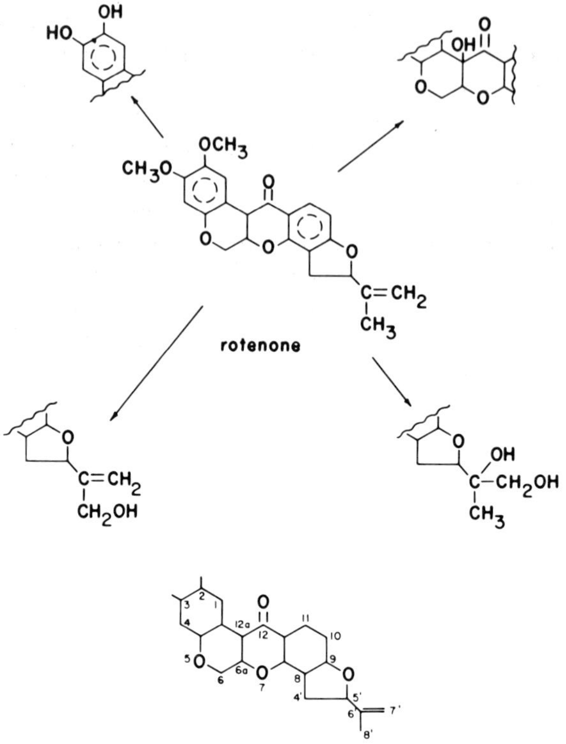

Fig. 12. Oxidation of rotenone (numbering according to Büchi *et al.*, 1961).

metabolic route of the carbamate carbofuran; the metabolite, 3-hydroxycarbofuran, is either conjugated directly or oxidized to a ketone (Metcalf *et al.*, 1968; Dorough, 1968; Ivie and Dorough, 1968).

Microsomal oxidations constitute major pathways in the metabolism of certain pyrethroids, particularly those which are esters of secondary alcohols (Fig. 11). A major reaction with chrysanthemates (allethrin and pyrethrin I) in rats and houseflies *in vivo* as well as in microsomal systems *in vitro* is the hydroxylation of the *trans*-methyl group of the isobutenyl chain in the acid moiety to form a hydroxymethyl derivative (Yamamoto *et al.*, 1969; Elliott *et al.*, 1972). The derivative may be conjugated with glucose (in houseflies) or further oxidized to a carboxylic acid via an aldehyde by a cyanide-sensitive oxidase. Similar oxidation of the *cis*-methyl group of allethrin is a minor pathway in houseflies. These studies, which were carried out with various labeled pyrethroids, established that metabolism of these esters does not involve significant amounts of hydrolysis at the central ester linkage.

Identification of rat urinary metabolites of these pyrethroids and a comparison with the *in vitro* metabolites indicate that microsomal oxidation occurs also on the side chain of the cyclopentenolone ring in the alcohol moiety (Elliott *et al.*, 1972). Thus oxidation of the *cis*-2′,4′-pentadienyl chain of the pyrethrins yields *in vivo* two types of diol, i.e., the *trans*-2′,5′-dihydroxypent-3′-enyl and the *cis*-4′,5′-dihydroxypent-2′-enyl derivatives, and a 4′-conjugate of the latter. These metabolites are likely to arise from a terminal epoxide. A similar diol is also a metabolite of allethrin in the rat. In addition, allethrin is hydroxylated at the methylene group of the allyl chain as well as at a methyl on the cyclopropane ring. Although several other polar metabolites are produced in houseflies, oxidation of the alcohol moiety appears less extensive than in rats.

Other synthetic pyrethroids such as tetramethrin (phthalthrin) and dimethrin (Fig. 17) are also metabolized by oxidation of the *trans*-methyl group in the chrysanthemic acid moiety (Yamamoto *et al.*, 1969). The oxidative pathway, however, may be of less significance *in vivo* for these esters of primary alcohols in view of the extensive hydrolysis *in vivo* (Miyamoto and Suzuki, 1973) and the demonstration of esterases active particularly toward *trans*-chrysanthemates of primary alcohols (Abernathy *et al.*, 1973).

Microsomal oxidation of this type is also important in the metabolism of rotenone in mammals, fish, and insects (Fukami *et al.*, 1967, 1969) (Fig. 12) in addition to the *O*-dealkylation reactions previously described. The major metabolite of [6a-^{14}C]rotenone produced by the rat liver microsomal system was identified by UV, IR, MS, NMR, and chemical analysis as 6′,7′-dihydro-6′,7′-dihydroxyrotenone. Further analysis of metabolites indicated three types of microsomal oxidations, i.e., dihydrodiol formation at the 6′,7′ positions of isopentenyl chain and hydroxylations at the 8′ position and at the 12a position. Metabolism both *in vivo* and *in vitro* involves various combinations of these oxidations. The diols are possibly produced via 6′,7′-epoxy intermediates (Unai *et al.*, 1973). Hydroxylation at the 12a position occurs mainly in the *cis* configuration relative to the 6a hydrogen to yield rotenolone I; the *trans* isomer, rotenolone II, is a minor metabolite. These hydroxylated metabolites can be further metabolized to more polar metabolites *in vitro* by soluble enzymes, whereas *in vivo* they are excreted either unchanged or as more polar metabolites including some conjugates.

4.1.3. Aromatic Hydroxylation

Information on the enzymatic mechanism of aromatic hydroxylation comes from basic studies of mixed-function oxidases. Although there are few examples of aromatic hydroxylation among insecticides, a discussion of this reaction is warranted in view of its far-reaching implications to P450-mediated microsomal oxidation.

The NIH shift (named after the National Institutes of Health, where it was discovered) has been established for aromatic hydroxylation using tritium- and deuterium-labeled compounds and is characteristic of aromatic hydroxylation by all mixed-function oxidases (Guroff *et al.*, 1967; Daly *et al.*, 1967). During such hydroxylation reactions, the hydrogen atom replaced by the hydroxyl group is not always expelled from the molecule, but may migrate to an adjacent position on the ring. The degree of hydrogen retention varies with different substrates. Substituents other than hydrogen (e.g., halogen) may behave in a similar manner. Such retention is not observed in the well-known electrophilic substitution reactions such as bromination, nitration, diazonium coupling, and Friedel–Crafts acylation and alkylation and in general does not occur in the so-called model reactions for enzymatic hydroxylation, such as the Hamilton system, the Fenton reagent, or the Udenfriend system. The only model agent that showed a small degree of retention was the trifluoroperacetic acid system. The NIH shift was explained by assuming cationoid intermediates which might in turn arise from the opening of intermediate epoxides (Fig. 13).

Evidence seems to confirm the earlier supposition that an epoxide (arene oxide) is an obligatory intermediate in the mixed-function hydroxylation of naphthalene (Fig. 14). Naphthalene-1,2-oxide was isolated as a metabolite of

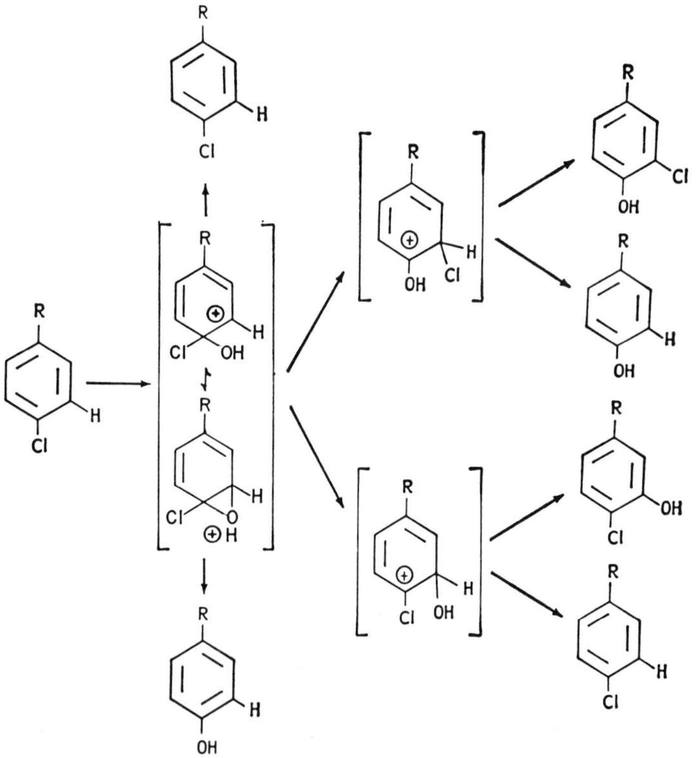

Fig. 13. NIH shift. From Guroff *et al.* (1967). Copyright 1967 by the American Association for the Advancement of Science.

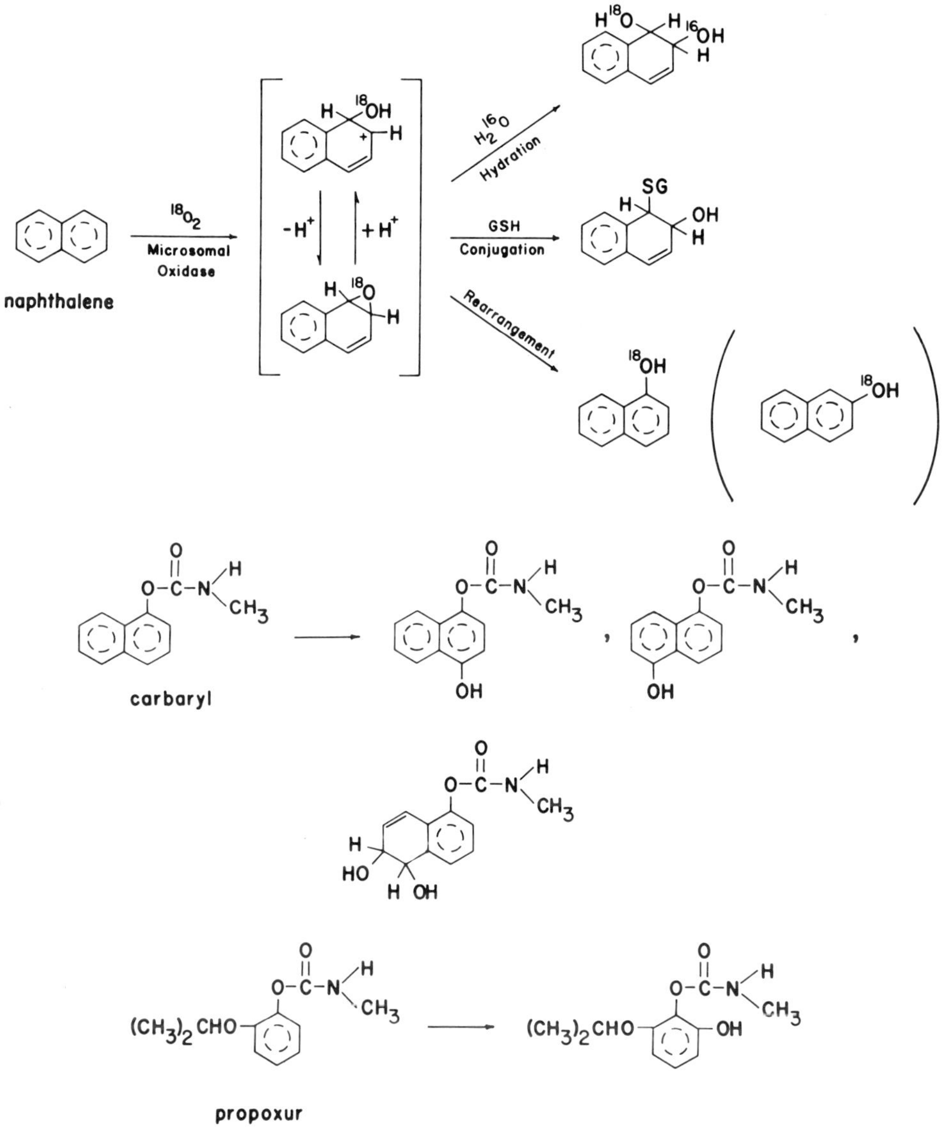

Fig. 14. Aromatic hydroxylation.

the rat liver microsomal oxidase system (Jerina *et al.*, 1969). Furthermore, synthetic naphthalene-1,2-oxide yields *in vitro* all major oxidative metabolites of naphthalene. Thus *trans*-1,2-dihydro-1,2-dihydroxynaphthalene, its glutathione conjugate, i.e., *S*-(1,2-dihydro-2-hydroxy-1-naphthyl)glutathione, 1-naphthol, and a trace amount of 2-naphthol were produced from both naphthalene and the epoxide. Hydration of the epoxide to dihydrodiol proceeds identically as in the case of naphthalene metabolism with respect to absolute stereochemistry and the source of the oxygen at the 2 position; the oxygen is derived from water. It appears, therefore, that naphthalene is converted by a microsomal oxidase to an epoxide which nonenzymatically rearranges almost exclusively to 1-naphthol; it may also be hydrated by a microsomal hydrase (Oesch *et al.*, 1972) or conjugated with glutathione by a soluble liver enzyme (Jerina *et al.*, 1969). Arene oxides are also implicated in other aromatic oxidations since dihydrodiols appear as urinary metabolites of benzene and halobenzenes as well as several polycyclic aromatic hydrocarbons (Jerina *et al.*, 1969). Synthetic benzene oxide is converted *in vitro* to phenol, dihydrodiol, catechol, and a glutathione conjugate (Jerina *et al.*, 1968). Moreover, the high retention of deuterium in the rearrangement of 3,4-[4-^{2}H]toluene oxide to 4-[3-^{2}H]hydroxytoluene is characteristic of the NIH shift. In spite of this evidence, direct hydroxylation of monocyclic arenes without the involvement of intermediate arene oxides has not been ruled out because no epoxide intermediate or dihydrodiol of any monocyclic arene has been isolated from an *in vitro* microsomal system.

Examples of aromatic hydroxylation of insecticides are rather limited, possibly because insecticides with aromatic nuclei often have other preferred sites for microsomal oxidation (Fig. 14, see also Figs. 5, 6, and 8). Aromatic hydroxylation, however, is the only pathway to initiate the biotransformation of naphthalene. It is also the major metabolic reaction of a naphthalene derivative, carbaryl (Dorough and Casida, 1964; Leeling and Casida, 1966; Oonnithan and Casida, 1968) (Fig. 14), where ring oxidation yields 4-hydroxy-1-naphthyl methylcarbamate, 5-hydroxy-1-naphthyl methylcarbamate, and 5,6-dihydro-5,6-dihydroxy-1-naphthyl methylcarbamate as major metabolites. Metabolism of propoxur in several strains of the housefly also involves ring hydroxylation to yield 5-hydroxypropoxur (Shrivastava *et al.*, 1969), which *in vivo* occurs mostly as a conjugate, probably the glucoside. Ring hydroxylation of DDT has also been suggested (Wilkinson, 1968; Agosin *et al.*, 1969).

4.1.4. Ester Oxidation of Organophosphates

Probably all organophosphorus insecticides with the P→S structure are desulfurated by microsomal oxidases of mammals and insects to their corresponding P→O analogues (Davison, 1955; Dauterman, 1971). The detached sulfur is apparently bound covalently to microsomal macromolecules *in vitro* (Nakatsugawa and Dahm, 1967) and *in vivo* (Poore and Neal, 1972), and is eventually excreted as inorganic sulfate (Nakatsugawa *et al.*, 1969*a,b*). Desulfuration, however, represents only part of the microsomal oxidation of these compounds. These esters are concurrently hydrolyzed to P→S acids and the corresponding leaving groups by the oxidase system (Nakatsugawa and

Dahm, 1967; Neal, 1967*a*; Nakatsugawa *et al.*, 1968, 1969*c*; Yang *et al.*, 1971*a*; Wolcott *et al.*, 1972; Wolcott and Neal, 1972; Motoyama and Dauterman, 1972). Since the P→O analogues are the actual toxic agents and the P → S acid derivatives are often the major degradation products *in vivo*, the balance between these two reactions is of critical importance in determining the action of phosphorothionate insecticides.

Experiments with $^{18}O_2$ and $H_2{}^{18}O$ have revealed that the P → O oxygen is derived from atmospheric oxygen, whereas the hydrolysis products contained no atmospheric oxygen but did contain oxygen from water as expected of usual hydrolysis. A common oxide-type intermediate has been proposed to account for the two concurrent reactions (McBain *et al.*, 1971*a*; Ptashne *et al.*, 1971). Presumably the intermediate either rearranges itself and releases a reactive sulfur atom or is hydrolyzed with the loss of the introduced oxygen atom (Fig. 15). Such a scheme is reminiscent of the aromatic hydroxylation mechanism discussed earlier.

Direct proof for the proposed organophosphorus intermediate, however, has not been obtained. Trifluoroperacetic acid, the only model agent that causes the NIH shift in aromatic hydroxylation (Guroff and Daly, 1967), produced paraoxon, diethyl phosphorothioic acid, and tetraethyl pyrophosphate from parathion, and dimethyl phosphorothioic acid from malathion

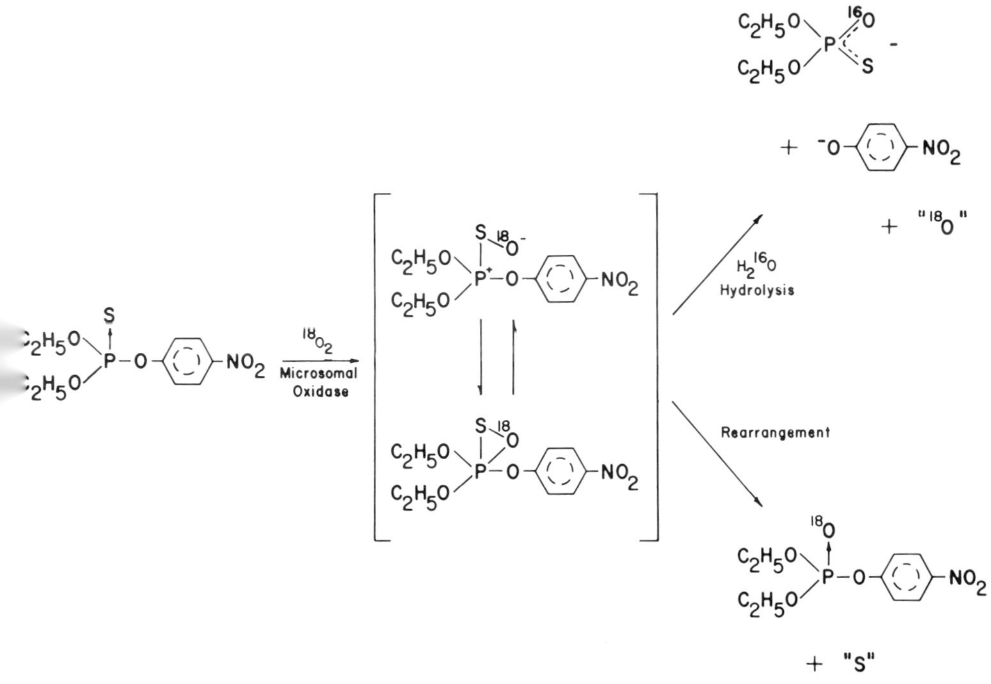

Fig. 15. Oxidative metabolism of parathion.

(Ptashne and Neal, 1972). Also, by use of another peracid, 3-chloroperbenzoic acid, an oxygenated product of Dyfonate was isolated and presumed to be the intermediate that decomposed to the hydrolysis products. Another, more transient intermediate was postulated to account for the formation of the P→O analogue of Dyfonate and the oxygenated Dyfonate (McBain *et al.*, 1971*b*). The identity of the isolated intermediate, however, has since been disputed (Wustner *et al.*, 1972).

It is not known whether the same enzyme is responsible for the two pathways. While the proposed reaction mechanism implies a single oxidase producing a common intermediate for the two pathways, differential inhibition of the two reactions by various inhibitors as well as differential response to inducing agents indicates the involvement of different enzymes (Neal, 1967*b*; Chapman and Leibman, 1971). It seems possible, however, that the two reactions are not due to separate oxidases, but to the different degree of coupling between a single oxidase and a hypothetical hydrolase responsible for the hydrolytic pathway. Such a situation is known for naphthalene hydroxylation by a reconstituted microsomal enzyme system of rat liver. The production of naphthol is increased at the expense of dihydrodiol production by 3,3,3-trichloropropene oxide, an inhibitor of the tightly coupled hydrase (Oesch *et al.*, 1972).

Oxidative hydrolysis of the type described above is not limited to P→S esters. A few P→O esters, including *n*-propyl paraoxon, paraoxon, and diazoxon, are also hydrolyzed by various microsomal oxidase systems (Nakatsugawa *et al.*, 1968; Lewis, 1969; Yang *et al.*, 1971*b*). If the proposed scheme for P→S esters applies to P→O esters as well, oxidation of a P→O ester should in part result in no net reaction. This parallels the NIH shift mechanism in which "no virtual reaction" is predicted (Guroff *et al.*, 1967) (Fig. 13).

Carboxyester groups may also be oxidatively hydrolyzed. Malaoxon β-monoacid has been identified as an *in vitro* oxidative metabolite of malaoxon in a microsomal system of resistant strains of houseflies (Welling *et al.*, 1974). The enzyme is specific to malaoxon; malathion is not a substrate. However, the β-monoacid could conceivably be produced either by *O*-deethylation or by a mechanism similar to that discussed for organophosphates.

Oxidative hydrolysis of carbamate insecticides has also been suggested (Douch and Smith, 1971*a*). However, the *in vitro* conditions under which this reaction would proceed are the same as those required to produce the *N*-demethyl derivatives. Since the latter may be extremely labile to hydrolysis (Fahmy *et al.*, 1966), it is not possible to determine whether carbamate ester cleavage occurs from direct oxidative hydrolysis or from indirect hydrolysis via *N*-demethylation.

4.1.5. Nitrogen and Thioether Oxidation

The oxidation of nitrogen to *N*-oxide is rare among insecticides. In fact, the only established case of *N*-oxidation is found in the metabolism of nicotine by a liver postmitochondrial supernatant system of several mammals (Fig. 7). Nicotine-1′-oxide is formed as a major metabolite in the presence of oxygen and NADPH (Booth and Boyland, 1970). Although the active form of schradan was once suspected to be an *N*-oxide, it was later identified as an *N*-hydroxymethyl derivative (O'Brien, 1960).

The *in vitro* requirements of this reaction for NADPH and oxygen suggest that the enzyme is a typical microsomal mixed-function oxidase involving cytochrome P450. The presumption, however, may not be valid in view of the existence of other microsomal mixed-function oxidases which catalyze *N*-oxidation under the same *in vitro* conditions but which are not P450 mediated. The *N*-oxidation of arylamines and heterocyclic nitrogen compounds apparently belongs in this category since it is not inhibited by carbon monoxide (Weisburger and Weisburger, 1971; Hlavica and Kiese, 1969). A flavoprotein (FAD) enzyme which catalyzes some of these reactions has been purified from pig liver and is distinct from NADPH-cytochrome *c* reductase and free of cytochromes, iron, and copper (Ziegler *et al.*, 1973).

The assumption that thioether oxidations (Fig. 16) are cytochrome P450-dependent may not be valid in all cases. Cytochrome P450 does catalyze the

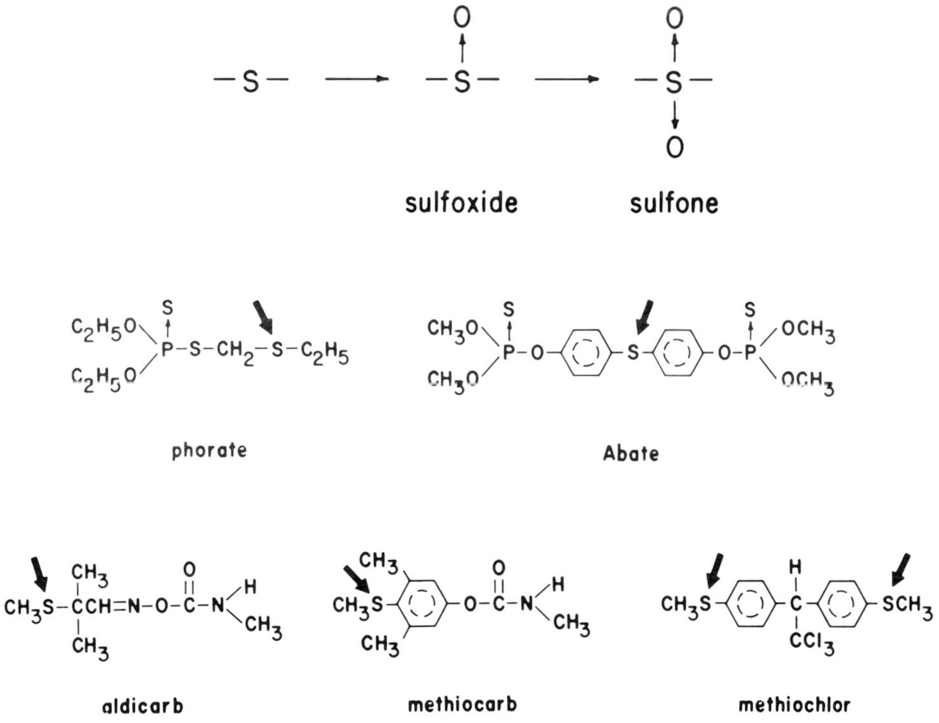

Fig. 16. Thioether oxidation.

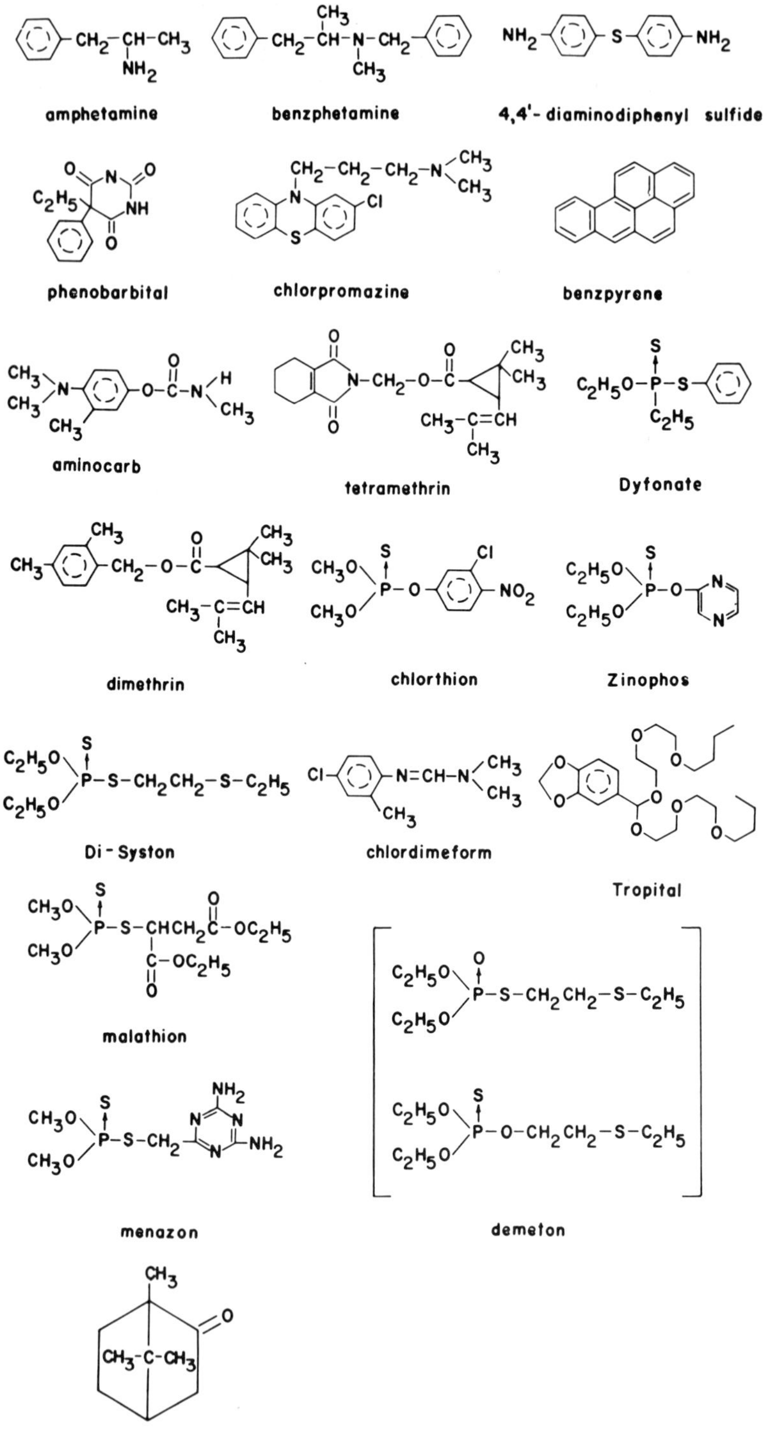

Fig. 17. Miscellaneous structures.

sulfoxidation of the drug chlorpromazine (Fig. 17) by guinea pig liver microsomes since the reaction is inhibited by a carbon monoxide (50%)–oxygen (10%) mixture. On the other hand, the sulfoxidation of 4,4′-diaminodiphenyl sulfide by a microsomes–NADPH system of guinea pig liver is insensitive to carbon monoxide (Gillette and Kamm, 1960; Gillette, 1966). A limited amount of information indicates that aldicarb sulfoxidation by rat liver enzymes is not confined to microsomes (Andrawes *et al.*, 1967; Oonnithan and Casida, 1968). The soluble fraction was somewhat more effective in converting aldicarb to its major *in vitro* metabolite, aldicarb sulfoxide.

Sulfoxidation is also a major metabolic route for other carbamates such as methiocarb (Oonnithan and Casida, 1968) and several organophosphates including demeton (March *et al.*, 1955), phorate (Bowman and Casida, 1958), Di-Syston (Bull, 1965) and Abate (Blinn, 1969). Sulfone derivatives, where identified, were only minor metabolites of both carbamates and organophosphates. Methiochlor, a DDT analogue, also undergoes thioether oxidations. Although the bis-sulfoxide and bis-sulfone of methiochlor are major *in vivo* metabolites of mice, only the mono- and bis-sulfoxides were identified as *in vitro* metabolites (Kapoor *et al.*, 1970) The characteristics of the enzymes responsible for these conversions have not been established.

4.2. Reaction Patterns

The preceding survey of microsomal oxidation reveals that many oxidative metabolites are in fact secondary products derived from the initial metabolites of enzymatic oxidation. For instance, a catechol isolated as an oxidative metabolite of piperonyl butoxide probably arises from the decomposition of the methylene-hydroxylated derivative. Likewise, the dihydrodiol of carbaryl results from the hydration of an intermediate epoxide. Thus the instability of some initial oxidation products can lead to a wide variety of metabolites, which belies the uniformity of oxidation itself. Brodie *et al.* (1958) were the first to recognize "hydroxylation" as the unifying feature of microsomal oxidation and today their proposal still seems valid despite discoveries of additional reaction types.

The identity of the hypothetical "hydroxyl," although still obscure, is suspected to be an electrophilic species (*cf.* Section 2.4.3). It appears that "hydroxylation" assumes three distinct reaction patterns dictated by the electronic nature of the target structure.

The first category is comprised of aliphatic hydroxylation and *O*-, *S*-, and *N*-alkyl hydroxylation (Figs. 5, 6, and 8). Although at first sight these reactions appear quite different, they are all initiated by the displacement of a hydrogen by a "hydroxyl." The diversity of the reaction products is directly attributable to the differences in the stability of the hydroxylated products, highly unstable intermediates being readily dealkylated.

Ullrich *et al.* (1972) have suggested that epoxide formation is characteristic of "hydroxyl" attack on a carbon–carbon π bond. Once an intermediate epoxide is formed, hydration to the corresponding dihydrodiol can occur. Hydration is apparent in the conversion of the epoxide dieldrin to aldrindiol, whereas hypothetical epoxides have been invoked to explain the dihydrodiols of rotenone and pyrethrins. In the case of the arene oxides, the high tendency to aromatize by rearrangement competes with enzymatic hydration. Hence naphthalene oxide yields both naphthol and dihydrodiol. A similar reaction pattern is evident in the "hydroxylation" of organophosphates. If, for example, parathion oxide is the actual intermediate, its rearrangement to paraoxon and hydrolysis to diethyl phosphorothioic acid and 4-nitrophenol is an exact parallel to the rearrangement and hydration (intramolecular hydrolysis) of naphthalene oxide. Therefore, the formation of cyclic oxides may be a common reaction pattern for "hydroxylation" of the π bonds of $P \rightarrow S$ and $P \rightarrow O$ structures as well as carbon-carbon double bonds (Figs. 9, 14, and 15).

Oxygenation of the unshared electron pair of nitrogen or sulfur comprises the third type of reaction pattern (Figs. 7 and 16). Although the reactions in this category are consistent with an electrophilic "hydroxyl" attack, cytochrome P450 has not been implicated in all cases. Other types of microsomal mixed-function oxidases or even nonmicrosomal oxidases may also be involved in these oxygenations.

Based on the preceding discussion the microsomal oxidation of insecticides is summarized as follows:

1. C-H hydroxylation
   a. *O*- and *S*-dealkylation
   b. *N*-alkyl hydroxylation, *N*-dealkylation
   c. aliphatic hydroxylation.
2. π-Bond oxygenation
   a. aliphatic epoxidation, dihydrodiol formation
   b. aromatic hydroxylation
   c. ester oxidation
3. Oxygenation at unshared electron pair
   a. nitrogen oxidation
   b. thioether oxidation

5. *Toxicological Consequences*

Microsomal oxidation is frequently an important determinant of insecticide toxicity. It may totally detoxify an insecticide in one step or it may convert a latent toxicant into an active metabolite. The *O*-deethylation of the organophosphate chlorfenvinphos is an example of direct detoxication since it yields an ionic metabolite with no anticholinesterase activity. The relative liver deethylase activities in the rat, mouse, rabbit, and dog (1 : 8 : 24 : 80) parallel the corresponding acute oral LD_{50} values of chlorfenvinphos (10, 100, 500, and >2000 mg/kg) (Donninger, 1971). The desulfuration of parathion and the *N*-methyl hydroxylation of schradan to potent cholinesterase inhibitors are classical cases of activation. Since the parent compounds themselves are practically inert as anticholinesterases, activation is obligatory to their lethal action. In fact, it was this phenomenon that initially drew attention to the microsomal oxidation of insecticides.

The great majority of microsomal oxidations, however, are less dramatic, and involve smaller degrees of detoxication and activation. Compared to the parent compounds, the *N*-hydroxymethyl derivatives of carbamates may display decreases in anticholinesterase activity of fourfold (propoxur), fivefold (Banol) and tenfold (Zectran and carbaryl). The *N*-methyl hydroxylation of aminocarb has no effect on toxicity (Oonnithan and Casida, 1968). Changes in the intrinsic toxicity (the ability of a given molecular species to attack the target) of an insecticide also depend on the pathways of oxidation. The relative concentrations required to inhibit mitochondrial respiration for rotenone, rotenolone I, 8′-hydroxyrotenone, 6′,7′-dihydroxyrotenone, and rotenolone II are 1 : 5 : 49 : 106+, respectively (Fukami *et al.*, 1969).

Activation products also exhibit a range with respect to intrinsic toxicity. Among carbamates, reactions which cause an increase in anticholinesterase activity include the 5-hydroxylation of propoxur (eightfold), sulfoxidation of methiocarb (tenfold), bis-*N*-demethylation of aminocarb (tenfold), mono- and bis-*N*-demethylation of Zectran (fivefold for both), and sulfoxidation of aldicarb (eightyfold) (Metcalf *et al.*, 1966; Oonnithan and Casida, 1968). For all the variation in the intrinsic toxicity of oxidative metabolites it is remarkable that microsomal oxidation typically serves animals as a detoxication mechanism.

It is evident from the preceding discussion that targets of insecticidal action may be exposed to many metabolites of varying toxicity as well as the parent compound. When the parent compound is itself toxic, the cause of eventual toxic effects is obscured because the biochemical lesion could potentially result from the action of parent compound, metabolites, or both. When obligatory activation is involved, toxicity is entirely due to the microsomal metabolites, and the degree of intoxication should reflect the degree of activation. Under these conditions, oxidase inhibitors should suppress toxicity and, in fact, cotreatment of houseflies with sesamex results in a moderate antagonism to parathion, methyl parathion, and chlorthion (Sun and Johnson, 1960).

However, factors involved in determining the eventual effect of microsomal oxidation are far more complex than casual interpretations might indicate and do not depend solely on the oxidative enzymes *per se*. The assumption that greater activation leads to greater intoxication cannot account for many experimental results. For instance, certain strains of houseflies resistant to organophosphates have a higher activation capacity than susceptible strains (Oppenoorth, 1971). Activation of parathion is about twice as high in the male rat as in the female (Neal, 1967*a*), whereas the higher oral LD_{50} for the male (7 mg/kg) than for the female (4 mg/kg) is a well-established fact (DuBois, 1971). A very similar situation exists for azinphosmethyl (Murphy and DuBois, 1957). Numerous studies have shown that inducers of hepatic microsomal oxidases reduce the toxicity of parathion and related organophosphates to rats and mice while enhancing activation (Ball *et al.*, 1954; Triolo and Coon, 1966; Bass *et al.*, 1972). These studies suggest that a higher total activation capacity in an animal need not necessarily result in a higher toxicant level at the target. In fact, the liver, with the highest activation capacity of all tissues, may not even release a significant amount of the active metabolite. The finding that hepatectomy causes no decrease in parathion toxicity (Diggle and Gage, 1951) seems to indicate that the liver is able to degrade nearly as much paraoxon as it produces. Neal (1972) has noted that a higher level of parathion activation in the liver should result in an accelerated detoxication of parathion when the capacity of the liver and serum to hydrolyze paraoxon exceeds the hepatic activation.

The importance of secondary metabolism is thus especially evident in obligatory activation owing to the unique "visibility" of the active metabolite. But, since many other microsomal oxidations do not necessarily abolish toxicity, secondary metabolism may generally constitute the definitive detoxication step. Yamamoto *et al.*, (1969) have suggested that the actual detoxication step for pyrethrins and allethrin may be the secondary conjugation of oxidative metabolites (alcohols) and/or the nonmicrosomal oxidation of the same metabolites to the corresponding carboxylic acids via aldehydes. Detoxication of carbamates and rotenone, which is initiated predominantly by microsomal oxidation, seems to be ensured by extensive secondary metabolism of the toxic oxidative metabolites. The high yield of conjugates and other water-soluble metabolites from these insecticides *in vivo* (Oonnithan and Casida, 1968; Dorough, 1968; Metcalf *et al.*, 1968; Miyamoto *et al.*, 1969; Shrivastava *et al.*, 1969; Fukami *et al.*, 1969) is probably due to the secondary metabolism of oxidative metabolites. The *N*-hydroxymethyl derivative of dicrotophos is also rapidly hydrolyzed to nontoxic metabolites in the rat (Bull and Lindquist, 1964).

Microsomal oxidases thus fulfill only a partial, albeit critical, function in the total detoxication system. While the consequences of microsomal oxidation are greatly influenced by secondary metabolism, the influence is not a simple function of the balance of the total enzyme activities for the two steps in an animal. Since the target tissue is usually not a major site of metabolism, the relative distribution and localization of enzymes for oxidative and secondary metabolism are important variables. Enzymes catalyzing the two steps may be located in separate tissues, in which case primary oxidative metabolites will enter the general circulation before undergoing further metabolism. On the other hand, both enzyme systems may be localized in the same cell or even within the same subcellular organelle, where their close coupling can promote the prompt disposal of all oxidative metabolites as soon as they are produced. Since both types of enzymes often coexist, varying degrees of coupling probably occur in many tissues.

One important consequence of coupling is that it may totally mask the intoxicating potential of microsomal oxidation in a particular tissue. Parathion activation in mammalian liver may approach such a situation. With effective coupling, oxidative metabolites are short-lived intermediates in the total detoxicative sequence of reactions, and hence their wide variation in toxicity is inconsequential. Indeed, it may be the tight coupling of microsomal oxidases with highly active enzymes of secondary metabolism that makes the liver a superior organ of detoxication in general.

The failure of such a cooperative arrangement in any tissue, however, can result in the exposure of vital systems to the contingent hazards of oxidative metabolites. In fact, the success of latent toxicants as insecticides is based on the accidental exploitation of such opportunities. Differential effectiveness of coupling among various animal strains and species may contribute to the selective toxicity of these insecticides. Ineffective coupling may also be associated with long-term health hazards of oxidative metabolites of insecticides. Such a possibility is indicated by studies on aromatic hydrocarbons (Oesch *et al.*, 1972; Keysell *et al.*, 1973) which suggest the importance of the coupled oxidase–hydrase system in relation to the hepatotoxicity and carcinogenicity of intermediate arene oxides.

6. Concluding Remarks

The near-omnipotence of microsomal oxidases probably reflects the combined substrate spectra of several forms of cytochrome P450, which may even include those which normally hydroxylate essential biochemicals. The occurrence of cytochrome P450 in lower as well as higher organisms alludes to the ubiquity of this enzyme system as a basic biochemical mechanism to deal with lipophilic substances. Most of our present knowledge of the oxidative system, however, has been obtained from surprisingly few species of mammals, insects, and microorganisms. Future comparative studies will undoubtedly include the neglected lower phyla of invertebrates. These studies should furnish us with a deeper insight into the basic role of the oxidases in animal physiology and their adaptive differentiation for the hydroxylation of foreign and essential compounds on the evolutionary scale. In view of the relevance of microsomal oxidases to the biodegradability of insecticides, comparative information would also be valuable in assessing potential side effects of insecticides in the ecosystem.

While many facets of microsomal oxidation remain to be elucidated, sufficient information is already available to provide an appropriate toxicological perspective. Clearly, microsomal oxidases play a central role in detoxication, but their detoxicative potential is fully attained only when they operate in close unison with the enzymes of secondary metabolism. By coupling these enzymes, nature has ensured an effective protection against normal levels of exposure to natural xenobiotics of acute or subacute toxicity. Nature, however, could not have selected for similar effective coupling against hazards that are realized long after the reproductive stage of animals. Consequently, in an environment replete with natural xenobiotics, man may face such long-term hazards as carcinogenesis from the oxidative metabolism of a large assortment of chemicals, of which insecticides are but a small part. Our understanding of oxidative detoxication which has been facilitated by the studies of insecticide toxicity will aid in assessing these hazards.

7. References

Abernathy, C. O., Neda, K., Engel, J. L., Gaughan, L. C., and Casida, J. E., 1973, Substrate-specificity and toxicological significance of pyrethroid-hydrolyzing esterases of mouse liver microsomes, *Pestic. Biochem. Physiol.* **3:**300.

Agosin, M., and Perry, A. S., 1974, Microsomal mixed-function oxidases, *in: The Physiology of Insecta*, Vol. V (M. Rockstein, ed.), pp. 537–596, Academic Press, New York.

Agosin, M., Michaeli, D., Miskus, R., Nagasawa, S., and Hoskins, W. M., 1961, A new DDT-metabolizing enzyme in the German cockroach, *J. Econ. Entomol.* **54:**340.

Agosin, M., Scaramelli, N., Gil, L., and Letelier, M. E., 1969, Some properties of the microsomal system metabolizing DDT in *Triatoma infestans*, *Comp. Biochem. Physiol.* **29:**785.

Alvares, A. P., Leigh, S., Kappas, A., Levin, W., and Conney, A. H., 1973, Induction of aryl hydrocarbon hydroxylase in human skin, *Drug Metab. Dispos.* **1:**386.

Amar-Costesec, A., Beaufay, H., Feytmans, E., Thinès-Sempoux, D., and Berthet, J., 1969, Subfractionation of rat liver microsomes, *in: Microsomes and Drug Oxidations*, (J. R. Gillette, A. H. Conney, G. J. Cosmides, R. W. Estabrook, J. R. Fouts, and G. J. Mannering, eds.), pp. 41–58, Academic Press, New York.

Andrawes, N. R., Dorough, H. W., and Lindquist, D. A., 1967, Degradation and elimination of temik in rats, *J. Econ. Entomol.* **60:**979.

Appleton, H. T., and Nakatsugawa, T., 1972, Paraoxon deethylation in the metabolism of parathion, *Pestic. Biochem. Physiol.* **2:**286.

Autor, A. P., Kaschnitz, R. M., Heidema, J. K., Van Der Hoeven, T. A., Duppel, W., and Coon, M. J., 1973, Role of phospholipid in the reconstituted liver microsomal mixed function oxidase system containing cytochrome P-450 and NADPH-cytochrome P-450 reductase, *Drug Metab. Dispos.* **1:**156.

Axelrod, J., 1954, Enzymatic demethylation of sympathomimetic amines, *Fed. Proc.* **13:**332.

Baldwin, M. K., and Robinson, J., 1969, Metabolism in the rat of the photoisomerization product of dieldrin, *Nature (London)* **224:**283.

Baldwin, M. K., Robinson, J., and Parke, D. V., 1970, Metabolism of endrin in the rat, *J. Agr. Food Chem.* **18:**1117.

Ball, W. L., Sinclair, J. W., Crevier, M., and Kay, K., 1954, Modification of parathion's toxicity for rats by pretreatment with chlorinated hydrocarbon insecticides, *Can. J. Biochem. Physiol.* **32:**440.

Baron, J., Hildebrandt, A. G., Peterson, J. A., and Estabrook, R. W., 1973, The role of oxygenated cytochrome P-450 and of cytochrome b_5 in hepatic microsomal drug oxidations, *Drug Metab. Dispos.* **1:**129.

Bass, S. W., Triolo, A. J., and Coon, J. M., 1972, Effect of DDT on the toxicity and metabolism of parathion in mice, *Toxicol. Appl. Pharmacol.* **22:**684.

Bend, J. R., Hook, G. E., and Gram, T. E., 1973, Characterization of lung microsomes as related to drug metabolism, *Drub Metab. Dispos.* **1:**358.

Benke, G. M., Wilkinson, C. F., and Telford, J. N., 1972, Micrósomal oxidases in a cockroach, *Gromphadorhina portentosa, J. Econ, Entomol.* **65:**1221.

Blinn, R. C., 1969, Metabolic fate of abate insecticide in rat, *J. Agr. Food Chem.* **17:**118.

Booth, J., and Boyland, E., 1970, The metabolism of nicotine into two optically-active stereoisomers of nicotine-1′-oxide by animal tissues *in vitro* and by cigarette smokers, *Biochem. Pharmacol.* **19:**733.

Bowman, J. S., and Casida, J. E., 1958, Further studies on the metabolism of thimet by plants, insects, and mammals, *J. Econ. Entomol.* **51:**838.

Brattsten, L. B., and Wilkinson, C. F., 1973, A microsomal enzyme inhibitor in the gut contents of the house cricket *(Acheta domesticus), Comp. Biochem. Physiol.* **45B:**59.

Brodie, B. B., 1956, Pathways of drug metabolism, *J. Pharm. Pharmacol.* **8:**1.

Brodie, B. B., Gillette, J. R., and LaDu, B. N., 1958, Enzymatic metabolism of drugs and other foreign compounds, *Ann. Rev. Biochem.* **27:**427.

Brooks, G. T., 1972, Pathways of enzymatic degradation of pesticides, *in: Environmental Quality and Safety* (F. Coulston, and F. Korte, eds.), pp. 106–164, Academic Press, New York.

Brooks, G. T., 1974 *Chlorinated Insecticides*, Vol. II, CRC Press, Cleveland.

Brooks, G. T., and Harrison, A., 1967, The metabolism of dihydrochlordene and related compounds by housefly (*M. domestica* L.) and pig liver microsomes, *Life Sci.* **6:**681.

Brooks, G. T., and Harrison, A., 1969*a*, Hydration of HEOD (dieldrin) and the heptachlor epoxides by microsomes from the livers of pigs and rabbits, *Bull. Environ. Contam. Toxicol.* **4:**352.

Brooks, G. T., and Harrison, A., 1969*b*, The oxidative metabolism of aldrin and dihydroaldrin by houseflies, housefly microsomes and pig liver microsomes and the effect of inhibitors, *Biochem. Pharmacol.* **18:**557.

Brooks, G. T., Harrison, A., and Cox, J. T., 1963, Significance of the epoxidation of the isomeric insecticides aldrin and isodrin by the adult housefly *in vivo, Nature (London)* **197:**311.

Brooks, G. T., Lewis, S. E., and Harrison, A., 1968, Selective metabolism of cyclodiene insecticide enantiomers by pig liver microsomal enzymes, *Nature (London)* **220:**1034.

Brooks, G. T., Harrison, A., and Lewis, S. E., 1970, Cyclodiene epoxide ring hydration by microsomes from mammalian liver and houseflies, *Biochem. Pharmacol.* **19:**255.

Büchi, G., Crombie, L., Godin, P. J., Kaltenbronn, J. S., Siddalingaiah, K. S., and Whiting, D. A., 1961, The absolute configuration of rotenone, *J. Chem. Soc.* **1961:**2843.

Bull, D. L., 1965, Metabolism of Di-Syston by insects, isolated cotton leaves, and rats, *J. Econ. Entomol.* **58:**249.

Bull, D. L., and Lindquist, D. A., 1964, Metabolism of 3-hydroxy-*N,N*-dimethylcrotonamide dimethyl phosphate by cotton plants, insects, and rats, *J. Agr. Food Chem.* **12:**310.

Bull, D. L., Lindquist, D. A., and Grabble, R. R., 1967, Comparative fate of the geometric isomers of phosphamidon in plants and animals, *J. Econ. Entomol.* **60:**332.

Burns, B., and Gurtner, G. H., 1973, A specific carrier for oxygen and carbon monoxide in the lung and placenta, *Drug Metab. Dispos.* **1:**374.

Camp, H. B., Fukuto, T. R., and Metcalf, R. L., 1969, Selective toxicity of isopropyl parathion. Metabolism in the housefly, honey bee, and white mouse, *J. Agr. Food Chem.* **17:**249.

Capdevila, J., Perry, A. S., Morello, A., and Agosin, M., 1973, Some spectral properties of cytochrome P-450 from microsomes isolated from control, phenobarbital- and naphthalene-treated houseflies, *Biochim. Biophys. Acta* **314:**93.

Casida, J. E., 1970, Mixed function oxidase involvement in the biochemistry of insecticide synergists, *J. Agr. Food Chem.* **18:**753.

Casida, J. E., Engel, J. L., Essac, E. G., Kamienski, F. X., and Kuwatsuka, S., 1966, Methylene-C^{14}-dioxyphenyl compounds: Metabolism in relation to their synergistic action, *Science* **153:**1130.

Chaplin, M. D., and Mannering, G. J., 1970, Role of phospholipids in the hepatic microsomal drug-metabolizing system, *Mol. Pharmacol.* **6:**631.

Chapman, S. K., and Leibman, K. C., 1971, The effects of chlordane, DDT, and 3-methylcholanthrene upon the metabolism and toxicity of diethyl-4-nitrophenyl phosphorothionate (parathion), *Toxicol. Appl. Pharmacol.* **18:**977.

Chedid, A., and Nair, V., 1972, Diurnal rhythm in endoplasmic reticulum of rat liver: Electron microscopic study, *Science* **175:**176.

Claude, A., 1938, A fraction from normal chick embryo similar to the tumor producing fraction of chicken tumor I, *Proc. Soc. Exp. Biol. Med.* **39:**398.

Claude, A., 1943, The constitution of protoplasm, *Science* **97:**451.

Claude, A., 1969, Microsomes, endoplasmic reticulum and interactions of cytoplasmic membranes, *in: Microsomes and Drug Oxidations* (J. R. Gillette, A. H. Conney, G. J. Cosmides, R. W. Estabrook, J. R. Fouts, and G. J. Mannering, eds.), pp. 3–39, Academic Press, New York.

Clemons, G. P., and Menzer, R. E., 1968, Oxidative metabolism of phosphamidon in rats and a goat, *J. Agr. Food Chem.* **16:**312.

Conney, A. H., and Kuntzman, R., 1971, Metabolism of normal body constituents by drug metabolizing enzymes in liver microsomes, *in: Handbook of Experimental Pharmacology*, Vol. XXVIII: *Concepts in Biochemical Pharmacology*, Part 2 (B. B. Brodie, and J. R. Gillette, eds.), pp. 401–421, Springer, Berlin.

Coon, M. J., Strobel, H. W., and Boyer, R. F., 1973, On the mechanism of hydroxylation reactions catalyzed by cytochrome P-450, *Drug Metab. Dispos.* **1:**92.

Cooper, D. Y., and Salhanick, H. A., 1973, *Multienzyme Systems in Endocrinology: Progress in Purification and Methods of Investigation*, New York Academy of Sciences, New York.

Cooper, D. Y., Levine, S., Narasimhulu, S., Rosenthal, O., and Estabrook, R. W., 1965, Photochemical action spectrum of the terminal oxidase of mixed function oxidase systems, *Science* **147:**400.

Cooper, D. Y., Schleyer, H., and Rosenthal, O., 1973, Chemistry of cytochrome P-450 purified from endocrine systems, *Drug Metab. Dispos.* **1:**21.

Dahm, P. A., and Nakatsugawa, T., 1968, Bioactivation of insecticides, *in: Enzymatic Oxidations of Toxicants* (E. Hodgson, ed.), pp. 89–110, North Carolina State University Press, Raleigh, N.C.

Dallner, G., Siekevitz, P., and Palade, G. F., 1966, Biogenesis of endoplasmic reticulum membranes. II. Synthesis of constitutive microsomal enzymes in developing rat hepatocyte, *J. Cell Biol.* **30:**97.

Daly, J., 1971, Enzymatic oxidation at carbon, *in: Handbook of Experimental Pharmacology*, Vol. XXVIII: *Concepts in Biochemical Pharmacology*, Part 2 (B. B. Brodie and J. R. Gillette, eds.), pp. 285–311, Springer, Berlin.

Daly, J., Guroff, G., Udenfriend, S., and Witkop, B., 1967, Hydroxylation-induced migrations of tritium in several substrates of liver aryl hydroxylases, *Arch. Biochem. Biophys.* **122:**218.

Dauterman, W. C., 1971, Biological and nonbiological modifications of organophosphorus compounds, *Bull. WHO* **44:**133.

Davidow, B., and Radomski, J. L., 1953, Isolation of an epoxide metabolite from fat tissues of dogs fed heptachlor, *J. Pharmacol. Exp. Ther.* **107:**259.

Davison, A. N., 1955, The conversion of schradan (OMPA) and parathion into inhibitors of cholinesterase by mammalian liver, *Biochem. J.* **61:**203.

Diggle, W. M., and Gage, J. C., 1951, Cholinesterase inhibition by parathion *in vivo, Nature (London)* **168:**998.

Dinamarca, M. L., Agosin, M., and Neghme, A., 1962, The metabolic fate of C^{14}-DDT in *Triatoma infestans*, *Exp. Parasitol.* **12:**61.

Donninger, C., 1971, Species specificity of phosphate triester anticholinesterases, *Bull. WHO* **44:**265.

Donninger, C., Hutson, D. H., and Pickering, B. A., 1972, The oxidative dealkylation of insecticidal phosphoric acid triesters by mammalian liver enzymes, *Biochem. J.* **126:**701.

Dorough, H. W., 1968, Metabolism of Furadan (NIA-10242) in rats and houseflies, *J. Agr. Food. Chem.* **16:**319.

Dorough, H. W., and Casida, J. E., 1964, Nature of certain carbamate metabolites of the insecticide, Sevin, *J. Agr. Food Chem.* **12:**294.

Douch, P. G. C., and Smith, J. N., 1971*a*, Metabolism of *m-tert.-butylphenyl N*-methylcarbamate in insects and mice, *Biochem. J.* **125:**385.

Douch, P. G. C., and Smith, J. N., 1971*b*, The metabolism of 3,5-di-*tert*.-butylphenyl *N*-methylcarbamate in insects and by mouse liver enzymes, *Biochem, J.* **125:**395.

DuBois, K. P., 1971, The toxicity of organophosphorus compounds to mammals, *Bull. WHO* **44:**233.

Elliott, M., Janes, N. F., Kimmel, E. C., and Casida, J. E., 1972, Metabolic fate of pyrethrin I, pyrethrin II, and allethrin administered orally to rats, *J. Agr. Food Chem.* **20:**300.

Ernster, L., and Orrenius, S., 1965, Substrate-induced synthesis of the hydroxylating enzyme system of liver microsomes, *Fed. Proc.* **24:**1190.

Ernster, L., and Orrenius, S., 1973, Dynamic organization of endoplasmic reticulum membranes, *Drug Metab. Dispos.* **1:**66.

Estabrook, R. W., and Cohen, B., 1969, Organization of the microsomal electron transport system, *in: Microsomes and Drug Oxidations* (J. R. Gillette, A. H. Conney, G. J. Cosmides, R. W. Estabrook, J. R. Fouts, and G. J. Mannering, eds.), pp. 95–109, Academic Press, New York.

Estabrook, R. W., Cooper, D. Y., and Rosenthal, O., 1963, The light reversible carbon monoxide inhibition of the steroid C21-hydroxylase system of the adrenal cortex, *Biochem. Z.* **338:**741.

Estabrook, R. W., Hildebrandt, A. G., Baron, J., Netter, K. J., and Leibman, K., 1971, A new spectral intermediate associated with cytochrome P-450 function in liver microsomes, *Biochem. Biophys. Res. Commun.* **42:**132.

Estabrook, R. W., Gillette, J. R., and Leibman, K. C. (eds.), 1973*a*, *Microsomes and Drug Oxidations*, Williams and Wilkins, Baltimore.

Estabrook, R. W., Matsubara, T., Mason, J. I., Werringloer, J., and Baron, J., 1973*b*, Studies on the molecular function of cytochrome P-450 during drug metabolism, *Drug Metab. Dispos.* **1:**98.

Fahmy, M. A. H., Metcalf, R. L., Fukuto, T. R., and Hennessy, D. J., 1966, Effects of deuteration, fluorination, and other structural modifications of the carbamyl moiety upon the anticholinesterase insecticidal activities of phenyl *N*-methylcarbamates, *J. Agr. Food Chem.* **14:**79.

Fujita, T., and Mannering, G. J., 1971, Differences in soluble P-450 hemoproteins from livers of rats treated with phenobarbital and 3-methylcholanthrene, *Chem.-Biol. Interact.* **3:**264.

Fukami, J., Yamamoto, I., and Casida, J. E., 1967, Metabolism of rotenone *in vitro* by tissue homogenates from mammals and insects, *Science* **155:**713.

Fukami, J., Shishido, T., Fukunaga, K., and Casida, J. E., 1969, Oxidative metabolism of rotenone in mammals, fish, and insects and its relation to selective toxicity, *J. Agr. Food Chem.* **17:**1217.

Gaudette, L. E., and Brodie, B. B., 1959, Relationship between the lipid solubility of drugs and their oxidation by liver microsomes, *Biochem. Pharmacol.* **2:**89.

Gillette, J. R., 1966, Biochemistry of drug oxidation and reduction by enzymes in hepatic endoplasmic reticulum, *Advan. Pharmacol.* **4:**219.

Gillette, J. R., and Kamm, J. J., 1960, The enzymatic formation of sulfoxides: The oxidation of chlorpromazine and 4,4′-diaminodiphenyl sulfide by guinea pig liver microsomes, *J. Pharmacol. Exp. Ther.* **130:**262.

Gillette, J. R., Conney, A. H., Cosmides, G. J., Estabrook, R. W., Fouts, J. R., and Mannering, G. J. (eds.), 1969, *Microsomes and Drug Oxidations,* Academic Press, New York.

Gillette, J. R., Davis, D. C., and Sasame, H. A., 1972, Cytochrome P-450 and its role in drug metabolism, *Ann. Rev. Pharmacol.* **12:**57.

Gordon, H. T., 1961, Nutritional factors in insect resistance to chemicals, *Ann. Rev. Entomol.* **6:**27.

Gram, T. E., 1971, Enzymatic *N*-, *O*-, and *S*-dealkylation of foreign compounds by hepatic microsomes, in: *Handbook of Experimental Pharmacology*, Vol. XXVIII: *Concepts in Biochemical Pharmacology*, Part 2 (B. B. Brodie and J. R. Gillette, eds.), pp. 334–348, Springer, Berlin.

Gram, T. E., and Fouts, J. R., 1968, Studies on the intramicrosomal distribution of hepatic enzymes which catalyze the metabolism of drugs and other foreign compounds, *in: Enzymatic Oxidations of Toxicants* (E. Hodgson, ed.), pp. 47–64, North Carolina State University, Raleigh, N.C.

Greim, H., Schenkman, J. B., Klotzbücher, M., and Remmer, H., 1970, The influence of phenobarbital on the turnover of hepatic microsomal cytochrome b_5 and cytochrome P-450 hemes in the rat, *Biochim. Biophys. Acta* **201:**20.

Gunsalus, I. C., Meeks, J. R., and Lipscomb, J. D., 1973, Cytochrome P-450_{cam} substrate and effector interactions, *Ann. N.Y. Acad. Sci.* **212:**107.

Guroff, G., and Daly, J., 1967, Quantitative studies on the hydroxylation-induced migration of deuterium and tritium during phenylalanine hydroxylation, *Arch. Biochem. Biophys.* **122:**212.

Guroff, G., Daly, J. W., Jerina, D. M., Renson, J., Witkop, B., and Udenfriend, S., 1967, Hydroxylation-induced migration: The NIH shift, *Science* **157:**1524.

Hamilton, G. A., Giacin, J. R., Hellman, T. M., Snook, M. E., and Weller, J. W., 1973, Oxenoid models for enzymic hydroxylations, *Ann. N.Y. Acad. Sci.* **212:**4.

Hathway, D. E. (senior reporter), 1970, *Foreign Compound Metabolism in Mammals*, Vol. I, The Chemical Society, London.

Hathway, D. E. (senior reporter), 1972, *Foreign Compound Metabolism in Mammals*, Vol. II, The Chemical Society, London.

Hayaishi, O., 1962, History and scope, *in: Oxygenases* (O. Hayaishi, ed.), pp. 1–29, Academic Press, New York.

Hildebrandt, A., and Estabrook, R. W., 1971, Evidence for the participation of cytochrome b_5 in hepatic microsomal mixed-function oxidation reactions, *Arch. Biochem. Biophys.* **143:**66.

Hlavica, P., and Kiese, M., 1969, *N*-Oxygenation of *N*-alkyl- and *N,N*-dialkylanilines by rabbit liver microsomes, *Biochem. Pharmacol.* **18:**1501.

Hodgson, E. (ed.), 1968, *Enzymatic Oxidations of Toxicants*, North Carolina State University, Raleigh, N.C.

Hodgson, E., and Casida, J. E., 1960, Biological oxidation of *N,N*-dialkyl carbamates, *Biochim. Biophys. Acta* **42:**184.

Hodgson, E., and Plapp, F. W., Jr., 1970, Biochemical characteristics of insect microsomes, *J. Agr. Food Chem.* **18:**1048.

Holtzman, J. L., Gram, T. E., Gigon, P. L., and Gillette, J. R., 1968, The distribution of the components of mixed-function oxidase between the rough and the smooth endoplasmic reticulum of liver cells, *Biochem. J.* **110:**407.

Hucker, H. B., Gillette, J. R., and Brodie, B. B., 1960, Enzymatic pathway for the formation of cotinine, a major metabolite of nicotine in rabbit liver, *J. Pharmacol. Exp. Ther.* **129:**94.

Imai, Y., and Sato, R., 1967, Studies on the substrate interactions with P-450 in drug hydroxylation by liver microsomes, *J. Biochem.* **62:**239.

Ishimura, Y., Ullrich, V., and Peterson, J. A., 1971, Oxygenated cytochrome P-450 and its possible role in enzymic hydroxylation, *Biochem. Biophys. Res. Commun.* **42:**140.

Ivie, G. W., and Dorough, H. W., 1968, Furadan-C^{14} metabolism in a lactating cow, *J. Agr. Food Chem.* **16:**849.

Jerina, D. M., Daly, J. W., and Witkop, B., 1968, The role of arene oxide-oxepin systems in the metabolism of aromatic substrates. II. Synthesis of 3,4-toluene-4-^{2}H oxide and subsequent "NIH shift" to 4-hydroxy toluene-3-^{2}H, *J. Am. Chem. Soc.* **90:**6523.

Jerina, D. M., Daly, J. W., Witkop, B., Zaltzman-Nirenberg, P., and Udenfriend, S., 1969, 1,2-Naphthalene oxide as an intermediate in the microsomal hydroxylation of naphthalene, *Biochemistry* **9:**147.

Jones, A. R., 1970, Further metabolites of hexamethylphosphoramide, *Biochem. Pharmacol.* **19:**603.

Jones, A. R., and Jackson, H., 1968, The metabolism of hexamethylphosphoramide and related compounds, *Biochem. Pharmacol.* **17:**2247.

Jori, A., DiSalle, E., and Santini, V., 1971, Daily rhythmic variation and liver drug metabolism in rats, *Biochem. Pharmacol.* **20:**2965.

Kamienski, F. X., and Casida, J. E., 1970, Importance of demethylenation in the metabolism *in vivo* and *in vitro* of methylenedioxyphenyl synergists and related compounds in mammals, *Biochem. Pharmacol.* **19:**91.

Kamin, H., and Masters, B. S. S., 1968, Electron transport in microsomes, *in: Enzymatic Oxidations of Toxicants* (E. Hodgson, ed.), pp. 5–26, North Carolina State University Press, Raleigh, N.C.

Kapoor, I. P., Metcalf, R. L., Nystrom, R. F., and Sangha, G. K., 1970, Comparative metabolism of methoxychlor, methiochlor, and DDT in mouse, insects, and in a model ecosystem. *J. Agr. Food Chem.* **18:**1145.

Kapoor, I. P., Metcalf, R. L., Hirwe, A. S., Lu, P. Y., Coats, J. R., and Nystrom, R. F., 1972, Comparative metabolism of DDT, methylchlor, and ethoxychlor in mouse, insects, and in a model ecosystem, *J. Agr. Food Chem.* **20:**1.

Kato, R., Takanaka, A., and Shoji, H., 1969, Inhibition of drug-metabolizing enzymes of liver microsomes by hydrazine derivatives in relation to their lipid solubility, *Jap. J. Pharmacol.* **19:**315.

Keysell, G. R., Booth, J., Grover, P. L., Hewer, A., and Sims, P., 1973, The formation of "K-region" epoxides as hepatic microsomal metabolites of 7-methylbenz[*a*]anthracene and 7,12-dimethylbenz[*a*]anthracene and their 7-hydroxymethyl derivatives, *Biochem. Pharmacol.* **22:**2853.

Khan, M. A. Q., Rosen, J. D., and Sutherland, D. J., 1969, Insect metabolism of photoaldrin and photodieldrin, *Science* **164:**318.

King, D. S., 1972, Ecdysone metabolism in insects, *Am. Zool.* **12:**343.

Klein, A. K., Dailey, R. E., Walton, M. S., Beck, V., and Link, J. D., 1970, Metabolites isolated from urine of rats fed ^{14}C-photodieldrin, *J. Agr. Food Chem.* **18:**705.

Knowles, C. O., and Sen Gupta, A. K., 1970, *N'*-(4-Chloro-*o*-tolyl)-*N,N*-dimethylformamidine-^{14}C (Galecron) and 4-chloro-*o*-toluidine-^{14}C metabolism in the white rat, *J. Econ. Entomol.* **63:**856.

Knowles, C. O., and Shrivastava, S. P., 1973, Chlordimeform and related compounds: Toxicological studies with house flies, *J. Econ. Entomol.* **66:**75.

Korte, F., and Arent, H., 1965, Metabolism of insecticides, IX (1) isolation and identification of dieldrin metabolites from urine of rabbits after oral administration of dieldrin, *Life Sci.* **4:**2017.

Krieger, R. I., and Wilkinson, C. F., 1970, An endogenous inhibitor of microsomal mixed-function oxidases in homogenates of the southern armyworm *(Prodenia eridania), Biochem. J.* **116:**781.

Krieger, R. I., and Wilkinson, C. F., 1971, The metabolism of 6,7-dihydroisodrin by microsomes and southern armyworm larvae, *Pestic. Biochem. Physiol.* **1:**92.

Krieger, R. I., Feeny, P. P., and Wilkinson, C. F., 1971, Detoxication enzymes in the guts of caterpillars: An evolutionary answer to plant defenses? *Science* **172:**579.

Ku, T. Y., and Dahm, P. A., 1973, Effect of liver enzyme induction on paraoxon metabolism in the rat, *Pestic. Biochem. Physiol.* **3:**175.

Kuriyama, Y., Omura, T., Siekevitz, P., and Palade, G. E., 1969, Effects of phenobarbital on the synthesis and degradation of the protein compounds of rat liver microsomal membranes, *J. Biol. Chem.* **244:**2017.

LaDu, B. N., Mandel, H. G., and Way, E. L. (eds.), 1971, *Fundamentals of Drug Metabolism and Drug Disposition,* Williams and Wilkins, Baltimore.

Lake, B. G., Hopkins, R., Chakraborty, J., Bridges, J. W., and Parke, D. V., 1973, The influence of some hepatic enzyme inducers and inhibitors on extrahepatic drug metabolism, *Drug Metab. Dispos.* **1:**342.

Leeling, N. N., and Casida, J. E., 1966, Metabolites of carbaryl (1-naphthyl methylcarbamate) in mammals and enzymatic systems for their formation, *J. Agr. Food Chem.* **14:**281.

Lewis, J. B., 1969, Detoxication of diazinon by subcellular fractions of diazinon-resistant and susceptible houseflies, *Nature (London)* **224:**917.

Lipscomb, J. D., and Gunsalus, I. C., 1973, Structural aspects of the active site of cytochrome P-450_{cam}, *Drug Metab. Dispos.* **1:**1.

Lu, A. Y. H., and Coon, M. J., 1968, Role of hemoprotein P-450 in fatty acid ω-hydroxylation in a soluble enzyme system from liver microsomes, *J. Biol. Chem.* **243:**1331.

Lu, A. Y. H., and Levin, W., 1972, Partial purification of cytochromes P-450 and P-448 from rat liver microsomes, *Biochem. Biophys. Res. Commun.* **46:**1334.

Lu, A. Y. H., Kuntzman, R., West, S., and Conney, A. H., 1971, Reconstituted liver microsomal enzyme system that hydroxylates drugs, other foreign compounds and endogenous substrates. I, *Biochem. Biophys. Res. Commun.* **42:**1200.

Lu, A. Y. H., West, S. B., Ryan, D., and Levin, W., 1973, Characterization of partially purified cytochromes P-450 and P-448 from rat liver microsomes, *Drug Metab. Dispos.* **1:**29.

Lucier, G. W., and Menzer, R. E., 1970, Nature of oxidative metabolites of dimethoate formed in rats, liver microsomes, and bean plants, *J. Agr. Food Chem.* **18:**698.

Lucier, G. W., and Menzer, R. E., 1971, Nature of neutral phosphorus ester metabolites of phosphamidon formed in rats and liver microsomes, *J. Agr. Food Chem.* **19:**1249.

Machin, A. F., Quick, M. P., Rogers, H., and Anderson, P. H., 1971, The conversion of diazinon to hydroxydiazinon in the guinea-pig and sheep, *Bull. Environ. Contam. Toxicol.* **6:**26.

Machin, A. F., Quick, M. P., Rogers, H., and Janes, N. F., 1972, An isomer of hydroxydiazinon formed by metabolism in sheep, *Bull Environ. Contam. Toxicol.* **7:**270.

Mailman, R. B., and Hodgson, E., 1972, The cytochrome P-450 substrate optical difference spectra of pesticides with mouse hepatic microsomes, *Bull. Environ. Contam. Toxicol.* **8:**186

Mannering, G. J., 1971, Microsomal enzyme systems which catalyze drug metabolism, *in: Fundamentals of Drug Metabolism and Drug Disposition* (B. N. LaDu, H. G. Mandel, and E. L. Way, eds.), pp. 206–252, Williams and Wilkins, Baltimore.

Mannering, G. J., Kuwahara, S., and Omura, T., 1974, Immunochemical evidence for the participation of cytochrome b_5 in the NADH synergism of the NADPH-dependent mono-oxidase system of hepatic microsomes, *Biochem. Biophys, Res. Commun.* **57:**476.

March, R. B., Metcalf, R. L., Fukuto, T. R., and Maxon, M. G., 1955, Metabolism of Systox in the white mouse and American cockroach, *J. Econ. Entomol.* **48:**355.

Martin, Y. C., and Hansch, C., 1971, Influence of hydrophobic character on the relative rate of oxidation of drugs by rat liver microsomes, *J. Med. Chem.* **14:**777.

Mason, H. S., 1957, Mechanisms of oxygen metabolism, *Advan. Enzymol.* **19:**79.

Masters, B. S. S., Nelson, E. B., Schacter, B. A., Baron, J., and Isaacson, E. L., 1973, NADPH-cytochrome *c* reductase and its role in microsomal cytochrome P-450-dependent reactions, *Drug Metab. Dispos.* **1:**121.

Matthews, H. B., and Matsumura, F., 1969, Metabolic fate of dieldrin in the rat, *J. Agr. Food Chem.* **17:**845.

Matthews, H. B., and McKinney, J. D., 1974, Dieldrin metabolism to *cis*-dihydroaldrindiol and epimerization of *cis*- to *trans*-dihydroaldrindiol by rat liver microsomes, *Drug Metab. Dispos.* **2:**333.

Matthews, H. B., McKinney, J. D., and Lucier, G. W., 1971, Dieldrin metabolism, excretion, and storage in male and female rats, *J. Agr. Food Chem.* **19:**1244.

Mazel, P., and Henderson, J. F., 1965, On the relationship between lipid solubility and microsomal metabolism of drugs, *Biochem. Pharmacol.***14:**92.

McBain, J. B., Yamamoto, I., and Casida, J. E., 1971*a*, Mechanism of activation and deactivation of Dyfonate (*O*-ethyl *S*-phenyl ethylphosphonodithioate) by rat liver microsomes, *Life Sci. (Part II)* **10:**947.

McBain, J. B., Yamamoto, I., and Casida, J. E., 1971*b*, Oxygenated intermediate in peracid and microsomal oxidations of the organophosphonothionate insecticide Dyfonate, *Life Sci. (Part II)* **10:**1311.

McIntosh, E. N., Mitani, F., Užgiris, V. I., Alonzo, C., and Salhanick, H. A., 1973, Comparative studies on mitochondrial and partially purified bovine corpus luteum cytochrome P-450, *Ann. N.Y. Acad. Sci.* **212:**392.

Menzer, R. E., and Casida, J. E., 1965, Nature of toxic metabolites formed in mammals, insects, and plants from 3-(dimethoxyphosphinyloxy)-*N,N*-dimethyl-*cis*-crotonamide and its *N*-methyl analog, *J. Agr. Food Chem.* **13:**102.

Metcalf, R. L., Fukuto, T. R., Collins, C., Borck, K., Burk, J., Reynolds, H. T., and Osman, M. F., 1966, Metabolism of 2-methyl-2-(methylthio)-propionaldehyde *O*-(methylcarbamoyl)-oxime in plant and insect, *J. Agr. Food Chem.* **14:**579.

Metcalf, R. L., Osman, M. F., and Fukuto, T. R., 1967, Metabolism of C^{14}-labeled carbamate insecticides to $C^{14}O_2$ in the house fly, *J. Econ. Entomol.* **60:**445.

Metcalf, R. L., Fukuto, T. R., Collins, C., Borck, K., El-Azia, S. A., Munoz, R., and Cassil, C. C., 1968, Metabolism of 2,2-dimethyl-2,3-dihydrobenzofuranyl-7 *N*-methylcarbamate (Furadan) in plants, insects, and mammals, *J. Agr. Food Chem.* **16:**300.

Mitani, F., Alvares, A. P., Sassa, S., and Kappas, A., 1971, Preparation and properties of a solubilized form of cytochrome P-450 from chick embryo liver microsomes, *Mol. Pharmacol.* **7:**280.

Miyamoto, J., and Suzuki, T., 1973, Metabolism of tetramethrin in houseflies *in vivo*, *Pestic. Biochem. Physiol.* **3:**30.

Miyamoto, J., Yamamoto, K., and Matsumoto, T., 1969, Metabolism of 3,4-dimethylphenyl *N*-methylcarbamate in white rats, *Agr. Biol. Chem.* **33:**1060.

Morello, A., 1964, Role of DDT-hydroxylation in resistance, *Nature (London)* **203:**785.

Morello, A., 1965, Induction of DDT-metabolizing enzymes in microsomes of rat liver after administration of DDT, *Can. J. Biochem.* **43:**1289.

Motoyama, N., and Dauterman, W. C., 1972, The *in vitro* metabolism of azinphosmethyl by mouse liver, *Pestic. Biochem. Physiol.* **2:**170.

Mücke, W., Alt, K. O., and Esser, H. O., 1970, Degradation of ^{14}C-labeled diazinon in the rat, *J. Agr. Food Chem.* **18:**208.

Murphy, S. D., and DuBois, K. P., 1957, Enzymatic conversion of the dimethoxy ester of benzotriazine dithiophosphoric acid to an anticholinesterase agent, *J. Pharmacol. Exp. Ther.* **119:**572.

Nair, V., and Casper, R., 1969, The influence of light on daily rhythm in hepatic drug metabolizing enzymes in rat, *Life Sci.* **8:**1291.

Nakatsugawa, T., and Dahm, P. A., 1967, Microsomal metabolism of parathion, *Biochem. Pharmacol.* **16:**25.

Nakatsugawa, T., and Nelson, P. A., 1972, Studies of insecticide detoxication in invertebrates: An enzymological approach to the problem of biological magnification, *in: Environmental Toxicology of Pesticides* (F. Matsumura, G. M. Boush, and T. Misato, eds.), pp. 501–524, Academic Press, New York.

Nakatsugawa, T., Ishida, M., and Dahm, P. A., 1965, Microsomal epoxidation of cyclodiene insecticides, *Biochem. Pharmacol.* **14:**1853.

Nakatsugawa, T., Tolman, N. M., and Dahm, P. A., 1968, Degradation and activation of parathion analogs by microsomal enzymes, *Biochem. Pharmacol.* **17:**1517.

Nakatsugawa, T., Tolman, N. M., and Dahm, P. A., 1969*a*, Degradation of parathion in the rat, *Biochem. Pharmacol.* **18:**1103.

Nakatsugawa, T., Tolman, N. M., and Dahm, P. A., 1969*b*, Metabolism of S^{35}-parathion in the house fly, *J. Econ. Entomol.* **62:**408.

Nakatsugawa, T., Tolman, N. M., and Dahm, P. A., 1969*c*, Oxidative degradation of diazinon by rat liver microsomes, *Biochem. Pharmacol.* **18:**685.

Neal, R. A., 1967*a*, Studies on the metabolism of diethyl 4-nitrophenyl phosphorothionate (parathion) *in vitro*, *Biochem. J.* **103:**183.

Neal, R. A., 1967*b*, Studies of the enzymic mechanism of the metabolism of diethyl 4-nitrophenyl phosphorothionate (parathion) by rat liver microsomes, *Biochem. J.* **105:**289.

Neal, R. A., 1972, A comparison of the *in vitro* metabolism of parathion in the lung and liver of the rabbit, *Toxicol. Appl. Pharmacol.* **23:**123.

Nebert, D. W., Considine, N., and Kon, H., 1973, Genetic differences in cytochrome P-450 during induction of mono-oxygenase activities, *Drug Metab. Dispos.* **1:**231.

Nelson, P. A., 1974, Aldrin epoxidation in *Lumbricus terrestris* L., M.S. thesis, SUNY College of Environmental Science and Forestry.

Nishibayashi, H., and Sato, R., 1968, Preparation of hepatic microsomal particles containing P-450 as sole heme constituent and absolute spectra of P-450, *J. Biochem. (Tokyo)* **63:**766.

Nishibayashi, H., Omura, T., Sato, R., and Estabrook, R. W., 1967, Comments on the absorption spectra of hemoprotein P-450, *in: Structure and Function of Cytochromes* (K. Okunuki, M. D. Kamen, and I. Sukuzu, eds.), pp. 658–665, University Park Press, Baltimore.

O'Brien, R. D., 1960, *Toxic Phosphorus Esters, Chemistry, Metabolism, and Biological Effects*, Academic Press, New York.

O'Brien, R. D., Kimmel, E. C., and Sferra, P. R., 1965, Toxicity and metabolism of famphur in insects and mice, *J. Agr. Food Chem.* **13:**366.

Oesch, F., Jerina, D. M., Daly, J. W., Lu, A. Y. H., Kuntzman, R., and Conney, A. H., 1972, A reconstituted microsomal enzyme system that converts naphthalene to *trans*-1,2-dihydroxy-1,2-dihydronaphthalene *via* naphthalene-1,2-oxide: Presence of epoxide hydrase in cytochrome P-450 and P-448 fractions, *Arch. Biochem. Biophys.* **153:**62.

Omura, T., and Sato, R., 1964, The carbon monoxide-binding pigment of liver microsomes, *J. Biol. Chem.* **239:**2370.

Omura, T., Sato, R., Cooper, D. Y., Rosenthal, O., and Estabrook, R. W., 1965, Function of cytochrome P-450 of microsomes, *Fed. Proc.* **24:**1181.

Omura, T., Siekevitz, P., and Palade, G. E., 1967, Turnover of constituents of the endoplasmic reticulum membranes of rat hepatocytes, *J. Biol. Chem.* **242:**2389.

Oonnithan, E. S., and Casida, J. E., 1968, Oxidation of methyl- and dimethylcarbamate insecticide chemicals by microsomal enzymes and anticholinesterase activity of the metabolites, *J. Agr. Food Chem.* **16:**28.

Oppenoorth, F. J., 1971, Resistance in insects: The role of metabolism and the possible use of synergists, *Bull. WHO* **44:**195.

Orrenius, S., Ellin, Å., Jakobsson, S. V., Thor, H., Cinti, D. L., Schenkman, J. B., and Estabrook, R. W., 1973, The cytochrome P-450-containing mono-oxygenase system of rat kidney cortex microsomes, *Drug Metab. Dispos.* **1:**350.

Palade, G. E., and Siekevitz, P., 1956, Liver microsomes, an integrated morphological and biochemical study, *J. Biophys. Biochem. Cytol.* **2:**171.

Papadopoulos, N. M., and Kintzios, J. A., 1963, Formation of metabolites from nicotine by a rabbit liver preparation, *J. Pharmacol Exp. Ther.* **140:**269.

Parke, D. V., 1968, *The Biochemistry of Foreign Compounds*, Pergamon Press, New York.

Perry, A. S., and Buckner, A. J., 1970, Studies on microsomal cytochrome P-450 in resistant and susceptible houseflies, *Life Sci. (Part II)* **9:**335.

Perry, A. S., Dale, W. E., and Buckner, A. J., 1971, Induction and repression of microsomal mixed-function oxidases and cytochrome P-450 in resistant and susceptible houseflies, *Pestic. Biochem. Physiol.* **1:**131.

Philpot, R. M., and Hodgson, E., 1971, Differences in the cytochrome P-450s from resistant and susceptible house flies, *Chem.-Biol. Interact.* **4:**399.

Poore, R. E., and Neal, R. A., 1972, Evidence for extrahepatic metabolism of parathion, *Toxicol. Appl. Pharmacol.* **23:**759.

Porter, K. R., 1961, The endopiasmic reticulum: Some current interpretations of its forms and functions, in: *Biological Structure and Function*, Vol. 1 (T. W. Goodwin, and O. Lindberg, eds.), pp. 127–155, Academic Press, New York.

Ptashne, K. A., and Neal, R. A., 1972, Reaction of parathion and malathion with peroxytrifluoroacetic acid, a model system for the mixed function oxidases, *Biochemistry* **11:**3224.

Ptashne, K. A., Wolcott, R. M., and Neal, R. A., 1971, Oxygen-18 studies on the chemical mechanisms of the mixed function oxidase catalyzed desulfuration and dearylation reactions of parathion, *J. Pharmacol. Exp. Ther.* **179:**380.

Radzialowski, F. M., and Bousquet, W. F., 1967, Circadian rhythm in hepatic drug metabolizing activity in the rat, *Life Sci.* **6:**2545.

Remmer, H., 1971, Enzyme induction phenomenon: Effects of vertebrate livers, *in: Pesticide Chemistry* Vol. II (A. S. Tahori, ed.), pp. 167–196, Gordon and Breach, New York.

Remmer, H., Estabrook, R. W., Schenkman, J., and Greim, H., 1968, Induction of microsomal liver enzymes, *in: Enzymatic Oxidations of Toxicants* (E. Hodgson, ed.), pp. 65–88, North Carolina State University Press, Raleigh, N.C.

Renson, J., Weissbach, H., and Udenfriend, S., 1965, On the mechanism of oxidative cleavage of aryl-alkyl ethers by liver microsomes, *Mol. Pharmacol.* **1:**145.

Richardson, A., Baldwin, M., and Robinson, J., 1968, Identification of metabolites of dieldrin (HEOD) in the faeces and urine of rats, *J. Sci. Food Agr.* **19:**524.

Sasame, H. A., Mitchell, J. R., Thorgeirsson, S., and Gillette, J. R., 1973, Relationship between NADH and NADPH oxidation during drug metabolism, *Drug. Metab. Dispos.* **1:**150.

Sato, R., Nishibayashi, H., and Ito, A., 1969, Characterization of two hemoproteins of liver microsomes, *in: Microsomes and Drug Oxidations* (J. R. Gillette, A. H. Conney, G. J. Cosmides, R. W. Estabrook, J. R. Fouts, and G. J. Mannering, eds.), pp. 111–132, Academic Press, New York.

Sato, R., Satake, H., and Imai, Y., 1973, Partial purification and some spectral properties of hepatic microsomal cytochrome P-450, *Drug Metab. Dispos.* **1:**6.

Schonbrod, R. D., and Terriere, L. C., 1971, Inhibition of housefly microsomal epoxidase by the eye pigment, xanthommatin, *Pestic. Biochem. Physiol.* **1:**409.

Sen Gupta, A. K., and Knowles, C. O., 1970, Galecron-^{14}C (*N'*-(4-chloro-*o*-tolyl)-*N,N*-dimethylformamidine) metabolism in the dog and goat, *J. Econ. Entomol.* **63:**951.

Shrivastava, S. P., Tsukamoto, M., and Casida, J. E., 1969, Oxidative metabolism of C^{14}-labeled Baygon by living house flies and by house fly enzyme preparations, *J. Econ. Entomol.* **62:**483.

Siekevitz, P., 1965, Origin and functional nature of microsomes, *Fed. Proc.* **24:**1153.

Stripp, B., Zampaglione, N., Hamrick, M., and Gillette, J. R., 1972, An approach measurement of the stoichiometric relationship between hepatic microsomal drug metabolism and the oxidation of reduced nicotinamide adenine dinucleotide phosphate, *Mol. Pharmacol.* **8:**189.

Strother, A., 1972, *In vitro* metabolism of methylcarbamate insecticides by human and rat liver fractions, *Toxicol. Appl. Pharmacol.* **21:**112.

Sun, Y.-P., and Johnson, E. R., 1960, Synergistic and antagonistic actions of insecticide–synergist combinations and their mode of action, *J. Agr. Food Chem.* **8:**261.

Taylor, J. M., Dehlinger, P. J., Dice, J. F., and Schimke, R. T., 1973, The synthesis and degradation of membrane proteins, *Drug Metab. Dispos.* **1:**84.

Triolo, A. J., and Coon, J. M., 1966, Toxicologic interactions of chlorinated hydrocarbon and organophosphate insecticides, *J. Agr. Food Chem.* **14:**549.

Tsukamoto, M., 1959, Metabolic fate of DDT in *Drosophila melanogaster*. I. Identification of a non-DDE metabolite, *Botyu-Kagaku* **24:**141.

Ullrich, V., and Diehl, H., 1971, Uncoupling of monooxygenation and electron transport by fluorocarbons in liver microsomes, *Eur. J. Biochem.* **20:**509.

Ullrich, V., Ruf, H. H., and Mimoun, H., 1972, Model systems for monooxygenases, *in: Biological Hydroxylation Mechanisms* (G. S. Boyd and R. M. S. Smellie, eds.), pp. 11–19, Academic Press, New York.

Unai, T., Cheng, H. M., Yamamoto, I., and Casida, J. E., 1973, Chemical and biological *O*-demethylation of rotenone derivatives, *Agr. Biol. Chem.* **37:**1937.

Warburg, O., 1926 Über die Wirkung des Kohlenoxyds auf den Stoffwechsel der Hefe, *Biochem. Z.* **177:**471.

Wattenberg, L. W., and Leong, J. L., 1971, Tissue distribution studies of polycyclic hydrocarbon hydroxylase activity, *in: Handbook of Experimental Pharmacology*, Vol. XXVIII: *Concepts in Biochemical Pharmacology*, Part 2 (B. B. Brodie and J. R. Gillette, eds.), pp. 422–430, Springer, Berlin.

Weisburger, J. H., and Weisburger, E. K., 1971, *N*-Oxidation enzymes, *in: Handbook of Experimental Pharmacology*, Vol. XXVIII: *Concepts in Biochemical Pharmacology*, Part 2 B. B. Brodie and J. R. Gillette, eds.), pp. 312–333, Springer, Berlin.

Welling, W., deVries, A. W., and Voerman, S., 1974, Oxidative cleavage of a carboxyester bond as a mechanism of resistance to malaoxon in houseflies, *Pestic. Biochem. Physiol.* **4:**31.

Wilkinson, C. F., 1968, Detoxication of pesticides and the mechanism of synergism, *in: Enzymatic Oxidations of Toxicants* (E. Hodgson, ed.), pp. 113–142, North Carolina State University Press, Raleigh, N.C.

Wilkinson, C. F., and Brattsten, L. B., 1972, Microsomal drug metabolizing enzymes in insects, *Drub Metab. Rev.* **1:**153.

Wilkinson, C. F., and Hicks, L. J., 1969, Microsomal metabolism of the 1,3-benzodioxole ring and its possible significance in synergistic action, *J. Agr. Food Chem.* **17:**829.

Willi, P., and Bickel, M. H., 1973, Liver metabolic reactions: Tertiary amine *N*-dealkylation, tertiary amine *N*-oxidation, *N*-oxide reduction, and *N*-oxide *N*-dealkylation. II. *N,N*-Dimethylaniline, *Arch. Biochem. Biophys.* **156:**772.

Williams, R. T., 1967, Comparative patterns of drug metabolism, *Fed. Proc.* **26:**1029.

Wilson, T. G., and Hodgson, E., 1972, Mechanism of microsomal mixed-function oxidase inhibitor from the housefly, *Musca domestica* L., *Pestic. Biochem. Physiol.* **2:**64.

Wolcott, R. M., and Neal, R. A., 1972, Effect of structure on the rate of the mixed function oxidase catalyzed metabolism of a series of parathion analogs, *Toxicol. Appl. Pharmacol.* **22:**676.

Wolcott, R. M., Vaughan, W. K., and Neal, R. A., 1972, Comparison of the mixed function oxidase-catalyzed metabolism of a series of dialkyl *p*-nitrophenyl phosphorothionates, *Toxicol. Appl. Pharmacol.* **22:**213.

Wong, D. T., and Terriere, L. C., 1965, Epoxidation of aldrin, isodrin, and heptachlor by rat liver microsomes, *Biochem. Pharmacol.* **14:**375.

Wustner, D. A., Desmarchelier, J., and Fukuto, T. R., 1972, Structure for the oxygenated product of peracid oxidation of Dyfonate® insecticide (*O*-ethyl *S*-phenyl ethylphosphonodithioate), *Life Sci. (Part II)* **11:**583.

Yamamoto, I., Kimmel, E. C., and Casida, J. E., 1969, Oxidative metabolism of pyrethroids in houseflies, *J. Agr. Food Chem.* **17:**1227.

Yang, R. S. H., Hodgson, E., and Dauterman, W. C., 1971*a*, Metabolism *in vitro* of diazinon and diazoxon in rat liver, *J. Agr. Food Chem.* **19:**10.

Yang, R. S. H., Hodgson, E., and Dauterman, W. C., 1971*b*, Metabolism *in vitro* of diazinon and diazoxon in susceptible and resistant houseflies, *J. Agr. Food Chem.* **19:**14.

Yu, S. J., and Terriere, L. C., 1971, Hormonal modification of microsomal oxidase activity in the housefly, *Life Sci.* **10:**1173.

Ziegler, D. M., McKee, E. M., and Poulsen, L. L., 1973, Microsomal flavoprotein-catalyzed *N*-oxidation of arylamines, *Drug Metab. Dispos.* **1:**314.

3

Cytochrome P450 Interactions

Ernest Hodgson and Laurence G. Tate

1. Introduction

1.1 Historical and General

Cytochrome P450 was first described, independently, by Klingenberg (1958) and Garfinkel (1958) as a carbon monoxide binding pigment located in the microsomal fraction of mammalian liver cells. The name is derived from the position of the peak in the optical difference spectrum of the carbon monoxide complex with the reduced cytochrome. In the almost two decades that have elapsed since that time, an enormous literature relative to this cytochrome has appeared. Although the general features of its distribution and function are now apparent, it has, in spite of all efforts, resisted extensive purification and characterization and its reaction mechanism is not clearly understood.

Cytochrome P450 is ubiquitous in distribution, having been demonstrated in at least 14 species of mammals, 17 species of insects, and one or more species of birds, reptiles, amphibians, fish, bacteria, and higher plants (Hodgson, 1974). In mammals it occurs in several but not all tissues, being found in liver, kidney, lung, small intestine, adrenal gland, and testis. Most of our information on the nature and role of the cytochrome in vertebrates is derived from that occurring in mammalian liver. In the Insecta, cytochrome P450 is less well studied, although a start has been made toward a detailed characterization of

---

Ernest Hodgson and Laurence G. Tate • Department of Entomology, North Carolina State University, Raleigh, North Carolina.

the cytochrome from the housefly, *Musca domestica* (Hodgson, 1974; Wilkinson and Brattsten, 1972). It has been demonstrated in fat body, Malpighian tubules, and midgut of insects, with the last generally appearing to have the highest concentration in many of the species studied.

In animal cells, cytochrome P450 is located in both rough and smooth endoplasmic reticulum, the former characterized by the presence of attached ribosomes. A microsomal preparation can be prepared by homogenization and differential centrifugation. Attempts to solubilize and purify cytochrome P450 have been partially successful, the principal problem being extreme lability, usually resulting in degradation to another pigment, cytochrome P420.

Cytochrome P450 is a mixed-function oxidase, its reaction mechanism involving the incorporation of one atom of molecular oxygen into the substrate while the other is reduced to water. The electrons involved are derived, via a flavoprotein reductase, from NADPH, making the enzyme one of a general class catalyzing the following type of reaction:

$$\text{NADPH} + \text{H}^+ + \text{AH}_2 + \text{0:0} \longrightarrow \text{NADP}^+ + \text{AHOH} + \text{H}_2\text{O}$$

The substrate specificity is variable, being narrow in the case of bacterial cytochrome P450 and wide in the case of mammalian liver cytochrome P450. In the latter case, many xenobiotics, including insecticides, are oxidized by a variety of reactions such as *N*-dealkylation, *O*-dealkylation, ring hydroxylation, desulfuration, and oxidative dearylation (Chapter 2). Cytochrome P450s from different mammalian tissues are to some extent tissue specific, at least to the extent that substrate specificity varies between tissues, causing the relative importance of the metabolism of endogenous substrates *vis-à-vis* xenobiotics to vary from one tissue to another.

The other electron transport system of the endoplasmic reticulum, that involving NADH, and NADH–cytochrome b_5 reductase and cytochrome b_5, may also be involved in the transfer of electrons to cytochrome P450, although this is still a matter of controversy (Chapter 2).

Since the original observations of Brown *et al.* (1954), the induction of mixed-function oxidase activity by xenobiotics has been much studied and it is now clear that it involves the appearance of qualitatively different forms of cytochrome P450. It is of interest that these forms may, in some species, be under genetic control.

1.2 Biochemistry and Physiology of Cytochrome P450

1.2.1. Methodology

Cytochrome P450 has been studied most extensively in microsomal preparations obtained by ultracentrifugation (e.g., 100,000–200,000*g* for 1 hr) of a postmitochondrial supernatant. Further fractionation—for example, into rough and smooth microsomes—can be achieved by centrifugation on continuous or discontinuous density gradients. Although a detailed review of homogenization and centrifugation techniques is inappropriate here, it should be borne in mind that, in order to preserve the integrity of the microsomal membranes and to prevent the degradation of cytochrome P450 to cytochrome P420, the optimum homogenization methods and media should be determined for each species and each tissue to be examined. Such parameters as the type of buffer, pH, and ionic strength are all of importance, as are the type of homogenizer, the time of homogenization, and the maintenance of a low temperature during the homogenization process. The use of glycerol as a protectant for cytochrome P450 is, generally speaking, unnecessary if the investigations are to be carried out on intact microsomes and other conditions are optimum.

The tool most used in the investigation of cytochrome P450 has been optical difference spectroscopy. Spectroscopy of microsomal preparations is complicated by the fact that turbid samples scatter a great deal of light and light scattering varies with the wavelength of the light. This means that changes in the absolute spectrum caused by the addition of a ligand are seen as variations in a sloping spectrum caused by the addition of a ligand are seen as variations in a sloping baseline, the principal component of which is not the pigments present in the preparation but rather the physical properties of the particles. Optical difference spectroscopy avoids the problems due to light scattering and other nonspecific absorption by recording only the change in light absorption caused by the addition of a ligand and not the absolute spectra themselves. This is accomplished by placing the microsomal suspension in both reference and sample cuvettes of a split-beam spectrophotometer, thus balancing these effects and permitting the recording of a flat baseline. The ligand under investigation is placed in the sample cuvette and on rerecording a difference spectrum is seen which shows only perturbations in the absolute spectrum and not the absolute spectrum itself. This principle is illustrated in Fig. 1.

The difference spectrum of greatest general importance is that of the carbon monoxide complex with the reduced cytochrome (Fig. 2). This shows a peak at or around 450 nm, with the degradation product cytochrome P420 appearing, if present, as a shoulder at 420 nm. The method for the demonstration of this spectrum, as described by Omura and Sato (1964*a*, *b*), includes saturation of both cuvettes with CO followed by reduction of the sample cuvette with dithionite. This method provides the basis for the estimation of cytochrome P450. The problems associated with the determination of the CO spectrum, particularly as they apply to insect preparations, have been discussed by Hodgson (1974) and Hodgson *et al.* (1974).

Type I and type II spectra (Schenkman *et al.*, 1967) (Fig. 3) are formed by the addition of various ligands to the oxidized form of cytochrome P450. Type I has a peak at 385 nm and a trough at 420 nm, while type II has a peak at about 430 nm and a trough at about 393 nm. Type I is formed by a large variety of different ligands and is believed to represent binding to a lipophilic site somewhat removed from the heme iron, while type II, formed primarily by

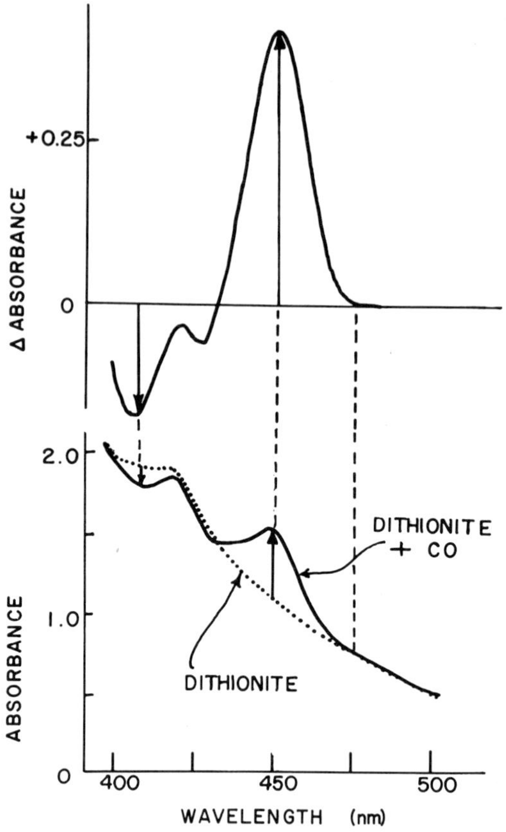

Fig. 1. Principle involved in difference spectroscopy. The lower graph represents the absolute spectra of cytochrome P450 before and after the addition of CO and dithionite, while the upper graph represents the difference spectrum obtained by adding microsomes to both cuvettes prior to the formation of the CO complex in the sample cuvette.

organic nitrogen compounds, is believed to represent binding to the heme iron. The modified type II (415–420 nm peak, 390 nm trough), caused primarily by alcohols, has also been called the reverse type I (Diehl *et al.*, 1970; Schenkman *et al.*, 1972; Wilson and Orrenius, 1972), since one hypothesis hold that it results from the displacement of an endogenous type I substrate in the sample cuvette. It is possible, however, that the modified type II spectrum represents the binding of a nucleophilic oxygen atom to the heme iron (Mailman *et al.*, 1974).

The type II spectrum formed by *n*-octylamine (Fig. 4) is of particular interest since it occurs in two forms, one with a double trough at 410 and 394 nm and the other with a single trough at 390 nm. These forms have been used in the characterization of qualitatively different cytochromes in mammals (Jefcoate *et al.*, 1969) and insects (Philpot and Hodgson, 1971*a*; Tate *et al.*, 1973).

Unusual spectral perturbations which complicate interpretation may be caused by nonlinear baseline changes due to turbidity differences between

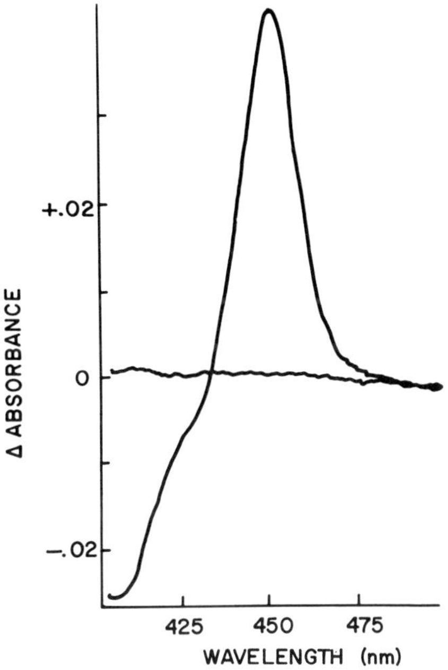

Fig. 2. CO difference spectrum obtained using mouse liver microsomes.

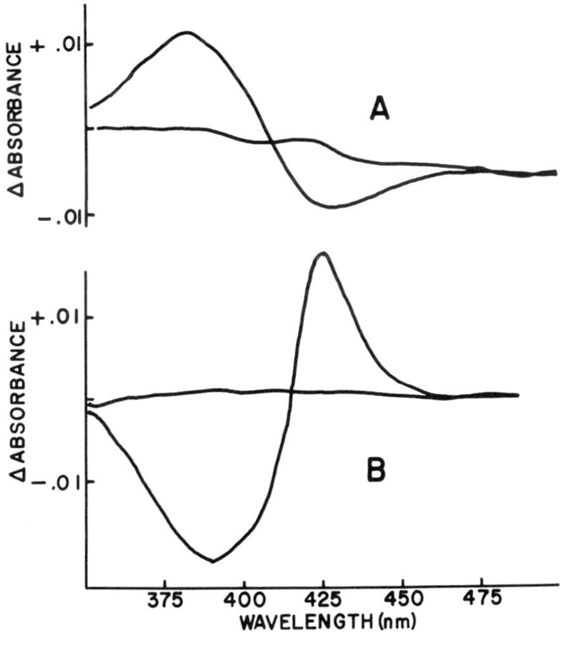

Fig. 3. Type I and type II spectra. A is a type I spectrum obtained with baygon and abdominal microsomes from Fc insecticide-resistant houseflies. B is a type II spectrum obtained with nicotine and mouse liver microsomes.

sample and reference cuvette, by mixed type I and type II interactions, by denaturation of cytochrome P450, and by native absorbance of added ligands. Guidelines have been established to simplify identification of these problems and interpret the resultant spectra (Hodgson *et al.*, 1974; Mailman *et al.*, 1974).

Ethyl isocyanide and type II spectra are the result of interactions with the reduced form of the cytochrome and characteristically have two peaks in the Soret region, at, or close to, 455 and 430 nm (Fig. 5). The size of the peaks is pH dependent and a pH equilibrium point can be calculated at which the two peaks are of equal magnitude. Changes in the pH equilibrium point have been used in studies of induction of mammalian hepatic cytochrome P450 to determine whether the induced cytochrome is qualitatively different from control cytochrome (Sladek and Mannering, 1966). Following the demonstration that similar spectra are formed by methylenedioxphenyl compounds, it was proposed that the term "type III" be used as a general designation for all pH-dependent double Soret spectra (Philpot and Hodgson, 1971*b*).

Electron paramagnetic resonance spectroscopy (EPR) has come into increasing use in recent years as a tool for the characterization of cytochrome P450 and for the investigation of its reaction mechanism, particularly since the observation that subtrate binding caused a change from a low-spin to a

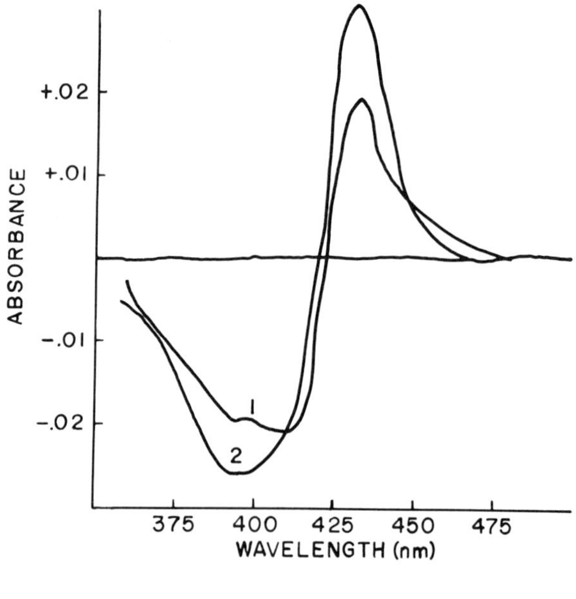

Fig. 4. *n*-Octylamine difference spectra obtained with abdominal microsomes from CSMA insecticide-susceptible houseflies (1) and Fc insecticide-resistant houseflies (2).

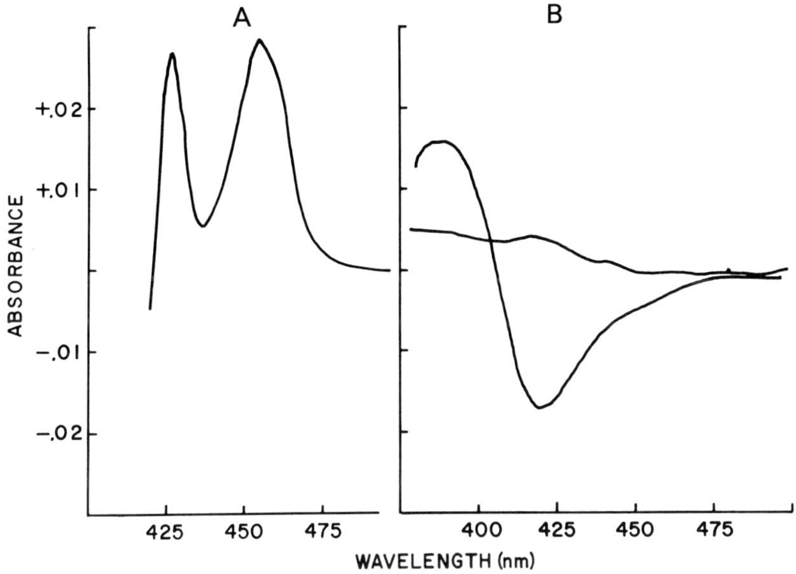

Fig. 5. Type III spectrum obtained with piperonyl butoxide and microsomes from mouse liver. A: Reduced spectrum. B: Spectrum formed with oxidized microsomes (type I).

high-spin form (T'sai *et al.*, 1970). As yet, this technique has not been used in the study of insecticide interactions with cytochrome P450.

Although solubilization and complete purification of cytochrome P450 have proved elusive, a number of methods for the preparation of partially purified cytochrome P450 have been described (Lu *et al.*, 1973; Sato *et al.*, 1973; Comai and Gaylor, 1973). They involve solubilization with detergents, with or without sonication, followed by a variety of well-known procedures which include ammonium sulfate precipitation, absorption on calcium phosphate gel, and DEAE-cellulose chromatography. Glycerol is usually added as it appears to stabilize cytochrome P450 to some extent.

These and similar methods have yielded preparations of high purity for which claims of homogeneity have been made (Imai and Sato, 1974; Levin *et al.*,1974; van der Hoeven *et al.*, 1974). Reconstitution studies are now feasible and should dramatically increase our knowledge of the function and reaction mechanism of cytochrome P450. Oxidation of insecticides by solubilized, reconstituted systems has not yet been demonstrated.

1.2.2. Role of Cytochrome P450 in Detoxication and Normal Metabolism

It has become abundantly clear that cytochrome P450 is the terminal oxidase in a multienzyme system responsible for the mixed-function oxidation of a wide variety of substrates, both xenobiotics and endogenous compounds. The former include not only drugs, pesticides, and environmental contaminants but also a wide variety of organic chemicals from almost every chemical class. The latter include steroid hormones, cholesterol, bile acids, fatty acids, thyroxine, bilirubin, and heme (Conney and Kuntzman, 1971). Since Chapter 2 deals with oxidation at the substrate level, this aspect will not be discussed further.

1.2.3. Mechanism of Action of Cytochrome P450 and Associated Enzymes

Although the mechanism by which cytochrome P450 and other enzymes of the microsomal mixed-function oxidase system act is not completely understood, many of its characteristics are known (Estabrook, 1971). The first step is the combination of the substrate with the oxidized form of the cytochrome to form a cytochrome P450–substrate complex (Fig. 6). This reduced, in the first of two reduction steps, by a reducing equivalent derived from NADPH via a flavoprotein reductase. This reductase has acquired the awkward name of NADPH-cytochrome *c* reductase, awkward because cytochrome *c* is not the normal acceptor in the microsomes but was used as an artificial acceptor during the early investigation of this enzyme. Whether or not another component exists between the reductase and cytochrome P450 is still largely conjectural, but it is established that lipid peroxidation occurs at this point in the electron transport chain. The reduced form of the cytochrome–substrate complex can react with carbon monoxide to form a carbon monoxide complex with a peak in the difference spectrum at or about 450 nm, or in the presence of molecular oxygen can form an oxygenated intermediate.

The next reduction step now occurs with the reducing equivalent, possibly derived from cytochrome b_5, reacting with the oxygenated intermediate to produce, via dismutation reactions not yet clearly understood, a hydroxylated product, water, and the oxidized form of the cytochrome. This mechanism is shown in Fig. 6; as indicated, the second electron can be derived from either NADH or NADPH. An alternative hypothesis (Cooper *et al.*, 1965) does not involve cytochrome b_5 but rather a second electron acceptor on cytochrome P450 such as an SH group.

1.2.4. Induction

Since the initial observations of Brown *et al.* (1954), it has become clear that the mixed-function oxidase activity of mammalian liver can be induced by a variety of xenobiotics, including a number of insecticides. The early work was summarized in an important review by Conney (1967). It was already clear at that time that all inducers did not induce the same enzyme activity. Some, such as phenobarbital, were found to have a fairly general effect, while others, such as the polycyclic hydrocarbons, induced a much narrower range of enzyme

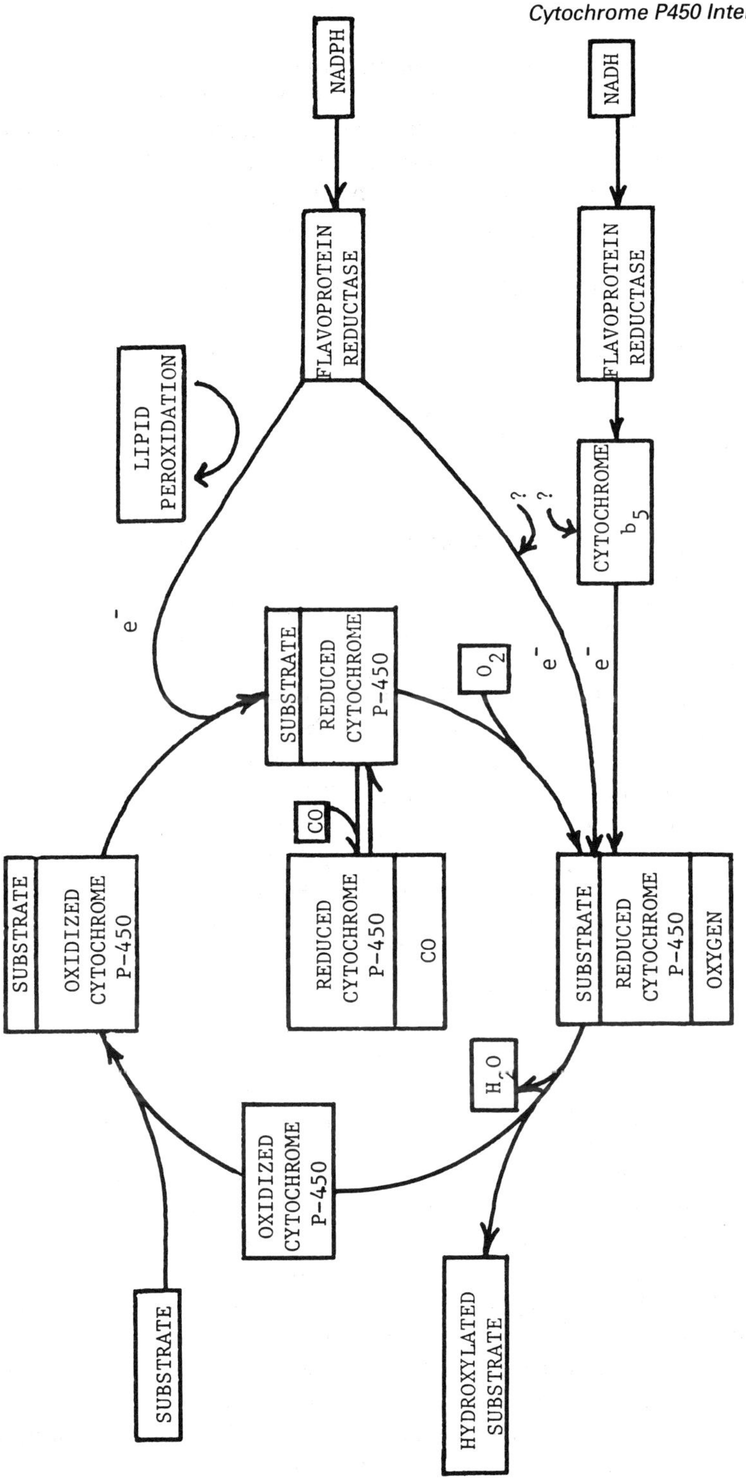

Fig. 6. Cytochrome P450 and the microsomal electron transport system.

activities. The latter also produced differences in the nature of the cytochrome P450 induced, including a shift in λ_{max} of the CO difference spectrum from 450 nm to 448 nm and a shift in the ethyl isocyanide pH equilibrium point (Sladek and Mannering, 1966). More recently (Lu *et al.*, 1973) it has become apparent, from studies of reconstituted enzyme systems, that these differences reside in the cytochrome P450 moiety of the microsomal electron transport chain. Induction in other mammalian organs is less well studied but is known to occur. It is also known to occur in insects and has been studied primarily in the adult housefly.

Many studies on induction, both *in vivo* and in cell culture (Gelboin, 1971), have been carried out. These studies indicate that induction occurs at the gene level, being first manifested by an increase in DNA-independent RNA synthesis.

2. Specific Interactions of Insecticides with Cytochrome P450

Type I interactions are caused by many compounds of different chemical and use classes. These include steroid hormones, drugs, and pesticides. Although most type I compounds are substrates for microsomal mixed-function oxidases, the relationship with type I binding is not a simple one. In general, K_s (concentration required for half-maximal spectrum development) is not equal to or correlated with K_m (concentration required for half-maximal enzyme activity) and no consistent correlations can be seen between the ability of a compound to act as a substrate and characteristics of its type I spectrum, including spectral size, peak or trough position, and rate of formation. Hexobarbital and aminopyrine are exceptions since in rat liver microsomes the K_s values of these compounds are equal to their K_ms (Schenkman *et al.*, 1967; Remmer *et al.*, 1969).

It should be noted that the presence of a type I binding spectrum, although it must indicate an interaction of some type, does not always indicate that the ligand is a substrate, nor is its absence necessarily an indication that metabolism will not occur. Although the insecticide mirex is not metabolized (Mehendale *et al.*, 1972), it gives rise to a distinct type I spectrum with both rat and mouse liver microsomes (Baker *et al.*, 1972) as well as microsomal preparations from sheep, rabbits, and insecticide-resistant (Fc strain) houseflies (Kulkarni *et al.*, 1975). On the other hand, many compounds which cause type I binding in resistant houseflies and mammals do not give rise to any detectable spectrum in CSMA insecticide-susceptible houseflies (Philpot and Hodgson, 1971*a*; Tate *et al.*, 1973; Kulkarni *et al.*, 1974*a,b*) but are known to be substrates for the microsomal mixed-function oxidases of this strain.

Following the earlier investigations of Mailman and Hodgson (1972) on the spectral binding of pesticides to mouse liver microsomes, Kulkarni *et al.* (1975) examined the binding spectra of numerous insecticides with hepatic microsomes from the rat, mouse, rabbit, and sheep as well as abdominal microsomes from insecticide-resistant (Fc) and -susceptible (CSMA) houseflies. While none gave rise to type I spectra with microsomes from susceptible houseflies, confirming earlier reports (Philpot and Hodgson, 1971*a*; Tate *et al.*, 1973), the following, with some exceptions due to species variation, gave rise to type I spectra in sheep, rat, mouse, and Fc housefly: the organochlorines *p,p′*-DDT, TDE, kelthane, methoxychlor, lindane, endrin, aldrin, dieldrin, chlordane, heptachlor, toxaphene, kepone, mirex, and endosulfan; the botanicals and related compounds pyrethrin, allethrin, bioallethrin, dimethrin, neopynamin, bioresmethrin, NRDC-119, NRDC-108, SBP-1382, cyclethrin, barthrin; the carbamates carbaryl, baygon, zectran, carbofuran, methomyl, mobam, dimetilan, temik, bux; the organophosphates diazinon, guthion, coumaphos, azodrin, malathion, ruelene.

Other investigations are sparse and relate to only a small number of species and insecticides. Thus type I interactions have been reported for DDT (Schenknan *et al.*, 1967), mirex and kepone (Baker *et al.*, 1972), and aldrin and dieldrin (Green and Stevens 1973). Roth and Neal (1972) reported parathion to form a type I spectrum with rabbit liver microsomes. Stevens *et al.* (1973) reported that in rat and mouse liver microsomes parathion, paraoxon, methyl parathion, malathion, disulfoton, and carbaryl all gave rise to more or less normal type I spectra. The lack of stability of the malathion and carbaryl spectra reported by these workers was not seen by Kulkarni *et al.* (1975). Mayer *et al.* (1973) reported that several juvenile hormone analogues gave type I spectral changes with rat liver microsomes.

Schenkman *et al.* (1967) suggested that type II spectra may result from the formation of a hemochrome with a nitrogenous base. Since that time, the type II interaction has been shown to occur with aromatic amines (T'sai *et al.*, 1970), with aliphatic amines and isocyanides (Jefcoate *et al.*, 1969), as well as with nonnitrogenous compounds such as steroids (Cooper *et al.*, 1971) and alcohols (Diehl *et al.*, 1970). Many of the nonnitrogenous compounds, however, are type IIA or modified type II in which the maxima and minima are at a lower wavelength. This spectrum, clearly different from the usual type II, is also referred to as the "reverse type I" since one hypothesis holds that it results from the displacement of an endogenous ligand (Schenkman *et al.*, 1972).

Studies using derivatives of pyridine, pyrrolidine, and piperdine as well as benzothiadiazoles, imidazoles, amides, amines, alkylnitriles, benzonitriles, phenols, alcohols, and other compounds together with microsomes from the mouse and from susceptible and resistant houseflies (Mailman *et al.*, 1974; Kulkarni *et al.*, 1974*a*) have attempted to define the chemical features required for type II binding. It appears that type II binding is caused by ligands in which sp^2 or sp^3 nonbonded electrons of nitrogen atoms are sterically accessible. Furthermore, compounds with other sterically accessible electrophilic atoms, such as the oxygen atoms of alcohols, may act in an analogous fashion and give rise to type II spectra, although at lower intensity and wavelength, due to the reduced electrophilicity. These correspond to the modified type II spectra, and their formation in this manner represents an alternative to the endogenous substrate hypothesis mentioned above.

Insecticides known to form type II spectra include the botanicals nicotine and anabasine and the phosphates zinophos and menazon; no examples have been described among the organochlorines or carbamates.

Certain unusual spectra, concentration-dependent changes, spectral shifts, and differences between spectra can be seen when the difference spectra of insecticides are compared. One example is the unusual spectrum seen with pyrethroids in housefly microsomes which is characterized by a peak at 415–418 nm and a trough at 445–447 nm; with mammalian liver microsomes, these compounds give typical type I spectra. Rotenone produces no detectable spectrum with rat, mouse, and housefly microsomes, but with sheep and rabbit gives a spectrum with a peak at 415–417 nm and a trough at 395 nm (Kuwatsuka, 1970; Kulkarni *et al.*, 1975). Schenkman *et al.* (1967) reported a modified type II spectrum for this insecticide.

Strain and species differences are not unusual and can be seen with both insecticidal and noninsecticidal ligands. The most dramatic example of a strain difference yet observed is the lack of a type I spectrum in abdominal microsomes from susceptible houseflies. Numerous ligands, including at least 36 insecticides, give clearly demonstrable type I spectra with such preparations from resistant houseflies of the Fc strain while failing to give any spectrally discernible interaction with microsomes from the susceptible CSMA strain. Species differences can be seen in the case of menazon, which gives a type II spectrum with mouse liver microsomes and a type I with microsomes from Fc flies. With rat liver microsomes and either menazon or dimethoate, a type I interaction is seen at low concentrations, a mixture of types I and II at intermediate concentrations, and a type II at saturating levels.

Table I. Ratios of Different Spectral Types Using Microsomal Preparations from Sheep, Rabbit, Rat, Mouse, and Resistant and Susceptible Houseflies

	Species					
	Sheep	Rabbit	Rat	Mouse	Housefly/Fc	Housefly/CSMA
Type I/type I						
DDT/kelthane	0.53	1.36	0.30	0.38	0.78	—
TDE/carbaryl	1.63	2.49	1.15	1.65	0.75	—
Type II/type II						
Nicotine/anabasine	1.30	0.79	1.32	1.13	1.72	1.39
Nicotine/zinophos	1.15	1.68	2.09	1.67	1.28	1.54
Type II/type I						
Nicotine/DDT	2.61	7.11	4.90	4.66	3.43	—
Zinophos/baygon	1.04	3.09	2.31	1.45	2.36	—

Within a single spectral type, variations between species in the ratios of spectral size for different compounds are quite common. This is shown in Table I.

These variations lead to the general conclusion that although the fundamental mechanisms of cytochrome P450 action may be the same throughout the animal kingdom, qualitative as well as quantitative differences are common. These differences can be seen between species and between different strains of the same species, and in the latter case are under genetic control. Concentration-dependent shifts almost certainly depend on the presence of more than one binding site with different affinities for the ligand in question, leaving open the question of whether these occur on the same or different cytochrome P450 molecules.

3. Interaction of Insecticide Synergists with Microsomal Mixed-Function Oxidases

3.1. General Introduction

Since the observation that sesame oil would synergize the action of pyrethrum (Eagleson, 1940), methylenedioxyphenyl (1,3-benzodioxole) derivatives have been used as insecticide synergists. The initial suggestion that these compounds functioned by inhibition of oxidative metabolism of insecticides was made by Sun and Johnson (1960) on the basis of *in vivo* tests on houseflies. Hodgson and Casida (1960, 1961) then demonstrated that piperonyl butoxide and sesamex inhibited the oxidative metabolism of carbamates by rat liver microsomes. A number of reviews of structure–activity relationships for synergism and the ability of synergists to inhibit microsomal mixed-function activity have appeared (e.g., Casida, 1970; Hewlett, 1960; Wilkinson, 1971).

More recently, new insights have been gained in to the interaction of methylenedioxyphenyl compounds with cytochrome P 450, and these have been reviewed at length (Hodgson and Philpot, 1974). Synergists of other chemical classes which interact with cytochrome P450 include arylalkylamines such as SKF-525A, Lilly 18947, and their derivatives, compounds containing acetylenic bonds, such as aryl-2-propenyl ethers and oxime ethers, organothiocyanates, *N*-(5-pentynyl)phthalimide, phosphate esters containing propynyl functions, phosphorothionates, 1,2,3-benzothiadiazoles, and certain imidazole derivatives (Hodgson and Philpot, 1974; Casida, 1970; Wilkinson, 1971).

When the available information on inhibition of microsomal mixed-function oxidases is compiled (Hodgson and Philpot, 1974), it is clear that this is a very general phenomenon. Over 50 compounds are known which inhibit the oxidation of almost as many substrates by microsomal preparations from numerous species of both vertebrates and invertebrates. It is also clear that kinetic studies do not provide a clear picture of the nature of the inhibition. Thus in the case of the methylenedioxyphenyl compounds it is variously described as competitive, "partly progressive," "partly competitive," "curvilinear," and noncompetitive. The lack of agreement is doubtless due, in part, to the difficulties inherent in working with particulate enzymes and substrates and inhibitors of limited solubility. The difficulties may be exaggerated by the use of nicotinamide, often added to protect NADPH by the inhibition of pyridine nucleotidase, since nicotinamide has been shown to be an inhibitor of microsomal oxidations. Addition of substrates and/or inhibitors in organic solvents may also be important, since solvent effects on these enzymes have been demonstrated (Anders, 1971).

The formation of a noncompetitive inhibitor complex when methylenedioxyphenyl compounds are incubated with NADPH-fortified microsomes, as described below, affects the inhibition kinetics. Franklin (1972*b*) has shown that the inhibition of ethylmorphine *N*-demethylation prior to complex formation is competitive and becomes noncompetitive after complex formation. The noncompetitive phase appears most important with respect to synergistic action. The problems caused by particulate enzymes, insoluble substrates and enzymes, nicotinamide, and solvent effects notwithstanding, the kinetic complexities observed with methylenedioxyphenyl compounds could be due in large part to the transition from competitive to noncompetitive inhibition during the course of the reaction.

3.2. Synergist Interaction with Cytochrome P450

Perry and Buckner (1970) and Matthews and Casida (1970) using houseflies and Matthews *et al.* (1970) using mice first demonstrated that treatment *in vivo* with piperonyl butoxide caused an apparent reduction in the cytochrome P450 level. Subsequently, a similar reduction was achieved *in vitro* by Philpot and Hodgson (1971*b*). Incubation of mouse hepatic microsomes with piperonyl butoxide and NADPH caused the appearance of a spectrally observable complex which blocks CO binding to a portion of the cytochrome, thus explaining the apparent reduction in level previously observed. The complex formed causes an optical difference spectrum with two peaks in the Soret region (Fig. 5), at 455 and 437 nm, which exist in a pH-dependent equilibrium. This has been termed a "type III" spectrum. In the absence of NADPH, piperonyl butoxide forms a typical type I spectrum in both mammals (Matthews *et al.*, 1970) and insects (Philpot and Hodgson, 1970).

Inhibition of CO binding is noncompetitive and corresponds to the formation of the type III spectrum. Maximum inhibition is about 50%, suggesting the presence of at least two forms of cytochrome P450, only one of which interacts with piperonyl butoxide to form the type III complex. These findings have been confirmed and extended by Franklin (1971, 1972*a*,*b*), who first suggested that a metabolite of piperonyl butoxide is responsible for the observed complex.

The demonstration that *in vivo* treatment of mice (Philpot and Hodgson, 1971*b*,*c*, 1972) and insects (Philpot and Hodgson, 1971*a*) with piperonyl butoxide results in a tightly bound cytochrome P450 complex that cannot be displaced by CO and withstands preparative procedures led Philpot and Hodgson (1971*b*) to suggest that this complex accounts for the apparent reduction in cytochrome P450 levels observed when animals are treated with methylenedioxyphenyl compounds (Philpot and Hodgson, 1971*b*,*c*; Škrinjarić-Spoljar *et al.*, 1971; Perry and Buckner, 1970; Matthews and Casida, 1970; Matthews *et al.*, 1970). That the K_s for piperonyl butoxide in the formation of the type III complex and the K_s for piperonyl butoxide in the inhibition of CO binding are the same confirms this hypothesis (Philpot and Hodgson, 1972).

In addition to the type III interaction described above, methylenedioxyphenyl compounds have been shown to give rise to a number of other difference spectra with reduced cytochrome P450 (Hodgson *et al.*, 1973), although a type I spectrum with oxidized microsomes, becoming type III on reduction with NADPH, is the most common. Compounds causing this combination of interactions include piperonyl butoxide, myristicin, isosafrole, dihydrosafrole, methylenedioxybenzene, sulfoxide, barthrin, methylenedioxybenzylidine aniline, 3,4-methylenedioxy-Ω-styrene, piperonylic acid, methylenedioxytoluene, and propyl isome. Although a few methylenedioxyphenyl compounds give type II spectra with oxidized microsomes, all of these, such as piperonyl alcohol (type II A), methylenedioxyphenylacetonitrile, and 3,4-methylenedioxyacetanilide, form a type III spectrum on incubation with NADPH-reduced microsomes.

Other methylenedioxyphenyl compounds, under a variety of reducing conditions, following the formation of a type III spectrum, give rise to a spectrum with a peak at 450 nm which resembles the CO spectrum in both ΔA_{max} and λ_{max}. These include such compounds as piperonylonitrile, 3,4-methylenedioxybromobenzene, and 3,4-methylenedioxy-1-bromo-6-methoxybenzene. It is not yet clear whether this peak is due to metabolically produced CO or to a molecular configuration which causes a similar perturbation of the spectrum.

One other unusual spectrum is seen with either *cis*- or *trans*-methylenedioxycyclohexane. On incubation with mouse liver microsomes and NADPH, a difference spectrum is formed with a single peak at 427 nm, corresponding exactly to one of the peaks of the double Soret type III spectrum. This does not appear to be the result of a different pH equilibrium point since the 455 nm peak does not appear when the pH is varied. Since these two compounds are even better inhibitors of the formation of the CO binding spectrum than piperonyl butoxide, these results do not appear to support the hypothesis that the 455 nm peak is associated with binding to the heme iron (Ullrich and Schnabel, 1973).

Examination of the difference spectra formed by methylenedioxyphenyl compounds and reduced cytochrome P450 from insecticide-susceptible and -resistant strains of houseflies (Kulkarni *et al.*, 1974*b*) reveals that, although some variations occur, essentially the same types of spectra are formed.

The cytochrome P450 difference spectra of synergists other than the methylendioxyphenyl compounds have not been studied in detail. Matthews and Casida (1970) showed that MGK 264, 2-(2,4,5-trichlorophenyl)propynyl ether, and 2-methylpropyl-2-propynylphenylphosphonate give type I spectra with mouse microsomes, while 5,6-dichlorobenzothiadiazole gives a type II spectrum at low concentrations which shifts to a mixed type II/type I as the concentration is increased. This last compound gives rise to an unusual spectrum, having a peak at 444–446 nm, with reduced microsomes (Gil, 1973). A new class of synergists, the substituted imidazoles, are type II ligands which bind very tightly to the cytochrome (Wilkinson *et al.*, 1972, 1974*a,b*), while one of them, isopropylphenylimidazole, is reported to yield a type III spectrum with reduced microsomes (Leibman and Ortiz, 1973).

3.3. Relationship of Spectra to Synergism

To verify a relationship between type III complex formation and synergism, it is necessary to establish that a direct correlation exists between the complex and inhibition, that the complex, once formed, is stable and unaffected by mixed-function oxidase reactions, and that the complex can be formed by various synergistically active methylenedioxyphenyl compounds in different species. Although not conclusive, the evidence presently available indicates that such a relationship exists.

Franklin (1972*b*) has shown that the inhibition of ethylmorphine *N*-demethylation by piperonyl butoxide *in vitro* is essentially competitive but changes to noncompetitive as the type III complex forms. In addition, if the type III complex is allowed to form by incubation of limiting amounts of piperonyl butoxide prior to the addition of ethylmorphine, the inhibition observed is proportional to the magnitude of the complex. Thus, for ethylmorphine, Franklin has clearly demonstrated a direct correlation between the type III complex and inhibition.

The type III compex has also been correlated with inhibition of the formation of type I and type II substrate difference spectra. It is of interest that inhibition of the formation of the type I spectrum (70–80%) is greater than that seen for the type II spectrum (30–40%) (Philpot and Hodgson, 1972) and that similar results have been obtained for the inhibition of the metabolism of several type I and type II substrates (Jaffe *et al.*, 1968; Škrinjarić-Spoljar *et al.*, 1971; Matthews *et al.*, 1970).

The type III interaction formed *in vitro* can also be observed *in vivo* (Philpot and Hodgson, 1971*c*), indicating the presence of a tightly coupled complex, and Franklin (1972*a*) has shown that the stability of the type III compex, once it is formed, is unaffected by oxidative and conjugative reactions, other mixed-function oxidase inhibitors, free radical scavengers, or mixed-function oxidase cofactors.

Although the foregoing discussion is concerned mainly with the effect of piperonyl butoxide in the rat and mouse, the formation of the type III complex (as indicated above) is not confined to this compound or these species. The magnitude of the interaction formed in the mouse is greater than that in the rat (Franklin, 1972*a*), which suggests an explanation for the greater inhibitory action of methylenedioxyphenyl compounds in the former species (Conney *et al.*, 1971). A similar conclusion can be drawn from the finding that the magnitude of the type III complex formed in microsomes from resistant houseflies is much greater than that seen in microsomes from susceptible flies (Philpot and Hodgson, 1971*a*).

3.4. Mechanism of Synergism

The following discussion of the mechanism of synergistic action is drawn, with only minor modifications, from that of Hodgson and Philpot (1974). The suggestion by Casida *et al.* (1966) that methylenedioxyphenyl compounds act as synergists because of their ability to serve as alternative substrates for the microsomal enzymes implies that the inhibition is strictly competitive. On the other hand, proposals by Hennessy (1965), Hansch (1968), and Ullrich and Schnabel (1973) that methylenedioxyphenyl compounds form carboxolium ions, free radicals, or carbanions, respectively, which interact with some component of the mixed-function oxidase system, all indicate that inhibition should be noncompetitive. In view of the complex formation between cytochrome P450 and methylenedioxphenyl compounds and the correlation between the formation of this complex and inhibition of the *N*-demethylation of ethylmorphine, it appears that the latter theories may have more merit. It should not be concluded, however, that methylenedioxyphenyl compounds cannot inhibit by acting as alternative substrates. Clearly, the extent to which inhibition occurs is dependent on the substrate being metabolized. This may be associated with the observation that the type III complex forms with only a portion of the available cytochrome, a finding which suggests the presence of at least two forms of cytochrome P450. The unreacted cytochrome can account for the enzyme activity remaining after maximum inhibition has developed (15% in the case of ethylmorphine) (Franklin, 1972*a*) as well as for the continued metabolism of methylenedioxyphenyl compounds after the maximum complex formation has occurred (Hodgson and Philpot, 1974). It is therefore possible that inhibition resulting from substrate competition takes place on that portion of the cytochrome not inhibited by formation of the type III complex. If inhibition of this type occurs, there is little to suggest that it would differ from that observed for many other substrates acting as competitive inhibitors but not as synergists.

Both Philpot and Hodgson (1971*a*) and Franklin (1972*a*) have proposed that the formation of the type III complex, which results in noncompetitive inhibition, is largely responsible for the synergistic activity of methylenedioxyphenyl compounds.

Microsomal mixed-function oxidases are known to metabolize the methylenedioxy group to the corresponding catechol and formate (Casida *et al.*, 1966; Kamienski and Casida, 1970; Wilkinson and Hicks, 1969; Kuwatsuka, 1970), and the suggestion has been made that catechols may play a role in the inhibition mechanism (Kuwatsuka, 1970; Desmarchellier *et al.*, 1973; Hennessy, 1965). In view of the ease with which catechols enter into biological oxidation–reduction reactions, the possibility that these compounds act at a point in the electron transport chain prior to cytochrome P450 (e.g., at the flavoprotein) and thus inhibit by a mechanism different from that of the methylenedioxy compounds should be investigated.

Any mechanism to account for the formation of the type III complex must be consistent with several established facts. The structural requirements for the methylenedioxyphenyl compounds with respect to synergistic activity are known, and several studies have shown that an unsubstituted methylene group at the 2 position of the ring is required for optimal activity. Substitution of the methylene hydrogens by methyl groups results in almost complete loss of activity, and significant deuterium effects have also been observed. (See Philpot and Hodgson, 1974, for appropriate references.) It seems clear, therefore, that the interaction between methylenedioxyphenyl compounds and cytochrome P450 involves a C—H bond of the methylene group at the 2 position of the ring. Further, the requirement for NADPH and oxygen suggests that a reaction with an activated oxygen complex of cytochrome P450 takes place.

The patterns of synergism observed with compounds containing numerous simple aromatic substituents were clearly not related to electronic effects (Wilkinson, 1967), and Hansch (1968) using steric, homolytic, and hydrophobic constants subsequently concluded that activity was related to their ability to produce free radicals via homolytic cleavage of a C—H bond of the methylene group. This type of substituent effect is known for homolytic reactions occuring with aromatic compounds (Williams, 1960).

Methylenedioxyphenyl compounds react with the Fenton's free radical generating system to inhibit the epoxidation of aldrin. This inhibition is competitive and has been related by Marshall and Wilkinson (1970, 1973) to the rate at which the methylenedioxyphenyl compounds are converted to their respective catechols. However, a low correlation was found between the model system inhibition and the $\sum\sigma$ constants calculated by Hansch (1968).

The proposal of Hennessy (1965) that methylenedioxyphenyl synergism is dependent on hydride ion transfer from the methylene group of the dioxole ring is supported by molecular orbital studies (Cloney and Scerr, 1968). This idea is consistent with C—H bond cleavage but requires some electrophilic agent to initiate the reaction. The effect of a peracetic acid epoxidation system, considered to generate the OH^+ ion, on several methylendioxyphenyl compounds has been investigated with negative results (Marshall and Wilkinson, 1973). No inhibition of epoxidation of modification of the methylenedioxyphenyl compounds was seen to occur in this system, suggesting that in model systems methylenedioxyphenyl compounds are subject to radical but not electrophilic ion attack.

The formation of a carbanion at the methylene group of methylenedioxyphenyl compounds has recently been proposed by Ullrich and Schnabel (1973). The interaction of the carbanion with the heme of cytochrome P450 is suggested as being the inhibitory mechanism. Evidence that methylenedioxyphenyl compounds undergo such a reaction is derived mainly from studies with fluorene, a nonmethylenedioxyphenyl compound with a methlyene bridge as the only reactive group. Incubation of this compound with microsomes plus NADPH and subsequent addition of dithionite result in a difference spectrum with a peak at 446 nm. An additional peak seen at 374 nm corresponds to the absorbance of the fluorenyl anion. The similarity between this spectrum and that produced by piperonyl butoxide under similar conditions is not at all clear. Indeed, it appears to be almost identical to that produced by metyrapone and WL 19255, a benzothiadiazole compound (Matthews *et al.*, 1970; Hildebrandt *et al.*, 1969), both of which produce difference spectra with peaks at 444 nm on addition to dithionite-reduced mouse microsomes. Also, these complexes do not inhibit CO binding and therefore appear to be reversible in nature. It appears that the structure–activity relationships correlating synergistic action with free radical stabilization or molecular orbital studies supporting the hydride transfer theory of Hennessy would be difficult to apply to the formation of a carbanion from methylenedioxyphenyl compounds.

One of the difficulties in assigning a mechanism to account for the synergistic action of methylenedioxyphenyl compounds is the paucity of knowledge concerning the active O_2 species participating in cytochrome P450 mixed-function oxidase reactions. The oxygen complex involved in the mixed-function oxidase reaction catalyzed by the copper-containing enzyme phenolase, however, is better understood. The formation of a reversible percupryl ion complex, $[Cu^+O_2]^{2+}$, has been implicated for phenolase by Bright *et al.* (1963). It is of interest that methylenedioxyphenyl compounds are known to inhibit this enzyme (Metcalf *et al.*, 1966). The partial inhibition of cytochrome P450 activity by superoxide dismutase described by Strobel and Coon (1971) suggests that $O_2\cdot^-$ is involved in mixed-function oxidase reactions carried out by the cytochrome P450 system. As suggested by Wilkinson (1971), the active oxygen species in this case, $[Fe^{3+}O\cdot^-]^{2+}$, could be considered analogous to that proposed for phenolase. If this is so, the mechanism of methylenedioxyphenyl inhibition of phenolase and that of cytochrome P450 reactions should be similar.

The evidence at present suggests that the synergistic activity of methylenedioxyphenyl compounds results from a stable inactive cytochrome P450 complex formed through the reaction of the methylene group at the 2 position with a cytochrome P450 activated oxygen complex. This reaction involves a cleavage of a C—H bond, resulting in the formation of highly reactive species which complexes with cytochrome P450. The following proposed mechanism is consistent with the above, although in view of the paucity of evidence it lacks detail and should be considered tentative.

The first step involves the binding of the methylenedioxyphenyl compound to oxidized cytochrome P450 to form a type I spectral complex:

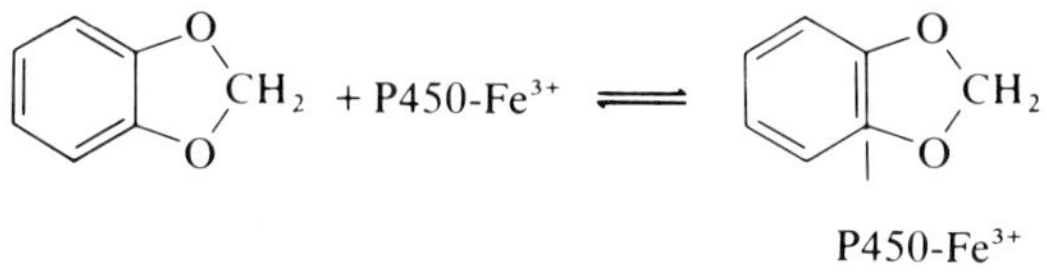

This is followed by the incorporation of one electron, from NADPH, and oxygen to form the following activated oxygen complex:

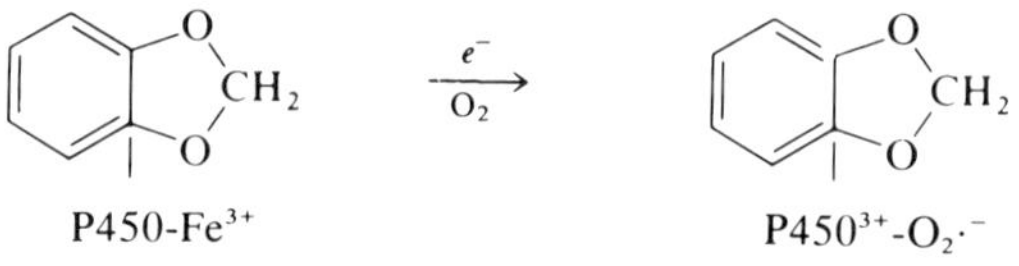

This complex abstracts a hydrogen from the methylenedioxy carbon, and in the presence of excess NADPH the following products are produced:

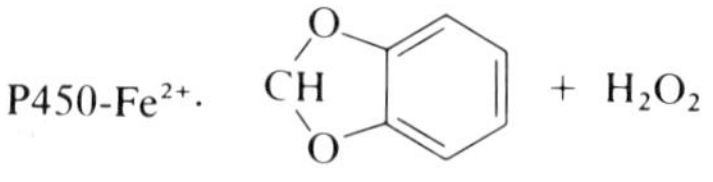

On complete oxidation of the NADPH, the ferrous form of the complex is formed.

Little is known concerning the mode of action of synergists, other than methylenedioxyphenyl compounds, which inhibit microsomal mixed-function oxidases. Although they may act by competitive inhibition, this is a viable mechanism only if they bind tightly to the cytochrome or if the enzyme-substrate complex is relatively stable. In the case of some of the substituted imidazoles, it is clear that the former condition exists since the K_s values are in the region of 1×10^{-6} to 1×10^{-5} M (Wilkinson *et al.*, 1972, 1974*a,b*). The possibility of some mode of action other than, or in addition to, competitive inhibition is raised by the demonstration of a type III spectrum for 1-phenylisopropylimidazole (Leibman and Ortiz, 1973) and the unusual spectrum of 5,6-dichlorobenzothiadiazole with reduced microsomes, which shows a peak at 446 nm (Matthews *et al.*, 1970; Gil, 1973).

4. *Induction of Cytochrome P450 by Insecticides*

4.1. *Induction in Mammals*

Many diverse organic compounds have been shown to be inducers of mixed-function oxidases in mammals. The characteristics and effects of the two general types of inducers have been reviewed extensively by Conney (1967). Polycyclic aromatic hydrocarbons induce specific alterations in enzyme activity with simultaneous qualitative changes in the spectral characteristics of cytochrome P450. The predominant type of induction, stimulated by compounds such as phenobarbital, is nonspecific and characterized by overall quantitative increases in all mixed-function oxidase components, including cytochrome P450. Insecticides are generally included in the category of nonspecific inducers. Most studies on pesticide induction of cytochrome P450 have been focused primarily on quantitative measurements using the CO difference spectra as the sole criterion.

The organochlorine pesticides are the most effective insecticides of mixed-function oxidase induction in mammals. Several chlorinated hydrocarbons such as chlordane, DDT, lindane, and mirex have been implicated in cytochrome P450 induction. Chlordane induces cytochrome P450 levels approximately twofold in the rat (Hart and Fouts, 1965; Mullen *et al.*, 1966; Lucier *et al.*, 1972). This is greater than the induction seen with either 3-methylcholanthrene or benzpyrene (Mullen *et al.*, 1966) but lower than the induction seen after phenobarbital administration (Hart and Fouts, 1965). In the squirrel monkey, hepatic microsomal cytochrome P450 levels were induced fivefold by chlordane (Cram *et al.*, 1965). In this instance, cytochrome P450 induction was greater than the simultaneous increase in other types of enzymatic activity. DDT is a good inducer of microsomal cytochrome P450. Treatment with DDT increased cytochrome P450 in hepatic microsomes of both rats (Bunyan *et al.*, 1970; Remmer *et al.*, 1968) and mice (Abernathy *et al.*, 1971*a,b*; Chhabra and Fouts, 1973). DDE, a metabolite of DDT, also induces microsomal cytochrome P450 in mammals (Bunyan *et al.*, 1970; Bunyan and Page, 1973). Bunyan *et al.* (1970) demonstrated in rats that DDE is a more effective inducer of hepatic microsomal cytochrome P450 than DDT. Mirex induces cytochrome P450 to twice control values at dietary levels of

1–25 ppm in rats and mice (Baker *et al.*,1972). This compound has the lowest threshold level for induction of cytochrome P450 reported for any insecticide, and at high dietary doses (100–250 ppm) it stimulates formation of P450 levels which are 6 times the control levels (Baker *et al.*, 1972).

Anticholinesterase insecticides are relatively poor inducers and may actually inhibit microsomal oxidative reactions under certain conditions (Conney, 1967; Stevens *et al.*, 1972). However, repeated exposure to relatively high doses below the LD_{50} may result in induction of cytochrome P450. Carbaryl administered orally at one-half the LD_{50} to mice for 3 days stimulated significant cytochrome P450 induction which paralleled increases in other microsomal oxidative parameters (Stevens *et al.*, 1973). The same study showed that feeding one-half LD_{50} doses of parathion, disulfoton, and carbaryl for 5 days resulted in increases in ethylmorphine-*N*-demethylase, NADPH-cytochrome *c* reductase, NADPH oxidase, and cytochrome P450.

Springfield *et al.* (1973) demonstrated induction by pyrethrum, a botanical insecticide, in mammals. Rats fed pyrethrum at 200 mg/kg for either 13 or 29 days exhibited significant increases in cytochrome P450 as well as other oxidative enzymes. It should be emphasized that the doses of nonorganohalide insecticides required for mixed-function oxidase induction are relatively high. Fouts (1973) has suggested that induction in some strains of mice is due to stress response rather than a direct effect of the inducer itself. The effects seen with anticholinesterases and pyrethrin may be similar in nature.

Methylenedioxyphenyl compounds, used as insecticide synergists, are inducers of cytochrome P450 and mixed-function oxidase activity in mammals (Hodgson and Philpot, 1974). *In vivo*, methylenedioxyphenyl compounds have a biphasic effect, a single dose initially inhibiting mixed-function oxidase activity (< 24–36 hr) and subsequently causing induction. Piperonyl butoxide, a compound widely used as a synergist, has been shown to induce hepatic cytochrome P450 levels in rats (Conney *et al.*, 1971; Wagstaff and Short, 1971) and mice (Matthews *et al.*, 1970; Škrinjarić-Spoljar *et al.*, 1971; Philpot and Hodgson, 1971*c*; Hodgson *et al.*, 1973). Isosafole induces cytochrome P450 levels in liver (Wagstaff and Short, 1971; Lake and Parke, 1971, 1972) and other extrahepatic tissues such as lung, kidney, gut, and spleen (Lake and Parke, 1972). Safrole (Parke, 1970; Wagstaff and Short, 1971; Gray *et al.*, 1972) and propyl isome (Philpot and Hodgson, 1971*c*; Hodgson *et al.*, 1973) are also inducers of cytochrome P450 in mammals.

Relatively few studies have been concerned with the qualitative effects on cytochrome P450 of pesticide induction in mammals. Several comparisons of enzymatic induction by pesticides and other inducers suggest qualitative differences may be apparent. Chlordane increases zoxazolamine hydroxylation more than that of either benzpyrene or 3-methylcholanthrene. However, benzpyrene hydroxylase and NADPH oxidase are induced similarly by all three inducers (Mullen *et al.*, 1966). In contrast to induction with phenobarbital or benzpyrene, a single dose of DDT in the rat causes aminopyrine demethylation and hexobarbital oxidation to remain higher than controls even after cytochrome P450 levels have returned to normal. This has been interpreted as a manifestation of molecular differences in the enzyme systems induced by these compounds (Remmer *et al.*, 1968).

Cytochrome P450 forms characteristic substrate-binding spectra which are often used in evaluating qualitative differences in cytochrome P450s such as those seen between control and phenobarbital- or 3-methylcholanthrene-induced microsomes. Details of the various types of spectra are discussed in a review by Hodgson (1974).

Pretreatment of the North Carolina Board of Health mouse strain with phenobarbital, benzpyrene, or DDT did not alter the ethyl isocyanide–cytochrome P450 difference spectrum, suggesting that a similar type of cytochrome P450 is induced in this strain (Abernathy *et al.*, 1971*b*). Ethyl isocyanide difference spectra in hepatic microsomes of mice induced by low doses of mirex were similar to those of controls, indicating induction of the same cytochrome (Baker *et al.*, 1972). On the other hand, when microsomes obtained from animals which had received high doses of mirex were compared to controls, the cytochrome P450/benzphetamine type I and/pyridine type II ratios were different. This indicates the possible induction of a qualitatively different cytochrome P450.

Some evidence suggests that cytochrome P450 induced by several methylenedioxyphenyl compounds is qualitatively different from that in controls. Administration of isosafrole and safrole to rats has been reported to induce a modified cytochrome P450 (Parke, 1970; Lake and Parke, 1972), and treatment of rats with isosafrole results in an increase in the binding of cytochrome P450 to both type I and type II substrates. At the same time, K_s values for type II substrates decrease while those for type I substrates do not (Lake and Parke, 1972).

Similarily, piperonyl butoxide has been reported to induce a quantatively different form of cytochrome P450. *In vivo* treatment of mice with piperonyl butoxide and propyl isome induces P450 levels after 36 hrs. However, in piperonyl butoxide induced animals, the ethyl isocyanide pH equilibrium point and type I to type II spectral binding ratios, both parameters indicative of qualitative changes, were significantly different from controls. In the propyl isome treated animals, no changes in these parameters were apparent (Philpot and Hodgson, 1971*c*; Hodgson *et al.*, 1973).

4.2. Induction in Insects

Pesticide induction of microsomal mixed-function oxidase activity has been demonstrated for several insect species, although detailed studies of cytochrome P450 induction have been largely confined to the housefly, *Musca domestica.* A broad spectrum of insecticides including DDT, cyclodienes, phosphoric acid esters, and juvenile hormone analogues are cytochrome P450 inducers in insects. Interpretation of inductive effects such as quantitative and qualitative characteristics of cytochrome P450 in these animals may be complicated by innate sex, age, or strain differences. For example, several uninduced resistant and susceptible housefly strains exhibit striking differences when characteristics of their cytochrome P450s are compared (Matthews and Casida, 1970; Philpot and Hodgson, 1971*a*; Perry *et al.*, 1971; Tate *et al.*, 1973).

Increased titers of cytochrome P450 have been reported after insecticide induction in several housefly strains. Dieldrin increases cytochrome P450 levels (Matthews and Casida, 1970) and oxidase activity (Yu and Terriere, 1971, 1972; Plapp and Casida, 1970) in the Orlando-R strain. DDT induces cytochrome P450 in the Orlando-R (Tate *et al.*, 1973) and Diazinon-R (Perry *et al.*, 1971) strains. Triorthocresylphosphate (TOCP), triphenylphosphate (TPP), and tributylphosphorotrithioate (DEF) increase cytochrome P448 (P450) levels in the Diazinon-R strain but have no effect on levels in the Malathion-R housefly strain (Perry *et al.*, 1971).

Terriere and Yu (1973) have demonstrated induction of cytochrome P450 in houseflies by juvenile hormone analogues. Cecropia JH and hydroprene (ethyl-3,7,11-triethyldodeca-2,4-dienoate) stimulated microsomal oxidation and increased cytochrome P450 levels by 31% in the Isolan-B strain.

Several other compounds are inducers of cytochrome P450 in insects. Phenobarbital and butylated hydroxytoluene are inducers of P450 in several housefly strains (Perry *et al.*, 1971). In the Fc strain, cytochrome P450 is also induced by naphthalene and phenobarbital (Capdevila *et al.*, 1973*a,b*). *In vivo* treatment with alkylbenzenes resulted in increases in mixed-function oxidase activity in gut, malpighian tubules, and fat body of the southern armyworm, *Prodenia eridania*. Increases in cytochrome P450 were concomitant with increases in other enzymatic parameters (Brattsten and Wilkinson, 1973).

Ascertaining qualitative alterations in chemically induced insect cytochrome P450 is confused by methodological differences utilized among various laboratories. Difference spectroscopy is generally used to measure variations in insect cytochrome P450s by comparisons of type I, type II, and type III binding spectra and CO spectral maxima. In uninduced mammals, cytochrome P450 characteristics are fairly uniform among strains. This makes detection of qualitatively different cytochromes easier. Several uninduced resistant housefly strains, however, possess cytochrome P450 which resembles in many respects that seen in 3-methylcholanthrene-induced mammals. This "resistant" cytochrome P450 has never been reported in any susceptible strain, whereas several resistant housefly strains with high oxidase activity may contain the "susceptible" cytochrome P450. Others may have characteristics of both types. Uninduced Diazinon-R (Perry and Buckner, 1970; Philpot and Hodgson, 1971*a*), Fc, and Dimethoate-R (Tate *et al.*, 1973) housefly strains contain the "resistant" type. This differs from the susceptible type of cytochrome P450 in several respects: (1) the CO maximum is shifted several nanometers to the blue; (2) type I binding is present; (3) type II binding is

Table II. Comparison of Cytochrome P450 Spectral Characteristics Between Resistant and Susceptible Housefly Strains

	Strain	
Spectrum	Susceptible[a]	Resistant[b]
P450 CO absorbance maxima (nm)	451–452	448–449
P450 titer/abdomen (% susceptible level)	100	150–200
Formation of type I difference spectrum	No	Yes
Formation of a single-trough type II *n*-octylamine spectrum	No	Yes
Magnitude of type III ethyl isocyanide 455 nm peak (% CO value)	100	60–70

[a]CSMA and *sbo*-S strains.
[b]Fc, Diazinon-R, and Dimethoate-R strains.

increased in relation to the CO spectrum; (4) the ethyl isocyanide spectrum is changed in magnitude (Table II).

In the Diazinon-R strain of housefly, *in vivo* administration of malathion, tropital, MGK-264, and three substituted phenyl-2-propynyl ether synergists produced a shift in the carbon monoxide spectrum from 448 (uninduced) to 450 nm. The same compounds did not affect microsomal cytochrome P450 (CO maximum 452) in the Malathion-R strain (Perry *et al.*, 1971). In Fc, a housefly strain previously reported to contain the "resistant" cytochrome P449 (Tate *et al.*, 1973), Capdevila *et al.* (1973*a,b*) have reported the induction of a qualitatively different cytochrome by naphthalene and phenobarbital. After induction, increased titers of cytochrome were measured which exhibited lower CO maxima, higher type I and type II ratios to the CO peak, and a different ethyl isocyanide equilibrium point than those in control microsomes. Apparently phenobarbital and naphthalene are specific inducers in certain insect strains, much like 3-methylcholanthrene in mammals.

There are contradictory data concerning insecticide induction of cytochrome P450 in other strains. The Orlando-R housefly strain normally has the "susceptible" type of cytochrome P450 (Tate *et al.*, 1973). Matthews and Casida (1970) reported a change in the ethyl isocyanide cytochrome P450 spectrum after topical application of dieldrin. More recently, Orlando-R flies fed 1000 ppm DDT showed no change in any qualitative cytochrome P450 characteristic, including ethyl isocyanide spectra (Tate *et al.*, 1973). Perhaps dieldrin and DDT induce by different mechanisms in this strain.

Treatment of *Prodenia eridania* with alkylbenzenes had no effect on the CO difference spectra maxima of cytochrome P450 (Brattsten and Wilkinson, 1973).

5. Cytochrome P450 and Insecticide Resistance

In the last several decades, many species of insects have acquired resistance to insecticides. Resistance in many instances is due at least in part to increased levels of mixed-function oxidase activity. Cytochrome P450 is the reactive focal point in the mixed-function oxidase system (Estabrook *et al.*, 1973). Since the discovery of cytochrome P450 in insects (Ray, 1967), there has been considerable interest in its role in the detoxication of xenobiotics by resistant insects. All comparative studies on cytochrome P450 in susceptible and resistant insects have been confined to houseflies.

Increased cytochrome P450 levels have been reported for a number of resistant housefly strains possessing high oxidase levels. Higher levels of cytochrome P450 as measured by the CO spectrum have been reported in the Diazinon-R (Folsom *et al.*, 1970; Perry and Buckner, 1970; Perry *et al.*, 1971; Philpot and Hodgson, 1971*a*), Fc (Matthews and Casida, 1970; Tate *et al.*, 1973), R-Baygon (Matthews and Casida, 1970; Tate *et al.*, 1973), Dimethoate-R (Tate *et al.*, 1973), Orlando-R (Matthews and Casida, 1970; Tate *et al.*, 1973), Malathion-R (Perry *et al.*, 1971), and Ronnel-R (Georghiou, 1969*a*). Although high titers of cytochrome P450 may be present in a resistant strain, there is not necessarily a correlation between this and mixed-function oxidase activity. For example, the housefly strains Fc,*bwb,stw* and R-Baygon,*bwb,ocra* have high NADPH-dependent oxidase activity (Plapp and Casida, 1969), yet their cytochrome P450 does not exceed the levels seen in susceptible strains. In no resistant strain assayed has more than a two-fold increase in cytochrome P450 been detected, whereas detoxication rates as measured by enzymatic assay are frequently several times the susceptible values. This fact has led some investigators to suggest that cytochrome P450 may not be the limiting factor in resistant mixed-function oxidase activity.

Several recent comparative studies have indicated the existence of qualitatively different cytochrome P450 in several resistant housefly strains. Characteristics of this resistant-strain cytochrome P450 as compared to the susceptible-strain cytochrome P450 are shown in Table II. There are several marked differences between the two forms. The "resistant" form differs in the following manner: (1) type I binding is present; (2) the CO maximum is shifted to a lower wavelength; (3) the magnitude of the type II spectrum is increased relative to the CO spectrum; (4) the 455 nm peak of the ethyl isocyanide type III spectrum is lowered with respect to the peak of the CO spectrum; (5) the octylamine type II spectrum has a single trough at 390 nm, in contrast to the "susceptible" P450 which binds with *n*-octylamine to form a double trough at 410 and 394 nm, respectively.

This qualitatively different cytochrome P450 was first described in the multiresistant Diazinon-R housefly strain. Perry and Buckner (1970) and Perry *et al.* (1971) observed that the CO maximum in microsomal preparations from this strain was several nanometers lower than that of other strains. Simultaneously, Philpot and Hodgson (1971*a*) demonstrated this lowered CO maximum wavelength as well as other qualitative differences between Diazinon-R and susceptible CSMA cytochrome P450s. High levels of cytochrome P450 similar to those seen in the Diazinon-R strain have been detected in the Fc and Dimethoate -R strains (Tate *et al.*, 1973). In the same study, the susceptible cytochrome P450 was detected in another Fc strain as well as R-Baygon and Orlando-R houseflies.

The significance of this qualitatively different cytochrome P450 in some housefly strains is unclear. No specific enzymatic reaction or resistant characteristic has been correlated with this cytochrome. NADPH-dependent oxidase levels in strains containing this cytochrome P450 are not substantially higher than those seen in resistant strains containing the susceptible type of cytochrome. However, high oxidase activity in strains having the resistant type of cytochrome can be correlated with the presence of type I binding, whereas other resistant cytochrome characteristics are not necessarily related to high oxidase activity. Microsomes from Fc_{Tex} and Fc,*bwb,stw* housefly abdomens, high in NADPH oxidase activity (Plapp and Casida, 1969), form binding spectra with type I substrates but lack any of the other resistant qualitative characteristics (Tate *et al.*, 1973). Evidence from genetic studies suggests that, in strains exhibiting type I binding, increased oxidase activity segregates on the same chromosome which confers type I binding regardless of the cytochrome P450 titer (Tate *et al.*, 1973, 1974).

While the genetics of resistance in insects has been studied extensively (Georghiou, 1969*b*), until recently no data were available on cytochrome P450 genetics in insects. The discovery of a qualitatively different form made possible a genetic analysis of cytochrome P450 and its relation to the genetics of resistance. High oxidase activity segregates with chromosome II and/or chromosome V in resistant houseflies (Hodgson and Plapp, 1970).

Tate *et al.* (1974) studied the genetics of cytochrome P450 in Diazinon-R, a strain of previously unknown genetic constitution, and Fc, one containing high-oxidase genes on chromosome V. Resistant strains were crossed with a susceptible strain carrying visible, recessive markers on chromosomes II, III, and V, the chromosomes generally responsible for increased mixed-function oxidase activity in houseflies. F_1 males were backcrossed to females of the marked susceptible strain. This procedure was utilized to eliminate crossing over in the F_1 generation. Each of the eight phenotype progenies resulting from the backcross contained specific combinations of resistant and susceptible chromosomes. These substrains were analyzed for cytochrome P450 qualitative characteristics, and it was found that those seen in the resistant strain are inherited as semidominants on chromosome II. On the other hand, in Fc, only type I binding segregates with chromosome V, the chromosome which was previously shown to confer high oxidase activity in that strain (Plapp and Casida, 1969). Chromosome II in Fc contains gene(s) which are quantitative in their effects on the expression of cytochrome P450. Other qualitative characteristics seen in the parent Fc strain could not be recovered in any resistant chromosome combinations in the backcross progeny, suggesting the possible involvement of background modifiers in expression of those phenotypes. These results indicate that in one strain, Fc, and possibly others, at least two or more loci are involved in phenotype expression of the resistant cytochrome P450.

More recent studies (Plapp *et al.*, 1975) of crossing over in the Diazinon-R strain have demonstrated that, of the various cytochrome P450 parameters tested, only the presence of type I binding is always linked to resistance. This emphasizes the fact that this characteristic is related to some enzymatic activity of importance in the resistance mechanisms and, moreover, that more than one gene is involved in the inheritance of different forms of cytochrome P450 in the housefly.

6. Conclusion

Cytochrome P450 is the key element of the microsomal mixed-function oxidase system which participates, in many tissues and organisms, in the detoxication of xenobiotics. Evolutionary considerations lead to the conclusion that this cytochrome has been adapted, in certain tissues of animals, for the detoxication of lipophilic dietary toxicants, of relatively low toxicity, which would otherwise accumulate to deleterious levels. Induction provides an additional level of flexibility in carrying out this function. As a result of the lack of specificity which is necessary to carry out such a function, this system also metabolizes synthetic organic compounds such as insecticides.

Insecticides can interact with cytochrome P450 in a variety of ways, either directly as substrates or inhibitors or indirectly as inducers. In many cases, insecticides or insecticide synergists can interact in more than one way, being first inhibitors of substrates and subsequently inducers.

It has been particularly difficult to ascribe mechanisms to these interactions or to discern structure–function relationships among the many substrates and inducers. Much of the difficulty lies in the experimental problems inherent in working with particulate enzymes and substrates and inhibitors of limited solubility in water. Optical difference spectroscopy, although a valuable technique, is an indirect measure of binding. Other problems are physiological, particularly the problem of how many cytochrome P450s there are in the uninduced mammal. Indirect evidence suggests that more than one exists in mammalian liver, and if this is the case it is important to know how they differ from one another and what their functions are.

The importance of cytochrome P450 cannot be seriously challenged: almost every insecticide used interacts with it in both target and nontarget organisms. The best hope for a more complete understanding appears to lie in improved purification techniques and reconstitution of simplified but still functional systems. Investigations of this type are now going on in several laboratories.

7. References

Abernathy, C. O., Hodgson, E., and Guthrie, F. E., 1971*a*, Structure activity relationships on the induction of hepatic microsomal enzymes in the mouse by 1,1,1-trichloro-2,2-bis-(*p*-chorophenyl)ethane (DDT) analogs, *Biochem. Pharmacol.* **20:**2385.

Abernathy, C. O., Philpot, R. M., Guthrie, F. E., and Hodgson, E., 1971*b*, Inductive effects of 1,1,1-trichloro-bis-(*p*-chlorophenyl)ethane (DDT), phenobarbital and benzpyrene on microsomal cytochrome P-450, ethyl isocyanide spectra, and metabolism, *in vivo* of zoxazolamine and hexobarbital in the mouse, *Biochem. Pharmacol.* **20:**2395.

Anders, M. W., 1971, Enhancement and inhibition of drug metabolism, *Ann. Rev. Pharmacol.* **11:**37.

Baker, R. C., Coons, L. C., Mailman, R. B., and Hodgson, E., 1972, Induction of hepatic mixed-function oxidases by the insecticide, Mirex, *Environ. Res.* **5:**418.

Brattsten, L. B., and Wilkinson, C. F., 1973, Induction of microsomal enzymes in the southern armyworm (*Prodenia eridania*), *Pestic. Biochem. Physiol.* **3:**393.

Bright, H. J., Wood, B. J. B., and Ingraham, L. L., 1963, Copper, tyrosinase, and the kinetic stability of oxygen, *Ann. N.Y. Acad. Sci.* **100:**965.

Brown, R. R., Miller, J. A., and Miller, E. C., 1954, The metabolism of methylated aminoazo dyes. 4. Dietary factors enhancing demethylation *in vitro, J. Biol. Chem.* **209:**211.

Bunyan, P. J., and Page, J. M. J., 1973, Some effects of 1,1-di(*p*-chlorophenyl)-2,2-dichloroethylene (DDE) and 1,1-di-(*p*-chlorophenyl)-2-chloroethylene (DDMU) in the rat and Japanese quail, *Chem.-Biol. Interact.* **6:**249.

Bunyan, P. J., Taylor, A., and Townsend, M. G., 1970, The effects of 1,1-di-(*p*-chlorophenyl)-2,2,2-trichloroethane (DDT) and 1,1-di-(*p*-chlorophenyl)-2,2-dichloroethylene (DDE) on hepatic microsomal oxidase in the rat and japanese quail, *Biochem. J.* **118:**51P.

Capdevila, J., Morello, A., Perry, A. S., and Agosin, M., 1973*a*, Effect of phenobarbital and naphthalene on some components of the electron transport system and hydroxylation activity of housefly microsomes, *Biochemistry* **12:**1445.

Capdevila, J., Perry, A. S., Morello, A., and Agosin, M., 1973*b*, Some spectral properties of cytochrome P-450 from microsomes isolated from control, phenobarbital and naphthalene-treated houseflies, *Biochim. Biophys. Acta* **314:**93–103.

Casida, J. E., 1970, Mixed-function oxidase involvement in the biochemistry of insecticide synergists, *J. Agr. Food Chem.* **18:**753.

Casida, J. E., Engel, J. L., Esaac, E. G., Kamienski, F. X., and Kuwatsuka, S., 1966, Methylene-C^{14}-dioxyphenyl compounds: Metabolism in relation to their synergest action, *Science* **151:**1130.

Chhabra, R. S., and Fouts, J. R., 1973, Stimulation of hepatic microsomal drug metabolizing enzymes in mice by 1,1,1-trichloro-2,2-bis (*p*-chlorophenyl)ethane (DDT) and 3,4-benzpyrene, *Toxicol. Appl. Pharmacol.* **25:**60.

Cloney, R. D., and Scherr, V. H., 1968, Molecular orbital study of hydride transferring ability of benzodiheterolines as a basis of synergistic activity, *J. Agr. Food Chem.* **16:**791.

Comai, K., and Gaylor, J. L., 1973, Existence and separation of three forms of cytochrome P-450 from rat liver microsomes. *J. Biol. Chem.* **248:**4947.

Conney, A. H., 1967, Pharmacological implications of microsomal enzyme induction, *Pharmacol. Rev.* **19:**317.

Conney, A. H., and Kuntzman, R., 1971, Metabolism of normal body constituents by drug metabolizing enzymes in liver microsomes, *in: Handbook of Experimental Pharmacology*, Vol. 28, (B. B. Brodie, J. R. Gillette, and H. S. Ackerman, eds.), pp. 401–421, Springer, New York.

Conney, A. H., Welch, R., Kuntzman, R., Chang, R., Jacobson, M., Munro-Faure, A. D., Peck, A. W., Bye., Poland, A., Poppers, P. J., Finster, M., and Wolff. J. A., 1971, Effects of environmental chemicals on the metabolism of drugs, carcinogens and normal body constituents in man, *Ann. N.Y. Acad. Sci.* **179:**155

Cooper, D. Y., Narashimhula, S., Rosenthal, O., and Estabrook, R. W., 1965, Spectral and kinetic studies of microsomal pigments, *in: Oxidase and Related Redox Systems* (T. E. King, H. S. Moses, and M. Morrison, eds.), pp. 838–860, Wiley, New York.

Cooper, D. Y., Schleyer, H., Levin, S. S., and Rosenthal, C., 1971, P-450 from adrenal-cortical mitochondria: Interaction with CO, substrates, iron sulfur protein and O_2, *Abst. 162nd Am. Chem. Soc. Natl. Meet.*

Cram, R. L., Juchau, M. R., and Fouts, J. R., 1965, Stimulation by chlordane of hepatic drug metabolism in the squirrel monkey. *J. Lab. Clin. Med.* **66:**906.

Desmarchellier, J., Krieger, R. O., Lee, P. W., and Fukuto, T. R., 1973, Carbaryl synergism by substituted 1,3-benzodioxoles and related compounds and inhibition of mixed-function oxidases, *J. Econ. Entmol.* **66:**631.

Diehl, H., Schadelin, J., and Ullrich, V., 1970, Studies on the kinetics of cytochrome P-450 reduction in rat liver microsomes, *Hoppe-Seylers Z. Physiol. Chem.* **351:**1359.

Eagleson, C., 1940, U.S. Patent 2,202,145.

Estabrook, R. W., 1971, Cytochrome P-450—Its function in the oxidative metabolism of drugs, *in:* Handbook of *Experimental Pharmacology,* Vol. 28, (B. B. Brodie, J. R. Gillette, and H. S. Ackerman, eds.), pp. 264–284, Springer, New York.

Estabrook, R. W., Gillette, J. R., and Leibman, K. C. (eds.), 1973, Second International Symposium on Microsomes and Drug Oxidations, *Drug Metab. Diposition* **1:**1.

Folsom, M. D., Hansen, L. G., Philpot, R. M., Yang, R. S. H., Dauterman, W. C., and Hodgson, E., 1970, Biochemical characteristics of microsomal preparations from diazinon-resistant and susceptible houseflies, *Life Sci.* **9:**869.

Fouts, J. R., 1973, Some selected studies on hepatic microsomal metabolizing enzymes–environmental interactions, *Drug Metab. Disposition* **1:**380.

Franklin, M. R., 1971, The enzymic formation of a methylenedioxyphenyl derivative exhibiting an isocyanide like spectrum with reduced cytochrome P-450 in hepatic microsomes, *Xenobiotica* **1:**581.

Franklin, M. R., 1972*a*, Piperonyl butoxide metabolism by cytochrome P-450: Factors affecting the formation and disappearance of the metabolite cytochrome P-450 complex, *Xenobiotica* **2:**517.

Franklin, M. R., 1972*b*, Inhibition of hepatic oxidative xenobiotic metabolism by piperonyl butoxide, *Biochem. Pharmacol.* **21:**3287.

Garfinkel, D., 1958, Studies on pig liver microsomes. 1. Enzymatic and pigment composition of different microsomal fractions. *Arch. Biochem. Biophys.* **77:**493.

Gelboin, H. V., 1971, Mechanisms of induction of drug metabolism enzymes, *in: Handbook of Experimental Pharmacology,* Vol. 23 (B. B. Brodie, J. R. Gillette, and H. S. Ackerman, eds.), pp. 264–284, Springer, New York.

Georghiou, G. P., 1969*a*, Isolation, characterization and resynthesis of insecticide resistance factors in the housefly *Musca domestica, Proc, 2nd Int. Cong. Pestic. Chem.* **2:**77.

Georghiou, G. P., 1969*b*,Genetics of resistance to insecticides in houseflies and mosquitos, *Exp. Parasitol.* **26:**224.

Gil, L. D., 1973, Structure–activity relationships of 1,2,3-benzothiadiazole insecticide synergists, Ph.D.dissertation, Cornell University.

Gray, T. J. B., Parke, D. V., Grasso, P., and Crampton, R. F., 1972, Biochemical and pathological differences in hepatic response to chronic feeding of safrole and butylated hydroxytoluene to rats, *Biochem. J.* **130:**91P.

Green, F. E., and Stevens, J. R., 1973, Species differences in aldrin and dieldrin interactions with components of the hepatic cytochrome P-450 dependent enzymes, *in: Pesticides and the Environment: A Continuing Controversy* (W. B. Deichman, ed.), International Medical Book Corp., New York.

Hansch, C., 1968, The use of homolytic, steric, and hydrophobic constants, in a structure–activity study of 1,3-benzodioxole synergists, *J. Med. Chem.* **11:**920.

Hart, L. C., and Fouts, J. R., 1965, Studies on the possible mechanism by which chlordane stimulates hepatic microsomal drug metabolism in the rat, *Biochem. Pharmacol.* **14:**263.

Hennessy, D. J., 1965, Hydride transferring ability of methylenedioxybenzenes as a basis of synergistic activity, *J. Agr. Food Chem.* **13:**218.

Hewlett, P. S., 1960, Joint action in insecticides, *Adv. Pest Control Res.* **3:**27.

Hildebrandt, A. G., Leibman, K. C., and Estabrook, R. W., 1969, Metyrapone interaction with hepatic microsomal cytochrome P-450 from rats treated with phenobarbital, *Biochem. Biophys. Res. Commun.* **37:**477.

Hodgson, E., 1974, Comparative studies of cytochrome P-450 and its interaction with pesticides, *in: Survival in Toxic Environments,* Academic Press, New York.

Hodgson, E., and Casida, J. E., 1960, Biological oxidation of *N,N*-dialkyl carbamates, *Biochim. Biophys. Acta* **42:**184.

Hodgson, E., and Casida, J. E., 1961, Metabolism of *N*: *N*-dialkyl carbamates and related compounds by rat liver, *Biochem. Pharmacol.* **8:**179.

Hodgson, E., and Philpot, R. M., 1974, Interaction of methylenedioxyphenyl (1,3-benzodioxole) compounds with enzymes and their effect, *in vivo,* on animals, *Drug Metabol. Revs.* **3:**231–301.

Hodgson, E., and Plapp, F. W., 1970, Biochemical characteristics of insect microsomes, *J. Agr. Food Chem.* **18:**1048.

Hodgson, E., Philpot, R. M., Baker, R. C., and Mailman, R. B., 1973, Effect of synergists on drug metabolism, *Drug Metab. Disposition* **1:**391.

Hodgson, E., Tate, L. G., Kulkarni, A. P., and Plapp, F. W., 1974, Microsomal cytochrome P-450: Characterization and possible role in insecticide resistance in *Musca domestica, J. Agr. Food Chem.* **22:**360–366.

Imai, Y., and Sato, R., 1974, A gel-electrophoretically homogeneous preparations of cytochrome P-450 from liver microsomes of phenobarbital-pretreated rabbits, *Biochem. Biophys. Res. Commun.* **60:**8.

Jaffe, H., Fujii, K., Sengupta, M., Guerin, M., and Epstein, S. S., 1968, *In vivo* inhibition of mouse liver microsomal hydroxylating systems by methylenedioxyphenyl insecticidal synergists and related compounds, *Life Sci.* **7:**1051.

Jefcoate, C. R. E., Gaylor, J. L., and Calabreze, R., 1969, Ligand interactions with cytochrome P-450. 1. Binding of primary amines, *Biochemistry* **8:**3455.

Kamienski, F. X., and Casida, J. E., 1970, Importance of demethylenation in the metabolism *in vivo* and *in vitro* of methylenedioxyphenyl synergists and related compounds in mammals, *Biochem. Pharmacol.* **19:**91.

Klingenberg, M., 1958, Pigments of rat liver microsomes, *Arch. Biochem. Biophys.* **75:**376.

Kulkarni, A. P., Mailman, R. B., and Hodgson, E., 1974*a*, Cytochrome P-450 difference spectra: Type II interactions in insecticide resistant and susceptible houseflies, *Drug Metab. Disposition* **2:**308–320.

Kulkarni, A. P., Mailman, R. B., and Hodgson, E., 1974*b*, Interaction of insecticide synergists with cytochrome P-450 from the housefly, *Musca domestica, Pestic. Biochem. Physiol.,* in press.

Kulkarni, A. P., Mailman, R. B., and Hodgson, E., 1975, Cytochrome P-450 optical difference spectra of insecticides: A comparative study, *J. Agr. Food Chem.,* **23:**177–183.

Kuwatsuka, S., 1970, Biochemical aspects of methylenedioxyphenyl compounds in relation to the synergistic action, *in: Biochemical Toxicology of Insecticides* (R. D. O'Brien and I. Yamamoto, eds.), pp. 131–144, Academic Press, New York.

Lake, B. G., and Parke, D. V., 1971, Interaction of safrole and isosafrole with hepatic microsomal haemoproteins, *Biochem. J.* **127:**23P.

Lake, B. G., and Parke, D. V., 1972, Induction of aryl hydrocarbon hydroxylase in various tissues of the rat by methylenedioxyphenyl compounds, *Biochem. J.* **130:**86P.

Leibman, K. C., and Ortiz, E., 1973, New potent modifiers of liver microsomal drug metabolism: 1-Arylimidazoles, *Drug Metab. Disposition* **1:**29.

Levin, W., Ryan, D., West, S., and Lu, A. Y. H., 1974, Preparation of partially purified, lipid-depleted cytochrome P-450 and reduced nicotinamide adenine dinucleotide phosphate–cytochrome *c* reductase from rat liver microsomes, *J. Biol. Chem.* **249:**1747.

Lu, A. Y. H., West, S. B., Ryan, D., and Levin, W., 1973, Characterization of partially purified cytochrome P-450 and P-448 from rat liver microsomes, *Drug Metab. Disposition* **1:**29.

Lucier, G. W., McDaniel, O. S., Williams, C., and Klein, R., 1972, Effects of chlordane and methylmercury on the metabolism of carbaryl and carbofuran in rats, *Pestic. Biochem. Physiol.* **2:**244.

Mailman, R. B., and Hodgson, E., 1972, The cytochrome P-450 substrate optical difference spectra of pesticides with mouse hepatic microsomes, *Bull. Environ. Contam. Toxicol.* **8:**186.

Mailman, R. B., Kulkarni, A. P., Baker, R. C., and Hodgson, E., 1974, Cytochrome P-450 difference spectra: Effect of chemical structure on type II spectra in mouse hepatic microsomes, *Drug Metab. Disposition* **2:**301–308.

Marshall, R. S., and Wilkinson, C. F., 1970, The epoxidation of aldrin by a modified Fentons reagent and its inhibition by substituted 1,3-benzodioxoles, *Biochem. Pharmacol.* **19:**2665.

Marshall, R. S., and Wilkinson, C. F., 1973, The interaction of insecticide synergists with nonenzymatic model oxidation systems, *Pestic. Biochem. Physiol.* **2:**425.

Matthews, H. B., and Casida, J. E., 1970, Properties of housefly microsomal cytochromes in relation to sex, strain, substrate specificity and apparent inhibition by synergist and insecticide chemicals, *Life Sci.* **9:**989.

Matthews, H. B., Škrinjarić-Špoljar, M., and Casida, J. E., 1970, Insecticide synergist interactions with cytochrome P-450 in mouse liver microsomes, *Life Sci.* **9:**1039.

Mayer, R. T., Wade, A. E., and Soliman, M. R. I., 1973, Juvenile hormone analogs as *in vitro* inhibitors of rat liver microsomal oxidases. *J. Agr. Food Chem.* **21:**360.

Mehendale, H. M., Fishbein, L., Fields, M., and Matthews, H. B., 1972, Fate of ^{14}C-Mirex in the rat and plants. *Bull. Environ. Contam. Toxicol.* **8:**200.

Metcalf, R. L., Fukuto, T. R., Wilkinson, C. F., Fahmy, M. H., El-Aziz, S. A., and Metcalf, E. R., 1966, Mode of action of carbamate synergists, *J. Agr. Food Chem.* **14:**555.

Mullen, J. O., Juchau, M. R., and Fouts, J. R., 1966, Studies of interactions of 3,4-benzpyrene, 3-methylcholanthrene, chlordane and methyltestosterone as stimulators of hepatic microsomal enzyme systems in the rat, *Biochem. Pharmacol.* **15:**137.

Omura, T., and Sato, R., 1964*a*, The carbon monoxide binding pigment of liver microsomes. 1. Evidence for its hemoprotein nature. *J. Biol. Chem.* **239:**2370.

Omura, T., and Sato, R., 1964*b*, The carbon monoxide binding pigment of liver microsomes, *J. Biol. Chem.* **239:**2379.

Parke, D. V., 1970, Mechanism and consequences of the induction of microsomal enzymes of mammalian liver, *Biochem. J.* **119:**53P.

Perry, A. S., and Buckner, A. J., 1970, Studies on microsomal P-450 in resistant and susceptible houseflies, *Life Sci.* **9:**335.

Perry, A. S., Dale, W. E., and Buckner, A. J., 1971, Induction and repression of microsomal mixed-function oxidases and cytochrome P-450 in resistant and susceptible houseflies, *Pestic. Biochem. Physiol.* **1:**131.

Philpot, R. M., and Hodgson, E., unpublished work cited in Hodgson, E., and Plapp, F. W., 1970, Biochemical characteristics of insect microsomes, *J. Agr. Food Chem.* **18:**1048.

Philpot, R. M., and Hodgson, E., 1971*a*, Differences in the cytochrome P-450s from resistant and susceptible houseflies, *Chem.-Biol. Interact.* **4:**399.

Philpot, R. M., and Hodgson, E., 1971*b*, A cytochrome P-450–piperonyl butoxide spectrum similar to that produced by ethyl isocyanide, *Life Sci.* **10:**503.

Philpot, R. M., and Hodgson, E., 1971*c*, The production and modification of cytochrome P-450 difference spectra by *in vivo* adminstration of methylenedioxyphenyl compounds, *Chem.-Biol. Interact.* **4:**185.

Philpot, R. M., and Hodgson, E., 1972, The effect of piperonyl butoxide concentration on the formation of cytochrome P-450 differences spectra in hepatic microsomes from mice, *Mol. Pharmacol.* **8:**204.

Plapp, F. W., and Casida, J. E., 1969, Genetic control of housefly NADPH dependent oxidases: Relation to insecticide chemical metabolism and resistance, *J. Econ. Entomol.* **62:** 1174.

Plapp, F. W., and Casida, J. E., 1970, Induction by DDT and dieldrin of insecticide metabolism by housefly enzymes, *J. Econ. Entomol.* **63:**1191.

Plapp, F. W., Jr., Tate, L. G., and Hodgson, E., 1975, Biochemical genetics of oxidative resistance to diazinon in the housefly. *Pestic. Biochem. Physiol.*, in press.

Ray, J. W., 1967, The epoxidation of aldrin by housefly microsomes and its inhibition by carbon monoxide, *Biochem. Pharmacol.* **16:**99

Remmer, H., Estabrook, R. W., Schenkman, J. B., and Greim, H., 1968, Induction of microsomal liver enzymes, *in: Enzymatic Oxidation of Toxicants* (E. Hodgson, ed.), pp. 65–88, North Carolina State University Press, Raleigh, N.C.

Remmer, H., Schenkman, J. B., and Greim, H., 1969, Spectral investigation on cytochrome P-450, *in: Microsomes and Drug Oxidation* (J. R. Gillette, A. H. Conney, G. J. Cosmides, R. W. Estabrook, J. E. Fouts, and G. J. Mannering, eds.), pp. 371–336, Academic Press, New York.

Roth, J. A., and Neal, R. A., 1972, Spectral studies of the binding of *O,O*-diethyl-*p*-nitrophenylphosphorothiote (parathion) to cytochrome P-450, *Biochemistry* **11:**955.

Sato, R., Sotake, H., and Imai, Y., 1973, Partial purification and some spectral properties of hepatic microsomal cytochrome P-450, *Drug Metab. Disposition* **1:**6.

Schenkman, J. B., Remmer, H., and Estabrook, R. W., 1967, Spectral studies of drug interaction with hepatic microsomal cytochrome, *Mol. Pharmacol.* **3:**113.

Schenkman, J. B., Cinti, D. L., Orrenius, S., Moldeus, P., and Kraschnitz, R., 1972, The nature of the reverse type I (modified type II) spectral change in liver microsomes, *Biochemistry* **11:**4243.

Škrinjarić-Špoljar, M., Matthews, H. B., Engel, J. L., and Casida, J. E., 1971, Response of hepatic microsomal mixed-function oxidases to various types of insecticide chemical synergists administered to mice, *Biochem. Pharmacol.* **20:**1607.

Sladek, N. E., and Mannering, G. J., 1966, Evidence for a new hemoprotein in hepatic microsomes from methylcholanthrene treated rats, *Biochem. Biophys. Res. Commun.* **24:**668.

Springfield, A. C., Carlson, G. P., and DeFeo, J. J., 1973, Liver enlargement and modification of hepatic microsome drug metabolism in rats by pyrethrum. *Toxicol. Appl. Pharmacol.* **24:**298.

Stevens, J. T., Stitzel, R. E., and McPhillips, J. J., 1972, Effects of anticholinesterase insecticides on hepatic microsomal metabolism, *J. Parmacol. Exp. Ther.* **181:**576.

Stevens, J. T., Green, F. E., Stitzel, R. E., and McPhillips, J. T., 1973, Effects of anticholinesterase insecticides on mouse and rat liver microsomal mixed-function oxidases, *in: Pesticides and the Environment* (W. B. Deichmann, ed.), pp. 498–501, Intercontinental Medical Book Corp., New York and London.

Strobel, H. W., and Coon, M. J., 1971, Effect of superoxide generation and dismutation on hydroxylation reactions catalyzed by liver microsomal cytochrome P-450, *J. Biol. Chem.* **246:**7826.

Sun, Y. P., and Johnson, E. R., 1960, Synergistic and antagonistic action of insecticide–synergist combinations and their mode of action, *J. Agr. Food Chem.* **8:**261.

Tate, L. G., Plapp, F. W., and Hodgson, E., 1973, Cytochrome P-450 difference spectra of microsomes from several insecticide-resistant and susceptible strains of the housefly, *Musca domestica* L., *Chem.-Biol. Interact.* **6:**237.

Tate, L. G., Plapp, F. W., and Hodgson E., 1974, Genetics of cytochrome P-450 in two insecticide resistant strains of the housefly, *Musca domestica, Biochem. Genet.* **11:**49.

Terriere, L. C., and Yu, S. J., 1973, Insect juvenile hormones: Induction of detoxifying enzymes in the housefly and detoxication by housefly enzymes, *Pestic. Biochem. Physiol.* **3:**96.

T'sai, R., Yu, A., Gunsalus, I. C., Peisach, J., Blumberg, W., Orme-Johnson, W. M., and Beinert, H., 1970, Spin state changes in cytochrome P-450_{cam} on binding of specific substrates, *Proc. Natl. Acad. Sci. U.S.A.* **66:**1157.

Ullrich, V., and Schnabel, K. H., 1973, Formation and binding of carbanions by cytochrome P-450 of liver microsomes, *Drug Metab. Disposition* **1:**176.

van der Hoeven, T. A., Haugen, D. A., and Coon, M. J. 1974, Cytochrome P-450 purified to apparent homogeneity from phenobarbital-induced rabbits liver microsomes: Catalytic activity and other properties. *Biochem. Biophys. Res. Commun.* **60:**569.

Wagstaff, D. J., and Short, C. R., 1971, Induction of hepatic microsomal hydroxylating enzymes by technical piperonyl butoxide and some of its analogs, *Toxicol. Appl. Pharmacol.* **19:**54.

Wilkinson, C. F., 1967, Penetration, metabolism, and synergistic activity with carbaryl of some simple derivities of 1,3-benzodioxide in the housefly, *J. Agr. Food Chem.* **15:**139.

Wilkinson, C. F., 1971, Insecticide synergists and their mode of action, *Proc. 2nd Int. I.U.P.A.C. Congr. Pestic. Chem.* **2:**117.

Wilkinson, C. F., and Brattsten, L. B., 1972, Micrósomal drug metabolizing enzymes in insects, *Drug Metabol. Rev.* **1:**153.

Wilkinson, C. F., and Hicks, L. J., 1969, Microsomal metabolism of the 1,3-benzodioxole ring and its possible significance in synergistic action, *J. Agr. Food Chem.* **17:**829.

Wilkinson, C. F., Yellin, T. O., and Hetnarski, K., 1972, Imidazole derivatives: A new class of microsomal enzyme inhibitors, *Biochem. Pharmacol.* **21:**2187.

Wilkinson, C. F., Hetnarski, K., Cantwell, G. P., and Di Carlo, F. J., 1974*a*, Structure–activity relationships in the effects of 1-alkylimidazoles on microsomal oxidation *in vitro* and *in vivo*, *Biochem. Pharmacol.* **23:**2377.

Wilkinson, C. F., Hetnarski, K., and Hicks, L. J., 1974*b*, Substituted imidazoles as inhibitors of microsomal oxidation and insecticide synergists, *Pestic. Biochem. Physiol.* **4:**299.

Williams, G. H., 1960, *Homolytic Aromatic Substitutions,* Pergamon Press, New York.

Wilson, B. J., and Orrenius, S., 1972, A study of the modified type II spectral change produced by the interaction of agroclavine with cytochrome P-450, *Biochim. Biophys. Acta* **261:**94.

Yu, S. J., and Terriere, L. C., 1971, Hormonal modification of microsomal oxidase activity in the housefly, *Life Sci.* **10:**1173.

Yu, S. J., and Terriere, L. C., 1972, Enzyme induction in the housefly: The specificity of the cyclodiene insecticides, *Pestic. Biochem. Physiol.* **2:**184.

4

Extramicrosomal Metabolism of Insecticides

W. C. Dauterman

1. Introduction

This chapter will be restricted to a discussion of the metabolic reactions of insecticides which are not mediated by either the mixed-function oxidases (Chapter 2) or the conjugation enzymes (Chapter 5). Emphasis will be placed on *in vitro* systems, particularly those in which the enzymes have been purified. The enzymes that catalyze the extramicrosomal metabolism of insecticides may be associated with specific subcellular fractions or may be more broadly distributed among several; in general, their natural biological functions are not well understood.

2. Phosphotriester Hydrolysis

The degradation of organophosphorus insecticides by a variety of hydrolases (O'Brien, 1960; Heath, 1961; Eto, 1974) almost certainly constitutes the most important mechanism by which these compounds are detoxified. The enzymes that catalyze the hydrolytic attack on the phosphorus ester or anhydride band may be referred to as phosphoric triester (phosphotriester) hydro-

W. C. Dauterman • North Carolina State University, Department of Entomology, Raleigh, North Carolina.

lases. Since triester hydrolysis results in the formation of a phosphorus-containing metabolite which is in an anionic form at neutral pH and therefore a poor cholinesterase inhibitor, the overall result of the reaction is the detoxication of the parent compound.

2.1. Types of Reactions

Organophosphorus compounds may undergo enzymatic hydrolysis at the ester or acid anhydride bond. The generalized reactions which are possible are shown as follows, where R is an alkyl group and X (the "leaving" group) may be a halide, another substituted phosphorus group linked by an anhydride bond, an alkoxy group, or aryloxy group:

$$\text{(a)}\quad (RO)_2\overset{\overset{\displaystyle S}{\|}}{P}\text{—}X + H_2O \rightarrow (RO)_2\overset{\overset{\displaystyle S}{\|}}{P}OH + HX$$

$$\text{(b)}\quad (RO)_2\overset{\overset{\displaystyle O}{\|}}{P}\text{—}X + H_2O \rightarrow (RO)_2\overset{\overset{\displaystyle O}{\|}}{P}OH + HX$$

$$\text{(c)}\quad (RO)_2\overset{\overset{\displaystyle S(O)}{\|}}{P}\text{—} + H_2O \rightarrow (RO)(HO)\overset{\overset{\displaystyle S(O)}{\|}}{P}\text{—}X + ROH$$

Reactions (a) and (b) lead to the formation of a dialkyl phosphorothioic acid and a dialkyl phosphoric acid, respectively, in addition to HX. In the case of (c), the metabolites formed are the desalkyl derivative and an alcohol.

Studies on the metabolism of organophosphorus insecticides have shown that these metabolites can be derived by a number of different enzymatic processes. Thus dialkyl phosphorothioic acids or dialkyl phosphoric acids, which were originally considered to be metabolites of phosphotriester hydrolases, can also be formed by mixed-function oxidases as well as glutathione transferases (Dauterman, 1971). This is also the case with the desalkyl derivatives. Consequently, the isolation and identification of these metabolites from *in vitro* or *in vivo* studies do not alone provide sufficient proof that a specific enzyme system is involved in their formation. In order to distinguish the involvement of a specific enzyme system, it is necessary to examine its subcellular localization, its cofactor requirements, and its response to inhibitors, and to effect a complete analysis and identification of the metabolites it yields from any particular substrate.

2.2. Source and Properties of Enzymes

2.2.1. Fluorohydrolases

The first report of the enzymatic hydrolysis of an organophosphorus insecticide was that of Mazur (1946), who showed that DFP and various analogues were hydrolyzed at the P—F bond by a number of rabbit tissue homogenates.

$$(i\text{-}C_3H_7O)\overset{\displaystyle O}{\overset{\|}{P}}\text{—}F + H_2O \rightarrow (i\text{-}C_3H_7O)_2\overset{\displaystyle O}{\overset{\|}{P}}\text{—}OH + HF$$

DFP

Since none of the organophosphorus insecticides is a fluoro compound, the fluorohydrolases (E.C. 3.8.2.1) will not be discussed in detail, but examples will be used to provide some general information pertinent to the phosphorus triester hydrolases that hydrolyze organophosphorus insecticides.

The fluorohydrolases (DFPases) have been reported to be present in almost every tissue and organism studied (Mounter *et al.*, 1955*b*) as well as in microorganisms (Mounter *et al.*, 1955*a*). All of the animal tissues evaluated could hydrolyze DFP, and the liver was particularly active. In all cases, the reaction was activated by Mn^{2+} and Co^{2+}, although the relative stimulation caused by these ions varied from tissue to tissue and indicated that more than one enzyme was involved (Mounter, 1955). The DFPases found in microorganisms exhibited considerable variation in activity, and based on the effects of Mn^{2+}, Co^{2+}, and Mg^{2+} at least 11 different patterns of activity indicating 11 different enzymes were observed. From the vast literature on DFPases it can be concluded that a large number of different enzymes exist and may be characterized according to their source (tissue and organism), ion sensitivity, substrate specificity, and ability to be activated by derivatives of imidazole or pyridine in the presence of Mn^{2+} (Mounter and Chanutin, 1954).

2.2.2. *Arylester Hydrolases*

a. Mammals. Aldridge (1953*a*) published a classical study in which he introduced the terms "A esterases" and "B esterases." The A esterases hydrolyzed *p*-nitrophenyl acetate faster than the butyrate and were not inhibited by paraoxon, while the B esterases were inhibited by 10^{-7} to 10^{-8} M paraoxon and hydrolyzed *p*-nitrophenyl butyrate faster than the acetate. In a subsequent study, he showed that an A esterase in mammalian serum hydrolyzed paraoxon to diethyl phosphoric acid and *p*-nitrophenol (Aldridge, 1953*b*) and that the activities of the A esterases in sera from nine mammalian species varied significantly. Rabbit serum had the highest A esterase activity while mouse serum had the lowest.

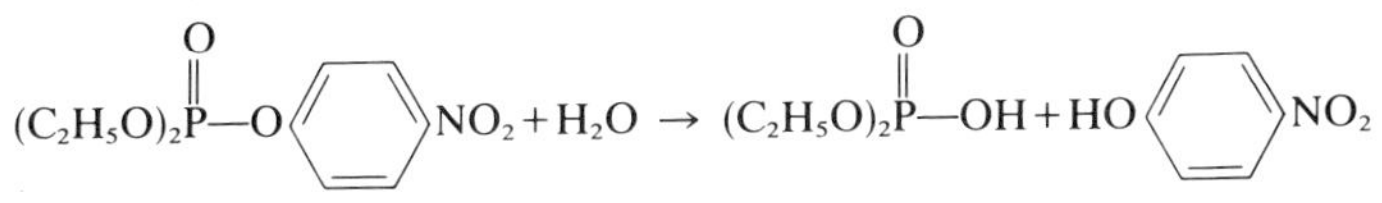

paraoxon

The paraoxon-hydrolyzing enzyme was not restricted to the sera but was found in a variety of mammalian tissues. With the rabbit, arylester hydrolase (E.C. 3.1.1.2) activity was highest in the serum, lower in the liver, and lowest in muscle. In the rat, the highest activity was found in the liver and the lowest in muscle. Based on these findings, it can be assumed that this enzyme is widely distributed in mammalian tissue.

Augustinsson and Heimburger (1954) studied the hydrolysis of organophosphorus compounds in rabbit plasma and found that tabun and DFP analogues as well as paraoxon and TEPP were hydrolyzed. Rabbit plasma did not degrade the phosphorothionates malathion, isosystox, systox, chlorothion, or diazinon.

Main (1960*a*) effected an approximately 385-fold purification of the enzyme hydrolyzing paraoxon (referred to as paraoxonase) from sheep serum. This is the only published method for the purification of this enzyme. The product was 80–95% pure based on electrophoretic evidence and had a molecular weight of 35,000–50,000. The K_m for paraoxon was fifteenfold greater for the purified fractions (4.2 mM) than for the serum (0.29 mM), suggesting the presence of an activator in the serum which was removed during purification.

The purified enzyme was inhibited 50% by 0.63 mM Ba^{2+} and 0.2 mM EDTA. Further evaluation of the enzyme indicated that some residual DFP-hydrolyzing activity was associated with the paraoxonase and that, based on its inhibition by Mn^{2+} ions, this activity was not due to the presence of a DFPase. The purified paraoxonase was capable of hydrolyzing tabun but not TEPP (Main, 1960*b*).

Kojima and O'Brien (1968) studied the enzymes in rat liver responsible for the metabolism of paraoxon. On the basis of different pH optima and ion sensitivities, three liver paraoxonases were found and were distributed in the soluble, microsomal, and mitochondrial fractions with relative total activities of 71 : 34 : 7 at pH 8.8 and 22 : 49 : 8 at pH 7.7. One of the enzymes in the soluble fraction was probably a glutathione transferase since desethyl paraoxon was a product and since dialysis—which would remove the endogenous cofactor glutathione (0.49 μmole/100 mg of liver, Johnson, 1965)—decreased the total activity by about 20%. The soluble fraction also degraded parathion, but this reaction was not necessarily catalyzed by the same enzyme(s) responsible for paraoxon degradation.

Becker and Barbaro (1964) studied the hydrolysis of *p*-nitrophenyl ethylphosphonate by rabbit plasma and found that the pH optimum was more alkaline than that reported for paraoxonase and that the enzyme was less readily inhibited by EDTA and Ba^{2+}. Using human serum as the enzyme source, Škrinjarić-Špoljar and Reiner (1968) concluded that paraoxon and its phosphonate analogue were hydrolyzed by different enzymes. Studies on the enzyme which hydrolyzed paraoxon in rabbit serum indicated that the reaction was not subject to either substrate or product inhibition (Lenz *et. al.*, 1973).

Diethyl *p*-aminophenyl phosphate and *p*-aminophenyl pinacolyl methylphosphonate were not substrates for the enzyme but were competitive inhibitors. It was concluded that paraoxonase was specific for phosphate esters and was not able to hydrolyze phosphonate esters.

The enzymatic hydrolysis of diazoxon was studied by Shishido and Fukami (1972) using tissue homogenates of rat and cockroach. A rat liver enzyme hydrolyzed diazoxon to diethyl phosphoric acid and 2-isopropyl-4-methyl-6-hydroxypyrimidine and was located in the microsomes.

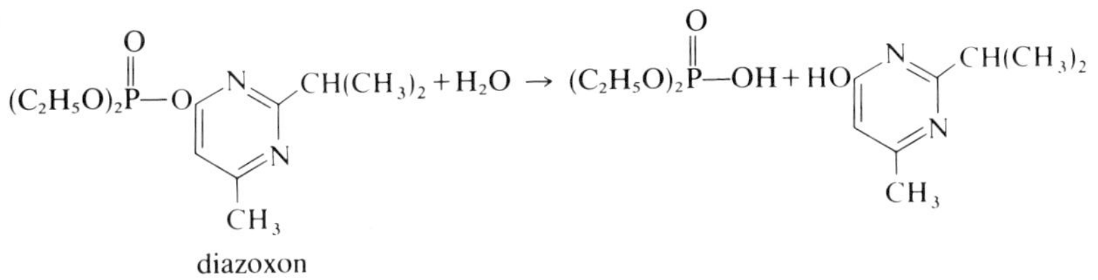

The activity of the enzyme was stimulated by Ca^{2+}, which also protected the enzyme from inactivation.

b. Insects. Krueger and Casida (1961) investigated the *in vitro* hydrolysis of a variety of organophosphorus insecticides in preparations from several insect and mammalian tissues. TEPP, DFP, and DDVP were readily hydrolyzed and the reaction was activated by 1 mM, Mn^{2+} and Co^{2+}. No activation or inhibition was observed at 1 mM with Ca^{2+}, Ba^{2+}, or Sr^{2+}. Also, no hydrolysis of several phosphorothionate insecticides evaluated was observed, indicating a certain specificity to the phosphates (rabbit serum paraoxonase).

It may be appropriate at this stage to discuss whether triester hydrolases are active toward both phosphorothionate and phosphate insecticides or only toward the latter. It has usually been assumed that the properties of an organophosphorus insecticide which allow it to inhibit cholinesterase (i.e., the electronegative character of the phosphorus atom) also determine its susceptibility to ester hydrolysis. However, the results of some workers have cast doubt on the validity of this assumption, which would dictate that phosphates but not phosphorothionates should be susceptible to attack. Matsumura and Hogendijk (1964) used [^{32}P]parathion as the substrate for the measurement of triester hydrolase activity during the partial purification of the enzyme from houseflies. The partially purified enzyme hydrolyzed parathion and diazinon to diethyl phosphorothioic acid but had little activity toward paraoxon. However, Welling *et al.* (1971) and Nakatsugawa *et al.* (1969) were unable to reproduce these results with [^{14}C]parathion or [^{35}S]parathion. In another study on insect hydrolases, Jarczyk (1966) investigated the hydrolases present in the midgut of lepidopterous caterpillars and partially purified two enzymes, one active toward phosphates and the other with a high activity to parathion analogues (P=S enzymes); thiophosphinic and thiophosphonic esters were also degraded by the latter. The P═O enzyme was activated by cysteine, DL-homocysteine, and GSH and was inhibited by Hg^{2+} and Cd^{2+}; the P═S enzyme was activated by Mn^{2+}.

Literature data on the substrate specificity of the phosphotriester hydrolases are confusing. The evidence for the mammalian hydrolases indicates that the phosphorothionates are not hydrolyzed, whereas data on the insect hydrolases indicate the possible existence of both "oxonases," which hydrolyze the phosphate esters, and "thionases," which directly hydrolyze thionoesters.

It is possible to roughly classify the phosphotriester hydrolases into groups, according to whether they are activated by calcium (group I) or manganese and cobalt (group II) (Table I). The first group appears to occur in mammalian but not in insect tissues and is responsible for the hydrolysis of paraoxon and diazoxon. The enzymes in this group appear to be quite specific for the phosphates and show little or no activity toward the phosphorothionates. Enzymes of the second group, Mn^{2+}-Co^{2+} activated hydrolases, appear to be present in both mammals and insects. In mammals they are responsible for the hydrolysis of DFP, tabun, and DDVP, while in insects paraoxon, DDVP, TEPP, and DFP as well as parathion and paraoxon analogues are degraded by this group. Further studies are required to clarify the substrate specificity and cofactor requirements of the phosphotriester hydrolases, with respect to their action on phosphorothionate and phosphate insecticides.

Table I. Activation by Ca and Mn-Co Ions of Phosphorotriester Hydrolases of Organophosphates

	Enzyme source[a] and references[b]	
Substrate	Ca	Mn-Co
Paraoxon	Rat liver (mit, sol., m) (1)	Housefly (h) (2)
Diazoxon	Rat liver (h) (3)	
DDVP	Rat liver (mit) (4)	Rat liver (sol.) (4)
		Housefly (h) (2)
TEPP		Housefly (h) (2)
P═S compounds		Lepidoptera (sol.) (5)
P═O compounds		Lepidoptera (sol.) (5)
DFP	Rat liver (insol.) (6)	Rat kidney (sol.) (6)
	Hog liver (insol.) (6)	Rat liver (sol.) (6)
		Hog kidney (sol.) (6)
		Hog liver (sol.) (6)
		Housefly (h) (2)

[a]Abbreviations: mit, mitochondria; sol., soluble fraction; m, microsomal fraction; h, homogenate; insol., insoluble fraction.

[b](1) Kojima and O'Brien (1968), (2) Krueger and Casida (1961), (3) Shishido and Fukami (1972), (4) Hodgson and Casida (1962), (5) Jarczyk (1966), (6) Mounter (1955).

2.2.3. O-Alkyl Hydrolases

O-Dealkylation or hydrolysis of the alkyl groups is another possible mechanism by which organophosphorus triesters may be degraded.

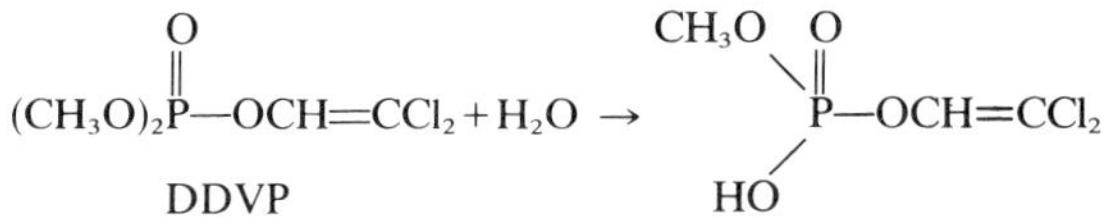

DDVP

Hodgson and Casida (1962) reported that DDVP was demethylated by an enzyme in the soluble fraction from rat liver, and since the reaction was not associated with the microsomes and had no requirement for NADPH it did not appear to result from oxidative *O*-dealkylation (Chapter 2). However, it is not clear whether sufficient amounts of endogenous GSH might have been present in the preparation to support *O*-dealkylation by the glutathione transferase (Chapter 5). One indication that the reaction might have been catalyzed by a hydrolase was that it was activated by Mn^{2+}, although Dicowsky and Morello (1971) showed that in the soluble fraction the degradation of DDVP to desmethyl DDVP was glutathione dependent. Nolan and O'Brien (1970) reported that [^{3}H]paraoxon was *O*-dealkylated in houseflies and resulted in the formation of labeled ethanol and derivatives. Since ethanol and not acetaldehyde or *S*-ethyl glutathione was isolated, it must be concluded that this reaction was mediated by a hydrolase.

2.2.4. Phosphodiester Hydrolases

The hydrolysis of phosphodiesters such as desmethyl DDVP (Hodgson and Casida, 1962) and desmethyl tetrachlorvinphos (Donninger *et al.*, 1971) has been reported to occur in the soluble fraction of rat liver and pig liver homogenates.

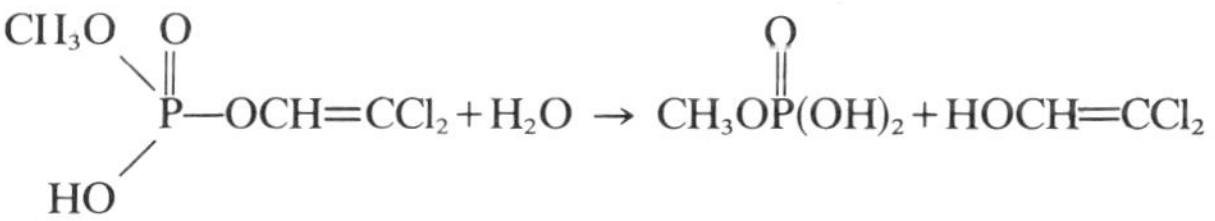

desmethyl DDVP

The enzyme from rat liver was purified 210-fold and had a pH optimum of 6.9. EDTA (10 mM) activated the enzyme, whereas phosphoric acid monoesters and phosphoric acid as well as high concentrations of Ca^{2+}, Zn^{2+}, and Fe^{3+} (1 mM) inhibited it. The enzyme appeared to be different from spleen phosphodiesterase since it would not hydrolyze desmethyl tetrachlorvinphos. Furthermore, the rat liver enzyme exhibited no activity toward the RNA "core," whereas snake venom and spleen phosphodiesterases hydrolyzed the "core" (Donninger *et al.*, 1971).

Considering the limited information available on hydrolytic dealkylation, further research is required to verify and determine the importance of this reaction in the metabolism of organophosphorus insecticides.

3. Carboxylester Hydrolysis (Organophosphorus Compounds)

Carboxylesterases (carboxylic ester hydrolase, E.C. 3.1.1.1), B esterases, or aliesterases catalyze the hydrolysis of both aliphatic and aromatic esters, but not choline esters. They are distinguished from lipases by their selective action on water-soluble rather than on water-insoluble substrates. Carboxylesterases have been demonstrated to be important in the metabolism of a number of types of insecticides. However, most of the available information is concerned with the organophosphorus insecticides malathion and acethion (O'Brien, 1960) and more recently with the pyrethroid insecticides (Abernathy and Casida, 1973).

3.1. Type of Reaction

The hydrolysis of malathion and acethion by carboxylesterases involves cleavage of the carboxyester group to form a water-soluble nontoxic product. With malathion the product formed is malathion monoacid (Cook and Yip, 1958), and with acethion the metabolite formed is acethion monoacid (O'Brien *et al.*, 1958).

$$\underset{\text{malathion}}{\begin{array}{l}(CH_3O)_2P(S)SCHOOC_2H_5 \\ \qquad\qquad\qquad\quad | \\ \qquad\qquad\qquad\quad CH_2COOC_2H_5\end{array}} \rightarrow \underset{\text{malathion } \alpha\text{-monoacid}}{\begin{array}{l}(CH_3O)_2P(S)SCHCOOH \\ \qquad\qquad\qquad\quad | \\ \qquad\qquad\qquad\quad CH_2COOC_2H_5\end{array}}$$

$$\underset{\text{acethion}}{C_2H_5O)_2P(S)SCH_2COOC_2H_5} \rightarrow \underset{\text{acethion monoacid}}{(C_2H_5O)_2P(S)SCH_2COOH}$$

Only one carbethoxy group of malathion was found to be hydrolyzed by the rat *in vivo* and by the purified rat liver carboxylesterase *in vitro*; the carboxylic acid produced was identified by NMR spectroscopy as the α-monoacid (Chen *et al.*, 1969). Welling and Blaakmeer (1971) also found that only one carbethoxy group of malathion was hydrolyzed but reported that both the α- and β-monoacids were produced by housefly homogenates and rat liver microsomes. It is not clear whether this discrepancy is due to differences in the strains of rats employed or whether it indicates the existence of more than one carboxylesterase in rat liver. The finding that only one carbethoxy group is cleaved is compatible with the early finding of Christmas and Lewis (1921) using hog liver esterase and diethyl succinate as the substrate. However, certain *in vivo* studies have indicated that malathion diacid is also a metabolite, and it is not yet clear whether this is generated enzymatically or whether it is formed nonenzymatically during the isolation of the metabolites.

3.2. Source and Properties of the Enzyme

The enzyme which hydrolyzes malathion is widely distributed in mammalian tissue, and has been found in liver, kidney, serum, lung, ileum, and spleen in the rat, mouse, guinea pig, and dog. In each species maximum activity was associated with the liver, the liver of dogs having particularly high activity (Murphy and DuBois, 1957). Similar findings with the rat were substantiated by identification of the carboxylic acid derivatives (Seume and O'Brien, 1960*a*), and the presence of the enzyme in human liver has also been demonstrated (Matsumura and Ward, 1966).

It is probable that the selectivity of malathion is directly related to the presence or absence of carboxylesterases in various species. Thus carboxylesterase activity is found to be low or absent in each of several insect species susceptible to malathion (Kojima, 1961) and is usually high in malathion-resistant strains or species (Motoyama and Dauterman, 1974).

Malathion is degraded by carboxylesterase activity in a soil microorganism *Tricoderma viride* (Matsumura and Boush, 1966), and a cell-free esterase catalyzing the reaction has recently been isolated from soil (Satyanarayana and Getzin, 1973). Unlike mammalian carboxylesterases, which hydrolyze both aliphatic and aromatic esters and triesters, the soil enzyme appears active only toward aromatic esters. Little or no information is available on plant esterases hydrolyzing malathion.

Main and Braid (1962) partially purified a rat liver esterase which was free from A esterases and cholinesterase. The enzyme cleaved only one of the carbethoxy groups of malathion and was characterized as a carboxylesterase. A series of nonphosphorus esters have been evaluated as substrates for the partially purified rat liver carboxylesterase, and Table II shows the K_m and V_{max}

Table II. Substrate Specificity of a Purified Preparation of Rat Liver Carboxylesterase[a]

Substrate	K_m (mM)	V_{max} (μ moles/mg protein/min)
Ethyl acetate	5.15	5.74
Ethyl propionate	2.86	19.6
Ethyl butyrate	3.9	42.6
Ethyl valerate	0.55	14.5
Diethyl malonate	3.57	37.0
Diethyl succinate	1.25	9.0
Diethyl adipate	0.47	19.2

[a]From Main and Dauterman (unpublished).

Table III. Activity of a Purified Preparation of Rat Liver Carboxylesterase Toward a Series of Malathion Homologues[a]

Compound	K_m (mM)	V_{max} (μmoles/mg/min)	Relative enzymatic half-life 0.695 (K_m/V_{max})
Carbmethoxy	0.506	0.52	0.676
Carbethoxy	0.179	1.81	0.069
Carb-*n*-propoxy	0.068	2.46	0.019
Carbisopropoxy	0.040	0.28	0.099
Carb-*n*-butoxy	0.022	2.97	0.005

[a] From Dauterman and Main (1966).

values for a series of mono- and diethylcarboxylic esters. The monoesters were good substrates, but the disubstituted esters had a higher affinity for the enzyme as indicated by their lower K_m values. With the monoesters the K_m decreased from acetate to valerate and this also occurred with the malonate to adipate series. In general, those members of the series having the longest acyl groups exhibited the highest affinity for the enzyme. Similar studies with dialkyl-substituted maleates and fumarates indicated that the enzyme could hydrolyze both geometric isomers of compounds containing a variety of alkyl substituents. There was no evidence that the enzyme was able to hydrolyze the second carbalkoxy group in any of the substrates evaluated.

Dauterman and Main (1966) studied the substrate specificity of rat liver carboxylesterase toward a series of malathion homologues. The results showed the carb-*n*-butoxy analogue to have the lowest K_m as well as the highest V_{max} values (Table III). Consequently, the longer-chained carbalkoxy compounds were better substrates and had a shorter half-life than the carbmethoxy analogue, a finding which agreed well with their *in vivo* toxicity.

When the optical isomers of malathion were used as substrates, the *d*-isomer bound better ($K_m = 0.084$ mM) to carboxylesterase than the *l*-isomer ($K_m = 0.205$ mM) (Hassen and Dauterman, 1968).

3.3. Inhibition of the Enzyme

Carboxylesterases or B esterases are susceptible to inhibition by a number of phosphates (Aldridge, 1953*a*; O'Brien, 1957). The enzyme which hydrolyzes malathion was shown by Cook *et al.* (1958) to be inhibited by the phosphate analogues of parathion, EPN, diazinon, chlorthion, and azinphos-methyl as well as systox and phosdrin. In all cases, the phosphate analogues were vastly better inhibitors of the carboxylesterase than the parent phosphorothionates. Murphy and DuBois (1957) also showed that EPN inhibited the enzymatic detoxication of malaoxon *in vitro*, and EPN administered *in vivo* decreased the detoxifying capacity of malaoxon by rat liver and serum.

It is worthwhile to note that although malaoxon is a substrate for the carboxylesterase (O'Brien 1957; Murphy and DuBois, 1957) the fact that it is a phosphate also makes it an inhibitor of the enzyme. Main and Dauterman (1967) reported that when malaoxon and carboxylesterase were mixed in solution two separate reactions occurred simultaneously. In one, malaoxon irreversibly inhibited carboxylesterase; in the other, malaoxon was a substrate for the carboxylesterase and was hydrolyzed. In the same study, the inhibition constants, k_i, of the carbalkoxy homologues of malaoxon were determined with respect to carboxylesterase activity (Table IV). The k_i values increased with increasing length of alkyl chain—a trend which, like that observed for the K_m values of the thiono analogues as enzyme substrates (Table III), probably reflects an increased affinity of the enzyme for the more lipophilic members of the series.

The carboxylesterases from malathion-susceptible and -resistant strains of mosquito (*Culex tarsalis*) were purified by DEAE column chromatography and found to have the same molecular weight (16,000) (Matsumura and Brown, 1963); properties such as K_m, energy of activation, and temperature coefficient were also similar in both strains and the oxygen analogues of EPN and *n*-propyl paraoxon inhibited the formation of malathion monoacid. It was concluded that in this case the interstrain difference was of a strictly quantitative nature.

On the other hand, Matsumura and Voss (1965) found a twentyfold difference in K_m to malathion in partially purified carboxylesterases from resistant and susceptible strains of the two-spotted spider mite, which indicated a qualitative difference between the two enzymes.

Plapp and coworkers evaluated a large number of organophosphorus compounds as synergists for malathion in resistant strains of houseflies and *Culex* mosquitoes (Plapp *et al.*, 1963; Plapp and Tong, 1966). The most active synergists were triphenyl phosphate, tributyl phosphorotrithiolate and its propyl analogue, and tributyl phosphorotrithioite. Saligenin cyclic phosphates (Eto *et al.*, 1965) and a number of noninsecticidal carbamates such as *o*-

Table IV. Bimolecular Reaction Constants (k_i) of a Series of Carbalkoxy Homologues of Malaoxon with Rat Liver Carboxylesterase[a]

Compound	k_i $(\text{M}^{-1}\ \text{min}^{-1}) \times 10^4$
Carbmethoxy	1.18
Carbethoxy	1.70
Carb-*n*-propoxy	8.26
Carbisopropoxy	3.23
Carb-*n*-butoxy	37.4

[a]From Main and Dauterman (1967).

chlorophenyl-*N,N*-dibutylcarbamate (Plapp and Valega, 1967) have also been shown to be excellent synergists for malathion against resistant insects.

4. Carboxylester Hydrolysis (Pyrethroids)

Early studies with the pyrethroids showed that they were rapidly metabolized and that certain synergists decreased the rate at which this occurred (Chamberlain, 1950; Zeid *et al.*, 1953; Winteringham *et al.*, 1955; Bridges, 1957; Hopkins and Robbins, 1957). It was generally assumed that hydrolysis of the ester linkage of the components of [^{14}C]allethrin and [^{14}C]pyrethrum constituted the major mechanism through which metabolism took place. These studies, however, failed to adequately identify the metabolites produced or the metabolic pathways involved, mainly because the radiolabeled compounds available at the time either were of low specific activity or were randomly labeled mixtures of isomers. It was not until procedures were developed for synthesizing stereochemically pure isomers of high specific activity that significant advances in metabolic studies were made with the pyrethroids (Nishizawa and Casida, 1965; Yamamoto and Casida, 1968; Elliott *et al.*, 1969; Elliott and Casida, 1972). Such studies both in insects (Yamamoto and Casida, 1966; Yamamoto *et al.*, 1969) and in mammals (Casida *et al.*, 1971) showed no evidence of hydrolysis of the cyclopropane ester bond and revealed that the major route of detoxication was via the mixed-function oxidases (see Chapter 2). More recently, Elliott *et al.* (1972) have reported the presence of an esterase in rat liver microsomes which hydrolyzes the methoxycarbonyl group of pyrethrin II and which may be responsible for the relatively low mammalian toxicity of this and related materials.

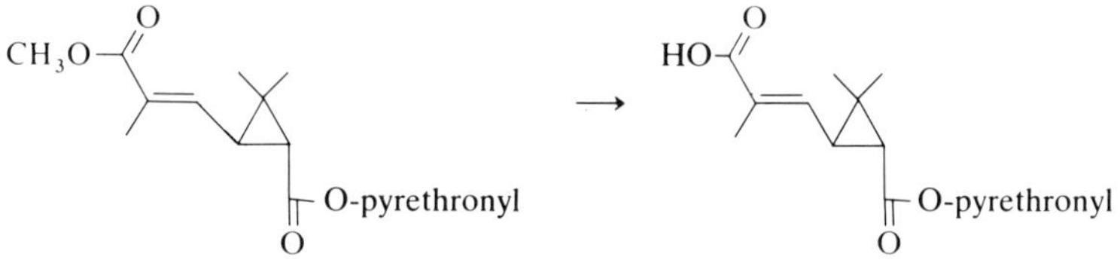

Abernathy and Casida (1973) have shown that the ester group of several of the newer synthetic primary alcohol chrysanthemates is readily cleaved by mouse liver microsomal esterases, although this does not happen to any great extent with the naturally occurring materials.

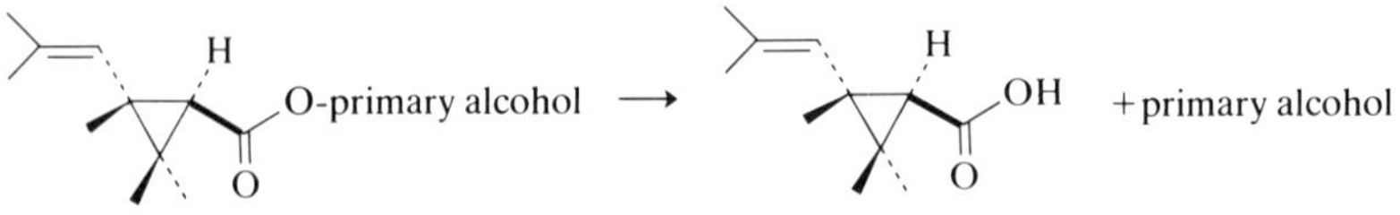

(+)-*trans*-chrysanthemate

The (+)-*trans* isomers were more rapidly hydrolyzed than the (+)-*cis* isomers and little or no hydrolysis occurred with secondary alcohol esters, such as *S*-allethrin (Abernathy *et al.*, 1973). With the acid moieties the *trans* isomers of chrysanthemates and ethano chrysanthemates were hydrolyzed 26- to 50-fold more rapidly than the corresponding *cis* isomers and the enzyme system involved was inhibited *in vitro* by paraoxon and *in vivo* by *S,S,S*-tributyl phosphorotrithioate. Consequently, depending on the pyrethroid involved either an esterase or a mixed-function oxidase or both may be rate limiting with respect to mammalian toxicity, the former being particularly important with certain (+)-*trans*-chrysanthemate insecticides of primary alcohols.

Although considerably less active than those in mouse liver, pyrethroid-hydrolyzing enzymes have also been found in insect preparations (Jao and Casida, 1974). Those obtained from the milkweed bug (*Oncopeltus fasciatus*) and larvae of the cabbage looper (*Trichoplusia ni*) cleaved the (+)-*trans* isomers of resmethrin and tetramethrin more rapidly than the corresponding (+)-*cis* compounds, although the isomer specificity was less pronounced with the enzymes from the housefly and cockroach (*Blattella germanica*) (Table V).

The V_{max} value for (+)-*trans*-resmethrin hydrolysis by the mouse liver esterase was more than thirtyfold greater than that for the insect enzymes, and the relative rate of hydrolysis of this compound in various preparations was in the order mouse liver ⋙ milkweed bug ≫ cockroach > cabbage looper > housefly. This clearly favors the mammal as far as selective toxicity is concerned.

The insect esterases were extremely sensitive to inhibition by 1-naphthyl-*N*-propylcarbamate relative to the mouse liver esterases, whereas *S,S,S*-tributyl phosphorotrithioate was more effective in inhibiting the mouse esterase than those from various insect sources. Esterase inhibitors may be useful as synergists in insect species where hydrolysis is rate limiting with respect to insecticidal action, but could also decrease the mammalian selectivity which the enzyme confers.

*Table V. Kinetic Parameters for Insect and Mouse Liver Esterases Hydrolyzing (+)-*trans*-Resmethrin*[a]

Enzyme source, acetone powder	K_m (M × 10^{-7})	V_{max} (pmoles/mg protein/min)
Insect		
Milkweed bug	25	66
Cockroach	83	58
Housefly	63	30
Looper	111	60
Mammal		
Mouse liver	125	2083

[a] Data from Jao and Casida (1974).

5. Carboxylamide Hydrolysis

A number of amide-containing organophosphate insecticides have been reported to be metabolized to their corresponding carboxylic acid derivatives (Dauterman *et al.*, 1959; Chamberlain *et al.*, 1961; Bull and Lindquist, 1964).

In vivo metabolism studies with dimethoate (Dauterman *et al.*, 1959) have suggested that the formation of the monocarboxylic acid is the direct result of amidase action.

5.1. Type of Reaction

$$(CH_3O)_2P(S)CH_2C(O)NHCH_3 \rightarrow (CH_3O)_2P(S)SCH_2COOH$$

dimethoate dimethoate acid

The dimethoate monoacid was initially isolated as a urinary metabolite and characterized by infrared spectroscopy by Dauterman *et al.* (1959). Subsequently Uchida *et al.* (1964) demonstrated the *in vitro* formation of dimethoate acid in various vertebrate tissues.

5.2. Source and Properties of the Enzyme

Uchida *et al.* (1964) studied the distribution of amidase activity in a number of rat tissues and found that dimethoate was hydrolyzed readily by liver, slightly by lung, muscle, and pancreas, and not at all by brain, spleen, and blood. With respect to the degradation of dimethoate by liver from a variety of species, highest hydrolytic activity was found in rabbit and sheep, medium activity in dog, rat, and steer, and low activity in pig, mouse, and guinea pig. Most of the hydrolytic activity (35%) was localized in both rat and sheep liver microsomes. Amidase activity was also implicated in the degradation of dimethoate by human liver (Uchida and O'Brien, 1967) and was considered to play an important role in the detoxication of this compound.

Chen and Dauterman (1971*a*) purified a carboxylamidase from sheep liver microsomes which catalyzed the hydrolysis of dimethoate as well as other *N*-monosubstituted and *N,N*-disubstituted amides. The enzyme was purified 59-fold by hydroxylapatite and benzyl-DEAE cellulose column chromatography. The purified enzyme has a pH optimum of approximately 9.0 and since it was not stimulated by divalent cations at a concentration of 10^{-6} M it appeared to be different from the arylamidase reported by Behal *et al.* (1968). The molecular weight of the carboxylamidase was 230,000–250,000 based on gel filtration, and it was classified as acylamide amidohydrolase (E.C. 3.5.1.4).

The enzyme was unable to hydrolyze the lower members of a series of *N*-methyl-substituted amides and had only a trace of activity to *N*-methyl butyramide (Table VI). The affinity of the enzyme for the *N*-methyl amides

Table VI. Substrate Specificity of Sheep Liver Amidase[a]

Substrate	K_m (M)	V_{max} (μ moles/mg/hr)
N-Methyl formamide	—	—
N-Methyl acetamide	—	—
N-Methyl butyramide	—	—
N-Methyl valeramide	5×10^{-3}	2.82
N-Methyl caproamide	6.6×10^{-4}	1.30
N-Methyl heptylamide	1.25×10^{-3}	2.14
N-Methyl caprylamide	3.7×10^{-3}	0.367
N-Ethyl caproamide	8.3×10^{-4}	1.71
N-Propyl caproamide	1.8×10^{-3}	3.42
N-Butyl caproamide	2.2×10^{-3}	0.087
N-Phenyl caproamide	2.5×10^{-3}	132.4
N,N-Dimethyl caproamide	1.9×10^{-4}	0.564
N,N-Diethyl caproamide	—	—

[a]From Chen and Dauterman (1971*a*).

Table VII. Relative Rate of Hydrolysis of Dimethoate Analogues by Sheep Liver Amidase[a]

Compound	Relative rate (%)
N-Alkyl analogues	
Dimethoate	100
N-Ethyl dimethoate	86
N-*n*-Propyl dimethoate	134
N-*n*-Butyl dimethoate	39
O,O-Diethyl dimethoate	66
O,O Diethyl *N*-ethyl dimethoate	38
O,O-Diethyl *N*-*n*-propyl dimethoate	40
O,O-Diethyl *N*-*n*-butyl dimethoate	27
N,N-Dialkyl analogues	
O,O-Diethyl *N,N*-dimethyl dimethoate	168
O,O-Diethyl *N,N*-diethyl dimethoate	0
Dialkoxy analogues	
Dimethoate	100
O,O-Diethyl dimethoate	66
O,O-Di-*n*-propyl dimethoate	0
O,O-Di-isopropyl dimethoate	0
Branched analogues	
Ethylene dimethoate	0
Branched methyl dimethoate	0

[a]The concentration of the compounds assayed was 5×10^{-4} M. The value 100 was assigned to dimethoate and the other figures are the relative values when compared to that of dimethoate. From Chen and Dauterman (1971*b*).

increased with increasing size of the acyl group, reached a maximum with *N*-methyl caproamide, and thereafter decreased. Of the two *N,N*-dimethyl compounds tested, only *N,N*-dimethyl caproamide was hydrolyzed.

The purified sheep liver amidase also hydrolyzed a series of dimethoate analogues (Chen and Dauterman, 1971*b*) (Table VII). Structural changes in the *O,O*-dialkoxy portion of the molecule as well as modifications in the branching of the alkyl chain affected the susceptibility to hydrolytic degradation by the amidase. With the *N,N*-dialkyl analogues, only the *N,N*-dimethyl compound was hydrolyzed. In this same study no correlation was found between the *in vitro* hydrolysis of the dimethoate analogues and their *in vivo* toxicity to houseflies.

5.3. Inhibition of the Enzyme

Seume and O'Brien (1960*b*) reported that dimethoate was potentiated in the mouse, housefly, and American cockroach by EPN, tri-*o*-cresyl phosphate, and tri-*o*-cresyl phosphorothionate; the degree of potentiation was less with insects than with the mouse. That the observed potentiation was a result of carboxylamidase inhibition was subsequently established by the effects of EPN on dimethoate metabolism in the mouse and guinea pig. EPN blocked dimethoate metabolism profoundly in the mouse (80%), less in the guinea pig (60%), and not at all in the housefly or milkweed bug (Uchida *et al.*, 1966).

Chen and Dauterman (1971*b*) found that when amide-containing phosphates such as azodrin, bidrin, and dimethoxon were incubated with sheep liver amidase no hydrolysis of any of these compounds occurred. Clearly they were not substrates of the enzyme, and indeed were inhibitory toward the hydrolysis of the substrate *N*-methyl caproamide. At concentrations of 1×10^{-4} M, dimethoxon inhibited the amidase 53%, bidrin 50%, and azodrin 27%; at 1×10^{-6} M, paraoxon caused complete inhibition of the enzyme.

It is therefore unlikely that the traces of monoacids reported in metabolic studies with amide-containing phosphates are formed by direct hydrolysis. They may, however, be produced by oxidative *N*-dealkylation followed by nonenzymatic hydrolysis of the unsubstituted amide.

It appears that carboxylamidases, like the carboxylesterases, can hydrolyze only phosphorothionate insecticides and are inhibited by the corresponding phosphate analogues.

6. Nitroreductases

One of the minor detoxication reactions involves the reduction of a number of nitro-containing organophosphorus compounds such as parathion, sumithion, and EPN.

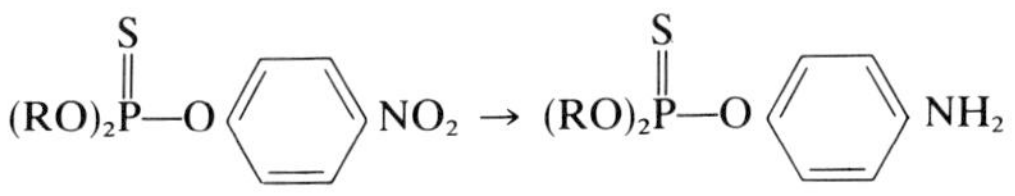

The reduction of the nitro group in parathion was first reported by Gardocki and Hazelton (1951), who found minute amounts of an aminophenol derivative in the urine of dogs treated with parathion. Pankaskie *et al.* (1952) *studied the degradation of parathion in cows and concluded that* "parathion must be hydrolyzed *in vivo* to *p*-nitrophenol, reduced to *p*-aminophenol, conjugated with glucuronic acid to an appreciable extent and then excreted in the urine as *p*-aminophenyl glucuronide." Cook (1957) reported that the nitro group in parathion was reduced by bovine rumen fluid. This was verified by Ahmed *et al.* (1958), who also showed that the formation of the amino derivatives of parathion and paraoxon greatly reduced their biological activity and toxicity. The reduction in bovin rumen fluid was the result of bacterial flora. Reductive degradation of parathion to aminoparathion was demonstrated with yeast which had been added to parathion-treated soil or water (Lichtenstein and Schulz, 1964) and with two species of *Rhizobium* (Mick and Dahm, 1970).

In vitro studies by Hitchcock and Murphy (1967) on the reduction of parathion, paraoxon, and EPN indicated that reductase activity was widespread in liver of mammalian, avian, and piscine species. Enzymatic activity was also present in mammalian kidneys, spleens, lungs, and erythrocytes, and also in avian kidneys. Reductase activity was uniformly distributed between various cell fractions of rat liver, and this has also been reported by a number of other investigators. NADPH and a high concentration of FAD (1.23 μmoles) were necessary as cofactors. Gillette (1971) questioned the findings of Hitchcock and Murphy (1967), stating "the finding that enzymes in tissues catalyze the reduction of nitro groups in the presence but not in the absence of high concentrations of flavins should, therefore, not be accepted as evidence for functional nitroreductases in those tissues." This is evident since animal tissues and microsomes contain less than 2.5×10^{-5} M of flavins. This might explain why Hitchcock and Murphy (1967) found excellent reductase activity in rat tissues but Ahmed *et al.* (1958) did not find appreciable quantities of amino derivatives as urinary metabolites of parathion in rats. At present, it would appear that reduction of nitro-containing organophosphate insecticides is important only in ruminants where intestinal bacteria are sufficiently active to reduce nitro compounds or and in other microorganisms associated with soil and water.

Lichtenstein and Fuhreman (1971) reported a nitroreductase in the soluble fraction of female houseflies which reduced parathion to aminoparathion. The reaction required NADPH but was not affected by the presence or absence of oxygen.

Studies with the Madagascar cockroach (*Gromphadorhina portentosa*) have established the presence of nitrobenzene reductase activity in several tissues, including the fat body, gut, and Malpighian tubules (Rose and Young, 1973); activity was also observed in whole housefly homogenates. In the cockroach tissues, activity was found in all subcellular fractions. The enzymes in the soluble fraction appeared to be NADH linked, whereas those in the microsomes were more dependent on NADPH. In all cases, activity was enhanced considerably by the addition of FAD, FMN, or riboflavin, and indeed a considerable amount of the observed activity appeared to result from the nonenzymatic reduction of the substrate by the reduced flavin (e.g., $FMNH_2$). The formation of the latter can apparently occur through the mediation of several enzymes in both the microsomal and soluble fractions of tissue homogenates, probably either NADH- or NADPH-linked flavoproteins such as the microsomal NADPH-cytochrome *c* reductase. The partial inhibition of nitrobenzene reductase activity by CO (which occurred more in the microsomal than soluble fraction) suggested that either cytochrome P450 or another CO-sensitive metalloprotein was also capable of flavin reduction. That the latter was in addition mediated through a nonenzymatic heat-stable factor was shown by the considerable activity observed following addition of FMN to preparations which had been boiled for 10 min or even autoclaved. Consequently, it appears that the true substrate for reductase activity is the flavin and that the reduction of the nitro compound *per se* occurs nonenzymatically. Similar results have recently been obtained with respect to the azo dye, azofuchsin, the reduction of which can be conveniently measured by a direct spectrophotometric method (Young, 1975). These have clearly shown that the azo compound is reduced by a stoichiometrically equivalent amount of FMN produced by pyridine nucleotide dependent flavin reductases. In *G. portentosa* midgut microsomes, FMN reduction is specifically NADPH linked, whereas in the soluble fraction it is mediated through systems requiring either NADPH or NADH.

7. Carbamate Hydrolysis

Enzymatic hydrolysis of the carbamate ester group of the insecticidal carbamates would initially appear as an obvious and potentially important route of detoxication of these compounds. In spite of this, the information in this area is very limited. Carbaryl has been shown to be hydrolyzed by a plasma albumin fraction from several mammalian and avian sources (Casida and Augustinsson, 1959) and the activity separated from cholinesterase as well as aliphatic and aromatic esterases. In a subsequent study, the same plasma albumin fraction was shown to hydrolyze the ethyl-, propyl-, and *i*-propylcarbamates of *p*-nitrophenol, whereas arylesterase, chymotrypsin, lipase, cholinesterase, pepsin, papain, and egg albumin showed no activity toward these esters (Casida *et al.*, 1960). Neither plasma albumin nor plasma arylesterase could readily hydrolyze the *N,N*-dialkylcarbamates, although *N,N*-dimethylcarbamoyl fluoride was hydrolyzed by enzyme(s) in human and rabbit plasma, possibly the same enzyme(s) which hydrolyzed DFP (Augustinsson and Casida, 1959). In rabbit liver and kidney, however, the hydrolysis of dimethylcarbamoyl fluoride and DFP involved different enzymes.

At the present time, it would appear that cleavage of the carbamate ester moiety is effected mainly by microsomal *N*-demethylation followed by hydrolysis of the unsubstituted carbamate (Douch and Smith, 1971) and that direct enzymatic hydrolysis is of only minor importance.

8. Epoxide Hydrases

Epoxide rings of certain arene and alkene compounds are hydrated enzymatically to give predominantly the corresponding *trans*-dihydrodiols. This reaction is responsible for the deactivation of certain labile epoxides, which may be responsible for carcinogenesis, and also for the detoxication of certain aromatic and olefinic xenobiotics (Jerina and Daly, 1974). The enzyme(s) which mediates this reaction, epoxide hydrase (E.C. 4.2.1.63), is found mainly in the microsomal fraction of mammalian liver homogenates (Oesch *et al.*, 1971); with styrene oxide as a substrate the relative activity in several species is in the order rhesus monkey > human = guinea pig > rabbit > rat > mouse (Oesch *et al.*, 1974). The highest epoxide hydrase activity in the guinea pig and rat was associated with the liver; low activity was found in the kidney, intestine, and lung; and no detectable activity was found in brain, heart, spleen, and muscle. A similar distribution pattern was reported by Stoming and Bresnick (1973) using 3-methylcholanthrene-11,12-oxide as a substrate. These workers also found that hepatic epoxide activity was barely detectable in fetuses and 1-day-old neonate rats but increased rapidly to adult levels within 25 days. Epoxide hydrase activity is induced on pretreatment of rats with phenobarbital or 3-methylcholanthrene, although the level of induction is different from that of the mixed-function oxidase system and under separate genetic control (Oesch, 1973).

Cleavage of the epoxide rings of certain cyclodiene insecticides and their analogues has also been demonstrated in several insect species, including the housefly (Brooks *et al.*, 1970), blowfly (*Calliphora erythrocephala*) and mealworm (*Tenebrio molitor*) (Brooks, 1973), southern armyworm (*Prodenia eridania*), and Madagascar roach (*Gromphadorhina portentosa*) (Slade *et al.*, 1975). At present, comparative studies have not been conducted with a large variety of insects using the same assay procedures. However, one might expect different levels of activity due to the close association of the epoxide hydrase and the monooxygenases (which are found in different amounts in insects) in the endoplasmic reticulum.

The mechanism of enzymatic epoxide ring hydration is still obscure, but activity is optimal at alkaline pH and probably involves a nucleophilic attack of the OH on the oxirane carbon. Nonenzymatic hydration become significant below pH 6.5.

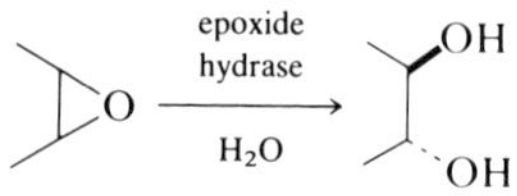

A relatively nonspecific epoxide hydrase has been purified from guinea pig liver (Oesch and Daly, 1971) and from human liver (Oesch, 1974) utilizing styrene oxide as the substrate. The enzyme was not dependent on the presence of metal ions and was not inhibited by SKF 525-A, piperonyl butoxide, and α-naphthoflavone but was strongly inhibited by 1,1,1-trichloropropane oxide (Oesch *et al.*, 1973). High concentrations of substrate inhibited the enzyme, whereas product diols had little effect (Oesch, 1973).

A number of simple alcohols such as cyclohexanol, 2-cyclohexen-1-ol, and glycidol were found to activate epoxide hydrase activity *in vitro*; certain imidazoles such as 1-(2-isopropylphenyl)imidazole and ketones such as α-tetralone also stimulated the enzyme. The partially purified epoxide hydrase displayed a sharp pH profile with an optimum at pH 9, and by SDS gel electrophoresis one major band with a molecular weight of 50,000 and several minor bands were detected. The hydrases purified from human liver and guinea pig liver had similar properties (Oesch *et al.*, 1974). An epoxide hydrase from rat liver microsomes has been partially purified but was contaminated with P450 and had low monooxygenase activity (Dansette *et al.*, 1974). Benzene oxide did not inhibit the hydration of styrene or naphthalene oxide in the most purified preparation, indicating the presence of at least two hydrases.

Evidence for the presence of more than one epoxide hydrase in hepatic microsomes is based on different purification factors (activity of the homogenate compared to purified preparations with benzene oxide and other oxide substrates), different ratios of hydrase activity toward various epoxides in preparations from different species, as well as differential stabilities.

Much of the interest in the pesticide area has centered on the role that epoxide hydrases play in the metabolism of two groups of compounds, the cyclodiene insecticides and the insect juvenile hormones (Menn and Beroza, 1972).

The detoxication of cyclodiene insecticides by hydration of the epoxide ring has been demonstrated with dieldrin, which is converted into the corresponding 6,7-*trans*-dihydroxydihydroaldrin. This has been shown to occur *in vivo* in the rabbit (Korte and Arent, 1965) and in larvae and adults of the mosquito (Oonnithan and Miskus, 1964; Tomlin, 1968). Studies by Brooks *et al.* (1970) indicated that housefly and rat liver microsomes had low hydrase activity to cyclodiene epoxide analogues, whereas pig and rabbit liver microsomes were slightly more active in the hydration of dieldrin and heptachlor epoxide (Brooks and Harrison, 1969). Of all the cyclodiene epoxide analogues evaluated, HEOM (1,2,3,4,9,9-hexachloro-6,7-epoxy-1,4,4a,5,6,7,8,8a-octahydro-1,4-methanonaphthalene) was most readily hydrated by the microsomes in all species investigated, and this compound has subsequently been used as a model substrate for measuring epoxide hydrase activity in a number of insect studies.

It was also reported that the enzymatic hydration of certain asymmetrical epoxides such as chlordene epoxide and HCE (1,2,3,4,9,9-hexachloro-*exo*-5,6-epoxy-1,4,4a,5,6,7,8,8a-octahydro-1,4-methanonaphthalene) was stereospecific, resulting in the hydration of mainly one enantiomer (Brooks *et al.*, 1968, 1970).

Studies by Matthews and McKinney (1974) indicated that 6,7-*cis*-dihydroxydihydroaldrin was produced as a metabolite *in vitro* from dieldrin in steady-state amounts while *trans*-diol formation steadily increased. Incubation of synthetic *cis*-diol with rat liver microsomes fortified with NADPH showed that the suspected epimerization reaction did indeed occur and that stereospecificity was introduced at some stage in the epimerization process, resulting in an optically active *trans*-diol. The enzyme responsible for this reaction had properties similar to those of the monooxygenase, but did not require oxygen, and appeared to be an epimerase. The epimerization was unidirectional since no *cis*-diol was formed on incubation of the *trans*-diol.

It is not clear whether the formation of *cis*-dihydrodiols proceeds via a dioxygenase-catalyzed reaction with a cyclic peroxide as an intermediate, as has been speculated for certain microbial dihydrodiol metabolites (Jerina *et al.*, 1971), or whether it is produced directly by epoxide hydration. Since the rate of diol formation with dieldrin is low compared to HEOM, it is possible that epoxide cleavage is mediated by two different enzyme systems which are present or absent in different species and react differently depending on the structure of the epoxide. Further studies are needed to clarify the mechanism of epoxide-ring openings with the cyclodiene epoxides.

The inactivation of juvenile hormone (JH) and its analogues in insects has been shown to involve hydrolysis of the ester group and hydration of the epoxide when terminal epoxide rings and ester groups are present (Slade and Zibitt, 1972; White, 1972) (Fig. 1). Slade and Wilkinson (1973) reported that certain JH analogues (JHA) interfered with the *in vitro* degradation of the natural [^{14}C]JH by inhibiting the enzymatic pathways responsible for its degradation. It was concluded that the action of many JHAs may be synergistic rather than intrinsically hormonal and that their morphogenetic activity results from a stabilization of the endogenous JH.

Using HEOM as a substrate for the measurement of epoxide hydrase activity present in pupal homogenates of the mealworm and blowfly, Brooks (1973, 1974) found a number of inhibitors of the insect HEOM epoxide hydrase, which included the cecropia hormone, piperonyl butoxide, sesoxane, SKF 525-A, 1,1,1-trichloropropane oxide, and a number of organophos-

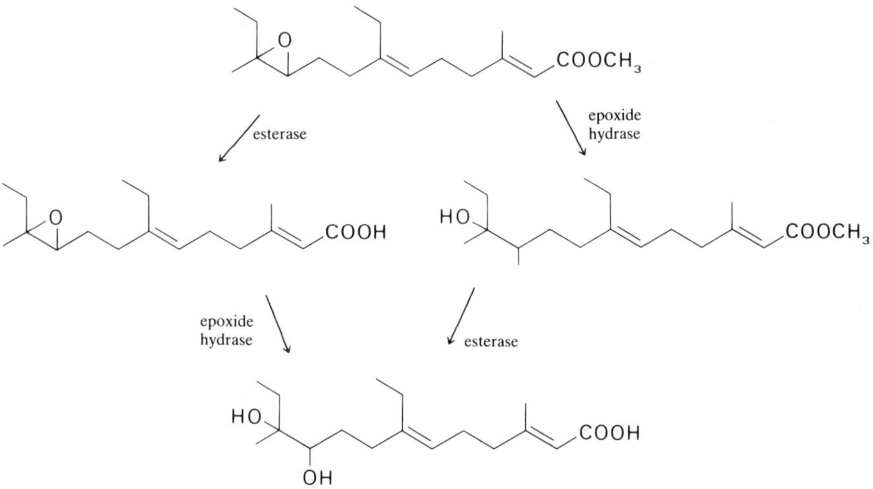

Fig. 1. Pathway of deactivation of cecropia JH in insects. From Slade and Wilkinson (1973).

phorus compounds. This contrasts with the results of Oesch (1973), who found that SKF 525-A and piperonyl butoxide had little effect on the styrene oxide hydrase of mammalian liver, and indicates a possible difference in the enzymes from the two sources.

Recent studies on some of the properties of insect HEOM epoxide hydrase have established a pH optimum of 9.0 for the enzyme from blowfly and southern armyworm and 8.1 for the hydrase from midgut microsomes of the Madagascar roach (Slade *et al.*, 1975). The pH optimum for the conversion of JH acid to the corresponding acid diol in the southern armyworm was 7.9, indicating a difference in the insect hydrases responsible for hydration of HEOM and JH. Hydrase activity from the blowfly was not affected by Mn^{2+}, Fe^{3+}, Fe^{2+}, Co^{2+}, Mg^{2+}. amd Ca^{2+} ions at 10^{-3} M but was inhibited 80% by 10^{-3} M Cu^{2+} ions. Metyrapone and 1-(2-isopropylphenyl)imidazole had no effect on the insect epoxide hydrases, whereas both compounds stimulated the mammalian styrene oxide hydrase.

The HEOM hydrase was inhibited *in vitro* by a variety of glycidyl ethers and various epoxides, of which 1,1,1-trichloropropane oxide was the most potent, as well as monooxygenase inhibitors, juvenile hormone analogues, and some organophosphorus compounds. Some representative compounds and their I_{50} values are given in Table VIII.

Other insecticides probably metabolized by epoxide hydrase are carbaryl and rotenone. This conclusion is based on isolation of the metabolite 5,6-dihydro-5,6-dihydroxy-1-naphthyl methylcarbamate from carbaryl (Leeling and Casida, 1966) and 6′,7′-dihydro-6′,7′,-dihydroxyrotenone from rotenone upon incubation with microsomes (Fukami *et al.*, 1969).

Table VIII. Effect of Various Compounds on the HEOM Hydrase in Tissue Preparation of Calliphora erythrocephala and Prodenia eridania[a]

	I_{50}(M)	
Compound	Blowfly	Southern army worm
3-Propargyloxyphenyl glycidyl ether	1.35×10^{-5}	—
2,2-bis[4-(2,3-Epoxypropyloxyl)phenyl] propane	1.4×10^{-5}	2.9×10^{-5}
1,1,1-Trichloro-2,3-epoxypropane	1.6×10^{-6}	8.0×10^{-6}
Styrene oxide	—	7.0×10^{-4}
Piperonyl butoxide	2.4×10^{-4}	1.4×10^{-5}
SKF 525-A	7.5×10^{-5}	9.5×10^{-6}
Tri-*o*-cresyl phosphate	4.4×10^{-5}	—
EPN	—	1.7×10^{-5}
Synthetic cecropia hormone (mixed isomers)	1.4×10^{-4}	—

[a]From Slade *et al.* (1975).

9. References

Abernathy, C. O., and Casida, J. E., 1973, Pyrethroid insecticides: Esterase cleavage in relation to selective toxicity, *Science* **179:**1235.

Abernathy, C. O., Ueda, K., Engel, J. L., Gaughan, L. C., and Casida, J. E., 1973, Substrate-specificity and toxicological significance of pyrethroid-hydrolyzing esterases of mouse liver microsomes, *Pestic. Biochem. Physiol.* **3:**300.

Ahmed, M. K., Casida, J. E., and Nichols, R. E., 1958, Significance of rumen fluid with particular reference to parathion. *J. Agr. Food Chem.* **6:**740.

Aldridge, W. N., 1953*a*, Serum esterases. 1. Two types of esterase (A and B) hydrolysing *p*-nitrophenyl acetate, propionate and butyrate, and a method for their determination, *Biochem. J.* **53:**110.

Aldridge, W. N., 1953*b*, Serum esterases. 2. An enzyme hydrolyzing diethyl *p*-nitrophenyl phosphate (E600) and its identity with A-esterase of mammalian sera, *Biochem. J.* **53:**117.

Augustinsson, K. B., and Casida, J. E., 1959, Enzymic hydrolysis of *N*:*N*-dimethylcarbamoyl fluoride, *Biochem. Pharmacol.* **3:**60.

Augustinsson, K. B., and Heimburger, G., 1954, Enzymatic hydrolysis of oranophosphorus compounds. IV. Specificity Studies, *Acta Chem. Scand.* **8:**1533.

Becker, E. L., and Barbaro, J. F., 1964, The enzymatic hydrolysis of *p*-nitrophenyl ethyl phosphonates by mammalian plasma, *Biochem. Pharmacol.* **13:**1219.

Behal, F. J., Little, G. H., and Klein, R. A., 1968, Arylamidase of liver, *Biochim. Biophys. Acta* **178:**118.

Bridges, P. M., 1957, Absorption and metabolism of [^{14}C] allethrin by the adult housefly, *Musca domestica* L., *Biochem. J.* **66:**316.

Brooks, G. T., 1973, Insect epoxide hydrase inhibition by juvenile hormone analogues and metabolic inhibitors, *Nature (London)* **245:**382.

Brooks, G. T., 1974, Inhibitors of cyclodiene epoxide ring hydrating enzymes of the blowfly, *Calliphora erythrocephala. Pestic. Sci.* **5:**177.

Brooks, G. T., and Harrison, A., 1969, Hydration of HEOD (dieldrin) and the heptachlor epoxide by microsomes from liver of pig and rabbits. *Bull. Environ. Contam. Toxicol.* **4:**352.

Brooks, G. T., Lewis, S. E., and Harrison, A., 1968, Selective metabolism of cyclodiene insecticide enantiomers by pig liver microsomal enzymes, *Nature* **220:**1034.

Brooks, G. T., Harrison, A., and Lewis, S. E., 1970, Cyclodiene epoxide ring hydration by microsomes from mammalian liver and houseflies, *Biochem. Pharmacol.* **19:**255.

Bull, D. L., and Lindquist, D. A., 1964, Metabolism of 3-hydroxy-*N,N*-dimethyl crotonamide dimethyl phosphate by cotton plants, insects, and rats, *J. Agr. Food Chem.* **12:**310.

Casida, J. E., and Augustinsson, K. B., 1959, Reaction of plasma albumin with 1-naphthyl *N*-methylcarbamate (the insecticide sevin) and certain other esters, *Biochim. Biophys. Acta* **36:**411.

Casida, J. E., Augustinsson, K. B., and Jonsson, G., 1960, Stability, toxicity and reaction mechanism with esterases of certain carbamate insecticides, *J. Econ. Entomol.* **53:**205.

Casida, J. E., Kimmel, E. C., Elliott, M., and Janes, N. F., 1971, Oxidative metabolism of pyrethrins in mammals, *Nature (London)* **230:**326.

Chamberlain, R. W., 1950, An investigation on the action of piperonyl butoxide with pyrethrum, *Am. J. Hyg.* **52:**153.

Chamberlain, W. F., Gatterdam, P. E., and Hopkins, D. E., 1961, The metabolism of P^{32}-labeled dimethoate in sheep, *J. Econ. Entomol.* **54:**733.

Chen, P. R. S., and Dauterman, W. C., 1971*a*, Alkylamidase of sheep liver, *Biochim. Biophys. Acta* **250:**216.

Chen, P. R. S., and Dauterman, W. C., 1971*b*, Studies on the toxicity of dimethoate analogs and their hydrolysis by sheep liver amidase, *Pestic. Biochem. Physiol.* **1:**340.

Chen, P. R., Tucker, W. P., and Dauterman, W. C., 1969, The structure of biologically-produced malathion monoacid, *J. Agr. Food Chem.* **17:**86.

Christman, A. A., and Lewis, H. B., 1921, Lipase studies. I. The hydrolysis of the esters of some dicarboxylic acids by the lipase of liver, *J. Biol. Chem.* **47:**495.

Cook, J. W., 1957, *In vitro* destruction of some organophosphate pesticides by bovine rumen fluid, *J. Agr. Food Chem.* **5:**859.

Cook, J. W., and Yip, G., 1958, Malathionase. II. Identity of a malathion metabolite, *J. Assoc. Off. Agr. Chem.* **41:**407.

Cook, J. W., Blake, J. R., Yip, G., and Williams, M., 1958, Malathionase. I. Activity and inhibition, *J. Assoc. Off. Agr. Chem.* **41:**399.

Dansette, P. M., Yagi, H., Jerina, D. M., Daly, J. W., Levin, W., Lu, A. Y. H., Kuntzman, R., and Conney, A. H., 1974, Assay and partial purification of epoxide hydrase from rat liver microsomes, *Arch. Biochem. Biophys.* **164:**511.

Dauterman, W. C., 1971, Biological and non-biological modifications of organophosphorus compounds, *Bull. WHO* **44:**133.

Dauterman, W. C., and Main, A. R., 1966, The relationship between acute toxicity and *in vitro* inhibition and hydrolysis of a series of carbalkoxy homologs of malathion, *Toxicol. Appl. Pharmacol.* **9:**409.

Dauterman, W. C., Casida, J. E., Knaak, J. B., and Kowalczyk, T., 1959, Metabolism of organophosphorus insecticides: Metabolism and residues associated with oral administration of dimethoate to rats and three lactating cows, *J. Agr. Food Chem.* **7:**188.

Dicowsky, L., and Morello, A., 1971, Glutathione-dependent degradation of 2,2-dichlorovinyl dimethyl phosphate (DDVP) by the rat, *Life Sci.* **10:**103.

Donninger, C., Nobbs, B. T., and Wilson, K., 1971, An enzyme catalyzing the hydrolysis of phosphoric acid diesters in rat liver. *Biochem. J.* **122:**51p.

Douch, P. G., and Smith, J. N., 1971, Metabolism of *m*-tert.-butylphenyl, *N*-methylcarbamate in insects and mice, *Biochem. J.* **125:**385.

Elliott, M., and Casida, J. E., 1972, Optically pure pyrethroids labeled with deuterium and tritium in the methylcyclopentenonyl ring, *J. Agr. Food Chem.* **20:**295.

Elliott, M., Kimmel, E. C., and Casida, J. E., 1969, ^{3}H-Pyrethrin I and -pyrethrin II: Preparation and use in metabolism studies, *Pyrethrum Post* **10:**3.

Elliott, M., Janes, N. F., Kimmel, E. C., and Casida, J. E., 1972, Metabolic fate of pyrethrin I, pyrethrin and allethrin administered orally to rats. *J. Agr. Food Chem.* **20:**300.

Eto, M., 1974, *Organophosphorus Pesticides: Organic and Biological Chemistry*, CRC press, Cleveland.

Eto, M., Oshima, Y., Kitakata, S., Tanaka, F., and Kojima, K., 1965, Studies on saligenin cyclic phosphorus esters with insecticidal activity. X. Synergism of malathion against susceptible and resistant insects, *Botyu-Kagaku* **31:**33.

Fest, C., and Schmidt, K. J., 1973, *The Chemistry of Organophosphorus Pesticides*, Springer, New York.

Fukami, J., Shishido, T., Fukuhaga, K., and Casida, J. E., 1969, Oxidative metabolism of rotenone in mammals, fish and insects and its relation to selective toxicity, *J. Agr. Food Chem.* **17:**1217.

Gardocki, J. F., and Hazelton, L. W., 1951, Urinary excretion of the metabolic products of parathion following intraveneous injection, *J. Ann. Pharm. Assoc. Sci. Ed.* **40:**491.

Gillette, J. R., 1971, Reductive enzymes, *in*: *Concepts of Biochemical Pharmacology*, Part 2 (B. B. Brodie and J. R. Gillette, ed.), pp. 349–361, Springer, New York.

Hassan, A., and Dauterman, W. C., 1968, Studies on the optically active isomers of *O,O*-diethyl malathion and *O,O*-diethyl malaoxon, *Biochem. Pharmacol.* **17:**1431.

Heath, D. F., 1961, *Organophosphorus Poisons*, Pergamon Press, New York.

Hitchcock, M., and Murphy, S. D., 1967, Enzymatic reduction of *O,O*-diethyl-(4-nitrophenyl)phosphorothioate, *O,O*-diethyl *O*-(4-*nitrophenyl)phosphate, and O*-ethyl *O*-(4-nitrophenyl)benzene thiophosphonate by tissues from mammals, birds and fishes, *Biochem. Pharmacol.* **16:**1801.

Hodgson, E., and Casida, J. E., 1962, Mammalian enzymes involved in the degradation of 2,2-dichlorvinyl dimethyl phosphate, *J. Agr. Food Chem.* **10:**208.

Hopkins, T. L., and Robbins, W. E., 1957, The adsorption, metabolism and excretion of ^{14}C-labeled allethrin by houseflies. *J. Econ. Entomol.* **50:**684.

Jao, L. T., and Casida, J. E., 1974, Insect pyrethroid-hydrolyzing esterases, *Pest. Biochem. Physiol.*, **4:**465.

Jarczyk, H. J., 1966, The influence of esterases in insects on the degradation of organophosphates of the E.605 series, *Pflanzenschutz-Nachr. Bayer* **19:**1.

Jerina, D. M., and Daly, J. W., 1974, Arene oxides: A new aspect of drug metabolism, *Science* **185:**573.

Jerina, D. M., Daly, J. W., Jefferey, A. M., and Gibson, D. T., 1971, *cis*-1,2-Dihydroxyl-1,2-dihydro-naphthalene: A bacterial metabolite from naphthalene, *Arch. Biochem. Biophys.* **142:**394.

Johnson, M. K., 1965, The influence of some aliphatic compounds on rat liver glutathione levels, *Biochem. Pharmacol.* **14:**1383.

Kojima, K., 1961, *Studies on the Selective Toxicity and Detoxication of Organophosphorus Compounds*, special report of the Ton Noyaku Co., Odawara, Japan.

Kojima, K., and O'Brien, R. D., 1968, Paraoxon hydrolyzing enzymes in rat liver, *J. Agr. Food Chem.* **16:**574.

Korte, F., and Arent, H., 1965, Metabolism of insecticides IX (1) isolation and identification of dieldrin metabolites from urine of rabbits after oral administration of dieldrin-^{14}C, *Life Sci.* **4:**2017.

Krueger, H. R., and Casida, J. E., 1961, Hydrolysis of certain organophosphates insecticides by housefly enzymes, *J. Econ. Entomol.* **54:**239.

Leeling, N. C., and Casida, J. E., 1966, Metabolites of carbaryl (1-naphthylmethylcarbamate) in mammals and enzymatic systems for their formation, *J. Agr. Food Chem.* **14:**281.

Lenz, D. E., Deguehery, L. E., and Holton, J. S., 1973, On the nature of the serum enzyme catalyzing paraoxon hydrolysis, *Biochim. Biophys. Acta* **321:**189.

Lichtenstein, E. P., and Fuhreman, T. W., 1971, Activity of an NADPH-dependent nitroreductase in houseflies, *Science* **172:**589.

Lichtenstein, E. P., and Schulz, K. R., 1964, The effects of moisture and microorganisms on the persistence and metabolism of some organophosphorus insecticides in soils with special emphasis on parathion, *J. Econ. Entomol.* **57:**618.

Main, A. R., 1960*a*, The purification of the enzyme hydrolysing diethyl *p*-nitrophenyl phosphate (paraoxon) in sheep serum, *Biochem. J.* **74:**10.
Main, A. R., 1960*b*, The differentiation of the A-type esterases in sheep serum, *Biochem. J.* **75:**188.
Main, A. R., and Braid, P. E., 1962, Hydrolysis of malathion by aliesterases *in vitro* and *in vivo, Biochem. J.* **84:**255.
Main, A. R., and Dauterman, W. C., 1967, Kinetics for the inhibition of carboxylesterase by malaoxon, *Can. J. Biochem.* **45:**757.
Matsumura, F., and Boush, G. M., 1966, Malathion degradation by *Trichoderma viride* and a *Pseudomonas* species, *Science* **153:**1278.
Matsumura, F., and Brown, A. W. A., 1963, Studies on carboxyesterase in malathion-resistant *Culex tarsalis, J. Econ. Entomol.* **56:**381.
Matsumura, F., and Hogendijk, C. J., 1964, The enzymatic degradation of parathion in organophosphate-susceptible and -resistant houseflies, *J. Agr. Food Chem.* **12:**447.
Matsumura, F., and Voss, G., 1965, Properties of partially purified malathion carboxylesterase of the two-spotted spider mite, *J. insect Physiol.* **11:**147.
Matsumura, F., and Ward, C. T., 1966, Degradation of insecticides by the human and the rat liver, *Arch. Environ. Health* **13:**257.
Matthews, H. B., and McKinney, J. D., 1974, Dieldrin metabolism to *cis*-dihydroaldrindiol and epimerization of *cis* to *trans* dihydroaldrindiol by rat liver microsomes, *Drug Metab. Disp.* **2:**333.
Mazur, A., 1946, An enzyme in animal tissues capable of hydrolyzing the phosphorus–fluorine bond of alkyl fluorophosphates, *J. Biol. Chem.* **164:**271.
Menn, J. J., and M. Beroza (eds.), 1972, *Insect Juvenile Hormones: Chemistry and Action,* Academic Press, New York.
Mick, D. L., and Dahm, P. A., 1970, Metabolism of parathion by two species of *Rhizobium, J. Econ. Entomol.* **63:**1155.
Motoyama, N., and Dauterman, W. C., 1974, The role of non-microsomal metabolism in organophosphorus resistance, *J. Agr. Food Chem.* **22:** 350.
Mounter, L. A., 1955, The complex nature of dialkylfluorophosphatases of hog and rat liver and kidney, *J. Biol. Chem.* **215:**705.
Mounter, L. A., and Chanutin, A., 1954, Dialkylfluorophosphatase of kidney. III. Studies of activation and inhibition of cofactors, *J. Biol. Chem.* **210:**219.
Mounter, L. A., Baxter, R. F., and Chanutin, A., 1955*a*, Dialkylfluorophosphatases of microorganisms, *J. Biol. Chem.* **215:**699.
Mounter, L. A., Dien, L. T. H., and Chanutin, A., 1955*b*, The distribution of dialkylfuoro phosphatase in the tissue of various species, *J. Biol. Chem.* **215:**691.
Murphy, S. D., and DuBois, K. P., 1957, Quantitative measurement of inhibition of the enzymatic detoxification of malathion by EPN (ethyl *p*-nitrophenyl thionobenzenephosphonate), *Proc. Soc. Expl. Biol. Med.* **96:**813.
Nakatsugawa, T., Tolman, N. M., and Dahm, P. A., 1969, Metabolism of S^{35}-parathion in the housefly, *J. Econ. Entomol.* **62:**408.
Nishizawa, Y., and Casida, J. E., 1965, Synthesis of *d-trans*-chrysanthemumic acid-1-^{14}C and its antipode on a semimicro scale. *J. Agr. Food Chem.* **13:**525.
Nolan, J., and O'Brien, R. D., 1970, Biochemistry of resistance to paraoxon in strains of houseflies, *J. Agr. Food Chem.* **18:**802.
O'Brien, R. D., 1957, Properties and metabolism in the cockroach and mouse of malathion and malaoxon, *J. Econ. Entomol.* **50:**159.
O'Brien, R. D., 1960, *Toxic Phosphorus Esters,* Academic Press, New York.
O'Brien, R. D., Thorn, G. D., and Fisher, R. W., 1958, New organophosphate insecticides developed on rational principles, *J. Econ. Entomol.* **51:**714.
Oesch, F., 1973, Mammalian epoxide hydrases: Inducible enzymes catalysing the inactivation of carcinogenic and cytotoxic metabolites derived from aromatic and olefinic compounds, *Xenobiotica* **3:**305.

Oesch, F., 1974, Purification and specificity of a human microsomal epoxide hydratese, *Biochem. J.* **139:**77.

Oesch, F., and Daly, J., 1971, Solubilization, purification and properties of a hepatic epoxide hydrase. *Biochim. Biophys. Acta* **227:**692.

Oesch, F., Jerina, D. M., and Daly, J. W., 1971, A radiometric assay for hepatic epoxide hydrase activity with (7-^{3}H) styrene oxide. *Biochim. Biophys. Acta* **227:**685.

Oesch, F., Jerina, D. M., Daly, J., and Rice, J., 1973, Induction, activation and inhibition of epoxide hydrase: An anomalous prevention of chlorobenzene-induced hepatotoxicity by an inhibitor of epoxide hydrase, *Chem.-Biol. Interact.* **6:**189.

Oesch, F., Thoenen, H., and Fahrländer, H., with technical assistance of Suda, K., 1974, Epoxide hydrase in human liver biopsy specimens: Assay and properties. *Biochem. Pharmacol.* **23:**1307.

Oonnithan, E. S., and Miskus, R., 1964, Metabolism of dieldrin C^{14} by dieldrin-resistant *Culex pipens quinquefasciatus* mosquitoes, *J. Econ. Entmol.* **57:**425.

Pankaskie, J. E., Fountaine, F. C., and Dahm, P. A., 1952, The degradation and detoxification of parathion in dairy cows, *J. Econ. Entomol.* **45:**51.

Plapp, F. W., Jr., and Tong, H. H. C., 1966, Synergism of malathion and parathion against resistant insects: Phosphorus esters with synergistic properties, *J. Econ. Entomol.* **59:**11.

Plapp, F. W., Jr., and Valega, T. M., 1967, Synergism of carbamate and organophosphate insecticides by noninsecticidal carbamates, *J. Econ. Entomol.* **60:**1094.

Plapp, F. W., Jr., Bigley, W. S., Chapman, G. A., and Eddy, G. W., 1963, Synergism of malathion against resistant houseflies and mosquitoes, *J. Econ. Entomol.* **56:**643.

Rose, H. A., and Young, R. G., 1973, Nitroreductases in the Madagascar cockroach, *Gromphadorhina portentosa, Pestic. Biochem. Physiol.* **3:**243.

Satyanarayana, T., and Getzin, L. W., 1973, Properties of a stable cell-free esterase from soil, *Biochemistry* **12:**1566.

Seume, F. W., and O'Brien, R. D., 1960*a*, Metabolism of malathion by rat liver tissue preparations and its modification by EPN, *J. Agr. Chem.* **8:**36.

Seume, F. W., and O'Brien, R. D., 1960*b*, Potentiation of the toxicity to insects and mice of phosphorothionates containing carboxyester and carboxyamide groups, *Toxicol. Appl. Pharmacol.* **2:**495.

Shishido, T., and Fukami, J., 1972, Enzymatic hydrolysis of diazoxon by rat tissue homogenates, *Pestic. Biochem. Physiol.* **2:**39.

Škrinjarić-Špoljar, M., and Reiner, E., 1968, Hydrolysis of diethyl-*p*-nitrophenyl phosphate and ethyl-*p*-nitrophenyl-ethyl phosphonate by human sera, *Biochim. Biophys. Acta* **165:**289.

Slade, M., and Wilkinson, C. F., 1973, Juvenile hormone analogs: A possible case of mistaken identity? *Science* **181:**672.

Slade, M., and Zibitt, C. H., 1972, Metabolism of *Cecropia* juvenile hormone in insects and in mammals, *In:* Insect Juvenile Hormones: Chemistry and Action (J. J. Menn and M. Beroza, eds.), P. 155, Academic Press, NewYork.

Slade, M., Brooks, G. T., Hetnarski, H. K., and Wilkinson, C. F., 1975, Inhibition of the enzymatic hydration of the epoxide HEOM in insects, *Pestic. Biochem. Physiol.*, **5:**35.

Stoming, T. A., and Bresnick, E., 1973, Gas chromatographic assay of epoxide hydrase activity with 3-methylcholanthrene-11-12-oxide, *Science* **181:**951.

Stoming, T. A., and Bresnick, E., 1974, Hepatic epoxide hydrase in neonatal and partially hepatectomized rats, *Cancer Res.* **34:**2810.

Tomlin, A. D., 1968, Trans-aldrin glycol as a metabolite of dieldrin in larvae of the southern house mosquito, *J. Econ. Entomol.*, **61:**855.

Uchida, T., and O'Brien, R. D., 1967, Dimethoate degradation by human liver and its significance for acute toxicity, *Toxicol. Appl. Pharmacol.* **10:**89.

Uchida, T., Dauterman, W. C., and O'Brien, R. D., 1964, The metabolism of dimethoate by vertebrate tissues, *J. Agr. Food Chem.* **12:**48.

Uchida, T., Zschintzsch, J., and O'Brien, R. D., 1966, Relation between synergism and metabolism of dimethoate in mammals and insects, *Toxicol. Appl. Pharmacol.* **8:**259.

Welling, W., and Blaakmeer, P. T., 1971, Metabolism of malathion in a resistant and susceptible strain of houseflies: Insecticide resistance, synergism, enzyme induction, *in: Proceedings of the Second International IUPAC Congress of Pesticide Chemistry*, Vol. II (A. S. Tahori, ed.), pp. 61–75, Gordon and Breach, New York.

Welling, W., Blaakmeer, P., Vink, G. J., and Voerman, S., 1971, *In vitro* hydrolysis of paraoxon by parathion resistant houseflies, *Pestic. Biochem. Physiol.* **1:**61.

White, A. F., 1972, Metabolism of the juvenile hormone analogue methyl farnesoate 10,11-epoxide in two insect species, *Life Sci.* **11:**201.

Winteringham, F. P. W., Harrison, A., and Bridges, P. M., 1955, Absorption and metabolism of [^{14}C] pyrethroids by the adult housefly, *Musca domestica* L., *Biochem. J.* **61:**359.

Yamamoto, I., and Casida, J. E., 1966, *O*-Demethyl pyrethrin II analogs from oxidation of pyrethrin I, allethrin, dimethrin and phthalthrin by housefly enzyme system, *J. Econ. Entomol.* **59:**1542.

Yamamoto, I., and Casida, J. E., 1968, Synthesis of ^{14}C-labelled pyrethrin I, allethrin, phthalthrin, and dimethrin on a submillimole scale, *Agr. Biol. Chem.* **32:**1382.

Yamamoto, I., Kimmel, E. C., and Casida, J. E., 1969, Oxidative metabolism of pyrethroids in houseflies, *J. Agr. Food Chem.* **17:**1227.

Young, R. G., 1975, Flavin-linked azoreductases of the Madagascar cockroach, *Gromphadorhina portentosa*, *Insect. Biochem.* **5:**in press.

Zeid, M. M. I., Dahm, P. A., Hein, R. E., and McFarland, R. H., 1953, Tissue distribution, excretion of $^{14}CO_2$ and degradation of radioactive pyrethrins administered to the American cockroach, *J. Econ. Entomol.* **46:**324.

5

Enzymatic Conjugation and Insecticide Metabolism

Raymond S. H. Yang

1. Introduction

Present knowledge concerning biochemical conjugations stemmed from the isolation and identification of conjugates from normal and pathological animal urine during the last century. As early as 1842, the formation of hippuric acid was conclusively established by Keller following self-administration of benzoic acid. During the latter part of the last century, conjugations such as the synthesis of sulfates, glucuronides, and ornithine conjugates were discovered, and reactions involving mercapturic acid formation, methylation, acetylation, and cyanide detoxication were recognized. Glutamine conjugation was demonstrated in man in 1914 and glucoside conjugation in plants and insects in 1938 and 1953, respectively (Smith and Williams, 1970; Williams, 1959, 1967). During the past two decades, considerable attention has centred around the biochemical conjugations of certain naturally occurring body constituents such as sterols, bile acids, steroid hormones, glycoproteins, and mucopolysaccharides. Consequently, a number of new conjugations have been discovered (Layne, 1970).

Since conjugation products are usually water soluble, readily excretable substances and their formation generally results in a decrease in toxicity, the

Raymond S. H. Yang • Institute of Comparative and Human Toxicology, Center of Experimental Pathology and Toxicology, Albany Medical College, Albany, New York 12208.

physiological significance of the conjugation mechanism was initially considered to be for the sole purpose of detoxication. The detoxication concept was widely accepted around the beginning of this century and as a consequence dominated scientific thinking in this area for many years (Smith and Williams, 1970). However, this hypothesis has been challenged from time to time by various investigators on the basis that it is inadequate to interpret several biochemical and toxicological observations. In 1928, A. J. Quick first expressed his doubts concerning the validity of the detoxication concept and suggested that conjugation processes could be viewed as normal metabolic reactions serving a dual role in intermediary metabolism and detoxication. This view subsequently was substantiated by Schachter and Marrian (1938) by their successful isolation of steroid sulfates from the urine of pregnant mares. Since then, the discovery of numerous conjugates of naturally occurring biochemical constituents has further emphasized the probable intrinsic physiological roles of various conjugation reactions. It is now generally recognized that in addition to their function in detoxication almost all conjugation mechanisms can act on natural substrates.

Although it is now almost one and a half centuries since the earliest studies of conjugations, this area of research remains a growing scientific discipline. Recent developments have served to modify earlier hypotheses. Thus the once prevalent but erroneous view that glucose conjugation is absent in mammals has recently been corrected, following the demonstration in several species of the formation of glucosides from both endogenous and foreign substrates (Gessner and Vollmer, 1969; Williamson *et al.*, 1969). The discovery that steroid sulfates are involved in the biosynthesis of hormones (Layne, 1970) has still further modified the classical detoxication concept.

The purpose of this chapter is to provide basic biochemical information concerning conjugation mechanisms and to consolidate the available information on conjugation reactions involving insecticides. Since much of the general biochemistry and enzymology involved in major conjugation processes is readily available elsewhere (Fishman, 1970; Hutson, 1970, 1972; Mandel, 1971; Parke, 1968; Smith, 1968; Williams, 1967), discussion in these areas has been kept to a minimum. However, attempts have been made to incorporate some of the latest advances in the general area of conjugation mechanisms.

2. Biochemistry of Selected Conjugation Mechanisms

On the basis of our present knowledge concerning the involvement of conjugation reactions in normal metabolic functions, the definition originally given by Parke (1968) must be slightly modified. Conjugations are biosyntheses in which natural or foreign compounds or their metabolites combine with readily available, endogenous conjugating agents (e.g., glucuronic acid, sulfate, acetyl, methyl, glycine) to form conjugates.

Since conjugations are biosynthetic processes, they are generally energy dependent and many are therefore directly or indirectly linked with high-energy compounds. In 1967, Williams classified conjugation reactions into two groups based on different forms of "activated intermediates." The first type involves the formation of an activated conjugation agent, whereas the second type involves the formation of an activated substrate (i.e., foreign or natural compounds).

$$\text{Type 1:}\quad \text{Conjugating agent} \xrightarrow{\text{energy}} \text{activated conjugating agent} \xrightarrow{\text{substrate}} \text{conjugated product}$$

$$\text{Type 2:}\quad \text{Substrate} \xrightarrow{\text{energy}} \text{activated substrate} \xrightarrow[\text{agent}]{\text{conjugating}} \text{conjugated product}$$

Type 1 reactions include such conjugations as methylation, acetylation, and the formation of glucuronides, glucosides, and sulfates. The sites at which these enzymatic reactions occur are distributed throughout the body, although the liver constitutes the principal site. Type 2 reactions consist of amino acid conjugations, which occur only in the liver and/or the kidney (Williams, 1967).

2.1. Glucuronic Acid Conjugation

Glucuronide formation is one of the most important conjugation mechanisms in animals. It has been studied extensively, and there are a number of comprehensive reviews on this subject (Dutton, 1966*a*; Miettinen and Leskinen, 1970; Williams, 1967).

2.1.1. Biochemical Mechanism

The mechanism of glucuronic acid conjugation involves two stages. First, an activated intermediate, uridine diphosphoglucuronic acid (UDPGA), is synthesized. Second, glucuronic acid is transferred from UDPGA to the substrate. The overall scheme may be shown as follows:

$$\text{D-glucose-1-P}_i + \text{UTP} \xleftrightarrow[\text{pyrophosphorylase}]{\text{UDPG}} \text{UDP-}\alpha\text{-D-glucose} + \text{PP}_i \qquad (1)$$

$$\text{UDP-}\alpha\text{-D-glucose} + 2\text{NAD} + \text{H}_2\text{O} \xrightarrow[\text{dehydrogenase}]{\text{UDPG}} \text{UDP-}\alpha\text{-D-glucuronic acid} + 2\text{NADH}_2 \qquad (2)$$

$$\text{UDP-}\alpha\text{-D-glucuronic acid} + \text{ROH} \xrightarrow[\text{glucuronyltransferase}]{\text{UDP}} RO\text{-}\beta\text{-D-glucuronic acid} + \text{UDP} + \text{H}_2\text{O} \qquad (3)$$

Reactions (1) and (2) are catalyzed by enzymes present in the nuclear and soluble fraction of the liver, respectively. The enzyme responsible for reaction (3), UDP glucuronyltransferase (UDP: glucuronate glucuronyltransferase;

E.C. 2.4.1.17), is located in the microsomal fraction. Glucuronide formation occurs mainly in the liver, although other organs and tissues such as kidney, intestines, and skin also possess enzyme activity.

A wide variety of chemicals can be conjugated with glucuronic acid, the most common functional groups involved being the hydroxyl, carboxyl, and amino moieties (Williams, 1967).

2.1.2. Distribution

Glucuronic acid conjugation is widely distributed among vertebrates. It occurs in mammals, marsupials, birds, amphibians, reptiles, and fish, although there appears to be some species variation (Williams, 1967). Examples of such variation can be found in the cat, the Gunn strain of Wistar rat, and some fish in which glucuronic acid conjugation is partially or totally defective (Mandel, 1971). In insects and plants, glucuronic acid conjugation is deficient or absent and appears to be replaced by glucose conjugation (Smith, 1968).

2.2. Glucose Conjugation

Until recently, glucose conjugation was regarded as a detoxication mechanism characteristic of plants, insects, and other invertebrates (Dutton, 1966*a*; Mandel, 1971; Parke, 1968; Smith, 1968; Williams, 1967). The underlying basis for this view undoubtedly stemmed from past failure to demonstrate glucoside formation in vertebrates and from its wide occurrence in plants and invertebrates. In 1969, Williamson *et al.* isolated a 17α-estradiol glucoside from the urine of rabbits receiving large doses of estrone benzoate, and this finding was confirmed *in vitro* by the same authors using rabbit liver microsomes fortified with uridine diphosphoglucose (UDPG). Meanwhile, Gessner and Vollmer (1969) reported the first successful transfer *in vitro* of glucose from UDPG to a foreign compound, 4-nitrophenol, by mouse liver microsomes. Since then, the formation of glucosides of various compounds including isoflavones, phenols, steroids, and bilirubin has been demonstrated *in vivo* or *in vitro* in different mammalian species (Collins *et al.*, 1970; Compernolle *et al.*, 1971; Fevery *et al.*, 1971, 1972; Gessner and Hamada, 1970; Gessner *et al.*, 1973; Heirwegh *et al.*, 1970, 1971; Kuenzle, 1970; Labow and Layne, 1972; Van Heyningen, 1971; Williamson *et al.*, 1972; Wong, 1971). These findings strongly suggest that glucose conjugation is not only of wide occurrence in mammals but also a biochemical process of considerable physiological significance.

2.2.1. Biochemical Mechanism

The detailed mechanism of glucose conjugation has not been fully elucidated, although by analogy to that of glucuronic acid conjugation it is probable that the following reaction sequence occurs:

$$\text{D-glucose-1-P}_i + \text{UTP} \xrightleftharpoons[\text{pyrophosphorylase}]{\text{UDPG}} \text{UDP-}\alpha\text{-D-glucose} + \text{PP}_i \quad (4)$$

$$\text{UDP-}\alpha\text{-D-glucose} + \text{ROH} \xrightarrow[\text{glucosyltransferase}]{\text{UDP}} \text{RO-}\beta\text{-D-glucose} + \text{UDP} + \text{H}_2\text{O} \quad (5)$$

The mammalian UDP glucosyltransferase appears to be located primarily in the liver microsomal fraction (Collins *et al.*, 1970; Labow and Layne, 1972; Gessner *et al.*, 1973), although steroid glucosyltransferase activity has been detected in rabbit kidney and large intestine (Collins *et al.*, 1970). The pH optimum for a phenol glucosyltransferase (UDP glucose: phenol β-glucosyltransferase; E.C. 2.4.1.35) in mouse liver was reported to be close to 7.0 (Gessner and Vollmer, 1969; Gessner *et al.*, 1973) and that of a rabbit liver steroid glucosyltransferase was found to be at 8.0 (Collins *et al.*, 1970). Both of these enzymes appeared to have broad substrate specificities and were in general less active than the corresponding glucuronyltransferases (Collins *et al.*, 1970; Gessner *et al.*, 1973).

In insects, relatively few studies have been conducted at the enzyme level. In contrast to the findings that mammalian glucosyltransferases are localized in the microsomes, work in insects has indicated considerable variation in subcellular distribution of the enzyme. In some instances, enzymatic activity is reported to be sedimented at 15,000–20,000*g* (Dutton, 1962; Mehendale and Dorough, 1972*a*; Trivelloni, 1964), whereas in the case of the tobacco hornworm (*Manduca sexta*) most of the activity appeared to be present in the high-speed (105,000*g*) supernatant (Mehendale and Dorough, 1972*a*). Since detailed subcellular fractionation studies involving marker enzymes and morphological examination have not been conducted with insect tissues, it is difficult to speculate on the patterns of localization which might exist. With respect to the tissue distribution of insect glucosyltransferase, the fat body appears to contain the highest enzyme titers (Gessner and Acara, 1968; Mehendale and Dorough, 1972*a*; Smith and Turbert, 1961), although Mehendale and Dorough (1972*a*) found that the midgut of the tobacco hornworm also constituted a suitable enzyme source. The tobacco hornworm midgut enzyme was active over a rather broad pH range toward the glucosylation of 1-naphthol and the optimal pH was estimated to be 8.5 (Mehendale and Dorough, 1972*a*).

2.2.2. Distribution

As indicated earlier, glucose conjugation occurs widely among insects and plants (Dutton, 1966*a*; Harbone, 1964; Pridham, 1964; Smith, 1968; Towers, 1964). It is also present in bacteria (Smith, 1964; Towers, 1964) and in invertebrates other than insects (Dutton, 1965, 1966*b*; Illing and Dutton, 1970). Among the mammalian species studied, glucose conjugation has been shown to occur in rabbits (Collins *et al.*, 1970; Labow and Layne, 1972;

Williamson *et al.*, 1969), mice (Gessner *et al.*, 1973; Gessner and Hamada, 1970; Gessner and Vollmer, 1969), rats (Fevery *et al.*, 1971, 1972; Heirwegh *et al.*, 1970), and humans (Kuenzle, 1970). On the other hand, Labow and Layne (1972) showed that sheep liver microsomal fraction while active in transferring glucuronic acid from UDPGA to isoflavones was inactive in the glucosylation of these compounds.

2.3. Sulfate Conjugation

Several comprehensive reviews concerning sulfate conjugation have been published (Dodgson and Rose, 1970; Roy, 1970; Roy and Trudinger, 1970), and the reader is referred to these for additional information.

2.3.1. Biochemical Mechanism

In common with other type 1 conjugation reactions, sulfate conjugation requires the biosynthesis of an active intermediate, 3′-phosphoadenosine-5′-phosphosulfate (PAPS), from which the sulfate group is transferred to either an endogenous or a foreign compound. The overall reaction sequence can be illustrated as follows:

$$ATP + SO_4^{2-} \underset{\text{adenylyl transferase}}{\overset{\text{ATP-SULFATE}}{\rightleftharpoons}} APS + PP_i \qquad (6)$$

$$APS + ATP \xrightarrow[\text{3'-phosphotransferase}]{\text{ATP-adenylyl sulfate}} PAPS + ADP \qquad (7)$$

$$ROH + PAPS \xrightarrow{\text{sulfotransferase}} ROSO_3H + PAP \qquad (8)$$

Reactions (6) and (7) entail the formation of PAPS from inorganic sulfate and 2 moles of ATP, through the intermediate adenosine-5′-phosphosulfate (APS). Enzymes responsible for both of these reactions are located in the soluble fraction of the cell. Reaction (8) involves the transfer of sulfate from PAPS to the substrate (ROH) to form the sulfate conjugate and 3′-phosphoadenosine-5′-phosphate (PAP). This reaction occurs with a very broad spectrum of natural and foreign substrates which include phenols, steroids, arylamines, chondroitin, choline, tyrosine methyl ester, luciferin, galactocerebroside, and heparin. It is known that there is a family of at least 12 sulfotransferases which catalyze the PAPS-dependent sulfate conjugation in various organisms (Florkin and Stotz, 1973). The mammalian sulfotransferases have been studied rather extensively, and the subject has been reviewed thoroughly by Dodgson and Rose (1970). In general, the enzymes are located in the soluble fraction of the cell and the liver appears to be the major source of enzyme activity. In insects, Yang and Wilkinson (1972) first demonstrated the presence of sulfotransferases in the gut tissues of the southern armyworm (*Prodenia eridania*). This enzyme system is active toward 4-nitrophenol as well as toward several naturally occurring mammalian, insect, and plant steroids, including cholesterol, α-ecdysone, and β-sitosterol. In a later study by the same authors (Yang and Wilkinson, 1973), similar enzyme systems were detected in seven other species of insects. The insect sulfotransferases are similar to the mammalian enzymes with respect to their cofactor requirements, pH optima, and subcellular localization.

2.3.2. Distribution

Sulfate conjugation is widely distributed in living organisms. It has been demonstrated in various mammals, birds, amphibians, mollusks, peripatus, and a large number of insects and other arthropods (Jordan *et al.*, 1970; Smith, 1968; Williams, 1967, 1971; Yang and Wilkinson, 1971, 1972, 1973) as well as in plants and microorganisms (Dodgson and Rose, 1970).

2.4. Glutathione Conjugation (Mercapturic Acid Formation)

The biochemistry of glutathione conjugation and mercapturic acid synthesis has been reviewed by a number of workers (Boyland and Chasseaud, 1969; Wood, 1970).

2.4.1. Biochemical Mechanism

As shown in the following scheme, mercapturic acid formation involves a four-step enzymatic process:

$$RX + GSH \xrightarrow{\text{GSH-S-transferase}} RSG + HX \tag{9}$$

$$RSG \xrightarrow{\gamma\text{-glutamyltransferase}} \text{R-Cys-Gly} + \text{glutamate} \tag{10}$$

$$\text{R-Cys-Gly} \xrightarrow{\text{peptidase}} \text{R—SCH}_2\underset{\substack{|\\ NH_2}}{\text{CH}}\text{COOH} + \text{glycine} \tag{11}$$

$$\text{R—SCH}_2\underset{\substack{|\\ NH_2}}{\text{CH}}\text{COOH} + \text{acetyl-CoA} \xrightarrow{\text{acetyltransferase}} \text{R—SCH}_2\underset{\substack{|\\ NHCOCH_3}}{\text{CH}}\text{COOH} + \text{CoA} \tag{12}$$

Initially, the substrate (RX) forms a conjugate (RSG) with glutathione (reaction 9). This glutathione conjugate is subsequently transformed to a cysteine conjugate through the stepwise enzymatic removal of a glutamyl residue (reaction 10) and a glycinyl residue (reaction 11). Finally, the cysteine conjugate is acetylated to give a mercapturic acid (reaction 12).

In recent years, research interest in this area has been focused mainly on the first reaction (glutathione conjugation). The reason for its popularity probably resides largely in the diversity of the substrates attacked and the corresponding multiplicity of the GSH-*S*-transferases involved. Furthermore, many toxicants, especially certain organophosphate insecticides, are detoxified readily by GSH-dependent transferase reactions. Boyland and Chasseaud (1969) identified several types of GSH-*S*-transferases and classified them according to their substrate specificities as GSH-*S*-aryl, -epoxide, -alkyl, -aralkyl, and -alkenetransferases. These enzymes are present in the soluble fraction of mammalian liver and/or kidney.

Reaction (10) is catalyzed by a γ-glutamyltransferase (glutathionase) which is most active in kidney and, to a lesser extent, in liver. Peptidases responsible for reaction (11) are present in kidney, liver, and pancreas. Acetylation is catalyzed by acetyl-CoA acetyltransferase present in the liver.

2.4.2. Distribution

The conjugation of foreign compounds with glutathione has been shown to occur in mammals, birds, reptiles, amphibians, fish, insects, and other invertebrates (Boyland and Chasseaud, 1969; Smith, 1968; Williams, 1971).

The ability to form mercapturic acid, on the other hand, is not as widely distributed, and there is evidence that the guinea pig, the pig, man, and the hen are defective in this respect (Williams, 1971). Some insect species have been shown to be capable of synthesizing mercapturic acids (Smith, 1968).

2.5. Phosphate Conjugation

Although the biosynthesis of phosphate esters is a common occurrence in intermediary metabolism, the conjugation of foreign compounds with phosphate is rarely encountered in nature (Hutson, 1970, 1972; Irving, 1970; Parke, 1968; Smith, 1968). Insects appear to be the major group of animals in which phosphate conjugation has been studied to any extent (Smith, 1968).

2.5.1. Biochemical Mechanism

Recently, Yang and Wilkinson (1973) demonstrated an active phosphotransferase in insects which catalyzed the phosphorylation of 4-nitrophenol in the presence of ATP and Mg^{2+}. This enzyme is present in the high-speed supernatant (100,000*g*) of gut tissue homogenates of the Madagascar cockroach (*Gromphadorhina portentosa*) and tobacco hornworm and of whole body homogenates of the housefly (*Musca domestica*). By analogy with other type 1 conjugations, it is possible that ATP may serve as the activated conjugating agent in the enzymatic phosphorylation of foreign compounds as shown in the following hypothetical reaction scheme:

$$\text{ROH} + \text{ATP} \xrightarrow[\text{Mg}^{2+}]{\text{phosphotransferase}} \text{ROPO}_3^{2+} + \text{ADP}$$

The physiological significance of this insect phosphotransferase is unknown at the present time. Since the Madagascar cockroach phosphotransferase could be induced by phenobarbital treatment (Gil *et al.*, 1974), it may have some toxicological implications.

2.5.2. Distribution

Among mammals, dogs and humans are to date the only two species in which phosphate conjugation has been demonstrated (Boyland *et al.*, 1961; Troll *et al.*, 1959, 1963). The possible formation of a carcinogenic phosphate conjugate of 2-acetylaminofluorene by rat liver enzymes has been suggested (King and Phillips, 1968, 1969). In insects, phosphate conjugation has been demonstrated either *in vivo* or *in vitro* in the housefly, blowfly (*Lucilia sericata*), fruit fly (*Drosophila melanogaster*), grass grub (*Costelytra zealandica*), Madagascar cockroach, and tobacco hornworm (Binning *et al.*, 1967; Darby *et al.*, 1966; Gil *et al.*, 1974; Heenan and Smith, 1967; Mitchell and Lunan, 1964; Yang and Wilkinson, 1973). There are also reports concerning the formation of phosphate conjugates in fungi and the primitive arthropod peripatus (Downing, 1962; Jordan *et al.*, 1970).

2.6. Glycine Conjugation

Aromatic and some aliphatic carboxylic acids are often conjugated with amino acids in various organisms (Hutson, 1970, 1972), the most widely occurring reaction involving the amino acid glycine.

2.6.1. Biochemical Mechanism

According to the reaction scheme presented below, glycine conjugation occurs in two stages, the first involving the activation of the substrate (RCOOH) through an enzyme system which requires ATP and coenzyme A (reactions 13 and 14) and the second the condensation of the activated substrate with glycine (reaction 15):

$$\text{RCOOH} + \text{ATP} \xrightarrow{\text{acyl synthetase}} \text{RCO-AMP} + \text{PP}_i \qquad (13)$$

$$\text{RCO-AMP} + \text{CoA-SH} \xrightarrow{\text{acyl thiokinase}} \text{RCO-S-CoA} + \text{AMP} \qquad (14)$$

$$\text{RCO-S-CoA} + \text{Glycine} \xrightarrow[\text{glycine } N\text{-acyl transferase}]{\text{acyl-CoA:}} \text{RCO-Gly} + \text{CoASH} \qquad (15)$$

Normally, these enzymatic reactions take place in the mitochondrial fraction of liver and kidney cells (Mandel, 1971), although rat intestinal preparations have also been shown to be active in glycine conjugation (Hutson, 1972).

2.6.2. Distribution

Glycine conjugation occurs in mammals, insects, amphibians, some birds, and reptiles. In certain species where glycine conjugation is absent, it appears to be replaced by conjugations involving other amino acids such as ornithine, arginine, and glutamine (Mandel, 1971; Smith, 1968; Williams, 1971).

2.7. Cyanide Detoxication (Thiocyanate Formation)

Readers are referred to a recent comprehensive review on cyanide detoxication (Westley, 1973).

2.7.1. Biochemical Mechanism

Inorganic cyanide ion is detoxified by conjugation with sulfur to form thiocyanate, a reaction catalyzed by the enzyme rhodanese:

$$SSO_3^{2-} + CN^- \xrightarrow{\text{rhodanese}} SO_3^{2-} + SCN^-$$

Rhodanese (thiosulfate: cyanide sulfurtransferase; E.C. 2.8.1.1) is present in the liver mitochondrial fraction of many species. Its molecular weight has been estimated to be 37,500 (dimer) and it has a rather broad substrate specificity. Sulfur donors include thiosulfate, thiosulfonates, persulfides, and polysulfides, and acceptors are cyanide, sulfite, organic sulfinites, thiols, dithiols, borohydride, and hydrosulfites (Westley, 1973).

2.7.2. Distribution

Rhodanese activity is widely distributed in different types of living organisms. It occurs in mammals, birds, fish, amphibians, and insects. It has also been detected in various plants and microorganisms (Smith, 1968; Westley, 1973).

3. Enzymatic Conjugation and Insecticide Metabolism

Once insecticides find their way into the living organisms, they are subjected to enzymatic oxidations, hydrolyses, reductions, and conjugations. With few exceptions, conjugations are preceded by other reactions, particularly by oxidations mediated by the microsomal mixed-function oxidases (Chapter 2). For this reason, conjugation reactions are usually considered as "secondary detoxication" mechanisms and have received little attention in studies on insecticide metabolism. Of the several types of conjugation reactions discussed, only the GSH-dependent transferases are capable of catalyzing the primary metabolic attack on insecticidal compounds. This can occur with organophosphates, organothiocyanates, and organochlorine compounds, and, since the products of the reactions are often less toxic than the parent compound, glutathione conjugation may be regarded as a "primary detoxication" process. As such it has received considerable attention in recent years.

In the following section, major groups of insecticides will be discussed with reference to conjugation mechanisms. Emphasis will be given to describing enzymatic aspects of the conjugation reactions and detailed descriptions of *in vivo* experiments will not be included. A number of recent reviews which cover some aspects of conjugation in relation to insecticide metabolism are available (Beynon *et al.*, 1973; Brooks, 1972; Bull, 1972; Dahm, 1970; Dauterman, 1971; Fukunaga *et al.*, 1969; Fukuto, 1972; Hollingworth, 1970, 1971; Hutson, 1970, 1972; Knaak, 1971; Mehendale and Dorough, 1972*b*; Smith, 1968).

3.1. Organophosphorus Insecticides

Among the numerous studies concerning insecticide metabolism, few deal directly with enzymatic conjugations, and evidence for the existence of certain conjugation mechanisms in a particular organism is often deduced solely from the nature of the metabolites excreted in *in vivo* studies. While the formation of conjugates such as glucuronide, glucoside, and sulfate clearly indicates the existence of the corresponding enzyme systems within the organism, this is not the case with products arising from the dealkylation and dearylation of some organophosphorus compounds.

As shown in Fig. 1, there are three possible enzymatic mechanisms which may effect the dealkylation of organophosphates: (1) the soluble GSH-*S*-alkyltransferase reported originally by Fukami and Shishido (1966), (2) the $NADPH_2$- and O_2-dependent microsomal mixd-function oxidases (Donninger *et al.*, 1966), and (3) the soluble hydrolytic enzyme reported by Kojima and O'Brien (1968). It is clear, therefore, that the *in vivo* isolation of a dealkylated organophosphate does not *per se* establish the enzymatic mechanism by which it was formed. A similar situation occurs in the case of dearylation of organophosphates (Fig. 2).

These examples illustrate the dangers involved in extrapolating from *in vivo* data and emphasize the importance of employing well-defined *in vitro*

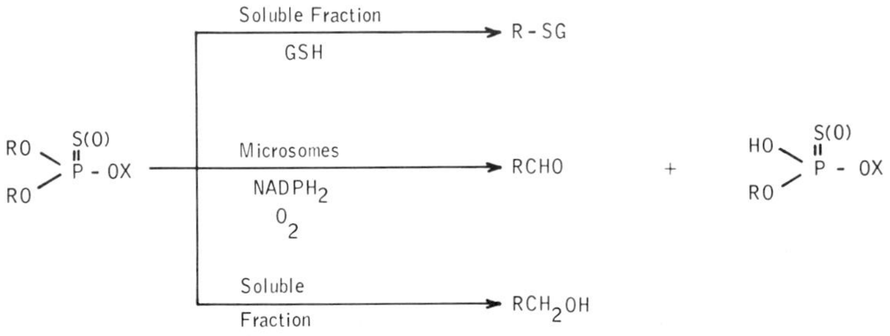

Fig. 1. Possible mechanisms of *O*-dealkylation of organophosphates (Donninger *et al.*, 1966; Fukami and Shishido, 1966; Kojima and O'Brien, 1968).

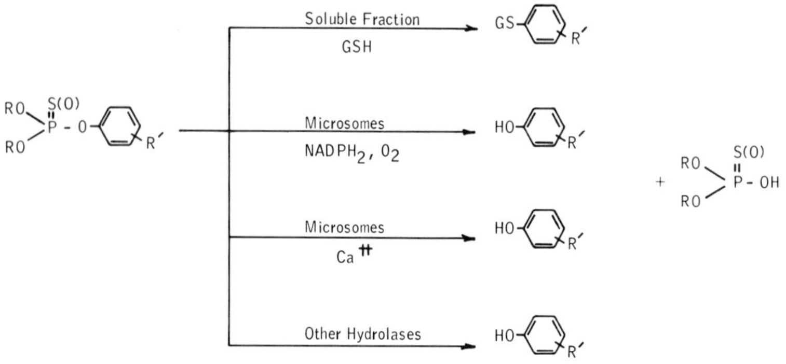

Fig. 2. Possible mechanisms of dearylation of organophosphates (Fukami and Shishido, 1966; Kojima and O'Brien, 1968; Nakatsugawa and Dahm, 1967; Neal, 1967; O'Brien, 1967).

systems coupled with *in vivo* studies to elucidate the enzymatic mechanisms responsible for the formation of a particular metabolite.

3.1.1. Glutathione Conjugation

GSH-dependent transferase reactions are characterized by a direct enzymatic attack on the substrate, and their importance in the metabolism of a large number of phosphorothionate and phosphate insecticides has now been demonstrated (Table I).

Research in this area was initiated in the early 1960s, when dealkylated metabolites were detected following the incubation of several organophosphorus compounds with high-speed supernatants of tissue homogenates from mammals and insects (Fukami and Shishido, 1963; Hodgson and Casida, 1962; Shishido and Fukami, 1963). Since the cofactor requirement of the enzyme system was not known at that time, studies were usually conducted either in the absence of any exogenous cofactors or in the presence of metal ions or a variety of pyridine nucleotides. The dealkylation activity observed under these conditions might well have been due to the presence of endogenous GSH. In one of the studies in which GSH was employed with phenyl mercuric acetate or *p*-chloromercuric benzoate, the stimulatory effect caused by addition of GSH was considered to result from the protection of the —SH groups of the enzyme (Fukami and Shishido, 1963). However, subsequent studies by Fukami and Shishido (1966) established a definite requirement for GSH in the dealkylation of methyl and ethyl parathion by a soluble enzyme in rat liver and insect tissues, and since then many insecticidal chemicals have been shown to be degraded by GSH-dependent transferases in various organisms (Table I). Among the most important findings, the following areas warrant special emphasis.

a. Mechanism of GSH-Dependent Dealkylation and Characteristics of Phosphoric Acid Triester-Glutathione-S-Alkyltransferase. In 1966, Fukami

Insecticidal chemicals	Organisms and tissues[a]	Enzyme preparations[b]	Cofactors required[c]	Methods of identification of metabolites[d]	References
		O-Demethylation			
Methyl parathion	Rat liver	$S_{54,500}$	—	PC, IEC	Fukami and Shishido (1963), Shishido and Fukami (1963)
	Rabbit liver	$S_{54,500}$	—	PC, IEC	
	Guinea pig liver	$S_{54,500}$	—	PC, IEC	
	Rice-stem borer	$S_{54,500}$	—	PC, IEC	
	Rat liver and other tissues	$S_{105,000}$, partially purified enzyme	GSH	PC, IEC	Fukami and Shishido (1966)
	Silkworm midgut, fat body, and others	$S_{105,000}$, partially purified enzyme	GSH	PC, IEC	
	Horn beetle larval midgut, fat body, and others	$S_{105,000}$, partially purified enzyme	GSH	PC, IEC	
	Mouse liver	$S_{145,000}$	GSH	TLC	Miyamoto *et al.* (1968*b*)
	Rat liver	S_{6000}	GSH	—	Rao and McKinley (1969)
	Guinea pig liver	S_{6000}	GSH	—	
	Chicken liver	S_{6000}	GSH	—	
	Rhesus monkey liver	S_{6000}	GSH	—	
Methyl paraoxon	Rat liver	$S_{54,500}$	—	PC, IEC	Shishido and Fukami (1963)
	American cockroach	$S_{54,500}$	—	PC, IEC	
	Mouse liver	$S_{145,000}$	GSH	TLC	Miyamoto *et al.* (1968*b*)
	Mouse, rat, human livers	$S_{117,000}$	GSH	PC, TLC, IEC, electrophoresis, chemical tests	Hollingworth (1969), Hollingworth *et al.* (1973)

Table I (cont.)

Insecticidal chemicals	Organisms and tissues[a]	Enzyme preparations[b]	Cofactors required[c]	Methods of identification of metabolites[a]	References
Sumithion	Rat liver	$S_{54,500}$	—	PC, IEC	Shishido and Fukami (1963)
	Rice-stem borer	$S_{54,500}$	—	PC, IEC	
	Rat liver	$S_{105,000}$ (desalted)	GSH	PC, IEC	Fukami and Shishido (1966)
	Mouse liver	$S_{145,000}$	GSH	TLC	Miyamoto *et al.* (1968*b*)
	Bollworm, tobacco budworm	$S_{(?)}$	GSH	—	Plapp (1973)
Sumioxon	Mouse liver	$S_{145,000}$	GSH	TLC	Miyamoto *et al.* (1968*b*), Hollingworth (1969)
	Mouse liver	$S_{177,000}$	GSH	—	
Dichlorvos	Rat liver	$S_{105,000}$	—	IEC	Hodgson and Casida (1962)
	Mouse liver	$S_{177,000}$	GSH	—	Hollingworth (1969)
	Rat liver	$S_{25,000}$ (dialyzed)	GSH	PC	Dicowsky and Morello (1971)
Tetrachlorvinphos	Rabbit liver	$S_{100,000}$ (dialyzed)	GSH	PC, TLC, isotope dilution analysis	Hutson *et al.* (1968, 1972)
	Mouse liver	$S_{100,000}$ (dialyzed)	GSH	PC, TLC, isotope dilution analysis	
	Rat liver	$S_{100,000}$ (dialyzed)	GSH	PC, TLC, isotope dilution analysis	
	Pig liver	$S_{100,000}$ (dialyzed)	GSH	PC, TLC, isotope dilution analysis	
	Pig liver	Partially purified enzyme	GSH	PC, TLC, isotope dilution analysis	

[illegible]	Rat liver	$S_{3000rpm}$	—	PC, IEC	Morikawa and Saito (1966)
	American cockroach	$S_{3000rpm}$	—	PC, IEC	
	Housefly	$S_{3000rpm}$	—	PC, IEC	
	Rice-stem borer	$S_{3000rpm}$	—	PC, IEC	
	Cabbage leaf	$S_{3000rpm}$	—	PC, IEC	
	Rice leaf	$S_{3000rpm}$	—	PC, IEC	
Azinphosmethyl	Rat liver	S_{6000}	GSH	—	Rao and McKinley (1969)
	Guinea pig liver	S_{6000}	GSH	—	
	Chicken liver	S_{6000}	GSH	—	
	Monkey liver	S_{6000}	GSH	—	
	Mite	$S_{100,000}$	GSH	PC, TLC	Motoyama *et al.* (1971)
	Housefly abdomen	$S_{100,000}$, partially purified enzyme	GSH	PC, TLC, IEC	Motoyama and Dauterman (1972*a*)
	Mouse liver	$S_{100,000}$ (dialyzed)	GSH	PC, TLC, IEC	Motoyama and Dauterman (1972*b*)
Azinphosmethyl oxygen analogue	Housefly abdomen	$S_{100,000}$	GSH	PC, TLC, IEC	Motoyama and Dauterman (1972*a*)
cis-Phosdrin	Mouse liver	$S_{100,000}$	GSH	PC	Morello *et al.* (1967, 1968)
		$S_{10,000}$ (dialyzed)	GSH	PC	
		$S_{45,000}$ (dialyzed)	GSH	PC	
Bromophos	Hog liver Mouse liver	$S_{30,000}$	GSH	TLC	Stenersen (1969, 1971)
Trichlorfon (dipterex)	Cotton leaf worm	Crude homogenate	—	PC, IEC	Zayed and Hassan (1965)
Ronnel	Rat liver	S_{6000}	GSH	—	Rao and McKinley (1969)
	Guinea pig liver	S_{6000}	GSH	—	
	Chicken liver	S_{6000}	GSH	—	
	Monkey liver	S_{6000}	GSH	—	
Malathion	Human liver Rat liver	Crude homogenate	—	TLC	Matsumura and Ward (1966)

Table I (cont.)

Insecticidal chemicals	Organisms and tissues[a]	Enzyme preparations[b]	Cofactors required[c]	Methods of identification of metabolites[d]	References
		O-Deethylation			
Chlorfenvinphos	Mouse liver	$S_{177,000}$	GSH	—	Hollingworth (1969)
	Rabbit liver	$S_{100,000}$	GSH	PC, TLC	Hutson *et al.* (1972)
	Pig liver	Partially purified enzyme	GSH	PC, TLC	
Parathion	Rat liver	$S_{54,500}$	—	PC, IEC	Shishido and Fukami (1963)
	Rice-stem borer	$S_{54,500}$	—	PC, IEC	
	Rat liver	$S_{105,000}$	GSH	PC, IEC	Fukami and Shishido (1966)
	Rat brain	$S_{105,000}$	GSH	PC, IEC	
	Rat kidney	$S_{105,000}$	GSH	PC, IEC	
	Horn beetle larvae midgut	$S_{105,000}$	GSH	PC, IEC	
	Rat liver	$S_{100,000}$	GSH	—	Kojima and O'Brien (1968)
	Rat liver	$S_{105,000}$	GSH	IEC	Nakatsugawa *et al.* (1969)
	Housefly abdomen	$S_{150,000}$	GSH	PC, IEC	Oppenoorth *et al.* (1972)
Paraoxon	Rat liver	$S_{127,000}$	GSH	PC, IEC	Kojima and O'Brien (1968)
	Mouse liver	$S_{177,000}$	GSH	—	Hollingworth (1969)
Diazinon	Housefly	$S_{100,000}$	GSH	PC	Lewis (1969)
Diazoxon	Housefly	$S_{100,000}$	GSH	PC	Lewis (1969)
Ethyl chlorothion	Housefly	$S_{100,000}$	GSH	—	Lewis and Lord (1969)
Ethyl dichlorvos	Mouse liver	$S_{177,000}$	GSH	—	Hollingworth (1969)
		Other *O*-Dealkylation			
Isopropyl paraoxon	Mouse liver	$S_{177,000}$	GSH	—	Hollingworth (1969)
		Dearylation			
Diazinon	Rat liver	$S_{105,000}$	GSH	PC, TLC, IEC, UV, IR	Fukami and Shishido (1966), Shishido *et al.* (1972)

	Cockroach fat body	$S_{105,000}$ (dialyzed)	GSH	PC, TLC, IEC, UV, IR	
	Housefly	$S_{100,000}$	GSH	PC	Lewis and Lord (1969)
	Housefly abdomen	$S_{100,000}$	GSH	IEC	Yang *et al.* (1971)
Diazoxon	Rat liver	$S_{105,000}$	GSH	PC, TLC, IEC, UV	Fukami and Shishido (1966)
	Cockroach				
	fat body	Homogenate	GSH	PC, TLC, IEC	Shishido *et al.* (1972)
	gut	Homogenate	GSH	—	
	muscle and cuticle	Homogenate	GSH	—	
	Housefly	$S_{100,000}$	GSH	PC	Lewis and Lord (1969)
	Housefly	$S_{100,000}$	GSH	IEC	Yang *et al.* (1971)
n-Propyl diazinon	Cockroach fat body	$S_{105,000}$ (dialyzed)	GSH	UV	Shishido *et al.* (1972)
Isopropyl diazinon	Cockroach fat body	$S_{105,000}$ (dialyzed)	GSH	UV	
Parathion	Rat liver	$S_{269,000}$ (desalted)	GSH	IEC	Nakatsugawa *et al.* (1969)
	Housefly	$S_{150,000}$	GSH	PC, IEC	Oppenoorth *et al.* (1972)
Paraoxon	Rat liver	$S_{176,000}$ (desalted)	GSH	TLC, electrophoresis, chemical tests	Hollingworth *et al.* (1973)
Methyl paraoxon	Rat liver	$S_{176,000}$ (desalted)	GSH	TLC, electrophoresis, chemical tests	

[a]Unless otherwise stated, an organism followed by no specification of tissues means the whole body is involved in the enzyme preparation.
[b]S means "supernatant"; the numbers refer to the *g* force.
[c]In studies where cofactors were added but served no known functions, it was considered that no cofactors were added.
[d]PC, paper chromatography; IEC, ion-exchange chromatography; TLC, thin-layer chromatography; UV, ultraviolet spectroscopy; IR, infrared spectroscopy.

and Shishido suggested that GSH might serve as an acceptor of the methyl group in the demethylation of methyl parathion. Their view was based on the suggested role of GSH in the metabolism of iodomethane in the rat (Johnson, 1966*a*,*b*). Later, Morello *et al.* (1967, 1968) and Hutson *et al.* (1968) confirmed this view by demonstrating chromatographically the formation of *S*-methylglutathione in the *in vitro* metabolism of *cis*-phosdrin and tetrachlorovinphos. Hollingworth (1969) also demonstrated the formation of *S*-methylglutathione in the *in vitro* demethylation of methyl paraoxon and its identity was established by a series of chromatographic, electrophoretic, and chemical tests. The involvement of GSH in the metabolism of organophosphates *in vivo* was shown by the depletion of GSH level in livers of mice dosed with Sumithion and Sumioxon and by the synergism of Sumithion toxicity by iodomethane (Hollingworth, 1969).

Hutson *et al.* (1972) have reported the detection of *S*-ethylglutathione chromatographically following the incubation of chlorofenvinphos with rabbit liver supernatant in the presence of GSH. Thus it is reasonable to assume that GSH may serve as an acceptor for ethyl or higher alkyl groups in the corresponding enzymatic dealkylation reactions.

In general, however, the enzyme is most active toward organophosphates of the dimethyl series. Although enzymatic dealkylation of diethyl and higher dialkyl organophosphates has been reported (Table 1), the reaction rates are usually quite low, often being only about 1/50 that of the demethylation reaction (Fukami and Shishido, 1966; Hollingworth, 1969; Hutson *et al.*, 1972). On the other hand, a high rate of deethylation of diazinon and diazoxon was reported by Lewis (1969) and by Lewis and Lord (1969), using a housefly soluble enzyme preparation fortified with GSH. Further, Nolan and O'Brien (1970) found deethylation to be a major pathway in the degradation *in vivo* of paraoxon in houseflies, and Appleton and Nakatsugawa (1972) have reported that monoethyl paraoxon is a major urinary metabolite of parathion in several strains of rats. Whether these discrepancies reflect variations among different strains of animals or are associated with the different substrates employed is not known.

The question concerning the number of alkyl groups of organophosphates removed during GSH-dependent dealkylation has been somewhat controversial. In 1969, Stenersen proposed a mechanism in which both methyl groups of bromophos were simultaneously cleaved by a GSH-dependent liver enzyme. However, this mechanism was not confirmed by other investigators and was later retracted by Stenersen (1971). It is now generally believed that only one alkyl group is transferred to GSH during the enzymatic process and that the second alkyl group is not removed until the insecticide molecule is further degraded (Bull, 1972).

The characterization and purification of the enzyme involved in the dealkylation of organophosphorus compounds were first attempted by Fukami and Shishido (1966). They achieved a fifteen- and tenfold purification of the GSH-*S*-alkyltransferase from rat liver and insect midgut tissues, respectively.

The partially purified enzymes from both sources required GSH for the reaction, and there appeared to be some quantitative and qualitative differences between the mammalian and the insect enzymes (Fukami and Shishido, 1966). More recently, Hutson *et al.* (1972) achieved a 45-fold purification of the GSH-*S*-alkyltransferase from the high-speed supernatant of pig liver. The enzyme was active toward a number of dialkyl organophosphates and was therefore referred to by these authors as "phosphoric acid triester-glutathione-*S*-alkyltransferase." Similar enzyme activities were also observed using liver preparations from mouse, rat, and rabbit. It was concluded that a single GSH-*S*-alkyltransferase with broad substrate specificity may be involved in the transfer of the methyl group from either iodomethane or dimethyl naphthyl phosphate.

b. GSH-Dependent Dearylation. Until quite recently, cleavage of the arylphosphate bond of many organophosphates was regarded as an exclusively hydrolytic process. In 1966, Fukami and Shishido first reported their preliminary findings concerning a GSH-dependent "hydrolysis" of diazinon and diazoxon by high-speed supernatants of rat liver and cockroach fat body homogenates. Subsequently GSH-dependent dearylation has been demonstrated with many different organophosphates using enzyme preparations from various animals (Hollingworth *et al.*, 1973; Lewis and Lord, 1969; Nakatsugawa *et al.*, 1969; Oppenoorth *et al.*, 1972; Shishido *et al.*, 1972; Yang *et al.*, 1971).

Elucidation of the mechanism involved in the GSH-dependent dearylation of organophosphates came from a key study by Shishido *et al.* (1972), who successfully isolated a glutathione conjugate following the incubation of diazinon or diazoxon with the high-speed supernatant of rat liver or cockroach fat body homogenate in the presence of GSH. Based on chromatographic and spectroscopic evidence, the conjugate was identified as *S*-(2-isopropyl-4-methyl-6-pyrimidinyl)glutathione. Thus the mechanism of GSH-dependent dearylation apparently involves the direct transfer of the aryl group of the organophosphate to GSH, and the enzyme responsible for this reaction is probably a GSH-*S*-aryltransferase. The subsequent isolation and identification of *S*-*p*-nitrophenylglutathione as a metabolite of parathion and its analogues (Hollingworth *et al.*, 1973) further substantiates this mechanism.

There appear to be some differences in the properties of the GSH-*S*-aryltransferase isolated in different laboratories. Shishido *et al.* (1972) reported that the dearylation of diazinon by a rat liver GSH-*S*-aryltransferase was most active around pH 6.0, while Hollingworth *et al.* (1973) reported that a similar rat liver enzyme preparation exhibited a broad pH optimum between 8.0 and 9.0 with respect to the dearylation of paraoxon. Whether or not there are a multiplicity of GSH-*S*-aryltransferases remains to be studied.

The interrelationship between GSH-*S*-alkyltransferase and GSH-*S*-aryltransferase with respect to their ability to attack different organophosphates is interesting. Motoyama and Dauterman (1972*a*) reported that while the high-speed supernatant of housefly abdomen homogenates exhibited both GSH-*S*-alkyltransferase and GSH-*S*-aryltransferase activities toward iodomethane and 3,4-dichloronitrobenzene, respectively, it was able to catalyze only the demethylation of azinphosmethyl. Hollingworth *et al.* (1973), using a rat liver preparation, found a competitive interaction between the two enzymes which depended on the nature of the alkyl and aryl moieties in the molecule. Thus they concluded that the rate of dearylation of methyl paraoxon was low because the GSH-*S*-alkyltransferase in the enzyme preparation degraded methyl paraoxon preferentially via demethylation. In the case of ethyl paraoxon, however, the dearylation mediated by GSH-*S*-aryltransferase was the predominant mechanism and very little deethylation activity was observed. When methyl paraoxon was incubated with a partially purified GSH-*S*-aryltransferase preparation which was relatively free of GSH-*S*-alkyltransferase activity, a high rate of dearylation was observed.

3.1.2. Other Conjugations

Other conjugation reactions include the formation of glucuronides, glucosides, and amino acid conjugates. Generally, these conjugations are preceded by oxidations, hydrolyses, or other reactions which are responsible for the primary attack on organophosphate compounds. However, there are exceptions where one or more of these conjugation reactions may occur directly with the organophosphorus insecticide (Bull, 1972). An interesting example was demonstrated by Bull and Ridgway (1969), who in the course of their studies on the metabolism of trichlorfon in animals and plants found that the insecticide was directly conjugated with glucuronic acid in the rat without prior biotransformation.

Table II represents a list of glucuronides, glucosides, and other conjugates formed in the metabolism of organophosphate insecticides by various organisms. In order to minimize possible confusion, only those studies in which the conjugates have been identified with some assurance have been included.

3.2. Carbamates

In the past, major emphasis in metabolic studies with the carbamate insecticides has been centered around hydrolytic and oxidative mechanisms. Although conjugates from various carbamates have been detected by many investigators in *in vivo* studies, the significance of conjugation reactions in the metabolism of carbamates has not been fully appreciated until recent years. In 1971, Mehendale and Dorough compiled experimental data from metabolic studies with several important carbamate insecticides and reported that as much as 78–91% of the parent compounds were conjugated in various forms by the rat within 48 hr after treatment. Some of the conjugates of carbamate insecticides are known to yield anticholinesterases on hydrolysis (Kuhr and

Table II. Formation of Glucoside, Glucuronide, and Other Conjugates in the Metabolism of Organophosphate Insecticides

Insecticidal chemicals	Organisms	Conjugates	References
Abate	Bean leaves	Glucosides of 4,4′-thiodiphenol, 4,4′-sulfinyldiphenol, 4,4′-sulfonyldiphenol	Blinn (1968)
Allied GC-6506[a]	Rat	Glucuronide or sulfate of substituted phenols	Bull and Stokes (1970)
	Tobacco budworm, cotton plants	Glucoside of substituted phenols	
Chlorfenvinphos	Rat, dog	1-(2,4-Dichlorophenyl)ethyl glucuronide; 2,4-dichlorophenyl-ethandiol glucuronide, glycine conjugate of 2,4-dichlorobenzoic acid	Hutson *et al.* (1967)
Dichlorvos	Rat	Dichloroethyl glucuronide	Casida *et al.* (1962)
Dicrotophos	Cotton	Glucoside of *N*-hydroxymethyl-dicrotophos	Bull and Lindquist (1964), Porter (1967)
Dimethoate	Rat, Cotton leaf worm	*N*-Methylcarbamoylmethyl glucuronide	Hassan *et al.* (1969), Zayed *et al.* (1970)
Monocrotophos	Cotton	*N*-Methylolmonocrotophos glucoside	Beynon *et al.* (1973), Lindquist and Bull (1967)
Tetrachlorvinphos	Rat	1-(2,4,5-Trichlorophenyl)ethyl glucuronide	Akintonwa and Hutson (1967)
	Rat, dog	2,4,5-Trichlorophenylethandiol glucuronide	
Trichlorofon	Rat	Glucuronide of trichlorofon	Bull and Ridgway (1969)
	Cotton leaf worm	Glucuronide of bisdemethyl trichlorfon	Hassan *et al.* (1965)

[a] Dimethyl-*p*-(methylthio)phenyl phosphate.

Casida, 1967), suggesting their formation from potentially toxic primary metabolites such as *N*-hydroxymethyl carbamates (Oonnithan and Casida, 1968). Consequently it is likely that at least some of these conjugation reactions are extremely important in enabling the organism to detoxify carbamate insecticides.

The involvement of conjugation mechanisms in carbamate metabolism in different organisms is summarized in Table III. Usually conjugation mechanisms act on carbamate metabolites arising from either hydrolytic cleavage or oxidation of the parent compound; only on rare occasions is direct conjugation of the parent carbamate observed (Knaak *et al.*, 1965; Mumma *et al.*, 1971). The major conjugates produced during carbamate metabolism are usually glucuronides, sulfates, and glucosides, and the formation of these three groups from carbaryl is summarized in Fig. 3.

3.3. Chlorinated Hydrocarbons

In recent years, considerable progress has been made in metabolic studies concerning chlorinated hydrocarbon insecticides. Many chemicals which were once regarded as metabolically inert have now been shown to be transformed into various metabolites in different organisms (Brooks, 1969, 1972). Of the known metabolites of chlorinated hydrocarbons, many bear hydroxy or other functional groups which are potential sites for conjugation. Unfortunately, little information is presently available concerning the involvement of conjugation mechanisms in the metabolism of this important group of insecticides.

The possible formation of conjugates (not identified) in rats treated with DDT was suggested as early as 1957 (Jensen *et al.*, 1957) and has subsequently been suggested by others (Heath and Vandekar, 1964; Metcalf *et al.*, 1970; Rowlands and Lloyd, 1969). There are, however, some cases where the formation of conjugates from chlorinated hydrocarbons has been demonstrated with reasonable assurance, and these are listed in Table IV.

Since there has been a considerable amount of controversy concerning the metabolism of γ-1,2,3,4,5,6-hexachlorocyclohexane (γ-HCH), particularly with respect to the initial stages of its biotransformation, special emphasis is given to this area of research in the following section. Furthermore, due to the possible involvement of DDT dehydrochlorinase or GSH-*S*-aryltransferase in the metabolism of γ-HCH, a brief account of the characteristics and properties of these enzymes is also given below.

3.3.1. Metabolism of γ-HCH

Based on the available information, the initial stage of metabolism of γ-HCH appeared to involve at least two possible mechanisms:

a. Monodehydrochlorination. Sternburg and Kearns (1956) first showed that γ-2,3,4,5,6-pentachlorocyclohex-1-ene (γ-PCCH) was an intermediate

Table III. Conjugates Formed in the Metabolism of Carbamate Insecticides

Insecticidal chemicals	Organisms	Conjugates	References
Banol	Rat	*N*-Glucuronide of Banol	Baron and Doherty (1967)
	Bean plant	Glucoside of hydroxylated Banol	Friedman and Lemin (1967)
	Bean plant	Glucoside of *N*-hydroxymethyl Banol	Kuhr and Casida (1967)
Baygon	Bean plant	Glucosides of *O*-depropyl, 4-hydroxy, and *N*-hydroxymethyl Baygon	Kuhr and Casida (1967)
	Housefly	Phosphate, glucoside, and sulfate of 5-hydroxy, *N*-hydroxymethyl, and *O*-depropyl Baygon[a]	Shrivastava *et al.* (1969)
Carbaryl	Cattle tick	Sulfate and/or glucoside of 1-napthol and 1,5-dihydroxy-naphthalene[a]	Bend *et al.* (1970)
	Guinea pig	1-Naphthyl methylcarbamate *N*-glucuronide (A)	Knaak *et al.* (1965)
		1-Naphthyl methylimidocarbonate *O*-glucuronide (B)	
		4-(Methylcarbamoyloxy)-1-naphthyl glucuronide (C)	
		4-(Methylcarbamoyloxy)-1-naphthyl sulfate (D)	
		1-Naphthyl glucuronide (E)	
		1-Naphthyl sulfate (F)	
	Rat	B, C, D, E, F	
	Man	E, F	
	Rat liver	B, C, E	
	Guinea pig liver	C, E	
	Pig	B, C, E	Knaak *et al.* (1968)
	Sheep	B, C, D, E, F	
	Man	C, E, F	
	Monkey	B, C, D	

Table III (cont.)

Insecticidal chemicals	Organisms	Conjugates	References
	Bean plant	Glucosides of 4-hydroxy, 5-hydroxy, *N*-hydroxymethyl, and dihydrodihydroxy carbaryl	Kuhr and Casida (1967)
	Rat liver	C, D, E, F, sulfate and glucuronide of 5,6-dihydro-1,5,6-trihydroxy-naphthalene, glucuronide of 5-hydroxy carbaryl	Leeling and Casida (1966)
	Rat liver	E, glucuronides of *N*-hydroxymethyl 5-hydroxy, and 6-hydroxy carbaryl	Mehendale and Dorough (1971)
	Bean, peas, pepper, corn plant	Glucosides of carbaryl, *N*-hydroxymethyl, 4-hydroxy, 5-hydroxy, 5,6-dihydro-5,6-dihydroxy carbaryl	Mumma *et al.* (1971)
	Chicken	E, F, sulfates of 4-hydroxy and 5-hydroxy carbaryl and others	Paulson *et al.* (1970)
Chevron RE 11775[b]	Rat	Glucuronides of 1- and 4-COOH-BPMC	Cheng and Casida (1973)
Furadan	Cotton leaf, salt marsh caterpillar larvae	Glucosides of 3-keto Furadan phenol, 3-hydroxy Furadan phenol, 3-hydroxy Furadan	Metcalf *et al.* (1968)
Landrin	Bean plant	Glucoside of (G) 3,4,5-trimethyl phenyl *N*-hydroxymethylcarbamate, (H) 4-hydroxymethyl-3,5-dimethyl phenyl methylcarbamate (I) 2,3,5-trimethylphenyl *N*-hydroxymethylcarbamate, (J) 3-hydroxymethyl-2,5-dimethylphenyl methylcarbamate, (K) 5-hydroxymethyl-2,3-dimethylphenyl methylcarbamate	Slade and Casida (1970)

	Housefly	Glucoside and/or sulfate of G, H, J, K,[a] and (L) 3-hydroxymethyl-4,5-dimethylphenyl methylcarbamate, (M) 3-carboxy-4,5-dimethylphenyl methylcarbamates	
	Mouse	Glucuronide and/or sulfate of H, L, M, 4-hydroxymethyl-3,5-dimethylphenyl, and 3,4,5-trimethylphenol[a]	
Meobal	Rat	Sulfate and glucuronide of 3,4-dimethylphenol, glucuronides of 3-hydroxymethyl-4-methylphenyl *N*-methylcarbamate, 3-methyl-4-carboxyphenol, 3-methyl-4-hydroxymethyl phenol, 3-hydroxymethyl-4-methyl phenol	Miyamoto (1970)
Mobam	Rat	4-Benzothienyl sulfate and glucuronides	Robbins *et al.* (1969)
	Goat, cow	4-Benzothienyl sulfate, 4-benzothienyl sulfate-1-oxide	Robbins *et al.* (1970)
SCNT[c]	Rat	Glucuronide and sulfate of the hydroxy thioacetimidate derivative	Hutson *et al.* (1971)

[a]The hydrolase preparation employed in the identification of conjugates contained two or more enzyme activities.
[b]Chevron RE 11775 = 3-(2-butyl)phenyl-*N*-benzenesulfenyl-*N*-methylcarbamate; BPMC = 3-(2-butyl)phenyl-*N*-methylcarbamate.
[c]SCNT = *S*-2-cyanoethyl-*N*-[(methylcarbamoyl)oxy]thioacetimidate.

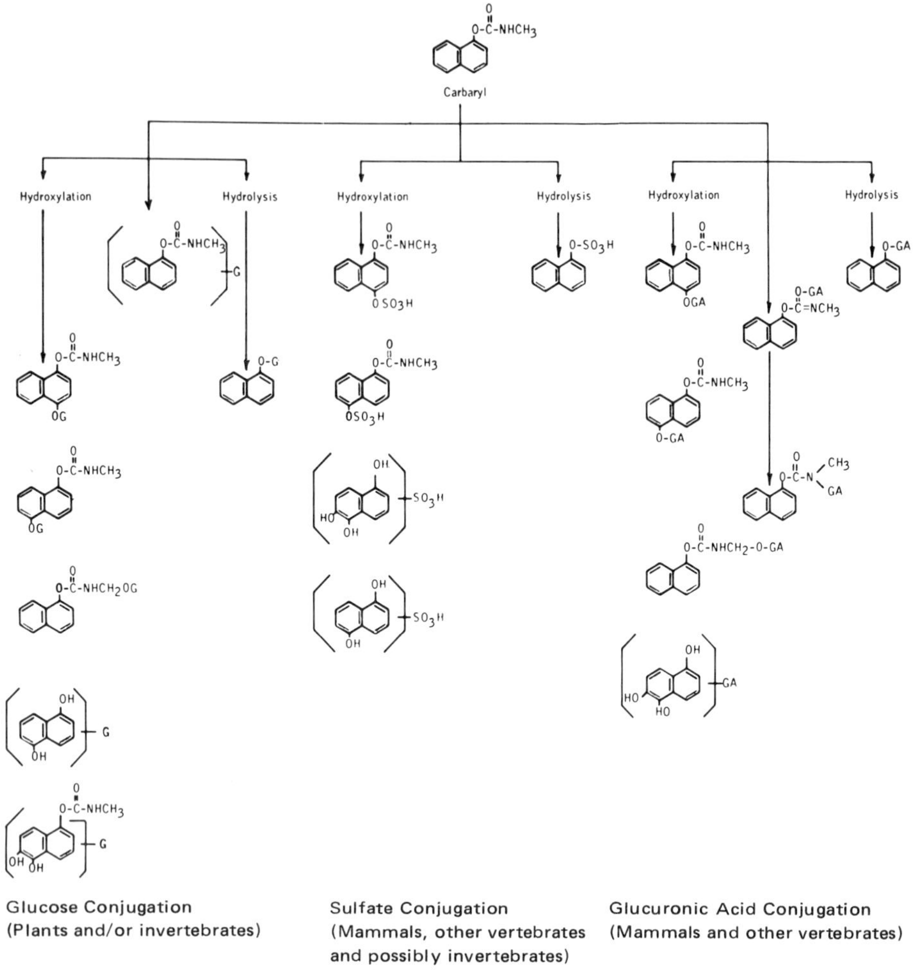

Fig. 3. Metabolic conjugation of carbaryl in animals and plants (Bend *et al.*, 1970; Knaak *et al.*, 1965, 1968; Kuhr and Casida, 1967; Leeling and Casida, 1966; Mehendale and Dorough, 1971; Mumma *et al.*, 1971; Paulson *et al.*, 1970). Abbreviations: GA, glucuronic acid; G, glucose.

of γ-HCH metabolism in houseflies. Subsequent studies (Bradbury and Standen, 1958) demonstrated that although γ-PCCH was formed in houseflies treated with γ-HCH the amount present never exceeded 3%, and in recent years evidence has been obtained indicating the formation of two additional isomers of γ-PCCH, which have been designated iso-PCCH and trito-PCCH (Freal and Chadwick, 1973; Reed and Forgash, 1968, 1970).

The enzymatic mechanism responsible for the monodehydrochlorination of γ-HCH has been the subject of much controversy. Ishida (1968) and Ishida and Dahm (1965*a*,*b*) reported that a partially purified housefly enzyme

Table IV. Conjugates Formed in the Metabolism of Chlorinated Hydrocarbon Insecticides

Insecticidal chemicals	Organisms	Conjugates	References
p,p'-DDT	Rat	A conjugated complex of DDA, serine, and aspartic acid	Pinto *et al.* (1965)
	Grain weevil	A complex glucoside of 3-hydroxy-4-chlorobenzoic acid, 4-hydroxybenzoic acid, and others	Rowlands and Lloyd (1969)
o,p'-DDT	Rat	Glycine and serine conjugates of *o,p'*-DDA	Feil *et al.* (1973)
Dieldrin (HEOD)	Rat liver, rabbit liver	Glucuronides of monohydroxylated dieldrin and *trans*-aldrindiol	Matthews and Matsumura (1969), Matthews *et al.* (1971)
α-HCH	Rat liver	Glutathione conjugate of an aromatic compound	Portig *et al.* (1973)
β-HCH	Mouse	Glucuronides and sulfates of 2,4,6-trichlorophenol, 2,4-dichlorophenol	Kurihara and Nakajima (1974)
γ-HCH	Housefly, tick, grass grub, locust	2,4-Dichlorophenylglutathione	Clark *et al.* (1966, 1969)
	Rat	Sulfates and glucuronides of 2,3,5-trichlorophenol and 2,4,5-trichlorophenol	Grover and Sims (1965)
	Mouse	Sulfates and glucuronides of 2,4-dichlorophenol and 2,4,6-trichlorophenol	Kurihara and Nakajima (1974)

preparation was active in catalyzing the dehydrochlorination of both γ-HCH and DDT in the presence of GSH. Their findings strongly implicated the involvement of DDT dehydrochlorinase in the initial metabolism of γ-HCH. Support for this view was provided by Sims and Grover (1965), who studied the enzymatic conjugation of various chlorocyclohexanes with GSH. While the housefly GSH-*S*-aryltransferase did not appear to be involved in the dehydrochlorination of γ-HCH, they found that appreciable amounts of Cl^- were released from γ-HCH by the housefly supernatant fraction without any enzymatic loss of GSH. Since GSH was not consumed in the reaction and no GSH conjugate of γ-HCH was detected, the involvement of DDT dehydrochlorinase appeared to be possible. Conflicting views were expressed from several different sources. Sternburg and Kearns (1956), while recognizing γ-PCCH as an intermediate in the metabolism of γ-HCH, claimed on the basis of few supporting data that DDT dehydrochlorinase was not involved in this reaction. Bradbury and Standen (1958) and Bridges (1959) questioned the importance of dehydrochlorination in the initial metabolism of γ-HCH since only small amounts of γ-PCCH was detected in houseflies treated with γ-HCH. However, their observation can be readily explained by the fact that γ-PCCH is rapidly degraded *in vivo* (Grover and Sims, 1965; Ishida and Dahm, 1965*a,b*; Reed and Forgash, 1970), rendering it a transitory intermediate. When attempts to trap ^{14}C-labeled γ-PCCH *in vivo* from [^{14}C] γ-HCH-treated houseflies failed (Bridges, 1959), it was attributed to the absence of dehydrochlorination of γ-HCH. This interpretation was later criticized by Clark *et al.* (1969) on the basis of uncertain *in vivo* conditions. Further evidence against the involvement of DDT dehydrochlorinase in the monodehydrochlorination of γ-HCH came from Clark *et al.* (1969), who employed selective inhibitors in studies of the metabolism of γ-HCH and γ-PCCH in various insect species. They demonstrated that bis-(*N,N*-dimethylaminophenyl)methane, a known DDT dehydrochlorinase inhibitor, while strongly inhibiting the metabolism of γ-PCCH in the blowfly (*Lucilia sericata*), was ineffective against the metabolism of γ-HCH in the same species. Thus these results suggest that the blowfly DDT dehydrochlorinase, instead of catalyzing the monodehydrochlorination of γ-HCH, is responsible for the metabolism of γ-PCCH—a mechanism previously suggested by Grover and Sims (1965) in a study of the metabolism of this compound. Whether there are some species variations in the metabolism of γ-HCH and γ-PCCH remains unknown.

b. Glutathione Conjugation. The second possible mechanism by which the initial metabolic attack on γ-HCH could occur may involve a GSH-dependent aryltransferase. In 1959, Bradbury and Standen showed that a housefly enzyme preparation, in the presence of GSH, converted γ-HCH to products which yielded thiophenols upon alkaline hydrolysis. They proposed a metabolic scheme for γ-HCH as follows:

$$C_6H_6Cl_6 + HSR \xrightarrow{-HCl} C_6H_6Cl_5SR \xrightarrow{H_2O} C_6H_3Cl_2SH + ROH + 3HCl$$

Since the compound HSR in the above scheme was suggested to be GSH, the possible formation of an intermediate *S*-pentachlorocyclohexyl glutathione ($C_6H_6Cl_5SR$) was indicated. In later studies by the same authors and by other investigators, the above hypothesis has been partially supported experimentally, although the presence of *S*-pentachlorocyclohexyl glutathione has never been demonstrated (Bradbury and Standen, 1960; Clark *et al.*, 1969).

Metabolic transformations of γ-HCH beyond the initial stage of monodehydrochlorination or glutathione conjugation may involve many different types of reactions, and various isomers of tetra-, tri-, and dichlorobenzenes as well as the corresponding chlorophenols have been shown to be formed *in vivo* from γ-HCH and/or γ-PCCH (Bradbury and Standen, 1959; Freal and Chadwick, 1973; Grover and Sims, 1965; Karapally *et al.*, 1973; Kurihara and Nakajima, 1974; Reed and Forgash, 1968, 1970). Some of these metabolites are reported to be excreted from animals as conjugates of glucuronic acid, sulfate, and glutathione (Clark *et al.*, 1966, 1969; Grover and Sims, 1965; Kurihara and Nakajima, 1974), whereas others are excreted unchanged. Although the precise sequence of metabolic transformations leading to the formation of these metabolites remains unknown, the involvement of microsomal mixed-function oxidases, UDP glucuronyltransferase, sulfotransferase, GSH-*S*-aryltransferase, and other enzymes appears quite certain.

3.3.2. DDT Dehydrochlorinase [1,1,1-Trichloro-2,2-bis-(p-chlorophenyl) ethane: Hydrogen-Chloride-Lyase; E.C. 4.5.1.1], GSH-S-Aryltransferase (Arylchloride: Glutathione S-Aryltransferase, E.C. 2.5.1.13), and "HCH-Metabolizing Enzymes"

During the last decade, there has been a considerable controversy concerning the true identities of the enzymes involved in the metabolism of DDT and γ-HCH. The terms "DDT dehydrochlorinase," "GSH-*S*-aryltransferase" and "HCH-metabolizing enzymes" (often erroneously referred to as "BHC-metabolizing enzymes") have appeared in some insecticide-related literature in a rather confusing manner. In certain cases, the experimental results appeared to suggest the existence of a nonspecific enzyme catalyzing each of the functions represented by the above designations (Ishida, 1968; Ishida and Dahm, 1965*b*). In other instances, DDT dehydrochlorinase and GSH-*S*-aryltransferase were clearly differentiated by various physicochemical methods (Balabaskaran, 1970; Balabaskaran and Smith, 1970; Goodchild and Smith, 1970). Table V summarizes the available information concerning the properties of DDT dehydrochlorinase and GSH-*S*-aryltransferase.

It may be concluded that DDT dehydrochlorinase is different from GSH-*S*-aryltransferase based on the following criteria: (1) different electrophoretic behavior (Goodchild and Smith, 1970), (2) different responses to specific inhibitors (Balabaskaran, 1970; Balabaskaran and Smith, 1970), (3) a different role of GSH in the reaction (Booth *et al.*, 1961; Dinamarca *et al.*, 1969, 1971; Lipke and Chalkley, 1962; Lipke and Kearns, 1959*a,b*), and (4) different

Table V. Properties of DDT Dehydrochlorinase (E.C. 4.5.1.1) and GSH-S-Aryltransferase (E.C. 2.5.1.13)

	E.C. 4.5.1.1	E.C. 2.5.1.13
Molecular weight	36,000[a] 120,000 (tetramer)[b]	36,000–40,000[c,d]
Type of protein	Lipoprotein[a]	—
pH optimum	7.4[a]	7.5–9.0[e]
K_m	5×10^{-7} M(DDT)[a]	2.5×10^{-3} M (3,4-Dichloronitrobenzene)[e]
Cofactor	GSH[a]	GSH[e]
Sulfhydryl groups	32/tetramer[f]	—
Disulfide bond	0[f]	—
Isoelectric point	~6.5[a]	—
Subcellular distribution	Soluble fraction[a]	Soluble fraction[e]
Enzyme source	Insect tissues[a,b,f]	Mammalian liver,[e,g] insect tissues and others

[a]Lipke and Kearns (1959*a*,*b*).
[b]Dinamarca *et al.* (1969).
[c]Ishida (1968).
[d]Goodchild and Smith (1970).
[e]Booth *et al.* (1961).
[f]Dinamarca *et al.* (1971).
[g]Boyland and Chasseaud (1969).

stability (Goodchild and Smith, 1970). In view of their very similar physical characteristics (Table V), it is obvious that the two enzymes may not readily be separated unless a rigorous enzyme purification scheme is followed. It is probable that much of the confusion between the enzymes involved in the metabolism of DDT and γ-HCH could be eliminated if highly purified homogeneous enzyme preparations were employed for further *in vitro* experiments.

The so-called HCH-metabolizing enzymes may be loosely referred to as an enzyme preparation which is capable of transforming various isomers of HCH and PCCH into hydrophilic compounds. It is apparent that such a general term may indicate the involvement of DDT dehydrochlorinase, GSH-*S*-aryltransferase, microsomal mixed-function oxidases, and conjugating enzymes.

3.4. Botanicals, Organothiocyanates, Insecticide Synergists, and Insect Hormones and Their Chemical Analogues

Among the many chemicals within the various groups listed in this section are some of the most widely used or most promising insecticides. As with other insecticides, the role of conjugation reactions in the metabolism of these compounds is relatively unexplored. In the following section, a brief account will be given of the metabolism of each group of chemicals with special reference to conjugation mechanisms. Conjugates formed in the metabolism of these chemicals are summarized in Table VI.

3.4.1. Botanicals

In recent years, the knowledge concerning the metabolism of pyrethroids and rotenoids has been greatly advanced, although as with most other groups of insecticides attention has been focused primarily on oxidative metabolism and relatively little attention given to conjugation. Miyamoto *et al.* (1968*a*) isolated several urinary glucuronides from rats treated with phthalthrin and tentatively identified one as the glucuronide of 3-hydroxycyclohexane-1,2-dicarboximide. In the following year, Yamamoto *et al.* (1969) demonstrated the formation in the housefly of glucosides of hydroxylated allethrin. They proposed that hydroxylation at the *trans*- and *cis*-methyl groups of the isobutenyl side chain in the acid moiety of allethrin followed by glucose conjugation was the major detoxication step. In later studies (Casida *et al.*, 1971; Elliott *et al.*, 1972), pyrethrins I and II were shown to be transformed by rats first into a common metabolite, pyrethrin ω_t-oic acid. This metabolite was subsequently modified at the *cis*-2′,4′-pentadienyl side chain to give a 4′,5′-diol, which was in turn conjugated with a *p*-methoxy aromatic acid at the 4′ position. The precise identity of the conjugating agent and the possible mechanism involved in this conjugation reaction were not discussed. In addition to the above conjugate, designated as pyrethrin metabolite C (PMC), glucuronide and sulfate conjugates of unidentified metabolites were probably present in the urine and feces as more polar metabolites (Elliott *et al.*, 1972).

Little or no information is presently available concerning the existence of conjugates of rotenoids or of their metabolites. Rotenone is known to be transformed oxidatively to various hydroxylated metabolites in mammals, fish, and insects (Fukami *et al.*, 1967, 1969), and it is likely that these constitute substrates for conjugation reactions. In this connection, it is possible that the large portion of unidentified water-soluble metabolites of rotenone observed *in vivo* and *in vitro* by Fukami *et al.* (1969) might in fact be conjugated products.

There is a general lack of information concerning the formation of conjugates of nicotinoids or of their metabolites, although there is a real possibility that this occurs.

3.4.2. Organothiocyanates

A number of organothiocyanates are useful in the control of household, veterinary, and agricultural pests, and their toxic action is believed to be caused by the *in vivo* liberation of hydrogen cyanide. The enzymatic mechanisms involved in this biotransformation have been studied by Ohkawa and Casida (1971) and Ohkawa *et al.* (1972), and GSH-*S*-transferases has been shown to play an important role. As shown in the following scheme (Ohkawa and Casida, 1971), the enzyme-mediated attack by GSH could occur either at the cyano group or at the thiocyanate sulfur, resulting in the liberation of hydrogen cyanide:

$$RCH_2SCN + GSH \nearrow [GSCN] + RCH_2SH;\quad [GSCN] \xrightarrow{GSH} HCN + GSSG$$

$$RCH_2SCN + GSH \searrow [RCH_2SSG] + HCN;\quad [RCH_2SSG] \xrightarrow{GSH} RCH_2SH + GSSG$$

This GSH-*S*-transferase-mediated reaction is unique in that it represents an intoxication process due to the formation of hydrogen cyanide.

3.4.3. Insecticide Synergists

Of the many types of insecticide synergists, the group of compounds containing the methylenedioxyphenyl moiety is the best known and most extensively studied (Casida, 1970; Wilkinson, 1971). In recent years, considerable attention has been given to the metabolism *in vivo* of the methylenedioxyphenyl compounds and various conjugates have been successfully isolated (Table VI). In addition to the formation of glucuronides, glucosides, and sulfates, amino acid conjugation may also be important in the metabolism of this group of insecticide synergists. Conjugation is usually preceded by oxidation at either the methylenedioxyphenyl moiety or the side chain, the latter often being a major route of detoxication. Consequently, conjugates formed from methylenedioxyphenyl compounds often contain the piperonylic acid moiety. An excellent example of this was reported by Esaac and Casida (1968), who demonstrated that living houseflies transform tropital, piperonyl alcohol, piperonal, and safrole to a common intermediate, piperonylic acid, which is then conjugated with alanine, glycine, glutamate, serine, or glutamine. The mechanism by which some of the amino acid conjugations take place are not clear, although the interconversion of amino acids was shown to occur (Esaac and Casida, 1968). The metabolism of a carbamate synergist, 2-propynyl 1-naphthyl ether, in mice and houseflies, was studied by Sacher *et al.* (1968). The formation of glucuronide and glucoside conjugates of naphtholic metabolites was demonstrated in mice and houseflies respectively.

3.4.4. Insect Hormones and Their Chemical Analogues

The two principal hormones involved in the growth and development of insects are the molting and the juvenile hormones. The possible involvement of conjugation reactions in the metabolism of insect hormones was not explored

Table VI. Conjugates Formed in the Metabolism of Botanicals, Organothiocyanates, Insecticide Synergists, and Insect Hormones and Their Chemical Analogues

Chemicals	Organisms	Conjugates	References
Botanicals			
Allethrin	Housefly	Glucoside of allethrin-ω_t-ol and allethrin-ω_c-ol	Yamamoto *et al.* (1969)
Pyrethrin I Pyrethrin II	Rat	Aromatic acid conjugate of pyrethrin ω_t-oic acid derivative	Casida *et al.* (1971), Elliott *et al.* (1972)
Phthalthrin	Rat	Glucuronide of 3-hydroxycyclohexane-1,2-dicarboximide and others	Miyamoto *et al.* (1968*a*)
Organothiocyanates			
Octyl thiocyanate	Mouse liver	Octyl mercaptan	Ohkawa and Casida (1971)
Benzyl thiocyanate	Mouse liver	Dibenzyl disulfide	
Insecticide synergists			
Isosafrole	Housefly	*N*-Piperonyl glycine (A) *N*-Piperonyl serine (B) *N*-Piperonyl glutamate (C)	Esaac and Casida (1969)
Safrole	Housefly	A, B, C, and *N*-piperonyl alanine (D), *N*-piperonyl glutamine (E)	Esaac and Casida (1968, 1969)
Piperonal	Housefly	A, B, C, D, E and β-glucoside of piperonyl alcohol	Esaac and Casida (1968, 1969)
Tropital	Housefly	A, B, C, D, E and β-glucoside of piperonyl alcohol	Esaac and Casida (1968, 1969)
	Mice, rat, hamster	Glycine conjugate and glucuronide of piperonylic acid	Kamienski and Casida (1970)
Piperonyl butoxide	Mice	Glycine conjugate of 6-propyl piperonylic acid	Kamienski and Casida (1970)
	Housefly	Glucoside of 6-propyl piperonylic acid, glucoside-6-phosphate of 6-propyl piperonylic acid	Esaac and Casida (1969)

Table VI (cont.)

Chemicals	Organisms	Conjugates	References
2,3-Methylene-dioxynaphthalene	Housefly	Glucoside of 1-hydroxy-2,3-methylene-dioxynaphthalene, 2-hydroxy-3-naphthyl glucoside, 2-hydroxy-3-naphthyl sulfate	Sacher *et al.* (1969)
	Mice	Glucuronide of 1-hydroxy-2,3-methylenedioxynaphthalene, 2-hydroxy-3-naphthyl glucuronide, 2-hydroxy-3-naphthyl sulfate	
2-Propynyl-1-naphthyl ether	Housefly	Glucoside of a naphtholic derivative	Sacher *et al.* (1968)
	Mice	1-naphthyl glucuronide	
Insect hormones and chemical analogues			
α-Ecdysone	Tobacco hornworm fat body	α-Glucoside of α-, β-, and 26-hydroxy-β-ecdysone	King (1972)
	Tobacco hornworm	Sulfated α-, β-, and 26-hydroxy-β-ecdysone	King (1972)
	Armyworm gut tissue	Sulfate of α-ecdysone	Yang and Wilkinson (1972)
20-Hydroxyecdysone	Blowfly fat body	α-Glucoside of 20-hydroxyecdysone	Heinrich and Hoffmeister (1970)
Ponasterone A	Blowfly fat body	α-Glucoside of ponasterone A	Heinrich and Hoffmeister (1970)
22,25-Bisdeoxyecdysone	Tobacco hornworm	Sulfates and/or glucosides of α-ecdysone, 22-deoxyecdysone, and 22,25-bisdeoxyecdysone	Kaplanis *et al.* (1972, 1974), Thompson *et al.* (1972, 1973)
	Housefly	β-Glucosides of tetrahydroxy, pentahydroxy steroids, and 22,25-bisdeoxyecdysone	Kaplanis *et al.* (1974), Thompson *et al.* (1972, 1973)
	Housefly	Sulfate of 22,25-bisdeoxyecdysone	Yang and Wilkinson (1973)
	Honeybee		
	worker	Sulfate of 22,25-bisdeoxyecdysone	Yang and Wilkinson (1973)
	drone	Sulfate of 22,25-bisdeoxyecdysone	Yang and Wilkinson (1973)
	Madagascar cockroach	Sulfate of 22,25-bisdeoxyecdysone	Yang and Wilkinson (1973)

	Cecropia moth	Sulfate of 22,25-bisdeoxyecdysone	Yang and Wilkinson (1973)
	Monarch butterfly	Sulfate of 22,25-bisdeoxyecdysone	Yang and Wilkinson (1973)
	Black swallowtail	Sulfate of 22,25-bisdeoxyecdysone	Yang and Wilkinson (1973)
	Armyworm	Sulfate of 22,25-bisdeoxyecdysone	Yang and Wilkinson (1973)
	Tobacco hornworm	Sulfate of 22,25-bisdeoxyecdysone	Yang and Wilkinson (1973)
Crecropia juvenile hormone	Fleshfly	Sulfate and/or glucoside of diol ester	Slade and Zibitt (1971, 1972)
	Armyworm	Sulfate of acid diol	Slade and Wilkinson (1974)
	Locust	Glucosides of diol ester, epoxy acid, and acid diol	White (1972)
Stauffer R 20458[a]	Rat	Sulfates and glucuronides of 4-acetylphenol and 4-hydroxybenzoic acid, glucuronide of 1-(4′-hydroxyphenoxy)-6,7-dihydroxy-3,7-dimethyl-2-octene, glucuronide of 1-(4′-carboxyphenoxy)-2,3,6,7-tetrahydroxy-3,7-dimethyloctene	Hoffman *et al.* (1973)

[a] 1-(4′-Ethylphenoxy)-6,7-epoxy-3,7-dimethyl-2-octene.

until quite recently. Heinrich and Hoffmeister (1970) first demonstrated the glucosylation *in vitro* of 20-hydroxyecdysone and ponasterone A by insect adipose tissues, and subsequently the formation of conjugates of ecdysone, juvenile hormones, and several synthetic analogues has been reported in various species (Ajami and Riddiford, 1973; Hoffman *et al.*, 1973; Kaplanis *et al.*, 1972, 1974; King, 1972; Slade and Wilkinson, 1974; Slade and Zibitt, 1971, 1972; Thompson *et al.*, 1972; White, 1972; Willig *et al.*, 1971; Yang and Wilkinson, 1972, 1973).

The physiological role of conjugation in the metabolism of insect growth hormones is of considerable interest. While conjugates of juvenile hormone metabolites are considered to be mainly excretory products (Slade and Wilkinson, 1974), the formation of conjugates from the molting hormones has been suggested by various investigators to have important physiological implications. Willig *et al.* (1971) demonstrated that isolated ring glands from blowfly larvae converted [^{14}C]cholesterol into molting hormones in the form of α-glucoside and ester derivatives and suggested that synthesis in the form of biologically inactive conjugates (prehormones) might be of importance for storage and transport. Such inactive prehormones could be reactivated later by hydrolysis (Willig *et al.*, 1971). A similar insect endocrine control mechanism was suggested by Yang *et al.* (1973) and Yang and Wilkinson (1972, 1973). These authors proposed that an insect sulfotransferase system acting in concert with arylsulfatase might represent a mechanism for the regulation of insect steroids in general and more specifically for the deactivation and activation of insect molting hormones (Yang *et al.*, 1973; Yang and Wilkinson, 1972, 1973). Their hypothesis was based on findings from various laboratories: (1) the isolation and characterization from the larvae of the southern armyworm of a sulfotransferase system which catalyzes the sulfation of α-ecdysone, 22,25-bisdeoxyecdysone, cholesterol, β-sitosterol, campesterol, and other animal and plant steroids (Yang and Wilkinson, 1972); (2) the wide occurrence of the sulfotransferase system in insects (Yang and Wilkinson, 1973); (3) the age dependency of arylsulfatase and sulfotransferase activities during larval development of the southern armyworm, arylsulfatase titer attaining a maximum during the molt and a minimum between molts and sulfotransferase following the inverse course (Yang *et al.*, 1973); (4) the isolation and identification of sulfate conjugates of cholesterol, β-sitosterol, and campesterol *in vivo* in tobacco hornworm (Hutchin and Kaplanis, 1969) and the reported function of these steroids as precursors in the biosynthesis of α-ecdysone in insects (Robbins *et al.*, 1971; Thompson *et al.*, 1972, 1973; Willig *et al.*, 1971); and (5) the periodic rise and fall in the titer of molting hormone at certain stages of insect development (Burdette, 1962; Shaaya and Karlson, 1965*a,b*; Kaplanis *et al.*, 1966).

The possible involvement of conjugation in the regulation, transport, and storage of insect molting hormones has also been independently suggested by other investigators (Thompson *et al.*, 1973; Kaplanis *et al.*, 1974).

4. Significance of Conjugation Mechanisms

4.1. Possible Intrinsic Physiological Roles and Toxicological Significance

Despite the recent advances in our understanding of conjugation mechanisms, little is yet known concerning the precise roles of these reactions in living organisms. Undoubtedly detoxication is one important function of conjugation since the generally high hydrophilicity of most conjugates allows those produced from potentially hazardous materials to be rapidly excreted from living organisms. Nevertheless, evidence is accumulating which indicates the participation of conjugation reactions in important metabolic functions (Dodgson and Rose, 1970; Layne, 1970; Roy, 1970; Wang and Bulbrook, 1968). The possibility that conjugation reactions can serve to temporarily inactivate endogenous compounds such as steroids by masking critical functional groups or by modifying the physicochemical properties (solubility) on which their physiological distribution depends is of considerable interest and importance. The presence of a multiplicity of conjugate hydrolases, often in different subcellular compartments from the conjugating enzymes (Dodgson and Rose, 1970; Van Lancker, 1970), provides an interesting basis for speculation on the possibility of regulatory mechanisms at the subcellular level, and it is conceivable that several enzyme pairs (β-glucuronidase–glucuronyl transferase, arylsulfatase–sulfotransferase, etc.) acting in concert could serve to maintain the critical balance between biologically active and inactive forms of a particular material. Indeed, such regulatory mechanisms have been suggested with respect to steroid hormones in mammals and in insects (Dodgson and Rose, 1970; Kaplanis *et al.*, 1974; Roy, 1970; Thompson *et al.*, 1973; Wang and Bulbrook, 1968; Willig *et al.*, 1971; Yang *et al.*, 1973; Yang and Wilkinson, 1972, 1973).

The potential toxicological significance of conjugation reactions may be considered from several different viewpoints. First, the intoxication processes such as the liberation of hydrogen cyanide from thiocyanates (Section 3.4.2) and the activation of *N*-hydroxy compounds (Section 4.2) are receiving increasing recognition and are of obvious toxicological importance. Especially in the latter case, an entirely new area of research in chemical carcinogenesis is just starting to be explored. In view of the vast number of chemicals which may be conjugated, it is entirely possible that new conjugates possessing significant biological activity will be discovered.

With respect to the role of conjugation in insecticide detoxication, the differential rate of dealkylation and dearylation of organophosphorus insecticides by insect and mammalian GSH transferases provides one example of the potential importance of conjugation in determining selective toxicity. Other conjugation reactions, although preceded by oxidations, hydrolyses, etc., may also be of considerable toxicological importance, and, in view of the significant biological activity often retained by these primary metabolites, could in some cases constitute the major step leading to detoxication. At present, little is known concerning the possible consequences to an organism of blocking or otherwise disrupting its conjugation systems. Hollingworth (1969) demonstrated the *in vivo* potentiation of Sumithion toxicity in mice following pretreatment with iodomethane, and Culver *et al.* (1970) reported that the metabolism of carbaryl in rat was reduced *in vivo* and *in vitro* in the presence of certain inhibitors of monoamine oxidase. Similar effects have been reported on the overall metabolism of carbaryl in rat liver preparation by lowering the level of UDPGA in the incubation system (Mehendale and Dorough, 1971). These findings suggest that detrimental effects might result if conjugation mechanisms were blocked and the overall detoxication capacity of the organism altered. Finally, the investigation of intrinsic physiological roles of conjugation mechanisms raises interesting possibilities regarding the exploration of biochemical targets for the design and development of new and selective insecticides.

4.2. Conjugation of N-Hydroxy Compounds and Chemical Carcinogenesis

During the last few years, the conjugation of *N*-hydroxy compounds has attracted considerable interest because of the discovery of the unexpectedly high reactivity of certain *N*-hydroxy conjugates toward nucleophilic centers in nucleic acids and proteins (Heidelberger, 1973; Irving, 1970, 1971; King and Phillips, 1968, 1969; Miller, 1970; Miller and Miller, 1966). Among the chemicals studied, one notable example is that of 2-acetylaminofluorene (AAF), where conjugation constitutes an activation reaction and gives rise to the ultimate carcinogen (Irving, 1970, 1971; Miller, 1970). The overall reaction sequence is shown in Fig. 4. AAF is first hydroxylated by the microsomal mixed-function oxidase system, and the resulting *N*-hydroxy-AAF (which is not reactive toward nucleophils) is subsequently conjugated with glucuronic acid, sulfate, or phosphate to form reactive electrophils. These conjugates are the ultimate carcinogens which react to form covalent bonds with critical cellular macromolecules, such as nucleic acid or protein, and eventually lead to carcinogenesis.

The above findings not only stress the inadequacy of the classical detoxication concept of conjugation but also provide us with an extremely stimulating area for further scientific exploration.

5. Conclusion

Despite the fact that enzymatic conjugations were among the first biochemical reactions to be studied, the foregoing discussion clearly indicates

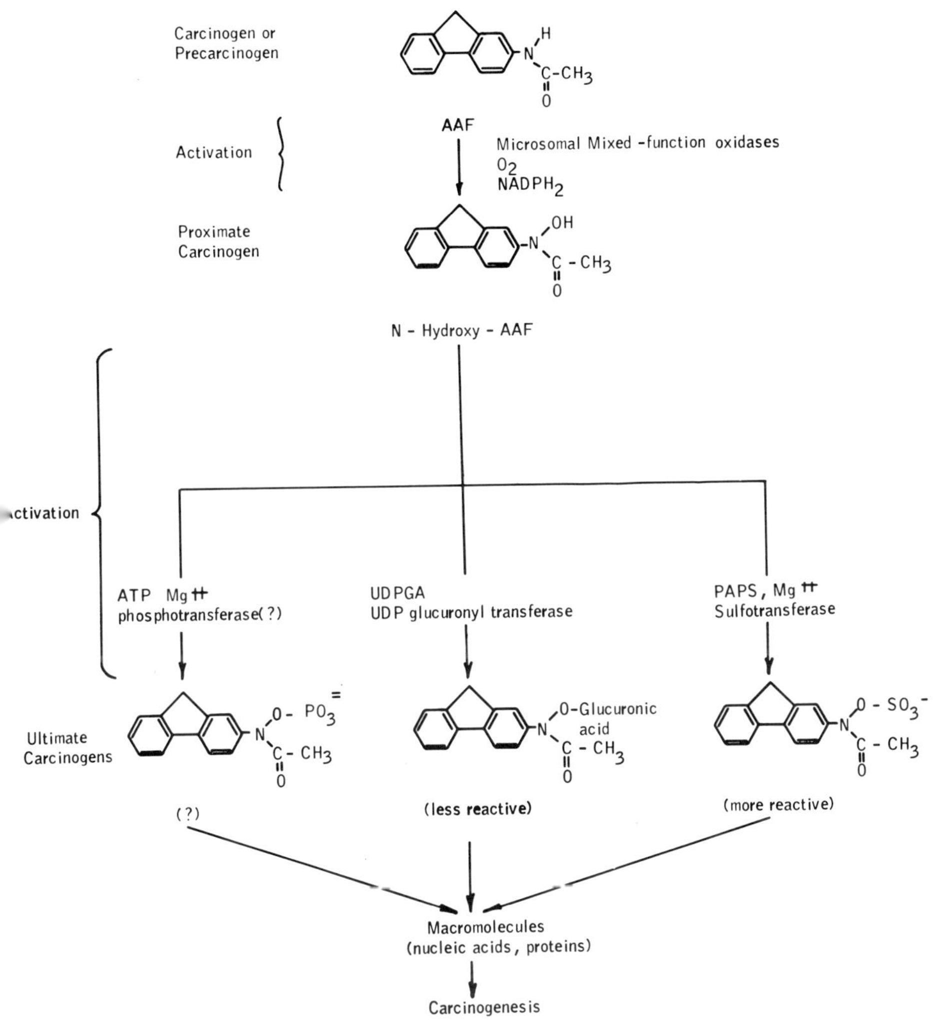

Fig. 4. Metabolic activation of 2-acetylaminofluorene and chemical carcinogenesis (Heidelberger, 1973; Irving, 1970, 1971; King and Phillips, 1968, 1969; Miller, 1970; Miller and Miller, 1966).

our current lack of understanding in this area. It is probable that future studies in which the enzymes are purified and characterized in different species will add considerably to our knowledge of mechanisms, substrate specificity, and the degree of coupling with other enzymes such as the mixed-function oxidases and hydrolases. Only then will we begin to understand the intrinsic physiological role of conjugation in the metabolism of endogenous substrates and its toxicological significance in the metabolisms of insecticides and other foreign compounds.

ACKNOWLEDGMENTS

I am grateful to Drs. J. L. Byard, F. Coulston, L. Golberg, and K. A. Pittman of this Institute for reviewing the manuscript, and to Ms. Rosie Kobryn for her skillful preparation of the manuscript.

6. References

Ajami, A. M., and Riddiford, L. M., 1973, Comparative metabolism of the cecropia juvenile hormone, *J. Insect Physiol.* **19:**635.

Akintonwa, D. A. A., and Hutson, D. H., 1967, Metabolism of 2-chloro-1-(2,4,5-trichlorophenyl) vinyl dimethyl phosphate in the dog and rat, *J. Agr. Food Chem.* **15:**632.

Appleton, H. T., and Nakatsugawa, T., 1972, Paraoxon deethylation in the metabolism of parathion, *Pestic. Biochem. Physiol.* **2:**286.

Balabaskaran, S., 1970, Insecticide metabolism and its inhibition in relation to resistance, *Proc. 23rd N.Z. Weed and Pest Control Conf.*, pp. 184–190.

Balabaskaran, S., and Smith, J. N., 1970, The inhibition of 1,1,1-trichloro-2,2-bis-(*p*-chlorophenyl)ethane (DDT) dehydrochlorinase and glutathione *S*-aryltransferase in grass-grub and housefly preparations, *Biochem. J.* **117:**989.

Baron, R. L., and Doherty, J. D., 1967, Metabolism and excretion of an insecticide (6-chloro-3,4-dimethylphenyl *N*-methylcarbamate) in the rat, *J. Agr. Food Chem.* **15:**830.

Bend, J. R., Holder, G. M., Protos, E., and Ryan, A. J., 1970, The metabolism of carbaryl in the cattle tick *Boophilus microplus* (Canestrini), *Aust. J. Biol. Sci.* **23:**361.

Beynon, K. I., Hutson, D. H., and Wright, A. N., 1973, The metabolism and degradation of vinyl phosphate insecticides, *Residue Rev.* **47:**55.

Binning, A., Darby, F. J., Heenan, M. P., and Smith, J. N., 1967, The conjugation of phenols with phosphate in grass grubs and flies, *Biochem. J.* **103:**42.

Blinn, R. C., 1968, The fat of *O,O,O′,O′*-tetramethyl *O,O′*-thiodi-*p*-phenylene phosphorothioate on bean leaves, *J. Agr. Food Chem.* **16:**441.

Booth, J., Boyland, E., and Sims, P., 1961, An enzyme from rat liver catalyzing conjugations with glutathione, *Biochem. J.* **79:**516.

Boyland, E., and Chasseaud, L. F., 1969, The rate of glutathione and glutathione *S*-transferase in mercapturic acid biosynthesis, *Advan. Enzymol.* **32:**173.

Boyland, E., Kinder, C. H., and Manson, D., 1961, Synthesis and detection of di-(2-amino-1-naphthyl) hydrogen phosphate, a metabolite of 2-naphthylamine in dogs, *Biochem. J.* **78:**175.

Bradbury, F. R., and Standen, H., 1958, The fate of γ-benzene hexachloride in resistant and susceptible houseflies, III, *J. Sci. Food Agr.* **9:**203.

Bradbury, F. R., and Standen, H., 1959, Metabolism of benzene hexachloride by resistant houseflies *Musca domestica, Nature (London)* **183:**983.

Bradbury, F. R., and Standen, H., 1960, Mechanism of insect resistance to the chlorohydrocarbon insecticides, *J. Sci. Food Agr.* **11:**92.

Bridges, R. G., 1959, Pentachlorocyclohexene as a possible intermediate metabolite of benzene hexachloride in houseflies, *Nature (London)* **184:**1337.

Brooks, G. T., 1969, The metabolism of diene-organochlorine (cyclodiene) insecticides, *Residue Rev.* **27:**81.

Brooks, G. T., 1972, Pathways of enzymatic degradation of pesticides, *in: Environmental Quality and Safety, Chemistry, Toxicology and Technology*, Vol. I (F. Coulston and F. Korte, eds.), pp. 106–164, Academic Press, New York.

Bull, D. L., 1972, Metabolism of organophosphorus insecticides in animals and plants, *Residue Rev.* **43:**1.

Bull, D. L., and Lindquist, D. A., 1964, Metabolism of 3-hydroxy-*N,N*-dimethyl crotonamide dimethyl phosphate by cotton plants, insects, and rats, *J. Agr. Food Chem.* **12:**310.

Bull, D. L., and Ridgway, R. L., 1969, Metabolism of trichlorfon in animals and plants, *J. Agr. Food Chem.* **17:**837.

Bull, D. L., and Stokes, R. A., 1970, Metabolism of dimethyl *p*-(methylthio)phenyl phosphate in animals and plants, *J. Agr. Food Chem.* **18:**1134.

Burdette, W. J., 1962, Changes in titer of ecdysone in *Bombyx mori* during metamorphosis, *Science* **135:**432.

Casida, J. E., 1970, Mixed-function oxidase involvement in the biochemistry of insecticide synergists, *J. Agr. Food Chem.* **18:**753.

Casida, J. E., McBride, L., and Niedermeier, R. P., 1962, Metabolism of 2,2-dichlorovinyl dimethyl phosphate in relation to residues in milk and mammalian tissues, *J. Agr. Food Chem.* **10:**370.

Casida, J. E., Kimmel, E. C., Elliott, M., and Janes, N. F., 1971, Oxidative metabolism of pyrethrins in mammals, *Nature (London)* **230:**326.

Cheng, H. M., and Casida, J. E., 1973, Metabolites and photoproducts of 3-(2-butyl) phenyl *N*-methylcarbamate and *N*-benzenesulfenyl-*N*-methylcarbamate, *J. Agr. Food Chem.* **21:**1037.

Clark, A. G., Hitchcock, M., and Smith, J. N., 1966, The metabolism of gammexane in flies, ticks and locusts, *Nature (London)* **209:**103.

Clark, A. G., Murphy, S., and Smith, J. N., 1969, The metabolism of hexachlorocyclohexanes and pentachlorocyclohexenes in flies and grass grubs, *Biochem. J.* **113:**89.

Collins, D. C., Williamson, D. G., and Layne, D. S., 1970, Steroid glucosides, enzymatic synthesis by a partially purified transferase from rabbit liver microsomes, *J. Biol. Chem.* **245:**873.

Compernolle, F., Van Hees, G. P., Fevery, J., and Heirwegh, K. P. M., 1971, Mass-spectrometric structure elucidation of dog bile azopigments as the acyl glucosides of glucopyranose and xylopyranose, *Biochem. J.* **125:**811.

Culver, D. J., Lin, T., and Dorough, H. W., 1970, The effect of monoamine oxidase inhibitors on carbaryl metabolism, *J. Econ. Entomol.* **63:**1369.

Dahm, P. A., 1970, Some aspects of the metabolism of parathion and diazinon, *in: Biochemical Toxicology of Insecticides* (R. D. O'Brien and I. Yamamoto, eds.), pp. 51–63, Academic Press, New York.

Darby, F. J., Heenan, M. P., and Smith, J. N., 1966, The absence of glucuronide conjugates from 1-naphthol dosed flies and grass grubs; detection of 1-naphthylphosphate, *Life Sci.* **5:**1499.

Dauterman, W. C., 1971, Biological and nonbiological modifications of organophosphorus compounds, *Bull. WHO* **44:**133.

Dicowsky, L., and Morello, A., 1971, Glutathione-dependent degradation of 2,2-dichlorovinyl dimethyl phosphate (DDVP) by the rat, *Life Sci.* **10:**1031.

Dinamarca, M. L., Saavedra, I., and Valdes, E., 1969, DDT-dehydrochlorinase. I. Purification and characterization, *Comp. Biochem. Physiol.* **31:**269.

Dinamarca, M. L., Levenbook, L., and Valdes, E., 1971, DDT-dehydrochlorinase. II. Subunits, sulfhydryl groups, and chemical composition, *Arch. Biochem. Biophys.* **147:**374.

Dodgson, K. S., and Rose, F. A., 1970, Sulfoconjugation and sulfohydrolysis *in: Metabolic Conjugation and Metabolic Hydrolysis*, Vol. I (W. H. Fishman, ed.), pp. 239–325, Academic Press, New York.

Donninger, C., Hutson, H. D., and Pickering, B. A., 1966, Oxidative cleavage of phosphoric acid triesters to diesters, *Biochem. J.* **102:**26.

Downing, D. F., 1962, The chemistry of the psychotomimetic substances, *Quart. Rev. (London)* **16:**133.

Dutton, G. J., 1962, The mechanism of *o*-aminophenyl glucoside formation in *Periplaneta americana*, *Comp. Biochem. Physiol.* **7:**39.

Dutton, G. J., 1965, Conjugation of phenols in molluscs: Demonstration of glucosyl- not glucuronyl-transfer from uridine diphosphate nucleotides, *Biochem. J.* **96:**36p.

Dutton, G. J., 1966*a*, The biosynthesis of glucuronides, *in*: *Glucuronic Acid, Free and Combined: Chemistry, Biochemistry, Pharmacology and Medicine* (G. J. Dutton, ed.), pp. 185–299, Academic Press, New York.

Dutton, G. J., 1966*b*, Uridine diphosphate glucose and the synthesis of phenolic glucosides by mollusks, *Arch. Biochem. Biophys.* **116:**399.

Elliott, M., Janes, N. F., Kimmel, E. C., and Casida, J. E., 1972, Metabolic fate of pyrethrin I, pyrethrin II, and allethrin administered orally to rats, *J. Agr. Food Chem.* **20:**300.

Esaac, E. G., and Casida, J. E., 1968, Piperonylic acid conjugates with alanine, glutamate, glutamine, glycine and serine in living houseflies, *J. Insect Phydiol.* **14:**913.

Esaac, E. G., and Casida, J. E., 1969, Metabolism in relation to mode of action of methylenedioxyphenyl synergists in houseflies, *J. Agr. Food Chem.* **17:**539.

Feil, V. J., Lamoureux, C. J. H., Styrvoky, E., Zaylskie, R. G., Thacker, E. J., and Holman, G. M., 1973, Metabolism of *o,p'*-DDT in rats, *J. Agr. Food Chem.* **21:**1072.

Fevery, J., Van Hees, G. P., Leroy, P., Compernolle, F., and Heirwegh, K. P.M., 1971, Excretion in dog bile of glucose and xylose conjugates of bilirubin, *Biochem. J.* **125:**803.

Fevery, J., Leroy, P., and Heirwegh, K. P. M., 1972, Enzymic transfer of glucose and xylose from uridine diphosphate glucose and uridine diphosphate xylose to bilirubin by untreated and digitonin-activated preparations from rat liver, *Biochem. J.* **129:**619.

Fishman, W. H., 1970, *Metabolic Conjugation and Metabolic Hydrolysis*, Vols. I and II, Academic Press, New York.

Florkin, M., and Stotz, E. H., 1973, *Comprehensive Biochemistry*, 3rd ed., Vol. 13, Elsevier, New York.

Freal, J., and Chadwick, R. W., 1973, Metabolism of hexachlorocyclohexane to chlorophenols and effect of isomer pretreatment on lindane metabolism in rat, *J. Agr. Food Chem.* **21:**424.

Friedman, A. R., and Lemin, A. J., 1967, Metabolism of 2-chloro-4,5-dimethylphenyl *N*-methylcarbamate in bean plants, *J. Agr. Food Chem.* **15:**642.

Fukami, J., and Shishido, T., 1963, Studies on the selective toxicities of organic phosphorous insecticides (III). The characters of enzyme system in cleavage of methyl parathion to desmethyl parathion in the supernatant of several species of homogenates (Part I), *Botyu-Kagaku* **28:**77.

Fukami, J., and Shishido, T., 1966, Nature of a soluble, glutathione-dependent enzyme system active in cleavage of methyl parathion to desmethyl parathion, *J. Econ. Entomol.* **59:**1338.

Fukami, J., Yamamoto, I., and Casida, J., 1967, Metabolism of rotenone *in vitro* by tissue homogenates from mammals and insects, *Science* **155:**713.

Fukami, J., Shishido, T., Fukunaga, K., and Casida, J. E., 1969, Oxidative metabolism of rotenone in mammals, fish and insects and its relation to selective toxicity, *J. Agr. Food Chem.* **17:**1217.

Fukunaga, K., Fukami, J., and Shishido, T., 1969, The *in vitro* metabolism of organophosphorus insecticides by tissue homogenates from mammal and insect, *Residue Rev.* **25:**223.

Fukuto, T. R., 1972, Metabolism of carbamate insecticides, *Drug Metab. Rev.* **1:**117.

Gessner, T., and Acara, M., 1968, Metabolism of thiols, *S*-glucosylation, *J. Biol. Chem.* **243:**3142.

Gessner, T., and Hamada, N., 1970, Identification of *p*-nitrophenyl glucoside as a urinary metabolite, *J. Pharm. Sci.* **59:**1528.

Gessner, T., and Vollmer, C. A., 1969, Glucosylation by mouse liver microsomes, *Fed. Proc.* **28:**545.

Gessner, T., Jacknowitz, A., and Vollmer, C. A., 1973, Studies of mammalian glucoside conjugation, *Biochem. J.* **132:**249.

Gil, D. L., Rose, H. A., Yang, R. S. H., Young, R. G., and Wilkinson, C. F., 1974, Enzyme induction by phenobarbital in the Madagascar cockroach, *Gromphadorhina portentosa*, *Comp. Biochem. Physiol.* **47B:**657.

Goodchild, B., and Smith, J. N., 1970, The separation of multiple forms of housefly 1,1,1-trichloro-2,2-bis-(*p*-chlorophenyl) ethane (DDT) dehydrochlorinase from glutathione *S*-aryltransferase by electrofocusing and electrophoresis, *Biochem. J.* **117:**1005.

Grover, P. L., and Sims, P., 1965, The metabolism of γ-2,3,4,5,6-pentachlorocyclohex-1-ene and γ-hexachlorocyclohexane in rats, *Biochem. J.* **96:**521.

Harborne, J. B., 1964, Phenolic glucosides and their natural distribution, *in: Biochemistry of Phenolic Compound* (J. B. Harborne, ed.), pp. 129–169, Academic Press, New York.
Hassan, A., Zayed, S. M. A. D., and Abdel-Hamid, F. M., 1965, Metabolism of organophosphorus insecticides. V. Mechanism of detoxification of dipterex in *Prodenia litura* F., *Biochem. Pharmacol.* **14:**1577.
Hassan, A., Zayed, S. M. A. D., and Bahig, M. R. E., 1969, Metabolism of organophosphorus insecticides. XI. Metabolic fate of dimethoate in the rat, *Biochem. Pharmacol.* **18:**2429.
Heath, D. F., and Vandekar, M., 1964, Toxicity and metabolism of dieldrin in rats, *Brit. J. Ind. Med.* **21:**269.
Heenan, M. P., and Smith, J. N., 1967, Conjugation of 1-naphthol and *p*-nitrophenol in flies: Formation of glucoside-6-phosphate, *Life Sci.* **6:**1753.
Heidelberger, C., 1973, Current trends in chemical carcinogenesis, *Fed. Proc.* **32:**2154.
Heinrich, G., and Hoffmeister, H., 1970, Bildung von Hormonglykosiden als Inaktivierungsmechanismus bei *Calliphora erythrocephala*, *Z. Naturforsch. B* **25:**358.
Heirwegh, K. P. M., Van Hees, G. P., Comperonlle, F., and Fevery, J., 1970, Excretion of bilirubin conjugated with glucose in dog bile, *Biochem. J.* **120:**17p.
Heirwegh, K. P. M., Meuwissen, J. A. T. P., and Fevery, J., 1971, Enzymic formation of β-D-monoglucuronide, β-D-monoglucoside and mixtures of β-D-monoxyloside and β-D-dixyloside of bilirubin by microsomal preparations from rat liver, *Biochem. J.* **125:**28p.
Hodgson, E., and Casida, J. E., 1962, Mammalian enzymes involved in the degradation of 2,2-dichlorovinyl dimethyl phosphate, *J. Agr. Food Chem.* **10:**208.
Hoffman, L. J., Ross, J. H., and Menn, J. J., 1973, Metabolism of 1-(4′-ethylphenoxy)-6,7-epoxy-3,7-dimethyl-2-octene (R 20458) in the rat, *J. Agr. Food Chem.* **21:**156.
Hollingworth, R. M., 1969, Dealkylation of organophosphorus esters by mouse liver enzymes *in vitro* and *in vivo*, *J. Agr. Food Chem.* **17:**987.
Hollingworth, R. M., 1970, The dealkylation of organophosphorus triesters by liver enzymes, *in: Biochemical Toxicology of Insecticides* (R. D. O'Brien and I. Yamamoto, eds.), pp. 75–92, Academic Press, New York.
Hollingworth, R. M., 1971, Comparative metabolism and selectivity of organophosphate and carbamate insecticides, *Bull. WHO* **44:**155.
Hollingworth, R. M., Alstott, R. L., and Litzenberg, R. D., 1973, Glutathione *S*-aryl transferase in the metabolism of parathion and its analogs, *Life Sci.* **13:**191.
Hutchin, R. F. N., and Kaplanis, J. N., 1969, Sterol sulfates in an insect, *Steroids* **13:**605.
Hutson, D. H., 1972, Mechanisms of biotransformation, *in: Foreign Compounds Metabolism in Mammals*, Vol. 1 (D. E. Hathway, senior reporter), pp. 314–395, The Chemical Society, London.
Hutson, D. H., 1972, Mechanisms of biotransformation, *in: Foreign Compounds: Metabolism in Mammals*, Vol. 2 (D. E. Hathway, senior reporter), pp. 328–397, The Chemical Society, London.
Hutson, D. H., Akintowa, D. A. A., and Hathway, D. E., 1967, The metabolism of 2-chloro-1-(2′,4′-dichlorophenyl) vinyl diethyl phosphate in the dog and rat, *Biochem. J.* **102:**133.
Hutson, D. H., Pickering, B. A., and Donninger, C., 1968, Phosphoric acid triester: Glutathione alkyl transferase, *Biochem. J.* **106:**20p.
Hutson, D. H., Hoadley, E. C., and Pickering, B. A., 1971, The metabolism of *S*-2-cyanoethyl-*N*-[(methylcarbamoyl)oxy]thio-acetimidate, an insecticidal carbamate, in the rat, *Xenobiotica* **1:**179.
Hutson, D. H., Pickering, B. A., and Donninger, C., 1972, Phosphoric acid triester: Glutathione alkyltransferase, a mechanism for the detoxification of dimethyl phosphate triesters, *Biochem. J.* **127:**285.
Illing, H. P. A., and Dutton, G. J., 1970, Observations on the biosynthesis of thioglucuronides and thioglucosides in vertebrates and molluscs, *Biochem. J.* **120:**16p.
Irving, C. C., 1970, Conjugates of *N*-hydroxy compounds, *in: Metabolic Conjugation and Metabolic Hydrolysis*, Vol. I (W. H. Fishman, ed.), pp. 53–119, Academic Press, New York.

Irving, C. C., 1971, Metabolic activation of *N*-hydroxy compounds by conjugation, *Xenobiotica* **1:**387.

Ishida, M., 1968, Comparative studies on BHC metabolizing enzymes, DDT dehydrochlorinase and glutathione *S*-transferases, *Agr. Biol. Chem.* **32:**947.

Ishida, M., and Dahm, P. A., 1965*a*, Metabolism of benzene hexachloride isomers and related compounds *in vitro*. I. Properties and distribution of the enzyme, *J. Econ. Entomol.* **58:**383.

Ishida, M., and Dahm, P. A., 1965*b*, Metabolism of benzene hexachloride isomers and related compounds *in vitro*. II. Purification and stereospecificity of housefly enzymes, *J. Econ. Entomol.* **58:**602.

Jensen, J. A., Cueto, C., Dale, W. E., Rothe, C. F., Pearce, G. W., and Mattson, A. M., 1957, DDT metabolites in feces and bile of rats, *J. Agr. Food Chem.* **5:**919.

Johnson, M. K., 1966*a*, Metabolism of iodomethane in the rat, *Biochem. J.* **98:**38.

Johnson, M. K., 1966*b*, Studies on glutathione *S*-alkyltransferase of the rat, *Biochem. J.* **98:**44.

Jordan, T. W., McNaught, R. W., and Smith, J. N., 1970, Detoxications in peripatus, sulphate, phosphate and histidine conjugations, *Biochem. J.* **118:**1.

Kamienski, F. X., and Casida, J. E., 1970, Importance of demethylenation in the metabolism *in vivo* and *in vitro* of methylenedioxyphenyl synergists and related compounds in mammals, *Biochem. Pharmacol.* **19:**91.

Kaplanis, J. N., Thompson, M. J., Yamamoto, R. T., Robbins, W. E., and Loloudes, S. J., 1966, Ecdysones from the pupa of the tobacco hornworm, *Manduca sexta*, *Steroids* **8:**605.

Kaplanis, J. N., Thompson, M. J., Dutky, S. R., Robbins, W. E., and Lindquist, E. L., 1972, Metabolism of [4-^{14}C]-22,25-bisdeoxyecdysone during larval development in the tobacco hornworm, *Manduca sexta*, *Steroids* **20:**105.

Kaplanis, J. N., Dutky, S. R., Robbins, W. E., and Thompson, M. J., 1974, The metabolism of the synthetic ecdysone analog, 1α-^{3}H-22,25-bisdeoxyecdysone in relation to certain of its biological effects in insects, *in: Invertebrate Endocrinology and Hormonal Heterophylly* (W. J. Burdette, ed.), pp. 161–175, Springer, New York.

Karapally, J. C., Saha, J. G., and Lee, Y. W., 1973, Metabolism of lindane-^{14}C in the rabbit: Ether-soluble urinary metabolites, *J. Agr. Food Chem.* **21:**811.

King, C. M., and Phillips, B., 1968, Enzyme-catalyzed reactions of the carcinogen *N*-hydroxy-2-fluorenylacetamide with nucleic acid, *Science* **159:**1351.

King, C. M., and Phillips, B., 1969, Enzymatic activation of *N*-hydroxy-2-fluorenylacetamide (N-OH-FAA): Structures of nucleic acid adducts, *Proc. Am. Assoc. Cancer Res.* **10:**46.

King, D. S., 1972, Metabolism of α-ecdysone and possible immediate precursors by insects *in vivo* and *in vitro*, *Gen. Comp. Endocrinol. Suppl.* **3:**221.

Knaak, J. B., 1971, Biological and nonbiological modifications of carbamates, *Bull. WHO* **44:**121.

Knaak, J. B., Tallant, M. J., Bartley, W. J., and Sullivan, L. J., 1965, The metabolism of carbaryl in the rat, guinea pig, and man, *J. Agr. Food Chem.* **13:**537.

Knaak, J. B., Tallant, M. J., Kozbelt, S. J., and Sullivan, L. J., 1968, The metabolism of carbaryl in man, monkey, pig, and sheep, *J. Agr. Food Chem.* **16:**465.

Kojima, K., and O'Brien, R. D., 1968, Paraoxon hydrolyzing enzymes in rat liver, *J. Agr. Food Chem.* **16:**574.

Kuenzle, C. C., 1970, Bilirubin conjugates of human bile: The excretion of bilirubin as the acyl glycosides of aldobiouronic acid, pseudoaldobiouronic acid and hexuronosylhexuronic acid, with a branched-chain hexuronic acid as one of the components of the hexuronosylhexuronide, *Biochem. J.* **119:**411.

Kuhr, R. J., and Casida, J. E., 1967, Persistent glycosides of metabolites of methylcarbamate insecticide chemicals formed by hydroxylation in bean plants, *J. Agr. Food Chem.* **15:**814.

Kurihara, N., and Nakajima, M., 1974, Studies on BHC isomers and related compounds. VIII. Urinary metabolites produced from γ- and β-BHC in the mouse: Chlorophenol conjugates, *Pestic. Biochem. Physiol.* **4:**220.

Labow, R. S., and Layne, D. S., 1972, The formation of glucosides of isoflavones and of some other phenols by rabbit liver microsomal fractions, *Biochem. J.* **128:**491.

Layne, D. S., 1970, New metabolic conjugates of steroids, *in: Metabolic Conjugation and Metabolic Hydrolysis*, Vol. I (W. H. Fishman, ed.), pp. 21–52, Academic Press, New York.

Leeling, N. C., and Casida, J. E., 1966, Metabolites of carbaryl (1-naphthyl methylcarbamate) in mammals and enzymatic systems for their formation, *J. Agr. Food Chem.* **14:**281.

Lewis, J. B., 1969, Detoxification of diazinon by subcellular fractions of diazinon-resistant and susceptible houseflies, *Nature (London)* **224:**917.

Lewis, J. B., and Lord, K. A., 1969, Metabolism of some organophosphorus insecticides by strains of housefly, *Proc. 5th Brit. Insecticide Fungicide Conf.*, pp. 465–471.

Lindquist, D. A., and Bull, D. L., 1967, Fate of 3-hydroxy-*N*-methyl-*cis*-crotonamide dimethyl phosphate in cotton plants, *J. Agr. Food Chem.* **15:**267.

Lipke, H., and Chalkley, J., 1962, Glutathione, oxidized and reduced, in some dipterans treated with 1,1,1-trichloro-2,2-di-(*p-chlorophenyl)ethane*, *Biochem. J.* **85:**104.

Lipke, H., and Kearns, C. W., 1959*a*, DDT dehydrochlorinase. I. Isolation, chemical properties and spectrophotometric assay, *J. Biol. Chem.* **234:**2123.

Lipke, H., and Kearns, C. W., 1959*b*, DDT dehydrochlorinase. II. Substrate and cofactor specificity, *J. Biol. Chem.* **234:**2129.

Mandel, H. G., 1971, Pathways of drug biotransformation: Biochemical conjugations, *in: Fundamentals of Drug Metabolism and Drug Disposition* (B. N. LaDu, H. G. Mandel, and E. L. Way, eds.), pp. 149–186, Williams and Wilkins, Baltimore.

Matsumura, F., and Ward, C. T., 1966, Degradation of insecticides by the human and the rat liver: Metabolic fate of carbaryl, malathion, and parathion, *Arch. Environ. Health* **13:**257.

Matthews, H. B., and Matsumura, F., 1969, Metabolic fate of dieldrin in the rat, *J. Agr. Food Chem.* **17:**845.

Matthews, H. B., McKinney, J. D., and Lucier, G. W., 1971, Dieldrin metabolism, excretion, and storage in male and female rats, *J. Agr. Food Chem.* **19:**1244.

Mehendale, H. M., and Dorough, H. W., 1971, Glucuronidation mechanisms in the rat and their significance in the metabolism of insecticides, *Pestic. Biochem. Physiol.* **1:**307.

Mehendale, H. M., and Dorough, H. W., 1972*a*, *In vitro* glucosylation of 1-naphthol by insects, *J. Insect Physiol.* **18:**981.

Mehendale, H. M., and Dorough, H. W., 1972*b*, Conjugative metabolism and action of carbamate insecticides, *in: Insecticide* (Proc. 2nd Int. IUPAC Congr. Pesticide Chem.), Vol. I (A. S. Tahori, ed.), pp. 15–28, Gordon and Breach, New York.

Metcalf, R. L., Fukuto, T. R., Collins, C., Borck, K., Abd El-Aziz, S., Munoz, R., and Cassil, C. C., 1968, Metabolism of 2,2-dimethyl-2,3-dihydrobenzofuranyl-7 *N*-methylcarbamate (Furadan) in plants, insects, and mammals, *J. Agr. Food Chem.* **16:**300.

Metcalf, R. L., Kapoor, I. P., Nystrom, R. F., and Sangha, G. K., 1970, Comparative metabolism of methoxychlor, methiochlor and DDT in mouse, insects and in a model ecosystem, *J. Agr. Food Chem.* **18:**1145.

Miettinen, T. A., and Leskinen, E., 1970, Glucuronic acid pathway, *in: Metabolic Conjugation and Metabolic Hydrolysis*, Vol. I (W. H. Fishman, ed.), pp. 157–237, Academic Press, New York.

Miller, E. C., and Miller, J. A., 1966, Mechanisms of chemical carcinogenesis: Nature of proximate carcinogens and interactions with macromolecules, *Pharmacol. Rev.* **18:**805.

Miller, J. A., 1970, Carcinogenesis by chemicals: An overview, *Cancer Res.* **30:**559.

Mitchell, H. K., and Lunan, K. D., 1964, Tyrosine-*O*-phosphate in *Drosophila*, *Arch. Biochem. Biophys.* **106:**219.

Miyamoto, J., 1970, A feature of detoxication of carbamate insecticide in mammals, *in: Biochemical Toxicology of Insecticides* (R. D. O'Brien and I. Yamamoto, eds.), pp. 115–130, Academic Press, New York.

Miyamoto, J., Sato, Y., Yamamoto, K., Endo, M., and Suzuki, S. I., 1968*a*, Biochemical studies on the mode of action of pyrethroidal insecticides. Part I. Metabolic fate of phthalthrin in mammals, *Agr. Biol. Chem.* **32:**628.

Miyamoto, J., Sato, Y., Yamamoto, K., and Suzuki, S., 1968*b*, Activation and degradation of Sumithion, methyl parathion and their oxygen analogs by mammalian enzymes *in vitro*, *Botyu-Kagaku* **33:**1.

Morello, A., Vardanis, A., and Spencer, E. Y., 1967, Differential glutathione-dependent detoxication of two geometrical vinyl organophosphorus (mevinphos) isomers, *Biochem. Biophys. Res. Commun.* **29:**241.

Morello, A., Vardanis, A., and Spencer, E. Y., 1968, Mechanism of detoxication of some organophosphorus compounds, the role of glutathione-dependent demethylation, *Can. J. Biochem.* **46:**885.

Morikawa, O., and Saito, T., 1966, Degradation of vamidothion and dimethoate in plants, insects and mammals, *Botyu-Kagaku* **31:**130.

Motoyama, N., and Dauterman, W. C., 1972*a*, *In vitro* metabolism of azinphosmethyl in susceptible and resistant houseflies, *Pestic. Biochem. Physiol.* **2:**113.

Motoyama, N., and Dauterman, W. C., 1972*b*, The *in vitro* metabolism of azinphosmethyl by mouse liver, *Pestic. Biochem. Physiol.* **2:**170.

Motoyama, N., Rock, G. C., and Dauterman, W. C., 1971, Studies on the mechanism of azinphosmethyl resistance in the predaceous mite, *Neoseiulus* (T.) *fallacis* (family: Phytoseiidae), *Pestic. Biochem. Physiol.* **1:**205.

Mumma, R. O., Khalifa, S., and Hamilton, R. H., 1971, Spectroscopic identification of metabolites of carbaryl in plants, *J. Agr. Food Chem.* **19:**445.

Nakatsugawa, T., and Dahm, P. A., 1967, Microsomal metabolism of parathion, *Biochem. Pharmacol.* **16:**25.

Nakatsugawa, T., Tolman, N. M., and Dahm, P. A., 1969, Degradation of parathion in the rat, *Biochem. Pharmacol.* **18:**1103.

Neal, R. A., 1967, Studies on the metabolism of diethyl 4-nitrophenyl phosphorothionate (parathion) *in vitro*, *Biochem. J.* **103:**183.

Nolan, J., and O'Brien, R. D., 1970, Biochemistry of resistance to paraoxon in strains of houseflies, *J. Agr. Food Chem.* **18:**802.

O'Brien, R. D., 1967, *Insecticides, Action and Metabolism*, Academic Press, New York.

Ohkawa, H., and Casida, J. E., 1971, Glutathione *S*-transferases liberate hydrogen cyanide from organic thiocyanates, *Biochem. Pharmacol.* **20:**1708.

Ohkawa, H., Ohkawa, R., Yamamoto, I., and Casida, J. E., 1972, Enzymatic mechanisms and toxicological significance of hydrogen cyanide liberation from various organothiocyanates and organonitriles in mice and houseflies, *Pestic. Biochem. Physiol.* **2:**95.

Oonnithan, E. S., and Casida, J. E., 1968, Oxidation of methyl and dimethylcarbamate insecticide chemicals by microsomal enzymes and anticholinesterase activity of the metabolites, *J. Agric. Food Chem.* **16:**28.

Oppenoorth, F. J., Rupes, V., ElBashir, S., Houx, N. W. H., and Voerman, S., 1972, Glutathione-dependent degradation of parathion and its significance for resistance in the housefly, *Pestic. Biochem. Physiol.* **2:**262.

Parke, D. V., 1968, *The Biochemistry of Foreign Compounds*, Pergamon Press, London.

Paulson, G. D., Zaylskie, R. G., Zehr, M. V., Portnoy, C. E., and Feil, V. J., 1970, Metabolites of carbaryl (1-naphthyl methylcarbamate) in chicken urine, *J. Agr. Food Chem.* **18:**110.

Pinto, J. D., Camien, M. N., and Dunn, M. S., 1965, Metabolic fate of *p,p'*-DDT [1,1,1-trichloro-2,2-bis(*p*-chlorophenyl)ethane] in rats, *J. Biol. Chem.* **240:**2148.

Plapp, F. W., Jr., 1973, Comparison of insecticide absorption and detoxification in larvae of the bollworm, *Heliothis zea*, and the tobacco budworm, *H. virescens*, *Pestic. Biochem. Physiol.* **2:**447.

Porter, P. E., 1967, Bidrin insecticide, *in: Analytical Methods for Pesticides, Plant Growth Regulators and Food Additives*, Vol. V (G. Zweig, ed.), p. 213, Academic Press, New York.

Portig, J., Kraus, P., Sodomann, S., and Noack, G., 1973, Biodegradation of alpha-hexachlorocyclohexane. I. Glutathione-dependent conversion to a hydrophilic metabolite by rat liver cytosol, *Naunyn-Schmiedeberg's Arch. Pharmacol.* **279:**185.

Pridham, J. B., 1964, The phenol glucosylation reaction in the plant kingdom, *Phytochemistry* **3:**493.

Quick, A. J., 1928, Quantitative studies of β-oxidation. I. The conjugation of benzoic acid and phenylacetic acid as the end-products from the oxidation of phenyl-substituted aliphatic acids, *J. Biol. Chem.* **77:**581.

Rao, S. L. N., and McKinley, W. P., 1969, Metabolism of organophosphorus insecticides by liver homogenates from different species, *Can. J. Biochem.* **47:**1155.

Reed, W. T., and Forgash, A. J., 1968, Lindane: Metabolism to a new isomer of pentachlorocyclohexene, *Science* **160:**1232.

Reed, W. T., and Forgash, A. J., 1970, Metabolism of lindane to organic-soluble products by houseflies, *J. Agr. Food Chem.* **18:**475.

Robbins, J. D., Bakke, J. E., and Feil, V. J., 1969, Metabolism of 4-benzothienyl *N*-methylcarbamate (Mobam) in rats, balance study and urinary metabolite separation, *J. Agr. Food Chem.* **17:**236.

Robbins, J. D., Bakke, J. E., and Feil, V. J., 1970, Metabolism of benzo[*b*]thien-4-yl methylcarbamate (Mobam) in dairy goats and a lactating cow, *J. Agr. Food Chem.* **18:**130.

Robbins, W. E., Kaplanis, J. N., Svoboda, J. A., and Thompson, M. J., 1971, Steroid metabolism in insects, *Ann. Rev. Entomol.* **16:**53.

Rowlands, D. G., and Lloyd, C. J., 1969, DDT metabolism in susceptible and pyrethrin-resistant *Sitophilus granarius* (L.), *J. Stored Prod. Res.* **5:**413.

Roy, A. B., 1970, Enzymological aspects of steroid conjugation, *in: Chemical and Biological Aspects of Steroid Conjugation* (S. Bernstein and S. Solomon, eds.), pp. 74–130, Springer, New York.

Roy, A. B., and Trudinger, P. A., 1970, *The Biochemistry of Inorganic Compounds of Sulfur*, Cambridge University Press, Cambridge.

Sacher, R. M., Metcalf, R. L., and Fukuto, T. R., 1968, Propynyl naphthyl ethers as selective carbamate synergists, *J. Agr. Food Chem.* **16:**779.

Sacher, R. M., Metcalf, R. L., and Fukuto, T. R., 1969, Selectivity of carbaryl-2,3-methylenedioxynaphthalene combination: Metabolism of the syngerist in two strains of houseflies and in mice, *J. Agr. Food Chem.* **17:**551.

Schachter, B., and Marrian, G. F., 1938, The isolation of estrone sulfate from the urine of pregnant mares, *J. Biol. Chem.* **126:**663.

Shaaya, E., and Karlson, P., 1965*a*, Der Ecdysontiter während der Insektenentwicklung. II. Die postembryonale Entwicklung der Schmeissfliege *Calliphora erythrocephala* Meig, *J. Insect Physiol.* **11:**65.

Shaaya, E., and Karlson, P., 1965*b*, Der Ecdysontiter während der Insektenentwicklung. IV. Die entwicklung der Lepidopteren *Bombyx mori* L. und *Cerura vinula* L., *Develop. Biol.* **11:**424.

Shishido, T., and Fukami, J., 1963, Studies on the selective toxicities of organic phosphorus insecticides (II). The degradation of ethyl parathion, methyl parathion, methyl paraoxon and sumithion in mammal, insect and plant, *Botyu-Kagaku* **28:**69.

Shishido, T., Usui, K., Sato, M., and Fukami, J., 1972, Enzymatic conjugation of diazinon with glutathione in rat and American cockroach, *Pestic. Biochem. Physiol.* **2:**51.

Shrivastava, S. P., Tsukamoto, M., and Casida, J. E., 1969, Oxidative metabolism of ^{14}C-labeled baygon by living houseflies and by housefly enzyme preparations, *J. Econ. Entomol.* **62:**483.

Sims, P., and Grover, P. L., 1965, Conjugations with glutathione: The enzymic conjugation of some chlorocyclohexenes, *Biochem. J.* **95:**156.

Slade, M., and Casida, J. E., 1970, Metabolic fate of 3,4,5- and 2,3,5-trimethylphenyl methylcarbamates, the major constituents in Landrin insecticide, *J. Agr. Food Chem.* **18:**467.

Slade, M., and Wilkinson, C. F., 1974, Degradation and conjugation of cecropia juvenile hormone by the southern armyworm (*Prodenia eridania*), *Comp. Biochem. Physiol.* **49B:**99.

Slade, M., and Zibitt, C. H., 1971, Metabolism of cecropia juvenile hormone in lepidopterans, *in: Chemical Releasers in Insects* (Proc. 2nd Int. IUPAC Congr. Pestic. Chem.), Vol. III (A. S. Tahori, ed.), pp. 45–58, Gordon and Breach, New York.

Slade, M., and Zibitt, C. H., 1972, Metabolism of cecropia juvenile hormone in insects and in mammals, *in: Insect Juvenile Hormones: Chemistry and Action* (J. J. Menn and M. Beroza, eds.), pp. 155–176, Academic Press, New York.

Smith, J. N., 1964, Comparative biochemistry of detoxification, *Comp. Biochem.* **6:**403.

Smith, J. N., 1968, The comparative metabolism of xenobiotics, *Advan. Comp. Physiol. Biochem.* **3:**173.

Smith, J. N., and Turbert, H. B., 1961, Enzymic glucoside synthesis in locusts, *Nature* (*London*) **189:**600.

Smith, R. L., and Williams, R. T., 1970, History of the discovery of the conjugation mechanisms, *in: Metabolic Conjugation and Metabolic Hydrolysis*, Vol. I (W. H. Fishman, ed.), pp. 1–19, Academic Press, New York.

Stenersen, J., 1969, Demethylation of the insecticide bromophos by a glutathione-dependent liver enzyme and by alkaline buffers, *J. Econ. Entomol.* **62:**1043.

Stenersen, J., 1971, Thin-layer chromatography of diesters and some monoesters of phosphoric acid, *J. Chromatog.* **54:**77.

Sternburg, J., and Kearns, C. W., 1956, Pentachlorocyclohexene, an intermediate in the metabolism of lindane by houseflies, *J. Econ. Entomol.* **49:**548.

Thompson, M. J., Svoboda, J. A., Kaplanis, J. N., and Robbins, W. E., 1972, Metabolic pathways of steroids in insects, *Proc. Roy. Soc. London Ser. B* **180:**203.

Thompson, M. J., Kaplanis, J. N., Robbins, W. E., and Svoboda, J. A., 1973, Metabolism of steroids in insects, *Advan. Lipid Res.* **11:**219.

Towers, G. H. N., 1964, Metabolism of phenolics in higher plants and microorganisms, *in: Biochemistry of Phenolic Compounds* (J. B. Harborne, ed.), pp. 249–294, Academic Press, New York.

Trivelloni, J. C., 1964, A study of the formation of β-glucosides in the locust (*Schistocerca cancellata*), *Enzymologia* **26:**329.

Troll, W., Belman, S., and Nelson, N., 1959, Aromatic amines. III. Note on bis(2-amino-1-naphthyl)phosphate, a urinary metabolite of 2-naphthylamine, *Proc. Soc. Exp. Biol. Med.* **100:**121.

Troll, W., Tessler, A. N., and Nelson, N., 1963, Bis(2-amino-1-naphthyl)phosphate, a metabolite of beta-naphthylamine in human urine, *J. Urol.* **89:**626.

Van Heyningen, R., 1971, Fluorescent glucoside in the human lens, *Nature* (*London*) **230:**393.

Van Lancker, J. L., 1970, Hydrolases and cellular death, *in: Metabolic Conjugation and Metabolic Hydrolysis*, Vol. I (W. H. Fishman, ed.), pp. 355–418, Academic Press, New York.

Wang, D. Y., and Bulbrook, R. D., 1968, Steroid sulphates, *Advan. Reprod. Physiol.* **3:**113.

Westley, J., 1973, Rhodanese, *Advan. Enzymol.* **39:**327.

White, A. F., 1972, Metabolism of juvenile hormone analogue methyl farnesoate 10,11-epoxide in two insect species, *Life Sci.* **11:**201.

Wilkinson, C. F., 1971, Effects of synergists on the metabolism and toxicity of anticholinesterases, *Bull. WHO* **44:**171.

Williams, R. T., 1959, *Detoxication Mechanisms*, 2nd ed., Chapman and Hall, London.

Williams, R. T., 1967, The biogenesis of conjugation and detoxication products, *in: Biogenesis of Natural Compounds* (P. Bernfeld, ed.), pp. 427–474, Pergamon Press, Oxford.

Williams, R. T., 1971, Species variation in drug biotransformations, *in: Fundamentals of Drug Metabolism and Drug Disposition* (B. N. LaDu, H. G. Mandel, and E. L. Way, eds.), pp. 187–205, Williams and Wilkins, Baltimore.

Williamson, D. G., Collins, D. C., Layne, D. S., Conrow, R. B., and Bernstein, S., 1969, Isolation of 17α-estradiol 17β-D-glucopyranoside from rabbit urine, and its synthesis and characterization, *Biochemistry* **8:**4299.

Williamson, D. G., Layne, D. S., and Collins, D. C., 1972, Steroid estrogen glycosides: Formation of glucosides and galactosides by human liver and kidney, *J. Biol. Chem.* **247:**3286.

Willig, A., Rees, H. H., and Goodwin, T. W., 1971, Biosynthesis of insect moulting hormones in isolated ring glands and whole larvae of *Calliphora*, *J. Insect Physiol.* **17:**2317.

Wong, K. P., 1971, Formation of bilirubin glucoside, *Biochem. J.* **125:**929.

Wood, J. L., 1970, Biochemistry of mercapturic acid formation, *in: Metabolic Conjugation and Metabolic Hydrolysis*, Vol. II (W. H. Fishman, ed.), pp. 261–299, Academic Press, New York.

Yamamoto, I., Kimmel, E. C., and Casida, J. E., 1969, Oxidative metabolism of pyrethroids in houseflies, *J. Agr. Food Chem.* **17:**1227.

Yang, R. S. H., and Wilkinson, C. F., 1971, Conjugation of *p*-nitrophenol with sulfate in larvae of the southern armyworm (*Prodenia eridania*), *Pestic. Biochem. Physiol.* **1:**327.

Yang, R. S. H., and Wilkinson, C. F., 1972, Enzymic sulphation of *p*-nitrophenol and steroids by larval gut tissues of the southern armyworm (*Prodenia eridania* Cramer), *Biochem. J.* **130:**487.

Yang, R. S. H., and Wilkinson, C. F., 1973, Sulphotransferases and phosphotransferases in insects, *Comp. Biochem. Physiol.* **46B:**717.

Yang, R. S. H., Hodgson, E., and Dauterman, W. C., 1971, Metabolism *in vitro* of diazinon and diazoxon in susceptible and resistant houseflies, *J. Agr. Food Chem.* **19:**14.

Yang, R. S. H., Pelliccia, J. G., and Wilkinson, C. F., 1973, Age-dependent arylsulphatase and sulphotransferase activities in the southern armyworm: A possible insect endocrine regulatory mechanism? *Biochem. J.* **136:**817.

Zayed, S. M. A. D., and Hassan, A., 1965, Metabolism of organophosphorus insecticides. I. Distribution and metabolism of dipterex in adult larvae of the cotton leaf worm (*Prodenia litura* F.), *Can. J. Biochem.* **43:**1257.

Zayed, S. M. A. D., Hassan, A., Fakhr, I. M. I., and Bahig, M. R. E., 1970, Metabolism of organophosphorous insecticides. X. Degradation of ^{14}C-dimethoate in the adult larva of the cotton leaf worm, *Biochem. Pharmacol.* **19:**17.

III.

Target Site Interactions

6

The Nervous System: Comparative Physiology and Pharmacology

D. L. Shankland

1. Nervous Integrity and Neurotoxins

Nature and the pesticide industry apparently have decided that the best way to poison an animal is through its nervous system. Of all the natural poisons known, by far the most lethal are nerve poisons. Tetanus and botulinus toxins, for instance, are lethal to mammals at about 1 ng/kg of body weight, making them about 10^6 times as toxic as sodium arsenite and HCN, which have "LD_{50}s" to mammals in the range 1–10 mg/kg. The "LD_{50}s" of tetrodotoxin, saxitoxin, and batrachotoxin are on the order of a few to about 15 μg/kg and that of black widow spider venom about 50 μg/kg. The minimal lethal dose of the neurotoxic venom of cobra to mammals is about 0.1–1.0 mg/kg, while the nonneurotoxic venoms of pit vipers, although highly toxic, are only one-tenth to one-hundredth as toxic (Evans, 1972; Bücherl and Buckley, 1971; van Heyningen, 1970; McCrone and Hatula, 1967; Halstead, 1965). As a corollary to this, by far the largest part of the insecticides manufactured are neurotoxic. This applies generally to the carbamates, organophosphates, and chlorinated hydrocarbons. On a decreasing scale of toxicity, however, these compounds begin where the venoms leave off, i.e., in the range of mg/kg.

The paramount importance of the nervous system to the functional integrity of animals makes it an extremely sensitive target for the action of

D. L. Shankland • Department of Entomology, Purdue University, West Lafayette, Indiana.

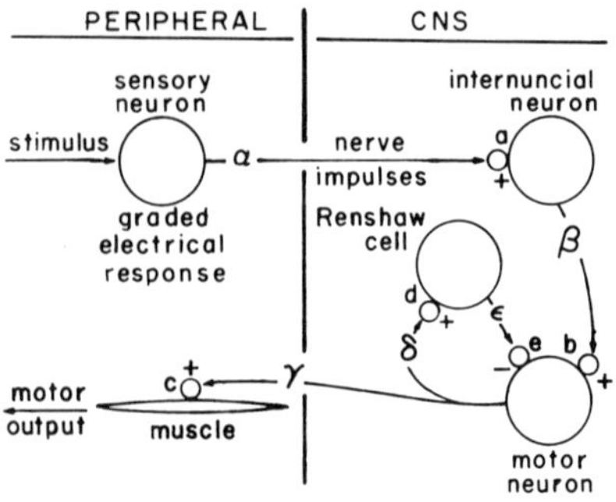

Fig. 1. A schematized reflex loop showing the variety of nervous processes susceptible to the actions of drugs and poisons. See text for explanation.

poisons. A discrete lesion at one site may be lethal to an animal through the far-reaching consequences of general disruption of nervous integrity. The complexity of even relatively simple nervous systems prohibits a comprehensive treatment here of the integrative mechanisms and great number of processes and sites on which poisons can act. The simplified reflex loop shown in Fig. 1, however, may provide some appreciation for those points. The figure represents a generalized scheme and includes elements of both vertebrate and invertebrate systems.

The input to the system is a stimulus of appropriate form, depending on the specific kind of sensory cell concerned. The sensory neuron transduces the energy form of the stimulus to an electrical signal having an amplitude proportional to the strength of the stimulus; i.e., it is graded as a function of stimulus strength. This initial process offers a site for drug or poison action. The inhibition of feeding in desert locust by azadirachtin, for example, appears to be an example of direct action on sensory cells (Butterworth and Morgan, 1971). The graded electrical response occurring in the sensory neuron is converted at the juncture of the cell body and the axon (afferent fiber α) to a pulse code of nerve impulses transmitted along the axon to the CNS. The impulses themselves are of constant amplitude, but they carry information by way of a frequency code. Such conduction occurs in axons throughout the nervous system and is the process on which the excitatory action of DDT and allethrin and the depressant action of tetrodotoxin are exerted. In the CNS, the afferent fiber terminates at an intermediate cell called an *internuncial neuron* but communicates with the latter by a chemical transmitter across a specialized junction between the cells called a synapse (*a* in Fig. 1). This chemically mediated transmission begins in the presynaptic terminal, where the impulses invade the axon terminal and cause the release of the chemical transmitter acetylcholine (ACh), for example. Recent evidence indicates that dieldrin (HEOD) or a metabolite (trans-aldrindiol) acts by causing too much ACh to be released in this process (Shankland and Schroeder, 1973; Akkermans *et al.*, 1974). The ACh released acts on the postsynaptic membrane and through another transducer process this chemical signal is converted into an electrical response, again graded as a function of the stimulus strength (i.e., the amount of ACh). The actions of nicotine, muscarine, carbamate and organophosphate insecticides, and many other drugs are exerted directly or indirectly on this postsynaptic transducer process. The internuncial neuron communicates with a motor neuron, again by means of a pulse code over an axon, β, and again the transmitter mediates the intercellular communication at the synapse, *b*. The excited motor neuron sends a pulse code over two pathways: (1) directly to the muscle over the efferent motor fiber, γ, and (2) via a collateral fiber, δ, to an inhibitory internuncial neuron. The signal to the muscle is mediated by ACh (in vertebrate muscle, but not in insects) at the nerve muscle junction, *c*. Junctional transmission at the motor neuron, *b*, and transmission along the efferent fiber, γ, and at the muscle, *c*, are each candidate processes for the selective action of nerve poisons. Transmission at *b* is sensitive to both muscarinic and nicotinic drugs, as well as carbamate and organophosphate insecticides. The motor fiber, γ, although usually less sensitive to DDT than the afferent fiber, α, shows the same kind of response to both DDT and allethrin. The nerve–muscle junction in mammals, *c*, is the site of action of botulinus toxin, decamethonium, nicotine, curariform drugs, nereistoxin, and a great many others. The inhibitory neuron in Fig. 1 represents a specialized neuron called a Renshaw cell in the CNS of vertebrates. The signal from the motor neuron is carried to the Renshaw cell at synapse *d* by ACh and is relayed as inhibitory feedback, called recurrent inhibition, over ε to the motor neuron. This inhibitory signal is mediated chemically at synapse, *e*, but the identity of the transmitter is not known. Strychnine and perhaps tetanus toxin act at *e* to block inhibition and in so doing produce the convulsive tonic spasms characteristic of these compounds.

Thus at every step in the flow of information through a nervous system there are processes which are especially sensitive to the action of some drug, and the action may be excitatory or depressant. A poison which causes hyperresponsiveness to a stimulus may, for example, be enhancing activity at any one of the various excitatory processes in a reflex loop, or it may be depressing the activity of an inhibitory process. Conversely, depression of reflex activity may be due to depression of any of the excitatory processes or enhancement of inhibitory ones.

The entire nervous system of even a lower animal, such as an insect, involves tens of thousands of cells in an extremely complicated communication network which is exquisitely integrated to produce functional integrity in virtually every activity. Higher animals possess nervous systems involving millions of cells with a commensurate increase in complexity. In any of the systems, however, the disruption of transmission at various discrete sites may so disrupt the integrity of the system as to cause death. In mammals, disruption of transmission at cholinergic synapses, for example, often causes failure of the respiratory system. In insects, not so dependent on active respiratory movements, the ultimate cause of death may be more complex, but it is not difficult to accept the disruption of nervous integrity as a causative factor.

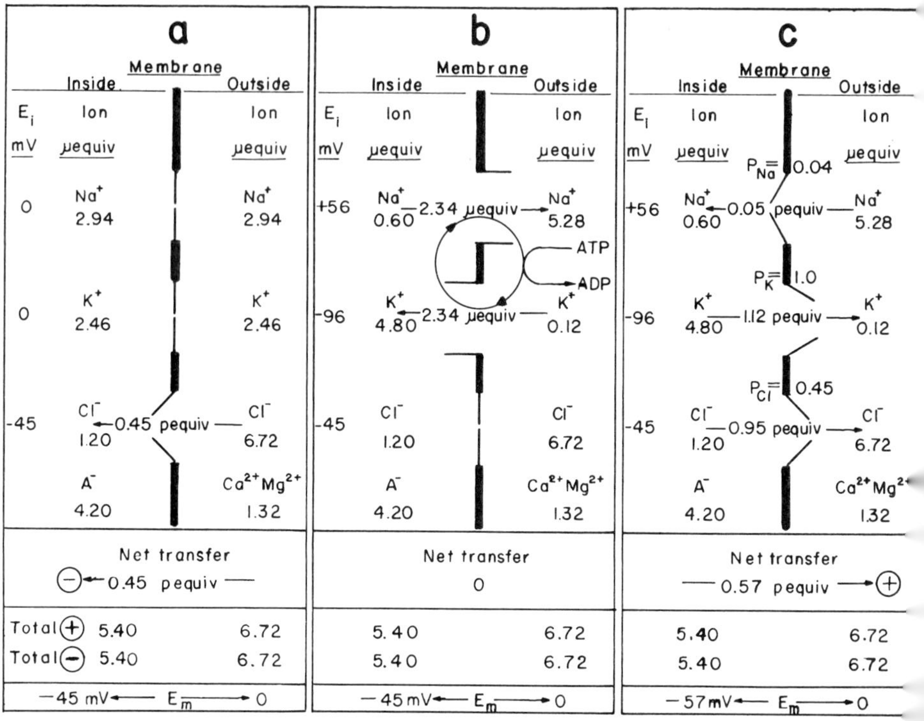

Fig. 2. Axonal membrane model based on the Hodgkin–Huxley (1952) membrane showing the roles of the ion pump and selective permeability of the membrane in the production of the resting potential. See text for explanation.

It is the purpose of this chapter to provide those who are not already intimately familiar with neurophysiology a background for understanding the neurological basis of the toxic mechanisms to be discussed in Chapters 7, 8, and 9.

2. *Electrophysiology of the Nervous System*

Many of the normal and pathological processes mentioned above can be studied through the use of electrophysiological techniques, and in fact they have figured importantly in the elucidation of the modes of action of a variety of insecticides and drugs.

If an electrode connected to a suitable voltage-measuring device is inserted into a live nerve cell, or any living cell for that matter, an electrical potential called the *membrane potential* will be recorded. It is consistently negative on the inside with respect to the outside. Burton (1973) discussed the origin and significance of this potential in animal cells, and it appears to be the secondary consequence of regulated distribution of different species of ions between the intra- and extracellular sites across the selectively permeable cell membrane. The primary need is for high internal potassium, low internal sodium, and, of course, an osmoregulatory capability. The osmolarity of cytoplasm is due partly to nondiffusible organic anions. Because of these internally fixed ions, if the plasma membrane were only passively permeable to small diffusible ions and water, the cell would swell due to osmosis whatever the osmolarity of an external salt solution. The swelling could be prevented by (1) a rigid cell wall, as in plants, (2) the active extrusion of water, at exorbitant energy cost, or (3) the active extrusion of ions. The last is found in animal cells as an active extrusion of sodium, calcium, and magnesium, often coupled with an active uptake of potassium. The active ion transport thus serves both to prevent swelling and to produce the required intracellular ionic environment. The combined effects of active transport, the net negative charge on intracellular organic molecules, and ion-selective permeability of the plasma membrane produce the electrical potential across the membrane, which in irritable cells, such as nerve and muscle, is called the *resting potential.*

Whereas in nonirritable tissues this electrical potential may have no known functional significance, in irritable tissues it is the basis for functionally significant potentials, one of the more obvious cases being the nerve impulse. Other kinds of electrical changes also occur, however, and electrophysiological techniques serve as powerful means of detecting and analyzing normal and pathological processes which produce them. A good understanding of the normal electrophysiological processes is prerequisite to understanding the modes of neurotoxic action of many insecticides.

2.1. Resting Potential

The explanation of resting potential used here has evolved through my own experience in attempting to make it clear to the neophyte. Alternative explanations with good illustrative support are available from Katz (1961, 1966) and Eccles (1965).

The resting potential of nerve and other irritable cells ranges from about 20 mV to 90 mV or so (Burton, 1973). The events producing the resting potential are illustrated in Fig. 2. The figure is based on the Hodgkin–Huxley (1952) model of squid giant axon but is not intended to represent all of its complexities. In Fig. 2, the membrane has an area of 1 cm^2 and separates 0.012 ml of inside axoplasm (the volume enclosed by 1 cm^2 of membrane of a cylindrical axon 500 μm in diameter) from an equal volume of outside interstitial fluid. This limitation on the outside volume, which is normally considered infinite, is imposed to permit the discussion of ionic gradients and fluxes to be based on numbers of equivalents rather than concentrations of ions. For the purpose at hand, no serious error is introduced by this. The gates in the membrane represent ion-specific channels which may open or close by varying amounts and thereby govern the permeability to the respective ion.

The axoplasm of squid giant axon contains about 50 mM Na^+ and 400 mM K^+ (Hodgkin, 1964). These cations are balanced by about 100 mM Cl^- and 350 mM of a variety of organic anions, the predominant one being isethionate ($HOC_2H_5SO_3^-$) at about 250 mM (Hodgkin, 1964). The organic ions cannot diffuse through the membrane. In Fig. 2b,c (2a is a hypothetical condition explained later), these molarities would represent the following amounts of the respective ions in μequiv: Na^+, 0.60; K^+, 4.80; Cl^-, 1.20; organic anions (A^-), 4.20. The ionic composition of the outside fluid, also derived from Hodgkin (1964), is shown in Fig. 2b,c in μequiv of each ion: Na^+, 5.28; K^+, 0.12; Ca^{2+} and Mg^{2+} combined, 1.32 (the membrane is impermeable to both ions); Cl^-, 6.72. The requirement for electrical neutrality throughout a bulk solution is met for each side of the membrane by the balance of cations and anions, shown near the bottom of Fig. 2.

The distribution of ions as they normally exist in resting squid axon is shown in Fig. 2c, but for purposes of explaining the development of the resting potential consider a hypothetical initial condition shown in Fig. 2a. Both Na^+ and K^+ are distributed across the membrane without concentration gradients, but the balance of cations and anions in the inside and outside solutions, respectively, is preserved. Presume that the membrane is temporarily impermeable to Na^+ and K^+, indicated by the closed gates. There is a concentration gradient of Cl^- from outside to inside, however, and the membrane is chloride permeable. The concentration gradient of Cl^- produces a diffusion pressure, and the permeability of the membrane allows inward diffusion to occur. Each Cl^- that diffuses inward carries with it one net negative charge of electricity. The impermeability of the membrane to Na^+ and K^+, hypothetically imposed here, would prevent any inward movement of cations to compensate for inward movement of Cl^-. As Cl^- diffuses inward, an excess of negative charge accumulates inside the membrane and opposes further diffusion. These two opposing forces, i.e., diffusion pressure and electrical potential, eventually come to equilibrium when the electrical potential is just sufficient to balance the diffusion pressure. This so-called equilibrium potential is defined by the Nernst equation, (1), which is discussed in textbooks of physical chemistry (*cf.* Adamson, 1969; Sheehan, 1961) and in its application to biological systems by K. Cole (1968), Davson (1964), Lakshminarayanaiah (1969), Plonsey (1969), Rodahl and Issekutz (1966), and Wilson (1972).

$$E_{Cl} = \frac{RT}{nF} \ln \frac{[Cl^-]_o}{[Cl^-]_i} \tag{1}$$

where E_{Cl} is the equilibrium potential of chloride in volts, R is the international gas constant, 8.314 joules/degree/mole, T is the temperature in °K, n is the number of unit electrical charges per ion, F is the faraday constant, 96,500 coulombs, and $[Cl^-]_o/[Cl^-]_i$ is the concentration gradient of Cl^- across the membrane from outside to inside.

Assuming a temperature of 30°C, converting to common logarithms, and expressing E_{Cl} in mV, equation (1) reduces to

$$E_{Cl} = 60 \log \frac{[Cl^-]_o}{[Cl^-]_i}$$

For the Cl^- gradient in Fig. 2, this equilibrium potential is 45 mV, inside negative. Under the special conditions of Na^+ and K^+ impermeability imposed on the model, this is also the voltage, E_m, that could be measured across the membrane. It would require a transfer of only about 1.0×10^{-12} equiv of Cl^- across 1 cm^2 of squid axon membrane to produce a potential of 100 mV (K. Cole, 1968; Lakshminarayanaiah, 1969). In Fig. 2a the 45 mV potential would therefore be produced by the net inward transfer of only 0.45 pequiv (0.45×10^{-12} equiv) of Cl^-. It bears emphasis that this excess internal Cl^- is confined to the cytoplasm–membrane interface and amounts to less than 0.00004% of the internal Cl^-. The amount transferred is so small that it has virtually no effect on the concentration of ions in the bulk solutions on either side of the membrane.

Beginning with these initial conditions (Fig. 2a), presume that an energy-requiring ion pump, resident in the membrane and drawing energy from ATP (Skou, 1965), begins to pump Na^+ outward and K^+ inward in a one for one exchange (Fig. 2b). The pump ultimately produces the gradients of K^+ and Na^+ across the membrane shown in Fig. 2b. If movement of Cl^- is prevented during this process (an artificial restriction imposed on this model), there will have been to this time no net transfer of electrical charge across the membrane, and the original potential of −45 mV due to Cl^- still prevails. But there are now concentration gradients of Na^+ and K^+, and it is possible to calculate by the Nernst equation what the equilibrium potential for each ion would be, and those values are shown as inside voltages of 56 mV and −96 mV for E_{Na} and E_K, respectively. Either of these would be produced as a measurable membrane potential only if the respective ion alone were allowed to diffuse across the membrane, down its concentration gradient, in the manner described above for Cl^-. Staverman (1952, 1954) has rigorously defined the factors controlling the diffusion of ions through membranes in various biological systems. In this discussion we need consider only three: (1) chemical potentials, referred to here as *diffusion pressures*, and equated to concentration gradients; (2) electrical potential gradients; and (3) membrane permeabilities, represented here by gate openings. Figure 2c shows how these factors interact in the production of the membrane resting potential.

Given the ionic gradients established in Fig. 2b, and assuming that the potassium gate is opened, K^+ would diffuse outward (Fig. 2c) because of both its concentration gradient and the membrane potential, E_m. As it diffuses, it carries positive charge outward, increasing the electrical potential across the membrane. This would force the outward diffusion of Cl^- simply because E_m

now exceeds E_{Cl} and would drive Cl^- up its concentration gradient. If the membrane were permeable to Na^+, that ion would enter into this process also, diffusing at a rate and direction which are functions of the Na^+ permeability of the membrane, the membrane potential, and its own concentration gradient. The way in which the three factors governing diffusion for each respective ion interact to attain a steady state is defined by the Goldman constant field equation (2), an extension of the Nernst equation, which is discussed in its application to the squid axon by Hodgkin and Katz (1949).

$$E_m = \frac{RT}{nF} \ln \frac{P_K[K^+]_i + P_{Na}[Na^+]_i + P_{Cl}[Cl^-]_o}{P_K[K^+]_o + P_{Na}[Na^+]_o + P_{Cl}[Cl^-]_i} \qquad (2)$$

where P is the relative permeability of each respective ion. Using the values of P from Hodgkin and Katz (1949) ($P_K : P_{NA} : P_{Cl} = 1.0 : 0.04 : 0.45$) and the respective ionic gradients from Fig. 2c, and reducing the constants as in equation (1), equation (2) predicts that the equilibrium potential for the entire system would be 57 mV, inside negative. This would be the potential which would be measurable across the membrane, and is the so-called resting potential.

It should be noted that the permeabilities used here are expressed as ratios and do not reflect absolute units of any kind. A relative permeability of 1.0 for K^+ does not mean that the membrane in the resting state is maximally permeable to that ion. This has important implications in the transient processes to be discussed later.

Here again, as shown in Fig. 2c, the amounts of the three ions involved in the fluxes to produce the membrane potential are too small to affect the concentrations of ions in the bulk solutions on either side of the membrane.

If, after the conditions illustrated in Fig. 2c were established, no more energy were put into the system, there would be leakage of Na^+ and Cl^- inward and K^+ outward. Eventually a Donnan equilibrium would be attained in which the distribution of the ions would differ from that shown in Fig. 2c as in Table I.

Table I. Comparison of Experimentally Determined Ion Distribution with That Predicted by a Donnan Equilibrium in the Membrane Model of Fig. 2c.

	Amounts of ions (μeq)			
	Inside		Outside	
Ion	Experimental[a]	Donnan[b]	Experimental[a]	Donnan[b]
Na^+	0.60	3.78	5.28	3.85
K^+	4.80	3.17	0.12	1.75
Cl^-	1.40	2.75	6.72	5.17

[a] Based on Hodgkin (1964) as per Fig. 2c.
[b] See Lakshminarayanaiah (1969) for a discussion of Donnan equilibria involving several species of permeant and nonpermeant ions.

The ion pump operates continuously to counteract these leakage currents, however, so the resting condition depicted in Fig. 2c is actually a dynamic equilibrium, and the maintenance of the characteristic resting membrane potential requires continual energy input.

Of course, in the living membrane the resting potential is not produced from an initial condition illustrated in Fig. 2a, nor is it produced by the discrete steps used here for illustrative purposes. All of the processes interact continually to maintain the normal conditions.

2.2. Transient Changes in Membrane Potential

The very basis of nerve excitability resides in the ability of the membrane to undergo active transient changes in the membrane potential. The transmission and processing of information in the nervous system involve a variety of such changes. The stimulus in Fig. 1, for example, elicits an electrical response from the sensory neuron which is graded as a function of stimulus strength. This sensory cell response is called a *generator potential*, and in any given sense cell it may extract only certain kinds of information from the stimulus. For example, a mechanoreceptor called the *PD organ* in the legs of certain crustaceans contains some cells that are purely tonic in that they respond continuously to maintained mechanical stimulation and others that are purely phasic in that they respond only to movement and not to maintained stimuli (Boettiger and Hartman, 1968). Similarly, the olfactory cells in the antennae of certain male silk worm adults, while exquisitely and selectively sensitive to female sex pheromone, respond only phasically to sustained stimulation. On the other hand, the generator potential of cells in the pit organ of *Locusta* antennae shows tonic response to continuous exposure to linoleic acid (Boeckh *et al.*, 1965). Even when the generator potential is purely tonic, however, as in this last case, the impulse code that it generates may contain phasic information. The initial high pulse rate produced by sudden exposure of *Locusta* pit organ to linoleic acid drops off to a lower level even if the stimulus is maintained at a constant intensity (Boeckh *et al.*, 1965). This phasitonic response contains information about both the rate of application of the stimulus and its continued presence. The nerve impulses are specifically adapted to the transmission of information from one part of the nervous system to another. At every synaptic junction, these all-or-none potentials lead to the generation of graded potentials (postsynaptic potentials). Nerve poisons alter many of these transient changes in various ways, and the mechanisms for the production of the graded and all-or-none responses are directly involved in their action.

2.2.1. Graded Responses

Given a resting state of the conditions depicted in Fig. 2c, and referring to equation (2), the membrane potential can be viewed as a compromise among the concentration gradients of the various permeant ions, and the relative permeability of the membrane to each of them. A change in permeability to one or more of the ions would allow ionic fluxes through the membrane, shifting the membrane potential to a new compromise value. Controlled and selective changes in permeability to ions form the basis of nerve membrane excitability.

Conventionally, ionic fluxes are discussed in terms of current, or the flow of electrical charge carried by the ions. Current is symbolized by I, and is measurable in amperes. Permeabilities are discussed in terms of conductances, and although precisely speaking conductance is not equal to permeability, in this discussion increased conductance can be taken simply as increased permeability. Conductance is symbolized by g, and is measurable in reciprocal ohms, or mhos. Conductance changes are caused by appropriate stimuli which vary with different kinds of neurons and different parts of the same neuron. The axonal membrane, for example, is responsive to galvanic shock (electrical stimulation) while the postsynaptic membrane of chemical synapses and the receptor membranes of sensory neurons are not. On the other hand, the postsynaptic membrane will respond to chemical stimuli, as will the receptor membrane of a variety of sensory neurons (Grundfest, 1965; Paintel, 1964). The response of the membrane, manifested as a potential change, is referred to as *electrogenesis* and is the consequence of the altered conductance, which is called a *transducer action* (Grundfest, 1961).

The effects of selective changes in conductance are shown in Fig. 3, the conditions of Fig. 2c being used as a basis. The resting membrane potential, $E_m(R)$, at −57 mV lies between E_{Cl} (−45 mV) and E_K (−96 mV) and is 113 mV more negative than E_{Na}. Under these conditions, a selective increase in K conductance, for example, called *K activation* (Grundfest, 1961) would allow an outward flow of potassium current (i.e., potassium ions), causing a shift in the membrane potential toward the potassium equilibrium potential (E_K). This, then, would be a hyperpolarization due to selective K activation. The amplitude of the change in the membrane potential would be a function of the amount of increase in K conductance, which would in turn be a function of the strength of a stimulus. The graded nature of the response is symbolized by the multiple arrowheads in Fig. 3.

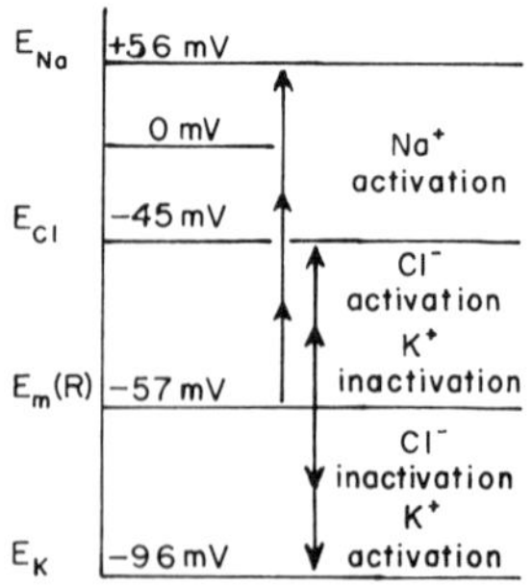

Fig. 3. Shifts in the membrane potential from the resting value $E_m(R)$ caused by selective increases (activation) and decreases (inactivation) in the permeability of the nerve membrane. See text for explanation.

In an analogous way, a selective increase in Cl conductance would cause a shift in E_m toward the chloride equilibrium potential, which in this case would be a depolarization due to selective Cl activation. Selective decreases in conductance of either of these ions would produce, respectively, depolarization due to K inactivation and hyperpolarization due to Cl inactivation.

Because the resting membrane potention in nerve is consistently more negative than E_{Na}, Na activation can be expected to produce only a depolarization. In the special case of nerve action potentials in axons, discussed below, Na activation is not graded but is an all-or-none process. In other electrogenic processes, however, at other sites in neurons it is graded, and often occurs in conjunction with K or Cl activation. When that happens, the resulting voltage change could be in either the hyperpolarizing or the depolarizing direction, depending on the relative changes in the respective conductances. Usually, however, in excitatory processes, Na activation and K activation are involved, and the membrane is depolarized.

These interacting forces then can produce depolarizations or hyperpolarizations which are graded or nongraded and transient or steady, depending on whether the transducer action is graded or not, and transient or steady. In cases of graded responses, the amplitudes are functions of stimulus strength. Nongraded events, like the nerve action potential, reach a single characteristic amplitude independent of stimulus strength, after the stimulus reaches a critical threshold.

2.2.2. The Nerve Action Potential

The nerve action potential has already been mentioned as the propagated all-or-none nerve impulse used in the transmission of pulse-coded information throughout the nervous system. Present understanding of the means of its production is due to the voltage clamp studies of Hodgkin and Huxley. These studies have been comprehensively reviewed by K. Cole (1968).

Figure 4 is based on the Hodgkin–Huxley studies and shows how transient changes in sodium conductance (g_{Na}) and potassium conductance (g_K) lead to inward and outward currents of Na^+ and K^+, respectively, producing a transient change in the membrane potential, which shows a prominant initial voltage wave form called a "spike" (Fig. 4c). Chloride is involved only passively, and chloride conductance does not change during spike production. Figure 4 shows the changes that would occur at a recording site as a propagated action potential passes a recording electrode as shown in the inset in Fig. 4c. (The events that will be described here are based on the resting condition shown in Fig. 2c.)

Spike production is initiated by Na activation, which is triggered by a depolarization of the membrane potential. The cause of this depolarization will be explained later. The depolarization must reach threshold, which is about 12–15 mV in squid axon. Once threshold has been reached, however, a

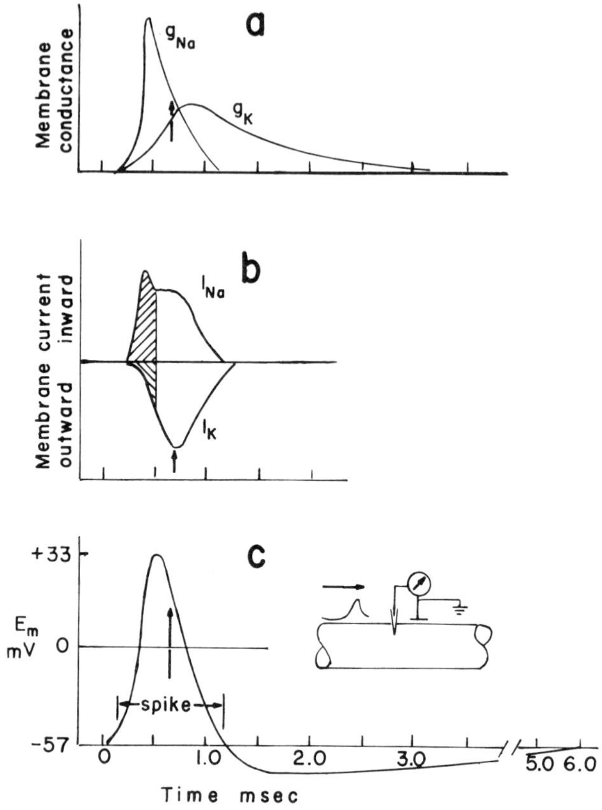

Fig. 4. Events occurring at a recording site in axonal membrane during the passage of an action potential showing (a) the time course of Na and K conductance changes, (b) Na and K currents which produce the action potential, and (c) the action potential itself, showing the portion called the spike. Modified from Hodgkin and Huxley (1952). See text for explanation.

nongraded Na activation occurs, indicated by the rising phase of the curve g_{Na} in Fig. 4a. This is followed immediately by Na inactivation, indicated by the descending portion of that same curve. Once triggered, this activation–inactivation cycle proceeds to completion no matter what the form of the stimulating voltage. It does not make any difference if the stimulating depolarization is a brief pulse, only a fraction of a millisecond in duration, or a prolonged voltage step. This matter is crucial to understanding the voltage clamp experiments discussed later in this chapter and in Chapter 9. Because there is a diffusion pressure of Na^+ from outside to inside (Fig. 2c), the increased g_{Na} allows a rapid surge of inward sodium current, I_{Na}, which reaches a peak slightly under 0.5 msec (Fig. 4b).

The Na current, in carrying positive charge inward, decreases or depolarizes the membrane potential, producing the rising phase of the action potential shown in Fig. 4c. This changed membrane potential initiates a process of K activation, producing an increase in g_K which lags behind g_{Na} (Fig. 4a). By this time, the membrane potential is no longer at the K^+ equilibrium potential, and there is an outward diffusion pressure of K^+ that is now unopposed by the membrane potential. The increased g_K facilitates an outward current of K^+ (Fig. 4b), which, in carrying positive charge outward, tends to repolarize the membrane. Note that the turning off of the Na current (by Na inactivation) is not sufficient in itself to account for the repolarization of the membrane. The latter requires the outward transfer of positive charge, accomplished by the K current. The total positive charge transferred inward by the Na current is a function of the area under the curve I_{Na} (Fig. 4b). Conversely, the total positive charge transferred outward is a function of the area under the curve I_K (Fig. 4b). In Fig. 4b, it can be seen that the shaded area under I_{Na} during the rising phase of the action potential is only partly countered by the smaller shaded area under I_K at the same time. The algebraic sum of the voltage changes produced by these two currents yields a maximum voltage of +33 mV (Fig. 4c) occurring at about 0.5 msec. The total excursion of E_m from the resting potential to the peak of the spike is 90 mV, and it is this voltage change which is used to express the amplitude of the action potential.

As I_K continues to increase with increasing g_K until about 0.7 msec, it begins to repolarize the membrane. But K conductance is inversely related to the membrane potential. That is, as the membrane is repolarized, the repolarization itself causes a reduction in g_K. The process can be likened to the refilling of a toilet water tank in which the rising water level closes a valve cutting off further flow into the tank. Note, however, that at the time the K current reaches its peak (arrow, Fig. 4b), there is still a significant Na conductance (arrow, Fig. 4a), and the membrane potential is well below the Na equilibrium potential (arrow, Fig. 4c). These conditions allow for a delay in the reduction of Na current shown by the inflection in the descending portion of the curve I_{Na} (Fig. 4b). The continued decrease in g_{Na}, however, ultimately turns off the Na current, leaving only K current during the final phase of repolarization. The membrane is actually hyperpolarized from about 1.2 msec to about 6 msec. Consideration of equation (2) and its implications as discussed in Section 2.2.1 will make the cause of this hyperpolarization apparent. The prolonged period of K activation producing the higher than normal g_K until about 3.5 msec (Fig. 4a) permits the membrane potential to more nearly approximate E_K. This is, then, a hyperpolarization due to K activation. At the time g_K reaches the resting level, the membrane is only slightly hyperpolarized, and E_m decays to the resting level over the next 2.5 msec or so. The decay is of the first order, and is defined by a time constant which is characteristic for the membrane.

The ionic currents of Fig. 4b represent currents flowing through the membrane. A close inspection of Fig. 4c will show that the membrane potential begins to change about 0.25 msec before the Na current begins to flow. This means that as a propagated nerve action potential approaches a recording electrode (inset, Fig. 4c) a small voltage change is detected before Na activation occurs. This early potential, referred to as the *foot* of the action potential, is, in fact, the cause of the depolarization of the membrane to the threshold required to trigger Na activation and spike production at the recording site. The foot itself and the distance ahead of the spike that it is detectable are functions of membrane impedance, and that matter will be discussed shortly. Before doing so, however, it is worth considering the magnitudes of the ionic fluxes that produce the action potential, and how the so-called ion pump is related to these events.

As was pointed out in reference to Fig. 2, it requires the net transfer of only about 1 pequiv of an ion to produce a change of 100 mV across 1 cm^2 of squid axon membrane. In the case of an action potential, the voltage change is about 90 mV. However, recall that over a good part of the duration of the action potential, there are both inward and outward currents flowing. Although there may be a net transfer of only 0.9 pequiv inward during depolarization and an equal amount outward during repolarization, the total inward and outward fluxes must be greater than that. In fact, it has been estimated that there is an entry of about 4 pequiv of Na^+ and a loss of an equal amount of K^+ per cm^2 per impulse in squid axon (Lakshminarayanaiah, 1969). This amounts to an uptake in the axoplasm of only about 0.0007% of the internal sodium and a loss of about 0.0001% of the internal potassium. Therefore, even if the ion pump were not working, the concentration gradients of Na^+ and K^+ that normally occur across the resting membrane could support the production of thousands of impulses before the gradients were seriously reduced.

Even in the resting membrane, however, the ion pump is continuously active, counteracting the slow leakage currents of sodium and potassium. Furthermore, it was shown by Hodgkin and Keynes (1956) that the rate of Na^+–K^+ exchange by the pump is controlled by the internal sodium concentration. Increased internal sodium resulting from periods of firing in the axon thus serves to accelerate the pump within seconds and hasten the restoration of normal resting ionic gradients. The ion pump is thus only indirectly related to the spike-generating process, and drugs which block the pump do not interfere directly with spike generation (Hodgkin and Keynes, 1955). Thus, although DDT has been reported to inhibit Na^-–K^- ATPase (Bratkowski and Matsumura, 1972; Koch, 1969), this action could not account for the marked unstabilizing effect of that compound on nerve membrane or for the typical syndrome of poisoning. In excised but otherwise untreated giant axons of cuttlefish (*Sepia*), Keynes (1951) determined that during repetitive stimulation at 100/sec the ratio of Na^+ influx to efflux was about 1.5 : 1. Apparently, then, even in the normal nerve the ion pump does not keep up with the influx of Na^+

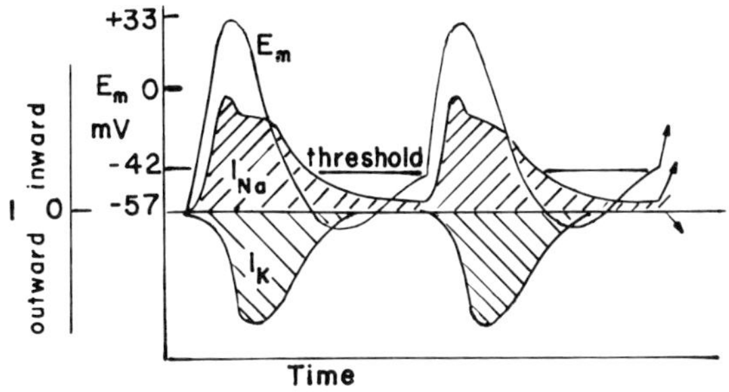

Fig. 5. The mechanism by which repetitive discharge occurs in DDT-poisoned nerves. See text for explanation.

during periods of high activity and must continue operating during quiet periods to restore normal resting conditions.

This is an appropriate place also to describe a pathological condition in spike production which results in repetitive discharge. This is a symptom produced by a variety of drugs and poisons, including DDT (Roeder and Weiant, 1948), the veratrine alkaloids (Shanes, 1958), and *p*-xylene (Sjodin and Mullins, 1958). If a drug acts to unduly prolong the Na current following an action potential, but the K current is not drastically altered, the situation shown in Fig. 5 occurs. The prolongation of the Na current beyond the termination of the K current depolarizes the membrane again after it has been repolarized by the K current. As soon as the depolarization reaches threshold, Na activation is triggered in the normal way, and another spike is produced. This process is self-perpetuating and may produce trains of hundreds of spikes in DDT-poisoned nerves (Eaton and Sternburg, 1964). The abnormal prolongation of Na current is the basis for the repetitive discharge caused by DDT, and probably also by allethrin (Narahashi, 1971), and appears to account for that same action by *p*-xylene.

2.2.3. Membrane Impedance and Cable Theory

The impedance properties of nerve membrane are often discussed in terms of cable or core conductor theory, and they are treated comprehensively in that way by K. Cole (1968), Plonsey (1969), and Taylor (1963). Briefly stated, impedance is a quantitative expression of the current–voltage relationships in membrane, and the cable properties of nerves determine how voltage changes spread electrotonically, or passively along the membrane.

The three elements of impedance are resistance, capacitance, and inductance. Whereas resistance can be measured under d.c. conditions, capacitance and inductance are measured under a.c. conditions, and a complete accounting of membrane impedance requires an involved analysis. Cable theory, as it

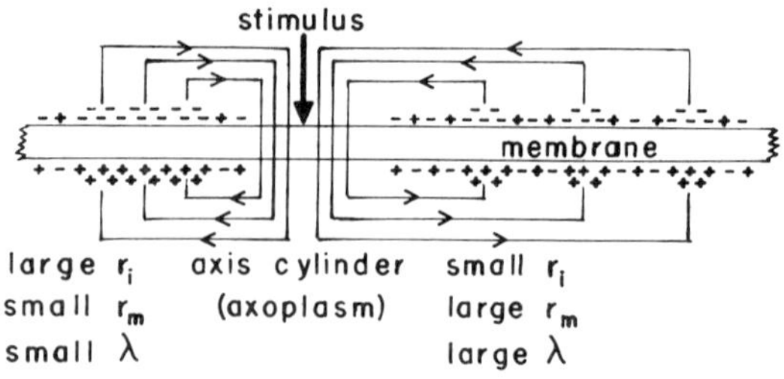

Fig. 6. Cable properties of nerve showing the relationships of membrane and axoplasm resistance (r_m and r_i, respectively) to the membrane space constant, λ. See text for explanation.

applies to nerves, requires a relatively simpler consideration of only resistance and capacitance.

If a stimulus, in the form of a depolarization, is imposed on axonal membrane, as in Fig. 6, the resulting current that flows inward through the membrane can turn either way along the axis cylinder. If the resistance of the cytoplasm or axoplasm to the flow of ions through it is relatively high, as on the left in Fig. 6, there will be an accumulation of positive charge on the inside of the membrane close to the site of the stimulus. The amount of current that flows into this region of accumulated charge is limited by the ability or capacity of the membrane surface to accommodate changes in ion density, and is called capacitive current. The capacity is, in electrical terminology, called *capacitance*, and is measured in units called *farads*, symbolized F. Biological membranes generally have a capacitance of about 1 $\mu F/cm^2$ (K. Cole, 1968). The accumulated charge will flow outward through the membrane at a rate that is inversely proportional to the resistance of the membrane. If it is low, as on the left in Fig. 6, the current will find ready egress from the axis cylinder.

Alternatively, if the axoplasm is of relatively low resistance, and the membrane highly resistant, as on the right in Fig. 6, the ionic current will flow relatively farther along the axis cylinder before flowing outward through the membrane. Obviously, then, the distance along the axon the current will flow, and consequently the spread of the potential change along the membrane due to that current, is a function of the relative resistance of the membrane and axoplasm. The precise relationship is given by cable theory in equation (3):

$$\lambda = r_m/r_i \tag{3}$$

where λ is the membrane space constant in cm, r_m is membrane resistance in Ωcm^2, and r_i is axoplasm (inside) resistance in Ω/cm. The space constant is the distance along the membrane from the current source at which the amplitude of the voltage change across the membrane is $1/e$th of the amplitude at the current source. Equation (4) is the mathematical statement of that relationship:

$$V_d = V_0\, e^{-d/\lambda} \tag{4}$$

where V_d is the amplitude of voltage change at any distance, d, from the current source, V_0 is the amplitude of voltage change at the current source, and d is distance in cm. If r_m is relatively low and r_i relatively high, λ is small, while it is large for the reciprocal relationship. Depending on the space constant, then, the Na current flowing inward at some point on an axon during an action potential will affect the membrane voltage at some distance, and the initial effect will be a depolarization of the membrane as the result of a capacitive current. Referring back to Fig. 4, then, the foot of the action potential in Fig. 4c, which precedes the beginning of the Na current (Fig. 4b), is due to this initial capacitive current induced by the approaching action potential. The change produced in E_m is what triggers the electrogenic process. Once triggered, the entire sequence of conductance changes responsible for production of the action potential occurs.

2.3. The Voltage Clamp

2.3.1. Rationale

The complex interplay of membrane conductance, current, and voltage in the production of a nerve action potential has already been described. Each of those factors is a function of the other two, and it would have been hopeless to attempt to determine the nature of their relationships without having experimental control over them.

The voltage clamp technique provides such control over membrane potential in that it allows the potential to be changed to any preset level and held there even though the voltage change triggers the transient processes which normally produce the action potential. Furthermore, since membrane currents are carried by ions, experimental manipulation of the ionic composition of the solution bathing the membrane provides control over specific ionic currents. And finally, because tetrodotoxin (TTX) is a specific blocker of sodium conductance in spike-producing membrane, the toxin affords experimental control over one of the two main components of membrane conductance. The voltage clamp used in conjunction with manipulation of the ionic environment of the axon and of membrane conductance thus becomes a powerful technique for investigations of both normal and pathological membrane processes.

The electronic details of the voltage clamp, which are available from Hodgkin *et al.* (1952), are not important at this point, but the principle on which it is based is. Ohm's law, as expressed in equation (5), defines the relationships of potential in volts (E), current in amperes (I), and resistance in ohms (R) in a simple resistive circuit. E is often referred to as "IR drop."

$$E = IR \tag{5}$$

Equation (5) predicts that if R changes, and if E is to be held constant, then I will have to change also. In axonal membrane, R is a function of both voltage and time: i.e., when the membrane is depolarized sufficiently to trigger spike generation, the membrane resistance is altered, first because of the changed membrane voltage and second because of a time-dependent process which controls Na activation and inactivation. The voltage clamp is a system which provides for changing the membrane potential to any desired level, and then, as R changes, supplying more or less current as needed to maintain a constant IR drop across the membrane. The current supplied in this way provides two important pieces of information: (1) its amplitude is an indication of membrane resistance, and (2) its direction, i.e., inward or outward through the membrane, combined with defined ionic environment of the membrane, provides information about which ions are carrying the current. For example, if in normal nerve a depolarization is followed by a large inward current which is eliminated when sodium is removed from the outside solution, then obviously the inward current was carried by sodium. Likewise, if an inward current following depolarization is eliminated by TTX, again it can be concluded that the inward current was due to sodium.

2.3.2. Graphic Presentation of Voltage Clamp Data

It was explained in Section 2.2.2 that a depolarization of axonal membrane triggers a cyclical change in Na conductance whether the depolarization is a brief pulse or sustained as in a voltage clamp. The depolarization also causes a change in K conductance, but that change occurs more slowly and is a relatively simpler function of membrane potential. Therefore, immediately after a depolarizing voltage step is imposed, a transient current reflecting the transient increase in sodium conductance occurs, and this is followed by a steady current which reflects the K conductance. The K conductance under these conditions is constant because, first, it is a function of membrane voltage and, second, a K activation–inactivation sequence does not occur as it does with Na conductance. If the current is displayed as a function of time on an oscilloscope, both the peak of the transient current and the steady-state current can be obtained. Figure 7a shows a family of typical (but not actual) oscilloscope traces obtained for each increment of a series of clamped depolarizations of an axonal membrane. In experiments of this kind, the membrane potential is initially set at some level, called the *holding potential*, such as the resting potential (e.g., Hodgkin *et al.*, 1952), or some more polarized level (e.g., Narahashi and Anderson, 1967). At time zero, a depolarizing voltage step is imposed and clamped across the membrane. If the step is of threshold amplitude or greater, the Na activation–inactivation cycle is triggered, during which Na current will flow in a direction and intensity governed by the clamped E_m, the concentration gradient of Na^+, and, of course, the Na conductance. In the example in Fig. 7, the initial holding potential is the resting potential $E_m(R)$ (−57 mV, as in Fig. 2c). As the voltage step is imposed, the membrane potential is clamped at selected increments of depolarization, as shown at the left in Fig. 7a. The oscillographic current records show a brief interruption, representing the capacitive current surge. The subsequent downward deflection indicates

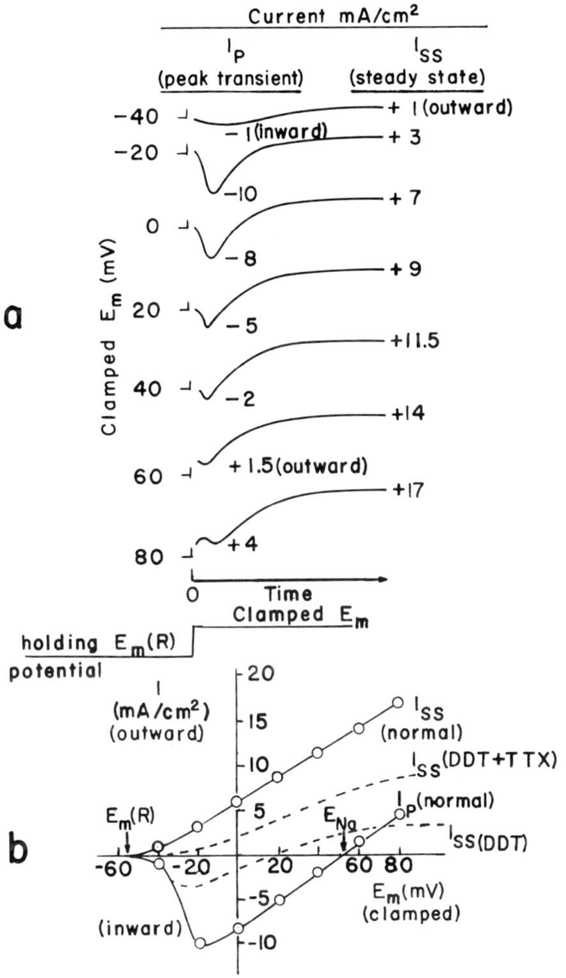

Fig. 7. Graphic presentation of voltage clamp data. (a) Simulated oscillographic records of membrane current during voltage clamp. (b) Plots of peak transient and steady-state currents obtained from (a), and after treatment with DDT and DDT plus TTX (tetrodotoxin). See text for explanation.

inward Na current which passes through a maximum or peak (I_p) as the Na conductance increases and then decreases in the familiar way. The later upward deflection represents outward K current which attains a steady state (I_{SS}) within a few milliseconds. These two values, I_p and I_{SS}, are plotted against the clamped E_m, as shown in Fig. 7b. The I_{SS} curve illustrates that any depolarization from the resting potential produces an outward K current. At small depolarizations, the slope of the curve is relatively less than at higher, because K conductance is increased by greater depolarizations. The I_p curve shows that as depolarizations increase they serve both to trigger Na activation and to impose a limit on the inward current. The reason for the limitation is apparent from the fact that when the clamped voltage reaches the Na equilibrium potential, E_{Na} (56 mV, *cf.* Fig. 2c), there is neither inward nor outward current. At that point, the process of Na activation–inactivation has occurred, of course, but because the diffusion pressure of Na^+ is exactly balanced by the clamped membrane potential no inward current flows. At greater depolarizations, the clamped E_m drives Na^+ outward, against its concentration gradient.

Pathological changes in membrane processes, caused, for example, by toxicants, become readily apparent in plots of I_{SS} and I_p. If, for example, Na conductance is pathologically prolonged, as it is in the case of DDT poisoning, the I_{SS} curve gives evidence of that, as shown by the dashed I_{SS} (DDT) curve in Fig. 7b. The curve indicates that the Na current is not turned off in the normal way and that the steady-state current ultimately attained is composed of both outward K current and inward Na current. I_{SS} is the algebraic sum of these two currents, resulting in depression of the I_{SS} curve. If TTX is added to the DDT-treated nerve, its specific blocking action on Na conductance eliminates the transient inward current entirely, and the I_{SS} curve appears as shown in Fig. 7b, curve I_{SS}(DDT+TTX). The depression of this curve below the normal I_{SS} curve indicates that the K conductance in this treated nerve is reduced. This suppression of K current and the prolongation of the Na conductance constitute two of the main effects of DDT on axonal membrane (Narahashi, 1971).

3. Junctional Transmission

The development of the concept of chemical mediation of synaptic transmission has been traced by Eccles (1964) from early theories through a broad range of compelling evidence which establishes it as an accepted fact. The evidence is derived from both vertebrates and invertebrates, and from interneuronal junctions in the central and peripheral nervous systems and from neuromuscular junctions. The largest part of it concerns cholinergic transmission, partly because the most thoroughly investigated synapse has been the vertebrate cholinergic neuromuscular junction. For that reason, and because the two following chapters deal specifically with cholinergic mechanisms, the discussion here will be on cholinergic synapses.

The architecture of synaptic machinery varies from simple contacts between cells to elaborate multibranched presynaptic structures forming many contacts with large dendritic trees on one or more postsynaptic cells (Bullock and Horridge, 1965; Palay and Chan–Palay, 1974).

In spite of those complexities, however, for the present purposes the synaptic junction can be depicted in simple form with emphasis on those

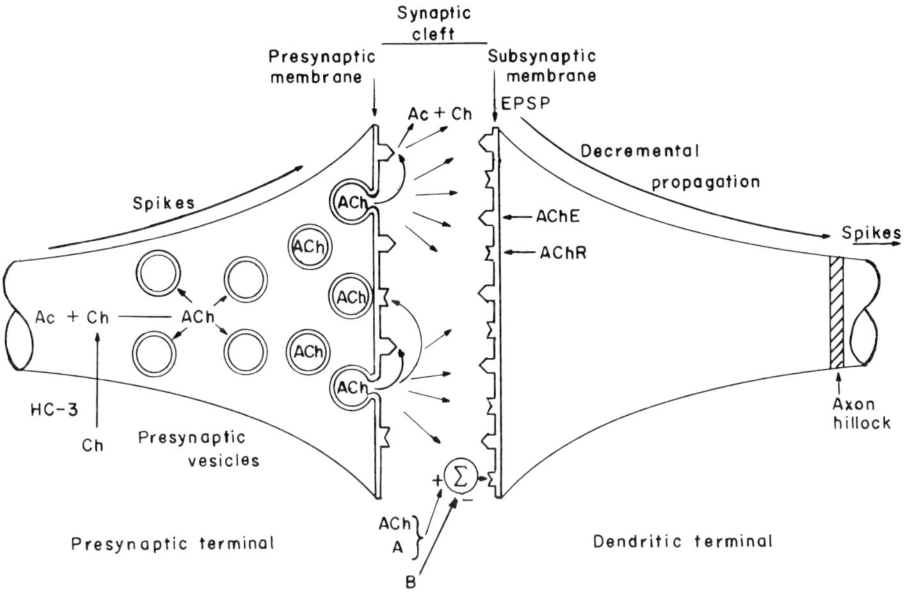

Fig. 8. Schematized cholinergic synapse showing elements of interneuronal and neuromuscular junctions. See text for explanation.

aspects of transmission which have relevant toxicological implications. Such a schematized cholinergic synapse is shown in Fig. 8. Although it is depicted as an interneuronal junction of the simplest sort, it contains the toxicologically relevant features of a neuromuscular synapse, also. Most of the basis for the discussion that follows is available from Eccles (1964) and McLennan (1970), and additional references will be cited only when necessary to supplement those works.

3.1. Synaptic Structure

The important structural features of the synapse include the presynaptic terminal with its specialized presynaptic membrane through which ACh is released from vesicles into the synaptic cleft. The vesicles have been well established as repositories of releasable stores of ACh (Whittaker, 1972). In vertebrates they range from 400 to 600 Å in diameter (Pappas and Waxman, 1972) and are found in about the same size range in invertebrates (Bullock and Horridge, 1965). The latter also possess vesicles as large as 2000 Å in diameter, but they are probably not cholinergic (Bullock and Horridge, 1965).

There is cytochemical evidence that presynaptic membrane in some synapses contains membrane-bound acetylcholinesterase (AChE) and pharmacological evidence for the presence of acetylcholine receptors (AChR) (Koelle, 1963). Both of these elements are shown in Fig. 8.

The pre- and postsynaptic membranes are separated by a synaptic cleft about 200–500Å wide (Akert *et al.*, 1972). There is evidence that in certain kinds of synapses the cleft contains some formed elements (De Robertis, 1966), but the exact nature of the cleft is still uncertain (Akert *et al.*, 1972). Whatever its makeup, no toxicological significance has been ascribed to any formed elements in the cleft except those at the membrane–cleft interface.

The presynaptic membrane is faced by an approximately equal area of specialized membrane on the postsynaptic side of the cleft called the subsynaptic membrane, which contains membrane-bound AChE and AChR. The latter is surrounded by nonspecialized membrane which extends to the axon hillock (Fig. 8), where it joins the axonal membrane.

3.2. Events in Transmission

3.2.1. Presynaptic Events

a. The Presynaptic Potential. Chemical transmission is initiated by the invasion of the presynaptic terminal by a depolarization in the form of a spike, or by electrotonic spread from a spike which stops short of the terminal membrane. In the generalized case (*cf.* Bishop, 1956), the depolarization resulting from each spike has a finite time course, decaying exponentially. The invading spikes cause the release of ACh into the synaptic cleft, and the amount released per unit time is a direct function of the frequency of incoming spikes, and not of the absolute voltage across the depolarized membrane (Auerbach, 1972). That is, a terminal which is normally depolarized by an incoming spike from say −70 V to −40 mV would release more transmitter than it would if it were initially partially depolarized and an incoming spike caused a depolarization from, say, −60 V to −40 mV.

b. Transmitter Release. Although the mechanism of transmitter release is not understood, some of its functional and pharmacological characteristics are known. The quantal nature of release, i.e., the release of packets of ACh of defined size, has been well established, and the evidence that the vesicles themselves contain the quantal units is convincing. The early evidence reviewed by Eccles (1964) has been supported and extended by more recent studies reviewed by McLennan (1970) and Pappas and Purpura (1972). The release of ACh from the vesicles through the presynaptic membrane appears to be by exocytosis, a process in which the presynaptic and vesicular membranes fuse, probably at specialized sites, with the subsequent formation of a hole from the lumen of the vesicle to the synaptic cleft (Akert *et al.*, 1972). Pfenninger *et al.* (1972) have discussed evidence that in ganglionic synapses in various vertebrates the exocytosis is reversible. That is, following release of the transmitter the synaptic vesicle closes and detaches from the presynaptic membrane. Ceccarelli *et al.* (1973) described a similar process in frog neuromuscular junctions. However, Heuser and Reese (1973) and Heuser *et al.* (1974), working with frog neuromuscular junctions, and Pysh and Wiley (1974), working with cat sympathetic ganglia, presented evidence for irreversible exocytosis. That is, the release of transmitter was followed by incorporation of the vesicular membrane into the presynaptic membrane with a consequent expansion of the terminal membrane and depletion of vesicles from the terminal. Subsequently, vesicles were reformed from cisternae which developed from the terminal membrane (Heuser and Reese, 1973).

From a toxicological standpoint, these conflicting reports are not entirely trivial. There is evidence reviewed by Hubbard and Quastel (1973) that the cyclical production of vesicles as described by Heuser and Reese (1973) is blocked by β-bungarotoxin and black widow spider venom by preventing their formation from the terminal plasma membrane. This issue does not bear heavily on insecticide action, however.

Estimates of the number of molecules of ACh per quantal unit at frog neuromuscular junction have varied from 10,000 to 60,000 (Auerbach, 1972), and in cerebral cortex of guinea pig and coypu from 1500 to 5000 (Whittaker and Sheridan, 1965). It has been estimated that about 150 quanta/impulse are released from amphibian and mammalian neuromuscular junctions (McLennan, 1970) and about 100/impulse in amphibian sympathetic ganglia (Nishi *et al.*, 1967).

In junctions capable of sustained transmission, the loss of transmitter from the immediate vicinity of the presynaptic membrane is compensated by its mobilization, involving migration of vesicles from more remote sites (Eccles, 1964; McLennan, 1970). Additional replacement is afforded through synthesis of new stores of ACh, a phase of transmitter turnover discussed below.

In a critical analysis of delay in transmission across the frog neuromuscular junction, Katz and Miledi (1965) found convincing evidence that at least 80% of the 0.5–2.6 msec required for transmission was due to the release mechanism. Synaptic delay at a variety of junctions in vertebrates and invertebrates is of this same order of magnitude (Eccles, 1964), and it is reasonable to suspect that in those junctions, too, the major portion of the delay is due to the release mechanism. Calcium is essential for ACh release as it is for many other neurosecretory processes (Rubin, 1970), and the calcium hypothesis as it applies to synaptic transmission has been discussed by Auerbach (1972). In essence, the hypothesis holds that presynaptic depolarization increases the permeability of the membrane to calcium. Because of its concentration gradient, calcium diffuses into the presynaptic terminal, where it facilitates the fusion of the vesicular and presynaptic membranes at specific release sites. Magnesium, which is known to inhibit the release of ACh, is presumed to compete with calcium at the release sites. Calcium is extruded from the terminal by an active process.

Koelle (1963) reviews cytochemical and pharmacological evidence that in the superior cervical ganglion of the cat the presynaptic membrane is the major site of junctional AChE and that it contains cholinoceptive sites (AChR). He also presents evidence that ACh released by afferent spikes acts on the cholinoceptive sites to cause additional release of ACh, thus amplifying transmitter release. It is hypothesized that the presynaptic AChE serves to prevent undue perpetuation of that process (Koelle, 1963). Inhibition of junctional AChE under these circumstances, then, would lead to excessive release of ACh and abnormally facilitated synaptic transmission.

3.2.2. Processes in the Synaptic Cleft

The postsynaptic phase of transmission begins with the binding of presynaptically liberated ACh to the receptor sites (AChR) incorporated in the subsynaptic membrane. This produces a transducer action which leads to a sequence of electrical events in the postsynaptic cell which will be discussed shortly.

A study by Katz and Miledi (1965) on frog neuromuscular transmission indicated that within a few milliseconds after the release of a quantum of ACh most of the molecules that were going to reach the AChR had done so. Evidence from several sources suggests that a significant number of molecules never combine with the receptor. There is no evidence that there are directive forces guiding ACh to the subsynaptic membrane or that the diffusion process in the cleft is very different from that expected in free solution (Katz and Miledi, 1965, 1973; Eccles and Jaeger, 1958). Salpeter and Eldefrawi (1973) estimated that a single presynaptic impulse liberated enough ACh into the cleft of rat diaphragm nerve muscle junction to produce a concentration there of about 10^{-5} M. In that same muscle estimates agree on about 8500 AChR and an approximate equal number of AChE esteratic sites per square micrometer of subsynaptic membrane (Salpeter *et al.*, 1972; Salpeter and Eldefrawi, 1973; Porter *et al.*, 1973). Furthermore, the AChR and esteratic sites have been estimated to each occupy about 20% (Porter *et al.*, 1973) to 60% or more (Fertuck and Salpeter, 1974) of the area of the subsynaptic membrane. Therefore, molecules of ACh diffusing freely in the cleft would have approximately equal probability of colliding with either a receptor or an esteratic site. Katz and Miledi (1973), in a consideration of the kinetics of the transmission process, estimated that the subsynaptic AChE ensures that few molecules of ACh survive after one or two receptor occupations. They further estimated, however, that even when AChE had been inhibited only about two-thirds of the ACh molecules in a quantal unit liberated at frog motor end plates ever came in contact with subsynaptic AChR. Those that did not presumably diffused out of the cleft before making contact. Eccles and Jaeger (1958) estimated that in synapses of simple configuration like those in frog muscles about half of the ACh in the cleft could diffuse out of the synaptic region within 0.2 msec.

These studies combine to suggest that normally only a fraction of the ACh released at the neuromuscular junction ever comes in contact with AChR. The inhibition of pre- and postsynaptic AChE, however, would prolong transmitter action by allowing repeated collision of ACh with the subsynaptic membrane, increasing the probability of receptor activation. Katz and Miledi (1973) suggest additionally that where AChE has been inhibited the diffusion of ACh from the cleft is delayed by the binding that occurs between ACh and AChR during repeated collision as the transmitter diffuses out of the cleft.

3.2.3. Postsynaptic Events

a. Receptor Activation and Electrogenesis. The activating or agonistic action of ACh on the AChR is presumed to involve some perturbation in the latter which results in changed membrane permeability to specific ions (Ehrenpreis *et al.*, 1969). In excitatory synapses, permeability (i.e., conductance) to both Na^+ and K^+ is increased (Grundfest, 1969). The potential changes to be expected from such increased conductances were discussed in Section 2.2.1. Depending on the relative and absolute changes in g_{Na} and g_K, the potential produced would fall somewhere between the equilibrium potentials for the two ions and is always (in excitatory synapses) a depolarization, called an *excitatory postsynaptic potential* or EPSP. Although it is the membrane conductance which is directly related to the degree of receptor activation, the EPSP is proportional to it and is often used in physiological and pharmacological studies as a parameter of receptor activation.

The EPSP produced at the subsynaptic membrane is transmitted decrementally along the postsynaptic membrane to the axon hillock at the juncture of the dendritic and axonal membrane (Fig. 8). The amount of decrement is a function of the space constant of the membrane, but in a given synapse the depolarization at the axon hillock is probably a constant proportion of the EPSP. Nondecrementally propagated action potentials arise at the axon hillock at a repetition rate which is a function of the amplitude of the EPSP.

The relationship between EPSP and spike repetition rate in a generalized synapse is shown in Fig. 9. Figure 9a shows EPSP of indefinite time course, and Fig. 9b shows the related spike repetition rate. The EPSP is characterized by a dynamic range within which spike activity is generated. Spikes are not generated by depolarizations less than some threshold or greater than some upper limit. At intermediate levels, the spike repetition rate is proportional to the amplitude of the EPSP. Synapses which do not transmit because the EPSP has exceeded its dynamic range are said to be under a *depolarizing block*. Figure 9c shows the appearance of an oscillogram produced by a single sweep of an oscilloscope beam covering the entire duration of an EPSP which exceeds its dynamic range. There may be, at various kinds of synapses, modification of this simple proportional relationship between EPSP and spike repetition rate. Adapting synapses, for example, may show reduced rates with prolonged presynaptic stimulation, even though the EPSP is maintained.

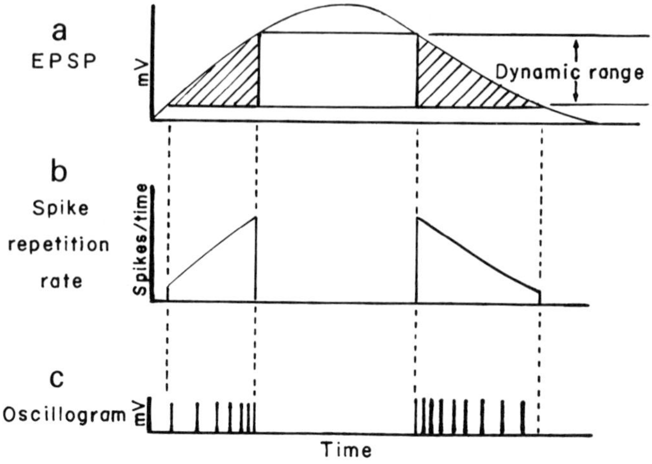

Fig. 9. Functional characteristics of an excitatory synapse showing the relationships between the excitatory postsynaptic potential (EPSP) and spike repetition rate in the postsynaptic fiber. See text for explanation.

These functional features of synapses have important toxicological implications, some of which will be discussed in detail in Section 4. It may be briefly stated here, however, that some drugs and poisons may produce both excessive synaptic excitation and depolarizing blocks by mimicking the action of the natural transmitter, as, for example, muscarine and nicotine mimic ACh. On the other hand, nondepolarizing blocks may be produced by drugs which block the action of the natural transmitter, as atropine and curare block ACh. Finally, still other poisons, such as the organophosphate and carbamate insecticides, produce excessive synaptic activity or depolarizing blocks indirectly, by inhibiting AChE, which normally limits the duration of action and accumulation of presynaptically liberated ACh.

b. Theories of Drug–Receptor Interaction: The transducer action produced in the receptor is the consequence of a specific interaction between the transmitter molecule, or in the general case an agonist molecule, and the receptor molecule itself. The exact nature of this interaction is not understood, but there are two predominant theories which purport to account for the kinetics of the process. Both theories concern drugs in general, and not just transmitter molecules.

The "occupation" theory has been comprehensively treated by Ariëns and Simonis (1964), and the "rate" theory was authored and described by Paton (1961). Ariëns and Simonis (1967) and Paton (1967) compared the two theories in their implications in experimental pharmacology. Only the essential elements of the theories will be considered here.

Occupation theory is based on the kinetic scheme shown in equations (6), (7), and (8):

$$[A]+[R] \underset{k_2}{\overset{k_1}{\rightleftharpoons}} [AR] \overset{\alpha}{\rightarrow} [AR^*] \tag{6}$$

$$P = \frac{[A]}{[A]+k_2/k_1} \tag{7}$$

$$E = \alpha p \tag{8}$$

in which [A] is the concentration of agonist at receptor site, [R] is the concentration of receptor sites, [AR] is the concentration of agonist–receptor complex, [AR*] is the concentration of activated agonist–receptor complex, k_1 is the association rate constant, k_2 is the dissociation rate constant, k_1/k_2 is the "affinity" of A for R, α is the "intrinsic activity" of A, p is the proportion of receptor sites occupied by A, and E is the quantitative expression of the biological response produced.

The theory holds that the agonist combines reversibly with the receptor, the ratio of the rate constants k_1/k_2 being a measure of the strength of the binding, or "affinity" (Ariëns and Simonis, 1964) of the drug for the receptor. Equation (7) defines the proportion of the receptors occupied by A. The activation of the receptor to R^* represents the transducer action, and the term [AR*] is considered to be quantitatively manifested by the biological response produced in the tissue, e.g., an EPSP. The ability of A to produce the activated state of R when bound to the receptor is called the *intrinsic activity* (Ariëns and Simonis, 1964) or *efficacy* (Stephenson, 1956) of A, and the magnitude of the response is defined by the product of the intrinsic activity (α) and the proportion of receptors occupied (equation 8). By definition, a drug capable of producing the maximal response of which the tissue is capable, no matter what its concentration, has an intrinsic activity of 1. A drug with both high affinity and unit intrinsic activity could produce the maximal response at low concentration. On the other hand, a drug of high affinity and zero intrinsic activity would combine strongly with the receptor without activating it and thus antagonize the effects of agonists. Such drugs are called antagonists, denoted as "B" in Fig. 8. Drugs having intermediate values of α may show both agonistic and antagonistic effects and are called *partial agonists.*

Rate theory is based on the kinetic scheme shown in equation (9), (10), and (11):

$$[A]+[R] \underset{k_2}{\overset{k_1}{\rightleftharpoons}} [AR^*] \tag{9}$$

$$\rho = \frac{k_2[A]}{[A]+k_2/k_1} \tag{10}$$

$$E = \phi\rho \tag{11}$$

where [A], [R], [AR*], k_1, k_2, and E have the same meaning as in occupation theory, and ρ is the rate of association of A with R and ϕ is a constant relating activity of chemical stimulation to the biological response.

In this scheme, the degree of receptor activation is a function of the rate at which agonist–receptor complexes are formed. In the equilibrium condition, maximal receptor activation, and consequently maximal biological effect, depends on continual and rapid association and dissociation of receptor and agonist. A drug which binds irreversibly with R is therefore, by definition, an antagonist, and one which is characterized by an intermediate rate of association would be comparable to the partial agonist of occupation theory.

The two theories do not predict different experimental results under equilibrium conditions, but do predict differences in transient events such as rate of onset of stimulant action, "fading" of response as equilibrium conditions are approached, and responses to successive doses of drugs. The points of difference between the theories and many of the problems of interpretation are discussed in considerable detail by Paton (1961, 1967) and Ariëns and Simonis (1967).

In both theories, termination of receptor activation occurs with the dissociation of the ACh–AChR complex. In the conventional view (Eccles, 1964; McLennan, 1970), this dissociation is mediated by the hydrolysis of ACh by AChE. Katz and Miledi (1973), however, hold that inhibition of AChE does not prolong the gating effect of ACh on the receptor and that the role of AChE is to prevent prolongation of transmitter action which would occur on repeated collision of ACh with AChR.

3.3. Transmitter Turnover

Whatever the mechanism of termination of transmitter action, ultimately ACh is hydrolyzed by AChE in the region of the synapse, liberating acetate and choline. Birks and MacIntosh (1957, 1961), on the basis of strong evidence, hypothesized that subsequently, choline is resorbed by the presynaptic terminal, where it is used to synthesize ACh to replenish releasable stores (Fig. 8). MacIntosh (1959) presented evidence that choline resorption is mediated by an active choline transport system which is competitively inhibited by the drug hemicholinium-3 (HC-3, Fig. 8).

This reuse of choline after the liberation of ACh is called transmitter turnover and has been shown to be important for sustained synaptic transmission. Working with cat superior cervical ganglion, Birks and MacIntosh (1961) found no reduction in postsynaptic output from prolonged administration of preganglionic stimuli of 20/sec, indicating no serious depletion of transmitter. That indication was supported by their estimates, based on bioassay of ACh in irrigation fluids, that the ganglion was releasing about 31 ng ACh/min and synthesizing about 29 ng/min.

McCandless *et al.* (1971), Čapek *et al.* (1971), and Esplin and Zablocka-Esplin (1971) used functional electrophysiological criteria to estimate transmitter turnover in various kinds of vertebrate cholinergic junctions. They determined the amount of ACh liberated per presynaptic volley as a fraction of the amount available for release and the rate of synthesis as a fraction of the amount of releasable ACh in the rested ganglion. Their studies supported the earlier conclusions of Birks and MacIntosh (1957) that releasable ACh is stored as a small immediately releasable fraction and a larger fraction called *depot ACh* mobilized after stimulation begins. Transmitter release was shown to be a first-order process, the amount released per presynaptic volley being a function of the amount of releasable stores present at the time. Replenishment was a zero-order process which proceeded at a fixed rate. Depending on the stores of ACh, and the synthesis rate, then, a synapse should be capable of continual transmission at or below a stimulation frequency at which release is balanced by synthesis.

4. Comparative Pharmacology

There are similarities in the form and function of nervous systems throughout the animal kingdom which account for a commonality of sensitivity to neuroactive compounds, while at the same time there are phylogenetically based differences which lead to variation in degree and/or loci of action of those same compounds. Similarities can be seen in the basic organization of nervous systems, in many processes such as electrogenesis and junctional transmission, and in biochemical entities such as chemical transmitters, receptors, and enzymes. And yet specializations imposed on these same features in different animals account for toxicologically significant differences among them.

4.1. Neurological Similarities and Drug Action

4.1.1. Structure and Organization

The quintessence of nervous systems is prescribed in the neuron doctrine, due mainly to Ramón y Cajal, and briefly stated by Bullock and Horridge (1965) thus: "all nervous systems consist in essence (whatever other nonnervous elements may be present) of distinctive cells called neurons, which are specialized for nervous functions and which produce prolongations and branches." From the functional standpoint, the important prolongations and branches are axons and dendrites. Axons are especially adapted for the transmission of information, often over considerable distances, as described in reference to Fig. 1. Dendrites are branches either from the cell body of a neuron or from an axon which are especially adapted for input. Input may be provided from other neurons by way of synapses, or in the case of sensory neurons by a suitable stimulus imposed on specialized sensory structures.

Among animals having central nervous systems, which includes flatworms and all higher forms, there is a basic organizational scheme. The CNS contains most of the motor and internuncial neuron cell bodies, while the peripheral system contains sensory cell bodies, various ganglia associated with the viscera, the body wall, and sensory and motor pathways, and the sensory and motor axons that make up the peripheral nerves. This organization provides the basis not only for reflex responses, present in the simplest animals, but also, with increasing complexity of the CNS in higher animals, for increasing central control over function. Even in mammals, however, some of the simplest reflexes, such as withdrawal from a noxious stimulus, persist in relatively uncomplicated form, requiring no input from the brain. Certain toxicological implications of this will be described below.

4.1.2. Electrogenesis and Drug Action

The basic similarity of the bioelectrical processes of neurons accounts for common responses among different animals to certain drugs and poisons. Thus, because the resting potential of neurons generally depends on high internal and low external potassium, they are generally depolarized by high external potassium. Similarly, action potentials in nerves and muscles of many animals are sodium dependent. Tetrodotoxin (TTX) and saxitoxin (STX) are specific blockers of Na activation in spike-producing membrane, and those toxins block axonal conductance in such diverse animals as marine worms, crustaceans, insects, cephalopods, and vertebrates (Evans, 1972). Interestingly, however, the nerves and muscles of certain newts and puffer fish, both sources of TTX, show remarkable tolerance to both toxins, even though action potentials in those tissues are sodium dependent (Evans, 1972). The basis for the tolerance is not known.

Batrachotoxin (BTX), from the Colombian poison arrow frog, on the other hand, causes an irreversible increase in sodium permeability in a variety of irritable membranes, so that the cells become depolarized and incapable of electrical responses (Albuquerque *et al.*, 1971). On axonal membrane, BTX, causing increased Na permeability, and TTX, causing reduced permeability, are antagonistic, and the toxic effects of both poisons are reduced when they are administered simultaneously (Albuquerque *et al.*, 1971).

The action of DDT on axonal membrane appears to be the same in invertebrates (Narahashi, 1971) and vertebrates (Hille, 1968) in producing instability and repetitive discharge. However, even within a single species, the sensitivity of different parts of the nervous system varies. Roeder and Weiant (1945) found, for example, that the sensory nerves in cockroach leg were more than 1000 times as sensitive to DDT as motor nerves in the same leg. Part of the DDT-poisoning syndrome in cockroaches is brought on by this differential sensitivity, in that repetitive discharge in sensory nerves causes reflex tremor in the motor system (Roeder and Weiant, 1948). In rats, although there is considerable brain involvement in DDT-poisoning syndrome, part of the typical DDT tremor is due to a reflex process exactly comparable to that in cockroaches (Shankland, 1964).

Shanes (1958) reviewed work on the veratrine alkaloids which gives further testimony to the similarity of axonal properties in diverse animals. Veratridine prolongs the negative afterpotential of spikes to the same extent in squid giant axons and nerves of crabs and frogs. Cevadine is less effective in this regard but is about equally effective in all three species. Veratrine, which contains a mixture of the two compounds, produces an intermediate effect, again comparable in all three animals.

4.1.3. Junctional Transmission and Drug Action

Chemically mediated junctional transmission appears in much the same form among diverse animals, and again is the basis for the action of some drugs against a broad spectrum of animal species. The quantum hypothesis of synaptic transmission was based on work with the cholinergic frog neuromuscular junction (del Castillo and Katz, 1954). Since that early work, the hypothesis has been extended by the demonstration that the presynaptic vesicles contain the quantal units of transmitter, and the extended hypothesis has been found to apply to cholinergic and noncholinergic junctions in central and peripheral sites in nervous systems of a great range of animal species (Gerschenfeld, 1973; Hall, 1972).

There appears to be a universal requirement for calcium in the release of neurotransmitter, and, in fact, in most neurosecretory processes (Rubin, 1970). Magnesium competes with calcium antagonistically in this process at many junctions, this fact apparently accounting for the depressant action of magnesium at such diverse sites as vertebrate neuromuscular junctions (Eccles, 1964), insect neuromuscular junctions (Smyth *et al.*, 1973), frog sympathetic ganglia (Blackman *et al.*, 1963), and insect central ganglia (Shankland *et al.*, 1971).

Black widow spider venom provides further evidence on the similarity of release mechanisms for various transmitters. This venom acts to greatly increase the spontaneous release of ACh from vertebrate central synapses (Simpson, 1974) and neuromuscular junctions (Hubbard and Quastel, 1973), from catecholamine-releasing terminals (Simpson, 1974), and from both inhibitory and excitatory insect neuromuscular junctions (Smyth *et al.*, 1973), which are, respectively, GABAnergic and probably glutaminergic (Usherwood, 1967).

The common features of transmitter release mechanisms, then, may account for the common effects of a drug on diverse junctions. Alternatively, the occurrence of the same transmitter in diverse animals may account for the common effects of a drug. For example, the antagonistic effects of picrotoxin on inhibitory systems in vertebrate CNS and in peripheral systems of invertebrates is well known (Curtis, 1963). The evidence for an inhibitory transmitter role of GABA in vertebrates is strong (Obata, 1972), and in invertebrates it is compelling (Gerschenfeld, 1973). The similar effects of picrotoxin at these different loci in such diverse animals appear almost certainly due to the common role of GABA in the two systems.

The widespread occurrence of cholinergic systems among animals is of more immediate relevance to the action of many organic insecticides. Florey (1961) reviewed biochemical evidence indicating the presence of ACh and/or AChE in all major groups of animals except coelenterates from one-celled forms to vertebrates. Crescitelli and Geissman (1962) and Kosterlitz (1967) reviewed pharmacological evidence on cholinergic transmission in coelenterates and higher animals, and, again, Coelenterata was the only major phylum showing no evidence of cholinergic transmission. However, the finding that ACh causes massive discharge of statocysts in hydra and that this action is blocked by the specific ACh antagonists *d*-tubocurarine and hexamethonium casts some doubt on even that exclusion (Van der Kloot, 1966). Of the animals with central nervous systems (flatworms and higher), the occurrence of cholinergic transmission in all of them has some basis and is well established in most of them (Gerschenfeld, 1973; Sakharov, 1970; Tauc, 1967).

Although cholinergic junctions may appear in various parts of any given nervous system, they do not necessarily appear in comparable parts in different systems. For example, there is evidence for cholinergic neuromuscular transmission in round and segmented worms, mollusks (Gerschenfeld, 1973), and vertebrates (Hubbard and Quastel, 1973). In crustaceans and insects, however, the evidence favors glutamate as the excitatory neuromuscular transmitter (Gerschenfeld, 1973). On the other hand, there is evidence for cholinergic transmission in the CNS of all animals from segmented worms to vertebrates, including crustaceans and insects (Gerschenfeld, 1973; Tauc, 1967). Additionally, in vertebrates the parasympathetic portion of the autonomic nervous system contains peripheral ganglia which are cholinergic. The peripheral ganglia of invertebrates, on the other hand, such as those associated with the stomatogastric system in crustaceans and insects, although not well defined pharmacologically, do not appear to be cholinergic (Davey, 1964).

In spite of the differences in location of cholinergic junctions among diverse animals, their presence accounts for the general effectiveness of many cholinergic drugs and poisons. A review by Whittaker (1963) of the bioassay systems for ACh illustrates this point in that they include venus clam heart, leech muscle, frog muscle, and pressor responses in cats. The toxicity of the anticholinesterase insecticides to both vertebrates and invertebrates is so well known as to need no documentation here. The curariform drugs antagonize cholinergic transmission peripherally in vertebrates (Karczmar, 1967) and centrally in insects (Flattum *et al.*, 1967; Larsen *et al.*, 1966; Shankland *et al.*,

1971). The drug hemicholinium-3, by blocking ACh turnover, blocks vertebrate neuromuscular junction (Birks and MacIntosh, 1957), vertebrate sympathetic ganglia (Birks and MacIntosh, 1961), and insect central ganglia (Shankland *et al.*, 1971).

The very occurrence of cholinergic transmission in any animal implies the presence of acetylcholine receptors (AChRs), and the common effectiveness of the cholinomimetic, nicotine, suggests some similarity in the properties of receptors in the two kinds of animals. In vertebrates, AChRs are classified on the basis of their location and pharmacological properties (Barlow, 1964). Muscarinic receptors are characteristically associated with smooth muscle, respond to muscarinic drugs such as muscarine, acetyl-β-methylcholine, and arecoline, and are blocked by atropine. Nicotinic receptors are of two types, one associated with neuromuscular junctions and one with autonomic ganglia. Both types are activated by nicotine, but muscle receptors are preferentially blocked by *d*-tubocurarine, and not by hexamethonium, while the reverse is true of ganglionic receptors. Karczmar (1967) points out that cholinergic transmission is essentially similar at neuromuscular junctions, at autonomic ganglia, and at certain CNS synapses, such as those between motor nerve collaterals and Renshaw cells (*d* in Fig. 1). He further explains that junctions once thought to be purely nicotinic are now known to show muscarinic responses, and the vertebrate skeletal neuromuscular junction is the only remaining nicotinic site presumably devoid of muscarinic receptors.

Even though invertebrates show responses to muscarinic and nicotinic drugs, Sakharov (1970) contends that the terms "muscarinic" and "nicotinic" do not apply to invertebrate receptors. However, Hubbard and Quastel (1973) describe species-related and site-related differences in the pharmacological characteristics of mammalian nicotinic receptors which indicate that even among those animals the term "nicotinic" does not define a single macromolecule. Furthermore, Kosterlitz (1967) describes pharmacological responses from vertebrates and invertebrates which suggest that intergrades between muscarinic and nicotinic receptors occur in both kinds of animals. Thus the use of muscarinic and nicotinic in reference to vertebrate AChR is probably an oversimplification and even in those animals might become less rigidly defined in time.

In any event, the receptors in insects show nicotiniclike and muscariniclike pharmacological properties. For example, in vertebrates, nicotinic receptors in both ganglia and muscles become desensitized on prolonged exposure to nicotine (Paton and Perry, 1953; Magazanik and Vyskočil, 1973). The so-called nicotinic block involves first a depolarization of the subsynaptic membrane and second, as desensitization occurs, repolarization to the normal resting potential. In the desensitized neuromuscular junction, where only nicotinic receptors occur, transmission is completely blocked. In autonomic ganglia, on the other hand, where muscarinic and nicotinic receptors are mixed, the desensitized ganglion, although nonresponsive to nicotine, is responsive to muscarinic drugs (Paton and Perry, 1953). The sixth abdominal ganglion of the American cockroach behaves in exactly this same way, in that it becomes desensitized on prolonged exposure to nicotine, and the desensitized ganglion is responsive to muscarinic drugs (Flattum and Shankland, 1971).

Furthermore, nicotinic receptors in vertebrate ganglia are relatively very sensitive to the antagonistic action of hexamethonium and insensitive to the depolarizing action of decamethonium (Barlow, 1964). Again, receptors in the sixth abdominal ganglion of the American cockroach are comparable in that they are sensitive to about 10^{-6} M hexamethonium but respond only weakly to 10^{-3} M decamethonium (Shankland *et al.*, 1971). Thus not only does this insect possess both muscariniclike and nicotiniclike receptors but also they are comparable to those in vertebrates, even to specific pharmacological characteristics.

From all of this, it is apparent that cholinergic junctional transmission endows animals generally with susceptibility to cholinergic drugs and poisons which act by a variety of mechanisms. Inhibitors of AChE, cholinomimetics, ACh antagonists, and drugs which block choline transport exert similar action at cholinergic junctions in widely diverse animals.

4.2. Neurological Dissimilarities and Drug Action

4.2.1. Structure and Organization

There are modifications imposed on the basic organizational scheme described in Section 4.1.1 which distinguish invertebrates from vertebrates (Bullock and Horridge, 1965). The axonal or fibrous portion and the nucleated portion of the CNS in all animals are divided into two zones, but the division is accomplished in distinctive ways in the two groups of animals. The ganglia in invertebrate CNS are composed of an outer rind of nucleated cell bodies and a central core of fibers and synapses. The core is separated into tracts and neuropil, the former being composed of axons en route, the latter being a plexus of axon terminals and dendrites and associated synapses. None of the conductive fibers enters the cell rind. The CNS of vertebrates, on the other hand, is characterized by two zones, referred to as *gray matter* and *white matter.* Although white matter is primitively peripheral to gray matter, the reverse arrangement is found in higher brain centers. Gray matter contains cell bodies of neurons as well as axonal and dendritic nerve endings and synapses. White matter is composed exclusively of axons en route. A corollary to the distinct arrangement in the two groups of animals is that synapses between axon terminals and cell bodies (i.e., axosomatic synapses) are common in vertebrates and rare in invertebrates. Synapses in the latter are generally of the axoaxonic or axodendritic types.

A further difference between the vertebrate and invertebrate CNS having profound pharmacological implications is the presence in the former of a blood–brain barrier. The barrier, which prevents the movement of a variety of electrolytes, colloids, vital dyes, and many drugs from the blood into the CNS, is a complex system of structural features and metabolic elements (i.e., enzymes and transport systems) and for that reason has been called by Lajtha (1962) the *brain barrier system.* It is generally conceded that such a barrier does not occur in invertebrates, but insects are an exception (Treherne and Pichon, 1972). In insects, too, the barrier involves both structural and metabolic elements and should be viewed as a barrier system. In vertebrates, the barrier does not protect the autonomic ganglia, and the associated cholinergic junctions are therefore readily accessible to a variety of drugs, including exogenously administered ACh. In insects, on the other hand, cholinergic junctions appear to be confined to the CNS, where they are protected by the barrier. This difference has accounted for a long-standing and striking pharmacological difference between insects on the one hand and other invertebrates and vertebrates on the other. Various invertebrate and vertebrate cholinergic systems show responses to exogenous ACh in concentrations equal to or less than about 10^{-7} M (Whittaker, 1963). The nerves of insects, however, most notably of the American cockroach, have been singularly insensitive to ACh, showing little or no response to concentrations as high as 10^{-3} M (Tauc, 1967). Various experimental approaches to circumvent the barrier system in that insect led finally to the demonstration that cholinergic junctions in the CNS were responsive to 10^{-7} M ACh, placing them in the range of sensitivity of other confirmed cholinergic systems (Shankland *et al.*, 1971).

O'Brien (1967) reviewed evidence that showed the CNS barrier system of aphid and cockroach to be comparable in resistance to penetration by large and polar molecules. The relatively greater toxicity of ionic anticholinesterase agents to vertebrates than to insects is due at least in part to the absence of the barrier system in vertebrate peripheral cholinergic ganglia and its presence in insect CNS.

A structural feature of nerves called *myelination* or *medullation* which occurs commonly in vertebrates and only rarely in invertebrates has been implicated by inference in a special sensitivity of certain animals to the action of certain organophosphorus compounds. Medullated nerves are characterized by layers of membrane or myelin tightly wrapped around individual axons forming a thick insulating sheath which is interrupted along its length by gaps called nodes of Ranvier. Through a phenomenon called *saltatory conduction,* thoroughly discussed by K. Cole (1968), this structure allows for very high speeds of nervous conduction. This affords nervous systems of higher animals many advantages related to the rapid transmission of information, including short response times and tight control over motor activity.

Davies (1963) has reviewed a kind of toxic action which was sometimes erroneously called simply "demyelination." The action was shown by only certain triaryl- and alkyl organophosphorus compounds. Chicken, cat, and man appear to be especially susceptible to the action, which develops after a week or two following poisoning. The symptoms are not at all related to anticholinesterase action and involve a paralysis of the upper and lower extremities. An early report on the histopathology of the condition referred to a specific demyelination. Later histological studies showed that degenerative changes occurred in both myelin sheaths and axons mainly in the sciatic nerve and in nerves of the spinal cord and medulla oblongata. Because the histopathology of the process involves more than demyelination, it has been called "delayed neurotoxic action" in more recent literature. Whereas "demyelination" was inaccurate, "delayed neurotoxic action" is lacking in precise meaning relevant to the disorder.

There is an extensive literature on nerve degeneration, and with it there has developed a special nomenclature. Seitelberger (1969) has reviewed a wide range of metabolic disorders which lead to various kinds of dystrophies and atrophies. Adams and Leibowitz (1968) have reviewed demyelinating diseases and their histopathologies. The pathological changes which occur as a result of nerve section or other experimental lesions have been reviewed by M. Cole (1968) and Lampert (1968). All of these specialties have led to priority positions for various descriptive terms of "atrophy", "dystrophy," and "degeneration." Since, according to Davies (1963), the delay in the appearance of symptoms in the paralyzing phosphate poisoning is truly characteristic of the process, the term *delayed axomyelin degeneration* would serve to identify the process without violating priorities in terminology. Whatever it is called, the peculiar distribution of especial susceptibility among chicken, cat, and man suggests that the basis for the action is not simply related to the presence of medullated nerves which are found commonly among all vertebrates, some of which have been shown to be resistant to this kind of poisoning.

4.2.2. Junctional Transmission

Probably the most frequently cited neurological difference between insects and vertebrates is in muscular innervation. The vertebrate system (*cf.* Cöers, 1960), which is cholinergic, is made up of motor units, the fibers of which are innervated at a single point, often about midway along their length. The fibers are capable of producing propagated action potentials, and although some gradation of contraction is accomplished by modulation of motor nerve discharge frequency it occurs mainly through recruitment of motor units. There is no peripheral inhibitory control in vertebrate muscle. Insect systems (Usherwood, 1967) involve muscle fibers which are multiply innervated, there being many terminals 30–60 μm apart on every muscle fiber. They are incapable of producing propagated action potentials. A single muscle fiber may be innervated by two or three kinds of nerve fibers. Fast and slow nerve fibers produce twitch and slow contractions, respectively, and are both probably glutaminergic. Inhibitory fibers may innervate the same muscle fiber, and they are GABAnergic.

Therefore, whereas vertebrate muscle is sensitive to a variety of cholinergic drugs (Hubbard and Quastel, 1973), insect muscle is not (Faeder *et al.*, 1970).

5. Conclusion

The integrative functions of nervous systems and the nervous processes involved show many similarities throughout the animal kingdom. Modern synthetic insecticides in the three major groups, organophosphates, carbamates, and chlorinated hydrocarbons, exert their neurotoxic action, whatever other action may occur, on basically similar processes in widely diverse animals. Where species-selective action occurs, it appears to be due primarily to differences other than those at the neurological target. O'Brien (1961) and Hollingworth (1971) have reviewed a variety of pharmacodynamic and metabolic differences among animals that appear to account for the most notable cases of selective action. The material reviewed here suggests that although the organization of nervous systems may vary greatly the basic neurological processes do not. On the contrary, the similarity in those processes among animals is striking and confers on them a common susceptibility to a great variety of neuroactive drugs and poisons, including insecticides.

6. References

Adams, C. W. M., and Leibowitz, S., 1968, The general pathology of demyelinating diseases, *in: The Structure and Function of Nervous Tissue*, Vol. 3 (G. H. Bourne, ed.), pp. 309–382, Academic Press, New York.

Adamson, A. W., 1969, *Understanding Physical Chemistry*, Benjamin, New York.

Akert, K., 1973, Dynamic aspects of synaptic ultrastructure, *Brain Res.* **49:**511.

Akert, K., Pfenninger, K., Sandri, C., and Moor, H., 1972, Freeze-etching and cytochemistry of vesicles and membrane complexes in synapses of the central nervous system, *in: Structure and Function of Synapses* (G. D. Pappas and D. P. Purpura, eds.), pp. 67–86, Raven Press, New York.

Akkermans, L. M., van den Bercken, J., van der Zalm, J. M., and van Straaten, H. W., 1974, Effects of dieldrin (HEOD) and some of its metabolites on synaptic transmission in frog motor endplate, *Pestic. Biochem. Physiol.* **4:**313.

Albuquerque, E. X., Daly, J. W., and Withrop, B., 1971, Batrachotoxin: Chemistry and pharmacology, *Science* **172:**995.

Ariëns, E. J., and Simonis, A. M., 1964, Drug receptor interaction, *in: Molecular Pharmacology* (E. J. Ariëns, ed.), pp. 119–393, Academic Press, New York.

Ariëns, E. J., and Simonis, A. M., 1967, Cholinergic and anticholinergic drugs, do they act on common receptors? *Ann. N.Y. Acad. Sci.* **144:**842.

Auerbach, A. A., 1972, Transmitter release at chemical synapses, *in: Structure and Function of Synapses* (G. D. Pappas and D. P. Purpura, eds.), pp. 137–159, Raven Press, New York.

Barlow, R. B., 1964, *Introduction to Chemical Pharmacology*, 2nd ed., Barnes and Noble, New York.

Birks, R., and MacIntosh, F. C., 1957, Acetylcholine metabolism at nerve endings, *Brit. Med. Bull.* **13:**157.

Birks, R., and MacIntosh, F. C., 1961, Acetylcholine metabolism of a sympathetic ganglion, *Can. J. Biochem.* **39:**787.

Bishop, G. H., 1956, The natural history of the nerve impulse, *Physiol. Rev.* **36:**376.

Blackman, G. J., Ginsburg, B. L., and Ray, C., 1963, Spontaneous synaptic activity in sympathetic ganglion cells of the frog, *J. Physiol. (London)* **167:**389.

Boeckh, J., Kaissling, K. E., and Schneider, D., 1965, Insect olfactory receptors, *Cold Spring Harbor Symp. Quant. Biol.* **30:**263.

Boettiger, E. G., and Hartman, H. B., 1968, Excitation of the receptor cells of crustacean PD organ, *in: Neurobiology of Invertebrates* (J. Salanka, ed.), pp. 381–390, Plenum Press, New York.

Bratkowski, T. A., and Matsumura, F., 1972, Properties of a brain adenosine triphosphatase sensitive to DDT, *J. Econ. Entomol.* **65:**1238.

Bücherl, W., and Buckley, E. E. (eds.), 1971, *Venomous Animals and Their Venoms,* Vol. 2, Academic Press, New York.

Bullock, T. H., and Horridge, G. A., 1965, *Structure and Function in the Nervous Systems of Invertebrates,* Vols. 1 and 2, Freeman, San Francisco.

Burton, R. F., 1973, The significance of ionic concentrations in the internal media of animals, *Biol. Rev. Camb. Phil. Soc.* **48:**195.

Butterworth, J. H., and Morgan, E. D., 1971, Investigation of the locust feeding inhibition of the seeds of the neem tree, *Azadirachta indica, J. Insect Physiol.* **17:**969.

Čapec, R., Esplin, D. W., and Salemaghaddum, S., 1971, Rates of transmitter turnover at the frog neuromuscular junction estimated by electrophysiological technique, *J. Neurophysiol.* **34:**831.

Ceccarelli, B., Hurlburt, W. P., and Mauro, A., 1973, Turnover of transmitter and synaptic vesicles at the frog neuromuscular junction, *J. Cell Biol.* **57:**499.

Cöers, C., 1960, Structural organization of the motor nerve endings in mammalian muscle spindles and other striated muscle fibers, *in: The Innervation of Muscle* (H. D. Bouman and A. L. Woolf, eds.), pp. 40–49, Wilkins Company, Baltimore.

Cole, K. S., 1968, *Membranes, Ions and Impulses,* University of California Press, Berkeley.

Cole, M., 1968, Retrograde degeneration of axon and soma in the nervous system, *in: The Structure and Function of Nervous Tissue,* Vol. 1 (G. H. Bourne, ed.), pp. 269–300, Academic Press, New York.

Crescitelli, F., and Geissman, T. A., 1962, Invertebrate pharmacology: Selected topics, *Ann. Rev. Pharmacol.* **2:**143.

Curtis, D. R., 1963, The pharmacology of central and peripheral inhibition, *Pharmacol. Rev.* **15:**333.

Davey, K. G., 1964, The control of visceral muscles in insects, *Advan. Insect Physiol.* **2:**219.

Davies, D. R., 1963, Neurotoxicity of organophosphorus compounds, *Handb. Exp. Pharmakol.* **15:**860.

Davson, H., 1964, *A Textbook of General Physiology,* Little, Brown, Boston.

del Castillo, J., and Katz, B., 1954, Quantal components of the end-plate potential, *J. Physiol. (London)* **124:**560.

De Robertis, E., 1966, Synaptic complexes and synaptic vesicles as structural and biochemical units of the central nervous system, *in: Nerve as a Tissue* (K. Rodahl and B. Issekutz, Jr., eds.), pp. 88–115, Harper and Row, New York.

Eaton, J. L., and Sternburg, J., 1964, Temperature and the action of DDT on the nervous system of *Periplaneta americana* (L.), *J. Insect Physiol.* **10:**471.

Eccles, J. C., 1964, *The Physiology of Synapses,* Springer, New York.

Eccles, J. C., 1965, The synapse, *Sci. Am.* **212:**56.

Eccles, J. C., and Jaeger, J. C., 1958, The relationship between the mode of operation and the dimensions of the junctional regions at synapses and motor end-organs, *Proc. Roy. Soc. London Ser. B* **148:**38.

Ehrenpreis, S., Fleisch, J. H., and Mittag, T. W., 1969, Approaches to the molecular nature of pharmacological receptors, *Pharmacol. Rev.* **21:**131.

Esplin, D. W., and Zablocka-Esplin, B., 1971, Rates of transmitter turnover in spinal monosynaptic pathway investigated by electrophysiological techniques, *J. Neurophysiol.* **34:**842.

Evans, M. H., 1972, Tetrodotoxin, saxitoxin, and related substances: Their applications in neurobiology, *Int. Rev. Neurobiol.* **15:**81.

Faeder, I. R., O'Brien, R. D., and Salpeter, M. M., 1970, A reinvestigation of evidence for cholinergic neuromuscular transmission in insects, *J. Exp. Zool.* **173:**187.

Fertuck, H. C., and Salpeter, M. M., 1974, Localization of acetylcholine receptor by ^{125}I-labeled α-bungarotoxin binding at mouse motor end plates, *Proc. Natl. Acad. Sci. U.S.A.* **71:**376.

Flattum, R. F., and Shankland, D. L., 1971, Acetylcholine receptors and the diphasic action of nicotine in the American cockroach, *Periplaneta americana* (L.), *Comp. Gen. Pharmacol.* **2:**159.

Flattum, R. F., Friedman, S., and Larsen, J. R., 1967, The effects of *d*-tubocurarine on nervous activity and muscular contraction in the house cricket *Acheta domesticus* (L.), *Life Sci.* **6:**1.

Florey, E., 1961, Comparative physiology of transmitter substances, *Ann. Rev. Physiol.* **23:**501.

Gerschenfeld, H. M., 1973, Chemical transmission in invertebrate central nervous systems and neuromuscular junctions, *Physiol. Rev.* **53:**1.

Grundfest, H., 1961, Ionic mechanisms in electrogenesis, *Ann. N.Y. Acad. Sci.* **94:**405.

Grundfest, H., 1965, Electrophysiology and pharmacology of different components of bioelectric transducers, *Cold Spring Harbor Symp. Quant. Biol.* **1:**14.

Grundfest, H., 1969, Synaptic and ephaptic transmission, *in: The Structure and Function of Nervous Tissue,* Vol. 2 (G. H. Bourne, ed.), pp. 463–491, Academic Press, New York.

Hall, Z. W., 1972, The storage, synthesis and inactivation of the transmitters acetylcholine, norepinephrine and gamma-aminobutyric acid, *in: Structure and Function of Synapses* (G. D. Pappas and D. P. Purpura, eds.), pp. 161–171, Raven Press, New York.

Halstead, B. W., 1965, *Poisonous and Venomous Marine Animals of the World*, Vol. 1, Government Printing Office, Washington, D.C.

Heuser, J. E., and Reese, T. S., 1973, Evidence for recycling of synaptic vesicle membrane during transmitter release at the frog neuromuscular junction, *J. Cell Biol.* **57:**315.

Heuser, J. E., Reese, T. S., and Landis, D. M. D., 1974, Functional changes in frog neuromuscular junctions studied with freeze-fracture, *J. Neurocytol.* **3:**109.

Hille, B., 1968, Pharmacological modifications of the sodium channels of frog nerve, *J. Gen. Physiol.* **51:**199.

Hodgkin, A. L., 1964, *The Conduction of the Nervous Impulse*, Charles C. Thomas, Springfield, Ill.

Hodgkin, A. L., and Huxley, A. F., 1952, A quantitative description of membrane current and its application to conduction and excitation in nerve, *J. Physiol. (London)* **117:**500.

Hodgkin, A. L., and Katz, B., 1949, The effect of sodium ions on the electrical activity of the giant axon of the squid, *J. Physiol. (London)* **108:**37.

Hodgkin, A. L., and Keynes, R. D., 1955, Active transport of cations in giant axons of *Sepia* and *Loligo*, *J. Physiol. (London)* **128:**28.

Hodgkin, A. L., and Keynes, R. D., 1956, Experiments on the injection of substances into giant axons by means of microsyringe, *J. Physiol. (London)* **131:**592.

Hodgkin, A. L., Huxley, A. F., and Katz, B., 1952, Measurement of current–voltage relations in the membrane of the giant axon of *Loligo*, *J. Physiol. (London)* **115:**424.

Hollingworth, R. M., 1971, Comparative metabolism and selectivity of organophosphate and carbamate insecticides, *Bull. WHO* **44:**155.

Hubbard, J. L., and Quastel, D. M. J., 1973, Micropharmacology of vertebrate neuromuscular transmission, *Ann. Rev. Pharmacol.* **13:**199.

Karczmar, A. G., 1967, Pharmacologic, toxicologic, and therapeutic properties of anticholinesterase agents, *in: Physiological Pharmacology*, Vol. 3 (W. W. Root and F. G. Hoffman, eds.), pp. 163–322, Academic Press, New York.

Katz, B., 1961, How cells communicate, *Sci. Am.* **205:**209.

Katz, B., 1966, *Nerve, Muscle, and Synapse*, McGraw-Hill, New York.

Katz, B., and Miledi, R., 1965, The measurement of synaptic delay, and the time course of acetylcholine release at the neuromuscular junction, *Proc. Roy. Soc. London Ser. B* **161:**483.

Katz, B., and Miledi, R., 1973, The binding of acetylcholine to receptors and its removal from the synaptic cleft, *J. Physiol.* (*London*) **231:**549.

Keynes, R. D., 1951, The ionic movements during nervous activity, *J. Physiol.* (*London*) **114:**119.

Koch, R. B., 1969, Inhibition of animal tissue ATPase by chlorinated hydrocarbon pesticides, *in: Chemical–Biological Interaction*, pp. 199–209, Elsevier, Amsterdam.

Koelle, G. B., 1963, Cytological distributions and physiological functions of cholinesterases, *Handb. Exp. Pharmakol.* **15:**187.

Kosterlitz, H. W., 1967, Effects of choline esters on smooth muscle and secretions, *in: Physiological Pharmacology*, Vol. 3 (W. S. Root and F. G. Hoffman, eds.), pp. 97–161, Academic Press, New York.

Lajtha, A., 1962, The "brain barrier system," *in: Neurochemistry* (K. A. C. Elliot, I. H. Page, and J. H. Quastel, eds.), pp. 399–430, Charles C. Thomas, Springfield, Ill.

Lakshminarayanaiah, N., 1969, *Transport Phenomena in Membranes*, Academic Press, New York.

Lampert, P. W., 1968, Fine structural changes of myelin sheaths in the central nervous system, *in: The Structure and Function of Nervous Tissue*, Vol. 1 (G. H. Bourne, ed.), pp. 187–204, Academic Press, New York.

Larsen, J. R., Miller, D. M., and Yamamoto, T., 1966, *d*-Tubocurarine chloride: Effect on insects, *Science* **152:**225.

MacIntosh, F. C., 1959, Formation, storage and release of acetylcholine at nerve endings, *Can. J. Biochem. Physiol.* **37:**343.

Magazanik, L. G., and Vyskočil, F., 1973, Desensitization at the motor endplate, *in: Drug Receptors: A Symposium* (H. P. Räng, ed.), pp. 105–119, University Park Press, Baltimore.

McCandless, O. L., Zablocka-Esplin, B., and Esplin, D., 1971, Rates of transmitter turnover in the cat superior cervical ganglion estimated by electrophysiological technique, *J. Neurophysiol.* **34:**817.

McCrone, J. D., and Hatula, R. J., 1967, Isolation and characterization of a lethal component from the venom of *Latrodectus mactans mactans*, *in: Animal Toxins* (F. E. Russell and P. R. Saunders, eds.), pp. 29–34, Pergamon Press, New York.

McLennan, H., 1970, *Synaptic Transmission*, 2nd ed., Saunders, Philadelphia.

Narahashi, T., 1971, Effects of insecticides on excitable tissues, *Advan. Insect Physiol.* **8:**1.

Narahashi, T., and Anderson, N. C., 1967, Mechanism of excitation block by the insecticide allethrin applied externally and internally to squid giant axons, *Toxicol. Appl. Pharmacol.* **10:**529.

Nishi, S., Soeda, H., and Koketsu, K., 1967, Release of acetylcholine from sympathetic preganglionic nerve terminals, *J. Neurophysiol.* **33:**114.

Obata, K., 1972, The inhibitory action of γ-aminobutyric acid, a probable synaptic transmitter, *Int. Rev. Neurobiol.* **15:**167.

O'Brien, R. D., 1961, Selective toxicity of insecticides, *Advan. Pest Contr. Res.* **4:**75.

O'Brien, R. D., 1967, Barrier systems in insect ganglia and their implications for toxicology, *Fed. Proc. Fed. Am. Soc.* **26:**1056.

Paintel, A. S., 1964, Effects of drugs on vertebrate mechanoreceptors, *Pharmacol. Rev.* **16:**341.

Palay, S. L., and Chan-Palay, V., 1974, *Cerebellar Cortex Cytology and Organization*, Springer, New York.

Pappas, G. D., and Purpura, D. P. (eds.), 1972, *Structure and Function of Synapses*, Raven Press, New York.

Pappas, G. D., and Waxman, S. G., 1972, Synaptic structure–morphological correlates of chemical and electronic transmission, *in: Structure and Function of Synapses* (G. D. Pappas and D. P. Purpura, eds.), pp. 1–43, Raven Press, New York.

Paton, W. D. M., 1961, A theory of drug action based on the rate of drug receptor combination, *Proc. Roy. Soc. London Ser. B* **154:**21.

Paton, W. D. M., 1967, Kinetic theories of drug action with special reference to the acetylcholine group of agonists and antagonists, *Ann. N.Y. Acad. Sci.* **144:**869.

Paton, W. D. M., and Perry, W. L. M., 1953, The relationship between depolarization and block in the cats superior cervical ganglion, *J. Physiol. (London)* **114:**43.

Pfenninger, K., Akert, K., Moor, H., and Sandri, C., 1972, The fine structure of freeze-fractured presynaptic membranes, *J. Neurocytol.* **1:**129.

Plonsey, R., 1969, *Bioelectrical Phenomena*, McGraw-Hill, New York.

Porter, C. W., Bernard, E. A., and Chu, T. M., 1973, Ultrastructural localization of cholinergic receptors at mouse motor endplate, *J. Membr. Biol.* **14:**383.

Pysh, J. J., and Wiley, R. G., 1974, Synaptic vesicle depletion and recovery in cat sympathetic ganglia electrically stimulated *in vivo:* Evidence for transmitter release by exocytosis, *J. Cell Biol.* **60:**365.

Rodahl, K., and Issekutz, B., Jr. (eds.), 1966, *Nerve as a Tissue*, Harper and Row, New York.

Roeder, K. D., and Weiant, E. A., 1945, The site of action of DDT in the cockroach, *Science* **103:**304.

Roeder, K. D., and Weiant, E. A., 1948, The effect of DDT on sensory and motor structures in the cockroach leg, *J. Cell. Comp. Physiol.* **32:**175.

Rubin, R. P., 1970, The role of calcium in the release of neurotransmitter substances and hormones, *Pharmacol. Rev.* **22:**389.

Sakharov, D. A., 1970, Cellular aspects of invertebrate neuropharmacology, *Ann. Rev. Pharmacol.* **10:**431.

Salpeter, M. M., and Eldefrawi, M. E., 1973, Sizes of endplate compartments, densities of acetylcholine receptor and other quantitative aspects of neuromuscular transmission, *J. Histochem. Cytochem.* **21:**769.

Salpeter, M. M., Plattner, H., and Rogers, A. W., 1972, Quantitative assay of esterases in endplates of mouse diaphragm by electron microscope autoradiography, *J. Histochem. Cytochem.* **20:**1059.

Seitelberger, F., 1969, General neuropathology of degenerative processes of the nervous system, *Neurosci. Res.* **2:**253.

Shanes, A. M., 1958, Electrochemical aspects of physiological and pharmacological action in excitable cells, *Pharmacol. Rev.* **10:**59.

Shankland, D. L., 1964, Involvement of spinal cord and peripheral nerves in DDT-poisoning syndrome in albino rats, *Toxicol. Appl. Pharmacol.* **6:**197.

Shankland, D. L., and Schroeder, M. E., 1973, Pharmacological evidence for a discrete neurotoxic action of dieldrin (HEOD) in the American cockroach, *Periplaneta americana* (L.), *Pestic. Biochem. Physiol.* **3:**77.

Shankland, D. L., Rose, J. A., and Donninger, C., 1971, The cholinergic nature of the cercal nerve–giant fiber synapse in the sixth abdominal ganglion of the American cockroach, *Periplaneta americana* (L.), *J. Neurobiol.* **2:**247.

Sheehan, W. F., 1961, *Physical Chemistry*, Allyn and Bacon, Boston.

Simpson, L. L., 1974, The use of neuropoisons in the study of cholinergic transmission, *Ann. Rev. Pharmacol.* **14:**305.

Sjodin, R. A., and Mullins, L. J., 1958, Oscillatory behavior of the squid axon membrane potential, *J. Gen. Physiol.* **42:**39.

Skou, J. C., 1965, Enzymatic basis for active transport of Na^+ and K^+ across cell membranes, *Physiol. Rev.* **45:**596.

Smyth, T., Jr., Greer, M. H., and Griffiths, D. J. G., 1973, Insect neuromuscular synapses, *Am. Zool.* **13:**315.

Staverman, A. J., 1952, Non-equilibrium thermodynamics of membrane processes, *Trans. Faraday Soc.* **48:**176.

Staverman, A. J., 1954, The physics of the phenomena of permeability, *Acta Physiol. Pharmacol. Neer.* **3:**522.

Stephenson, R. P., 1956, A modification of receptor theory, *Brit. J. Pharmacol.* **11:**379.

Tauc, L., 1967, Transmission in invertebrate and vertebrate ganglia, *Physiol. Rev.* **47:**521.

Taylor, R. E., 1963, Cable theory, *in: Physical Techniques in Biological Research*, Vol. 6 (W. L. Nastuk, ed.), pp. 219–262, Academic Press, New York.

Treherne, J. E., and Pichon, Y., 1972, The insect blood–brain barrier, *Advan. Insect Physiol.* **9**:257.

Usherwood, P. N. R., 1967, Insect neuromuscular mechanisms, *Am. Zool.* **7**:553.

Van der Kloot, W. G., 1966, Comparative pharmacology, *Ann. Rev. Pharmacol.* **6**:175.

van Heyningen, W. E., 1970, General characteristics, *in: Microbial Toxins*, Vol. 1 (S. J. Ajl and T. C. Montie, eds.), pp. 1–28, Academic Press, New York.

Whittaker, V. P., 1963, Identification of acetylcholine and related esters of biological origin, *Handb. Exp. Pharmakol.* **15**:1.

Whittaker, V. P., 1972, The use of synaptosomes in the study of synaptic and neural membrane function, *in: Structure and Function of Synapses* (G. D. Pappas and D. P. Purpura, eds.), pp. 87–100, Raven Press, New York.

Whittaker, V. P., and Sheridan, M. N., 1965, The morphology and acetylcholine content of isolated cerebral synaptic vesicles, *J. Neurochem.* **12**:363.

Wilson, J. A., 1972, *Principles of Animal Physiology*, Macmillan, New York.

7

Acetylcholinesterase and Its Inhibition

R. D. O'Brien

1. The Biology of Acetylcholinesterase

Acetylcholinesterase (AChE) (3.1.1.7) belongs to the very large group of enzymes called *hydrolases*, defined as enzymes which split substrates by the introduction of the elements of water. The hydrolases which split acetylcholine are called *cholinesterases*; they catalyze the reaction

$$\underset{\text{acetylcholine}}{CH_3COOCH_2CH_2\overset{+}{N}(CH_3)_3} \xrightarrow{H_2O} \underset{\text{acetate}}{CH_3COO^-} + H^+ + \underset{\text{choline}}{HOCH_2CH_2\overset{+}{N}(CH_3)_3} \tag{1}$$

The reaction is reversible, but under usual conditions it lies far to the right.

There are two kinds of cholinesterase. The most important is usually called *acetylcholinesterase* (AChE); other names for it are "true cholinesterase" or "specific cholinesterase" or such source names as "erythrocyte cholinesterase." Special features of this enzyme are that acetylcholine is its best substrate and that it shows excess substrate inhibition—i.e., if one increases the substrate concentration progressively the rate of substrate hydrolysis increases up to some rather high substrate concentration (typically $10^{-2.5}$ M), after which further increases lead to *less* hydrolysis (Fig. 1). The second kind of cholinesterase is best called *butyrylcholinesterase* (3.1.1.8); other names for it are "pseudocholinesterase" or "nonspecific cholinesterase" or such source names

R. D. O'Brien • Section of Neurobiology and Behavior, Cornell University, Ithaca, New York.

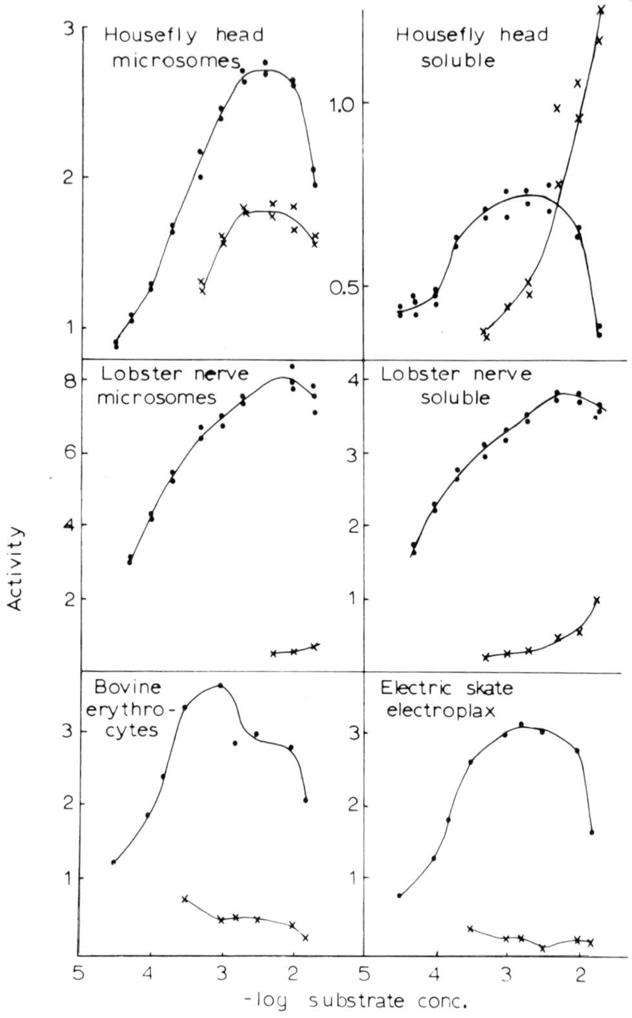

Fig. 1. Substrate dependence of AChEs. Substrate is acetylcholine (• • •) or butyrylcholine (×××). All show excess substrate inhibition for acetylcholine, although optimal substrate concentration varies. For housefly only, substantial butyrylcholine hydrolysis occurs, either because of the existence of butyrylcholinesterase in the head or because of an unusual property of this AChE. From Soeda *et al.* (1975).

as "plasma cholinesterase." Sometimes (very confusingly) it is called simply "cholinesterase." Its special features are that butyrylcholine is its best substrate and that it does not exhibit excess substrate inhibition. AChE and butyrylcholinesterase may also be distinguished by their susceptibility to various inhibitors. Thus butyrylcholinesterase of horse serum is 11,300 times more sensitive to iso-OMPA than AChE of horse erythrocytes, and for mipafox the ratio is 4200. The ratio, however, varies a good deal in different species (Aldridge, 1953).

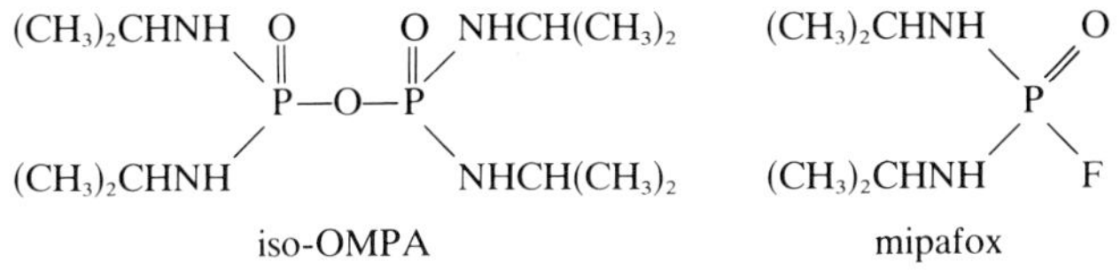

In vertebrates, both kinds of cholinesterase are common. AChE is found only in erythrocytes and nervous and muscle tissue. If the AChE of certain strategic locations (discussed below) is blocked, the animal dies. By contrast, butyrylcholinesterase is found in numerous tissues, including blood plasma and liver as well as nervous tissue, and it can be inhibited with no apparent ill effect to the animal; its inhibition in plasma can be a useful index of the course of poisoning by anticholinesterases.

In insects, butyrylcholinesterase has not been found, but AChE is plentiful in nervous tissue. As will be discussed later, AChE in the neuromuscular junction of the vertebrate is one of the important targets in poisoning by anticholinesterases. It is therefore critical to note that most studies of the neuromuscular junction in insects have shown clearly that there is no cholinesterase there and that transmission at the synapse does not involve acetylcholine; it probably involves glutamate instead. Most studies have centered upon the insect leg muscle. Thus Harlow (1958) showed the insensitivity of the locust leg nerve–muscle preparation to anticholinesterases, and this was confirmed for the cockroach *Periplaneta americana* (O'Connor *et al.*, 1965). Studies with more refined techniques have shown similar results for the leg nerve preparation of the cockroaches *Periplaneta americana* and *Gromphadorhinda portentosa* and the locust *Locusta migratoria*, as well as for the larval trunk muscle of the silkworm *Samia cecropia* (Faeder *et al.*, 1970). The absence of AChE has also been shown histochemically in the neuromuscular junction of the assassin bug *Rhodnius* (Wigglesworth, 1958). One extraordinary exception has been reported by Booth and Lee (1971), who found that the cricket has AChE in "virtually all muscles of the body" and in this way differs from the other insects included in this study, i.e., cockroach, milkweed bug, tobacco hornworm, and honeybee. With this exception, however, it appears that the AChE in insects (and probably other arthropods) is confined to the central nervous system.

It was formerly believed that a given enzyme in a given tissue was comprised of a single kind of macromolecule. It is now recognized, however, that an increasingly large number of enzymes exist in a variety of forms which have a common catalytic property but which are separable by physical techniques, principally electrophoresis. AChE from a variety of sources has been shown to be separable in this way. We shall call these different forms *isozymes*, although in some ways it would be better to refer to them as "multiple molecular forms" until it has been clearly shown that each form is under separate genetic control rather than their being merely aggregates of one another or possibly even artifacts of preparation. One of the earliest demonstrations of the existence of AChE isozymes was that of Bernsohn *et al.* (1963) working with the enzymes in vertebrate brain. Later work with rat retina indicated a potential difference in the physiological importance of the AChE isozymes, for it was shown that of the ten isozymes detected one had an unusually short half-life (about 3 hr) as judged by the fact that new enzyme was synthesized after DFP treatment. No resynthesis of the other isozymes was detected (Davis and Agranoff, 1968).

In the case of houseflies, it has been shown that there are four AChE isozymes in the head and three quite different ones in the thorax (Tripathi *et al.*, 1973; Tripathi and O'Brien, 1973*a*). The isozymes differed measurably in their sensitivity to inhibition by anticholinesterases *in vitro*; for instance, they varied over a 2.3-fold range in their inhibition by malaoxon. More importantly, they varied a great deal in their inhibition when the insect was poisoned by various organophosphates applied to the tip of the abdomen. Figure 2 shows that during poisoning by an LD_{50} dose of paraoxon applied to the tip of the abdomen none of the head isozymes was severely inhibited, and the three thoracic isozymes differed profoundly in their behavior. By comparing the sensitivities of the isozymes to four different organophosphates, it was possible to conclude that the inhibition of thoracic isozyme V was the most significant in poisoning; i.e., the extent of its inhibition correlated most closely with the lethal effects of the inhibitor.

2. Acetylcholinesterase as a Target

There is little doubt that most organophosphates and carbamates kill vertebrates by inhibiting their AChE. The ultimate cause of death is virtually always asphyxiation, but the cause of the asphyxiation varies according to the animal species and the compound. Four factors have been explored: constriction of the bronchi, which lead air into the lungs; lowering of blood pressure; blockade of the neuromuscular junctions of the diaphragm and the intercostal muscles; and failure of the respiratory center of the brain. The classic study in this case is that of DeCandole *et al.* (1953), who explored the action of seven organophosphates in nine mammalian species. Failure of the respiratory center of the brain was the commonest cause of death and in monkey was virtually the sole factor involved. In the cat broncial constriction was very important, and in the rabbit failure of the diaphragmatic muscles was a significant factor. Because all of these effects ultimately lead to asphyxiation, artificial respiration can be an important factor in therapy of organophosphate poisoning.

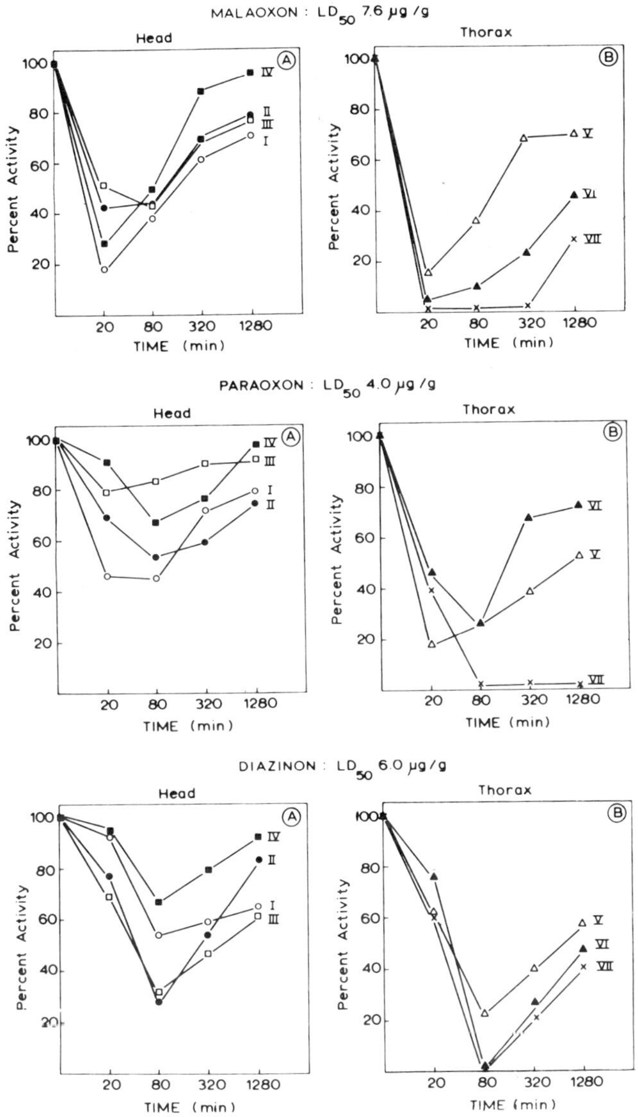

Fig. 2. Effect of poisoning by LD_{50} doses of three organophosphates on housefly AChE isozymes of head (left) and thorax (right). Top to bottom: malaoxon, paraoxon, diazinon. From Tripathi and O'Brien (1973*a*).

When we turn to the situation in insects, the picture is not nearly as clear. Although the majority of findings support the view that these compounds kill by inhibition of AChE, there are a number of paradoxical observations. These were reviewed in detail by O'Brien (1960), who concluded that the best available evidence favored the view that inhibition of AChE was involved in the poisoning process. Although the opposite conclusion was reached in a subsequent review (Chadwick, 1963), there has been in the last decade a rather widespread acceptance of the view that AChE is indeed the true target. This is not because there has been any ultimate proof or because all the earlier paradoxes have been removed. It is simply that none of the alternative possibilities (such as the involvement of other hydrolases) has ever been well supported, and it would appear that in poisoned insects cholinesterase is the enzyme which is most inhibited (O'Brien, 1961). Unfortunately, it is still completely unknown how inhibition of the enzyme leads to the ultimate death of the insect. A rather detailed analysis of some of the physiological events following anticholinesterase poisoning has been provided by Miller and Kennedy (1972), but it seems unlikely that the causal chain leading to death will be known in the near future.

3. Inhibition of Acetylcholinesterase

We will consider here the details of the mechanism by which the anticholinesterases inhibit the enzyme and comment briefly on the structural factors which promote anticholinesterase activity. The disputes of former years have now died away, and there is full agreement about the overall mechanism of inhibition. It is now recognized that organophosphates and carbamates are in fact substrates for AChE, but with some rather unusual properties. AChE hydrolyzes acetylcholine by first combining with it to form an enzyme–substrate complex. The acetylcholine then acetylates the enzyme, and the acetylated enzyme is subsequently hydrolyzed in a third step to restore the original enzyme. If we use A for the acetyl group and X for the choline, the reaction can be formulated as follows:

$$\mathrm{E+AX} \underset{k_{-1}}{\overset{k_{+1}}{\rightleftharpoons}} \mathrm{EAX} \xrightarrow{k_2} \mathrm{EA} \xrightarrow{k_3} \mathrm{E+A} \quad (+\,\mathrm{X}) \tag{2}$$

reversible complex (EAX) acetylated enzyme (EA)

The affinity with which the substrate binds to the enzyme in the first step is often described by the dissociation constant, $K_d = k_{-1}/k_{+1}$. The smaller K_d is, the greater the affinity. After complex formation, there is an acetylation step, characterized by a rate constant k_2, and ultimately a deacetylation step, defined by the rate constant k_3, which regenerates the free enzyme. All of the steps are extremely fast. The slowest one in vertebrates and the one which therefore controls the rate of the overall reaction is probably deacetylation, which is typically completed in about 100 μsec (Lawler, 1961). In the housefly head enzyme, however, the k_2 (acetylation) step is the slowest (Hellenbrand and Krupka, 1970).

The organophosphates and carbamates react in an absolutely analogous way. Thus the organophosphates first form a complex with the enzyme, then they phosphorylate it, and then there is a dephosphorylation step which is usually extremely slow, and which controls the rate of the overall reaction; it can take months, as we shall see. Similarly, the carbamates first form a complex, then they carbamylate the enzyme, and decarbamylation subsequently occurs with a half-time in the order of 20 min. Under most experimental conditions, the form of the inhibited enzyme is the phosphorylated one in the case of organophosphates and the carbamylated one in the case of the carbamates. Indeed, only by performing experiments under rather special conditions can one have substantial amounts of the enzyme–inhibitor complex present. It was for this reason that it was many years before the existence of such a complex was satisfactorily demonstrated.

Most of the good anticholinesterases will, at concentrations of about 10^{-6} M, satisfactorily inhibit much of the enzyme in experimentally convenient times, in the order of 5–50 min. We will first discuss the kinetics under these conditions and then show how work with higher inhibitor concentrations leads to more complicated, but more valuable information.

In the case of most organophosphates, for which the k_3 (phosphorylation) step is extremely slow, one can allow various inhibitor concentrations (I) to react with the enzyme for appropriate times (t) and then add a large surplus of substrate to terminate inhibition, and construct plots as in Fig. 3. This shows

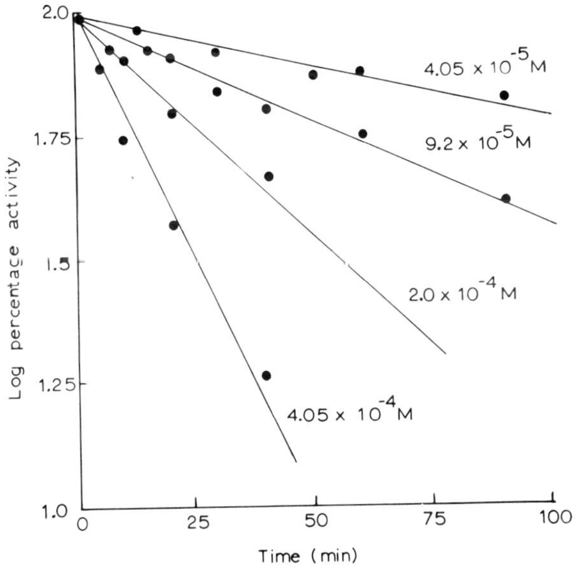

Fig. 3. Rate of inhibition of erythrocyte AChE by various concentrations of diethyl *p*-chlorophenyl phosphate. From Aldridge and Davison (1952*b*).

that the log of the residual percent of enzyme activity (P) is proportional to time. One can readily show (O'Brien, 1960) that

$$\log_{10} P = 2 - \frac{k_i I}{2.303} \cdot t \tag{3}$$

where k_i is what we should now call the bimolecular reaction constant (Main, 1964). This equation describes what is called *first-order reaction kinetics*; i.e., the rate depends only on the concentration of one component, the inhibitor. The slope of each line of Fig. 3 is given by $k_i I/2.303$ and can therefore be used to calculate k_i, which is a useful measure of the overall reactivity of the inhibitor. We shall later see that it has two components.

Fairly often one sees in the literature the term pI_{50}, which is the negative logarithm of the I_{50}, the inhibitor concentration which (for any given incubation time) gives 50% inhibition. For example, one can see from Fig. 3 that 9.2×10^{-5} M inhibits about 50% in about 60 min, so for $t = 60$ min that compound has an approximate I_{50} of 9.2×10^{-5} M and a pI_{50} of log 9.2×10^{-5} M $= 4.04$. From equation (3), when $P = 50$ one can calculate that $k_i = 0.695/I_{50}t$, so in principle the I_{50} (*if t* is provided) has the same information as the k_i. The only drawback is that when k_i is reported the investigator is giving direct assurance that plots such as Fig. 3 are linear. When they are *not* linear, it is still easy to find a concentration giving 50% inhibition at some time t, and this is a kind of I_{50}, but a kind from which no k_i can be deduced, because the kinetics are too complex to be described by a single rate constant. For example, reactivation (the k_3 step) may be significant, or there may be different isozymes present with different reactivities. Consequently, it is far more useful and informative to measure and report k_i rather than I_{50} or pI_{50}.

Now let us consider organophosphates more thoroughly. Equation (3) (which turns out to fit the data for most organophosphates) was derived on the assumption that the reaction proceeds as follows:

$$\text{PX} + \text{E} \xrightarrow{k_i} \text{PE} + \text{X} \tag{4}$$

where E is enzyme, P is the dialkyl phosphoryl group (in most cases), and X is the other or "leaving" group, e.g., p-nitrophenyl in the case of paraoxon:

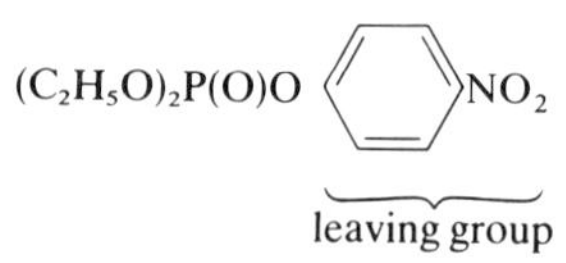

Reaction (4) yields first-order kinetics if PX is far more plentiful than E, which is the usual case. But to portray the reaction more accurately one has to note that the inhibitor first has to "sit down on" the enzyme before it phosphorylates it (it has to form a complex, PXE) and that some dephosphorylation of PE can also occur. Consequently, we should replace (4) by

$$\begin{array}{c} \mathrm{PX+E} \underset{}{\overset{K_d}{\rightleftharpoons}} \mathrm{PXE} \xrightarrow{k_2} \mathrm{PE} \xrightarrow{k_3} \mathrm{P+E} \\ \qquad\qquad\qquad\quad + \\ \qquad\qquad\qquad\quad \mathrm{X} \end{array} \tag{5}$$

where (as in the case of equation 2) K_d is a dissociation constant which is an inverse measure of affinity, k_2 (sometimes called k_p) is a phosphorylation constant, and k_3 is a dephosphorylation or reactivation constant. Fortunately, in the case of most organophosphates, k_3 is small enough to be ignored in brief experiments. Its size depends on the enzyme and on P, not at all on X, which has left prior to the k_3 step. We shall discuss it below.

Main (1964) showed that $k_i = k_2/K_d$; in other words, overall potency (measured by k_i) is made up of an affinity term (K_d) and a phosphorylation term (k_2). Since factors exist which improve k_2 and worsen K_d (and *vice versa*), it is not surprising that analyses of structure–activity relationships may often fail, if only k_i is examined, rather than its constituent parts.

For some years, people doubted that equation (5) could be verified experimentally. The problem was that, as we shall see, the K_d values for phosphates are often about 10^{-4} M. This means that if one added 10^{-4} M phosphate to enzyme, in the first instant 50% of the enzyme would be converted to PXE. But the phosphorylation reaction is usually so fast (i.e., k_2 is so large) that a second or two later all the PXE would be converted to PE. Under usual experimental conditions, one works with much less inhibitor, because it is convenient to study the course of inhibition over many minutes, and so very little PXE is formed, and the reaction behaves in accordance with equation (4). In order to work under conditions in which measurable PXE is present, and thus K_d and k_2 can be measured, strategems are needed to study the whole reaction within a few seconds or less.

Main and Iverson (1966) designed special reaction vessels by which one could react enzyme and inhibitor together for periods of 1–10 sec and then add excess substrate and assay at leisure. Hart and O'Brien (1973) used a standard recording spectrophotometer and added inhibitor to an ongoing reaction of enzyme and substrate. Subsequently (Hart and O'Brien, 1974) they used the same theory along with a stopped-flow method, in which enzyme is rapidly mixed with inhibitor plus substrate and the time course of the reaction is followed on an oscilloscope, from periods of milleseconds up to 2 min, according to choice.

In the Main–Iverson method, one can show that

$$1/I = k_i\left(\frac{r}{2.3\Delta \log v}\right) - 1/K_d \tag{6}$$

A plot of $1/I$ against $t/2.3\Delta \log v$ has a slope of k_i and a y intercept of $-1/K_d$; from K_d and k_i one calculates $k_2 = k_i K_d$. In the Hart–O'Brien method, the

Table I. Values of K_d and k_2 for Inhibition of AChE at pH 7.4–7.6[a]

	T (°C)	K_d (mM)	k_2 (min^{-1})
Organophosphates			
DFP $(isoPrO)_2P(O)F$	5	1.6	11.9
Amiton $(EtO)_2P(O)SCH_2CH_2NEt_2$	25	0.0072	6.7
$(EtO)_2P(O)SCH_2CH_2NMe_2$	5	0.18	126
Paraoxon $(EtO)_2P(O)OC_6H_4NO_2$	5	0.36	42.7
Methyl paraoxon $(MeO)_2P(O)OC_6H_4NO_2$	5	0.88	50.2
Carbamates			
Eserine	25	0.0033	10.8
Substituted phenyl methylcarbamates			
3-Trifluoromethyl	38	1.33	3.00
4-Chlorophenyl	38	3.27	1.33
1-Naphthyl	38	0.0106	1.33
3,5-Diisopropylphenyl dimethylcarbamate	38	0.008	1.31

[a]From summaries by Aldridge and Reiner (1972).

substrate, whose concentration is $[S]$ and whose Michaelis constant is K_m, is present throughout the reaction; and one of the ways of analyzing the data utilizes equation (6) but with $I(1-\alpha)$ in place of I, where $\alpha=[S]/(K_m+[S])$.

Table I shows a number of cases where K_d and k_2 have been measured for organophosphates. Note that both vary greatly from one compound to another.

Next let us consider carbamates. Although the earlier literature was full of disagreement, and must be read with great skepticism, it is now universally agreed that carbamates react precisely as do organophosphates, with the exception that the decarbamylation (k_3) step is fast enough that it must be reckoned with. Where C is the methyl-carbamyl group and X is the leaving group,

$$\mathrm{CX+E} \overset{K_d}{\rightleftharpoons} \mathrm{CXE} \overset{k_2}{\longrightarrow} \underset{\mathrm{X}}{\underset{+}{\mathrm{CE}}} \overset{k_3}{\longrightarrow} \mathrm{C+E} \tag{7}$$

As a result of the faster k_3 step, a plot like that of Fig. 3 for phosphates shows for carbamates a quite different form (Fig. 4). Even at fairly high inhibitor concentrations, the amount of inhibition (defined as formation of carbamylated enzyme) does not go to completion but achieves a steady-state typical of that inhibitor concentration. It has been shown (O'Brien, 1968) that the level of the steady state is determined in part by the Michaelis constant, K_m, of the enzymatic hydrolysis of the carbamate. It should be noted that because of the nonlinearity of plots such as Fig. 4 it is particularly dangerous to try to use equation (3) to compute k_i from the I_{50} as measured in experiments lasting more than a few minutes. For instance, after about 20 min the apparent I_{50}

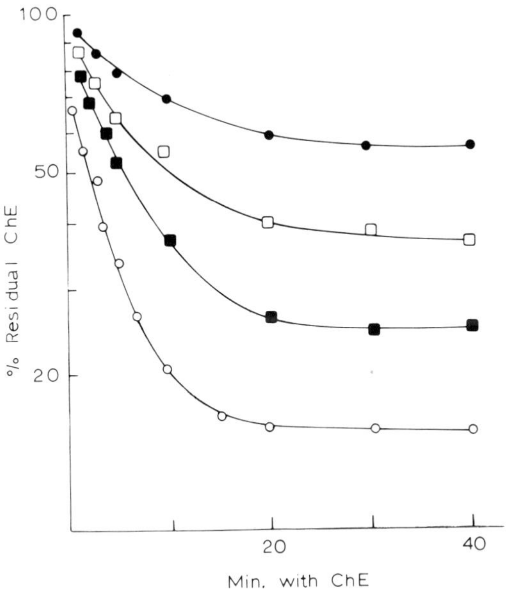

Fig. 4. Progress of inhibition of bovine erythrocyte AChE at 38°C by the following molar concentrations of carbaryl (sevin). Top to bottom: 5×10^{-7}, 10^{-6}, 2.5×10^{-6}, 5×10^{-6}. From O'Brien *et al.* (1966).

from Fig. 4 is time independent, so clearly one cannot use the formula $k_i = 0.695/I_{50}t$.

The Main or Hart–O'Brien approach described above for phosphates can be used with carbamates to measure K_d and k_2, and examples are given in Table I. It is essential to work with short reaction times, preferably of a few seconds, and then the influence of the k_3 step is negligible.

4. *Enzyme Sites and Inhibitor Structure*

4.1. *The Catalytic Site*

It has already been pointed out that the mechanism by which organophosphates and carbamates react with AChE is precisely analogous to that of "true" substrates. In fact, all these compounds are substrates, but phosphates and carbamates are exceptionally bad substrates because of their low k_3 rates. This enables them to "tie up" so much of the enzyme that they are inhibitors of the hydrolysis of "true" substrates, such as acetylcholine.

The above reactions are therefore analogous, and although they all involve ultimate reaction with a single catalytic site we shall see later that different compounds utilize different binding sites prior to their reaction with that catalytic site. The catalytic site, often called the *esteratic site*, is almost certainly the hydroxyl group of the amino acid serine or $HOCH_2CH(NH_2)COOH$. More precisely, it is one particular serine of the enzyme, whose properties are uniquely modified by the effects of neighboring amino acids (free serine is far less reactive to acylation than the serine of the catalytic site). It is the OH group of this special serine which is acylated by acetylcholine, phosphorylated by inhibitory organophosphates, and carbamylated by inhibitory carbamates.

One question we must consider (see Sections 4.4 and 4.5 below) is what factors influence the rate with which bound inhibitors react with the serine (i.e., in the k_2 step). A different question, to be discussed in Sections 4.2 and 4.3, is what factors influence the affinity with which inhibitors bind to the enzyme (i.e., the K_d step) prior to reaction with the serine. Since the overall potency of inhibition may be described by k_i (if we set aside the k_3 or reactivation step) and since $k_i = k_2/K_d$, it is clear that equal attention must be paid to k_2 and to K_d. The situation is therefore quite unlike that found in reactions of the form $a \rightarrow b \rightarrow c \rightarrow d$, in which one step only (the slowest one) controls the overall reaction.

4.2. Binding Sites

4.2.1. The Anionic Site

Although all substrates and inhibitors of AChE react with the same catalytic site, they may bind (prior to reaction) to a whole variety of binding sites. The early literature implies that there is only one kind of binding site, the so-called anionic site which is probably responsible for binding the $N^+(CH_3)_3$ group of acetylcholine. Although it seems very implausible today, it was frequently argued that when a substituent such as an isopropyl group on a phenyl ring was found to promote potency the group "must bind to the anionic site." Nowadays we realize that a hydrophobic group such as isopropyl would not bind to the polar region of the anionic site. Instead, the picture which one should have is of a catalytic serine which, like all amino acids in proteins, is surrounded by a veritable sea of other amino acid residues, any one (or group) of which can potentially act as a binding site for a matching substrate or inhibitor. Indeed, the recent discovery of a mutant AChE from a resistant strain of houseflies which binds acetylcholine almost normally, but which has its affinity for organophosphates and carbamates reduced by up to 500-fold, strongly suggests that the latter bind to a site or sites quite distinct from the anionic site (Tripathi and O'Brien, 1973*b*; Tripathi, 1976).

4.2.2. The Hydrophobic Site

Hydrophobic binding is a relatively recently discovered phenomenon. It arises because nonpolar (hydrophobic) regions of any molecule dissolved in water enforce the ordering of water molecules in their immediate vicinity, in order to minimize interference with the water molecules' attractive interactions with each other. Now consider a hydrophobic portion or patch on an enzyme; if we add a small molecule such as an organophosphate, bearing a nonpolar group such as an alkyl chain, —$CH_2CH_2CH_3$, it too forces orderliness in its surrounding water. But if the alkyl chain comes alongside the hydrophobic patch, it can lie up against the enzyme's patch without repulsion, and therefore the total amount of ordered water needed to surround these two nonpolar sites is reduced. A saving in total orderliness (entropy) is therefore realized and constitutes the energy with which the "hydrophobic bond" is endowed. Typically each methylene group that is involved in such a "bond" contributes 0.6 kcal to ΔF. Thus a propyl group, —$CH_2CH_2CH_3$, if it found a matching hydrophobic patch, could contribute $3 \times 0.6 = 1.8$ kcal. From the relationship $\Delta F = 1.34 \log K_d$ (at 25°C), one can calculate that each "bond" decreases K_d, and thus increases affinity, by a factor of 2.68. Consequently a fully effective propyl group would increase affinity $2.68^3 = 19$ times.

Although hydrophobic binding has been amply demonstrated for butyrylcholinesterase (see review by Kabachnik *et al.*, 1970), the evidence for AChE is not extensive. Although it was reported that simple trialkyl phosphates could be potent inhibitors simply because of hydrophobic binding (Bracha and O'Brien, 1970), this work was invalidated when it was shown that compounds prepared by these routes were contaminated with potent impurities (Gazzard *et al.*, 1974; Hart and O'Brien, 1974*b*). Furthermore, in phosphonates of the structure

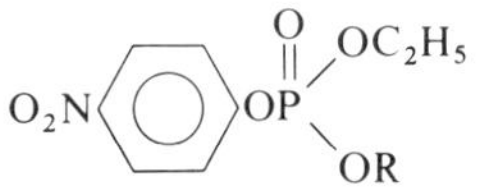

the progressive extension of R (from C_3 to C_{10}) gave progressive *loss* of activity against AChE, even though activity against chymotrypsin *increased* progressively from C_3 to C_7 (Becker *et al.*, 1963). Thus arguments for the presence of a hydrophobic site on AChE depend on the following sorts of indirect evidence. (a) In several dozen aromatic methylcarbamates, the effect of ring substituents is approximately additive (Metcalf and Fukuto, 1965), and in many cases these additive effects fit quite closely to an improvement of threefold per added methylene (O'Brien, 1971). (b) In 24 aromatic methylcarbamates, it was necessary to consider variations in hydrophobicity as well as in charge-transfer complex formation ability in order to account for the variations in affinity for AChE (Hetnarski and O'Brien, 1975*b*). (c) When fluorescent inhibitors, derived from 5-dimethylaminonaphthalene sulfonamide, bind to AChE, they change their absorption maximum in a way that indicates that they occupy a nonpolar binding site (Mayer and Himel, 1972).

4.2.3. Site(s) for Charge-Transfer Complex (CTC) Formation

When two molecules are of such shapes that they can approach one another closely, and if one, the donor, can easily lose an electron (i.e., it has a

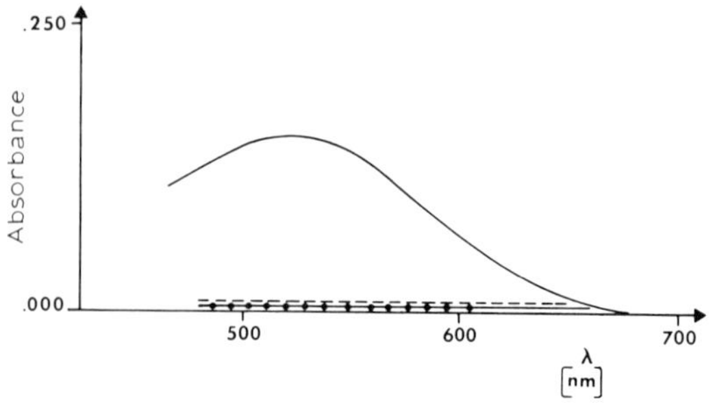

Fig. 5. Absorption curves for the donor, tetracyanoethylene, at 5×10^{-2} M (– – –); for the acceptor, *p*-methoxybenzyl methylcarbamate, alone at 5×10^{-3} M (–●–●–); and for their mixture (——). Solvent was 1,2-dichloroethane. The strong absorption of the mixture shows formation of a complex. From Hetnarski and O'Brien (1973).

low ionization potential) and one, the acceptor, can easily gain an electron (i.e., it has a high electron affinity), then a CTC may be formed, the presence of new bonding being demonstrated by the appearance of a new absorption band. For example, a good electron acceptor, tetracyanoethylene $(CN)_2CH{=}CH(CN)_2$, can form CTC with most aromatic carbamates. Figure 5 shows that alone neither tetracyanoethylene nor *p*-methoxybenzyl methylcarbamate absorbs between 300 and 500 nm but that after mixing they give a new absorption spectrum with a peak at 365 nm. Since the carbamates act as donors, their ability to form a CTC is improved by ring substituents such as CH_3 or NH_2, which provide electrons to the ring, and is worsened by those that withdraw electrons; *p*-nitrophenyl methylcarbamate, for instance, cannot form a CTC with tetracyanoethylene.

There is indirect evidence that aromatic hydrocarbons (such as benzene and naphthalene) can inhibit AChEs by such CTC formation and that most aromatic carbamates owe much of their affinity for the enzyme (*not*, of course, their carbamylating activity) to the CTC formed by their aromatic moieties. The evidence in both cases is based on parallelisms between variations in the ability of series of inhibitors to form CTC with model acceptors (Hetnarski and O'Brien, 1972, 1973, 1975*a,b*) and variations in their ability to combine with the enzyme.

The formation of CTC is helpful in understanding one aspect of a well-known anomaly. In substituted phenyl phosphates, electron-withdrawing substituents such as NO_2, Cl, or CH_3O improve k_i and electron-pushing substituents such as NH_2 have the reverse effect through effects involving the k_2 step (see Section 4.4). Since the reactions are mechanistically analogous, a similar effect would be expected with the substituted phenyl carbamates and it has long been a puzzle as to why with these compounds electron-withdrawing substituents either worsen or have little effect on inhibitory activity (Kolbezen *et al.*, 1954). It now appears that in phenyl carbamates (a) CTC formation is a major factor, and is improved by electron-pushing and worsened by electron-withdrawing substituents; these effects only modify K_d, and (b) the carbamylation (k_2) step is surprisingly insensitive to substituents (see Section 4.5).

We still do not know if CTCs are formed by phenyl organophosphates; a complicating factor is that the "lone pair" electrons of the P=O could be a better electron donor than the aromatic ring and far less sensitive to ring substituents.

It should be stressed that the evidence for CTC formation with AChE is still indirect; spectral evidence is necessary to prove it, and because only one of the thousands of amino acids in the enzyme is affected, such evidence may be hard to get.

4.2.4. An Indophenyl Binding Site

It has been demonstrated that treatment of AChE with several agents affects its reaction with indophenyl compounds in one way and its reaction with all other compounds so far studied in a different way. For instance, treatment of AChE with an alkylating agent such as the aziridinium compound MCP, or with the diazotizing compound Tdf, alters the enzyme so that it loses its reactivity toward such different agents as the substrates 1-naphthyl acetate, phenyl acetate, and acetylcholine or the inhibitors carbaryl, eserine, and Tetram (the oxalate of *O,O*-diethyl *S*-diethylaminoethyl phosphorothiolate) but increases its reactivity toward indophenyl acetate (Purdie and McIvor, 1966; O'Brien, 1969; Chiu and O'Brien, 1971).

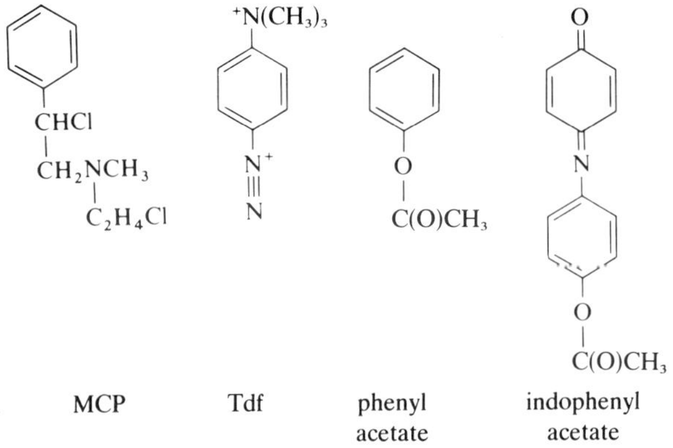

The evidence suggested that the difference between these responses is caused by different locations of the binding site for indophenyl esters as compared to the other agents. Supporting evidence was later provided by the demonstration that compounds such as acetylcholine, phenyl acetate, and 1-naphthyl acetate all inhibit the hydrolysis of acetylthiocholine by AChE competitively but are noncompetitive inhibitors of the hydrolysis of indophenyl acetate. Furthermore, the reactivities for AChE of indophenyl dimethylcarbamate and indophenyl diethyl phosphate, unlike those of other carbamates and phosphates, are unaffected by diazotization of AChE with Tdf (Chiu and O'Brien, 1971). These findings all support the view that there is a special indophenyl binding site in AChE.

4.3. Allosteric Sites

The binding sites described so far are all so close to the neighboring catalytic site (a few Ångstroms away) that compounds which occupy them can have direct access to the catalytic site. In recent years, many enzymes have been found to have allosteric sites; i.e., sites far removed from the active site, whose occupancy, however, modifies the reactivity of the active site by long-range effects, presumably mediated by changes in the configuration of the enzyme.

One almost completely unambiguous kind of evidence that compels one to postulate allosteric site effects is the finding that certain agents can increase the reactivity of the active site for other substrates or inhibitors. The effects of MCP and Tdf on the hydrolysis of indophenyl acetate by AChE are just two of the numerous examples involving AChE. Tetramethylammonium accelerates hydrolysis of phenyl acetate (Roufogalis and Thomas, 1968); gallamine accelerates carbamylation (Kitz *et al.*, 1970); decamethonium and related compounds activate sulfonylation by methanesulfonyl fluoride (Belleau *et al.*, 1970).

It is therefore almost certain that AChE is an allosteric enzyme, and although the properties and locations of the allosteric site(s) are presently unknown their exploration will undoubtedly occur in the coming few years.

4.4. Factors Affecting Phosphorylation Rates

It has been stated above that the overall inhibitory potency of phosphates as measured by k_i is made up of an affinity term (K_d), whose features have just been discussed and a phosphorylation term (k_2) whose characteristics we will now discuss. The earlier literature described most of the facts which we shall describe as factors affecting k_i, but we are now reasonably safe in assuming that they really affected k_2 primarily and consequently affected $k_i = k_2/K_d$.

A long series of persuasive studies, stretching back to Aldridge and Davison (1952*a*), can be summarized as follows. Organophosphates phosphorylate AChE by virtue of an electrophilic attack of the P on a serine hydroxyl of AChE. Hence a requirement for potency is that an electron-withdrawing substituent be attached to the P in order to give it sufficient electrophilic character, e.g., the *p*-nitrophenyl group in paraoxon:

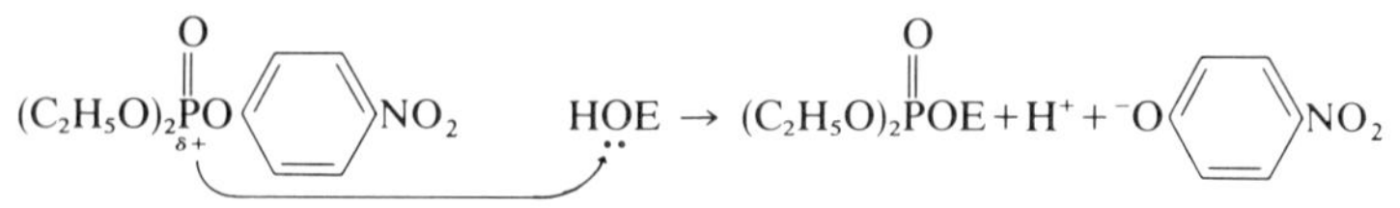

This generalization has been expressed in several ways, all showing that within any sufficiently comparable series in which one substituent is varied the more electrophilic the substituent the more potent the inhibitor. The electrophilicity of the substituent, or the electrophilicity which it imparts to the P, has been measured in several different ways. The classic approach of Aldridge and Davison (1952*a*) dealt with variously substituted phenyl groups attached to

$$(C_2H_5O)_2\overset{\overset{\displaystyle O}{\|}}{P}O—$$

and the measure was the sensitivity of the P to attack by OH^-, i.e., the sensitivity of the organophosphates to alkaline hydrolysis. This sensitivity nicely paralleled the anticholinesterase potency of the series (Fig. 6). A similar relation has been shown with related phosphonates (Fukuto and Metcalf, 1959).

Other measures of electrophilic character are the acidity of the acid derived from the substituent, e.g., HONO_2 (Ketelaar, 1953), the infrared stretching frequency of the P—O—substituent bond, or, in the case of

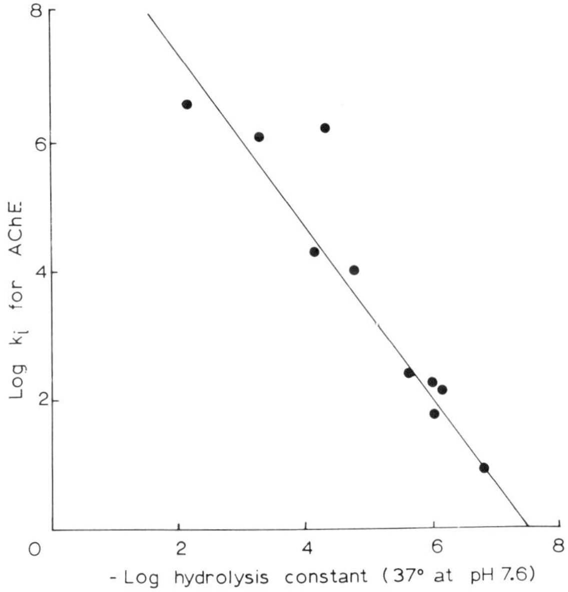

Fig. 6. Relation between hydrolyzability and anticholinesterase activity for ten substituted diethyl phenyl phosphates. Redrawn from Aldridge and Davison (1952*a*).

phenyl compounds, the Hammett σ constant of the substituent in the phenyl ring (Fukuto and Metcalf, 1956) (see Chapter 11). Thus the optimization of k_2 consists in large part in maximizing the electrophilic character of the substituent. Excessive electrophilic character can be counterproductive, however, by making the phosphate excessively sensitive to hydrolysis even at neutral pH. Thus the insertion of one chlorine atom into schradan increased its anticholinesterase activity 10^5 times and reduced its half-life at neutral pH from 10 years to 2 hr; the incorporation of three chlorines reduced its half-life to 3 min and abolished anticholinesterase activity (Spencer and O'Brien, 1953).

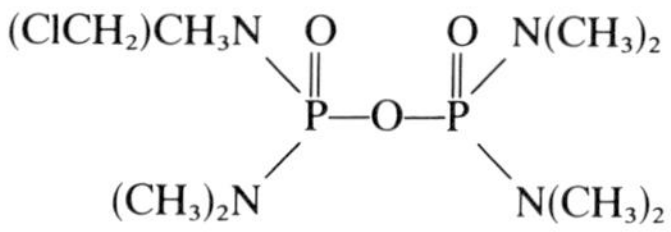

monochloro schradan

The above account explains to a considerable extent why P(S) compounds, such as parathion, are (in the pure form) such poor inhibitors of AChE. The P of P(S) has far less electrophilic character than the P of P(O) and is consequently far less sensitive to alkaline hydrolysis and far less potent against AChE. Some of the consequences are as follows: (a) P(S) compounds (phosphorothionates) are poor inhibitors but are greatly activated by metabolic desulfuration or by isomerization, e.g., where X is *p*-nitrophenyl:

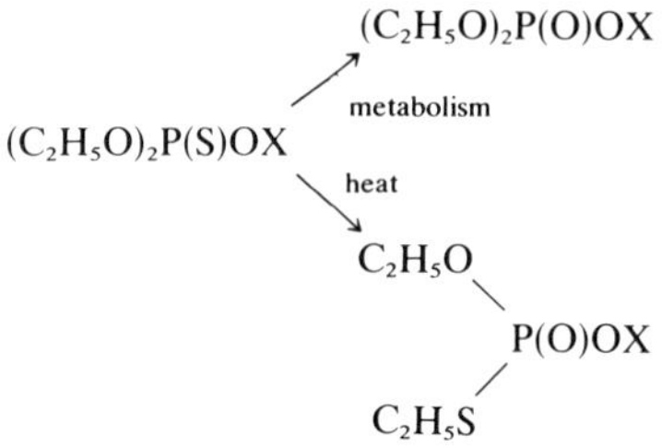

poor inhibitor potent inhibitors

(b) Metabolic changes which result in the formation of O^- derivatives greatly reduce the electrophilic character of the P and virtually abolish anticholinesterase activity (Aharoni and O'Brien, 1968). For instance, the *O*-demethylation of dimethoxon reduces the anticholinesterase activity 665,000-fold:

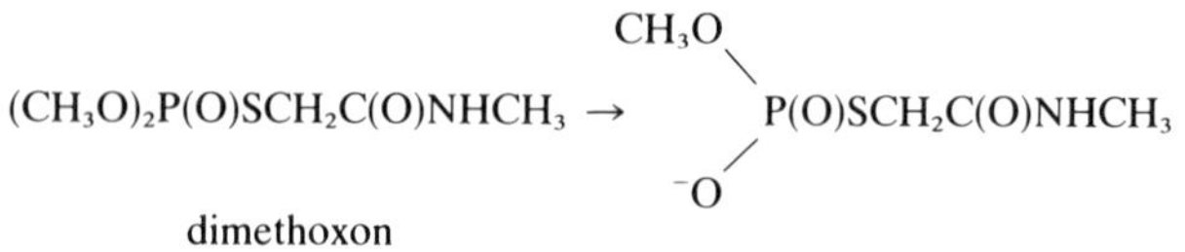

dimethoxon

The average reduction from seven such desalkylations was 158,000-fold. In a few rare cases the reduction was much less, probably because the desalkylation product formed an internal salt and thus reduced the electronic consequences of the introduction of O^-. Thus desethyl Amiton had 1/179 of the activity of its parent, Amiton:

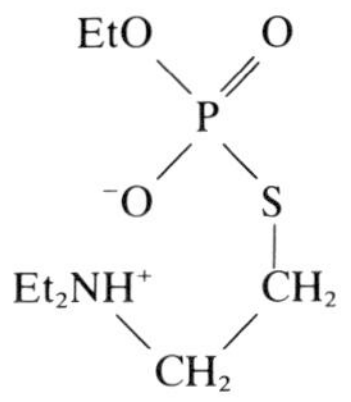

desethyl Amiton

4.5. Factors Affecting Carbamylation Rates

In view of the parallelism that has been stressed between organophosphate and carbamate inhibition, the reader has every reason to expect that the k_2 (carbamylation) step would have the same sensitivity to the electrophilic character of substituents as in the case of phosphates. This phenomenon has been much studied in phenyl methylcarbamates. But in spite of the fact that electrophilic substituents in the phenyl ring increase alkaline hydrolysis in the expected direction, the k_2 values for methylcarbamates are virtually insensitive to such substitution. Dimethylcarbamates do show the expected positive relation between σ and k_i (and hence with k_2, because $k_i = k_2/K_d$ in short experiments and K_d is relatively constant), the correlation coefficient being 0.78, and between alkaline hydrolyzability and k_i, the correlation coefficient being 0.87 (O'Brien *et al.*, 1966; Hetnarski and O'Brien, 1975*b*). In spite of the puzzling insensitivity of the k_2 of methylcarbamates to σ, there is little doubt that the overall reaction between phenyl methylcarbamates and AChE is just like that with OH^-. The only speculation advanced to explain the paradox (O'Brien *et al.*, 1966) was borrowed from a similar situation in acetanilide hydrolysis and proposes that the overall k_2 reaction proceeds through two steps, the first with a positive dependence on σ and the second with a negative one, e.g., where R is a phenyl group:

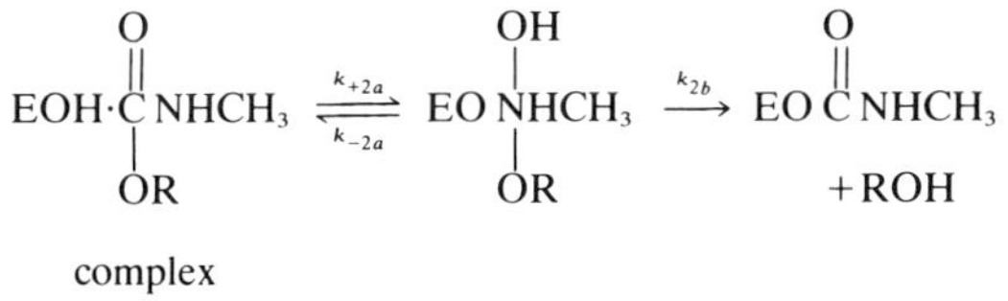

As for the differences between methylcarbamates and dimethylcarbamates, it is well known that the former hydrolyze much faster than the latter, the ratio of rates reaching 2×10^6 in some cases (O'Brien *et al.*, 1966), probably because of special mechanism in methylcarbamates involving deprotonation of the NH and cleavage to methyl isocyanate (Dittert and Higuchi, 1963). Such a facilitation might also occur in the reaction with AChE.

5. Recovery from Inhibition

We shall discuss first the spontaneous recovery of phosphorylated or carbamylated AChE in the absence of promoting agents. In the case of organophosphates, the k_3 (recovery) step is quite slow. Of course, since the leaving group has left, recovery depends only on the nature of the dialkyl phosphoryl moiety. For rabbit erythrocyte AChE at 37°C, the dimethyl phosphates give an enzyme which recovers with a half-life of about 80 min, and the diethyl phosphates give an enzyme which recovers with a half-life of about 500 min (Aldridge and Davison, 1953). By contrast, diisopropyl phosphates give an enzyme with virtually no recovery. Other cholinesterases vary a good deal: thus rat serum butyrylcholinesterase recovers from dimethyl phosphate inhibition with a half-life of over 200 h (Davies *et al.*, 1960). Housefly head AChE recovers well from poisoning *in vivo*, but *in vitro* there is no recovery from inhibition by any phosphate so far studied (van Asperen and Dekhuijzen, 1958; Mengle and O'Brien, 1960).

In carbamates the k_3 step is relatively fast; again the rate depends only on the nature of the carbamyl moiety, which means in most cases the methylcarbamyl or dimethylcarbamyl groups. The rate for a variety of AChEs and experimental conditions does not vary greatly with methylcarbamates (see summary in Aldridge and Reiner, 1972). Thus half-lives (in minutes) have been reported as follows: 19 for bovine erythrocytes (pH 7, 38°C), 24 for housefly head (pH 8, 30°C), 26 for bee head (pH 8, 30°C), 38 for electric eel (pH 7, 25°C). But for butyrylcholinesterase from serum the k_3 step is slower, with half-lives of 2.6 and 3 hr for horse and human (pH 7.4, 25°C).

With dimethylcarbamates, the rates vary a good deal according to enzyme. For AChEs, values are 27 min for electric eel (pH 7, 25°C), 56 min for bovine erythrocytes (pH 8, 25°C), and 4 hr for housefly head (pH 7, 25°C). For butyrylcholinesterase from serum, the k_3 is remarkably slow, with half-lives of 17 and 3.5 hr for horse and human (pH 7.4, 25°C).

In many cases, the k_3 step can be greatly accelerated by certain catalysts. This can be achieved not only *in vitro* but also in living animals, so that these catalysts are valuable therapeutic agents in cases of poisoning. They are all nucleophilic reagents and act by attacking the phosphorus or carbamyl carbon and substituting themselves for the enzyme. Just as the enzyme, in the inhibition step, displaced the leaving group of the inhibitor ($E + PX \rightarrow EP + X$), so the nucleophilic reagent A displaces the enzyme ($A + EP \rightarrow EA + P$). The EA form is unstable and rapidly splits to restore the free enzyme.

One of the first reagents in which the displacement reaction was shown was hydroxylamine. It is a very weak reactivator of organophosphate-inhibited eel AChE; thus after 5 hr of contact 1.2 M hydroxylamine permitted 88% recovery of AChE inhibited by TEPP (diethyl phosphoric anhydride) in comparison with 10% spontaneous recovery in the absence of hydroxylamine (Wilson, 1951). It also has a small effect on carbamylated eel enzyme, accelerating the k_3 step sevenfold for the methylcarbamyl enzyme and twofold for the dimethylcarbamyl enzyme (Wilson *et al.*, 1961).

Since 1955, many new reactivators have been described which are far superior to hydroxylamine. One commonly used as a therapeutic agent is the oxime 2-PAM (pyridine-2-aldoxime methiodide), and a close relative, TMB4, is a potent reactivator of phosphorylated AChE:

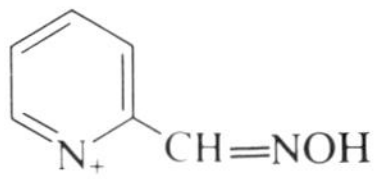

2-PAM

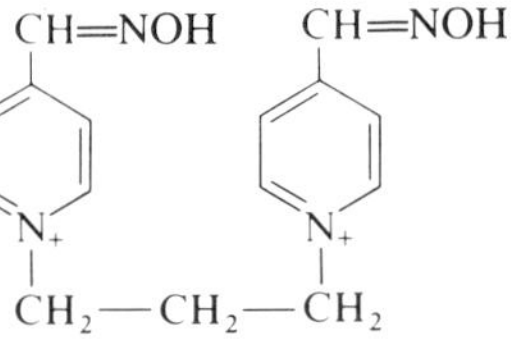

TMB4

Although TMB4 is more potent as a reactivator, it is not as safe as 2-PAM and so does not enjoy widespread use. The therapeutic use of these agents will be discussed below.

Surprisingly enough, 2-PAM does not catalyze the reactivation of AChE which has been methylcarbamylated (O'Brien, 1968) or dimethylcarbamylated (Wilson *et al*,. 1961), even though (as mentioned above) hydroxylamine does have some small effect. Both agents will reactivate organophosphate-inhibited housefly head AChE, in spite of its lack of spontaneous recovery (Mengle and O'Brien, 1960).

6. Aging

For organophosphates only, recovery from inhibition is sometimes complicated by the phenomenon of aging, defined as a progressive conversion of the inhibited enzyme to a form which cannot be reactivated by oximes. For example, Hobbiger (1955) showed that whereas diethyl phosphoryl butyrylcholinesterase which had been prepared 10 min earlier could be 90% reactivated by an oxime, less than 10% could be reactivated if the enzyme was allowed to "age" for 24 hr.

The basis of the aging reaction is now known to be a dealkylation reaction:

$$\mathrm{EOP(O)(OR)_2} \rightarrow \mathrm{EOP(O)}\begin{matrix}\diagup \mathrm{OR} \\ \diagdown \mathrm{O^-}\end{matrix}$$

In other words, there is a spontaneous loss of one alkyl group from the dialkyl phosphorylated enzyme, creating an anionic phosphate moiety which is insensitive to oximes and other nucleophiles. The best evidence is that of Berends *et al.* (1959), who showed that there was identity between the rate at which horse serum butyrylcholinesterase inhibited by radioactive DFP or $(i\text{PrO})_2\text{P(O)F}$ lost its ability to be reactivated by 2-PAM ("aged") and the rate of increase in the amount of radioactive $(i\text{Pr})(\text{HO})\text{P(O)O}^-$ which was found attached to the enzyme.

The rate of aging depends on the enzyme and the alkyl groups on the phosphoryl portion. A few compounds give almost "instant aging": for bovine erythrocyte AChE inhibited by the "nerve gas" Soman (pinacolyl methylphosphoryl fluoride), the half-life of aging was 2.3 min at 37°C (Fleisher and Harris, 1965). And if the organophosphate is anionic in the first place [a rare situation: one unusual example is desethyl Amiton, $(\text{HO})(\text{C}_2\text{H}_5\text{O})\text{P(O)SCH}_2\text{CH}_2\text{N}(\text{C}_2\text{H}_5)_2$] the inhibited enzyme is completely insensitive to reactivation (Aharoni and O'Brien, 1968). Poisoning by such compounds is therefore refractory to oxime therapy. But these are both special cases. Results with more typical compounds were reported by Hobbiger (1957) for brain and erythrocyte AChE of mice; the diethyl phosphorylated enzyme spontaneously changed to an aged form with a half-life of 36 hr, and the diisopropyl phosphorylated enzyme with a half-life of 4 hr. Extensive data from more recent studies are summarized by Aldridge and Reiner (1972).

7. Therapy for Poisoning by Anticholinesterases

Poisoning by organophosphates and carbamates leads to the accumulation of excessive amounts of acetylcholine in strategic locations, due to its inadequate breakdown by AChE. Attempts to reverse the poisoning of vertebrates involve two quite different principles: counteracting the excess acetylcholine by an antagonist, usually atropine, and restoring the effectiveness of the AChE, usually with 2-PAM. Neither procedure is very effective with insects (Mengle and O'Brien, 1960), probably because the lesion is always in the ganglia of the central nervous system, and ionic and ionizable compounds such as 2-PAM and atropine penetrate poorly into insect ganglia (Eldefrawi and O'Brien, 1967).

Excessive acetylcholine has the effect of overexciting synapses and subsequently blocking them by maintaining the acetylcholine receptor in an open (ion–conducting) configuration (see Chapter 8 for a detailed discussion of the receptor). Atropine competes with acetylcholine for the active sites of the receptor, but the atropine–receptor complex does not "open" the receptor. Consequently, atropine can antagonize the excitatory action of acetylcholine and can compensate for the elevated levels of acetylcholine caused by AChE inhibition. It should be noted, however, that there are at least two kinds of acetylcholine receptors: "nicotinic," found at the junction of nerves with keletal muscles, and "muscarinic," found at junctions of nerves with glands nd smooth muscles and probably also in the central nervous system. Atropine as affinity only for muscarinic receptors. Consequently it will relieve only the ymptoms caused by overexcited muscarinic junctions. These include lac-imation, salivation, urination, myosis (constriction of the pupil), and robably central effects, e.g., on the respiratory center of the brain. 'hey exclude skeletal muscle twitching or paralysis, which involves nicotinic eceptors.

Atropine is a tertiary-nitrogen base with a pK_a of about 9.3; consequently, t pH 9.3 it is 50% ionized and pH 7.4 it is 98.8% ionized. Although most of he atropine is ionized at physiological pH, and the ionized form penetrates ery poorly into brain, the fraction which is un-ionized penetrates readily. As it iffuses into the brain, the reservoir of ionized form in the blood reequilibrates o produce more un-ionized form, so that eventually the concentration in brain nay equal that in blood. Consequently, atropine is effective against central ervous system muscarinic receptors as well as peripheral muscarinic recep-ors.

The mode of action of 2-PAM was described above. It restores phos-horylated AChE by dephosphorylating it. Because it cannot decarbamylate he carbamylated enzyme, it is useless as a therapeutic agent in carbamate oisoning and in fact enhances the toxicity of carbaryl to dogs (Carpenter *et al.*, 961). Because its action is on the AChE and not the receptor, it does not listinguish between muscarinic and nicotinic synapses. It is a fully ionic ompound (unlike atropine) and consequently has virtually no effect upon entral synapses.

Because atropine can help the poisoned muscarinic synapses (central and eripheral) and 2-PAM can help peripheral synapses (muscarinic and icotinic), the two agents can be used together in the treatment of poisoning, nd the combination is usually better than either one alone in cases of rganophosphate poisoning.

3. References

Aharoni, A. H., and O'Brien, R. D., 1968, The inhibition of acetylcholinesterases by anionic organophosphorus compounds, *Biochemistry* **7:**1538.

Aldridge, W. N., 1953, The differentiation of true and pseudocholinesterase by organophosphorus compounds, *Biochem. J.* **53:**62.

Aldridge, W. N., and Davison, A. N., 1952*a*, The inhibition of erythrocyte cholinesterase by tri-esters of phosphoric acid. 2. Diethyl *p*-nitrophenyl thiophosphate (E605) and analogues, *Biochem. J.* **52:**663.

Aldridge, W. N., and Davison, A. N., 1952*b*, The inhibition of erythrocyte cholinesterase by tri-esters of phosphoric acid. 1. E600 analogues, *Biochem. J.* **51:**62.

Aldridge, W. N., and Davison, A. N., 1953, The mechanism of inhibition of cholinesterases by organophosphorus compounds, *Biochem. J.* **55:**763.

Aldridge, W. N., and Reiner, E., 1972, *Enzyme Inhibitors as Substrates*, North-Holland/American Elsevier, Amsterdam and New York.

Becker, E. L., Fukuto, T. R., Boone, B., Canham, D. C., and Boger, E., 1963, The relationship of enzyme inhibitory activity to the structure of *n*-alkylphosphonate and phenylalkyl phosphonate esters, *Biochemistry* **2**:72.

Belleau, B., Ditullio, V., and Tasai, Y.-H., 1970, Kinetic effects of leptocurares and pachycurares on the methanesulfonylation of acetylcholinesterase, *Mol. Pharmacol.* **6**:41.

Berends, F., Posthumus, C. H., van der Sluys, I., and Deierkauf, F. A., 1959, The chemical basis of the "aging process" of DFP inhibited pseudocholinesterase, *Biochim. Biophys. Acta* **34**:576.

Bernsohn, J., Barron, K. D., and Hedrick, M. T., 1963, Some properties of isozymes of brain acetylcholinesterase, *Biochem. Pharmacol.* **12**:761.

Booth, G. M., and Lee, A. H., 1971, Distribution of cholinesterases in insects, *Bull. WHO* **44**:91.

Bracha, P., and O'Brien, R. D., 1970, Hydrophobic bonding of trialkyl phosphates and phosphorothiolates to acetylcholinesterase, *Biochemistry* **9**:741.

Carpenter, C. P., Weil, C. S., Palm, P. E., Woodside, M. W., Nair, J. H., III, and Smyth, H. F., Jr., 1961, Mammalian toxicity of 1-naphthyl-*N*-methylcarbamate (sevin insecticide), *J. Agr. Food Chem.* **9**:30.

Chadwick, L. E., 1963, Actions on insects and other invertebrates, *in: Handbuch der experimentallen Pharmakologie, Ergaenzungswerk*, Vol. 15, pp. 741–798, Springer, Berlin.

Chiu, Y. C., and O'Brien, R. D., 1971, Separate binding sites on acetylcholinesterase for indophenyl and other esters, *Pestic. Biochem. Physiol.* **1**:434.

Davies, D. R., Holland, P., and Rumens, N. J., 1960, Relation between the chemical structure and neurotoxicity of alkyl organophosphorus compounds, *Br. J. Pharmacol.* **15**:271.

Davis, G. A., and Agranoff, B. W., 1968, Metabolic behavior of isozymes of acetylcholinesterase, *Nature (London)* **220**:277.

DeCandole, C. A., Douglas, W. W., Evans, C. L., Holms, R., Spencer, K. E. V., Torrance, R. W., and Wilson, K. M., 1953, The failure of respiration in death by anticholinesterase poisoning, *Br. J. Pharmacol.* **8**:466.

Dittert, L. W., and Higuchi, T., 1963, Rates of hydrolysis of carbamate and carbonate esters in alkaline solution, *J. Pharm. Sci.* **52**:852.

Eldefrawi, M. E., and O'Brien, R. D., 1967, Permeability of the abdominal nerve cord of the American cockroach, *Periplaneta americana* (L.), to quaternary ammonium salts, *J. Exp. Biol.* **46**:1.

Faeder, I. R., O'Brien, R. D., and Salpeter, M. M., 1970, A re-investigation of evidence for cholinergic neuromuscular transmission in insects, *J. Exp. Zool.* **173**:187.

Fleisher, J. N., and Harris, L. W., 1965, Dealkylation as a mechanism for aging of cholinesterase after poisoning with pinacolyl methylphosphonofluoridate, *Biochem. Pharmacol.* **14**:641.

Fukuto, T. R., and Metcalf, R. L., 1956, Structure and insecticidal activity of some diethyl substituted phenyl phosphates, *J. Agr. Food Chem.* **4**:930.

Fukuto, T. R., and Metcalf, R. L., 1959, The effect of structure on the reactivity of alkylphosphonate esters, *J. Am. Chem. Soc.* **81**:372.

Gazzard, M. F., Sainsbury, G. L., Swanson, D. W., Sellers, D., and Watts, P., 1974, The anticholinesterase ability of diethyl *S n*-propyl phosphorothiolate: Errors caused by the presence of an active impurity, *Biochem. Pharmacol.* **23**:751.

Harlow, P. A., 1958, The action of drugs on the nervous system of the locust (*Locusta migratoria*), *Ann. Appl. Biol.* **46**:55.

Hart, G. J., and O'Brien, R. D., 1973, Recording spectrophotometric method for determination of dissociation and phosphorylation constants for the inhibition of acetylcholinesterase by organophosphates in the presence of substrate, *Biochemistry* **12**:2940.

Hart. G. J., and O'Brien, R. D., 1974, Stopped-flow studies of the inhibition of acetylcholinesterase by organophosphates in the presence of substrate, *Pestic. Biochem. Physiol.* **4**:239.

Hart, G. J., and O'Brien, R. D., 1975, Trialkyl phosphates and phosphorothiolates—Lack of hydrophobic interaction with acetylcholinesterase, *Biochem. Pharmacol.* **24**:540.

Hellenbrand, K., and Krupka, R. M., 1970, Kinetic studies on the mechanism of insect acetylcholinesterase, *Biochemistry* **9**:4665.

Hetnarski, B., and O'Brien, R. D., 1972, The role of charge-transfer complex formation in the inhibition of acetylcholinesterases by aromatic carbamates, *Pestic. Biochem. Physiol.* **2:**132.

Hetnarski, B., and O'Brien, R. D., 1973, Charge-transfer in cholinesterase inhibition: Role of the conjugation between carbamyl and aryl groups of aromatic carbamates, *Biochemistry* **12:**3883.

Hetnarski, B., and O'Brien, R. D., 1975*a*, The charge-transfer constant; a new substituent constant for structure–activity relationships, *J. Med. Chem.* **18:**29.

Hetnarski, B., and O'Brien, R. D., 1975*b*, Electron-donor and affinity constants and their application to the inhibition of acetylcholinesterase by carbamates, *J. Agr. Food Chem.* **23:**709.

Hobbiger, F., 1955, Effect of nicotinhydroxamic acid methiodide on human plasma cholinesterase inhibited by organophosphates containing a dialkylphosphate group, *Br. J. Pharmacol.* **10:**356.

Hobbiger, F., 1957, Protection against the lethal effects of organophosphates by pyridine-2-aldoxime methiodide, *Br. J. Pharmacol.* **12:**438.

Kabachnik, M. I., Brestkin, A. P., Godovikov, N. N., Michelson, M. J., Rozengart, E. V., and Rozengart, V. I., 1970, Hydrophobic areas on the active surface of cholinesterases, *Pharmacol. Rev.* **22:**355.

Ketelaar, J. A. A., 1953, Chemical structure and insecticidal activity of organic phosphorus compounds, *Trans. 9th Int. Congr. Entomol. 1951*, **2:**318.

Kitz, R. J., Braswell, L. M., and Ginsburg, S., 1970, On the question: Is acetylcholinesterase an allosteric protein? *Mol. Pharmacol.* **6:**108.

Kolbezen, M. J., Metcalf, R. L., and Fukuto, T. R., 1954, Insecticidal activity of carbamate cholinesterase inhibitors, *J. Agr. Food Chem.* **2:**864.

Lawler, H. C., 1961, Turnover time of acetylcholinesterase, *J. Biol. Chem.* **236:**2296.

Main, A. R., 1964, Affinity and phosphorylation constants for the inhibition of esterases by organophosphates, *Science* **144:**992.

Main, A. R., and Iverson, F. I., 1966, Measurement of the affinity and phosphorylation constants governing irreversible inhibition of cholinesterases by di-isopropyl phosphorofluoridate, *Biochem. J.* **100:**525.

Mayer, R. T., and Himel, C. M., 1972, Dynamics of fluorescent probe–cholinesterase reactions, *Biochemistry* **11:**2082.

Mengle, D. C., and O'Brien, R. D., 1960, The spontaneous and induced recovery of the fly brain cholinesterase after inhibition by organophosphates, *Biochem. J.* **75:**201.

Metcalf, R. L., and Fukuto, T. R., 1965, Effect of chemical structure on intoxication and detoxication of phenyl *N*-methylcarbamates in insects, *J. Agr. Food Chem.* **13:**220.

Miller, T., and Kennedy, J. M., 1972, Flight motor activity of houseflies as affected by temperature and insecticides, *Pestic. Biochem. Physiol.* **2:**206.

O'Brien, R. D., 1960, *Toxic Phosphorus Esters*, Academic Press, New York.

O'Brien, R. D., 1961, Esterase inhibition in organophosphorus poisoning of houseflies, *J. Econ. Entomol.* **54:**1161.

O'Brien, R. D., 1968, Kinetics of the carbamylation of cholinesterase, *Mol. Pharmacol.* **4:**121.

O'Brien, R. D., 1969, Binding sites of cholinesterases—Alkylation by an aziridinium derivative, *Biochem. J.* **113:**713.

O'Brien, R. D., 1971, The design of organophosphate and carbamate inhibitors of cholinesterases, *in: Drug Design* (E. J. Ariens, ed.), pp. 162–212, Academic Press, New York.

O'Brien, R. D., Hilton, B. D., and Gilmour, L., 1966, The reaction of carbamates with cholinesterase, *Mol. Pharmacol.* **2:**593.

O'Connor, A. K., O'Brien, R. D., and Salpeter, M. M., 1965, Pharmacology and fine structure of peripheral muscle innervation in the cockroach *Periplaneta americana*, *J. Insect Physiol.* **11:**1351.

Purdie, J. E., and McIvor, R. A., 1966, The properties of acetylcholinesterase modified by interaction with the alkylating agent *N,N*-dimethyl-2-phenylaziridinium ion, *Biochim. Biophys. Acta* **128:**590.

Roufogalis, B. D., and Thomas, J., 1968, The acceleration of acetylcholinesterase activity at low ionic strength by organic and inorganic cations, *Mol. Pharmacol.* **4:**181.

Soeda, Y., Eldefrawi, M. E., and O'Brien, R. D., 1975, Lobster axon acetylcholinesterase: A comparison with acetylcholinesterases of bovine erythrocytes, housefly head and *Torpedo* electroplax, *Comp. Biochem. Physiol.* **50C:**163.

Spencer, E. Y., and O'Brien, R. D., 1953, Enhancement of anticholinesterase activity in octamethylpyrophosphoramide by chlorine, *J. Agr. Food Chem.* **1:**716.

Tripathi, R. K., 1976, Relation of acetylcholinesterase sensitivity to cross-resistance of a resistant housefly strain to organophosphates and carbamates, *Pestic. Biochem. Physiol.*, in press.

Tripathi, R. K., and O'Brien, R. D., 1973*a*, Effects of organophosphates *in vivo* upon acetylcholinesterase isozymes from housefly head and thorax, *Pestic. Biochem. Physiol.* **2:**418.

Tripathi, R. K., and O'Brien, R. D., 1973*b*, Insensitivity of acetylcholinesterase as a factor in resistance to the organophosphate Rabon^R in houseflies, *Pestic. Biochem. Physiol.* **3:**495.

Tripathi, R. K., Chiu, Y. C., and O'Brien, R. D., 1973, Reactivity *in vitro* towards substrates and inhibitors of acetylcholinesterase isozymes from electric eel electroplax and housefly brain, *Pestic. Biochem. Physiol.* **3:**55.

van Asperen, K., and Dekhuijzen, H. M., 1958, Quantitative analysis of the kinetics of cholinesterase inhibition in tissue homogenate of mice and houseflies, *Biochim. Biophys. Acta* **28:**603.

Wigglesworth, V. B., 1958, The distribution of esterase in the nervous system and other tissues of insect *Rhodnius prolixus*, *Quart. J. Microsc. Sci.* **99:**441.

Wilson, I. B., 1951, Acetylcholinesterase. XI. Reversibility of tetraethyl pyrophosphate inhibition, *J. Biol. Chem.* **190:**111.

Wilson, I. B., Harrison, M. A., and Ginsburg, S., 1961, Carbamyl derivatives of acetylcholinesterase, *J. Biol. Chem.* **236:**1498.

8

The Acetylcholine Receptor and Its Interactions with Insecticides

Amira Toppozada Eldefrawi

1. Introduction

The beginning of this century saw the development of two important principles of neurobiology: that the nerve cell (neuron) is the building unit of the nervous system and that neurons usually communicate with one another by chemicals. Direct electrical communication between neurons is rare, because even at their closest point of approach they are insulated from each other by a gap (about 200–250 Å) filled with extracellular fluid. When an electrical signal arrives at the end of a neuron, packets of a small molecule, called *neurotransmitter*, are released from the neuron into the gap. These chemicals act as messengers and interact with a component of the membrane of the receiving cell, whether it is another neuron, muscle, or gland. This component is called the *receptor* and the point of communication between the cells is called the *synapse*. Different chemicals have been identified as neurotransmitters, and one of these is acetylcholine (ACh). The synapse in this case is said to be *cholinergic* (Fig. 1). Such synapses are located in the animal brain, ganglia, and nerve–gland junctions, and also in nerve–muscle junctions in vertebrates.

Amira Toppozada Eldefrawi • Section of Neurobiology and Behavior, Cornell University, Ithaca, New York.

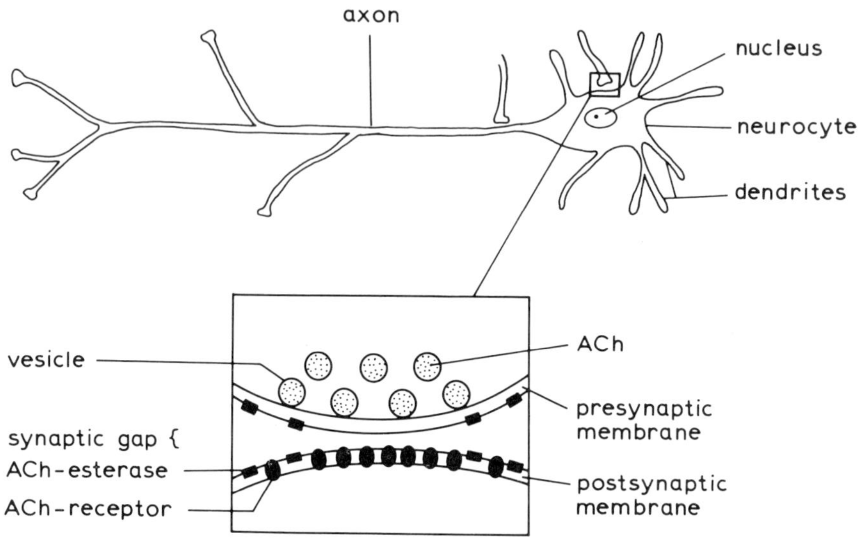

Fig. 1. A schematic diagram representing a neuron, with an enlarged inset of a cholinergic synapse.

Based on electrophysiological, pharmacological, and recently biochemical data, it is postulated that the binding of a neurotransmitter to its receptor causes a change in the three-dimensional structure (i.e., conformation) of the latter. As a consequence, there is an increase of ion fluxes down electrochemical gradients through "gates" or "channels" in the membrane. These gates are regulated by molecules called *ionophores*, *gate proteins*, or *ionic conductance modulators.* It is not yet known whether the channel molecule is part of, or closely associated with, the molecule that carries the neurotransmitter binding site (i.e., the receptor). Whether each channel is opened by the action of a single or several transmitter molecules on one receptor or one transmitter on several receptor molecules is not yet established. The mechanism of this chemoelectrical transduction is also not yet understood. It is conceivable that in the case of the ACh receptor it is coupled directly to a gate protein, or indirectly by means of an enzyme, which amplifies the message by producing a second messenger in a manner analogous to coupling of hormone receptors with nucleotide cyclases. In fact, several recent reports suggest that one type of ACh receptor is coupled to guanyl cyclase, since application of ACh or agonists to brain tissue increases the cellular levels of cyclic GMP (Illiano *et al.*, 1973). The cyclic nucleotide produced could conceivably phosphorylate the "gate protein," thereby opening the channel.

Some synapses are excitatory, so that the opening of gates leads to influx of Na^+ and efflux of K^+, causing localized depolarization of the postsynaptic membrane. If there are many transmitter–receptor interactions the depolarization is greater, and if it reaches a certain level an action potential is initiated and propagated along the cell. Other synapses are inhibitory, preventing an excitation that may occur, by hyperpolarizing the postsynaptic membrane through increased flux of Cl^- and perhaps K^+. Since the same transmitter may act in both kinds of synapses, it is plausible that what determines whether a synapse is excitatory or inhibitory is the nature of the ion gate protein coupled to the receptor, with the latter serving as the discriminator for recognition of the appropriate chemical signals.

The action of the transmitter is terminated by its dissociation from the receptor (because of the nature of its reversible binding) and by a concomitant reduction in transmitter concentration in the synaptic gap. This reduction may be caused by its hydrolysis by an enzyme, such as acetylcholinesterase (AChE) in cholinergic synapses, by its reabsorption into the neuron as in adrenergic synapses, or by its diffusion.

The majority of insecticides are poisons of the nervous system, and most of these attack one target, AChE. However, this enzyme is but one of the regulatory proteins involved in cholinergic transmission. Others are the ACh-synthesizing enzyme choline acetyltransferase, the ACh receptor, and the proteins involved in the selective absorption of ACh into vesicles or of choline from the synaptic cleft into the neuron. Of these, the ACh receptor is in fact the main target for a few insecticides and may act as a secondary target for others.

It is essential to know the molecular properties of ACh receptors in order to understand their interactions with insecticides. Since the only ACh receptors that have so far been purified are those of electric organs of fish, the emphasis in this chapter will be on the nature of these receptors. However, whenever possible, the properties of the ACh receptors of insects will be compared and contrasted with those of vertebrates. In addition, the known effects of insecticides on vertebrate and insect ACh receptors will be discussed.

2. Electrophysiological Studies on the Acetylcholine Receptor

2.1. The ACh Receptor as a Pharmacological Entity

The earliest notion of a *receptor* was that of Ehrlich (1913), who suggested the presence of highly specialized chemical groupings in the protozoan protoplasm upon which chemotherapeutic drugs acted.

Long before the transmitter role of ACh was known or even the presence of ACh in excitable tissues was established, vertebrate physiologists knew that various poisons and drugs either stimulated or inhibited nerves and muscles (Bernard, 1857). Based on their response to drugs, these cholinergic synapses were classified into muscarinic ones (those activated by muscarine and blocked by atropine, found in smooth muscles and glands) and nicotinic ones (those activated by nicotine and inhibited by *d*-tubocurarine in skeletal muscle or hexamethonium in autonomic ganglia). Langley in 1906 proposed the term *receptive substance* to describe a component of the skeletal muscle cell which reacted with nicotine and *d*-tubocurarine to cause or block contraction of the muscle. Later, starting with the experiments of Dale (1914) and Loewi (1921), ACh was proven to be the transmitter in these junctions and the pharmacologically defined "receptive substance" was recognized as a physiological entity, the ACh receptor.

The introduction of microelectrodes and iontophoresis made possible the study of synapses of the central nervous system (CNS) of vertebrates and invertebrates. Nicotinic as well as muscarinic synapses were found (Crawford and Curtis, 1966; Crawford *et al.*, 1966; Flattum and Shankland, 1971). However, whereas neuromuscular synapses in vertebrates are cholinergic, those of insects are not (Faeder *et al.*, 1970; Faeder and Salpeter, 1970).

The term *ligand* will be used throughout this chapter to designate ions or small molecules that bind to macromolecules through any kind of bonds. Thus a ligand can be a drug, an ion, or a neurotransmitter. The ligands that act like the transmitter, causing depolarization of the postsynaptic membrane in an excitatory synapse, are called *agonists* and those that block transmission are called *antagonists.* Under special conditions (e.g., high or persistent dosages), agonists may block transmission yet the postsynaptic membrane remains depolarized. This phenomenon is called *desensitization.*

2.2. The ACh Receptor as a Biochemical Entity

Evidence for the existence of the ACh receptor has been based largely on pharmacological and electrophysiological data. Because of its ability to recognize and bind molecules and its nature as a regulatory molecule, Nachmansohn (1955) postulated that the ACh receptor might be a protein. Years later this was proved to be true. The early evidence was obtained from electrophysiological studies, where active-site-directed reagents were used to learn more about the biochemical nature of, and functional groups in, the nicotinic neuromuscular ACh receptor. The importance of a disulfide bond for the function of the ACh receptor was revealed by the fact that reduction with 1,4-dithiothreitol (DTT) blocked the response to applied agonists in an electric organ (Karlin and Bartels, 1966) and frog sartorius muscle (Lindstrom *et al.*, 1973). Upon reoxidation of the disulfide bond, the ACh receptor became functional again. However, reoxidation could be prevented by alkylation of the produced sulfhydryl groups with any of a number of quaternary ammonium maleimide compounds, thus irreversibly inactivating the receptor. Estimates vary with regard to the location of this important disulfide bond on the ACh receptor. In the electric organ of the eel it is suggested to be 9–12 Å away from the anionic site (Karlin *et al.*, 1970), but in the receptor of the frog sartorius muscle one report suggests it to be a few Ångstroms away from the anionic site (Ben-Haim *et al.*, 1973), and another suggests it to be far from it (Lindstrom *et al.*, 1973).

One or more free sulfhydryl groups seem also to play an important role in the functioning of the ACh receptor. Exposure of the electric organ of the electric eel to the sulfhydryl reagent *p*-chloromercuribenzoate inhibits its depolarization by ACh or carbamylcholine (Karlin and Bartels, 1966). The opposite effect is found in the ACh receptor of the frog sartorius muscle, since several sulfhydryl reagents (oxidants, organic mercurials, and heavy metal ions) cause depolarization and muscle contraction, an effect that is also blocked by *d*-tubocurarine (Sobrino and del Castillo, 1972). These observations suggest the presence of a functional sulfhydryl group close to the active site of the ACh receptor. A diazotizable side chain also seems to be present, because exposure to *p*-trimethylammonium benzenediazonium fluoroborate, which forms a covalent bond with tyrosine, histidine, and lysine side chains of proteins, blocks depolarization of the electric organ by applied agonists (Changeux *et al.*, 1969).

It is obvious that the data one can obtain from electrophysiological and pharmacological experiments, although important, are limited. Only elaborate biochemical studies on pure ACh receptors can provide us with the detailed information on its chemical nature, structure, and molecular properties that is necessary for understanding its regulatory role. Since ACh receptors were purified only in the last 3 years, such information is just now becoming available and will be discussed in Section 4 of this chapter.

3. Studies on the ACh Receptor in Cell-Free Preparations

Physiologically, the presence of the ACh receptor in excitable tissues is inferred from the changes that occur in membrane potential or the ensuing muscle contraction that results from its binding of ACh (or agonists), or blockade of the response by antagonists. In a cell-free preparation, none of these changes can be monitored. Thus a major difficulty encountered in the purification of ACh receptors is the correct identification of the ACh receptor *in vitro*. Another is the need for large quantities of tissues, for unlike an enzyme whose activity is easily detectable *in vivo* or *in vitro* by virtue of its catalytic nature, the only means of identifying the ACh receptor is by its reversible binding with ACh and specific drugs or the almost irreversible binding of certain neurotoxins. This binding should fulfill certain criteria before it is certain that the binding macromolecule is the ACh receptor. The most important criterion is that the binding should be blocked by drugs that are known to interact *in situ* with the ACh receptor. The concentration of the binding macromolecules should parallel the estimated concentration of ACh receptors in the tissue tested. Thus there should be more binding in tissues rich in cholinergic synapses and binding should be absent in those known to be noncholinergic. In vertebrates, the neurotransmitter receptors are expected to be found in membrane fractions and not in nuclear or soluble ones. Final proof can be obtained only after purification of the ACh receptors, by reconstituting them, and by using them to produce antibodies which in turn can be radiolabeled and used in autoradiography to locate the receptors *in situ*. By explaining the techniques used and the data obtained, it will become clear how these difficulties have been overcome.

3.1. Identification of ACh Receptors

3.1.1. Use of Equilibrium Dialysis

Several methods such as equilibrium dialysis, centrifugation, and gel filtration have been used to measure the reversible binding of radiolabeled ligands to macromolecules (O'Brien *et al.*, 1974). In the widely used method of equilibrium dialysis (Fig. 2), a small volume (e.g., 1 ml) of tissue preparation is placed in a dialysis bag tied at both ends. It is made of membranes, such as cellophane, which allow diffusion of inorganic ions and small molecules such as cholinergic ligands. The bag is suspended into a large volume (e.g., a hundred-fold the volume of the tissue sample) of buffered solution (such as Krebs phosphate Ringer, pH 7.4 and 0.2 μ ionic strength), which contains a known concentration of the radiolabeled ligand. The system is shaken, preferably at 4°C, until equilibrium is reached, then equal samples are taken from the contents of the dialysis bag and the bath, and their radioactivities are counted. By subtracting the count of the bath sample from that of the sample from bag content, a difference is obtained which if significant represents the number of binding sites occupied at equilibrium. If a nonradiolabeled specific cholinergic ligand (agonist or antagonist) is added to the dialysis bath containing the radiolabeled one, the former competes with the latter for the ACh receptor sites, consequently reducing the amount of radiolabeled ligand bound. Thus by this method one can study the drug specificity of the binding macromolecules, which should mirror their drug sensitivities, *in vivo*.

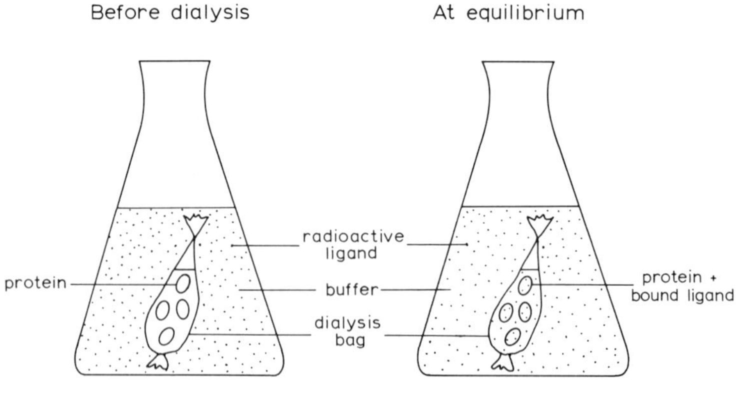

Fig. 2. Diagramatic representation of equilibrium dialysis.

Since we are dealing with charged molecules, the dialysis medium should be a buffer of high ionic strength, to guard against possible Donnan effects and artifactual binding. Because the tissue homogenate contains many proteins, and most proteins are negatively charged at physiological pH, and because the cholinergic ligands used are positively charged, they will undoubtedly bind to many proteins. What distinguishes the ACh receptor from other proteins should thus be its high affinity for specific ligands resulting from multiple bond formation. Thus, whereas the nonspecific binding has a dissociation constant (K_d) in the 10^{-3} M range, the binding of specific cholinergic ligands to the ACh receptor is characterized by K_d values of 10^{-9}–10^{-5} M.

An important criterion for checking the validity of the technique for *in vitro* identification of ACh receptors is the correct estimation of the concentration of receptor sites in the tissue. By measuring the degree of binding at several ligand concentrations and extrapolating the data in a double reciprocal plot, it is possible to calculate the maximum concentration of binding sites. If this value is the same for several specific cholinergic ligands, but much higher for another, one would tend to conclude that the latter is binding not only to the ACh receptor but to other molecules as well, unless of course the receptor has a different and larger number of binding sites for this ligand. An excellent example of this is seen in the binding of various ligands to receptors in the 100,000 *g* 1-hr supernatant of housefly head homogenates. In this system, the concentration of the binding sites for muscarone, nicotine, atropine, and decamethonium ranges from 2.2 to 3.2 nmoles per gram of heads, whereas in the case of dimethyl *d*-tubocurarine it is found to be 23 nmoles per gram of heads (Eldefrawi *et al.*, 1971*d*).

The validity of the equilibrium dialysis technique for identification of ACh receptors was established when values for the concentration of receptor sites obtained by this method were confirmed by techniques using the irreversible binding of specific chemicals. For example, in the electric organ of the electric ray *Torpedo marmorata* a concentration of 1 nmole muscarone or ACh binding sites per gram of tissue obtained by equilibrium dialysis (O'Brien *et al.*, 1969; Eldefrawi *et al.*, 1971*a*) was confirmed by the results of studies on the almost irreversible binding of α-bungarotoxin (α-BGT) (Miledi *et al.*, 1971). In the case of the electric eel, *Electrophorus electricus*, a concentration of 0.02–0.03 nmole per gram was obtained by the affinity label reagent *N*-ethylmaleimide (Karlin *et al.*, 1970) and also by equilibrium dialysis of cholinergic ligands (Eldefrawi *et al.*, 1971*c*; Suszkiw, 1973). In mammalian brain, the concentration of the muscarinic ACh receptor was calculated by equilbrium dialysis to be 3 pmoles per gram of mouse brain (i.e., 302 pmoles per gram of protein) (Schleifer and Eldefrawi, 1974) and 400 pmoles per gram protein of rat cortex using the irreversible binding of *N*-2′-chloroethyl-*N*-[2″,3″-$^3H_2$]propyl-2-aminoethyl benzilate (Hiley *et al.*, 1972). On the other hand, the concentration of nicotinic receptors was estimated by equilibrium dialysis to be 3 pmoles per gram of mouse brain (Schleifer and Eldefrawi,

1974), and by α-BGT binding to be 3.4 and 2.1 pmoles per gram of rat and guinea pig brains, respectively (Salvaterra and Moore, 1973).

3.1.2. Use of Neurotoxins

The identification and purification of ACh receptors of the nicotinic neuromuscular type of electric organs were greatly enhanced by the discovery of the highly specific binding of certain α-neurotoxins found in venoms of the krait *Bungarus multicinctus* and the cobra *Naja naja* (Lee, 1972; Lester, 1971). These neurotoxins are large polypeptides (molecular weight 6800 and 8000 daltons for the cobra toxin and α-BGT, respectively) that are highly basic (p*I* 9.3–9.5). The binding of cobra toxin is considered reversible ($t_{1/2}$ = 160 min) (Klett *et al.*, 1973), yet is much slower than that of ACh ($t_{1/2}$ < 1 msec) (Katz and Miledi, 1973). In contrast, the binding of α-BGT is considered almost irreversible. With such tight specific binding, it was feasible to use the radiolabeled toxins as affinity labels to identify the ACh receptor and also to tag and follow the receptor–toxin complex during purification (Miledi *et al.*, 1971; Potter and Molinoff, 1972). Their use also permitted the localization of ACh receptors by autoradiography (Hartzell and Fambrough, 1973). On the other hand, although very specific for the ACh receptor of skeletal muscles *in situ*, α-BGT was also found to label soluble proteins (Porter *et al.*, 1973), axons (Denburg *et al.*, 1972), and possibly the gate protein (Albuquerque *et al.*, 1973).

3.2. ACh Receptors of Electric Organs

The electric organs of fish are powerful bioelectric generators made up of sequentially arranged electric cells, each of which is called an *electroplax*. These organs are embryologically similar to skeletal muscle, but their cells are modified for the production of electricity. The electric organs of two types of fish have been exclusively and extensively used in ACh receptor research. One fish is the electric eel, *Electrophorus electricus*, which is a freshwater fish found in the Amazon, and produces up to 600 V. Its 5000–6000 electric cells are lined ventrally along the length of its body, making up approximately 60% of its body weight. The electroplax is innervated along only one of its membranes and is chemically as well as electrically excitable. The other kind of fish is the electric ray; examples are *Torpedo marmorata* and *T. Ocellata* (Fig. 3) in the Mediterranean and *T. californica* in the Pacific. The electric ray is a flattened disc with a tail. The electric organ makes up about 25% of its body weight, has fewer cells (about 500), and produces less voltage (about 20–60 V) than that of the electric eel. The innervated membrane of *Torpedo* electroplax has many more synapses than that of the eel and it is excitable chemically but not electrically. As a consequence, the concentrations of AChE and ACh receptor are thirty- to fiftyfold higher in *Torpedo* than in *Electrophorus* (Eldefrawi *et al.*, 1971*a,c*; O'Brien *et al.*, 1972).

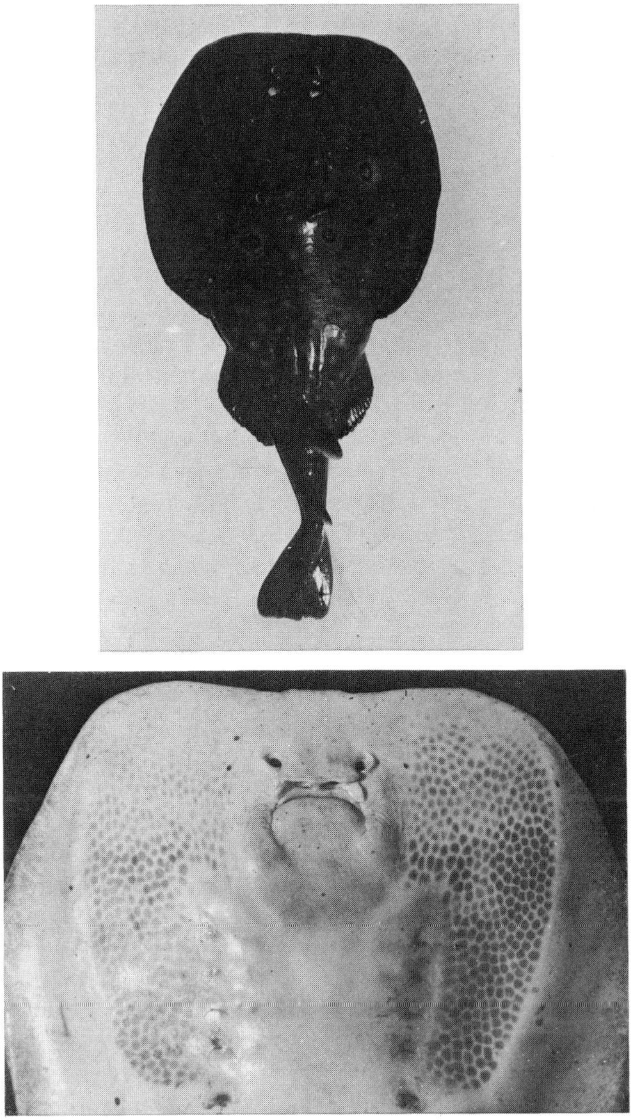

Fig. 3. Dorsal and ventral views of *Torpedo ocellata*. The honeycomblike structure, visible through the ventral skin, is the electric organ, which extends all the way to the dorsal skin. The organ of this species is as rich in ACh receptors as that of *T. marmorata* and *T. californica* (Eldefrawi, unpublished).

One advantage of working with ACh receptors of electric organs is th homogeneity of the receptor molecules, all being of the neuromuscula nicotinic type. This is in contrast to those of the brain, where there are not onl receptors for several transmitters, but the cholinergic ones are of two or thre kinds. A second advantage of working with electric organs is the high content o ACh receptors, which facilitates their identification and purification. The AC receptor of *Torpedo* electroplax makes up about 1% of its total protein and i about 200- to 300-fold more concentrated than that of mouse brain (Eldefraw and Eldefrawi, 1973*a*; Schleifer and Eldefrawi, 1974). *Electrophorus* electro plax offers a unique advantage in the wealth of information available on it pharmacology obtained from electrophysiological studies of the monocellula electroplax preparation (Nachmansohn, 1971). For these reasons, the AC receptors of electric organs were the first neurotransmitter receptors to b isolated and characterized.

The affinities of the ACh receptors of electric organs for cholinergi ligands ($K_d = 10^{-8}$–10^{-6} M, obtained by direct measurement of radiolabele ligand binding) are 1–2 orders of magnitude higher than those obtained fron electrophysiological experiments (Eldefrawi *et al.*, 1971*a,c,e*). Although suc high affinities were at first suspected by some of being artifacts, they have nov been confirmed by evidence obtained from other methods, such as th displacement of a fluorescent ligand from membrane-bound *Torpedo* receptor (Cohen and Changeux, 1973). The discrepancy between the affinity value measured by electrophysiological and biochemical methods may be accounte for by the fact that in the former the responses measured *in situ* reflect not onl the receptor affinity for the ligand but also the ability of the ligand to depolariz or block. Furthermore, electrophysiological calculations are based on assump- tions such as the presence on the ACh receptor molecule of a single binding sit for agonists and antagonists and the absence of any cooperative interactions Only lately, after isolation of the pure receptor, has it been realized that thes may not be totally correct.

A most interesting finding in this context is that nicotine, in addition to it desensitizing effect at very high concentrations, exhibits a dual action on frog sartorius muscle end plates (Wang and Narahashi, 1972). Thus at lov concentrations nicotine was found to decrease the end plate potential through suppression of its sensitivity to ACh without causing depolarization, and a higher concentrations it depolarized the postsynaptic membrane. It is inter- esting to speculate that the blocking and depolarizing actions of nicotine at low and higher concentrations, respectively, are related to the two affinites observed in the receptor of *Torpedo* electroplax (Eldefrawi *et al.*, 1971*c*).

3.3. ACh Receptors of Mammalian and Insect Brains

The most interesting tissue to study is the brain, but it is also the most difficult because of its complexity. Cholinergic transmission in the CNS of mammals and insects is well documented, but our knowledge of CNS pharmacology lags behind our understanding of the peripheral nervous system, particularly in the case of insect brain. None of these ACh receptors has yet been purified, but their *in vitro* identification has been accomplished through binding studies with various radiolabeled cholinergic ligands.

It is clear that major differences exist between the ACh receptors of mouse and housefly brains with respect to ligand affinity, concentration, solubility, and pharmacology. The K_d values of mouse brain receptors for nicotine, pilocarpine, and ACh are in the 7–23 × 10^{-9} M range, whereas in the housefly brain they are in the $10^{-7} - 10^{-5}$ M range. The concentrations of ACh receptors are 7.0 pmoles and 3.2 nmoles per gram of mouse brain and housefly head, respectively. The ACh receptor of mouse brain is membrane bound (Schleifer and Eldefrawi, 1974), as is that of electric organs, whereas much of the ACh receptor protein of the housefly brain is found in the 100,000 *g* 1-hr supernatant (Eldefrawi and O'Brien, 1970). The ACh receptors of housefly heads are extractable without detergents, yet precipitate on centrifugation at 105,000 *g* for 5 hr, possibly because of their easy aggregation (Aziz and Eldefrawi, 1973). The ability to work with the housefly receptors in the absence of detergents is a great asset, since these materials have been found to interfere with all aspects of the molecular studies of electroplax receptors.

The pharmacology of the ACh receptors of housefly heads is different from those of mouse brain and electric organs. The ACh receptors of *Torpedo* electroplax are typically nicotinic in their response to drugs, exhibiting low affinities for atropine and pilocarpine (Table I). On the other hand, the housefly receptors are both muscarinic and nicotinic, for not only is their muscarone binding blocked by nicotine and atropine but also their binding of nicotine is blocked by atropine and *vice versa* (Eldefrawi *et al.*, 1971*d*). In addition, unlike *Torpedo* receptors, those of the housefly are blocked by a few drugs of no known cholinergic effect: bretylium, amphetamine, tyramine, and hordenine (Eldefrawi and O'Brien, 1970). Recently, they were also shown to have high affinities for procaine and other local anesthetics, but their binding of [^{3}H]decamethonium was unaffected by α-bungarotoxin or α-cobra toxin (Eldefrawi, unpublished). In the absence of electrophysiological data on the effect of drugs on housefly head receptors, it is not possible to conclude whether these observed pharmacological characteristics are specific for central ACh receptors (compared to peripheral ones), whether they are specific for insect receptors, or whether they are due to the impurity of the suspected housefly head ACh receptors. Since the results of *in vitro* binding studies with mouse brain ACh receptors reveal them to be either nicotinic or muscarinic in almost equal amounts (Schleifer and Eldefrawi, 1974), and electrophysiological studies of cercal nerve giant fiber synapse of the American cockroach show these ACh receptors to be either nicotinic or muscarinic (Flattum and Shankland, 1971), one may conclude that the mixed character of the housefly receptor may be characteristic of an insect brain receptor.

One disturbing aspect of these putative ACh receptors of housefly head is their low affinity for ACh (K_i by blockade of [^{3}H]nicotine is 1.6 × 10^{-5}M) (Aziz

Table I. Pharmacological Profiles of the ACh Receptors of Torpedo Electroplax, Housefly Heads (from Eldefrawi and O'Brien, 1970), and Mouse Brain (from Schleifer and Eldefrawi, 1974)[a]

	Percent blockade		
Drug	*Torpedo* electroplax	Housefly head	Mouse brain
Nicotine[b]	59	39	—
Anabasine	—	—	50
Gallamine[b]	58	36	52
d-Tubocurarine[c]	79	54	32
ACh[d]	82	76	—
Decamethonium[b]	77	65	52
Atropine	21	72	49
Pilocarpine	0	84	37

[a] The ACh receptors of the first two are recognized in these experiments by their binding of [^{3}H]muscarone at 10^{-6} M, and the drugs are used at 10^{-4} M unless otherwise noted. In the case of the mouse brain ACh receptors, these are recognized by their binding of [^{3}H]ACh at 10^{-8} M after inhibition of cholinesterases with pyridostigmine, and the competing drugs are used at 10^{-6} M.
[b] Drugs used at 10^{-5} M to block the binding of the ACh receptors of housefly heads.
[c] Tubocurarine is used at 10^{-5} M to block the *Torpedo* receptors and at 10^{-4} M to block the housefly ones.
[d] ACh is used at 4×10^{-5} M after inhibition of endogenous cholinesterases with paraoxon.

and Eldefrawi, 1973). It is, of course, possible that the ACh receptors of insects have a lower affinity for ACh than those of vertebrates, and this is in harmony with the finding in brains and nerve cords of some insect species of concentrations of ACh that are approximately 2 orders of magnitude higher than those found in the vertebrate central nervous system (Metcalf and March, 1950; Treherne, 1966). Indirect evidence supporting the correct identity of the insect cholinergic binding molecules as ACh receptors is their similar concentration in the housefly head extract to that of AChE (Eldefrawi, unpublished). A ratio of 1 : 3 for ACh receptor : AChE was found in membrane preparations from electric organs (O'Brien *et al.*, 1969) and mouse brain (Schleifer and Eldefrawi, 1974). The resort to such indirect evidence demonstrates the uncertainty which exists when the receptors of a tissue of unknown pharmacology are studied. Since there is some information on the *in situ* pharmacology of the nerve cord of the American cockroach, it may be a better tissue to use for the study of receptors, if enough tissue is available. As long as the number of insect species whose ACh receptors have been studied biochemically remains so pitifully small (in fact, essentially one), no generalization can be made.

4. Purification and Molecular Properties of Acetylcholine Receptors

4.1. Purification

In order to study the molecular properties of ACh receptors, they must be solubilized from the membrane and purified. Some membrane proteins are weakly bound and can be readily displaced by hypotonic washes, strong salt solutions, or sonication (e.g., AChE of electric organs and vertebrate brain). Detergents are essential for solubilization of the ACh receptors of electric organs, but unfortunately they result in the formation of stable receptor–detergent complexes which complicate studies of the structure of the receptor protein. This is one good reason why it is more advantageous to study receptors of insects and crustaceans, which seem to be more easily solubilized without detergents.

Conventional biochemical techniques used in the purification of enzymes proved inadequate for the ACh receptors of electric organs. Column chromatography separates a few proteins from the ACh receptor, but in the process there is an equivalent loss of receptor activity (i.e., binding of ACh) so that there is no enrichment of the specific binding (i.e., active sites per milligram of protein). Isoelectric focusing purifies the receptor to a much higher degree, but because of denaturation at the low pH of its isoelectric point there is only a 6.3-fold increase in specific binding (Eldefrawi and Eldefrawi, 1972).

Recently, a simple and efficient technique called *affinity chromatography* or *affinity adsorption* (Cuatrecasas and Anfinsen, 1971) has been successfully used in the purification of AChE (Rosenberry *et al.*, 1972) and insulin receptors (Cuatrecasas, 1972). In it, a specific ligand is covalently attached to a solid support, such as Sepharose, and is either packed into a column and used as in chromatography or added directly to a solution of the appropriate protein mixture. In the latter case, the affinity gel selectively adsorbs the desired protein from the solution. Nonspecifically adsorbed proteins are easily desorbed by salt solutions, and the desired one is desorbed by a solution of a specific ligand.

Affinity adsorption is a rapid technique and has been successfully applied to the purification of ACh receptors of electric organs of several ligands which were covalently attached to Sepharose (Karlsson *et al.*, 1972; Schmidt and Raftery, 1973; Biesecker, 1973; Olsen *et al.*, 1972; Klett *et al.*, 1973, Eldefrawi and Eldefrawi, 1973*a*). The best proved to be the neurotoxin from cobra venom, which reversibly binds the nicotinic ACh receptor with a very high affinity. The use of affinity gel in a batch method produced an ACh receptor of higher purity and with less AChE than by affinity chromatography. The specific activity of the receptor obtained by the former method is 7.8–11.3 nmoles per milligram of protein and the ratio of the remaining active sites of

AChE to ACh receptor is 1 : 20,000 (Eldefrawi and Eldefrawi, 1973*a*), compared to reported ratios of 1 : 100 (Meunier *et al.*, 1974) and 1 : 2000 (Klett *et al.*, 1973).

The protocol of ACh receptor purification by the batch method is as follows: The 1% Triton-solubilized proteins are incubated with the affinity gel for 2 hr, followed by filtration and incubation of the gel for 30 min with phosphate Ringer solution containing 0.1% Triton. The mixture is filtered and the step repeated, followed by two incubations of the gel with 1 M sodium chloride in Ringer and two filtrations. Finally, the ACh receptor is desorbed by incubation with 1 M carbamylcholine for 5–16 hr and recovered in the filtrate. Extensive dialysis removes most of the carbamylcholine from the solution of pure receptor. An advantage of this technique is that over 99.9% of the Triton used for solubilization of the receptor is removed in the process. By extraction of the ACh receptor in [^{3}H] Triton X-100, the number of detergent molecules remaining attached to the pure receptor was calculated to be 16 per ACh-binding subunit (mol wt about 90,000), and this number was further reduced to eight molecules of Triton by an extra dialysis overnight at room temperature (Edelstein *et al.*, 1974).

It is important to know the method used for purification of a macromolecule, because it could influence its size or chemical nature. Such variations were noticed in the AChE extracted from *Electrophorus* electroplax by homogenization as compared to those isolated by tryptic treatment or by the action of endogenous proteases (Rieger *et al.*, 1973; Powell *et al.*, 1973). So far, ACh receptors have been purified only by affinity gels, but the specific ligand attached to the gel has varied and so has the ligand used to desorb the receptor. Even minor changes have been found to affect the receptor. Thus desorption of the ACh receptor by 10^{-2} M benzoquinonium produced a denatured ACh receptor that could bind α-BGT but not its own transmitter, ACh, whereas its desorption by another agonist, carbamylcholine, produced a functional ACh receptor (Eldefrawi and Eldefrawi, 1973*a*).

4.2. Molecular Nature of the ACh Receptor

4.2.1. Chemical Character

The ACh receptor is a protein, and its amino acid composition has been determined. As shown in Table II, the amino acid content of the ACh receptor of electric organs is remarkably similar to that of the AChE of this tissue. The basic amino acids of either molecule make up approximately 11–12 moles %, and the acidic amino acids make up about 19–26 moles % of both molecules, with a higher percentage of aspartic than glutamic acid. The relatively high content of acidic amino acids is reflected in the low isoelectric point (4.5–4.8) reported for the ACh receptor of electric organs (Raftery *et al.*, 1971; Biesecker, 1973; Eldefrawi and Eldefrawi, 1972, 1973*a*). In this property it is

Table II. Amino Acid Composition (Expressed as Moles %) of the ACh Receptors of Electric Organs of Torpedo marmorata and Electrophorus electricus, Compared to the Different Analyses of Amino Acids of AChE of Electrophorus

	ACh receptor			AChE of *Electrophorus*		
	Torpedo	*Electrophorus*				
Amino acid	Eldefrawi and Eldefrawi (1973*a*)	Klett *et al.* (1973)	Meunier *et al.* (1974)	Leuzinger and Baker (1967)	Dudai *et al.* (1972)	Rosenberry *et al.* (1972)
Lysine	6.1	5.7	6.3	4.3	4.8	4.6
Histidine	2.1	2.8	2.5	2.3	2.1	2.3
Arginine.	3.5	5.3	4.2	5.4	5.0	5.2
Aspartic acid	11.8	14.4	9.8	10.8	12.6	13.1
Threonine	6.3	7.0	6.0	4.3	4.1	4.5
Serine	7.1	7.7	8.2	6.9	6.8	6.8
Glutamic acid	10.7	12.8	9.0	9.4	11.1	10.4
Proline	6.2	7.3	6.7	8.1	7.0	5.9
Glycine	6.4	7.4	4.8	7.7	8.8	8.7
Alanine	6.0	7.3	5.4	5.5	7.4	6.2
Half-cystine	2.0	1.8	1.7	1.1	0.9	1.6
Valine	5.5	10.8	6.9	7.0	6.9	7.1
Methionine	1.7	2.5	3.4	3.0	1.3	2.7
Isoleucine	5.2	8.0	8.1	3.7	4.0	3.8
Leucine	9.3	13.2	10.7	9.0	8.2	8.6
Tyrosine	3.6	5.0	3.8	3.8	2.9	3.6
Phenylalanine	4.4	4.4	5.1	5.3	5.1	5.3
Tryptophan	2.1	0.0	2.4	2.0	—	2.0

similar to most other "intrinsic" or "integral" membrane proteins which penetrate into the lipid bilayer of membranes and have low polarity (Capaldi and Vanderkooi, 1972).

The three analyses of the receptors of *Torpedo* and *Electrophorus* differ slightly in the percentages of several amino acids, most notably tryptophan. These differences may be due to the different species or to varying degrees of purification of the preparations employed. The inability to detect tryptophan in one preparation may have resulted from interference caused by the detergent present. In fact, the ACh receptor molecule, when excited at 290 nm, fluoresces strongly with a peak similar to that of tryptophan (336 nm) (Eldefrawi *et al.*, 1975).

There is a good possibility that the ACh receptor of electric organs is a conjugated protein. Some report it as a glycoprotein based on the finding of simple sugars and hexosamines in the purified receptor (Raftery *et al.*, 1973) and its precipitation by concanavalin A (Changeux, 1974). In the membrane-bound or solubilized state, the ACh receptor is reported to be phospholipoprotein because of the partial blockade of ACh binding by the receptor after its exposure to phospholipases (Eldefrawi *et al.*, 1971*c,e,* 1972), but analysis of lipid phosphorus in the pure ACh receptor indicates that, if present, it would be less than 1%. These discrepancies should soon be resolved. In contrast, the housefly head ACh receptors seem not to be phospholipoproteins (Eldefrawi and O'Brien, 1970), which may be the reason for their easy solubilization simply by homogenization.

4.2.2. Molecular Weight and Shape

It is very difficult to determine the molecular weight of the vertebrate ACh receptor because of its need for a detergent to remain in a soluble form even after purification. Molecular weight values of about 450,000 determined by column chromatography (Meunier *et al.*, 1972; Raftery *et al.*, 1973) appear to be overestimations (Hucho and Changeux, 1973), while those obtained by centrifugation in sucrose gradients (Meunier *et al.*, 1972) may be underestimations because of the buoyancy caused by the detergent. The receptor molecules are too large to adequately penetrate disc gels when electrophoresis is performed in the absence of detergents. To overcome this problem, the ACh receptor was fixed with glutaraldehyde (Biesecker, 1973) or suberimidate (Hucho and Changeux, 1973), and the molecular weights, determined by sodium dodecyl sulfate (SDS) gel electrophoresis, were found to be 260,000 and 230,000, respectively. Again this method may give underestimations as it did for cytochrome b_5 and other hydrophobic membrane proteins (Spatz and Strittmater, 1973).

The success in reducing the concentration of Triton, during purification of the ACh receptor, to eight molecules per subunit carrying one ACh-binding site permitted the use of sedimentation velocity and equilibrium to determine its molecular weight (Carroll *et al.*, 1973; Edelstein *et al.*, 1975). In the presence of such low Triton concentrations, the ACh receptor molecules apparently existed in aggregates of 330,000 and 660,000 or 1,300,000. When 0.1% Triton was added, molecules of only the lower molecular weight of 330,000 were present.

Since the molecular weight corresponding to one ACh-binding site was calculated to be 83,000–112,000 (Eldefrawi and Eldefrawi, 1973*a*), yet the protein was dissociated with SDS into one major subunit of 46,000 and possibly three other subunits, the ACh receptor molecule of 330,000 might consist of several protomers (i.e., the unit carrying one ACh-binding site). Each protomer would be made of one 46,000 subunit and another one. If the molecular weight of the receptor were 660,000 and the protomer 110,000, then the molecule would consist of six protomers. A more plausible alternative is that the molecular weight 330,000 consists of four of the protomers of approximately 80,000 mol. wt. These uncertainties should be resolved in the near future.

The subunit structure of the ACh receptor was confirmed by electron microscopy. The negatively stained pure receptor molecules were doughnut-like in shape, having three to six subunits with a central electron-dense core (Cartaud *et al.*, 1973; Eldefrawi *et al.*, in press). Hexagons of similar size and shape were found in the innervated membrane of *Torpedo* electroplax from which the ACh receptor was extracted (Cartaud *et al.*, 1973; Nickel and Potter, 1973). An interesting aspect is that similar hexagonal arrays were observed in the luminal surface of rat bladder (Hicks and Ketterer, 1970), as well as in "gap junctions" of liver (Benedetti and Emmelot, 1968) and goldfish brain (Zampighi and Robertson, 1973), where no ACh receptors are expected to be found since transmission is not chemical but electrical. It is plausible that the hexagonal array is important for regulation of ion fluxes in special membranes, and in cholinergic postsynaptic membranes the molecules are modified so as to have ACh-binding sites.

Correlation of the molecular weight with the surface dimensions of the ACh receptor molecules led to the suggestion that the molecule should be elongated (Nickel and Potter, 1973).

4.2.3. Cooperative Interactions

Electrophysiological experiments on *Electrophorus* electroplax (Karlin, 1967; Changeux and Podleski, 1968) and vertebrate muscles (Ariëns, 1964) gave sigmoidal dose–response curves which indicated positive cooperativity. To put it simply, this is the case where binding of a molecule to one subunit induces three-dimensional changes in another subunit so that it becomes more receptive to a second molecule. These electrophysiological data could be due to interactions at the binding stage or to effects later on in the chain of events leading to depolarization. At first, when binding of cholinergic ligands to the ACh receptor was determined over a limited range of ligand concentrations, there was a straight-line relationship between the reciprocals of amount bound and concentration. By expanding the ligand concentrations tested, deviations from Menten kinetics were observed. The availability of a purified ACh receptor of *Torpedo marmorata* with very high specific binding allowed measurements of its binding of [^{3}H]ACh at very low concentrations (8×10^{-10} 10^{-8} M); positive cooperativity of ACh binding was observed under these conditions (Eldefrawi and Eldefrawi, 1973*b*). Positive cooperativity was also exhibited for the binding of *d*-tubocurarine and decamethonium (Eldefrawi, unpublished). At higher ligand concentrations (10^{-7}–10^{-5} M), a lower affinity for binding was observed in the pure receptor as well as in the membrane-bound or solubilized *Torpedo* receptors from aged tissue (Eldefrawi *et al.*, 1971*a*, 1972, 1975). The high and low affinities may represent binding to two different noninteracting sites on the ACh receptor or to sites on two different molecules, but both are nicotinic and totally blocked with α-neurotoxins (O'Brien and Gibson, 1974). Alternatively, one may be carried by the postsynaptic ACh receptors and one by an extrajunctional receptor. Electrophysiological data have led to the suggestion that two such populations of ACh receptors occur in frog (Feltz and Mallart, 1971) and of glutamate receptors in locust muscles (Cull-Candy and Usherwood, 1973). The purification procedures used might not distinguish between these similar macromolecules. Another possibility is that, in the process of purification, groups of a few molecules at the active site might become oxidized or reduced, or the receptor hydrolysed, thus reducing its affinity for ligands. A plausible explanation for the reduced affinity at higher ligand concentrations is the occurrence of negative cooperativity between the receptor subunits (Eldefrawi and Eldefrawi, 1973*b*). This is the situation where, in a two-subunit protein, binding of the first molecule induces a conformational change that makes binding to the second active site more difficult. Insulin receptors on cultured lymphocytes and liver plasma membranes have recently been shown to exhibit negative cooperative interactions (DeMeyts *et al.*, 1973).

We may thus conclude that the ACh receptor molecule, possibly made of six subunits, exhibits positive cooperativity at low ligand concentrations. If the ACh receptor turns out to exhibit negative cooperativity as well, it would not be unique in exhibiting both kinds of cooperativities in its binding of ligands. There are few other examples, one of which is yeast glyceraldehyde-3-phosphate dehydrogenase in its binding of nicotinamide-adenine dinucleotide (Cook and Koshland, 1970). Such cooperativities would be advantageous for the ACh receptor, amplifying the effect of low concentrations of ACh and desensitizing the ACh receptor at high ACh concentrations and thus providing it with mechanisms for its regulatory role.

5. Interactions of Insecticides with ACh Receptors

5.1. Anti-ACh Receptors

Since several macromolecules interact with ACh in its role as a transmitter, it would be expected that these macromolecules should have some similarities. This is underscored by the effect of some cholinergic drugs on more than one of these macromolecules. A good example is *d*-tubocurarine, which not only blocks the nicotinic receptor but also affects the presynaptic release of ACh and its hydrolysis by AChE. The ability of a drug to interact with these different targets will depend largely on its relative affinity for each of the macromolecules concerned. The macromolecule having the highest affinity for each drug presumably becomes its major target. Among currently used insecticides, only two groups interact with the ACh receptor or possibly the gate proteins coupled to it. The first group includes materials whose target is the receptor and/or gate proteins, and the second includes insecticides which have a different primary target (e.g., AChE) but which may interact with the ACh receptor under special circumstances. These functions of insecticides were discovered initially through electrophysiological studies and subsequently confirmed by biochemical experiments *in vitro* on ACh receptors. Equivalent studies have not yet been done on pure ACh receptors.

5.1.1. Nicotine and Analogues

Nicotine is a well-established agonist, which at low concentrations stimulates nicotinic junctions and at high concentrations depresses them. It is interesting that its action as a blocker was the one first discovered over 100 years ago (Bernard, 1857). Electrophysiological data provided evidence that the nicotinic ACh receptor is the primary target of nicotine in vertebrates (Volle and Koelle, 1969) as well as insects (Roeder and Roeder, 1939; Flattum and Shankland, 1971).

Because nicotinic synapses in insects occur only in the central nervous system, where there are barriers against the penetration of hydrophilic and charged molecules (Eldefrawi and O'Brien, 1966, 1967*a,b*; Toppozada and O'Brien, 1967; Eldefrawi *et al.*, 1968), quaternarized molecules have low toxicities to insects. Yamamoto *et al.* (1968) found an apparent correlation between the structures of many nicotinoids and 2-pyridylmethylamines, their toxicities to houseflies, and their inhibition of AChE. It was clear, however, that the target for these compounds was not AChE, because the toxic concentrations of these compounds were far below those required for enzyme inhibition; K_i values were $1–100 \times 10^{-3}$ M (Fujita *et al.*, 1970). These data suggested the possible interaction of nicotine with another macromolecule, possibly the ACh receptor, with steric and electronic characteristics similar to those of AChE. Both optical isomers of nicotine (or nornicotine) were toxic to insects, but their relative toxicities varied with the insect species concerned (Soeda and Yamamoto, 1969). The essential requirement for insect toxicity was established to be the 3-pyridylmethylamine with a basic (but not highly so) amino nitrogen (Yamamoto *et al.*, 1962, 1968; Kamimura *et al.*, 1963).

With the availability of an insect tissue extract containing proteins exhibiting the pharmacological binding characteristics of ACh receptors (Eldefrawi *et al.*, 1970), it was possible to study directly the effect of nicotinoids on insect ACh receptors. A similar number of receptors in the housefly head extract bound [^{3}H]nicotine or [^{3}H] muscarone, and when the effect of the presence of several nicotinoids on the binding of these ligands to their receptors was determined it was found that only the toxic compounds blocked the binding (Table III). The correlation observed between toxicity and blockade of binding was poor, possibly because toxicity is determined not only by binding to the ACh receptors but also by other factors such as metabolism and penetration through membrane barriers.

5.1.2. Nereistoxin and Cartap

Nereistoxin (Fig. 4) was isolated from a marine annelid, *Lumbriconereis heteropoda* (Nitta, 1934), and proved to be toxic to many insects, particularly the plant-chewing species (Sakai, 1964). Among several synthesized derivatives of nereistoxin, the ones toxic to insects are the 4-alkylamino-1,2-dithiolanes and the 2-dimethylamino-1,3-propane dithiols (Sakai, 1966*b*).

Table III. Relationship Between the Blockade of the Binding of [^{3}H]Muscarone to the ACh Receptors of Housefly Heads by Two Concentrations of Nicotinoids and Their Toxicities to Houseflies[a]

Drug	Percent blockade of [^{3}H]muscarone binding (10^{-6} M)		Toxicity to housefly LD_{50}(μg/fly)
	10^{-5} M	10^{-4} M	
Nicotine	50	102	5
Anabasine	25	98	4
3-Pyridylmethyl dimethylamine	45	92	16
3-Pyridylmethyl diethylamine	61	97	11
N,N-Diethylnicotinamide	0	0	>100
N-(3-Pyridylmethyl) morphine	0	0	>100

[a] From Eldefrawi *et al.*, (1970).

Cartap (4-*N,N*-dimethylamino-1,2-dithiolane) (Fig. 4) is a synthetic insecticide believed to be metabolized *in vivo* to nereistoxin (Sakai and Sato, 1971).

Judging from its effect on frog rectus abdominis and sartorius muscles (Sakai, 1966*a*; Deguchi *et al.*, 1971), rat diaphragm (Chiba *et al.*, 1967), and the cockroach sixth abdominal ganglion (Sakai, 1967; Bettini *et al.*, 1973), the toxicity of nereistoxin has been attributed to its blockade of cholinergic transmission. Its action is not due to inhibition of AChE, because it has a low affinity for the enzyme (Sakai, 1966*b*). Nereistoxin reduces the amount of transmitter released from the presynapse and also reduces the sensitivity of the postsynaptic membrane to applied ACh. It has been suggested that its major action is to inhibit the mechanisms that increase conductances of the post-

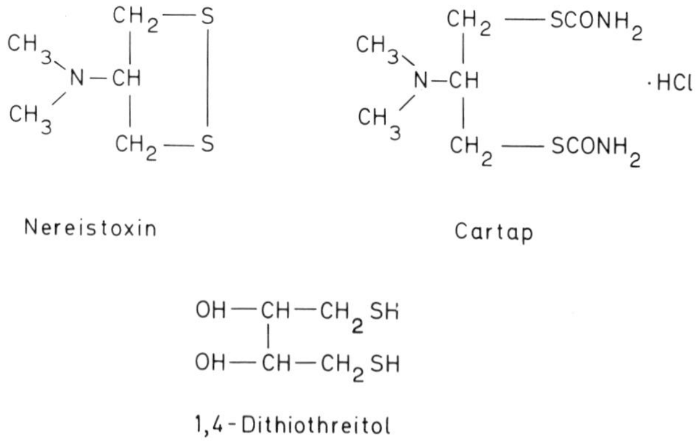

Fig. 4. Nereistoxin, cartap, and 1,4-dithiothreitol.

synaptic membrane to sodium and potassium ions (Deguchi *et al.*, 1971). Whether its action is on the ACh receptor itself or on the gates controlling ionic permeability is not yet known. It is interesting to note the similarity in structure between nereistoxin and 1, 4-dithiothreitol (DTT), another chemical that is very effective in inhibiting the ACh receptor. One can speculate that nereistoxin may be reduced to a compound similar to DTT, which may be the one attacking the receptor. Such a compound would also be expected to attack other kinds of receptors which have an essential S—S bond. In fact, nereistoxin has been found to act not only at nicotinic junctions but also at those which are muscarinic. It decreased mammalian heart rate and increased salivary gland excretion, and both effects were antagonized by atropine (Nitta, 1941). Because of the pharmacological specificity of mammalian ACh receptors, the action of nereistoxin at both nicotinic and muscarinic junctions leads to the suggestion that nereistoxin may not bind to the active sites of these receptors.

5.2. Anticholinesterases

5.2.1. Electrophysiological Studies

It is widely accepted that the mode of action of organophosphates and carbamates is inhibition of AChE, which results in the retardation of ACh hydrolysis and consequently its prolonged action at cholinergic synapses. As a consequence, the few results of several earlier electrophysiological studies demonstrating reversible intereactions of the anticholinesterases, which were independent of their enzyme-inhibiting action, received only little attention. For example, in several vertebrate neuromuscular preparations in which cholinesterases had been totally inhibited, the addition of eserine, neostigmine (Miquel, 1946; Randall, 1950), or edrophonium (Cohen and Posthumus, 1955) caused muscle contraction. Also, DFP, TEPP, parathion, sarin, and eserine (Koppanyi and Karczmar, 1951; Van der Meer and Meeter, 1956) potentiated the action of applied ACh, succinylcholine (a slowly hydrolyzable ester), and even nonhydrolyzable agonists such as decamethonium, choline, and edrophonium (Zaimis, 1951; Hutter, 1952; Cohen and Posthumus, 1955). In other words, the anticholinesterases in these experiments were acting as receptor agonists. In several other studies on vertebrates, anticholinesterases acted as antagonists. In one, DFP reduced muscle contraction and its effect was reversed without noticeable recovery of cholinesterase activity, while in another TEPP was inhibitory even in the presence of a great deal of AChE activity (McNamara *et al.*, 1954).

In a more recent study on the monocellular preparation of *Electrophorus* electroplax where membrane depolarization was measured, low concentrations of the organophosphates DFP, paraoxon, or phospholine (5×10^{-5} to 2×10^{-4} M) potentiated the action of ACh, as expected from their inhibition of AChE. However, when high concentrations ($1\text{–}8 \times 10^{-3}$ M) of these drugs were

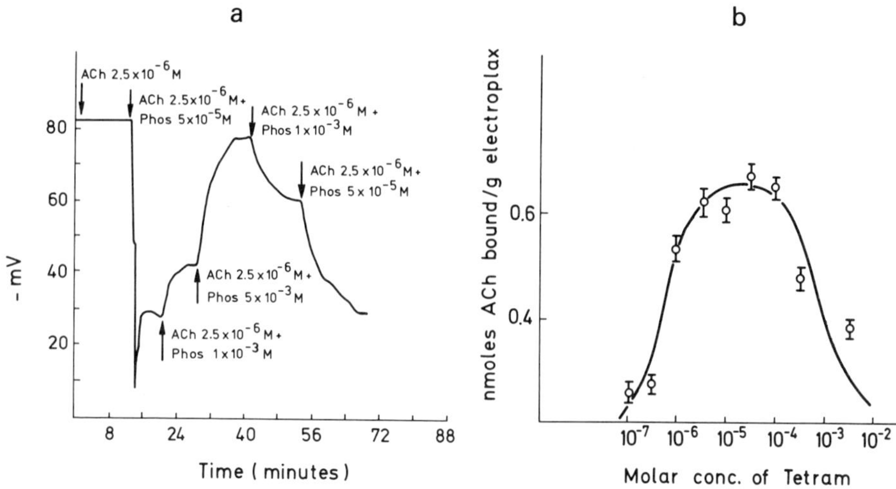

Fig. 5. Effect of organophosphates on ACh receptors *in situ* and *in vitro*. (a) Receptor inhibition by high concentrations of phospholine in the monocellular preparation of eel electroplax. ACh at 2.5×10^{-6} M causes a biphasic depolarization only in the presence of the organophosphate phospholine at 50×10^{-6}M. Increasing the concentration of phospholine to 1×10^{-3} M or higher results in repolarization, which is reversed by decreasing the phospholine concentration. Reprinted from Bartels and Nachmansohn (1969) by courtesy of Academic Press. (b) Effect of increasing tetram concentration on the binding of ACh (at 2.5×10^{-7} M) to membrane-bound ACh receptors of *Torpedo marmorata*. The blocking effect of high tetram concentrations on ACh binding is reversible when tetram is removed by dialysis. The circles and vertical bars represent the means and standard deviations for three experiments, five samples each. Reprinted from Eldefrawi *et al.* (1971*b*) by courtesy of Academic Press.

applied with ACh, they reversibly repolarized the membrane, acting as inhibitors of the ACh receptor (Bartels and Nachmansohn, 1969) (Fig. 5). In another study using the voltage clamp technique on frog sartorius muscle, DFP was again found to be inhibitory, lowering the amplitude and reducing the decay time of the end-plate current; these effects are the opposite to those expected from excess ACh (Kuba *et al.*, 1973). The DFP effects were also reversible by washing off excess DFP, and remained as long as all of the AChE was still inhibited. Assuming that the interaction of ACh with its receptor is a reversible two-step process, the first leading to the formation of an intermediate inactive complex and the second to the active depolarizing conformation (del Castillo and Katz, 1957; Katz and Thesleff, 1957), and that the latter step is coupled with the opening of ionic channels, the effects of DFP were interpreted as an interaction with the receptor–gate protein complex and not with the receptor alone (Kuba *et al.*, 1973).

These observations demonstrate that several compounds recognized primarily as anticholinesterases can also interact with the ACh receptor or the receptor–gate protein complex, acting either as activators or as inhibitors depending on the compound, its concentration, and the tissue preparation used. As in the case of nereistoxin, the use of biochemical techniques should further elucidate the mode of action of these anticholinesterases.

5.2.2. Biochemical Studies

Biochemical studies on the membrane-bound ACh receptors of electric organs showed that high concentrations of DFP (and also tetram and guthoxon) could inhibit almost all of the enzyme without affecting receptor interactions (Eldefrawi *et al.*, 1971*b*). The ACh receptor of *Electrophorus* was more sensitive to blockade by those organophosphates than that of *Torpedo*. With the latter, increasing concentration of teram up to 5×10^{-4} M had no effect on the binding of muscarone to the ACh receptor but increased the binding and retention of ACh because of its inhibition of AChE. Because of the extreme sensitivity of the receptors of the housefly head extract to these organophosphates, but their low affinity for R-16661, this drug was used to inhibit AChE in the preparation before attempting to study the binding of ACh to its receptors (Aziz and Eldefrawi, 1973).

The *in vitro* binding of anticholinesterases to the ACh receptor of *Torpedo* is not restricted to organophosphates. Neostigmine and pyridostigmine acted like the organophosphates in blocking the binding of ACh to the *Torpedo* receptor (Eldefrawi *et al.*, 1972). Since this occurs only at high concentrations and is competitive, it is possibly mediated only by an electrostatic attraction between the anionic site of the receptor and these positively charged carbamates. An interesting finding was that although edrophonium was slightly less active than the two carbamates on AChE its K_i toward ACh binding to the ACh receptor was 1–3 orders of magnitude smaller (Seifert and Eldefrawi, 1974). Such a high affinity for the ACh receptor supports the earlier electrophysiological observations on the effect of edrophonium as an agonist (Riker, 1953).

6. Concluding Remarks

Of the macromolecules involved in cholinergic transmission, AChE is the target for most insecticides in use today, the ACh receptor is the main target for nicotine and analogues, and dieldrin may affect the presynaptic release of ACh (Shankland and Shroeder, 1973). The interest generated in the area of neurotransmitter receptors from various disciplines such as pharmacology, toxicology, neurochemistry, neurophysiology, biochemistry, and membrane and cell biology has resulted in a sudden expansion of receptor research into new and more interesting areas. There is ongoing research on the chemical composition, functional groups, and amino acid sequence of ACh receptors. These molecules are being photographed and their shapes and sizes determined under various conditions. Immunological and crystallographic studies are in progress as well as attempts to reconstitute the receptor into synthetic lipid membranes. Fluorescent probes are also being used to do fast kinetic studies on the interactions of ligands with receptors. ACh receptors from various sources are being compared, and differences between their drug sensitivities have been observed. Although a great deal of receptor research is currently being conducted, it is obvious that much more is needed before we can develop a variety of effective and selective receptor poisons.

In addition to the ACh receptor, other neurotransmitter receptors are suspected to be present in insects. Examples are dopamine receptors in salivary gland cells of the American cockroach (House *et al.*, 1973), serotonin receptors in Malpighian tubule muscle (Crowder and Shankland, 1972), adrenergic receptors in the light organ of the firefly, *Photinus pyralis* (Smalley, 1965), as well as glutamate receptors in insect neuromuscular junctions. Therefore, there is no shortage of possible targets for insecticides. What is needed for the development of new generations of insecticides is more basic information on insect neurochemistry and neuropharmacology, which currently lags far behind that for vertebrates.

ACKNOWLEDGMENT

Experimental work from this laboratory on insect acetylcholine receptors is supported by U.S. Public Health Service Research Grant ES 00901.

7. References

Albuquerque, E. X. Barnard, E. A. Chiu, T. H., Lapa, A. J., Dolly, J. O., Jansson, S. E., Daly, J., and Witkop, B., 1973, Acetylcholine receptor and ion conductance modulator sites at the murine neuromuscular junction: Evidence from specific toxin reactions, *Proc. Natl. Acad. Sci. U.S.A.* **70:**949.

Ariens, E. J., 1964, *in: Molecular Pharmacology*, Vol. 1 (E. J. Ariens, ed.), p. 119, Academic Press, New York.

Aziz, S. A., and Eldefrawi, M. E., 1973, Cholinergic receptors of the central nervous system of insects, *Pestic. Biochem. Physiol.* **3:**168.

Bartels, E., and Nachmansohn, D., 1969, Organosphosphate inhibitors of acetylcholine-receptor and -esterase tested on the electroplax, *Arch. Biochem. Biophys.* **133:**1.

Benedetti, E. L., and Emmelot, P., 1968, Hexagonal array of subunits in tight junctions separated from isolated rat liver plasma membranes, *J. Cell Biol.* **38:**15.

Ben-Haim, D., Landau, E. M., Silman, I., 1973, The role of a reactive disulphide bond in the function of the acetylcholine receptor at the frog neuromuscular junction, *J. Physiol. (London)* **234:**305.

Bernard, C., 1857, *Lecons sur les Effects des Substances Toxiques et Medicamenteuses*, Bailliere, Paris.

Bettini, S., D'Ajello, V., and Maroli, M., 1973, Cartap activity on the cockroach nervous and neuromuscular transmission, *Pestic. Biochem. Physiol.* **3:**199.

Biesecker, G., 1973, Molecular properties of the cholinergic receptor purified from *Electrophorus electricus, Biochemistry* **12:**4403.

Capaldi, R. A., and Vanderkooi, G., 1972, The low polarity of many membrane proteins, *Proc. Natl. Acad. Sci. U.S.A.* **69:**930.

Carroll, R. C., Eldefrawi, M. E., and Edelstein, S. J., 1973, Studies on the structure of the acetylcholine receptor from *Torpedo marmorata, Biochem. Biophys. Res. Commun.* **55:**864.

Cartaud, J., Benedetti, E. L., Cohen, J. B., Meunier, J. C., and Changeux, J. P., 1973, Presence of a lattice structure in membrane fragments rich in nicotinic receptor protein from the electric organ of *Torpedo marmorata, FEBS Letters* **33:**109.

Changeux, J. P., 1974, The cholingeric receptor protein: Functional properties and its role in the regulation of developing synapses *in: Cell Surface Development* (A. A. Moscana, ed.), Wiley, New York.

Changeux, J. P., and Podleski, T. R., 1968, On the excitability and cooperativity of the electroplax membrane, *Proc. Natl. Acad. Sci. U.S.A.* **59:**944.

Changeux, J. P., Podleski, T., and Meunier, J. C., 1969, On some structural analogies between acetylcholinesterase and the macromolecular receptor of acetylcholine, *J. Gen. Physiol.* **54:**225.

Chiba, S., Sajo, Y., Takeo, Y., Yui, T., and Aramaki, Y., 1967, Nereistoxin and its derivatives, their neuromuscular blocking and convulsive actions, *Jap. J. Pharmacol.* **17:**491.

Cohen, J. B., and Changeux, J. P., 1973, Interaction of a fluorescent ligand with membrane-bound cholinergic receptor from *Torpedo marmorata, Biochemistry* **12:**4855.

Cohen, J. A., and Posthumus, C. H., 1955, The mechanism of action of anticholinesterase, *Acta Physiol. Pharmacol. Neerl.* **4:**17.

Cook, R. A., and Koshland, D. E. Jr., 1970, Positive and negative cooperativity in yeast glyceraldehyde 3-phosphate dehydrogenase, *Biochemistry* **9:**3337.

Crawford, J. M., and Curtis, D. R., 1966, Pharmacological studies on feline Betz cells, *J. Physiol. (London)* **186:**121.

Crawford, J. M., Curtis, D. R., Voorhoeve, P. E., and Wilson, V. J., 1966, Acetylcholine sensitivity of cerebral neurones in the cat, *J. Physiol. (London)* **186:**139.

Crowder, L. A., and Shankland, D. L., 1972, Pharmacology of the Malpighian tubule muscle of the American cockroach *Periplaneta americana, J. Insect Physiol.* **18:**929.

Cuatrecasas, P., 1972, Affinity chromatography and purification of the insulin receptor of liver cell membranes, *Proc. Natl. Acad. Sci. U.S.A.* **69:**1277.

Cuatrecasas, P., and Anfinsen, C. B., 1971, Affinity chromatography, *Ann. Rev. Biochem.* **40:**259.

Cull-Candy, S. G., and Usherwood, P. N. R., 1973, Two populations of L-glutamate receptors on locust muscle fibres, *Nature New Biol.* **246:**62.

Dale, H. H., 1914, The action of certain esters and ethers of choline, and their relation to muscarine, *J. Pharmacol. Exp. Ther.* **6:**147.

Deguchi, T., Narahashi, T., and Haas, H. G., 1971, Mode of action of nereistoxin on the neuromuscular transmission in the frog, *Pestic, Biochem. Physiol.* **1:**196.

del Castillo, J., and Katz, B., 1957, Interaction at end-plate receptors between different choline derivatives, *Proc. Roy. Soc. London Ser. B* **146:**369.

DeMeyts, P., Roth, J., Neville, D. M., Jr., Gavin, J. R., III, and Lesniak, M. A., 1973, Insulin interactions with its receptors: Experimental evidence for negative cooperativity, *Biochem. Biophys. Res. Commun.* **55:**154.

Denburg, J. L., Eldefrawi, M. E., and O'Brien, R. D., 1972, Macromolecules from lobster axon membranes that bind cholinergic ligands and local anesthetics, *Proc. Natl. Acad. Sci. U.S.A.* **69:**177.

Dudai, Y., Silman, I., Kalderon, N., and Blumberg, S., 1972, Purification by affinity chromatography of acetylcholinesterase from electric organ tissue of the electric eel subsequent to tryptic treatment, *Biochim. Biophys. Acta* **268:**138.

Edelstein, S. J., Beyer, W. B., Eldefrawi, A. T., and Eldefrawi, M. E., 1975, Molecular weight of the acetylcholine receptors of electric organs and the effect of Triton X-100, *J. Biol. Chem.* **250:**6101.

Ehrlich, P., 1913, Chemotherapeutics: Scientific principles, methods and results, *Lancet* **2:**445.

Eldefrawi, M. E., and Eldefrawi, A. T., 1972, Characterization and partial purification of the acetylcholine receptor from *Torpedo* electroplax, *Proc. Natl. Acad. Sci. U.S.A.* **69:**1776.

Eldefrawi, M. E., and Eldefrawi, A. T., 1973*a*, Purification and molecular properties of the acetylcholine receptor from *Torpedo* electroplax, *Arch. Biochem. Biophys.* **159:**362.

Eldefrawi, M. E., and Eldefrawi, A. T., 1973*b*, Cooperativities in the binding of acetylcholine to its receptor, *Biochem. Pharmacol.* **22:**3145.

Eldefrawi, M. E., and O'Brien, R. D., 1966, Permeability of the abdominal nerve cord of the American cockroach to fatty acids, *J. Insect Physiol.* **12:**1133.

Eldefrawi, M. E., and O'Brien, R. D., 1967*a*, Permeability of the abdominal nerve cord of the American cockroach, *Periplaneta americana* (L.), to aliphatic alcohols, *J. Insect Physiol.* **13:**691.

Eldefrawi, M. E., and O'Brien, R. D., 1967*b*, Permeability of the abdominal nerve cord of the American cockroach, *Periplaneta americana* (L.), to quaternary ammonium salts, *J. Exp. Biol.* **46:**1.

Eldefrawi, A. T., and O'Brien, R. D., 1970, Binding of muscarone by extracts of housefly brain: Relationship to receptors for acetylcholine, *J. Neurochem.* **17:**1287.

Eldefrawi, M. E., Toppozada, A., Salpeter, M. M., and O'Brien, R. D., 1968, The location of penetration barriers in the ganglia of the American cockroach, *Periplaneta americana* (L.), *J. Exp. Biol.* **48:**325.

Eldefrawi, M. E., Eldefrawi, A. T., and O'Brien, R. D., 1970, Mode of action of nicotine in the housefly, *J. Agr. Food Chem.* **18:**1113.

Eldefrawi, M. E., Britten, A. G., and Eldefrawi, A. T., 1971*a*, Acetylcholine binding to *Torpedo* electroplax: Relationship to acetylcholine receptors, *Science* **173:**338.

Eldefrawi, M. E., Britten, A. G., and O'Brien, R. D., 1971*b*, Action of organophosphates on binding of cholinergic ligands, *Pestic. Biochem. Physiol.* **1:**101.

Eldefrawi, M. E., Eldefrawi, A. T., and O'Brien, R. D. 1971*c*, Binding sites for cholinergic ligands in a particulate fraction of *Electrophorus* electroplax, *Proc. Natl. Acad. Sci. U.S.A.* **68:**1047.

Eldefrawi, M. E., Eldefrawi, A. T., and O'Brien, R. D., 1971*d*, Binding of five cholinergic ligands to housefly brain and *Torpedo* electroplax, *Mol. Pharmacol.* **7:**104.

Eldefrawi, M. E., Eldefrawi, A. T., Gilmour, L. P., and O'Brien, R. D., 1971*e*, Multiple affinities for binding of cholinergic ligands to a particulate fraction of *Torpedo* electroplax, *Mol. Pharmacol.* **7:**420.

Eldefrawi, M. E., Eldefrawi, A. T., and Wilson, D. B., 1975, Tryptophan and cysteine residues of the acetylcholine receptors of *Torpedo* species. Relationship to binding of cholinergic ligands, *Biochemistry* **14:**4304.

Eldefrawi, M. E., Eldefrawi, A. T., Seifert, S., and O'Brien, R. D., 1972, Properties of Lubrol-solubilized acetylcholine receptor from *Torpedo* electroplax, *Arch. Biochem. Biophys.* **150:**210.

Feader, I. R., and Salpeter, M. M., 1970, Glutamate uptake by a stimulated insect nerve muscle preparation, *J. Cell Biol.* **46:**300.

Faeder, I. R., O'Brien, R. D., and Salpeter, M. M., 1970, A re-investigation of evidence for cholinergic neuromuscular transmission in insects, *J. Exp. Zool.* **173:**187.

Feltz, A., and Mallart, A., 1971, An analysis of acetylcholine response of junctional and extrajunctional receptors of frog muscle fibres, *J. Physiol. (London)* **218:**85.

Flattum, R. F., and Shankland, D. L., 1971, Acetylcholine receptors and the diphasic action of nicotine in the American cockroach, *Periplaneta americana* (L.), *Comp. Gen. Pharmacol.* **2:**159.

Fujita, T., Yamamoto, I., and Nakajima, M., 1970, Analysis of the structure–activity relationship of nicotine-like insecticides using substituent constants, *in: Biochemical Toxicology of*

Insecticides (R. D. O'Brien and I. Yamamoto, eds.), p. 21, Academic Press, New York.

Hartzell, H. C., and Fambrough, D. M., 1973, Acetylcholine receptor production and incorporation into membranes of developing muscle fibers, *Dev. Biol.* **30:**153.

Hicks, R. M., and Ketterer, B., 1970, Isolation of the plasma membrane of the luminal surface of rat bladder epithelium, and the occurrence of a hexagonal lattice of subunits both in negatively stained whole mounts and in sectioned membranes, *J. Cell Biol.* **45:**542.

Hiley, C. R., Young, J. M., and Burgen, A. S. V., 1972, Labelling of cholinergic receptors in subcellular fractions from rat cerebral cortex, *Biochem. J.* **126:**86P.

House, C. R., Ginsborg, B. L., and Silinsky, E. M., 1973, Dopamine receptors in cockroach salivary gland cells, *Nature New Biol.* **245:**63.

Hucho, F., and Changeux, J. P., 1973, Molecular weight and quaternary structure of the cholinergic receptor protein extracted by detergents from *Electrophorus electricus* electric tissue, *FEBS Letters* **38:**11.

Hutter, O. F., 1952, Effect of choline on neuromuscular transmission in the cat, *J. Physiol. (London)* **117:**241.

Illiano, G. Tell, G. P. E., Siegel, M. I., and Cuatrecasas, P., 1973, Guanosine 3′ : 5′-cyclic monophosphate and the action of insulin, *Proc. Natl. Acad. Sci. U.S.A.* **70:**2443.

Kamimura, H., Matsumoto, A., Miyazaki, Y., and Yamamoto, I., 1963, Studies on nicotinoids as an insecticide. IV. Relation of structure of toxicity of pyridylmethylamines, *Agr. Biol. Chem.* **27:**684.

Karlin, A., 1967, On the application of "a plausible model" of allosteric proteins to the receptor for acetylcholine, *J. Theor. Biol.* **16:**306.

Karlin, A., and Bartels, E., 1966, Effects of blocking sulfhydryl space groups and of reducing disulfide bonds on the acetylcholine-activated permeability system of the electroplax, *Biochim. Biophys. Acta* **126:**525.

Karlin, A., Prives, J., Deal, W., and Winnik, M., 1970, Counting acetylcholine receptors in the electroplax, *Ciba Foundation Symposium on Molecular Properties of Drug Receptors* (R. Porter and M. O'Connor, eds.), p. 247, Churchill, London.

Karlsson, E., Heilbronn, E., and Widlund, L., 1972, Isolation of the nicotinic acetylcholine receptor by biospecific chromatography on insolubilized *Naja naja* neurotoxin, *FEBS Letters* **28:**107.

Katz, B., and Miledi, R., 1973, The characteristics of "end-plate noise" produced by different depolarizing drugs, *J. Physiol. (London)* **230:**707.

Katz, B., and Thesleff, S., 1957, A study of the 'desensitization' produced by acetylcholine at the motor end-plate, *J. Physiol. (London)* **138:**63.

Klett, R. P., Fulpius, B. W., Cooper, D., Smith, M., Reich, E., and Possani, L. D., 1973, The acetylcholine receptor. I. Purification and characterization of a macromolecule isolated from *Electrophorus electricus, J. Biol. Chem.* **248:**6841.

Koppanyi, T., and Karczmar, A. G., 1951, Contribution to the study of the mechanism of action of cholinesterase inhibitors, *J. Pharmacol. Exp. Ther.* **101:**327.

Kuba, K., Albuquerque, E. X., and Barnard, E. A., 1973, Diisopropylfluorophosphate: Suppression of ionic conductance of the cholinergic receptor, *Science* **181:**853.

Langley, J. N., 1906, On the reaction of cells and of nerve endings to certain poisons, chiefly as regards the reaction of striated muscles to nicotine and curare. *J. Physiol. (London)* **20:**223.

Lee, C. Y., 1972, Chemistry and pharmacology of polypeptide toxins in snake venoms, *Ann. Rev. Pharmacol.* **12:**265.

Lester, H., 1971, Cobra toxin's action on nicotinic acetylcholine receptors, *J. Gen. Physiol.* **57:**255.

Leuzinger, W., and Baker, A. L., 1967, Acetylcholinesterase. I. Large-scale purification, homogeneity, and amino acid analysis, *Proc. Natl. Acad. Sci. U.S.A.* **7:**446.

Lindstrom, J. M., Singer, S. J., and Lennox, E. S., 1973, The effects of reducing and alkylating agents on the acetylcholine receptor of frog sartorius muscle, *J. Membr. Biol.* **11:**217.

Loewi, O., 1921, Ueber humorale Uebertragbarkeit der Herznervenzirkung, *Arch. Ges. Physiol. Pfluegers* **189:**239.

McNamara, B. P., Martha, E. F., Bergner, A. D., Robinson, E. M., Bender, C. W., and Willis, J. H., 1954, Studies on the mechanism of action of DFP and TEPP, *J. Pharmacol. Exp. Ther.* **110:**232.

Metcalf, R. L., and March, R. B., 1950, Properties of acetylcholine esterases from the bee, the fly and the mouse and their relation to insecticide action, *J. Econ. Entomol.* **43:**670.

Meunier, J. C., Olsen, R. W., and Changeux, J. P., 1972, Studies on the cholinergic receptor protein from *Electrophorus electricus*: Effect of detergents on some hydrodynamic properties of the receptor protein in solution, *FEBS Letters* **24:**63.

Meunier, J. C., Sealock, R., Olsen, R., and Changeux, J. P., 1974, Purification and properties of the cholinergic receptor protein from *Electrophorus electricus* electric tissue, *Eur. J. Biochem.* **45:**371.

Miledi, R., Molinoff, P., and Potter, L. T., 1971, Isolation of the cholinergic receptor protein of *Torpedo* electric tissue, *Nature (London)* **229:**557.

Miquel, O., 1946, The action of physostigmine, di-isopropyl fluorophosphate and other parasympathomimetic drugs on the rectus muscle of the frog, *J. Pharmacol.* **88:**67.

Nachmansohn, D., 1955, Metabolism and function of the nerve cell, *Harvey Lect. (1953/1954)* **49:**57.

Nachmansohn, D., 1971, Proteins in bioelectricity: Acetylcholine-esterase and -receptor, *in: Principles of Receptor Physiology* (W. R. Loewenstein, ed.), p. 18, Springer, New York.

Nickel, E., and Potter, L. T., 1973, Ultrastructure of isolated membranes of *Torpedo* electric tissue, *Brain Res.* **57:**508.

Nitta, S., 1934, Ueber Nereistoxin, einen giftigen Bestandteil von *Lumbriconereis heteropoda* (Eunicidae), *Yakagaku Zasshi* **54:**648.

Nitta, S., 1941, Pharmakologische Untersuchung des Nereistoxins, das vom Verf. im Körper des *Lumbriconereis heteropoda* (Isome) isoliert wurde, *Tokyo Igaku Zaschi* **55:**285.

O'Brien, R. D., and Gibson, R. E., 1974, Two forms of acetylcholine receptor in *Torpedo marmorata* electroplax, *Arch. Biochem. Biophys.* **165:**681.

O'Brien, R. D., Gilmour, L. P., and Eldefrawi, M. E., 1969, A muscarone-binding material in electroplax and its relation to the acetylcholine receptor. II. Dialysis assay, *Proc. Natl. Acad. Sci. U.S.A.* **65:**438.

O'Brien, R. D., Eldefrawi, M. E., and Eldefrawi, A. T., 1972, Isolation of acetylcholine receptors, *Ann. Rev. Pharmacol.* **12:**19.

O'Brien, R. D., Eldefrawi, M. E., and Eldefrawi, A. T., 1974, Techniques in isolation of acetylcholine receptors, *in: Methods in Neurochemistry* (R. Fried, ed.), Dekker, New York, in press.

Olsen, R. W., Meunier, J. C., and Changeux, J. P., 1972, Progress in the purification of the cholinergic receptor protein from *Electrophorus electricus* by affinity chromatography, *FEBS Letters* **28:**96.

Porter, C. W., Chiu, T. H., Weickowski, J., and Barnard, E. A., 1973, Types and locations of cholinergic receptor-like molecules in muscle fibres, *Nature New Biol.* **241:**3.

Potter, L. T., and Molinoff, P. B., 1972, Isolation of cholinergic receptor proteins *in: Perspectives in Neuropharmacology* (S. H. Snyder, ed.), p. 9, Oxford University Press, New York.

Powell, J. T., Bon, S., Rieger, F., and Massoulié, J., 1973, *Electrophorus* acetylcholinesterase: A glycoprotein; molecular weight of its subunits, *FEBS Letters* **36:**17.

Raftery, M. A., Schmidt, J., Clark, D. G., and Wolcott, R. G., 1971, Demonstration of a specific α-bungarotoxin binding component in *Electrophorus electricus* electroplax membranes, *Biochem. Biophys. Res. Commun.* **45:**1622.

Raftery, M. A., Schmidt, J., Martinez-Carrion, M., Moody, T., Vandlen, R., and Duguid, J., 1973, Biochemical studies on *Torpedo californica* acetylcholine receptors, *J. Supramol. Struct.* **1:**360.

Randall, L. O., 1950, Anticurare action of phenolic quaternary ammonium salts, *J. Pharmacol.* **100:**83.

Rieger, F., Bon, S., and Massoulie, J., 1973, Phospholipids in "native" *Electrophorus* acetylcholinesterase, *FEBS Letters* **37:**12.

Riker, W. F., Jr., 1953, Excitatory and anti-curare properties of acetylcholine and related quaternary ammonium compounds at the neuromuscular junction, *Pharmacol. Rev.* **5:**1.

Roeder, K. D., and Roeder, S., 1939, Electrical activity in the isolated ventral nerve cord of the cockroach. 1. The action of pilocarpine, nicotine, eserine and acetylcholine, *J. Cell. Comp. Physiol.* **14:**1.

Rosenberry, T. L., Chang, H. W., and Chen, Y. T., 1972, Purification of acetylcholinesterase by affinity chromatography and determination of active site stoichiometry, *J. Biol. Chem.* **247:** 1555.

Sakai, M., 1964, Studies on the insecticidal action of nereistoxin, 4-*N,N*-dimethylamino-1,2-dithiolane. I. Insecticidal properties, *Jap. J. Appl. Entomol. Zool.* **8:**324.

Sakai, M., 1966*a*, Studies on the insecticidal action of nereistoxin, 4-*N,N*-dimethylamino-1,2-dithiolane. III. Antagonism to acetylcholine in the contraction of rectus abdominis muscle of frog, *Boytyu-Kagaku* **21:**61.

Sakai, M., 1966*b*, Studies on the insecticidal action of nereistoxin, 4-*N,N*-dimethylamino-1,2-dithiolane. IV. Role of the anticholinesterase activity in the insecticidal action to housefly, *Musca domestica L.* (Diptera: Muscidae), *Appl. Entomol. Zool.* **1:**73.

Sakai, M., 1967, Studies on the insecticidal action of nereistoxin, 4-*N,N*-dimethylamino-1,2-dithiolane. V. Blocking action on the cockroach ganglion, *Botyu-Kagaku* **32:**21.

Sakai, M., and Sato, Y., 1971, Metabolic conversion of the nereistoxin-related compounds into nereistoxin as a factor of their insecticidal action, *in: Abst. 2nd Int. Congr. Pestic. Chem.*, Tel Aviv, February.

Salvaterra, P. M., and Moore, W. J., 1973, Binding of [^{125}I]-α-bungarotoxin to particulate fractions of rat and guinea pig brain, *Biochem. Biophys. Res. Commun.* **55:**1311.

Schleifer, L. S., and Eldefrawi, M. E., 1974, Identification of the nicotinic and muscarinic acetylcholine receptors in subcellular fractions of mouse brain, *Neuropharmacology* **13:**53.

Schmidt, J., and Raftery, M. A., 1973, Purification of acetylcholine receptors from *Torpedo californica* electroplax by affinity chromatography, *Biochemistry* **12:**852.

Seifert, S. A., and Eldefrawi, M. E., 1974, Affinity of myasthenia drugs to acetylcholinesterase and acetylcholine receptor, *Biochem. Med.* **10:**258.

Shankland, D. L., and Shroeder, M. E., 1973, Pharmacological evidence for discrete neurotoxic action of dieldrin (HEOD) in the American cockroach, *Periplaneta americana* (L.), *Pestic. Biochem. Physiol.* **3:**77.

Smalley, K. N., 1965, Adrenergic transmission in the light organ of the firefly, *Photinus pyralis, Comp. Biochem. Physiol.* **16:**467.

Sobrino, J. A., and del Castillo, J., 1972, Activation of the cholinergic end-plate receptors by oxidizing reagents, *Int. J. Neurosci.* **3:**251.

Soeda, Y., and Yamamoto, I., 1969, Studies on nicotinoids as insecticides. VIII. Physiological activities of the optical isomers of nicotinoids, *Botyu-Kagaku* **34:**57.

Spatz, L., and Strittmater, P., 1973, A form of cytochrome b_5 that contains an additional hydrophobic sequence of 40 amino acid residues, *J. Biol. Chem.* **248:**793.

Suszkiw, J. B., 1973, Quantitation of acetylcholinesterase and acetylcholine-binding sites in excitable membrane fragments from electric eel, *Biochim. Biophys. Acta* **318:**69.

Toppozada, A., and O'Brien, R. D., 1967, Permeability of the ganglia of the willow aphid, *Tuberolachnus salignus*, to organic ions, *J. Insect Physiol.* **13:**941.

Treherne, J. E., 1966, *in: The Neurochemistry of Arthropods*, p. 156, Cambridge University Press, Cambridge.

Van der Meer, C., and Meeter, E., 1956, The mechanism of action of anticholesterases. II. The effect of diisopropylfluorophosphonate (DPF) in the isolated rat phrenic nerve–diaphragm preparation, *Acta Physiol. Pharmacol. Neerl.* **4:**472.

Volle, R. L., and Koelle, G. B., 1969, Ganglionic stimulating and blocking agents, *in: The Pharmacological Basis of Therapeutics*, 3rd ed. (L. S. Goodman and A. Gilman, eds.), p. 578, Macmillan, New York.

Wang, C. M., and Narahashi, T., 1972, Mechanisms of dual action of nicotine on end-plate membranes, *J. Pharmacol. Exp. Ther.* **182:**527.

Yamamoto, I., Kamimura, H., Yamamoto, R., Sakai, S., and Goda, M., 1962, Studies on nicotinoids as an insecticide. I. Relation of structure to toxicity, *Agr. Biol. Chem.* **26:**709.

Yamamoto, I., Soeda, Y., Kamimura, H., and Yamamoto, R., 1968, Studies on nicotinoids as an insecticide. VII. Cholinesterase inhibition by nicotinoids and pyridylalkylamines—Its significance to mode of action, *Agr. Biol. Chem.* **32:**1341.

Zaimis, E. J., 1951, The action of decamethonium on normal and denervated mammalian muscle, *J. Physiol. (London)* **112:**176.

Zampighi, G., and Robertson, J. D., 1973, Fine structure of the synaptic discs separated from the goldfish medulla oblongata, *J. Cell Biol.* **56:**92.

9

Effects of Insecticides on Nervous Conduction and Synaptic Transmission

Toshio Narahashi

1. Introduction

The intoxication of an organism with insecticide involves a variety of steps and reactions (Narahashi, 1971*a*). The uptake of insecticide is the first step to occur, and a number of factors such as lipid solubility and vapor pressure of insecticide are related to this process (Chapter 1). The insecticide that has entered the body is then transported to various organs and may undergo a variety of biotransformations in which it is either converted into a more potent compound or degraded to one which is relatively nontoxic (Chapters 2, 4, and 5). The active form of the insecticide eventually reaches its target site and exerts effects characteristic of the insecticide and the tissue concerned.

Since the nervous system constitutes the major, and in many cases the sole, target site for the majority of insecticides which have been developed, an ability to measure their effects on this system is critical to establishing the precise mechanism by which they act.

In spite of the established importance of the nervous system as a target site, studies in this field have not attained the same level of popularity as several other areas of research on insecticide action and disposition. One obvious reason for this lies in the fact that the electrophysiological approach, which

Toshio Narahashi • Department of Physiology and Pharmacology, Duke University Medical Center, Durham, North Carolina.

remains the most straightforward and powerful manner of studying the mode of action of neuropoisons, requires more specialized training, expertise, and equipment than most other types of research.

To avoid unnecessary duplication, this chapter assumes that the reader is already equipped with a knowledge and understanding of the basic concepts of neurophysiology and neuropharmacology. These are covered in some detail in Chapter 6 of this book, and only a few points particularly pertinent to an interpretation of insecticidal action will be included here.

2. Rationale

2.1. Nerve Excitation and Conduction

At rest, the nerve membrane potential is close to the potassium equilibrium potential defined by the Nernst equation for potassium ions. When stimulated by application of an outward electrical current across the nerve membrane, the membrane permeability (conductance) to sodium ions is increased in a regenerative manner, so that the membrane potential approaches the sodium equilibrium potential; this constitutes the rising phase of the action potential. The increased sodium permeability decreases shortly thereafter, and the potassium permeability increases beyond that occurring at the resting level. This is the falling phase of the action potential which eventually brings the membrane potential back to its original resting level. As a result of these permeability changes, sodium ions enter and potassium ions leave the cell according to their electrochemical gradients.

Any change in the action potential caused by an insecticide may be accounted for in terms of these permeability changes in the nerve membrane. Thus the action potential may be suppressed by an inhibition of the increase in sodium permeability, which in turn is brought about either by a direct inhibition of the sodium activation mechanism or by a decrease in the membrane potential on which the sodium activation depends. Slowing of the falling phase of the action potential or increase in negative (depolarizing) afterpotential may be brought about by an inhibition of the increase in potassium permeability, a slowing of the decrease in sodium permeability, or a combination of both. Such an increase in negative afterpotential may induce repetitive afterdischarges, as described in Chapter 6.

It should be emphasized that these changes in the ionic permeability of the nerve membrane occur as a result of the opening and closing of ionic channels ("gates") in the membrane and that these processes are totally independent of the supply of metabolic energy. Metabolic energy comes into play after excitation, when it is required to operate the ionic pump responsible for extruding internal sodium ions and reabsorbing external potassium ions across the membrane. Thus metabolic energy plays a role in the recovery of a nerve fiber after excitation and subsequently in maintaining the proper ionic concentration gradients across the nerve membrane at rest. It is the metabolism-independent permeability parameters that are primarily affected by most neurotoxic agents including insecticides. In addition, the conduction of an impulse along a nerve fiber is mediated by an electrical current flowing across the nerve membrane in a manner totally independent of metabolic energy. Consequently, measurements of ionic permeabilities are extremely important in elucidating the mechanism of action of insecticides on the nerve membrane.

2.2. Synaptic and Neuromuscular Transmission

In discussing the synaptic effects of insecticides, only chemical synapses will be considered since although electrical synapses occur in certain nervous systems they are generally less vulnerable to the influence of neuroactive agents.

When an impulse arrives at a nerve terminal, a transmitter substance is released and stimulates the postsynaptic membrane by increasing ionic permeabilities (conductances). As a result, the postsynaptic membrane is depolarized (excitatory postsynaptic potential) and a local circuit current flowing across the extrajunctional membrane produces an action potential. The transmitter substance may be acetylcholine, norepinephrine, L-glutamate, γ-aminobutyric acid, or some other chemical depending on the type of synapse and the animal concerned. The permeability increase in the postsynaptic membrane may be with respect to sodium, potassium, chloride, or other ions depending on the kind of postsynaptic cell and organism. A portion of the released transmitter may be rapidly hydrolyzed by a specific enzyme such as acetylcholinesterase. In the case of an inhibitory synapse, the postsynaptic response may be manifested as a transient hyperpolarization (inhibitory postsynaptic potential), and the net result is not the production of an action potential but the inhibition of synaptic transmission mediated by the excitatory presynaptic fiber. A comprehensive account of synaptic transmission is provided in Chapter 6.

The processes involved in synaptic transmission are considerably more complicated than those in nerve conduction. For the skeletal neuromuscular junction of a vertebrate, the following steps are involved: (1) the transmitter substance acetylcholine is synthesized and stored in vesicles in the presynaptic neuron; (2) it is released by a depolarization of the presynaptic nerve terminal, normally by an action potential; (3) the released acetylcholine interacts with the receptor in the postsynaptic (end-plate) membrane, causing permeability increases to sodium and potassium; (4) the increased permeabilities of the postsynaptic membrane cause the membrane to depolarize, which in turn produces an action potential from the extrajunctional muscle membrane; (5) part of the released transmitter is quickly hydrolyzed by acetylcholinesterase.

In view of the several reactions involved in synaptic transmission, studies of the mechanism of action of drugs require numerous approaches involving electrophysiological, biochemical, and histological techniques. However, since impulse transmission is a fast process, lasting for only a matter of milliseconds, studies employing routine chemical methods are usually inappropriate and electrophysiological techniques are required to trace the reactions involved. It should be pointed out that these techniques must be integrated at a high level to elucidate the detailed mechanism of action of insecticides.

2.3. Two Ways of Using the Electrophysiological Approach

Any experimental technique can become a very powerful tool depending on how it is utilized; those involved in the electrophysiological approach to studies of insecticidal action are no exception. Two basic types of approach are employed. One of these, which is particularly useful in elucidating the cellular and molecular mechanism of insecticidal action, is through a careful comparative analysis of the characteristics of the electrical signals recorded from poisoned and normal nerves. Thus, for example, by detailed analysis of the action potential produced in the presence of an insecticide it is often possible to speculate on the ionic mechanism of action (e.g., effects on sodium and/or potassium conductance), and this speculation can often be proved or disproved by direct measurements of the ionic permeability of the nerve membrane. The results of such studies may ultimately provide some notion of how the insecticide interacts with the ionic channels of the membrane.

The other major way in which electrophysiology has proved useful is in providing an extremely sensitive indicator for comparing the activities of a series of related insecticide analogues. This is extremely important in structure–activity studies since the use of electrophysiological techniques allows the intrinsic activity of a compound with its target receptor to be measured without the added complication of factors such as cuticle penetration and metabolism usually attendant on other techniques. Similarly, the technique can be used to compare the effects of a single insecticide on different types of nerves, nerves from different species or strains of organisms, or nerves under a variety of different conditions (temperature, etc.). Thus the electrical response of a nerve to an insecticide may provide valuable information in studies of insect resistance to insecticides and selective toxicity.

This chapter is concerned mainly with the mechanism of action of insecticides at the membrane level, and emphasis will be placed on the effects of insecticides on membrane ionic permeabilities which account for changes in excitability.

3. DDT

DDT has long been known to produce hyperactivity, ataxia, and convulsions in insects and vertebrates. When recordings are made from various nerves of the DDT-intoxicated insect, multiple and recurrent discharges of impulses can be observed (Becht, 1958; Eaton and Sternburg, 1964; Gordon and Welsh,

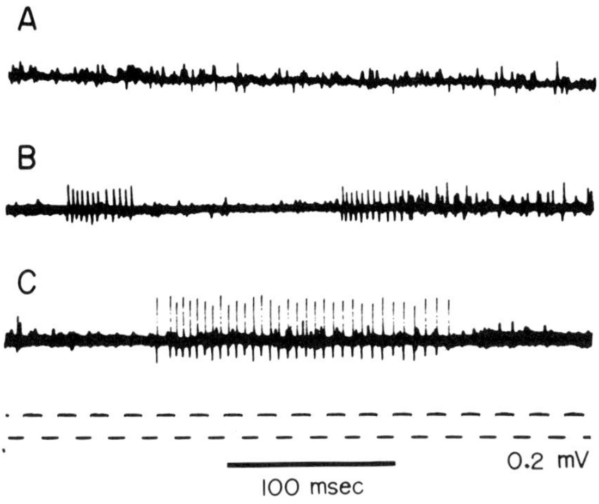

Fig. 1. Trains of impulses from the sensory cells of the cockroach leg 10 min after injection of 2.8×10^{-6} M DDT into the leg. A: Before injection. B and C: After injection. From Narahashi (1966).

1948; Lalonde and Brown, 1954; Narahashi, 1964; Narahashi and Yamasaki, 1960*b*; Roeder and Weiant, 1946, 1948, 1951; Shanes, 1949*a,b*, 1951; Welsh and Gordon, 1947; van den Bercken, 1968; Yamasaki and Ishii, 1952*a*, 1954*a,b,c*; Yamasaki and Narshashi, 1958*b*, 1962) (Fig. 1). In 1949, Shanes reported that the negative afterpotential of the crab nerve was increased in amplitude after treatment with DDT (Shanes, 1949*b*). However, since the action potentials were recorded from multifiber preparations by means of external electrodes, he himself was not convinced of the results and thought that the observed delay in the falling phase of the action potential might be due to superimposed multiple afterdischarges. Shortly after, we confirmed the increase in negative afterpotential by DDT using the cockroach nerve. In studies on the effect of DDT on the synaptic transmission of the last abdominal ganglion of the cockroach, the postsynaptic afterdischarges derived from individual giant axons clearly showed an augmented negative afterpotential (Yamasaki and Ishii, 1952*b*). Intracellular recordings from the cockroach giant axon also demonstrated this point, as is shown in Fig. 2 (Yamasaki and Narahashi, 1957; Narahashi and Yamasaki, 1960*b*).

3.1. Microelectrode Analyses of DDT Action

The action potential recorded by an intracellular microelectrode from the cockroach giant axon is composed of three phases: (1) a spike potential of

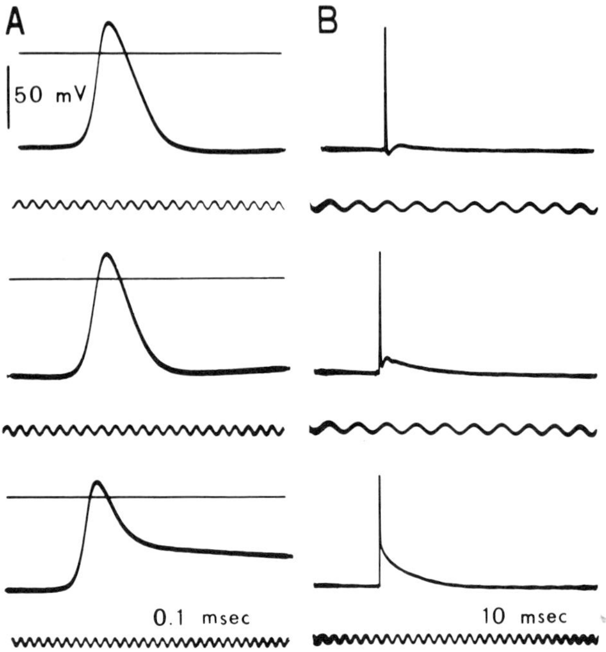

Fig. 2. Effects of 1×10^{-4} M DDT on the action potential of the cockroach giant axon. A (from top to bottom): Before, 38 min after, and 90 min after treatment with DDT. The horizontal lines refer to zero potential level. B: As in A, but with a slower sweep. From Narahashi and Yamasaki (1960).

about 100 mV in amplitude exhibiting an overshoot beyond 0 mV level, (2) an undershoot or positive phase of a few millivolts in amplitude immediately following the spike, and (3) a negative (depolarizing) afterpotential of a few millivolts in amplitude following the positive phase. There is some confusion about the terminology, because with intracellular microelectrode recording the positive phase and negative afterpotential are in fact a negative and a positive deflection, respectively. The traditional terminology derives from extracellular recording, where the spike deflects in the negative direction. The positive phase has been interpreted as being due to an increase in membrane potassium permeability, whereas the negative afterpotential is considered to result from a transient accumulation of potassium ions in the narrow space between the nerve membrane and Schwann cell membranes (Narahashi and Yamasaki, 1960*a*). Repetitive stimuli cause an increasing accumulation of potassium, thereby building up the negative afterpotentials (Fig. 3).

The negative afterpotentials augmented by DDT (Fig. 2) behave in a different manner during repetitive activity (Narahashi and Yamasaki, 1960*b*).

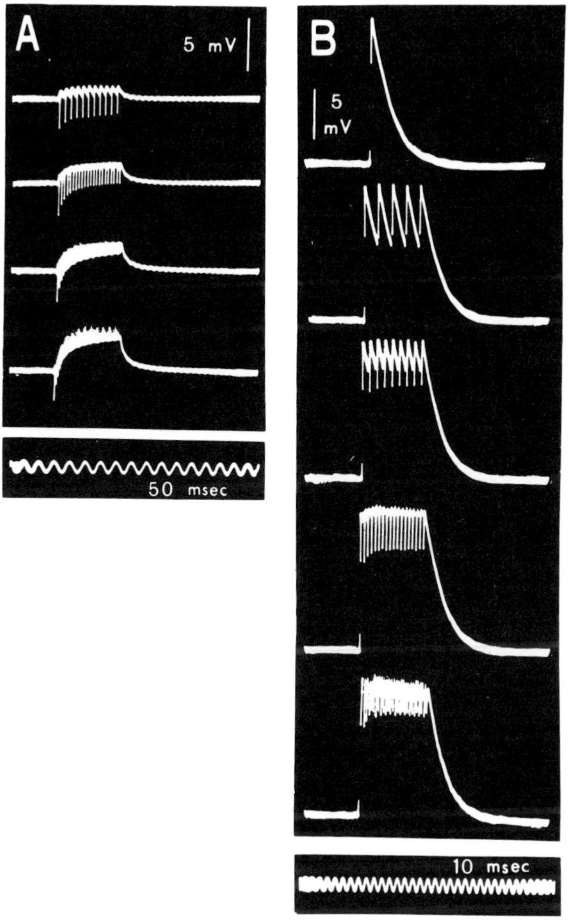

Fig. 3. Afterpotentials during repetitive stimuli at different frequencies in the normal (A) and DDT-poisoned (B) cockroach giant axons. The spike potentials are too large to be recorded. The frequencies of stimulation are, from top to bottom in A, 50, 100, 150, and 200 cps, and, in B, single stimulus, 50, 100, 200, and 300 cps. From Narahashi and Yamasaki (1960*a,b*).

No additions of negative afterpotentials are seen during repetitive excitation, and the peak levels of the negative afterpotentials are maintained almost constant (Fig. 3). This excludes the possibility that the augmentation of negative afterpotential by DDT is caused by an increased accumulation of potassium ions outside the nerve membrane.

It should be noted that during the course of *in vitro* DDT intoxication of nerve there is a period when a single stimulus induces repetitive afterdischarges (Narahashi and Yamasaki, 1960*b*). This occurs when the negative afterpotential begins to increase and is due to the sustained depolarizing stimulation. It is of interest that the DDT-poisoned axon eventually gives rise to spontaneous action potentials (Narahashi and Yamasaki, 1960*c*).

3.2. Voltage Clamp Analyses of DDT Action

The increase in negative afterpotential caused by DDT has been analyzed in detail by means of the voltage clamp technique using lobster giant axons (Narahashi and Haas, 1967, 1968) and frog nodes of Ranvier (Hille, 1968). As described before, this is the most powerful and straightforward method for studying the mechanism of action of drugs on nerve membranes in terms of ionic permeabilities.

Figure 4 shows families of membrane currents associated with step depolarization of various magnitudes in lobster giant axons before and during application of DDT. Before application of DDT, each of the membrane currents associated with depolarizations to the levels indicated on the right-hand side of the recordings is composed of a transient sodium current which is followed by a steady-state potassium current. The sodium currents flow inward across the nerve membrane (downward deflection) with small to moderate

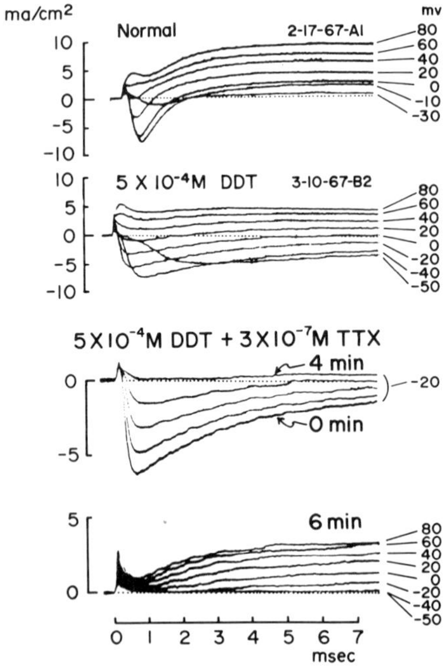

Fig. 4. Families of membrane currents associated with step depolarizations to the membrane potentials indicated under voltage clamp conditions in a normal lobster axon (2-17-67-A1) and in an axon treated with 5×10^{-4} M DDT and with DDT and 3×10^{-7} M tetrodotoxin (TTX) (3-10-67-B2). The third set of records shows changes in membrane current during the course of TTX action. Dotted lines refer to the zero baseline. From Narahashi and Haas (1967).

depolarizations, but flow outward with greater depolarizations. The potassium currents, however, always flow outward (upward deflection). After exposure of the axon to 5×10^{-4} M DDT, one drastic change occurs—the steady-state currents flow inward at certain membrane potentials.

The inward steady-state currents cannot be carried by potassium ions, because the electrochemical gradient for potassium across the nerve membrane is in the opposite direction. Therefore, we must seek other ions as carriers of these currents. One obvious candidate is sodium, since the electrochemical gradient is in the right direction and the sodium conductance of the membrane is transiently increased during stepwise depolarization. In order to identify the ion species responsible for the steady-state current in the DDT-poisoned axon, tetrodotoxin (TTX) can be used as a convenient tool. TTX is a poison from the ovary and liver of the puffer fish and blocks the conductance of the membrane to sodium without having any effect on potassium conductance (Narahashi *et al.*, 1964; Narahashi, 1972). When TTX is applied to the DDT-poisoned axon, the transient current associated with a step depolarization to -20 mV membrane potential is abolished in 4 min, while the steady state inward current is converted into a small outward current (Fig. 4). Thus the membrane current that has disappeared in the presence of TTX (current at 0 min minus current at 4 min in Fig. 4) represents a sodium current, and the membrane current still remaining in the presence of TTX (current at 4 min) represents a potassium current. It can be seen that the sodium current, after an initial transient flow, is maintained at the steady state. The control experiment in which TTX is applied to the normal axon (not illustrated) clearly shows that the sodium current usually terminates in 1–2 msec. DDT therefore slows the falling phase of the sodium current (sodium inactivation) to a considerable extent. Since sodium inactivation is one of the two factors that are responsible for the falling phase of the action potential, the DDT-induced slowing should cause an increase in negative afterpotential, as has actually been observed.

The bottom set of Fig. 4 showing a family of membrane currents recorded 6 min after application of TTX to the DDT-poisoned axon illustrates another important effect of DDT. From these recordings, it can be seen that the amplitude of the steady-state current, which is carried only by potassium ions, is greatly reduced compared with the control record at the top (note the difference in scale on the ordinate axis). Thus in addition to its action in slowing sodium inactivation DDT suppresses the potassium permeability increase, which is the other factor responsible for the falling phase of the action potential. It is a combination of both of these effects which causes the overall increase in negative afterpotential observed in the presence of DDT.

3.3. Effects of DDT on Neurons and Synaptic Transmission

The effects of DDT on spontaneous discharges in the nerve cord and certain sensory cells have been studied by a number of investigators (see review by Narahashi, 1971*a*). In the isolated nerve cord of the cockroach, the frequency of discharge of spontaneous impulses is increased after exposure to DDT (Yamasaki and Ishii, 1954*a*), and these often appear in the form of trains of impulses. The sensory cells are also affected by DDT, and the campaniform sensilla in the cockroach trochanter is particularly sensitive (Lalonde and Brown, 1954; Roeder and Weiant, 1946, 1948, 1951; Yamasaki and Ishii, 1954*a*,*b*). Many of the spontaneously occurring single impulses are converted into trains of impulses after application of DDT.

Synaptic transmission is also facilitated by DDT (Dresden, 1949; Yamasaki and Ishii, 1952*b*). However, studies on the effect of DDT on synaptic and neuromuscular transmission are complicated by its effects on conduction in the presynaptic nerve fiber and the resulting production of repetitive discharges. The mechanism of facilitation of synaptic transmission by DDT remains to be explored, although both repetitive discharges in the presynaptic element and augmentation of the negative afterpotential in the postsynaptic element are undoubtedly two contributing factors.

4. Pyrethroids

Pyrethroids initially stimulate nerve to produce repetitive discharges and subsequently cause paralysis (see reviews by Narahashi, 1971*a*,*b*). These effects are produced in crayfish and cockroach nerve cord preparations which contain ganglia and synapses (Schallek and Wiersma, 1948; Camougis and Davis, 1971; Burt and Goodchild, 1971*a*,*b*; Lowenstein, 1942) as well as in nerve fiber preparations without synapses (Welsh and Gordon, 1947; Narahashi, 1962*a*,*b*; Yamasaki and Ishii, 1952*a*). The stimulating action of pyrethroids is generally much more pronounced than that of DDT. The exact sites of action of pyrethroids at synapses are unknown, and it remains to be seen whether the repetitive discharges they produce in the nerve fibers can account completely for those observed across synapses.

4.1. Microelectrode Analyses of Allethrin Action

At a low concentration (1×10^{-6} M), allethrin increases the amplitude of negative afterpotential and prolongs its duration in the cockroach giant axon (Narahashi, 1962*a*,*b*). When the amplitude attains a certain critical level, repetitive afterdischarges are induced by a single stimulus. However, the repetitive discharges induced by allethrin are highly dependent on temperature, as will be described later. At a higher concentration (3×10^{-6} M), allethrin eventually suppresses the action potential (Fig. 5) and slightly depolarizes the membrane. Partial recovery of the action potential occurs slowly after washing with allethrin-free medium, but it is never complete. Thus it is clear that the increased negative afterpotential is responsible for repetitive afterdischarges. The mechanism underlying this increase has been studied by voltage clamp techniques.

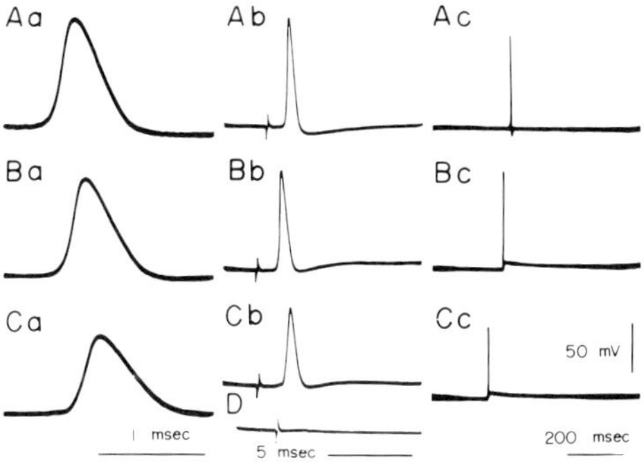

Fig. 5. Effects of 3×10^{-6} M allethrin on the action potential of the cockroach giant axon. Aa–Ac, before application of allethrin; Ba–Bc, 2.5 min after application; Ca–Cc, 4 min; D, 10.5 min. From Narahashi (1962*a*).

4.2. Voltage Clamp Analyses of Allethrin Action

Voltage clamp experiments have been carried out with squid and crayfish giant axons (Narahashi, 1969, 1971*b*; Narahashi and Anderson, 1967; Murayama *et al.*, 1972; Wang *et al.*, 1972). All of the three permeability parameters responsible for production of the action potential are affected by allethrin. The transient sodium current is suppressed in amplitude, and at certain membrane potentials it is followed by an inward steady-state current. The ionic components of the steady-state current can be separated by the use of TTX as in the case of DDT or by the measurements of tail sodium currents associated with step repolarizations at various times.

An example of such an experiment is shown in Fig. 6. Before application of allethrin, the transient sodium current is terminated in 1–2 msec, as can be measured as the difference between the currents before and after application of TTX (not shown). After exposure to allethrin, however, the sodium current is maintained over a period of about 8 msec. The potassium current, which is

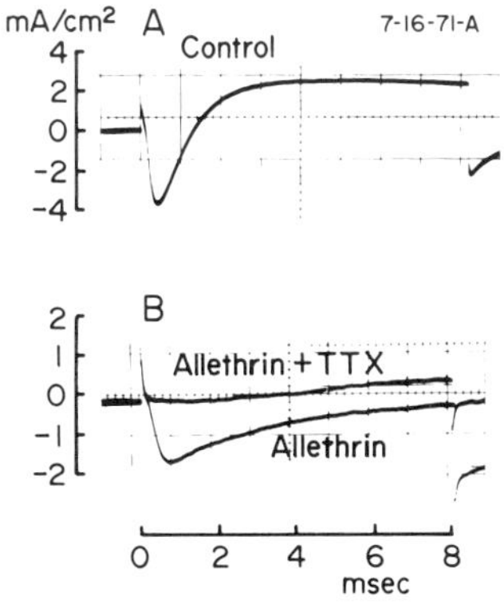

Fig. 6. Membrane currents from a voltage-clamped squid axon (7-16-71-A) before (A) and after (B) application of 2×10^{-5} M allethrin, and after application of both allethrin and 3×10^{-7} M tetrodotoxin (TTX) (B). The difference between two records in B represents the sodium current, which is greatly prolonged. From Wang *et al.* (1972).

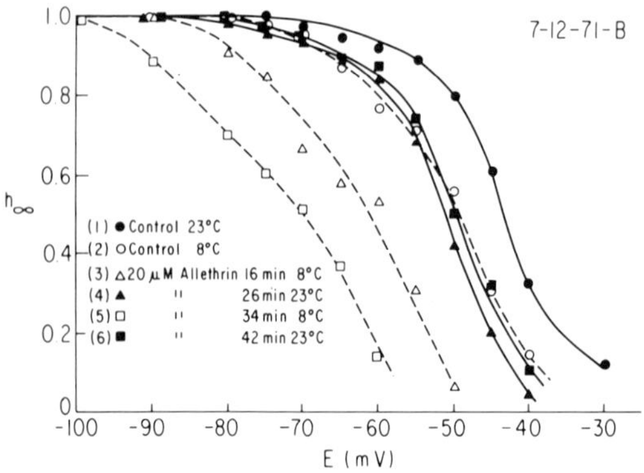

Fig. 7. Steady-state sodium inactivation curves from a voltage-clamped squid axon (7-12-71-B) before and after application of 2×10^{-5} M allethrin at 23°C and 8°C. The ordinate represents the peak amplitude of sodium current associated with the test pulse in a value relative to its maximum, and the abscissa represents the membrane potential of the conditioning pulse. Numbers in parentheses show the order of measurements. From Wang *et al.* (1972).

represented by the steady-state current remaining after addition of TTX, is suppressed by allethrin. Thus allethrin inhibits the sodium permeability increase, thereby suppressing the action potential, and inhibits both the sodium inactivation and the potassium activation, thereby augmenting and prolonging the negative afterpotential. Sodium inactivation appears to be the process most sensitive to the action of allethrin, which is in agreement with the observation that the negative afterpotential is increased and repetitive discharges are induced at very low concentrations of the insecticide.

The steady-state sodium inactivation curve is affected by allethrin, being shifted in the direction of hyperpolarization (Fig. 7). This change also contributes to the suppression of the action potential, because in the presence of allethrin the degree of sodium activation is decreased at membrane potentials near the resting level. The shift of the steady-state sodium inactivation is expected from the Hodgkin–Huxley equations describing the steady-state condition (h_∞) and the time constant (τ_h) of sodium inactivation:

$$h_\infty = \alpha_h/(\alpha_h + \beta_h) \tag{1}$$

$$\tau_h = 1/(\alpha_h + \beta_h) \tag{2}$$

where α_h and β_h are the rate constants (Hodgkin and Huxley, 1952). The fact that τ_h is lengthened by allethrin indicates changes in α_h or β_h or both, so that h_∞ is also expected to change.

4.3. Temperature Dependency of Allethrin Action

In the cockroach giant axon, the effects of allethrin depend on temperature in a somewhat complicated manner (Narahashi, 1962*a*). Repetitive discharges show a positive temperature dependency and are abolished on decreasing the temperature below a certain critical level (about 26°C). However, the allethrin-induced block has a negative temperature coefficient and is significantly potentiated, for example, by decreasing the temperature from 28°C to 12°C.

The mechanism involved in the temperature-dependent block has been studied by voltage clamp techniques with squid giant axons (Wang *et al.*, 1972), and it has been established that the effects of allethrin that are directly responsible for conduction block are augmented by lowering the temperature. The suppression of the transient sodium conductance, the shift of the sodium conductance curve along the potential axis in the direction of depolarization, and the shift of the steady-state sodium inactivation curve in the direction of hyperpolarization (Fig. 7) are all more pronounced at low temperature (8°C) than at high temperature (23°C).

The temperature dependency of the allethrin-induced repetitive discharges in the squid axon is more complicated (Narahashi, 1974). There is a temperature range (22–28°C) where repetitive discharges can be produced most intensively, and repetitive responsiveness is decreased as the temperature is elevated beyond 28°C or lowered below 22°C. Preliminary experiments have shown that the threshold depolarization for production of action potential is increased slightly as the temperature is raised from 13°C to 28°C. The amplitude of negative afterpotential is also increased but to a greater extent. Therefore, the latter exceeds the former at temperatures ranging from 22°C to 28°C, inducing repetitive afterdischarges. As the temperature is raised beyond 28°C, the threshold depolarization is further increased, but the negative afterpotential is decreased and prevents the occurrence of repetitive afterdischarges. A combination of these two temperature-dependent parameters can consequently account for the observed temperature dependency of repetitive responsiveness, although the ionic mechanism involved in this phenomenon remains to be studied.

4.4. *Effects of Pyrethroids on Synaptic Transmission*

Although as previously described the pyrethroids have a profound stimulating action on the spontaneous discharges of the central nerve cord of the cockroach (Camougis and Davis, 1971; Burt and Goodchild, 1971*a*,*b*), their precise site of action in the ganglia remains unknown. Pyrethroids may facilitate synaptic transmission by increasing the release of the transmitter substance from nerve terminals, by sensitizing the postsynaptic membrane to the transmitter, by inducing repetitive discharges in the nerve fibers of nerve terminals, or by a combination of these effects. Microelectrode experiments with appropriate synapse preparations are clearly needed to locate the site of action of pyrethroids in the synaptic region.

5. Nicotine

For a long time, nicotine has been known to stimulate and paralyze the ganglia, and indeed because of this action Langley and his associates utilized nicotine to study the innervation of the autonomic nervous system (Langley, 1901; Langley and Dickinson, 1889). Despite such time-honored ganglionic action, two unique actions of nicotine have been recently unveiled.

5.1. Nerve Blocking Action

Nicotine has recently been found to have a fairly potent local anesthetic action when applied directly to the internal membrane surface of the squid giant axon (Frazier *et al.*, 1973). When applied externally at a concentration of 1×10^{-3} M, nicotine has no effect on ionic conductances. When applied internally, however, nicotine suppresses the potassium conductance increase without having much effect on the sodium conductance increase. At a concentration of 1×10^{-2} M, internal application of nicotine almost completely abolishes both conductance increases, and external application at this concentration results in a 30–40% reduction of the potassium conductance with little effect on the sodium conductance. Thus it is clear that nicotine has a greater effect on the potassium conductance mechanism than on the sodium mechanism. Judging from its higher internal potency, the nicotine receptor is presumably located on or near the internal membrane surface and is more easily accessible from inside than from outside. The internal site of action of nicotine is reminiscent of that of local anesthetics (Narahashi *et al.*, 1970; Frazier *et al.*, 1970; Narahashi and Frazier, 1971).

5.2. Dual Action on Neuromuscular Junction

It has been known for some time that nicotine depolarizes the end-plate membrane of the vertebrate skeletal muscle and that the depolarization is followed by a repolarization in the continuous presence of nicotine. The acetylcholine receptor of the end-plate membrane remains blocked during the depolarization–repolarization process (Thesleff, 1955). The block after repolarization is often called desensitization. However, there are some observations with mammals which cannot be accounted for in terms of depolarization or desensitization block, and careful microelectrode experiments have revealed an additional effect of nicotine on the end-plate membrane (Wang and Narahashi, 1972).

At low concentrations in the order of 5×10^{-6} M, nicotine suppresses the end-plate potential (EPP) of the frog sartorius muscle without causing a sizable depolarization (Fig. 8). This effect is reversible after washing with drug-free medium, and is dependent on the concentration. The transient end-plate depolarization produced by iontophoretic application of acetylcholine (ACh) is

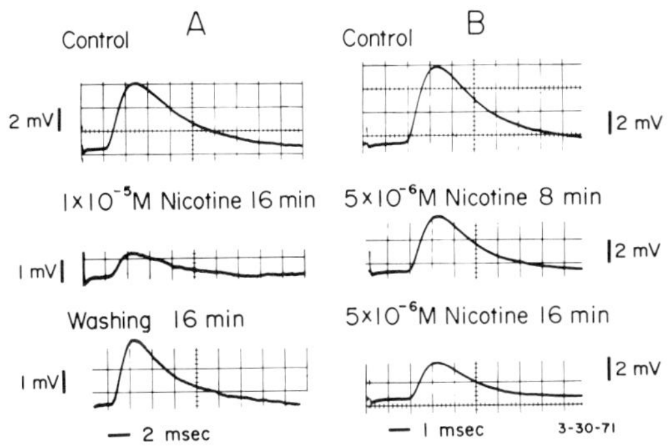

Fig. 8. End-plate potentials from the frog sartorius muscle (3-30-71) before and during application of nicotine and after washing with drug-free Ringer's solution. The muscle was treated with glycerol to block excitation–contraction coupling and soaked in a Ringer's solution containing 10 mM Mg^{2+} to decrease transmitter release. From Wang and Narahashi (1972).

also suppressed by nicotine effectively. At concentrations 1 order of magnitude higher (5×10^{-5} M), the end-plate membrane is depolarized and then partially repolarized in the continuous presence of the drug (Fig. 9). Thus nicotine has a dual action on the end-plate membrane with the two dose–response curves separated by a factor of about 10 (Fig. 10).

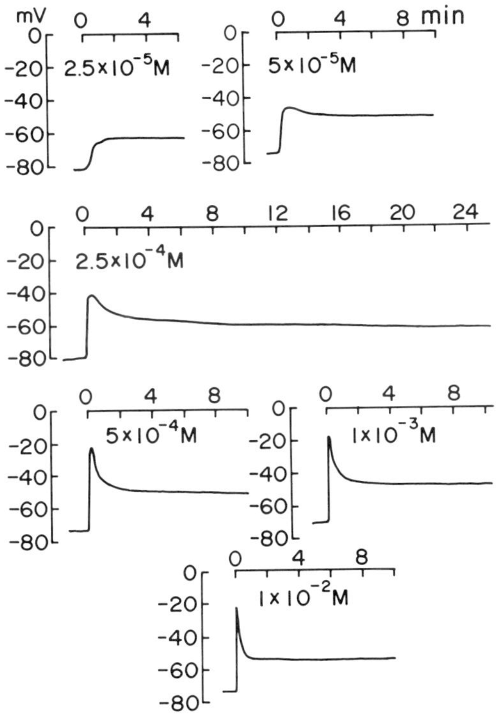

Fig. 9. Effects of various concentrations of nicotine on the resting membrane potential of the end plate of the frog sartorius muscle. From Wang and Narahashi (1972).

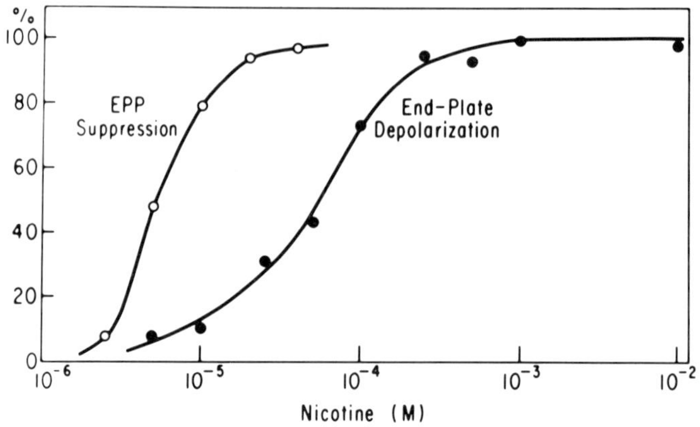

Fig. 10. Dose–response curves for the effect of nicotine in suppressing the end-plate potential (EPP) and in depolarizing the end-plate membrane in the frog sartorius muscle. The EPP suppression is corrected for that due to small depolarizations.

The ionic conductances of the end-plate membrane involved in these nicotine actions have been studied by means of the voltage clamp technique. With low concentrations of nicotine, both sodium and potassium components of the end-plate current (EPC) are equally suppressed (Fig. 11). The indiscriminate inhibitory action of nicotine on both ionic conductances is in contrast to that of procaine, which affects each component differently (Deguchi and Narahashi, 1971; Gage and Armstrong, 1968; Maeno, 1966; Maeno *et al.*, 1971; Kordaš, 1970).

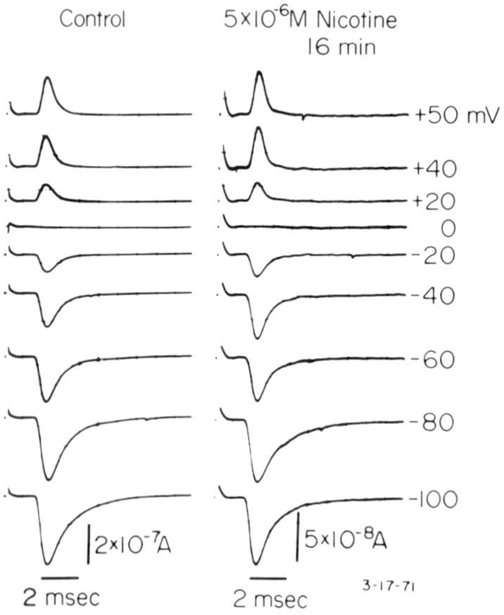

Fig. 11. End-plate currents recorded from the frog sartorius end plate (3-17-71) voltage-clamped at various membrane potentials indicated on the right before and after application of 5×10^{-6} M nicotine. Note different current calibrations. From Wang and Narahashi (1972).

At higher depolarizing concentrations, prolonged application of nicotine increases both sodium and potassium conductances of the end-plate membrane without nerve stimulation (Wang and Narahashi, 1972). These conductance increases are responsible for the initial depolarization. Following this, the potassium conductance returns to its original level while the sodium conductance decreases partially but remains slightly elevated. The latter accounts for the residual depolarization in the continuous presence of nicotine. It should be noted that although nicotine suppresses the ACh-induced increases in sodium and potassium conductances to the same extent it also intrinsically increases each conductance component in a different way. These results suggest some difference between the site where nicotine blocks ACh action and that where nicotine itself interacts to cause conductance increases. This may provide a clue to further identification and characterization of the ACh and nicotine receptors in the end-plate or postsynaptic membranes.

6. Nereistoxin and Cartap

Nereistoxin (NTX) is the active ingredient of the toxin obtained from the annelid *Lumbriconereis heteropoda*. The chemical structure has been identified as 4-*N*,*N*-dimethylamino-1,2-dithiolane (Fig. 12) by Okaichi and Hashimoto (1962). Studies of NTX and its derivatives have led directly to the development of the insecticide cartap (see Sakai, 1969; Narahashi, 1973).

Sakai and his associates have studied the mode of action of NTX and its derivatives on the cockroach ganglia and frog rectus abdominis (Sakai, 1966, 1967, 1969; Chiba *et al.*, 1967) and have concluded that NTX suppresses the sensitivity of the nicotinic postsynaptic membrane to ACh.

More detailed studies with microelectrodes and voltage clamp have been carried out using frog sartorius muscle (Deguchi *et al.*, 1971; Narahashi, 1973). NTX blocks neuromuscular transmission without causing a large depolarization of the end-plate membrane. An EPP can be recorded from the end-plate region (Fig. 13). The quantal content of EPP is not significantly affected. The transient end-plate depolarization induced by iontophoretic application of ACh is suppressed by NTX (Fig. 14). It is clear that NTX acts primarily on the postsynaptic membrane. Voltage clamp experiments have shown that both sodium and potassium conductances of the end-plate membrane are equally suppressed, and consequently NTX may be categorized as a nondepolarizing, curare-type neuromuscular blocking agent.

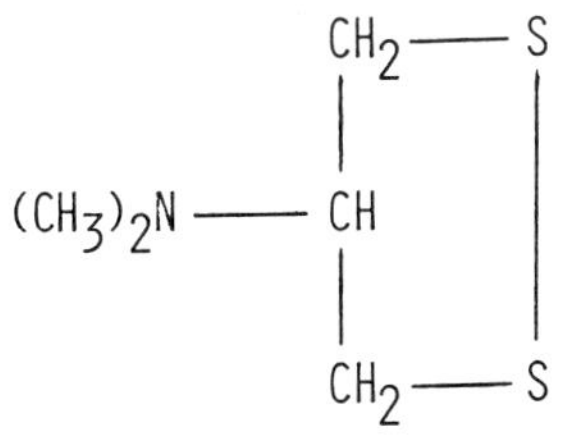

Fig. 12. Structure of nereistoxin.

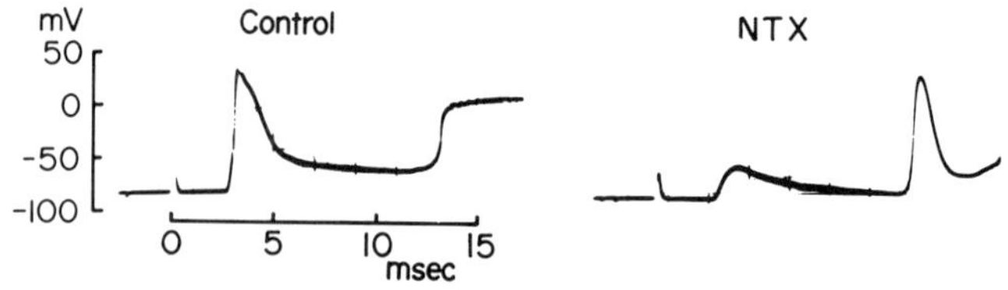

Fig. 13. Action potential and end-plate potential recorded from the end plate of the frog sartorius muscle before and 18 min after application of 1×10^{-4} M nereistoxin (NTX). Stimulation was applied to the nerve. In NTX, an end-plate potential was followed by an action potential produced at the other end plate of the same muscle fiber and propagating to the point of recording. From Deguchi *et al.* (1971).

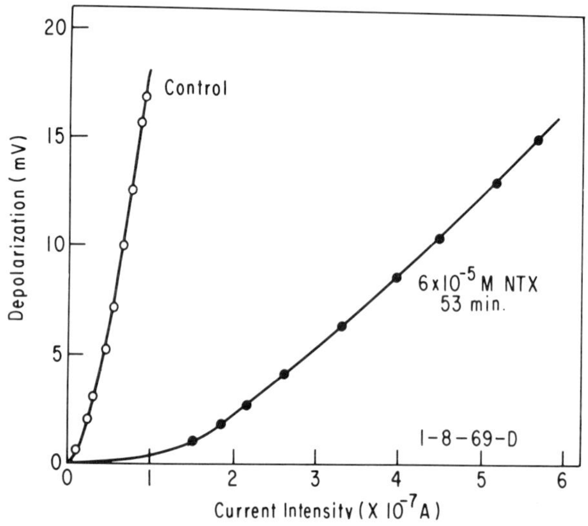

Fig. 14. Effects of 6×10^{-5} M nereistoxin (NTX) on the end-plate depolarization induced by iontophoretic application of acetylcholine. Frog sartorius muscle (1-8-69-D). The peak depolarization is plotted against the intensity of current applied to acetylcholine pipette. From Deguchi *et al.* (1971).

In addition to its cholinergic blocking action on the nicotinic receptors, NTX stimulates certain muscarinic receptors in mammals, causing an increase in motility of the gastrointestinal tract and uterus, an increase in secretion of the salivary and lacrimal glands, and a constriction of the pupil (Nitta, 1941).

These effects are antagonized by atropine. It remains to be seen whether the cholinergic blocking and cholinomimetic action of NTX are exerted at different dose levels.

Another interesting feature of NTX action is its antagonism by certain sulfhydryl reagents (Nagawa *et al.*, 1971). It appears that the disulfide bond of NTX plays a critical role in blocking the cholinergic receptor on the end-plate membrane since the insecticidal potency of several dithiol esters of NTX depends on their ability to be converted into NTX in insects (Sakai and Sato, 1971).

7. Cyclodienes

Early studies showed that dieldrin facilitates the synaptic transmission across the sixth abdominal ganglion of the cockroach (Yamasaki and Narahashi, 1958*a*). A single stimulus applied to the presynaptic nerve induces prolonged afterdischarges in the postsynaptic neurons. However, subsequent experiments with more purified samples of dieldrin did not confirm the above observation, and it was suspected that the initial positive results were due to impurities contained in the dieldrin sample (see Narahashi, 1971*a*). It takes an hour or longer after treatment with dieldrin for noticeable synaptic afterdischarges to be produced (Shankland and Schroeder, 1973). Despite the failure to observe rapid and direct effects with purified dieldrin, insects poisoned with dieldrin certainly show symptoms of poisoning characteristic of nervous disorders.

7.1. Effects on Synapses and Sensory Receptors

Studies with several dieldrin analogues and metabolites (Fig. 15) have thrown new light on the neurotoxic mechanism of action of dieldrin (Wang *et al.*, 1971). When applied to the metathoracic ganglion of the cockroach, purified dieldrin has no immediate effect on the synaptic transmission from the giant axons in the abdominal nerve cord to the motoneurons innervating the leg muscles. It takes 60–90 min before synaptic afterdischarges are observed. In contrast, aldrin *trans*-diol causes synaptic afterdischarges in less than 5–10 min. Photodieldrin and photoaldrin are intermediate in terms of the latency, producing the effect in 20–30 min.

Spontaneous discharges from the sensory cells of the cockroach leg are increased in frequency after dieldrin is directly injected in the leg. The latency is very long (about 45 min). However, aldrin *trans*-diol increases the frequency in about 2.5 min after injection. Photodieldrin and photoaldrin are again intermediate in this respect, increasing the frequency in 9–10 min.

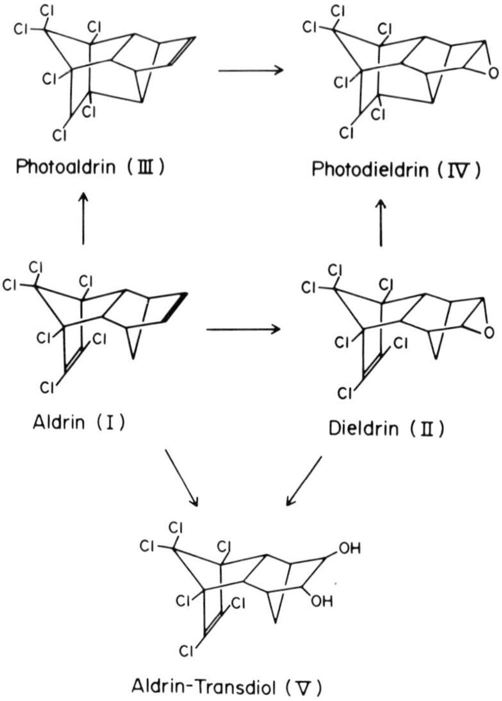

Fig. 15. Structures of aldrin, dieldrin, and their metabolites and derivatives. Wang *et al.* (1971).

7.2. Effects on Axonal Membranes

More recent experiments with squid giant axons give additional support to the fact that aldrin *trans*-diol is more effective than dieldrin on nerve (van den Bercken and Narahashi, 1974). Dieldrin itself has little or no effect on the action potential and membrane ionic currents. Aldrin *trans*-diol (5×10^{-5} M) suppresses the action potential and inhibits the transient sodium current (Fig. 16). The steady-state potassium current is also inhibited only to a small extent. These effects are not reversible after washing of the axon with drug-free medium.

7.3. Effects on Neuromuscular Transmission

The frequency and amplitude of spontaneous miniature end-plate potentials of the frog muscle are not affected by dieldrin. However, aldrin *trans*-diol causes a rapid and marked increase in their frequency and a decrease in their amplitude (Akkermans *et al.*, 1974). Spontaneous transmitter release is obviously stimulated, and the quantal content is also increased transiently. Aldrin *trans*-diol has a postsynaptic effect, also, decreasing the sensitivity of the end-plate membrane to ACh. Thus aldrin *trans*-diol exerts multiple actions on both presynaptic and postsynaptic elements, effects which are lacking with dieldrin itself. Shankland and Schroeder (1973) also suggested presynaptic action of dieldrin in the cockroach ganglion.

7.4. Discussion

Aldrin *trans*-diol has been found in houseflies, rats, rabbits, and pigs as a metabolite of dieldrin (Matthews and Matsumura, 1969; Brooks and Harrison,

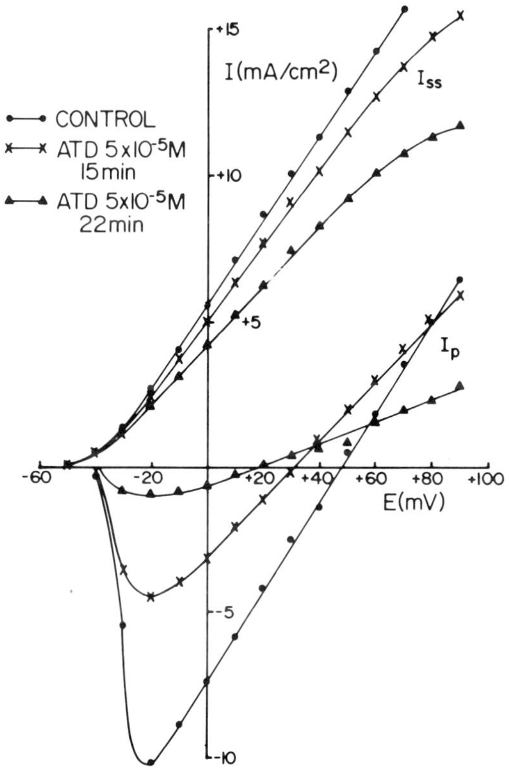

Fig. 16. Current–voltage relations for peak sodium current (I_p) and steady-state potassium current (I_{ss}) before, 15 min after, and 22 min after application of 5×10^{-5} M aldrin *trans*-diol (ATD) to the voltage-clamped squid giant axon (7-13-73A). From van den Bercken and Narahashi (1974).

1969; Brooks *et al.*, 1970; Korte and Arent, 1965; Korte and Kochen, 1966; Korte, 1967). No measurement has yet been made on the content of aldrin *trans*-diol in the brain or other parts of the nervous system in either mammals or insects poisoned with dieldrin. Despite the lack of such direct evidence, the present neurophysiological data strongly support the hypothesis that aldrin *trans*-diol is an active form of dieldrin. Further experiments are under way in our laboratory to document this hypothesis.

8. Summary and Conclusion

In order to elucidate the precise mode of action of insecticides, studies at the target site are essential because they are not complicated by factors such as penetration through the cuticle and metabolic activation or detoxication. With the exception of a few groups of insecticides such as the organophosphates, carbamates, and rotenone, no true target sites have yet been isolated *in vitro*. For many other insecticides such as DDT, the pyrethroids, and the cyclodienes, the target sites are known to be located in nerve and muscle membranes and are presumably on the proteins and/or phospholipids that control the gating mechanism of the ionic channels. Electrophysiological techniques have proved powerful tools for the study of these insecticides and have been quite successful in characterizing their mechanism of action on nerve membranes.

DDT acts on the nerve membrane. It augments and prolongs the negative afterpotential that follows the spike, and its action is characterized by the production of repetitive afterdischarges following a single stimulus. The increase in negative afterpotential has been demonstrated to be due to inhibition of both sodium inactivation and potassium activation during excitation. DDT has a profound negative temperature coefficient of action with respect to repetitive discharges.

Allethrin exerts four actions on the nerve membrane. It slightly depolarizes the membrane, increases the negative afterpotential, induces repetitive afterdischarges, and eventually blocks the action potential. The negative afterpotential is due to inhibitions of both sodium inactivation and potassium activation. Conduction block is primarily due to an inhibition of the sodium activation and a shift of the steady-state sodium inactivation curve in the direction of hyperpolarization. These effects are augmented by lowering the temperature, accounting for a large negative temperature dependency of the allethrin block of nerve.

When applied to the internal surface of the nerve membrane, nicotine blocks nervous conduction through inhibition of both sodium and potassium activation; it also exerts a dual action on the end-plate membrane. At low concentrations it suppresses the end-plate sensitivity to acetylcholine, thereby blocking neuromuscular transmission, and at higher concentrations it depolarizes and then partially repolarizes the end-plate membrane through increases and decreases of the sodium and potassium conductances.

Nereistoxin blocks neuromuscular transmission by suppressing the end-plate sensitivity to acetylcholine without causing much depolarization.

Dieldrin stimulates synapses and sensory receptors and induces the formation of multiple discharges after a long latency. Aldrin *trans*-diol, a metabolite of dieldrin, exerts the same effects much more quickly, and inhibits the sodium and potassium activation mechanisms of nerve membranes. It is possible that aldrin *trans*-diol is one of the active forms of dieldrin.

ACKNOWLEDGMENTS

Some of the studies quoted in this chapter were supported by NIH Grant NS06855. Unfailing secretarial assistance by Mrs. Frances Bateman and Mrs. Delilah Munday is deeply appreciated.

9. References

Akkermans, L. M. A., van den Bercken, J., van der Zalm, J. M., and van Straaten, H. W. M., 1974, Effects of dieldrin (HEOD) and some of its metabolites on synaptic transmission in the frog motor end-plate, *Pestic. Biochem. Physiol.* **4:**313.

Becht, G., 1958, Influence of DDT and lindane on chordotonal organs in the cockroach, *Nature (London)* **181:**777.

Brooks, G. T., and Harrison, A., 1969, Hydration of HEOD (dieldrin) and the heptachlor epoxides by microsomes from the liver of pigs and rabbits, *Bull. Environ. Contam. Toxicol.* **4:**352.

Brooks, G. T., Harrison, A., and Lewis, S. E., 1970, Cyclodiene epoxide ring hydration by microsomes from mammalian liver and houseflies, *Biochem. Pharmacol.* **19:**255.

Burt, P. E., and Goodchild, R. E., 1971*a*, The site of action of pyrethrin I in the nervous system of the cockroach *Periplaneta americana* L., *Entomol. Exp. Appl.* **14:**179.

Burt, P. E., and Goodchild, R. E., 1971*b*, The spread of topically-applied pyrethrin I from the cuticle to the central nervous system of the cockroach *Periplaneta americana*, *Entomol. Exp. Appl.* **14:**255.

Camougis, G., and Davis, W. M., 1971, A comparative study of the neuropharmacological basis of action of pyrethrin, *Pyrethrum Post* **11:**7.

Chiba, S., Saji, Y., Takeo, Y., Yui, T., and Aramaki, Y., 1967, Nereistoxin and its derivatives, their neuromuscular blocking and convulsive actions, *J.J. Pharmacol.* **17:**491.

Deguchi, T., and Narahashi, T., 1971, Effects of procaine on ionic conductances of end-plate membranes, *J. Pharmacol. Exp. Ther.* **176:**423.

Deguchi, T., Narahashi, T., and Haas, H. G., 1971, Mode of action of nereistoxin on the neuromuscular transmission in the frog, *Pestic. Biochem. Physiol.* **1:**196.

Dresden, D., 1949, *Physiological Investigations into the Action of DDT*, Drukkerij en Uitgeverij, G. W. Van der Wiel and Co., Arnhem, Netherlands.

Eaton, J. L., and Sternburg, J., 1964, Temperature and the action of DDT on the nervous system of *Periplaneta americana* (L), *J. Insect Physiol.* **10:**471.

Frazier, D. T., Narahashi, T., and Yamada, M., 1970, The site of action and active form of local anesthetics. II. Experiments with quaternary compounds, *J. Pharmacol. Exp. Ther.* **171:**45.

Frazier, D. T., Sevcik, C., and Narahashi, T., 1973, Nicotine: Effect on nerve membrane conductances, *Eur. J. Pharmacol.* **22:**217.

Gage, P. W., and Armstrong, C. M., 1968, Miniature end-plate currents in voltage-clamped muscle fibres, *Nature (London)* **218:**363.

Gordon, H. T., and Welsh, J. H., 1948, The role of ions in axon surface reactions to toxic organic compounds, *J. Cell. Comp. Physiol.* **31:**395.

Hille, B., 1968, Pharmacological modifications of the sodium channels of frog nerve, *J. Gen. Physiol.* **51:**199.

Hodgkin, A. L., and Huxley, A. F., 1952, A quantitative description of membrane current and its application to conduction and excitation in nerve, *J. Physiol.* (*London*) **117:**500.

Kordaš, M., 1970, The effect of procaine on neuromuscular transmission, *J. Physiol.* (*London*) **209:**689.

Korte, F., 1967, Metabolism of ^{14}C-labelled insecticides in microorganisms, insects, and mammals, *Botyu-Kagaku* **32:**46.

Korte, F., and Arent, H., 1965, Metabolism of insecticides 1X(1). Isolation and identification of dieldrin metabolites from urine of rabbits after oral administration of dieldrin-^{14}C, *Life Sci.* **4:**2017.

Korte, F., and Kochen, W., 1966, Insektizide in Stoffwechsel. XII. Isolierung und Identifizierung von Metaboliten des Aldrom-^{14}C aus dem Urin von Kaninchen, *Med. Pharmacol. Exp.* **15:**409.

Lalonde, D. I. V., and Brown, A. W. A., 1954, The effect of insecticides on the action potentials of insect nerve, *Can. J. Zool.* **32:**74.

Langley, J. N., 1901, On the stimulation and paralysis of nerve-cells and of nerve ending. Part I, *J. Physiol.* (*London*) **27:**224.

Langley, J. N., and Dickinson, W. L., 1889, On the local paralysis of peripheral ganglia, and on the connexion of different classes of nerve fibres with them, *Proc. Roy. Soc.* **46:**423.

Lowenstein, O., 1942, A method of physiological assay of pyrethrum extract, *Nature* (*London*) **150:**760.

Maeno, T., 1966, Analysis of sodium and potassium conductances in the procaine end-plate potential, *J. Physiol.* (*London*) **183:**592.

Maeno, T., Edwards, C., and Hashimura, S., 1971, Difference in effects on end-plate potentials between procaine and lidocaine as revealed by voltage-clamp experiments, *J. Neurophysiol.* **34:**32.

Matthews, H. B., and Matsumura, F., 1969, Metabolic fate of dieldrin in the rat, *J. Agr. Food Chem.* **17:**845.

Murayama, K., Abbott, N. J., Narahashi, T., and Shapiro, B. I., 1972, Effects of allethrin and *Condylactis* toxin on the kinetics of sodium conductance of crayfish axon membranes, *Comp. Gen. Pharmacol.* **3:**391.

Nagawa, Y., Saji, Y., Chiba, S., and Yui, T., 1971, Neuromuscular blocking actions of nereistoxin and its derivatives and antagonism by sulfhydryl compounds, *Jpn. J. Pharmacol.* **21:**185.

Narahashi, T., 1962*a*, Effect of the insecticide allethrin on membrane potentials of cockroach giant axons, *J. Cell. Comp. Physiol.* **59:**61.

Narahashi, T., 1962*b*, Nature of the negative after-potential increased by the insecticide allethrin in cockroach giant axons, *J. Cell. Comp. Physiol.* **59:**67.

Narahashi, T., 1964, Insecticide resistance and nerve sensitivity, *Jpn. J. Med. Sci. Biol.* **17:**46.

Narahashi, T., 1966, Mode of action of insecticides, *Kagaku To Seibutsu* (*Chemistry and Biology*) **4:**134.

Narahashi, T., 1969, Mode of action of DDT and allethrin on nerve: Cellular and molecular mechanisms, *Residue Rev.* **25:**275.

Narahashi, T., 1971*a*, Effects of insecticides on excitable tissues, *in: Advances in Insect Physiology*, Vol. 8 (J. W. L. Beament, J. E. Treherne, and V. B. Wigglesworth, eds.), pp. 1–93, Academic Press, New York.

Narahashi, T., 1971*b*, Mode of action of pyrethroids, *Bull. WHO* **44:**337.

Narahashi, T., 1972, Mechanism of action of tetrodotoxin and saxitoxin on excitable membranes, *Fed. Proc.* **31:**1124.

Narahashi, T., 1973, Mode of action of nereistoxin on excitable tissues, *in*: *Marine Pharmacognosy: Action of Marine Biotoxins at the Cellular Level* (D. F. Martin and G. M. Padilla, eds.), pp. 107–126, Academic Press, New York.

Narahashi, T., 1974, Nerve membrane as a target of pyrethroids, in: *3rd Int. Congr. Pestic. Chem.*, Helsinki, Abst. No. 349.

Narahashi, T., and Anderson, N. C., 1967, Mechanism of excitation block by the insecticide allethrin applied externally and internally to squid giant axons, *Toxicol. Appl. Pharmacol.* **10:**529.

Narahashi, T., and Frazier, D. T., 1971, Site of action and active form of local anesthetics, *in: Neurosciences Research*, Vol. 4 (S. Ehrenpreis and O. C. Salnitzky, eds.), pp. 65–99, Academic Press, New York.

Narahashi, T., and Haas, H. G., 1967, DDT: Interaction with nerve membrane conductance changes, *Science (N.Y.)* **157:**1438.

Narahashi, T., and Haas, H. G., 1968, Interaction of DDT with the components of lobster nerve membrane conductance, *J. Gen. Physiol.* **51:**177.

Narahashi, T., and Yamasaki, T., 1960*a*, Mechanism of the after-potential production in the giant axons of the cockroach, *J. Physiol. (London)* **151:**75.

Narahashi, T., and Yamasaki, T., 1960*b*, Mechanism of increase in negative after-potential by dicophane (DDT) in the giant axons of the cockroach, *J. Physiol. (London)* **152:**122.

Narahashi, T., and Yamasaki, T., 1960*c*, Behaviors of membrane potential in the cockroach giant axons poisoned by DDT, *J. Cell. Comp. Physiol.* **55:**131.

Narahashi, T., Moore, J. W., and Scott, W. R., 1964, Tetrodotoxin blockage of sodium conductance increase in lobster giant axons, *J. Gen. Physiol.* **47:**965.

Narahashi, T., Frazier, D. T., and Yamada, M., 1970, The site of action and active form of local anesthetics. I. Theory and pH experiments with tertiary compounds, *J. Pharmacol. Exp. Ther.* **171:**32.

Nitta, S., 1941, Pharmakalogische Untersuchung des Nereistoxins, das vom Verf. im Körper des *Lumbriconereis heteropoda* (Isome) isoliert wurde, *Tokyo Igaku Zasshi* **55:**285.

Okaichi, T., and Hashimoto, Y., 1962, The structure of nereistoxin, *Agr. Biol Chem.* **26:**224.

Roeder, K. D., and Weiant, E. A., 1946, The site of action of DDT in the cockroach, *Science (N.Y.)* **103:**304.

Roeder, K. D., and Weiant, E. A., 1948, The effect of DDT on sensory and motor structures in the cockroach leg, *J. Cell. Comp. Physiol.* **32:**175.

Roeder, K. D., and Weiant, E. A., 1951, The effect of concentration, temperature and washing on the time of appearance of DDT-induced trains in sensory fibers of the cockroach, *Ann. Entomol. Soc. Am.* **44:**372.

Sakai, M., 1966, Studies on the insecticidal action of nereistoxin, 4-*N,N*-dimethylamino-1,2-dithiolane. III. Antagonism to acetylcholine in the contraction of rectus abdominis muscle of frog, *Botyu-Kagaku* **31:**61.

Sakai, M., 1967, Studies on the insecticidal action of nereistoxin, 4-*N,N*-dimethylamino-1,2-dithiolane. V. Blocking action on the cockroach ganglion, *Botyu-Kagaku* **32:**21.

Sakai, M., 1969, Nereistoxin and cartap: Their mode of action as insecticides, *Rev. Plant Protection Res.* **2:**17.

Sakai, M., and Sato, Y., 1971, Metabolic conversion of the nereistoxin related compounds into nereistoxin as a factor of their insecticidal action, *in: Abstr. 2nd Int. Congr. Pestic. Chem.*, Tel Aviv, February.

Schallek, W., and Wiersma, C. A. G., 1948, The Influence of various drugs on a crustacean synapse, *J. Cell. Comp. Physiol.* **31:**35.

Shanes, A. M., 1949*a*, Electrical phenomena in nerve. I. Squid giant axons, *J. Gen. Physiol.* **33:**57.

Shanes, A. M., 1949*b*, Electrical phenomena in nerve. II. Crab nerve, *J. Gen. Physiol.* **33:**75.

Shanes, A. M., 1951, Electrical phenomena in nerve. III. Frog sciatic nerve, *J. Cell. Comp. Physiol.* **38:**17.

Shankland, D. L., and Schroeder, M. E., 1973, Pharmacological evidence for a discrete neurotoxic action of dieldrin (HEOD) in the American cockroach, *Periplaneta americana* (L.), *Pestic. Biochem. Physiol.* **3:**77.

Thesleff, S., 1955, The mode of neuromuscular block by acetylcholine, nicotine, decamethonium and succinylcholine, *Acta Physiol. Scand.* **34:**218.

van den Bercken, J., 1968, The action of DDT and dieldrin on nerves and muscles of *Xenopus laevis, Meded. Rijksfak, Landbouw-Wetenschappen Gent.* **33:**1241.

van den Bercken, J., and Narahashi, T., 1974, Effects of aldrin-transdiol, a metabolite of the insecticide dieldrin, on nerve membrane, *Eur. J. Pharmacol.* **27:**255.

Wang, C. M., and Narahashi, T., 1972, Mechanism of dual action of nicotine on end-plate membranes, *J. Pharmacol. Exp. Ther.* **182:**427.

Wang, C. M., Narahashi, T., and Yamada, M., 1971, The neurotoxic action of dieldrin and its derivatives in the cockroach, *Pestic. Biochem. Physiol.* **1:**84.

Wang, C. M., Narahashi, T., and Scuka, M., 1972, Mechanism of negative temperature coefficient of nerve blocking action of allethrin, *J. Pharmacol. Exp. Ther.* **182:**442.

Welsh, J. H., and Gordon, H. T., 1947, The mode of action of certain insecticides on the arthropod nerve axon, *J. Cell. Comp. Physiol.* **30:**147.

Yamasaki, T., and Ishii, T. (Narahashi, T.), 1952*a*, Studies on the mechanism of action of insecticides (IV). The effects of insecticides on the nerve conduction of insect, *Oyo-Kontyu* **7:**157.

Yamasaki, T., and Ishii, T. (Narahashi, T.), 1952*b*, Studies on the mechanism of action of insecticides (V). The effects of DDT on the synaptic transmission in the cockroach, *Oyo-Kontyu* **8:**111.

Yamasaki, T., and Ishii, T. (Narahashi, T.,), 1954*a*, Studies on the mechanism of action of insecticides (VII). Activity of neuron soma as a factor of development of DDT symptoms in the cockroach, *Botyu-Kagaku* **19:**1. English translation (1957), *in: Japanese Contributions to the Study of the Insecticide-Resistance Problem*, pp. 140–154, published by Kyoto University for WHO.

Yamasaki, T., and Ishii, T. (Narahashi, T.), 1954*b*, Studies on the mechanism of action of insecticides (VIII). Effects of temperature on the nerve susceptibility to DDT in the cockroach *Botyu-Kagaku* **19:**39. English translation (1957), *in: Japanese Contributions to the Study of the Insecticide-Resistance Problem*, pp. 155–162, published by Kyoto University for WHO.

Yamasaki, T., and Ishii, T. (Narahashi, T.), 1954*c*, Studies on the mechanism of action of insecticides (IX). Repetitive excitation of the insect neuron soma by direct current stimulation and effects of DDT, *Jpn. J. Appl. Zool.* **19:**16. English translation (1957), *in: Japanese Contributions to the Study of the Insecticide-Resistance Problem*, pp. 163–175, published by Kyoto University for WHO.

Yamasaki, T., and Narahashi, T., 1957, Intracellular microelectrode recordings of resting and action potentials from the insect axon and the effects of DDT on the action potential: Studies on the mechanism of action of insecticides (XIV), *Botyu-Kagaku* **22:**305.

Yamasaki, T., and Narahashi, T., 1958*a*, Nervous activity as a factor of development of dieldrin symptoms in the cockroach: Studies on the mechanism of action of insecticides (XVI), *Botyu-Kagaku* **23:**47.

Yamasaki, T., and Narahashi, T., 1958*b*, Resistance to houseflies to insecticides and the susceptibility nerve to insecticides: Studies on the mechanism of action of insecticides (XVII), *Botyu-Kagaku* **23:**146.

Yamasaki, T., and Narahashi, T., 1962, Nerve sensitivity and resistance to DDT in houseflies, *Jpn. J. Appl. Entomol. Zool.* **6:**293.

10

Insecticides as Inhibitors of Respiration

Jun-ichi Fukami

1. Introduction

Under aerobic conditions, the energy required by the tissue cell is provided mainly by respiration, which refers to oxidation of the biological fuel molecules by molecular oxygen. The many chemical steps involved in respiration and in the subsequent conservation of the derived energy in the form of ATP are catalyzed by numerous species of enzymes. These do not occur in soluble form in the cytoplasm of the cell but are located exclusively in the mitochondria, which are considered to be the energy-generating sites of the cell.

Because of the exceptional demands placed on it, the flight muscle of the insect has a very different energy requirement from that of the muscles of other living organisms. To accommodate these requirements, the final products of glycolysis in the insect flight muscle differ from those in the vertebrate muscle and consequently the respiration processes in these two tissues are quite different.

Emphasis in this chapter will be on description of the differences between insects and mammals and on the processes by which energy derived from respiration is utilized via the respiratory chain of the insect and stored as energy-rich phosphate bonds during the process of oxidative phosphorylation.

Jun-ichi Fukami • Laboratory of Insect Toxicology, The Institute of Physical and Chemical Research, Wako, Saitama, Japan.

The interactions between insecticides and the respiratory system of the insect will also be discussed.

2. Biochemistry of the Respiratory System

2.1. The Respiratory Enzyme System

2.1.1. Functional Metabolism in Mitochondria

The number of wing beats of an insect, which is equivalent to the number of flight muscle contractions, reaches several hundred per second and the rate of respiration during active flight exceeds that of the resting state by a factor of about 100 (Sacktor, 1965; Weis-Fogh, 1964). Consequently, a great deal of energy is consumed during insect flight and the mechanism by which this is generated has attracted the attention of a large number of researchers.

Several reviews concerning the respiratory processes occurring in the insect flight muscle have appeared (Sacktor, 1961, 1965, 1974; Chefurka, 1965; Harvey and Haskell, 1966), most of which emphasize the biochemical properties of the mitochondria.

a. Oxidation of α-Glycerophosphate. In the flight muscle mitochondria (sarcosomes) of a variety of insect species, TCA cycle intermediates are oxidized at rates similar to those observed in the muscles of other living organisms. This corresponds to the basal respiration rate in flight muscle of insects and is far less than that required for active flight. For example, early experiments by Davis and Fraenkel (1940) with the blowfly (*Lucilia sericata*) revealed that with the initiation of flight some individuals consumed oxygen at a rate of about 3000 μl/min/g, elevating their basal rates approximately a hundredfold. In the skeletal muscle of the vertebrate, the lactic acid produced by glycolysis is subsequently oxidized completely to carbon dioxide and water during respiration. However, in the insect flight muscle little or no oxidation of lactic acid occurs and the question is raised of what is the source of the energy required for muscle contraction during flight.

The answer in part lies in the fact that lactic acid is not the major end product of glycolysis in the insect flight muscle. It was found that in the thoracic muscle of the cockroach (*Periplaneta americana*) lactic acid is a quantitatively unimportant product of the anaerobic breakdown of glycogen. Most of the glycogen broken down undergoes a dismutation at the level of triosephosphate, while the reduced product of this dismutation, α-glycerophosphate, accumulates in the tissue; its amount corresponds well to one-half of the glycogen consumed (Kubišta, 1958). This immediately suggests that there is some special relationship between α-glycerophosphate and insect flight (Fig. 1).

The α-glycerophosphate which is produced by the insect flight muscle cell may originate either from the catabolism of phospholipids or from dihydroxyacetone phosphate (DHAP) derived from the classical glycolytic pathway of Embden-Meyerhof. In vertebrate muscle, α-glycerophosphate is not formed

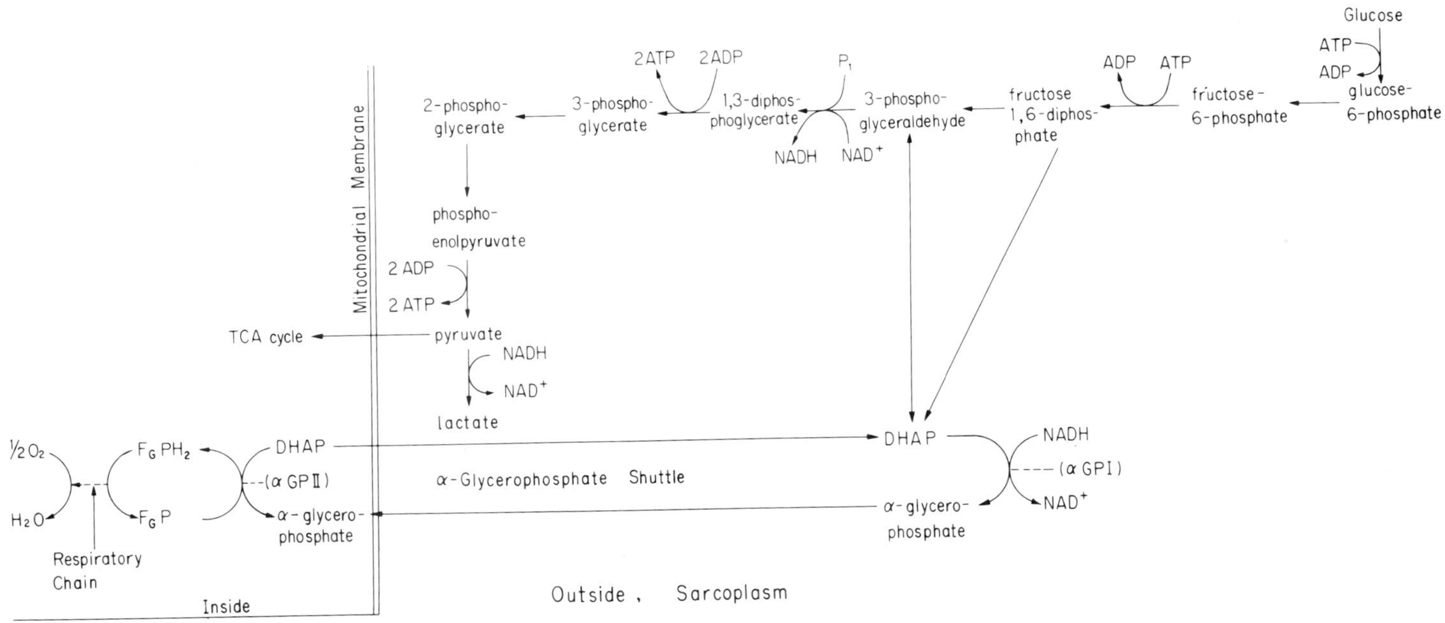

Fig. 1. Glycolysis and the α-glycerophosphate shuttle in insect flight muscle. Abbreviations: αGPII, α-glycerophosphate dehydrogenase in mitochondria; F_GP, mitochondrial flavoprotein; DHAP, dihydroxyacetone phosphate; α-GPI, NAD-linked α-glycerophosphate dehydrogenase in sarcoplasm. Modified from Sacktor (1974).

since the NADH formed during the glycolytic oxidation of 3-phosphoglyceraldehyde is rapidly reoxidized to NAD during the subsequent conversion of pyruvate to lactate by lactic dehydrogenase. In the insect flight muscle, this reaction is minimized since little if any lactic dehydrogenase is present. Instead, in the presence of NADH, DHAP is rapidly reduced to α-glycerophosphate by the extremely active α-glycerophosphate dehydrogenase (αGPI) which occurs in the sarcoplasm of the flight muscle (Estabrook and Sacktor, 1958; Chance and Sacktor, 1958; Marquardt and Brosemer, 1966). The equilibrium constant of this reaction favors the accumulation of α-glycerophosphate, which crosses the mitochondrial membrane and is reoxidized to DHAP by the active mitochondrial α-glycerophosphate dehydrogenase (αGPII); the equilibrium of the latter reaction favors the formation of DHAP.

This system, referred to as the α-glycerophosphate shuttle (Fig. 1), is important since it represents a cyclic mechanism whereby reducing equivalents from NADH, produced by the sarcoplasmic enzymes in the soluble fractions of the cell, are made available to the respiratory chain in the mitochondria.

That a system of this type is required is evident from the observation that the rate of NADH oxidation by insect flight muscle mitochondria is comparatively slow. In contrast, the rate of oxidation of α-glycerophosphate is extremely fast since it is readily accessible to the mitochondrial respiratory chain; α-glycerophosphate is therefore acting as an efficient carrier of reducing equivalents from the cytoplasm of the cell to the respiratory chain in the mitochondria.

In vertebrates, the oxidation of α-glycerophosphate by (αGPII) is low in liver mitochondria but occurs to a limited extent in muscle and brain mitochondria. In the flight muscle of the insect, α-glycerophosphate is the substrate for which mitochondria have the highest activity.

Chance and Sacktor (1958) discussed the relative physiological roles of different substrates in flight muscle mitochondria of the housefly (*Musca domestica*) and concluded that α-glycerophosphate was the only known substrate oxidized at a rate sufficient to account for the high respiratory activity during flight. This is supported by other investigators (Zebe and McShan, 1957; Kubišta, 1957; Sacktor and Cochran, 1958). It is remarkable that the rates of oxidation of pyruvate (Birt, 1961; Fukami, 1961; Cochran, 1963) and several TCA cycle intermediates are so much lower than that of α-glycerophosphate since in vertebrate mitochondria it is usually found that succinate and α-ketoglutarate promote maximal respiratory activity. Indeed, the fact that α-glycerophosphate dehydrogenase (αGPII) activity can exceed the rate of succinate oxidation by a factor exceeding 10 suggests that, on a kinetic basis, the TCA cycle may not play a significant role in flight muscle metabolism (Chance and Sacktor, 1958).

The possibility exists that during periods of relative inactivity the α-glycerophosphate shuttle is in an inhibited state and becomes activated only to provide the high respiratory activity required during flight. Under conditions of reduced activity, the TCA cycle may provide the necessary level of basal respiratory activity. Activation of the α-glycerophosphate shuttle may result from the accumulation of either substrate or divalent cations such as magnesium released during nervous stimulation of the muscle (Estabrook and Sacktor, 1958).

A theory involving metal chelation has been proposed to explain the activation process. EDTA inhibits the α-glycerophosphate dehydrogenase (αGPII) activity of mitochondria isolated from the flight muscle of the housefly (*Musca domestica*) and this inhibition is reversed by addition of 5 mM Mg^{2+}, Ca^{2+}, and a number of other bivalent metal ions (Estabrook and Sacktor, 1958). From these results, it was suggested that in the resting condition endogenous chelating agents, equivalent in their action to EDTA, exist in mitochondria and prevent the oxidation of α-glycerophosphate. When flight muscle is induced to contract by nerve stimulation, divalent cations, especially Ca^{2+}, are released and in addition to activating the myosin ATPase reverse the inhibition of mitochondrial α-glycerophosphate dehydrogenase (αGPII) by the chelating agent. As a result, α-glycerophosphate is rapidly oxidized and produces the large amount of ATP which is used as the energy source for insect flight. This suggestion has received extensive confirmation through the work of Hansford and Chappell (1967). These workers showed that α-glycerophosphate dehydrogenase (αGPII) of isolated flight muscle mitochondria from thoraces of the blowfly (*Calliphora vomitoria*) is markedly stimulated by low levels of Ca^{2+}. The dehydrogenase in intact and sonicated mitochondria shows allosteric kinetics and Ca^{2+} acts by lowering the K_m for substrate. In the presence of a constant low level of Ca^{2+}, achieved by using EDTA–Ca^{2+} buffer, the oxidation of α-glycerophosphate (αGPII) was stimulated three- to fourfold on addition of ADP and phosphate. From these results, it was deduced that Ca^{2+} released on initiation of contraction in flight muscle plays a dual physiological role first in activating the myosin ATPase and second in allowing rapid rates of α-glycerophosphate oxidation and therefore ATP synthesis. The control of mitochondrial α-glycerophosphate dehydrogenase (αGPII) depends both on the lowering of the K_m by Ca^{2+} and on the allosteric nature of the enzyme kinetics. At about 2 mM α-glycerophosphate, which is the physiological level in the flight muscle of blowfly (*Phormia regina*), a tenfold increase in uncoupled and ADP-stimulated oxidation occurred on increasing the Ca^{2+} concentration from 10^{-8} to 10^{-5} gm-ion/liter (Sacktor and Wormser-Shavit, 1966). Donnellan and Beechey (1969) also studied the role of Ca^{2+} in flight muscle mitochondria isolated from *Sarcophaga barbata, Pieris brassicae, Apis mellifera, Schistocerca gregaria*, and *Musca domestica* in relation to mitochondrial α-glycerophosphate dehydrogenase (αGPII). Their results confirmed that Ca^{2+} plays a key role in the oxidation of α-glycerophosphate by flight muscle mitochondria and that low levels of Ca^{2+} effect the oxidation by both mitochondria and submitochondrial particles through heterotropic allosteric interaction. The α-glycerophosphate acts as a homotropic effector in the absence of Ca^{2+} ion. EDTA particles effectively oxidize α-glycerophosphate independently of the Ca^{2+} concentrations, whereas calcium particles show a marked dependence on Ca^{2+}.

b. Oxidation of Pyruvate and Other TCA Cycle Intermediates. Whether the oxidation of α-glycerophosphate provides the only source of energy available to the insect during flight remains controversial, and some results suggest that at least in some species pyruvate and proline may also be involved.

Thus, in contrast to the results described earlier, Van den Bergh and Slater (1962) showed that the end products of glycolysis in the thoracic muscle of housefly (*Musca domestica*) are not only α-glycerophosphate but also pyruvate. In fact, they confirmed Chefurka's (1958) finding that pyruvate is produced to the same extent as α-glycerophosphate during glycolysis in thoracic muscle homogenates of the cockroach (*Periplaneta americana*). Subsequently, α-glycerophosphate, pyruvate, and pyruvate plus malate have been shown to be oxidized in the mitochondria of the thoracic muscle of the housefly, and their P/O and respiratory control ratios are quite high. Van den Bergh and Slater (1962) concluded from these results that pyruvate as well as α-glycerophosphate can be used as a substrate to provide energy to the flight muscle.

The P/O and respiratory control ratios of other TCA cycle intermediates are quite low and suggest that these are not utilized to any significant extent. Possibly the major reason for this is that unlike α-glycerophosphate and pyruvate, which readily gain entry into the mitochondria, TCA cycle intermediates are almost completely excluded from passage through the inner mitochondrial membrane. The importance of membrane permeability is clearly indicated by the fact that following sonic disruption of the mitochondria several TCA cycle intermediates can be rapidly oxidized (Van den Bergh and Slater, 1962; Van den Bergh, 1964); the respiratory rate with TCA cycle intermediates as a result of this treatment is sometimes increased well over tenfold.

It may be deduced, therefore, that because of the membrane barrier TCA cycle intermediates are maintained at high concentrations within the mitochondria and are responsible for the high metabolic turnover observed. Consequently, it appears that TCA cycle intermediates are utilized during insect flight.

Hansford (1971) has shown that mitochondria from the flight muscle of the cicada (*Magicicada septendecim*) oxidize pyruvate and α-glycerophosphate at high rates, implying that these are the products of glycolysis in the tissue. Palmitoyl-L-carnitine is also oxidized, but at a low rate which appears to rule out any major role in supporting flight. In contrast to their failure to oxidize other TCA cycle intermediates, the rapid oxidation of 2-oxoglutamate and L-glutamate by these mitochondria is interesting. It has been suggested that utilization of L-glutamate and 2-oxoglutamate in the cicada may be related to the observed autolysis of the flight muscle to support egg-laying in this short-lived adult. In this connection, it has been suggested that the TCA cycle intermediates are needed almost exclusively for respiratory function (as precursors of oxaloacetate for catabolism of pyruvate) and not for their participation in various biosynthetic process outside the mitochondria, as occurs in tissues such as the mammalian liver (Sacktor, 1974).

c. Fatty Acid Oxidation. The energy sources utilized by flying insects are of two major types (Van den Bergh, 1967). In general, insects such as the locust and the butterfly which can fly for extended periods of time utilize fats as the major source of energy during flight. In contrast, insects such as the housefly and bee which usually fly for much shorter periods rely more on carbohydrate as their main source of flight energy.

The mitochondria of insects such as the locust (*Locusta migratoria*) which can fly for long periods of time differ from those of housefly (*Musca domestica*) (Van den Bergh, 1967). Noteworthy differences from those of the housefly are in the rate of respiration and phosphorylation with L-glutamate as substrate, the fact that they can oxidize only for very short periods in the absence of added TCA cycle intermediates, their ability to oxidize carnitine esters of fatty acids (but not the free fatty acid) (Bode and Klingenberg, 1964), and the substantial amount of endogenous substrate present in locust mitochondria. Respiratory control by omission of phosphate or phosphate acceptor can be demonstrated with pyruvate plus malate and with L-glutamate as substrate. With α-glycerophosphate as the substrate, respiratory control is weaker and less reproducible. The dragonfly (*Pantala flavescens*), which is also capable of sustained flight, uses fat as its major energy source, and the oxidation of fatty acids plays an important role in the flight of this insect. It is possible that the flight muscle of the dragonfly has sufficiently high reserves of carnitine to maintain the oxidation of palmitic acid required for sustained flight, and carnitine release may increase with a rise in body temperature of the insect (Kallapur and George, 1973).

d. Proline Oxidation. The significance of the oxidation of α-glycerophosphate and pyruvate in energy production during insect flight has already been discussed. However, isolated mitochondria of the blowfly (*Phormia regina*) rapidly lose their capacity to oxidize pyruvate. This loss can be reversed by proline but not by TCA cycle intermediates or glutamate (Sacktor and Childress, 1967). These findings, combined with the rapid oxidation of proline, suggest that, *in vivo*, flight muscle mitochondria may be deficient in the TCA cycle intermediates and that these are generated from proline. On the initiation of flight, the concentration of proline in the flight muscle rapidly decreases (Sacktor and Wormser-Shavit, 1966). It has been found that isolated flight muscle mitochondria of the blowfly (*Calliphora vomitoria*) oxidize proline at a rate that can account for the utilization of this amino acid during flight (Sacktor and Childress, 1967). Proline is converted to L-glutamate as follows by a two-step oxidative reaction through the intermediate Δ^1-pyrroline-5-carboxylate, and the L-glutamate thus formed can generate TCA cycle intermediates by further mitochondrial oxidation.

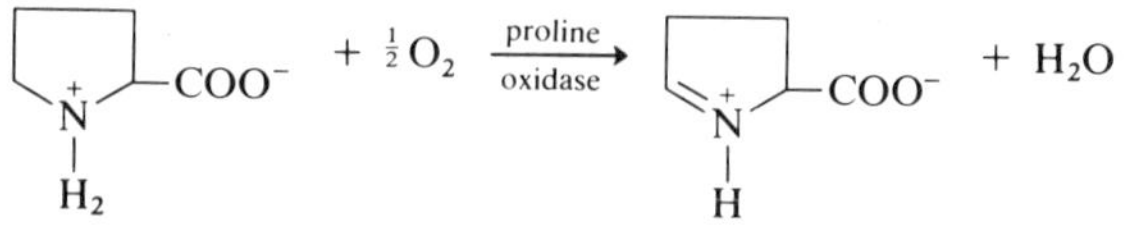

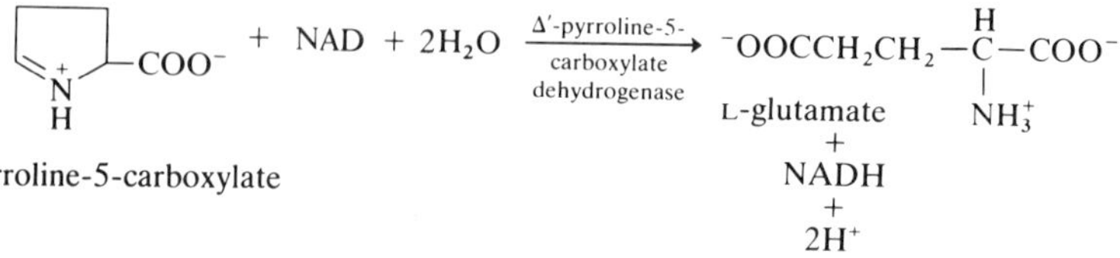

Δ¹-pyrroline-5-carboxylate

Proline significantly enhances the rate of pyruvate oxidation, but no stimulation is observed with either L-glutamate or malate, probably because of their inability to penetrate the mitochondrial membrane (Sacktor and Childress, 1967; Van den Bergh and Slater, 1962). These findings suggest that proline enhances the rate of pyruvate metabolism by penetrating the mitochondrial barrier, forming the intramitochondrial precursors of oxaloacetate and thus enabling the complete oxidation of pyruvate via the TCA cycle at a maximal rate (Sacktor and Childress, 1967). As a result of these studies, it has been concluded that pyruvate is not immediately oxidized on the initiation of flight since it cannot enter the TCA cycle until proline penetrates the mitochondrial membrane and provides the necessary TCA cycle intermediates. In the flight muscle of the Colorado potato beetle (*Leptinotarsa decemlineata*), proline is metabolized much faster than in other insects and is coupled completely with oxidative phosphorylation. Furthermore, the rate of pyruvate oxidation is directly related to the rate at which proline is utilized, clearly establishing proline as an important respiratory substrate in the Colorado potato beetle flight muscle (De Kort *et al.*, 1973). Hansford and Sacktor (1970) established that the oxidation of proline by flight muscle mitochondria of the blowfly (*Phormia regina*) is stimulated by ADP in the presence of uncoupling agents. The rate of oxidation of proline, in the presence of ADP, is enhanced additionally by a high level of inorganic phosphate, although in the absence of ADP the level of phosphate is of little significance.

Pyruvate also increases the affinity of proline dehydrogenase for its substrate (Hansford and Sacktor, 1970), and since on initiation of flight there is a considerable increase in the concentration of pyruvate in the flight muscle (Sacktor and Wormser-Shavit, 1966), this suggests an attractive mechanism for stimulating proline oxidation at the dehydrogenase level. The rate of proline oxidation is dependent on the absolute level of ADP (Hansford and Sacktor, 1970).

2.1.2. Role of NAD-Linked Isocitric Dehydrogenase

It is also of interest to note that the NAD-linked isocitric dehydrogenase of mitochondria from the aphid (*Myzus persicae*) is greatly activated by ADP.

NAD-linked isocitric dehydrogenase activity in mitochondria of alatiform larvae is much higher than that of apteriform larvae, suggesting some association of this enzyme with flight. On the other hand, the activity of malic dehydrogenase is essentially the same in the two morphs of the aphid. These results indicate that the NAD-linked isocitric dehydrogenase may also play a regulatory role in respiration and is further evidence for the importance of the TCA cycle in insect flight (Zahavi and Tahori, 1972).

2.2. Respiratory Chain

The respiratory chain is commonly known as the mitochondrial electron transport system. It receives electrons (reducing equivalents) from the TCA cycle, the fatty acid cycle, or NAD- or flavin-linked enzyme systems and transfers them through a series of redox intermediates to oxygen. The energy produced during the process is trapped in the form of the high-energy phosphate bond of ATP and eventually utilized for mechanical or chemical energy. The respiratory chain is an extremely complex enzyme system and is of critical importance to the functioning of living organisms. A schematic representation of the respiratory enzyme system of mammalian liver mitochondria is shown in Fig. 2. The major oxidation–reduction components that occur in the respiratory chain in mitochondria are (a) NAD or NADP dehydrogenase (pyridine nucleotide linked dehydrogenases) which require either NAD or NADP as coenzyme, (b) flavin-linked enzymes which contain flavin adenine dinucleotide

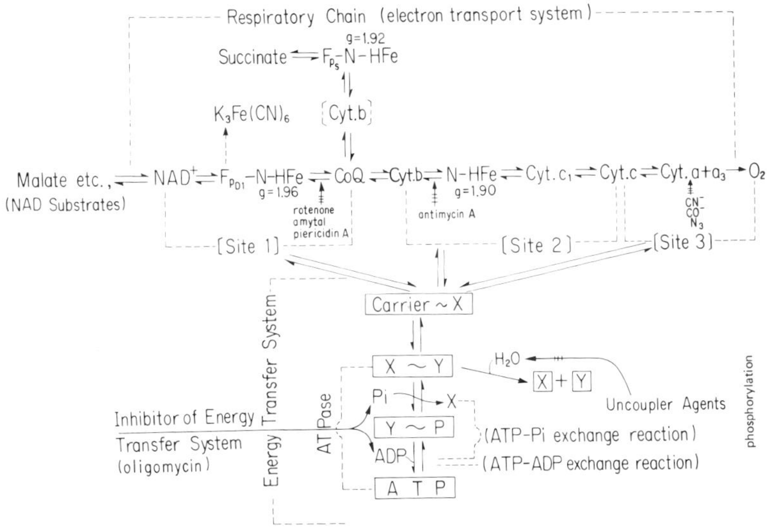

Fig. 2. Respiratory enzyme system of mammalian liver mitochondria. Abbreviations : F_{PD_1}, NADH dehydrogenase; F_{P_s}, succinate dehydrogenase; N-HFe, specific protein containing nonheme iron. See text for explanation. Modified from Lardy and Ferguson (1969).

(FAD) or flavin mononucleotide (FMN) as their prosthetic groups, (c) the cytochromes, which contain an iron–porphyrin ring system, and (d) the lipid-soluble quinone. The cytochrome system functions very well in insect flight muscle and its physiological properties have been investigated in detail by Chance and Sacktor (1958), using mitochondria of housefly (*Musca domestica*) flight muscle.

Fukami (1961) has investigated the cytochrome pigments of the meso- and metathoracic muscles of the cockroach (*Periplaneta americana*) following their reduction with sodium hydrosulfite; the absorption bands were examined by the opal glass transmission method at room temperature (Fig. 3). The α bands of the cytochromes occur at the usual position (*a*, 603 nm; *b*, 562 nm; *c*, 551 nm), and in the Soret region the absorption bands of cytochromes *b* and *c* can be distinguished at 429 and 419 nm, respectively. Absorption bands at the same locations are also observed in the presence of various substrates such as succinate, α-glycerophosphate, L-glutamate, and NADH (Fig. 4). Similar results were obtained in spectral studies of the electron transport system of flight muscle mitochondria of the housefly (*Musca domestica*) (Chance and Sacktor, 1958; Estabrook and Sacktor, 1958) in the presence of succinate, α-glycerophosphate, or NADPH or following nonenzymatic reduction with sodium dithionite or ascorbic acid.

Spectra similar to those reported for the flight muscle mitochondria of the housefly (*Musca domestica*) and cockroach (*Periplaneta americana*) have subsequently been recorded in preparations from flight, coxal, femoral, and sound-producing muscle from over 20 species belonging to the orders Orthoptera, Odonata, Hemiptera, Coleoptera, and Lepidoptera (Fukami, 1957; Fukami and Nakatsugawa, 1961; Bruce and Banks, 1973).

From these studies, it has been concluded that the respiratory chain in insect flight muscle mitochondria is very similar to that of rat liver mitochondria (Fig. 2).

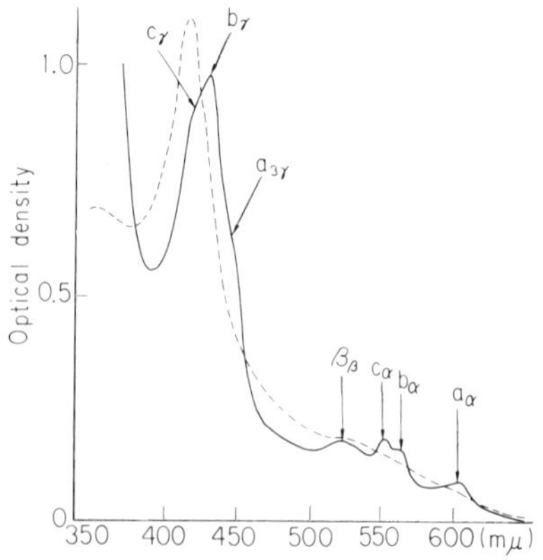

Fig. 3. Absorption spectrum of muscle mitochondria of the cockroach (*Periplaneta americana*) showing absorption maxima of a_α, b_α, c_α, β_β, and Soret ($a_{3\gamma}$, b_γ, c_γ) bands of the cytochromes as measured by the opal glass transmission method (Fukami, 1961). The reaction mixture consisted of 50 μM phosphate buffer (pH 7.4), 1 μM ATP, 5 μM $MgCl_2$, 0.01 μM cytochrome *c*, 0.3 ml of water, and 1 ml mitochondrial suspension (0.73 mg N/ml) in a total volume of 3.3 ml. Temperature 16°C. – – –, Oxidized; ——, reduced with sodium hydrosulfite.

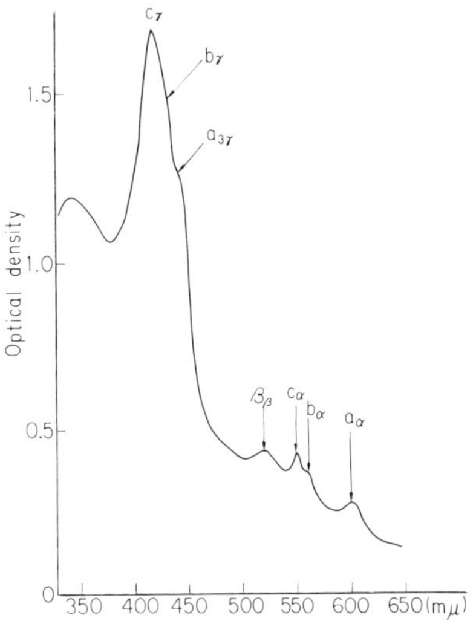

Fig. 4. Absorption spectrum of muscle mitochondria of cockroach (*Periplaneta americana*) in the presence of α-glycerophosphate, showing absorption maxima of a_α, b_α, c_α, β_β, and Soret ($a_{3\gamma}$, b_γ, c_γ) bands of the cytochromes (Fukami, 1961). Composition of reaction mixture was similar to that described in the caption of Fig. 3 except for the addition of 0.3 ml of 0.5 M α-glycerophosphate. Temperature 17°C.

As described above, cockroach muscle mitochondria exhibit the characteristic α absorption bands of cytochromes *a*, *b*, and *c* at 603, 562, and 551 nm, respectively, and appropriate Soret bands at 445, 429, and 419 nm. This agrees with the findings of Estabrook and Sacktor (1958), except for the α band at 551 nm. This band seems to be attributable to cytochrome *c*. Estabrook and Sacktor (1958) also pointed out the presence of a trough at 470 nm attributable to flavoprotein, but Fukami (1961) was not able to confirm this. In the ultraviolet region, absorption increases slowly toward 320 nm without any obvious peak at 340 nm to indicate the presence of reduced pyridine nucleotide. This is reportedly due to interference at this wavelength by cytochromes which are present in the tissue in concentrations much greater than those of the pyridine nucleotides (Estabrook and Sacktor, 1958). Using other techniques of difference spectroscopy, Klingenberg and Bücher (1959) clearly showed the peak of pyridine nucleotide in flight muscle mitochondria of the locust (*Locusta migratoria*).

Following purification, the amino acid compositions of various cytochrome pigments in the flight muscle of housefly (*Musca domestica*) larvae have been compared with those of beef heart muscle. Ultracentrifugal studies have established that beef heart cytochrome *b* is made up of aggregates, whereas the purified preparation of larval cytochrome *b* is monomeric (Ohnishi, 1966). The cytochrome b_5 *present during diapause of the insect* was purified from housefly larvae and shown to be a solubilized form of the cytochrome b_5 which in vertebrates usually exists bound to the mitochondrial membrane (Okada, 1973). The complete amino acid sequence of cytochrome *c* from the flight muscle of a saturniid moth (*Samia cynthia*) has also been established. This protein is functionally and structurally homologous to the mammalian-type cytochrome *c* with the exception that it has a nonacetylated

NH_2-terminal residue, has 107 rather than 104 residues, and carries an arginine rather than a lysine at position 13, immediately before the first thioether-bound cysteine. These characteristics appear to be common to invertebrate cytochrome *c*, including that isolated from baker's yeast. In other primary structural features, however, the saturniid moth cytochrome *c* resembles the vertebrate rather than the yeast protein, as might be expected from the phylogenetic relationship of fungi, invertebrates, and vertebrates (Chan and Margoliash, 1966).

Chance and Sacktor (1958) demonstrated the presence of roughly equal concentrations of the various respiratory cytochromes in housefly flight muscle mitochondria, and Klingenberg and Bücher (1959) subsequently showed that the relative cytochrome concentrations in flight muscle mitochondria of the locust (*Locusta migratoria*) closely resembled those in mammalian heart muscle mitochondria. However, the ratio of pyridine nucleotide to cytochrome appears to be strikingly different between insect and mammalian mitochondria. Thus in contrast to liver mitochondria, where the ratio of pyridine nucleotide to cytochrome *a* is 40 : 1, that in housefly flight muscle mitochondria is very low (Chance and Sacktor, 1958) and is reported to be only 4 : 1 in locust mitochondria (Klingenberg and Bücher, 1959). It should be noted, however, that the amount of pyridine nucleotide per gram mitochondrial protein does not differ appreciably between insect flight muscle and mammalian mitochondria (Klingenberg *et al.*, 1959); the difference in the ratio of the two components results mainly from the much greater cytochrome content in the insect flight muscle.

The relatively low pyridine nucleotide to cytochrome ratio in insect flight muscle mitochondria might be a further reflection of the importance of α-glycerophosphate as an initial energy source for insect flight since the oxidation of α-glycerophosphate does not require NAD as a cofactor (Fig. 1). Instead, reducing equivalents resulting from the oxidation of α-glycerophosphate by the mitochondrial αGPII are passed directly to the respiratory cytochromes via the flavoprotein and ATP is produced for muscle contraction. The P:O ratio with α-glycerophosphate is experimentally determined to be from 1.0 to 1.7, close to the integral value of 2, which is that of NAD-unlinked substrates. The rate of oxidation of α-glycerophosphate in isolated mitochondria is sufficient to account for the rate of respiration during flight.

On the other hand, since the oxidation of pyruvate by housefly flight muscle mitochondria yields usable energy at a rate ($Q_{O_2} \times P{:}O$) almost 3 times that of α-glycerophosphate (Gregg *et al.*, 1960), it would appear premature to conclude that α-glycerophosphate is the principal substrate activating the respiratory chain during flight. It is of course entirely possible that under physiological conditions the insect uses α-glycerophosphate to initiate flight and then utilizes pyruvate for energy production once flight is under way.

In general, the pyridine nucleotide linked dehydrogenases are specific for either NAD or NADP. Of the prominent dehydrogenases in insect mitochondria, isocitrate, malate, dihydrolipoyl, L-β-hydroxyacyl-CoA, and D-β-hydroxybutyrate dehydrogenases require NAD. A second isocitrate dehydrogenase requires NADP (Sacktor, 1974). In mammalian mitochondria, and presumably also in insect mitochondria, L-glutamic dehydrogenase can react with either NAD or NADP (Lehninger, 1970).

The flavin-linked dehydrogenases contain either flavin adenine dinucleotide (FAD) or flavin mononucleotide (FMN) as prosthetic groups. Oxidation–reduction of FAD or FMN is usually pictured as a simultaneous transfer of two reducing equivalents, but there is evidence that the reaction occurs in two separate one-electron steps (Lehninger, 1970). The most important flavin-linked dehydrogenases are (a) NADH dehydrogenase, which catalyzes the transfer of electrons from NADH to some unknown acceptor, possibly a nonheme iron protein, in the respiratory chain and (b) succinic dehydrogenase, α-glycerophosphate dehydrogenase, etc. (Lehninger, 1970). In mammalian tissues, the prosthetic group of NADH dehydrogenase is FMN, while the other dehydrogenases mentioned above contain FAD (Lehninger, 1970). Ernster and Navazio (1958) and Ernster *et al.* (1960, 1962) have found in the soluble fraction of rat liver a flavoenzyme, named "DT-diaphorase," which catalyzes the reduction of cytochrome *c* by either NADH or NADPH in the presence of a cofactor such as vitamin K_3 (menadione). DT-diaphorase is also present in rat liver mitochondria (Conover and Ernster, 1962), and Quagliariello *et al.* (1966) have suggested that, in addition to vitamin K_3, 3-hydroxyanthranilic acid can possibly act as a cofactor. Chino (1963) has shown the presence of an NADH/NADPH cytochrome *c* reductase in the soluble fraction of eggs of the silkworm (*Bombyx mori*) and suggests that xanthommatin may be required as a cofactor for the reaction. Subsequent experiments (Chino and Harano, 1966) revealed that dialysis of the enzyme caused a loss of activity and that this could be restored by addition of boiled soluble fraction or addition of 3-hydroxykynurenine or its dimer, xanthommatin, both of which are known to occur in the silkworm (*Bombyx mori*) egg (Chino, 1963; Osanai, 1966). Recently, a new flavoenzyme has been partially purified from the soluble fraction of the egg of the silkworm (*Bombyx mori*). The enzyme catalyzes the reduction of cytochrome *c* equally well in the presence of either NADH or NADPH and requires an aminophenol such as 3-hydroxykynurenine or xanthommatin as a cofactor; 2,6-dichlorophenolindophenol is a direct electron acceptor without cofactor. The enzyme is also found in the soluble fraction of rat liver (Harano and Chino, 1971), but several lines of evidence tend to rule out its possible identity with the DT-diaphorase reported by Ernster and Navazio (1958) and Ernster *et al.* (1960, 1962). Osanai (1966) has concluded that the flavoprotein which he has detected in the soluble fraction of silkworm eggs plays an important role in respiration during diapause. The appearance of FAD when the eggs of silkworm hatch suggests the possible formation of a normal respiratory system containing cytochrome *c*, which takes the place of this flavin-respiration during diapause.

Although the cytochrome system is the ultimate acceptor of electrons from the flavin-linked dehydrogenases, there is considerable evidence that both NAD-linked dehydrogenases and succinic dehydrogenase either contain or are closely associated with specific proteins called generically "nonheme iron proteins" because the iron they contain is present in some form other than heme (Lehninger, 1970).

Donnellan *et al.* (1970) studied the oxidation of α-glycerophosphate by flight muscle mitochondria isolated from the flesh fly (*Sarcophaga barbata*). The α-glycerophosphate–cyanoferrate oxidoreductase system in these mitochondria is antimycin A insensitive, whereas the corresponding NADH oxidoreductase is extremely sensitive to this respiratory chain inhibitor. It has been concluded from these and other data that a nonheme iron protein is not involved in the oxidation of α-glycerophosphate (αGPII) by the blowfly (*Sarcophaga bullata*).

Coenzyme Q is a benzoquinone derivative with a large isoprenoid side chain, varying in size from 6 to 10 units. In the course of a survey of the presence of coenzyme Q in a large number of animals, plants, and microorganisms, Lester and Crane (1959) became aware of the fact that coenzyme Q is not a single compound but exists in the form of several closely related homologues. Substantial amounts of coenzyme Q occur in most animal tissues studied, and in general the concentrations found are consistent with a possible role of coenzyme Q in electron transport. Thus, in keeping with its higher respiratory activity, cardiac tissue is always found to contain more coenzyme Q than skeletal muscle. Insect tissues such as those from whole houseflies and whole cabbage butterflies (*Pieris rapae*) also contain large amounts of coenzyme Q. In general, coenzyme Q_{10} is the homologue most commonly found in higher animals, but the presence of coenzyme Q_9 in the walleyed pike shows that this is not the case in all vertebrates. Further work will be necessary to establish whether coenzyme Q_9 is typical of insect tissues. Kröger and Klingenberg (1966) concluded that the redox reactions of ubiquinone in mitochondria from rat heart are consistent with its role as an electron carrier in the respiratory chain, and it is considered by many to act as a fat-soluble molecule shuttling between the flavoproteins and cytochromes in the lipid phase of the mitochondrial membrane. Coenzyme Q has been found in mitochondria of the blowfly (*Phormia regina*) flight muscle, where it undergoes rapid enzymatic reduction when α-glycerophosphate is present as substrate (Sacktor, 1961).

2.3. Oxidative Phosphorylation

With the exception of bacteria and photosynthetic organisms, the energy required to maintain living organisms is supplied from oxidative phosphorylation of mitochondria. Oxidative phosphorylation is a coupled reaction consisting of two complex enzyme systems, the respiratory chain (electron transport system) and the energy transfer system (phosphorylation), which utilizes redox energy liberated from the respiratory chain for the synthesis of ATP (Fig. 2).

The exact mechanism by which oxidation in the mitochondria is coupled with the esterification of ADP by inorganic phosphate to form ATP is still not known.

Oxidative phosphorylation is the process by which redox energy liberated during the passage of electrons from NADH to molecular oxygen via the respiratory chain is utilized for the synthesis of ATP (Fig. 2). The precise mechanism by which respiratory free energy is coupled with the esterification of ADP by inorganic phosphate (P_i) to form ATP is not yet fully understood and is under active investigation. As shown in Fig. 2, there are at least three sites at which phosphorylation can occur in rat liver mitochondria. One of these lies between NAD^+ and coenzyme Q, a second between cytochromes *b* and *c*, and the third between cytochrome *c* and oxygen. It has been postulated that at each of these coupling sites a high-energy carrier (presumably a protein) is generated and that the energy is transferred through other intermediates (X and Y in Fig. 2), ultimately resulting in the synthesis of ATP from ADP and P_i (Lehninger, 1970). Evidence supporting the general scheme shown in Fig. 2 comes from the influence on oxidative phosphorylation of several inhibitors and the established existence of at least four partial reactions in the ATP-forming process. It is probable that the coupling process is similar at each of the phosphorylation sites.

Even in actively respiring mitochondria, phosphorylation does not occur in the absence of ADP and P_i. Under physiological conditions, the absence of P_i is rare and since other factors are less important ADP is considered to be the key substance in respiratory control. Whenever a function which requires energy (ATP) is carried out in the cell, ADP is inevitably produced and more ATP is generated. If the ATP produced is not consumed, respiration is repressed by the absence of ADP. This phenomenon was initially observed by Lardy and Wellman (1952) and can be observed *in vitro* only in carefully prepared intact mitochondria which are defined as being "tightly coupled." Using the oxygen electrode, Chance and Williams (1955*a*,*b*, 1956) identified and characterized six respiratory states in intact mitochondria (Table I). These are now commonly used to describe the physiological condition of isolated mitochondria and to simplify and standardize the analysis of ADP as a metabolic regulator.

Polarographic methods using the oxygen electrode have now largely supplanted the traditional manometric procedures used to study oxidative phorphorylation. Such methods not only provide a continuous record of the time course of the reaction but also allow the convenient addition of several reagents during the course of a single incubation. The procedures employed have been reviewed by Estabrook (1967) and Hagihara (1965), but a brief description will be included here to illustrate how several parameters can be determined.

A suspension of coupled mitochondria in a suitable medium exhibits a low rate of respiration (oxygen uptake) in the absence of added substrate or ADP (Fig. 5). This is the endogenous respiration rate (state 1) due to the presence of

Table I. Respiratory Control at Steady States in Isolated Mitochondria[a]

Characteristics[b]	Oxygen	ADP level	Substrate level	Respiration rate	Rate-limiting component
State 1. Partially starved	Excess	Low	Low, endogenous	Slow	Phosphate acceptor
State 2. Starved	Excess	High	Approaching 0	Slow	Substrate
State 3. Active	Excess	High	High	Fast	Respiratory chain
State 4. Acceptorless	Excess	Low	High	Slow	Phosphate acceptor
State 5. Anaerobic	None	High	High	0	Oxygen
State 6.	Excess	Low	High	Slow	Respiratory chain

[a] From Chance and Williams (1955*a*).
[b] State 1: Respiration due to the presence of endogenous substrates in mitochondria. State 2: Respiration of endogenous substrates (state 1) but in the presence of added ADP as an acceptor for phosphorylation. State 3: Active respiration coupled with phosphorylation in the presence of exogenous substrate and ADP. State 4: Controlled respiration in the presence of high substrate concentrations but after all the exogenous ADP has been phosphorylated to ATP. State 5: Occurring under anaerobic condition but in the presence of high levels of substrate and ADP. State 6: The inhibited state following the repeated addition of small amounts of Ca^{2+} to mitochondria in state 4.

small amounts of substrate in the preparation. The addition of succinate causes an increase in respiration, but this is limited by the availability of ADP (state 2). When ADP is added, however, respiration coupled with phosphorylation causes a large increase in oxygen uptake (state 3), which continues until the ADP again becomes rate limiting and respiration returns to a low (state 4) level. The amount of oxygen uptake in state 3 is proportional to the amounts of ADP which are added to the incubation medium (Fig. 5).

The procedure by which such data can be quantitated is shown in Fig. 6, which gives actual values obtained with a mitochondrial preparation from muscle of the thorax and leg femurs of the American cockroach (*Periplaneta americana*) (Matsuda and Fukami, 1972).

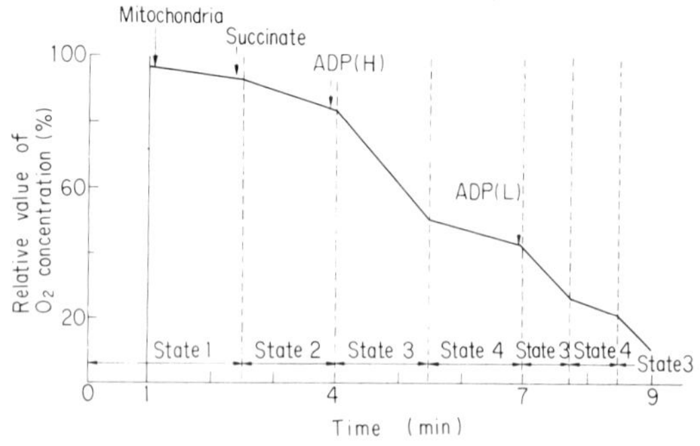

Fig. 5. Steady-state respiration of succinate in rat liver mitochondria in the absence (state 4) and presence of (state 3) high (H) and low (L) concentrations of ADP added at the points indicated by the arrows. Modified from Hagihara (1974).

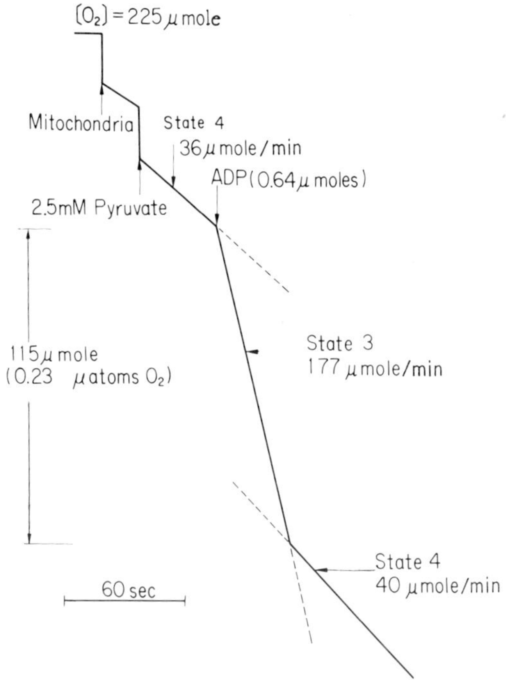

Fig. 6. Measurement of ADP : O ratio and respiratory control in a typical mitochondrial respiration experiment (Matsuda and Fukami, 1972) with mitochondria isolated from the thoracic and femur muscles of male American cockroach (*Periplaneta americana*). See text for explanation.

Prior to the experiment, the recorder is calibrated and chart divisions per μatom of O_2 are calculated from the volume of the incubation medium and the solubility of oxygen under the temperature conditions used. Since it is usually assumed that electron transfer occurs in two-electron steps, and the complete reduction of oxygen to water involves a four-electron change, it is usual to express the rates and amounts of oxygen uptake in μatoms of O_2 (μmoles $O_2 \times 2$).

In the example in Fig. 6, the addition of 0.64 μmole of ADP increased the rate of respiration with pyruvate from 40 to 177 μmoles of O_2/min. Since the respiratory control ratio is defined as the ratio of the respiration in the presence of the acceptor (ADP) (state 3) to that after the added ADP has been converted to ATP (state 4) (Chance and Williams, 1955*a*), the ratio in the example is 177/40 or 4.4.

Since O_2 uptake during the rapid phase (state 3) of the reaction is proportional to the amount of ADP added and the conversion of ADP to ATP is essentially quantitative, the efficiency of oxidative phosphorylation can be estimated (Chance and Williams, 1955*a*). In the example in Fig. 6, addition of 0.64 μmole of ADP results in an accelerated rate of respiration which consumes 0.23 μ atom of O_2. The ADP : O ratio or P : O ratio (P : $2e^-$) is 0.64/0.23 or 2.78. In general, mitochondria with high respiratory control ratios also exhibit high P : O ratios.

P:O ratios can also be determined by measuring the incorporation of ^{32}P-labeled orthophosphate into ATP by means of the hexanal column chromatography technique (Hagihara and Lardy, 1960; Hagihara, 1965).

Comparative studies have established that oxidative phosphorylation in isolated insect muscle mitochondria is usually less efficient than in those from vertebrate liver or heart. The results in Table II show that although respiratory activity in American cockroach muscle mitochondria is high with several substrates, a respiratory control ratio of greater than 1.0 was obtained only with pyruvate (R.C. 4.9, P:O 2.9). In similar experiments with rat liver mitochondria, R.C. values greater than 1.0 were measured with pyruvate, succinate, and α-ketoglutarate (Matsuda and Fukami, 1972); Ilivicky *et al.* (1967) have reported similar results.

Respiratory control ratios obtained with several other insect mitochondrial preparations during the oxidation of pyruvate or α-glycerophosphate have been reported by Sacktor (1974), and in the case of the blowfly (*Phormia regina*) P:O values during pyruvate oxidation approach the theoretical limit of 3 (Childress and Sacktor, 1966; Bulos *et al.*, 1972).

The yield of oxidative phosphorylation in insect flight muscle mitochondria can often be increased by the addition of bovine serum albumin (Sacktor, 1953, 1954; Lewis and Slater, 1954) or other protein (Rees, 1954) to the incubation medium. This effect is most obvious with mitochondria from the thoracic muscle of the housefly, in which no phosphorylation is observed until bovine serum albumin is added. Serum albumin has no effect on oxidative phosphorylation in fresh mammalian liver mitochondria but has a pronounced effect in protecting against the decreased phosphorylation normally associated

Table II. Respiratory Control and ADP:O Ratios with Several Substrates[a]

	American cockroach muscle				Rat liver			
	Respiratory activity[b]				Respiratory activity[b]			
	State 3	State 4	R.C.	ADP:O	State 3	State 4	R.C.	ADP:O
Pyruvate	227	46	4.9	2.9	15	13	1.2	1.8
Citrate	8	8	1.0		5	5	1.0	
α-Ketoglutarate	77	77	1.0		19	19	1.3	1.4
Succinate	195	195	1.0		69	27	2.6	1.9
Fumarate	21	21	1.0					
Malate	19	19	1.0					
Glutamate	35	35	1.0		18	18	1.0	
α-Glycerophosphate	215	215	1.0		13	13	1.0	

[a]From Matsuda and Fukami (1972).
[b]mμ atoms O/mg protein/min. Incubation conditions were as shown in Fig. 6.

with the "aging" of mitochondria (Pullman and Racker, 1956). This has been attributed to the ability of albumin to bind an uncoupling substance, a hemoprotein called "mitochrome," which is liberated from mitochondria during the aging process. The active component of mitochrome has been shown to be extractable with organic solvents and has been identified as a mixture of fatty acids (Hülsmann *et al.*, 1960). Of several saturated fatty acids evaluated, myristic acid exhibited the greatest activity. Activity diminished progressively in compounds with longer or shorter chains and was enhanced by introduction of a *cis* (but not *trans*) unsaturated bond (Pressman and Lardy, 1956).

Van den Bergh and Slater (1962) confirmed the observation of Gregg *et al.* (1960) with mitochondria isolated from the thoraces of houseflies that the oxidation of α-glycerophosphate and pyruvate was accompanied by high P : O ratios and did not require the addition of serum albumin to the reaction medium. The low P : O ratios found with all other substrates were substantially increased in the presence of serum albumin and with the exception of that for α-oxaloglutarate approached those reported for mammalian preparations. Remmert and Lehninger (1959) have described a "releasing" factor, derived from mitochondria, which is capable of transforming mitochondrial respiration from the "tightly coupled" for the "loosely coupled" state. Many authors have shown the presence in both animal (Lehninger and Remmert, 1959) and insect (Wojtczak and Wojtczak, 1959) mitochondria of an endogenous uncoupling agent or agents. The actions of both the releasing and uncoupling factor(s) are prevented by addition of serum albumin to the incubation medium, and with some insect preparations this increases oxygen and/or phosphorus uptake (Lewis and Slater, 1954; Wojtczak and Wojtczak, 1959). Sacktor *et al.* (1958) have reported an absolute requirement for a high level of serum albumin for phosphorylation in flight muscle mitochondria of the housefly and suggest that this may reflect the presence of uncoupling factors in their preparation. It was concluded that the effect of serum albumin in the insect preparation is very similar to that observed in "aged" mammalian mitochondria. However, in contrast to mammalian preparations, even freshly prepared flight muscle mitochondria show a requirement for bovine serum albumin. This suggests that insect mitochondria release the phosphorylation inhibitor very quickly or that they are extremely sensitive to the uncoupler.

Human and bovine serum and plasma, and to a smaller degree β-lactoglobulin, stimulate oxidative phosphorylation and the ATP–P_i exchange reaction in mitochondria from larvae of the wax moth (*Galleria mellonella*), the cause being removal by these proteins of an uncoupling agent. It has been demonstrated that like that from mammals the fraction showing uncoupling activity contains fatty acids, and palmitic, stearic, oleic, linoleic, and linolenic acids have been identified by paper chromatography (Wojtczak and Wojtczak, 1959).

Van den Bergh and Slater (1960) also showed that addition to the reaction medium of the soluble fraction from housefly thoracic muscle increased the respiration rate ten- to twentyfold, approaching that measured *in vivo*. Under these conditions, respiratory control with ADP was established and high P : O ratios could be obtained in the absence of serum albumin. It was established that the supernatant effect was due to the formation in the soluble fraction of a rapidly oxidizable substrate and that the low respiration rates, the lack of respiratory control, and the low P : O ratios observed with most substrates were caused by permeability barriers which prevented their entry into mitochondria.

Slater (1960) offered three possible explanations to account for the discrepancy between the high oxygen consumption of the housefly *in vivo* and the low rate of oxygen uptake with isolated mitochondria (sarcosomes) when oxidizing a variety TCA cycle intermediates: (a) the respiratory enzyme systems of the isolated mitochondria are damaged during isolation; (b) the isolated mitochondria are not readily permeable to intermediates of the TCA cycle (Lewis and Slater, 1954); (c) a large part of the respiration in insect mitochondria *in vivo* occurs by a pathway different from that of the TCA cycle. The existence of a permeability barrier is clearly demonstrated by the fact that sonication of mitochondria often causes over a tenfold increase in the respiratory rate with TCA cycle intermediates. To date, no physiological function for this low permeability has been established. Since the TCA cycle intermediates play an important catalytic role in pyruvate oxidation (a decrease in their concentration leads to a decreased rate of pyruvate oxidation), it has been suggested that the true physiological function of the permeability barrier in the mitochondria may be the prevention of the outward leakage of TCA cycle intermediates from the mitochondria.

3. Interaction of Insecticides with the Respiratory System

3.1. Insecticides as Respiratory Inhibitors

Initial information indicating whether an insecticide is acting primarily as an inhibitor of the respiratory enzyme system can be obtained by investigating factors such as the rate of heartbeat, oxygen consumption, and the electrical properties of muscle and nerve in poisoned insects. Rotenone will serve as a typical example.

Early studies on the effects of rotenone on insects established that it depressed oxygen consumption and decreased the rate of heartbeat (Orser and Brown, 1951). The primary biochemical lesion causing these effects was not known, although Tischler (1936) had suggested that the primary action of rotenone was associated with inhibition of the insect's ability to utilize oxygen. It was reasonable to assume that if this were correct it would occur by inhibition of the respiratory enzyme system of the insect and that such inhibition would cause a disturbance in the physiological function of one or more organs (Fukami, 1954). The results of subsequent experiments (Fukami, 1954, 1956) established that the insect was paralyzed and that its rate of respiration (oxygen uptake) decreased slowly in a manner which paralleled the intensity of the symptoms. Electrophysiological experiments with nerve and muscle from poisoned insects revealed an initial decrease in spontaneous discharge in the central nervous system followed by a conduction block; the ability of muscle to contract was progressively blocked after rotenone treatment (Fukami, 1954, 1956). These effects are similar to those observed with various respiratory inhibitors such as potassium cyanide and antimycin A and the insecticide Lethane 60, all of which are known to inhibit the respiratory enzymes. Failure to observe any indication of enhanced nervous activity in the early stages of rotenone poisoning suggested little or no direct action on the nervous system, and it was assumed that the primary action of rotenone was the inhibition of cell respiration in nerve and muscle tissue (Fukami, 1955, 1956).

The poisoning symptoms of insecticides such as pyrethrin, nicotine, DDT, and the organophosphorus compounds and carbamates differ greatly from those of rotenone. Oxygen consumption and rate of heartbeat increase substantially in insects poisoned by these insecticides, and the occurrence of vigorous convulsions during the initial stages of intoxication clearly suggests that their primary action is on the nervous system. There are no reports indicating that pyrethrin, nicotine, or DDT acts primarily as an inhibitor of the respiratory enzyme system either *in vivo* or *in vitro* (Fukami, 1956).

In general, inhibitors of the respiratory enzyme system can be classified into three types: (a) inhibitors of the respiratory chain (electron transport system) (type I), (b) inhibitors of oxidative phosphorylation (type II), and (c) inhibitors of the energy transfer system (phosphorylation) (type III). The inhibitory interactions of insecticides representing each of these types of materials are described in the following section.

3.2. In Vitro Mode of Action of Insecticides as Respiratory Enzyme Inhibitors

3.2.1. Inhibitors of the Respiratory Chain

Inhibitors of the respiratory chain may be classified into four major groups according to the site at which they exert their effect (Fig. 2). These are (a) inhibitors acting between NAD^+ and coenzyme Q, (b) the inhibitors of succinate oxidation, (c) inhibitors acting between cytochromes *b* and c_1, and (d) cytochrome *c* oxidase inhibitors.

These inhibitors completely block mitochondrial respiration, and, since oxidative phosphorylation is coupled with the respiratory chain, ATP is not produced in their presence. In general, inhibitors of the respiratory chain do not act directly on the phosphorylation process. The site at which inhibition occurs can be studied with the oxygen electrode, as shown in Fig. 7, which indicates the effect of antimycin A on succinate oxidation before (a) and after (b) the addition of ADP. It can clearly be seen that antimycin A causes an immediate and complete block of respiration in both cases. When cytochrome *c*

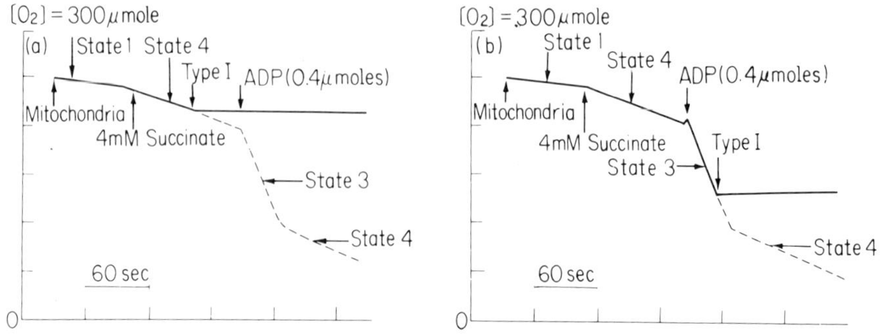

Fig. 7. Effect of antimycin A (a type I inhibitor) on state 4 (a) and state 3 (b) respiration of succinate (Hagihara, 1974). The broken line indicates the rate of respiration in the absence of the inhibitor. See text for explanation.

is reduced with ascorbic acid in a system containing a respiratory inhibitor acting between cytochromes *b* and c_l, the respiratory block can be circumvented and the production of ATP at this point (site 2) becomes completely coupled again. If appropriate respiratory chain donors and acceptors are selected, such experiments are possible with other kinds of inhibitors and can contribute significantly to our understanding of the sites at which phosphorylation occurs as well as the site of action of various inhibitors. If sufficiently high concentrations of respiratory inhibitors are employed, respiration at states 1, 2, 3, and 4 can be completely inhibited (Fig. 7).

a. Inhibitors Acting Between NAD^+ and Coenzyme Q. Amytal, progesterone, and methyleneglycol have been described as respiratory chain inhibitors acting between NAD^+ and coenzyme Q, but it is now known that they also inhibit the energy transfer system at site 1. Inhibitory substances which specifically block the respiratory chain at this site are the botanical insecticide rotenone and an insecticidal antibiotic piericidin A.

As previously discussed, the symptoms of rotenone poisoning (blockage of nervous conduction and muscle contraction) are similar to those caused by other types of respiratory inhibitors. It was initially concluded that rotenone inhibited respiration in nerve and muscle tissue and that, at least in part, its action was due to the inhibition of L-glutamate oxidation (Fukami, 1956; Fukami and Tomizawa, 1956). Structure–activity studies with a series of rotenone derivatives showed a good correlation among the level of L-glutamate oxidation, insecticidal activity, and the blockage of nerve conduction (Fukami *et al.*, 1959). In addition, while studying the inhibition of oxygen uptake of gill filament by rotenone, Lindahl and Öberg (1961) indicated using rat liver mitochondria that rotenone blocks a link in the respiratory chain, which is situated at the diaphorase level.

Rotenone exhibits a considerable degree of selective toxicity, and its high toxicity to insects and low toxicity to mammals are difficult to rationalize simply on the basis of its inhibitory effects on L-glutamate oxidation. Studies to determine the effect of rotenone on respiration and oxidative phosphorylation in cockroach (*Periplaneta americana*) muscle mitochondria showed that at concentrations of 10^{-5} to 10^{-7} M it caused a remarkable inhibition of oxygen uptake in the presence of α-ketoglutarate, L-glutamate, malate, fumarate, oxaloacetate, and citrate but did not inhibit the oxidation of succinate, α-glycerophosphate, and *p*-phenylendiamine. These results clearly indicated that rotenone was acting as a specific inhibitor of a pyridine nucleotide linked oxidase (Fukami, 1961). Since the inhibition of L-glutamate oxidation by rotenone was not reversed by addition of large amounts of NAD, it appeared that inhibition of the NAD-linked enzyme system was not due to removal of NAD from the mitochondrial structure but resulted from a direct inhibitory interaction with the enzyme protein itself. NADH oxidase activity of muscle mitochondria is increased by addition of cytochrome *c* and is decreased in the presence of rotenone or antimycin A; dehydrorotenone, which is without insecticidal activity, has no effect on NADH oxidase activity. Like rotenone, antimycin A inhibits the oxidation of α-ketoglutarate and L-glutamate, but in contrast to rotenone it also blocks α-glycerophosphate and succinate oxidation. This indicates that whereas antimycin A inhibits both NAD-linked and NAD-unlinked respiration, rotenone blocks only the NAD-linked system.

Further information on the inhibitory effects of rotenone and antimycine A was obtained from difference spectra of mitochondrial preparations using cockroach (*Periplaneta americana*) muscle mitochondria containing the inhibitors and L-glutamate, α-glycerophosphate, and NADH under various conditions. Difference spectra of mitochondrial suspensions in the presence of substrate were investigated, following treatment with antimycin A or rotenone. As is shown in Fig. 8a, the characteristic absorption bands of cytochrome *b* appeared at 430 and 562 nm in suspensions treated with antimycin A. This corresponds to the difference spectra obtained with mitochondria of ascites cells and with flight muscle mitochondria (sarcosomes) of the housefly treated with antimycin A (Chance and Hess, 1959; Chance and Sacktor, 1958). Under similar conditions, no absorption bands are observed following the addition of rotenone (Fig. 8b).

The effect of antimycin A on the absorption spectrum of the reduced cytochromes of mitochondrial suspensions in the presence of α-glycerophosphate was examined by the opal glass transmission method, and the results are shown in Fig. 9. The absorption band of reduced cytochrome *c* cannot be observed immediately but appears gradually with time. This obser-vaton agrees with that of Chance and Hess (1959), who reported that the absorption band is observed following treatment with antimycin A, under anaerobic conditions. It appears that some impurity in the preparation activates the oxidation of endogenous substrates and causes the reduction of cytochrome *c* since no spectra are observed in preparations thoroughly washed to remove all endogenous substrate. Difference spectra of these purified mitochondrial preparations in the presence of L-glutamate and rotenone

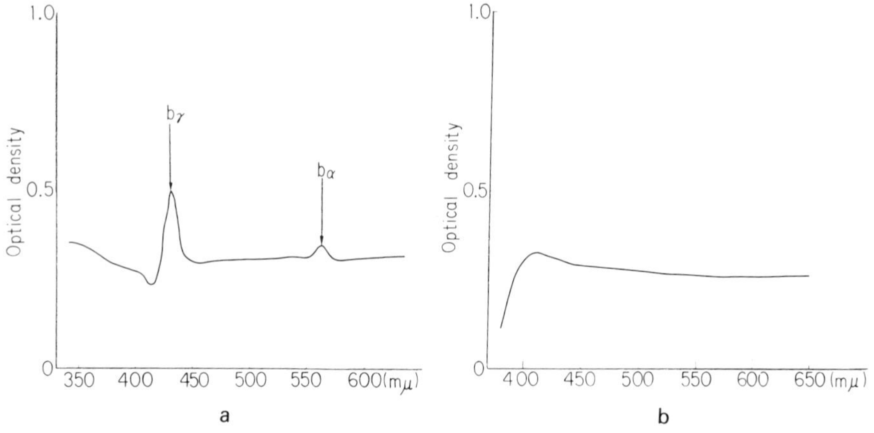

Fig. 8. Difference spectra of mitochondria in the presence of (a) antimycin A (1.6 μg) and (b) rotenone (5×10^{-5} M) (Fukami, 1961). In the test cuvette was 2.7 ml of reaction mixture containing 1 ml of mitochondrial suspension (1.68 mg N/ml). The control cuvette was identical except for the addition of antimycin A or rotenone, and in the latter case both cuvettes contained 0.3 ml of a 0.5 M NADH solution. The reaction mixture was as described in Fig. 3; temperature 17°C.

show the characteristic absorption bands of cytochromes *a*, a_3, *b*, and *c* between 630 and 330 nm (Fukami, 1961) (Fig. 10).

At concentrations of 10^{-7} to 10^{-6} M, rotenone depresses both oxygen uptake and the formation of ATP, although it is usually considered that the effects of rotenone on oxidative phosphorylation occur indirectly through its inhibitory action in the respiratory chain.

From the results described above, it was concluded that the inhibitory effects of rotenone are exerted at some point on the respiratory chain between NADH and cytochrome *b* (Fukami, 1961; Lindahl and Öberg, 1961). On the addition of NADH to submitochondrial particles in which NADH oxidation is

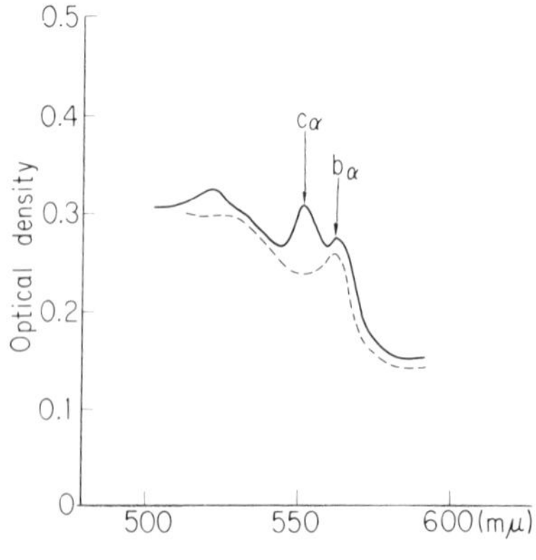

Fig. 9. Effect of antimycin A on the absorption spectrum of a reduced mitochondrial suspension in the presence of α-glycerophosphate (opal glass transmission method, Fukami, 1961). In the cuvette was 2.7 ml of reaction mixture containing 1 ml of mitochondrial suspension (1.52 mg N/ml), 0.3 ml of 0.5 M α-glycerophosphate, and 0.3 ml of antimycin A (1–6 μg). Temperature 17°C. ——, Control; – – –, antimycin A treated. The reaction mixture was similar to that described in Fig. 3.

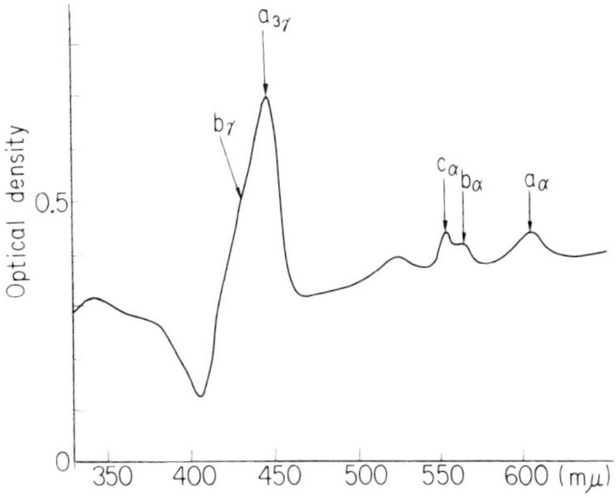

Fig. 10. Difference spectrum of rotenone in purified mitochondrial suspensions containing L-glutamate (Fukami, 1961). In the test cuvette was 2.7 ml of reaction mixture containing 1 ml of mitochondrial suspension (1.52 mg N/ml), 0.3 ml of 0.5 M L-glutamate, and rotenone (5×10^{-5} M). The control cuvette was identical except for the addition of rotenone. Temperature 19°C. The reaction mixture was similar to that described in Fig. 3.

blocked by rotenone, the $g = 1.94$ signal associated with NADH dehydrogenase appears to be essentially the same as in untreated preparations (Gutman *et al.*, 1970). However, appearance of the NADH-induced ion signal of succinic dehydrogenase and of cytochromes *b* and c_1 is blocked in the presence of rotenone. It can therefore be deduced that rotenone blocks NADH oxidation on the O_2 side of the nonheme iron of NADH dehydrogenase (Gutman *et al.*, 1970).

Other studies have been made on the binding of [^{14}C] rotenone to submitochondrial particles (electron transport particles or ETPs) prepared from beef heart mitochondria, although a major problem in such studies is to differentiate between binding at the enzymatically active site (specific binding) and nonspecific binding at other loci (Horgan *et al.*, 1968*a*) (Fig. 11). Rotenone binds vigorously not only to the specific binding site directly associated with its inhibitory action on respiration but also to nonspecific site(s) unrelated to its

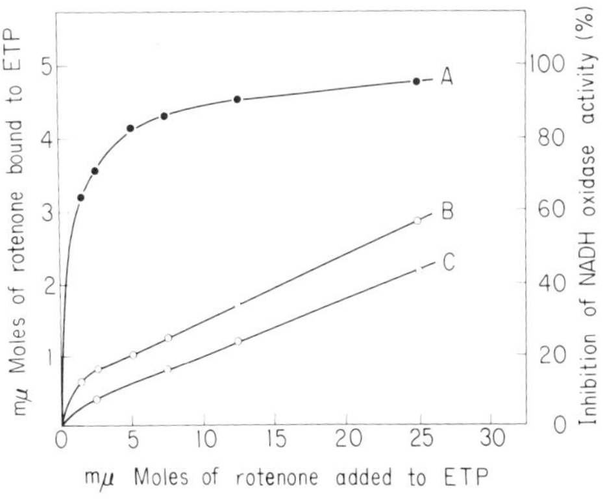

Fig. 11. Inhibitory action of rotenone toward NADH oxidase activity (curve A) and the binding of [^{14}C]rotenone to submitochondrial particles (ETPs) (25 mg) before (curve B) and after (curve C) their preincubation with unlabeled rotenone. From Horgan *et al.* (1968*a*).

mode of action. Nonspecific binding can be reduced by treatment of the particles with rotenone in the presence of bovine serum albumin. The specific binding sites of inhibited particles can be extracted into an organic solvent such as anhydrous acetone but are not affected by bovine serum albumin (Horgan *et al.*, 1968*a*). In this way, rotenone can be recovered intact from its specific binding sites in inhibited particles.

Although specific rotenone binding to submitochondrial particles occurs in an approximately 1:1 molar ratio with the NADH dehydrogenase content of mitochondria, the extraction of phosphorylating or nonphosphorylating particles labeled with [^{14}C]rotenone results in almost complete separation of [^{14}C]rotenone from this enzyme (Horgan *et al.*, 1968*a*,*b*). In addition, rotenone displaces a portion of the endogenous ubiquinone involved in electron transport from NADH but not from succinate (Gutman *et al.*, 1971). It therefore appears that the inhibition of NADH-linked respiration by rotenone occurs through specific binding to a component (possibly a lipoprotein) of the respiratory chain between NADH dehydrogenase and coenzyme Q and not to NADH dehydrogenase itself (Fig. 2).

In the course of a screening program to discover new types of insecticides from microorganisms, two materials, piericidins A and B, were isolated from *Streptomyces mobaraensis* and their chemical structures elucidated (Tamura *et al.*, 1963). These compounds were found to have a broad spectrum of insecticidal and miticidal activity, and to the peach aphid (*Myzus persicae*) and spider mite (*Tetranychus telarius*) are as toxic as rotenone and other miticides (Takahashi *et al.*, 1968).

Hall *et al.* (1966) found that piericidin A was a powerful inhibitor of the mitochondrial respiratory chain in beef heart mitochondria and suggested that its insecticidal activity might result directly from this activity. The inhibitory activity of piericidin A toward the aerobic oxidation of NADH-linked substrates was similar to that of rotenone. In contrast to rotenone, however, piericidin A also inhibited succinate oxidation at high concentrations (Jeng *et al.*, 1968). The strong structural resemblance of piericidin A to coenzyme Q (Fig. 12) and the fact that the inhibition of succinate oxidation by piericidin A could be reversed by addition of coenzyme Q strongly suggested that piericidin

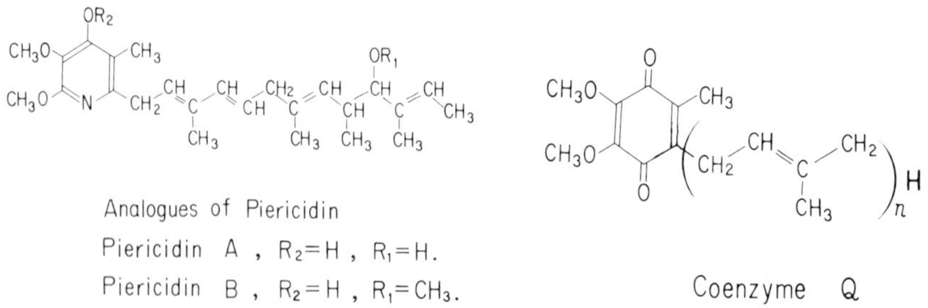

Fig. 12. Chemical structures of analogues of piericidin and coenzyme Q.

A was acting as a competitive inhibitor of coenzyme Q. This suggestion was strengthened by the finding that piericidin A irreversibly inhibited NADH oxidation at a concentration of 3.6×10^{-5} μmoles/mg protein and inhibited succinate oxidation reversibly only at extremely high concentrations (Jeng *et al.*, 1968).

Studies on muscle mitochondria of the cockroach (*Periplaneta americana*) have further established that piericidins A and B inhibit NADH oxidation at approximately the same concentrations as rotenone, and their similar patterns of activity strongly suggest that they are acting at the same respiratory site as rotenone (Mitsui *et al.*, 1969). That these compounds do indeed bind to and compete for a common site is supported by the fact that preincubation of ETP with unlabeled rotenoids, amytal, and piericidin A abolishes the subsequent binding of [^{14}C]rotenone. The binding affinity of piericidin A to ETP appears to be higher than that of the other compounds since neither rotenone nor amytal can displace [^{14}C] piericidin A from the site, and unlike the situation with rotenone the inhibition of NADH oxidation by piericidin A is not reversed by washing with BSA. Piericidin A readily displaces bound rotenone from the site, and its higher binding affinity possibly explains its higher oral toxicity to the mouse (rotenone LD_{50} 135 mg/kg, piericidin A LD_{50} 20 mg/kg). Piericidin A binding does not appear to involve covalent binding, since the radiocarbon bound to the site is released by protein denaturants, phospholipase A, and proteolytic enzymes (Horgan *et al.*, 1968*b*).

In conclusion, it appears that piericidin, rotenone, and amytal all exert their primary inhibitory action at a common respiratory site on the oxygen side of NADH dehydrogenase and not between NADH and the flavoprotein (Fig. 2). Although much of the work to elucidate rotenone's mode of action has been carried out with vertebrate mitochondria, it is likely that the results are fully applicable to its action on insects.

b. Inhibitors of Succinate Oxidation. Biochemical studies of the enzymes involved in succinate oxidation, isolated and purified from heart muscle, show that the active sites of the enzymes contain not only FMN but also from two to four iron atoms, two SH groups, and a disulfide bond (Keilin and King, 1960). Consequently, all substances such as chelating agents (8-hydroxyquinoline, *o*-phenanthroline, α,α-dipyridyl) and SH reagents which react with these groups inhibit one or more of these enzymes. DDT at 10^{-3} M has also been reported to inhibit succinate oxidation in the yellow mealworm (*Tenebrio molitor*) and housefly, but as will be discussed later this is not considered to be a primary lesion of DDT (Anderson *et al.*, 1954; Barsa and Ludwig, 1959).

c. Inhibitors Acting Between Cytochromes b and c_1. Many chemicals, particularly antibiotics and narcotics, have long been known to inhibit the respiratory chain between cytochromes *b* and c_1. As a result, this particular site on the chain has been considered to possess certain unusual properties, although it is not yet known whether the inhibitors bind directly to specific components of the chain or whether the site is functionally unstable and therefore particularly susceptible to the action of inhibitors.

One of the best known and most frequently used compounds interacting at this site is the antibiotic antimycin A, which is secreted by *Streptomyces griseus.* This compound is one of several exhibiting high inhibitory specificity at the site between cytochromes *b* and c_1 and is active at a 1:1 molar ratio with the cytochrome; at higher concentrations, it also inhibits oxidative phosphorylation. Antimycin A has insecticidal activity to the housefly and almond moth (*Cadra*), but it also exhibits high toxicity to fish (Hosotsuji1956) and has not been of any practical importance in agriculture.

Venom of the Australian bulldog ant (*Myrecia gulosa*) inhibits oxygen consumption in housefly flight muscle mitochondria in the presence of either succinate or α-glycerophosphate as substrate. The site of inhibition would appear, therefore, to be at some point on the respiratory chain common to both succinate and α-glycerophosphate, and may be between cytochrome *b* and cytochrome c_1. However, the possibility that the active component of the venom is a general flavoprotein poison has not been excluded (Lilian and Ilse, 1970).

d. Inhibitors of Cytochrome c Oxidase. Since cytochrome *c* oxidase is the terminal oxidase of the respiratory chain, its inhibition stops the flow of electrons down the respiratory chain and each component of the chain is in the reduced state under aerobic conditions. The actual site at which inhibition of this enzyme occurs remains largely unknown, although recently the enzyme has been isolated and its mechanism has been described at the molecular level. Most of the inhibitors of cytochrome *c* oxidase are materials which combine chemically with the heme moiety of the enzyme. Thus CO, NO, N_3^-, and CN^- are known to combine with the sixth ligand of the heme iron at which molecular oxygen is bound, and CN^- and S can interact with the formyl groups of the side chains of heme moiety (Fig. 2).

The insecticide malathion has been reported to inhibit succinate oxidation *in vitro* and cytochrome *c* oxidase *in vivo* during advanced stages of poisoning (Brown and Brown, 1956; O'Brien, 1956; McAllan and Brown, 1960); at high concentration *in vitro* (but not *in vivo*), parathion also inhibits cytochrome *c* oxidase (Brown and Brown, 1956). DDT, pyrethrin, and parathion have no effect on respiration in cockroach muscle either *in vivo* or *in vitro* (Fukami, 1956), and as will be discussed later the primary action of these insecticides is not associated with the inhibition of cytochrome *c* oxidase. Hydrogen cyanide has been used for many years as an insect fumigant, and several compounds which are converted *in vivo* to cyanide also have high insecticidal activity. In each case, the mode of action appears to be an interruption of the respiratory chain at cytochrome a_3 (Dixon and Webb, 1964).

Recent studies have established that the glutathione-*S*-transferases in mouse liver and houseflies are able to liberate hydrogen cyanide from a variety of organothiocyanates, including Lethane 384, and it is probable that this is directly associated with the insecticidal action of these compounds. The enzyme from mouse liver catalyzes the attack by glutathione at the thiocyanate sulfur and results in the liberation of hydrogen cyanide, oxidized glutathione, and the mercaptan moiety of the organothiocyanate. Since cyanide levels in the brain appear to be more important for toxicity than those in the liver, the degree of inhibition of nerve cytochrome *c* oxidase may be the ultimate cause of death in animals poisoned with organothiocyanates (Ohkawa *et al.*, 1972).

The resistance of California red scale (*Aonidiella aurantii*) to hydrocyanic acid is a matter of practical as well as theoretical interest. The inhibitory effect of cyanide on respiration and oxygen consumption is brought about by the inactivation of metal-containing respiratory enzymes through ligand interaction between the cyanide and the metal. The most usual site at which this occurs is the iron of the metalloporphyrin of cytochrome *c* oxidase. It is possible, therefore, that the cyanide-resistant respiration in resistant scale insects is associated with the activity of some metal-free respiratory enzyme such as an autooxidizable flavoprotein which would not be impaired by cyanide at moderate concentrations (Yust and Shelden, 1952).

3.2.2. Uncoupling Agents

Under normal conditions, the mitochondrial respiratory chain does not function independently but is closely linked to and controlled by the reactions associated with oxidative phosphorylation. Thus, as has been described, the addition of a respiratory substrate to an intact ("coupled") mitochondrial preparation increases the rate of respiration (state 3) until the supply of ADP and P_i is exhausted through conversion to ATP (Table I and Fig. 2). In the presence of an agent such as 2,4-dinitrophenol, the close coupling between the respiratory chain and phosphorylation is broken, respiratory control is lost, and electron transport along the chain occurs at a maximal rate without formation of ATP (Fig. 13). Compounds such as 2,4-dinitrophenol are therefore called "uncouplers."

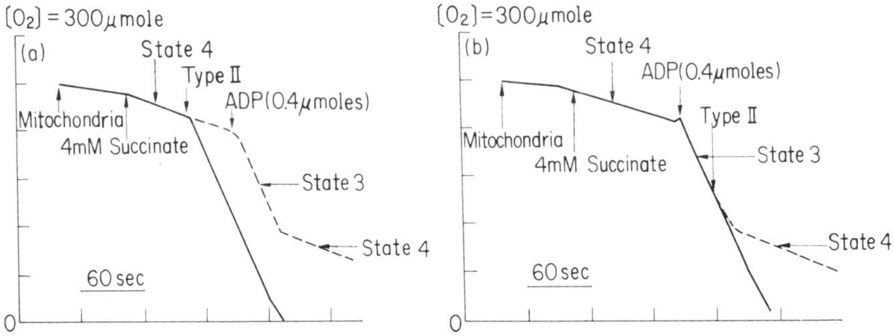

Fig. 13. Effects of the uncoupling agent 2,4-dinitrophenol (a type II inhibitor) on the state 4 (a) and state 3 (b) respiration of succinate (Hagihara, 1974). The broken line indicates the rate of respiration in the absence of the inhibitor. See text for explanation.

One of the remarkable characteristics of uncoupling agents is their ability to accelerate the decomposition of ATP by enhancing ATPase activity. In order to explain the mechanism by which energy is passed from the respiratory chain to the high-energy phosphate bond of ATP, two intermediate coupling factors (X and Y) have been proposed. Factor X is considered to trap the redox energy released from the respiratory chain and to react with Y to form a carrier $X \sim Y$. This in turn reacts with P_i to produce another intermediate $Y \sim P$, which passes on the high-energy phosphate to ADP to yield ATP (Fig. 2). It is probable, therefore, that uncoupling agents such as 2,4-dinitrophenol act by promoting the rapid hydrolysis of one or more of these high-energy phosphorylation intermediates and thereby preventing the formation of ATP.

In general, uncoupling agents contain both anionic and lipophilic groups in their structure. They include a variety of nitrophenol analogues, halogenated compounds, antibiotics, unsaturated fatty acids, and other compounds.

Dinitrophenols such as the miticides DNOC (4,6-dinitro-*o*-cresol) and DNCHP (2,4-dinitro-6-cyclohexylphenol) are well-established uncouplers of oxidative phosphorylation and their activity is enhanced by a low pK_a and high degree of lipid solubility (Parker, 1958). *In vitro*, uncoupling activity also depends on the affinity of the compound for protein, the pH of the reaction mixture, and the species and organ from which the mitochondria are obtained (Lewis and Slater, 1954). Serum albumin inhibits the uncoupling action of 2,4-dinitrophenol, possibly by competitive binding.

Since many organophosphorus insecticides have nitro-containing aromatic moieties incorporated into their structures and since these are released as phenols during metabolism, it was of interest to investigate the possible uncoupling action of some of these compounds. Sumithion and methyl parathion, which are metabolized to 4-nitro-*m*-cresol and 4-nitrophenol, respectively, inhibited state 4 respiration in cockroach muscle mitochondria at high concentration but had no effect in rat liver mitochondria. These insecticides also inhibited oxidative phosphorylation at high concentration (Matsuda and Fukami, 1972). The compounds 2,4-dinitrophenol, 4-nitro-*m*-cresol and 3-nitro-*p*-cresol were toxic to the cockroach (*Periplaneta americana*) *in vivo*, stimulated state 4 respiration, and inhibited oxidative phosphorylation in cockroach muscle and rat liver mitochondria. The mode of action of 4-nitro-*m*-cresol as an uncoupler is assumed to be the same as that of 2,4-dinitrophenol *in vivo*. Mononitrophenols are generally less active uncouplers than the dinitrophenols, and 2-nitrophenol is inactive; 4-nitrophenol and 4-nitro-*m*-cresol are of approximately equal potency (Matsuda and Fukami, 1972).

The best known of the halogenated uncouplers is PCP (pentachlorophenol), which is widely used as a herbicide in rice fields. PCP is strongly bound to mitochondria (Weinbach, 1954; Weinbach *et al.*, 1963) and is an uncoupler of oxidative phosphorylation in mitochondria from rat liver and cockroach muscle at low concentrations (Matsuda and Fukami, 1972).

Salicylanilides have recently been shown to possess important fungicidal, bacteriostatic, cestocidal, and molluscicidal properties, and 5-chloro-2′-

chloro-4′-nitrosalicylanilide has been developed for the control of schistosomiasis. Whitehouse (1964) demonstrated that salicylanilide and *N*-salicyloyl anthranilate are uncouplers of oxidative phosphorylation in rat liver mitochondria, and investigations with housefly thoracic mitochondria have subsequently revealed that several salicylanilide derivatives are among the most potent uncoupling agents yet described.

Although the mode of action of these compounds is still obscure, they appear to possess certain common structural features. Inspection of molecular models reveals the presence of strong electron-withdrawing centers, such as NO_2, CN, or CF_3, located within a certain distance of a halogenated aryl ring. These features, combined with the enhanced inhibitory effect on oxidative phosphorylation and the related ATP–P_i exchange reaction resulting from incorporation into the molecule of additional bulky groups such as naphthyl, biphenyl, or *tert*-butyl, suggest that inhibition results from binding and interaction at a specific active site on the enzyme surface. The similar inhibitory potencies of the salicylanilides toward the ATP–P_i exchange reaction in both housefly and rat liver mitochondria also illustrate the essential similarity of oxidative phosphorylation in vertebrates and invertebrates (Williamson and Metcalf, 1967).

In order to establish the mode of action of substituted 2,4-dinitrophenols, 2-trifluoromethylbenzimidazoles, salicylanilides, carbonyl cyanide, phenylhydrazone, and other compounds, studies have been made to determine their selectivity as uncouplers *in vitro*, their effects *in vivo* on the mitochondrial enzymes, and the relationship between their uncoupling potency and toxicity. In some of the results reported (e.g., for 6-*sec*-butyl-1-isopropylcarbonate, 2,4-dinitro-6-(1-methylheptyl)phenylcrotonate, and 5,6-dichloro-1-phenylcarbonyl-2-trifluoromethylbenzimidazole), it is probable that the actual uncoupler or inhibitor involved is not the compound administered but a metabolite. By appropriate combinations of *in vivo* and *in vitro* studies, it is possible to determine whether the original compound or a metabolite is responsible for the uncoupling action or inhibition observed (Ilivicky and Casida, 1969; Casida, 1973). In general, mitochondria from mouse liver appear to be less sensitive to the action of uncoupling agents than those from mouse brain or from insect tissues (Ilivicky and Casida, 1969).

DDT and parathion exhibit no inhibitory activity toward oxidative phosphorylation in homogenates of locust flight muscle either *in vitro* or *in vivo* (Tomizawa and Fukami, 1956). However, the *in vivo* stimulation of respiration by DDT and lindane in the cockroach resembles that caused by 2,4-dinitrophenol and its analogues (Ela *et al.*, 1970), and certain similarities between the effects of these two compounds are worth recalling. DDT-treated adult houseflies show decreased levels of ATP and increased levels of inorganic phosphate (Winteringham *et al.*, 1960) and *in vitro* studies indicate that, like 2,4-dinitrophenol, DDT abolishes oxidative phosphorylation (Sacklin *et al.*, 1955; Gonda *et al.*, 1957; Gregg *et al.*, 1960) and the ATP–P_i exchange reaction in a particulate fraction from the mosquito (*Aedes aegypti*) (Avi-Dor and Gonda, 1959); unlike 2,4-dinitrophenol, however, DDT has no effect on Mg^{2+}-stimulated ATPase activity. Since, like oligomycin, DDT abolishes the 2,4-dinitrophenol-stimulated ATPase activity in insect mitochondria (Gregg *et al.*, 1960), it is possible that *in vivo* it may be acting as an inhibitor of oxidative phosphorylation rather than an uncoupler (Ela *et al.*, 1970). Whether these effects are of any significance in DDT toxicity is not known since it is generally believed that the primary lesion of this insecticide is in the nervous system. It is, however, believed that the inhibition of oxidative phosphorylation and stimulation of mitochondrial ATPase activity in mitochondria from cockroach (*Blatella germanica*) and rat liver contribute to the toxicity of the acaricide and larvicide chlordimeform (Abo-Khatwa and Hollingworth, 1972).

3.2.3. Inhibitors of the Energy Transfer System

Reports concerning inhibitors of this type have appeared only relatively recently. Hollunger (1955) reported that guanidine inhibits only respiration coupled with phosphorylation (state 3) and has no effect on the respiration of uncoupled mitochondrial particles. Subsequently, oligomycin was found to inhibit energy transfer (Lardy *et al.*, 1958) and is the best-known compound of this type. As shown in Fig. 14, oligomycin exhibits quite different properties from those of uncoupling agents and causes little or no inhibition of state 4 respiration. Respiration in the presence of oligomycin is not stimulated by addition of ADP but increases remarkably following the addition of 2,4-dinitrophenol. As shown in Fig. 14, the addition of oligomycin to mitochondria respiring in state 3 (after addition of ADP) causes an immediate inhibition of respiration to the level of state 4. The characteristics of compounds inhibiting energy transfer are that (a) they inhibit state 3 respiration and ATP synthesis, (b) they inhibit latent ATPase activity stimulated by uncoupling agents, (c) they

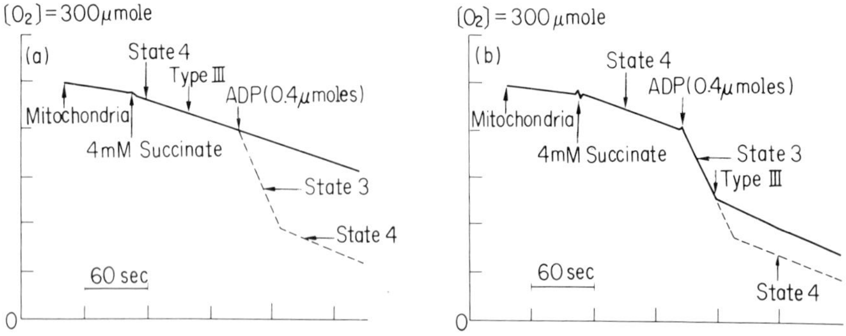

Fig. 14. Effects of oligomycin (a type III inhibitor) on the state 4 (a) and state 3 (b) respiration of succinate (Hagihara, 1974). The broken line indicates the rate of respiration in the absence of inhibitor. See text for explanation.

inhibit the ATP–P_i exchange reaction, and (d) they inhibit the ATP–ADP exchange reaction.

Of several types of compounds with a mode of action similar to that of oligomycin, the organotins are the only ones of any insecticidal importance. Organotins are becoming increasingly important as miticidal and biocidal agents, and of these the best known are the trialkyltins. These materials act directly on the energy conservation processes in mitochondria of both insects and mammals, optimal potency occurring with compounds containing a total of six to 18 carbon atoms (Piper and Casida, 1965). The affinity constants for the binding of trimethyltin and triethyltin to the membrane of rat liver mitochondria have been measured using ^{113}Sn (Aldridge and Street, 1970), and the ratio of these constants was found to be similar to the ratio of their inhibitory activities toward oxidative phosphorylation. A protein fraction capable of binding organotins has been isolated from guinea pig liver, and a histidine residue of this protein was linked with its binding properties (Rose and Lock, 1970).

3.2.4. Others

Organofluorine compounds such as fluoroacetate have long been recognized as insecticides and act by a process termed "lethal synthesis" (O'Brien, 1967). In this process, the fluoroacetate is metabolized to fluoroacetyl-CoA, which is subsequently condensed with oxaloacetate to form fluorocitrate. The fluorocitrate thus formed is a potent inhibitor of aconitase and the TCA cycle is blocked. The lesion is characterized by an accumulation of citrate and results in a depression of the energy acquired through the cycle.

Many organofluorine compounds which are metabolized to fluoroacetate (e.g., fluoroacetamide derivates and the miticide monofluorooxalic acid) are also potent respiratory inhibitors and exert their action through a similar mechanism (Drummel and Kun, 1969). It has been established that these compounds block citrate utilization by irreversible binding to the tricarboxylate carrier of intact rat mitochondria (Eanes *et al.*, 1972).

3.3. In Vivo Inhibition of Respiratory Enzymes by Insecticides

The interaction of insecticides with a variety of respiratory enzymes *in vitro* has been described. However, even when a particular insecticide is known to inhibit an enzyme system *in vitro*, it is still necessary to determine whether this inhibition constitutes its primary biochemical lesion in the living organism. Consequently, studies to determine whether or not *in vitro* inhibition data correlate with the observed *in vivo* toxicity become important in establishing the primary action of insecticides. Once the primary action of an insecticide has been established, the relationship between chemical structure and insecticidal activity of any group of compounds can be discussed not simply in terms of chemical structure vs. toxicity (which represents the net result of a series of complex toxicological interactions) but in terms of specific molecular parameters associated with its target site interaction.

There are several approaches which may be used to determine whether the ability of an insecticide to inhibit a specific enzyme *in vitro* is of physiological significance *in vivo* and directly associated with its toxic action. Studies to correlate the degree of enzyme inhibition *in vitro* with the presence or absence of specific chemical or physicochemical features of the insecticide provide valuable information on the nature and mechanism of the interaction. However, it is necessary to correlate the *in vitro* activities of a series of compounds with their *in vivo* toxicity before any firm conclusions can be made. This is often complicated by the involvement *in vivo* of factors such as penetration and metabolism, but considerable success has been achieved in some cases, as in the relationship between the toxicity and anticholinesterase activity of organophosphorus insecticides (Chapter 7).

Barsa and Ludwig (1959) reported that DDT at 10^{-3} M has a specific effect on cyanide-sensitive respiration, in that it inhibits only cytochrome *c* oxidase and succinate oxidation but has no effect on alcohol, glucose, lactic, α-glycerophosphate, or isocitric dehydrogenases in homogenates of the yellow mealworm (*Tenebrio molitor*) and housefly. However, the fact that DDE and other nontoxic DDT analogues also inhibit the succinate oxidation in housefly homogenates led Anderson *et al.* (1954) to conclude that this inhibition was not a prime factor in the mode of action of DDT. Fukami (1956) has also presented evidence to show that DDT is not an *in vivo* inhibitor of the respiratory enzyme system in the cockroach at the time of paralysis. O'Brien (1956, 1957) reached a similar conclusion concerning the inhibition of succinate and pyruvate oxidation by malathion. *In vitro* studies strongly suggest that the primary biochemical lesion of rotenone is located between NADH dehydrogenase and coenzyme Q and not between NADH and the flavoprotein. However, it is not yet known whether these *in vitro* results accurately reflect the primary action of rotenone in all living organisms. Rotenone has been shown to inhibit L-glutamate oxidation in guinea pig brain mitochondria (Fukami and Tomizawa, 1958). In the mammalian brain, L-glutamic acid is closely associated with nerve function and is present in a concentration higher than that found in other organs and tissues; it is the only amino acid to be oxidized in the brain during respiration. Although the importance of L-glutamic acid in cell respiration in the insect is not known, it is possible that it plays an important regulatory role. Consequently, the inhibition of L-glutamate oxidation by rotenone might be of major importance in the respiratory inhibition observed in insect nerve as well as in mammalian brain (Fukami *et al.*, 1970) and may be responsible for the block of nerve function which is a major cause of paralysis in insects treated with rotenone (Fukami, 1956; Fukami and Tomizawa, 1956; Yamasaki and Narahashi, 1957). Studies with rotenone and several derivatives have revealed a close correlation among the *in vitro* inhibition of L-glutamate oxidation, the blockage of nerve conduction, and insecticidal activity in various insect species and in no case was there high insecticidal activity in the absence of enzyme inhibition. This strongly suggests that the inhibition of L-glutamate oxidation plays a major role in killing the insect (Fukami and Tomizawa, 1958; Fukami *et al.*, 1970). Rotenone is metabolized to 8′-hydroxyrotenone, 6′,7′-dihydro-6′,7′-dihydroxyrotenone, and various other products when incubated with the microsomal mixed-function oxidase system of mammalian liver, fish liver, and insect tissue (Fukami *et al.*, 1969). The metabolic pathway for rotenone is the same in each of the animals studied, and the mixed-function oxidases initiate the detoxication reactions and limit the persistence of rotenone. These reactions appear to be detoxication mechanisms, as shown by direct assays with authentic rotenone metabolites (as well as unidentified metabolite mixtures) for potency as NADH oxidase inhibitors and as toxicants for mice (Fukami *et al.*, 1969). In view of the high structural specificity of rotenoids, with respect to both toxicity and NADH oxidase inhibition (Fukami *et al.*, 1959; Burgos and Redfearn, 1965), it is not surprising that the structural modifications which occur in the rotenone molecule during metabolism cause a reduction in biological activity. It appears fairly certain that the *in vivo* toxicity of rotenone can be fully explained on the basis of its inhibitory action on the respiratory chain.

Several attempts have been made to relate the toxicity of uncoupling agents such as the substituted 2,4-dinitrophenols, 2-trifluoromethylbenzimidazoles, salicylanilides, carbonylcyanides, and phenylhydrazones with their *in vitro* and *in vivo* effects on the mitochondrial enzymes (Williamson and Metcalf, 1967; Ilivicky and Casida, 1969; Muraoka and Terada, 1972; Corbett and Wright, 1970). There is considerable variation in the sensitivity of mitochondria from different species to the action of these uncouplers. Although it appears that the acute toxicity of the compounds results directly from their uncoupling action, particularly that occurring in the brain, this is not necessarily the case with regard to chronic or long-term exposures. It appears almost certain that the actual uncoupling agent is a metabolite rather than the parent compound.

3.4. Selective Toxicity

In contrast to most other groups of insecticides, very few studies have been made to determine the basis for the selective toxicity of insecticides acting as respiratory inhibitors. Oxygen consumption in cyanide-resistant strains of insects is much less susceptible to inhibition by HCN and H_2S than that of susceptible strains and the resistant strains are more susceptible to anoxia. It has been suggested that respiration in cyanide-resistant insects might be largely independent of cytochrome *c* oxidase and depend instead on an autooxidizable flavoprotein (Yust and Sheldon, 1952). Kurland and Schneiderman (1959) have shown that in diapausing saturniid pupae the insensitivity of the respiration to HCN and CO results from a great excess of cytochrome *c* oxidase over cytochrome *c*.

The basis for the selective toxicity exhibited by rotenone and piericidin A might be due to the presence in some species of a quinone which allows electrons to bypass the inhibited site on the respiratory chain. Conover and Ernster (1960) showed that in rat liver mitochondria the inhibition of NADH-linked substrate oxidation by rotenone and amytal could be reversed in the presence of vitamin K_3 and concluded that the latter bypassed the rotenone-sensitive site by way of enzyme DT-diaphorase. As shown in Table III, this effect was observed with both rotenone and piericidin A in rat liver mitochondria. In sharp contrast, respiration in cockroach (*Periplaneta americana*) muscle mitochondria was not restored by vitamin K_3 after inhibition with rotenone or piericidin A. Other quinones such as vitamin K_1, benzoquinone, CoQ_9, and CoQ_{10} have no effect in rat liver mitochondria. Similar results have been reported in mitochondria obtained from cockroach midgut and honeybee (*Apis mellifera*) muscle (Mitsui *et al.*, 1969). Recently, vitamin K_3 has been found to restore respiratory activity in carp liver mitochondria inhibited by rotenone (Fukami and Mitsui, unpublished).

An enzyme catalyzing the oxidation of NADH and NADPH and utilizing an artificial electron acceptor DCPIP (2,6-dichlorophenolindophenol) was first found in the soluble fraction of mammalian tissue by Ernster and Navazio (1958) and its properties were reported by Ernster *et al.* (1962). The enzyme occurs in mitochondria, and its relationship to mitochondrial electron transport and its similarity to the vitamin K reductase described by Martius and Märki

Table III. Bypass of Rotenone and Piericidin A Sensitive Sites of Respiratory Chain by Quinones[a]

Additions	Cockroach muscle respiration (O_2-μatoms)	Rat liver respiration (O_2-μatoms)	P:O
None	8.46	7.59	2.21
Piericidin A	0.65	1.62	0
Vitamin K_1	7.41	—	
Piericidin A, vitamin K_1	0.76	2.01	
Vitamin K_3	8.64	6.77	2.17
Piericidin A, vitamin K_3	1.07	8.20	1.39
1,4-Benzoquinone	7.58	8.05	
Piericidin A, 1,4-benzoquinone	0.68	1.21	
Coenzyme Q_9	8.35		
Piericidin A, coensyme Q_9	0.57	1.88	
Coenzyme Q_{10}	8.09		
Piericidin A, coenzyme Q_{10}	0.52	1.80	
Rotenone	1.08	1.00	0
Rotenone, vitamin K_3	1.21	7.50	1.40

[a]The substrate was α-ketoglutarate; the final concentrations of materials added were piericidin A, 10^{-6} M, rotenone, 10^{-6} M, and each quinone, 10^{-5} M. From Mitsui *et al.* (1970).

(1957) have been investigated in some detail (Conover and Ernster, 1962, 1963; Conover *et al.*, 1963). Enzyme levels are highest in the soluble fraction of rat liver, although it occurs also in other subcellular fractions. NADH-diaphorase activity is generally higher than NADPH-diaphorase activity in all subcellular fractions except the soluble. In cockroach muscle, however, there is little difference in NADH- and NADPH-diaphorase activities in the various subcellular fractions, although they are substantially lower than those in rat liver.

Further studies are required to establish whether DT-diaphorase is generally less active in insects than in mammals and whether it has any physiological significance in relation to the selective toxicity of rotenone to mammals, fish, and insects (Fukami *et al.*, 1970).

3.5. *Inhibitors of the α-Glycerophosphate Shuttle*

The extremely high activities of the enzymes involved in α-glycerophosphate metabolism (αGPI and αGPII) in the cytoplasmic and mitochondrial fractions of insect flight muscle are of special significance as potential insecticide targets in view of their important physiological function. In the insect flight muscle, the activity of αGPI is approximately 100- to 130-fold higher than that of the lactic dehydrogenase of vertebrate skeletal muscle; indeed, the activity of lactic dehydrogenase in the insect flight muscle is only 1–2% of that occurring in vertebrate muscle (Zebe and McShan, 1957).

In attempts to compare the biochemical properties of the insect and mammalian α-glycerophosphate dehydrogenase (αGPI), the enzyme has been purified and crystallized from rabbit muscle and from the thoraces (including legs and wings) of honeybees (Brosemer and Marquardt, 1966). The enzymes differed with respect to both electrophoretic and immunological properties and had different structural properties, particularly with respect to amino acid sequence and *C*-terminal composition (Marquardt and Brosemer, 1966). Similar studies with the mitochondrial enzyme (αGPII) showed only quantitative differences between insects and mammals.

Because of these enzymatic differences and because of the extreme physiological importance to the insect of the α-glycerophosphate shuttle, this system has been considered as a promising target around which to develop new types of selective insecticides. Several types of compounds have been evaluated as potential inhibitors. Cinammic acid and its derivatives inhibit the α-glycerophosphate dehydrogenase (αGPI) of insect flight muscle cytoplasm (O'Brien *et al.*, 1965). However, most of these compounds inhibit the αGPI of housefly flight muscle *in vitro* only at concentrations of about 10^{-4} M and show little toxicity *in vivo*. Significant *in vitro* inhibition at concentrations lower than 5×10^{-5} M is considered necessary to elicit any lethal effects *in vivo*. No attempts have yet been made to develop inhibitors for the mitochondrial enzyme (αGPII).

Despite the present lack of success in this area, the α-glycerophosphate shuttle remains an attractive target site against which to design new selective insecticides, and it will undoubtedly receive further attention in the future.

4. Conclusion

It is evident from the foregoing discussion that several types of compounds are effective inhibitors of respiration. Because of the basic uniformity of cell respiration in different organisms, most of these compounds have a broad spectrum of toxicity to many different species. There are, however, several leads which might prove useful in the development of compounds with some degree of selective toxicity.

Rotenone, for example, exhibits a low toxicity to mammals, and no phytotoxicity and yet is highly toxic to insects and fish. To date, studies on the binding of rotenone to its respiratory target site(s) have been effected only with preparations from mammalian mitochondria. Obviously there is a need to conduct similar studies with preparations from insects and fish to establish whether the selectivity of rotenone results primarily from basic differences in the target site of these species. If this proved to be the case, it might be possible to develop new derivatives with a low binding affinity for the mammalian target.

Additional studies are also required to provide a clearer understanding of the inhibitory interactions occurring at other respiratory sites, and there is every reason to suppose that despite its ubiquitous nature respiration can still provide several targets of potential importance in the development of new and more effective insecticides.

5. References

Abo-Khatwa, N., and Hollingworth, R. M., 1972, Chlorodimeform: The relation of mitochondrial uncoupling to toxicity in the German cockroach, *Life Sci. (Part II)* **11:**1181.

Aldridge, W. N., and Street, R. W., 1970, Oxidative phosphorylation: The specific binding of trimethyltin and triethyltin to rat liver mitochondria, *Biochem. J.* **118:**171.

Anderson, A. D., March, R. B., and Metcalf, R. L., 1954, Inhibition of the succinoxidase system of susceptible and resistant houseflies by DDT and related compounds, *Ann. Entomol. Soc. Am.* **47:**595.

Avi-Dor, -Dor, Y., and Gonda, O., 1959, Studies on the adenosine triphosphate–phosphate exchange and the hydrolysis of adenosine triphosphate catalyzed by a particulate fraction from the mosquito, *Biochem J.* **72:**8.

Barsa, M. C., and Ludwig, D., 1959, Effects of DDT on the respiratory enzymes of the mealworm, *Tenebrio molitor* (L.), and of the housefly, *Musca domestica* (L.), *Ann. Entomol. Soc. Am.* **52:**179.

Birt, L. M., 1961, Flight-muscle mitochondria of *Lucilia cuprina* and *Musca domestica*: Estimation of the pyridine nucleotide content and of the response of respiration to adenosine diphosphate, *Biochem. J.* **80:**623.

Bode, C., and Klingenberg, M., 1964, Carnitine and fatty acid oxidation in mitochondria of various organs, *Biochim. Biophys. Acta* **84:**93.

Brosemer, R. W., and Marquardt, R. R., 1966, Insect extramitochondrial glycerophosphate dehydrogenase. II. Enzymic properties and amino acid composition of the enzyme from honey bee (*Apis mellifera*) thoraces, *Biochim Biophys. Acta* **128:**464.

Brown, R. E., and Brown, A. W. A., 1956, The effects of insecticidal poisoning on the level of cytochrome oxidase in the American cockroach, *J. Econ. Entomol.* **49:**675.

Bruce, A. L., and Banks, W. M., 1973, Metabolism of muscle of cockroach *Blaberus gigantens, Ann. Entomol. Soc. Am.* **66:**1209.

Bulos, B., Shukla, S., and Sacktor, B., 1972, Bioenergetic properties of mitochondria from flight muscle of aging blow flies, *Arch. Biochem. Biophys.* **149:**461.

Burgos, J., and Redfearn, E. R., 1965, The inhibition of mitochondrial reduced nicotinamide-adenine dinucleotide oxidation by rotenoids, *Biochim. Biophys. Acta* **110:**475.

Casida, J. E., 1973, Insecticide biochemistry, *Ann. Rev. Biochem.* **42:**259.

Chan, S. K., and Margoliash, E., 1966, Properties and primary structure of the cytochrome *c* from the flight muscles of the moth, *Samia cynthia, J. Biol. Chem.* **241:**335.

Chance, B., and Hess, B., 1959, Metabolic control mechanisms. I. Electron transfer in the mammalian cell, *J. Biol. Chem.* **234:**2402.

Chance, B., and Sacktor, B., 1958, Respiratory metabolism of insect flight muscle. II. Kinetics of respiratory enzymes in flight muscle sarcosomes, *Arch. Biochem. Biophys.* **76:**509.

Chance, B., and Williams, G. R., 1955*a*, Respiratory enzymes in oxidative phosphorylation. III. The steady state, *J. Biol. Chem.* **217:**409.

Chance, B., and Williams, G. R., 1955*b*, Respiratory chain and oxidative phosphorylation. IV. The respiratory chain, *J. Biol. Chem.* **217:**429.

Chance, B., and Williams, G. R., 1956, The respiratory chain and oxidative phosphorylation, *Advan. Enzymol.* **17:**65.

Chefurka, W., 1958, On the importance of α-glycerophosphate dehydrogenase in glycolyzing insect muscle, *Biochim. Biophys. Acta* **28:**660.

Chefurka, W., 1965, Some comparative aspects of the metabolism of carbohydrate in insect, *Ann. Rev. Entomol.* **10:**345.

Childress, C. C., and Sacktor, B., 1966, Pyruvate oxidation and the permeability of mitochondria from blowfly flight muscle, *Science* **154:**268.

Chino, H., 1963, Respiratory enzyme system of the *Bombyx* silkworm egg in relation to the mechanism of the formation of sugar alchols, *Arch. Biochem. Biophys.* **102:**400.

Chino, H., and Harano, T., 1966, The significance of 3-hydroxykynurenine for the NADH, NADPH-cytochrome *c* reductase which exists in soluble fraction, *Seikagaku* **38:**675.

Cochran, D. G., 1963, Respiratory control in cockroach-muscle mitochondria, *Biochim. Biophys. Acta* **78:**393.

Conover, T. E., and Ernster, L., 1960, By-pass of the amytal-sensitive site of the respiratory chain in mitochondria by means of vitamin K_3, *Acta Chem. Scand.* **14:**1840.

Conover, T. E., and Ernster, L., 1962, DT diaphorase. II. Relation to respiratory chain of intact mitochondria, *Biochim. Biophys. Acta* **58:**189.

Conover, T. E., and Ernster, L., 1963, DT diaphorase. IV. Coupling of extramitochondrial reduced pyridine nucleotide oxidation to mitochondrial respiratory chain, *Biochim. Biophys. Acta* **67:**268.

Conover, T. E., Danielson, L., and Ernster, L., 1963, DT diaphorase. III. Separation of mitochondrial DT diaphorase and respiratory chain, *Biochim. Biophys. Acta* **67:**254.

Corbett, J. R., and Wright, B. J., 1970, Uncoupling of oxidative phorphorylation in intact mites and in isolated mite mitochondria by a new acaricide, 5,6-dichloro-l-phenyoxycarbonyl-2-trifluromethylbenzimidazole, *Biochem. J.* **118:**50.

Davis, R. A., and Fraenkel, G., 1940, The oxygen consumption of flies during flight, *J. Exp. Biol.* **17:**402.

De Kort, C. A. D., Bartelink, A. K. M., and Schuurmans, R. R., 1973, The significance of L-proline for oxidative metabolism in the flight muscles of the Colorado beetle, *Leptinotarsa decemlineata, Insect Biochem.* **3:**11.

Dixon, M., and Webb, E. C., 1964, *Enzymes*, Longmans Green, London.

Donnellan, J. F., and Beechey, R. B., 1969, Factors, affecting the oxidation of glycerol-l-phosphate by insect flight muscle mitochondria, *J. Insect Physiol.* **15:** 367.

Donnellan, J. F., Barker, M. D., Wood, J., and Beechey, R. B., 1970, Specificity and locale of the L-3-glycerophosphate-flavoprotein oxidoreductase mitochondria isolated from the flight muscle of *Sacophaga barbata, Biochem. J.* **120:**467.

Dummel, R. J., and Kun, E., 1969, Studies with specific enzyme inhibitors. XII. Resolution of DL-erythro-fluorocitric acid into optically active isomers, *J. Biol. Chem.* **244:**2966.

Eanes, R. Z., Skilleter, D. N., and Kun, E., 1972, Inactivation of the tricarboxylate carrier of liver mitochondria by (−)-erythrofluorocitrate, *Biochem. Biophys. Res. Commun.* **46:**1618.

Ela, R., Chefurka, W., and Robinson, J. B., *In vivo* glucose metabolism in the normal and poisoned cockroach, *Periplaneta americana, J. Insect Physiol.* **16:**2137.

Ernster, L., and Navazio, F., 1958, Soluble diaphorase in animal tissues, *Acta Chem. Scand.* **12:**595.

Ernster, L., Ljunggren, M., and Danielson, L., 1960, Purification and some properties of a highly dicumarol-sensitive liver diaphorase, *Biochem. Biophys. Res. Commun.* **2:**88.

Ernster, L., Danielson, L., and Ljunggren, M., 1962, DA diaphorase. I. Purification from the soluble fraction of rat-liver cytoplasm, and properties, *Biochim. Biophys. Acta* **58:**171.

Estabrook, R. W., 1967, Mitochondrial respiratory control and the polarographic measurement of ADP:O ratios, *in: Methods in Enzymology*, Vol. 10 (R. E. Estabrook, and M. F. Pullman, eds.), p. 41, Academic Press, New York.

Estabrook, R. W., and Sacktor, B., 1958, α-Glycerophosphate oxidase of flight muscle mitochondria, *J. Biol. Chem.* **233:**1014.

Fukami, J., 1954, Effect of rotenone on the succinoxidase system in the muscle of the cockroach, *Jpn. J. Appl. Zool.* **19:**29.

Fukami, J., 1955, Effect of rotenone on respiration in the muscle of the cockroach, *Periplaneta americana* L., *Jpn. J. Appl. Zool.* **19:**148.

Fukami, J., 1956, Effect of some insecticides on the respiration of insect organs, with special reference to the effects of rotenone, *Botyukagaku* **21:**122

Fukami, J., 1957, Studies on red and white muscle of insect, *Ins. Insect Contrib., Kyoto Univ. WHO*, p. 217.

Fukami, J., 1961, Effect of rotenone on the respiratory enzyme system of insect muscle, *Bull. Natl. Inst. Agr. Sci. C* **13:**33.

Fukami, J., and Nakatsugawa, T., 1961, Studies on red and white muscles of insects with special reference to spectrophotometric observation of cytochromes in muscles, *Bull. Natl. Inst. Agr. Sci. C* **13:**47.

Fukami, J., and Tomizawa, C., 1956, Effects of rotenone on the L-glutamic oxidase system in the insect, *Botyu-Kagaku* **21:**129.

Fukami, J., and Tomizawa, C., 1958, The effects of rotenone and its derivatives on the respiration of brain in guinea pig, *Botyu-Kagaku* **23:**205.

Fukami, J., Nakatsugawa, T., and Narahashi, T., 1959, The relation between chemical structure and toxicity in rotenone derivatives, *Jpn. J. Appl. Entomol. Zool.* **3:**259.

Fukami, J., Shishido, T., Fukunaga, K., and Casida, J. E., 1969, Oxidative metabolism of rotenone in mammals, fish and insects and its relation to selective toxicity, *J. Agr. Food. Chem.* **17:**1217.

Fukami, J., Mitsui, T., Fukunaga, and Shishido, T., 1970, The selective toxicity of rotenone between mammals, fish and insects, *in: Biochemical Toxicology of Insecticides* (R. D. O'Brien and I. Yamamoto, eds.), pp. 159–178, Academic Press, New York.

Gonda, O., Traub, A., and Avi-Dor, Y., 1957, The oxidative activity of a particulate fraction from mosquitos, *Biochem. J.* **67:**487.

Gregg, C. T., Heisler, C. R., and Remmert, L. F., 1960, Oxidative phosphorylation and respiratory control in housefly mitochondria, *Biochim. Biophys. Acta* **45:**561.

Gutman, M., Singer, T. P., Beinert, H., and Casida, J. E., 1970, Reaction sites of rotenone piericidin A, and amytal in relation to the nonhem ion components of NADH dehydrogenase *Proc. Natl. Acad. Sci. U.S.A.* **65:**763.

Gutman, Coles, C. J., Singer, T. P., and Casida, J. E., 1971, On the functional organization of the respiratory chain at the dehydrogenase–coenzyme Q junction, *Biochemistry* **10 (11):**2036.

Hagihara, B., 1965, Measurement of respiration with polarography, *Protein Nucleic Acid Enzyme* **10:**1689.

Hagihara, B., 1974, The outline of mitochomdria, *in: Mitochondria* (B. Hagihara, ed.), pp. 1–104, Asakura Shoten, Tokyo.

Hagihara, B., and Lardy, H. A., 1960, A method for the separation of orthophosphate from other phosphate compounds, *J. Biol. Chem.* **235:**889.

Hall, C., Wu, M., Crane, F. L., Takahashi, N., Tamura, S., and Folkers, K., 1966, Piericidin A: A new inhibitor of mitochondrial electron transport, *Biochem. Biophys. Res. Commun.* **25:**373.

Hansford, R. G., 1971, Some properties of mitochondria isolated from the flight muscle of periodical cicada, *Magicicada septendecim, Biochem. J.* **121:**771.

Hansford, R. G., and Chappell, J. B., 1967, The effect of Ca^{2+} on the oxidation of glycerol phosphate by blowfly mitochondria, *Biochem. Biophys. Res. Commun.* **27:**686.

Hansford, R. G., and Sacktor, B., 1970, The control of the oxidation of proline by isolated flight muscle mitochondria, *J. Biol. Chem.* **245:**991.

Harano, T., and Chino, H., 1971, A new diaphorase from *Bombyx* silkworm eggs—Cytochrome *c* reductase activity mediated with xanthommatin, *Arch. Biochem. Biochem. Biophys.* **146:**467.

Harvey, W. R., and Haskell, J. A., 1966, Metabolic control mechanisms in insect, *Advan. Insect Physiol.* **3:**133.

Hollunger, G., 1955, Guanidines and oxidative phosphorylations, *Acta pharmacol. Toxicol.* **11:** Suppl. No. 1, 84 pp.

Horgan, D. J., Singer, T. P., and Casida, J. E., 1968*a*, Studies on the respiratory chain-linked reduced nicotine-amide adenine dinucleotide dehydrogenase. XII. Binding sites of rotenone, piericidin A, and amytal in the respiratory chain, *J. Biol. Chem.* **243:**834.

Horgan, D. J., Ohno, H., Singer, T. P., and Casida, J. E., 1968*b*, Studies on the respiratory chain-linked reduced nicotine-amide adenine dinucleotide dehydrogenase. IV. Interactions of piericidin with the mitochondrial respiratory chain, *J. Biol. Chem.* **243:**5967.

Hosotsuji, T., 1956, Japanese Patent S 36-147.

Hülsmann, W. C., Elliott, W. B., and Slater, E. C., 1960, The nature and mechanism of action uncoupling agents present in mitochrome preparations, *Biochim. Biophys. Acta* **39:**267.

Ilivicky, J., and Casida, J. E., 1969, Uncoupling action of 2,4-dinitrophenols, 2-trifluoromethylbenzimidazoles and certain other pesticide chemicals upon mitochondria from different sources and its relation to toxicity, *Biochem. Pharmacol.* **18:**1389.

Ilivicky, J., Chefurka, W., and Casida, J. E., 1967, Oxidative phosphorylation and sensitivity to uncouplers of housefly mitochondria: Influence of isolation medium, *J. Econ. Entomol.* **60:**1404.

Jeng, M., Hals, C., Crane, F. L., Takahashi, S., Tamura, S., and Folkers, K., 1968, Inhibition of mitochondrial electron transport by piericidin A and related compounds, *Biochemistry* **7:**1311.

Kallapur, V. L., and George, C. J., 1973, Fatty acid oxidation by the flight muscle of the dragonfly, *Pantala flavescens, J. Insect Physiol.* **19:**1035.

Keilin, D., and King, T. E., 1960, Effect of inhibitors on the activity of soluble succinic dehydrogenase and on the reconstitution of the succinic dehydrogenase cytochrome system from its components, *Proc. Roy. Soc.* **152B:**163.

Klingenberg, M., and Bücher, T., 1959, Flugmuskelmitochondrien aus *Locusta migratoria* mit Atmungskontrolle, *Biochem. Z.* **331:**312.

Klingenberg, M., Slenczka, W., and Ritt, E., 1959, Vergleichende Biochemie der Pyridinucleotid-System in Mitochondria Verschiedener Organe, *Biochem. Z.* **332:**47.

Kröger, A., and Klingenberg, M., 1966, On the role of ubiquinone in mitochondria. II. Redox reaction of ubiquinone under the control of oxidative phosphorylation, *Biochem. Z.* **344:**317.

Kubišta, V., 1957, Inorganic phosphate and the rate of glycolysis in insect muscle, *Nature (London)* **180:**549.

Kubišta, V., 1958, Anaerobic Glykolyse in den Insectenmuskeln, *Biochem. Z.* **330:**315.

Kurland, C. G., and Schneiderman, H. A., 1959, The respiratory enzymes or diapausing silkworm pupae: A new interpretation of carbon monoxide-insensitive respiration, *Biol. Bull.* **116:**136.

Lardy, H., and Ferguson, S. M., 1969, Oxidative phosphorylation in mitochondria, *Ann. Rev. Biochem.* **38:**991.

Lardy, H. A., and Wellman, H., 1952, Oxidative phosphorylation: Role of inorganic phosphate and acceptor systems in control of metabolic rates, *J. Biol. Chem.* **195:**215.

Lardy, H. A., Johnson, D., and McMurray, W. C., 1958, Antibiotics as tools for metabolic studies. I. Survey of toxic antibiotics in respiratory, Phosphorylative and glycolytic systems, *Arch. Biochem. Biophys.* **78:**587.

Lehninger, A. L., 1970, *Biochemistry*, Worth Publishers, New York.

Lehninger, A. L., and Remmert, L. F., 1959, An endogenous uncoupling and swelling agent in liver mitochondria and its enzymic formation, *J. Biol. Chem.* **234:**2459.

Lester, R. L., and Crane, F. L., 1959, The natural occurrence of coensyme Q and related compounds, *J. Biol. Chem.* **234:**2169.

Lewis, S. E., and Slater, E. C., 1954, Oxidative phosphorylation in insect sarcomes, *Biochem. J.* **58:**207.

Lilian, M. E., and Ilse, D., 1970, An inhibitor of mitochondrial respiration in venom of the australian bull dog ant, *Myrecia gulosa, J. Insect Physiol.* **16:**1531.

Lindahl, P. E., and Öberg, K. E., 1961, The effect of rotenone on respiration and its point of attack, *Exp. Cell. Res.* **23:**228.

Marquardt, R. R., and Brosemer, R. W., 1966, Insect extramitochondrial glycerophosphate dehydrogenase. I. Crystallization and physical properties of the enzyme from honeybee (*Apis mellifera*) thoraces, *Biochim. Biophys. Acta* **128:**454.

Martius, C., and Märki, F., 1957, Ueber, Phyllochinon-Reductase, *Biochem. Z.* **329:**450.

Matsuda, M., and Fukami, J., 1972, Preliminary survey of effects of phenols on the oxidative phosphorylation in the American cockroach muscle mitochondria, *Appl. Entomol. Zool.* **7:**27.

McAllen, J. W., and Brown, A. W. A., 1960, The effect of insecticides on transamination in the American cockroach, *J. Econ. Entomol.* **53:**166.

Mitsui, T., Fukami, J., Fukunaga, K., Sagawa, T., Takahashi, N., and Tamure, S., 1969, Studies on piericidin. I. Effect of piericidin A and B on mitochondrial electron transport in insect, *Botyu-Kagaku* **34:**126.

Mitsui, T., Fukami, J., Fukunaga, K., Takahashi, N., and Tamura, S., 1970, Studies on piericidin: Antagonistic effect of vitamin K_3 on the respiratory chain of insects and mammals in the presence of piericidin A, *Agr. Biol. Chem.* **34:**1101.

Muraoka, S., and Terada, H., 1972, 3,5-Di-*tert*-butyl-4-hydroxybenzylidene-malononitrile: New powerful uncoupler of respiratory-chain phosphorylation, *Biochim. Biophys. Acta* **275:**271.

O'Brien, R. D., 1956, The inhibition of cholinesterase and succinoxidase by malathion and its isomer, *J. Econ. Entomol.* **49:**484.

O'Brien, R. D., 1957, The effect of malathion and its isomer on carbohydrate metabolism of the mouse, cockroach and housefly, *J. Econ. Entomol.* **50:**79.

O'Brien, R. D., 1967, *Insecticides: Action and Metabolism,* Academic Press, New York.

Obrien, R. D., Cheng, L., and Kimmel, E. C., 1965, Inhibition of the α-glycerophosphate shuttle in housefly flight muscle, *J. Insect Physiol.* **11:**1241.

Ohkawa, H., Ohkawa, R., Yamamoto, I., and Casida, J. E., 1972, Enzyme mechanisms and toxicological significance of hydrogen cyanide liberation from various organothiocyanates and organonitriles in mice and houseflies, *Pestic. Biochem. Physiol.* **2:**95.

Ohnishi, K., 1966, Studies on cytochrome *b*. III. Comparison of cytochrome *B*'s from beef heart muscle and larvae of the housefly, *J. Biochem.* **59:**17.

Okada, Y., 1973, *The Studies of Cytochrome,* pp. 165–176, University of Tokyo Press, Tokyo.

Orser, W. B., and Brown, A. W. A., 1951, The effect of insecticides on the heart beat of *Periplaneta, Can. J. Zool.* **29:**54.

Osanai, M., 1966, The pigment of silkworm's egg, *Jpn. Zool. Mag.* **71:**381.

Parker, V. H., 1958, Effect of nitrophenols and halogenophenols on the enzymic activity of rat-liver mitochondria, *Biochem. J.* **69:**306.

Piper, G. R., and Casida, J. E., 1965, Housefly adenosine triphosphatases and their inhibition by insecticidal organotin compounds, *J. Econ. Entomol.* **58:**392.

Pressman, B. C., and Lardy, H. A., 1956, Effect of surface active agents on the latent ATPase of mitochondria, *Biochim. Biophys. Acta* **21:**458.

Pullman, M. E., and Racker, E., 1956, Spectrophotometric studies of oxidative phosphorylation, *Science* **123:**1105.

Quagliariello, E., Palmieri, F., Alifano, A., and Papa, S., 1966, 3-Hydroxyanthranilic acid-mediated respiration in the inhibited respiratory chain, *Biochim. Biophys. Acta* **113:**482.

Rees, K. B., 1954, Aerobic metabolism of the muscle of *Locusta migratoria, Biochem. J.* **58:**196.

Remmert, L. F., and Lehninger, A. L., 1959, A mitochondrial factor producing "loose-coupling" of respiration, *Proc. Natl. Acad. Sci. U.S.A.* **45:**1.

Rose, M. S., and Lock, E. A., 1970, The interaction of triethyltin with a component of guinea-pig liver supernatant: Evidence for histidine in the binding sites, *Biochem. J.* **120:**151.

Sacklin, J. A., Terriere, L. C., and Remmert, L. F., 1955, Effect of DDT on enzymatic oxidation and phosphorylation, *Science* **122:**377.

Sacktor, B., 1953, Investigations on the mitochondria of the housefly, *Musca domestica* L., *J. Gen. Physiol.* **36:**371.

Sacktor, B., 1954, Investigations on the mitochondria of housefly, *Musca domestica* L. III. Requirements for oxidative phosphorylation. *J. Gen. Physiol.* **37:**343.

Sacktor, B., 1961, The role of mitochondria in respiratory metabolism of flight muscle, *Ann. Rev. Entomol.* **6:**103.

Sacktor, B., 1965, Energetics and metabolism of muscular contraction, *in: Physiology of Insecta*, Voll. II (M. Rockstein, ed.), pp. 484–580, Academic Press, New York.

Sacktor, B., 1974, Biological oxidations and energetics in insect mitochondria, *in: Physiology of Insecta*, Vol. IV (M. Rockstein, ed.), pp. 271–353, Academic Press, New York.

Sacktor, B., and Childress, C. C., 1967, Metabolism of proline in insect flight muscles and its significance in stimulating the oxidation of pyruvate, *Arch. Biochem. Biophys.* **120:**583.

Sacktor, B., and Cochran, D. G., 1958, The respiratory metabolism of insect flight muscle. I. Manometric studies of oxidation and concomitant phosphorylation with sarcosomes, *Arch. Biochem. Biophys.* **74:**266.

Sacktor, B., and Wormser-Shavit, E., 1966, Regulation of metabolism in working muscle *in vivo*. I. Concentrations of some glycolytic, tricarboxylic acid cycle, and amino acid intermediates in insect flight muscle during flight, *J. Biol. Chem.* **241:**624.

Sacktor, B., O'Neil, J. J., and Cochran, D. G., 1958, The requirement of serum albumin in oxidative phosphorylation of flight muscle mitochondria, *J. Biol. Chem.* **233:**1233.

Slater, E. C., 1960, *in: The Structure and Function of Muscle*, Vol. 2 (G. Bourne, ed.), p. 105, Academic Press, New York.

Takahashi, N., Suzuki, A., Kimura, Y., Miyamoto, S., Tamura, S., Mitsui, T., and Fukami, J., 1968, Isolation, structure and physiological activities of piericidin B: Natural insecticide produced by a streptomyces, *Agr. Biol. Chem.* **32:**1115.

Tamura, S., Takahashi, N., Miyamoto, S., Mori, R., Suzuki, S., and Nagatsu, J., 1963, Isolation and physiological activities of piericidin A, a natural insecticide produced by streptomyces, *Agr. Biol. Chem.* **27:**576.

Tischler, N., 1936, Studies on how derris kills insects, *J. Econ. Entomol.* **28:**215.

Tomizawa, C., and Fukami, J., 1956, Biochemical studies on the action of insecticides. II. The oxidative phosphorylation in the flight muscle of *Locusta migratoria* and the influences of insecticides, *Oyo-Kontyu* **12:**1.

Van den Bergh, S. G., 1964, Pyruvate oxidation and the permeability of housefly sarcosomes, *Biochem. J.* **93:**128.

Van den Bergh, S. G., 1967, Insect mitochondria, *in: Methods in Enzymology*, Vol. 10 (R. E. Estabrook and M. F. Pullman, eds.), p. 117, Academic Press, New York.

Van den Bergh, S. G., and Slater, E. C., 1960, The respiratory activity and respiratory control of sarcosomes isolated from the thoracic muscle of the housefly, *Biochim. Biophys. Acta* **40:**176.

Van den Bergh, S. G., and Slater, E. C., 1962, The respiratory activity and permeability of housefly sarcosomes, *Biochem. J.* **82:**362.

Weinbach, E. C., 1954, The effect of pentachlorophenol on oxidative phosphorylation, *J. Biol. Chem.* **210:**545.

Weinbach, E. C., Sheffield, H., and Garbers, J., 1963, Restoration of oxidative and morphological integrity to swollen, and uncoupled rat mitochondria, *Proc. Natl. Acad. Sci. U.S.A.* **49:**561.

Weis-Fogh, T., 1964, Biology and physics of locust flight. VIII. Lift and metabolic rate of flying insects, *J. Exp. Biol.* **41:**257.

Whitehouse, H. W., 1964, Biochem. Pharmocol. **13:**319.

Williamson, R. L., and Metcalf, R. L., 1967, Salicylanilides: A new group of active uncouplers of oxidative phosphorylation, *Science* **158:**1694.

Winteringham, F. P. W., Hellyer, G. C., and McKay, M. A., 1960, Effects of the insecticides DDT and dieldrin on phosphorus metabolism of the adult housefly *Musca domestica, Biochem. J.* **76:**543.

Wojtczak, L., and Wojtczak, A. B., 1959, The action of serum albumin on oxidative phosphorylation in insect mitochondria, *Biochim. Biophys. Acta* **31:**297.

Yammasaki, T., and Narahashi, T., 1957, Effects of oxygen lack, metabolic inhibitors, and DDT on the resting potential of insect nerve. Studies on the mechanism of action of insecticides. XII, *Botyu-Kagaku* **22:**259.

Yust, H. R., and Shelden, F. F., 1952, A study of the physiology of resistance to hydrocyanic acid in the California red scale, *Ann. Entomol. Soc. Am.* **45:**220.

Zahavi, M., and Tahori, A. S., 1972, Activity of mitochondrial NAD-linked isocitric dehydrogenase in alatiform and apteriform larve of *Myzus persicae, J. Insect Physiol.* **18:**608.

Zebe, E. C., and McShan, W. H., 1957, Lactic and α-glycerophosphate dehydrogenase in insect, *J. Gen. Physiol.* **40:**779.

11

Physicochemical Aspects of Insecticidal Action

T. R. Fukuto

1. Introduction

Enormous advances have been made during the past three decades in the development of organic chemicals as insect control agents. During this period, beginning approximately after the discovery of the insecticidal properties of DDT, countless numbers of new compounds of widely varying structures have been synthesized and evaluated for insecticidal activity. For the large part, the approaches taken to discover new insecticides have been empirical, i.e., through systematic analogue synthesis and screening. While such an approach has resulted in the unveiling of many useful compounds for insect control, it has been less than satisfying to individuals having the propensity to rationalize biological activity in terms of fundamental chemical and biochemical concepts. This is not meant to minimize the value of the empirical approach, for without it the establishment of meaningful relationships between chemical structure and biological activity would not be possible.

Intoxication of insects (and mammals) by an insecticide is dependent on a number of factors. The poisoning process is undoubtedly quite complex, but the principal events which control intoxication probably are (a) penetration and translocation to the site of action, (b) metabolic activation or degradation,

T. R. Fukuto • Department of Entomology, University of California, Riverside, California.

and (c) target site interaction. While the properties of the toxicant which control penetration, translocation, and target site interaction often are amenable to physicochemical analysis, metabolic transformations are on the whole less readily accounted for. For this reason, in most cases insecticidal activity cannot be quantitated on the basis of physicochemical concepts alone. However, in certain cases—e.g., with carbamate insecticides—metabolism may be minimized by the use of adjuvants (synergists) and in these instances a reasonable account of whole animal response can be made by the application of physical organic theory. Although successful prediction of the insecticidal activity of randomly selected chemicals is beyond our present state of capability, it is true that within restricted classes of compounds, particularly those with established modes of action, the tools provided by physical organic chemistry are of considerable value in relating biological activity to chemical structure. The principal thrust of this chapter will be centered on the application of physicochemical parameters in relating insecticidal action to chemical structure.

2. Physicochemical Parameters

The design of new insecticides through a rational approach is a standing goal among scientists engaged in the development of insecticides. One of the most fruitful tools for this purpose is the application of linear free energy parameters for correlation of structure with biological response. Before dealing with specific cases where these parameters have been used, it is appropriate to present a brief review of the various parameters which have been developed in assessing changes in chemical structure, i.e., substituent effects with biological activity.

2.1. σ and Related Constants

Substituent effects which have proved useful in relating biological activity to chemical structure may be divided into three broad categories: electronic, steric, and hydrophobic. Hammett's σ constant (Hammett, 1940; Jaffe, 1953), the first parameter developed to assess electronic effects, found its origin in the linear relationship observed between the ionization constants of *meta*- and *para*-substituted benzoic acids and the reactivity of the corresponding benzoate esters. The effect of aromatic ring substituents—placed in positions remote from the reaction center—on chemical reactivity may be estimated by the Hammett equation:

$$\log k/k_0 = \rho\sigma \qquad (1)$$

where in a given reaction k and k_0 are the rate (or equilibrium) constants for the substituted and unsubstituted compounds and ρ is a constant which depends on the nature of the reaction. The substituent constant σ provides an estimate of the electronic effect transmitted by a substituent through the aromatic system to the reaction center.

The Hammett equation indicates that rate or equilibrium constants of reactions involving substituted benzoic acid derivatives are related to σ, and therefore σ may be used for estimation of reactivity. The application of the equation is not restricted to benzoic acid derivatives and applies to other aromatic systems as well, e.g., phenols and anilines.

Since its introduction, the Hammett equation has been applied to a variety of reactions involving substituted aromatic compounds. However, because of the many cases in which the original σ values failed to provide satisfactory correlations, there has been a proliferation of alternative sets of substituent constants (Swain and Lupton, 1968). Resonance interaction between the substituent and the reaction center appears to be one of the principal causes for the failure of the original σ constants. For example, in the case of substituted phenols and anilines, direct conjugation is possible between electron-attracting *p*-substituents and the electron-rich hydroxyl or amino moiety. For reactions involving these compounds, the σ_p^- constant is normally used (Cohen and Jones, 1963; Cram, 1965). On the other hand, σ_p^+ constants (Brown and Okamoto, 1958) find use in the opposite situation, where electron-donating substituents conjugate with an electron-deficient center. These have been obtained from data based on solvolysis rates of substituted cumyl chlorides.

2.2. Polar and Steric Substituent Constants

For aliphatic compounds, the free energy parameters introduced by Taft (1956) for estimation of polar and steric effects have been useful in structure–activity correlations. These were derived from alkaline and acidic hydrolysis rate constants of esters of the general structure X—COOR according to

$$\sigma^* = [\log (k/k_0)_B - \log (k/k_0)_A]/2.48 \qquad (2)$$

where σ^* is defined as the polar substituent constant for the substituent X, and k_0 and k are rate constants for the hydrolysis of the unsubstituted (X = methyl) and substituted esters, respectively. The subscripts A and B denote reactions carried out under acidic and basic conditions, and 2.48 is the proportionality factor necessary to place σ^* on the same relative scale as σ. The same equation may be used to calculate σ constants for ortho substituents (σ_0) in aromatic compounds.

Estimation of steric effects attributable to the substituent X is provided by the steric substituent constant, E_s. This constant may be calculated from

$$E_s = \log (k/k_0)_A \qquad (3)$$

$$E_s = \log (k/k_0)_B - \rho^*\sigma^* \qquad (4)$$

where ρ^* is the reaction constant analogous to ρ in the Hammett equation. In reactions where steric (and also resonance) effects are negligible, i.e., E_s is zero, equation (4) reduces to (5), identical in form to the Hammett equation:

$$\log (k/k_0)_B = \rho^* \sigma^* \tag{5}$$

Derivation of the above equation (2–5) is based on three assumptions: (a) the relative free energy of a reaction may be treated as the sum of independent contributions from polar, steric, and resonance effects, (b) steric and resonance effects are the same under basic and acidic reaction conditions, and (c) polar effects are substantially greater in basic than in acid media. Although the validity of these assumptions has been questioned (Shorter, 1970), σ^* and E_s have been of proven usefulness in assessing substituent effects in a variety of chemical reactions, including those associated with biological systems.

2.3. Oil–Water Partition, the π Constant

The free energy parameters described above provide an assessment of the effect of various substituents on the chemical reactivity of a molecule. In situations where chemical reactivity happens to be the underlying basis for biological activity, e.g., the inhibition of acetylcholinesterase (Fukuto, 1971), the parameters σ, σ^*, and E_s, used individually or in combination, often suffice for the estimation of substituent effects on activity. These parameters, however, have little value in accounting for effects attributable to the physical properties of the molecule. Since it was indicated earlier that whole animal response to biologically active materials depends to some degree on the ability of the material to penetrate into the animal and translocate to the site of action, a parameter which estimates substituent effects on the movement of chemicals in biological systems is desirable before reasonable prediction of activity can be made. A parameter which has been used successfully for this purpose is based on the lipophilic properties of organic compounds as estimated by their oil–water partition coefficients (Hansch, 1969; Hansch and Dunn, 1972). Of particular usefulness has been the π constant, a parameter developed by Hansch and coworkers (Hansch and Fujita, 1964; Fujita *et al.*, 1964), which when used in conjunction with other free energy parameters has led to remarkable improvement in the correlation of biological activity with structure.

The π constant finds its origin in early studies by Meyer (1899), Overton (1901), and others (Meyer and Hemmi, 1935), who demonstrated the existence of a direct relationship between oil–water partition coefficients of organic molecules and isonarcosis in tadpoles. Subsequently, Ferguson (1939), in an effort to rationalize biological activity on thermodynamic grounds, demonstrated general correlation between activity and the logarithm of a number of physical constants, including partition coefficients. The logarithm of octanol–water partition coefficients ($\log P$) has had wide applicability in quantitative studies on structure–activity relations, and a large variety of congeneric materials possessing biological activity have been shown to follow

$$pC = a \log P + b \tag{6}$$

where C is the amount of material required to effect a fixed biological response, usually expressed in terms of molar concentration or moles per unit weight, P is the octanol–water partition coefficient, and a and b are constants (Hansch and Dunn, 1972). Evidently, equation (6) may be applied to correlation of activity at all levels of biological systems, including enzymes, membranes, cells, and the whole animal. Since P, in essence, describes the lipophilic character of an organic compound, it has often been used to assess hydrophobic effects in the interaction of organic compounds with biological tissue.

The substituent constant, π, analogous to the Hammett σ constant, was developed (Hansch and Fujita, 1964) in order to place the properties associated with lipophilic (or hydrophobic) character on a comparative basis. It is defined by

$$\pi = \log P_X / P_H \tag{7}$$

where P_H is the octanol–water partition coefficient of the parent or standard compound (i.e., the substituent X is hydrogen) and P_X is the corresponding value for the compound containing the substituent X. π is a free energy related constant and provides a measure of the relative free energy change resulting from the movement of the substituted compound from one phase to another. Since for a given series of compounds π is always related to the unsubstituted parent compound, it becomes, in effect a substituent constant and is dependent only on the nature of the substituent.

According to the Hansch model, a biologically active molecule applied to an external phase proceeds by a random walk process through the internal or cellular phases and eventually reaches the site of action. This applies to enzymes, cells, membranes, and whole animals, although, obviously, the process becomes increasingly more complex at the higher levels of organization. In moving from one biological environment to another, the molecule presumably undergoes multiple partitions between lipophilic and hydrophilic phases as it passes through the biological system en route to the target site. Since π is based on the oil (octanol)–water partitioning character of a compound, it provides for the estimation of the effect of change in structure on penetration and translocation. Further, according to the model, once the molecule reaches the site of action it is expected to interact with the site, either through binding forces or by chemical reaction, or possibly by a combination of both. Hydrophobic bonding is an important contributing factor to the binding of chemicals to biological tissue and occurs when two or more nonpolar groups come into contact. The formation of the bond is attributable to the entropy change associated with the change in water structure surrounding the groups (Némethy, 1967). Again, π (or $\log P$), because of its proven usefulness in estimating changes in lipophilic character, has found applicability as a parameter for estimation of hydrophobic bonding. Thus binding effects at the molecular level may be estimated by the constant π.

2.4. Multiple Parameter Analysis

As stated earlier, biological response to a foreign chemical may depend on a number of factors associated with the chemical and physical properties of the molecule. From the physicochemical viewpoint, i.e., disregarding metabolic and related effects, biological response may be considered a function of the electronic, steric, and hydrophobic properties of the molecule:

$$\text{biological response (BR)} = f\ (\text{hydrophobic}) + f\ (\text{electronic}) + f\ (\text{steric}) \quad (8)$$

This relationship has been placed on a qualitative basis by Hansch (1969), and is represented in general terms by

$$\log \text{BR} = k_1\pi^2 + k_2\pi + k_3\sigma + k_4E_s + k_5 \quad (9)$$

Coefficients k_1, k_2, k_3, and k_4 provide the weight which each contributing parameter contributes to biological response. In practice, it is possible by means of multiple regression analysis to determine the contribution from each of the free energy parameters. The π^2 term in equation (9) indicates that biological response is parabolically related to π, implying that molecules which are highly hydrophilic or lipophilic will have difficulty in reaching and interacting with the active site and that a value of π exists where maximum response may be expected.

In addition to the various free energy constants described above, other parameters have been used in the correlation of biological activity and chemical structure. While these have not found much applicability in relating physicochemical properties of organic compounds to insecticidal action, they have been useful in analogous studies with drugs. These include such molecular properties as polarizability, dipole moments, molar refractivity, molar volume, and van der Waal's forces (Wilkinson, 1973; Cammarata and Martin, 1970).

3. Structure–Activity Correlations

3.1. Organophosphorus Esters

The toxicity of organophosphorus and carbamate esters to insects and mammals is associated with the inhibition of acetylcholinesterase (AChE), an enzyme whose normal function is to catalyze the hydrolysis of acetylcholine (ACh). Physiologically, ACh may be regarded as one of the neurotransmitting agents in the nervous systems of animals. Inhibition of the AChE–ACh enzyme system therefore represents the biochemical lesion responsible for the toxic action of phosphate and carbamate esters (Chapter 7). The inhibition of AChE by organophosphorus and carbamate insecticides has been demonstrated beyond reasonable doubt to be the result of an actual chemical reaction between the enzyme and ester as exemplified below with a phosphate ester

(Fukuto, 1971):

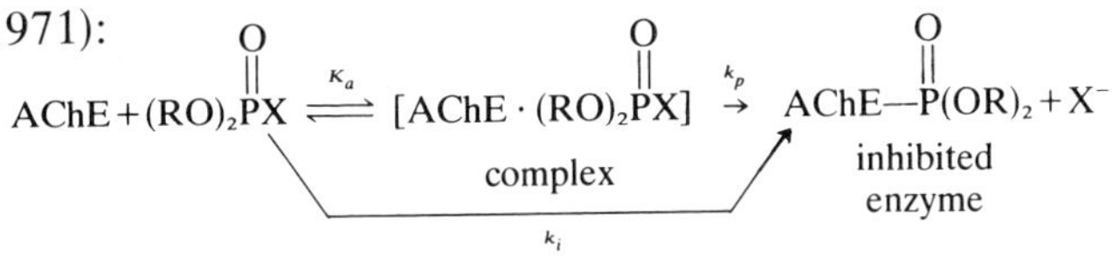

where K_a is the affinity (dissociation) constant between complex and reactants, k_p is the phosphorylation constant, and k_i is the overall bimolecular inhibition constant and is equal to k_p/K_a (Main, 1964). Since K_a is a measure of the tendency for the enzyme–inhibitor complex to dissociate, it is generally regarded as a binding or affinity constant and should be dependent on the structural and steric properties of the molecule. On the other hand, the effect of change in the reactivity of the ester has been associated with the phosphorylation constant, k_p (Hollingworth *et al.*, 1967). The bimolecular rate constant for inhibition, k_i, is dependent on the values of K_a and k_p.

3.1.1. Reactivity and AChE Inhibition

Investigations on the relation between chemical structure of organophosphorus esters and the inactivation of AChE have revealed that anticholinesterase activity depends to a large extent on the chemical reactivity of the ester. For example, the inhibition of erythrocyte AChE by a series of diethyl-substituted phenyl phosphates has been shown to occur in a bimolecular manner and a linear relationship has been observed between the bimolecular inhibition constant k_i and alkaline hydrolysis rates (Aldridge and Davison, 1952). Further, the inhibition of housefly head AChE (HFAChE) by a series of diethyl-substituted phenyl phosphates has been related to the reactivity of these esters as estimated by Hammett's σ constants and shifts in *P-O*-phenyl stretching frequencies in the infrared region (Fukuto and Metcalf, 1956). In general, esters with electron-withdrawing substituents, i.e., substituents with large positive σ values, were decidedly more potent as inhibitors of AChE and a linear relationship was observed between anticholinesterase activity and σ values.

As a rule, the chemical reactivity of the phosphorus atom is the single most important property which determines the anticholinesterase activity of an organophosphorus ester. This is supported by the estimation of charge density distribution between the phosphorus atom and atoms adjacent to it by application of molecular orbital theory (Kier, 1971). Calculation of charge densities in several organophosphorus insecticides reveals that a large positive charge is required at the phosphorus atom for high anticholinesterase activity (Pullman and Valdemoro, 1960). It is well known that parathion, a phosphorothionate ester, is virtually devoid of anticholinesterase activity, while its phosphate analogue, paraoxon, is a potent inhibitor. The toxicity of parathion is attributed to its metabolic conversion to paraoxon in the animal (Chapter 2). Molecular orbital calculations show that the positive charge associated with the phosphorus atom of paraoxon is substantially greater than that of parathion.

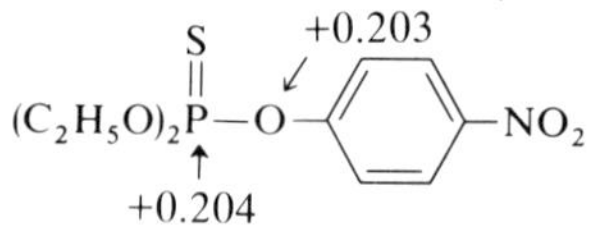

parathion

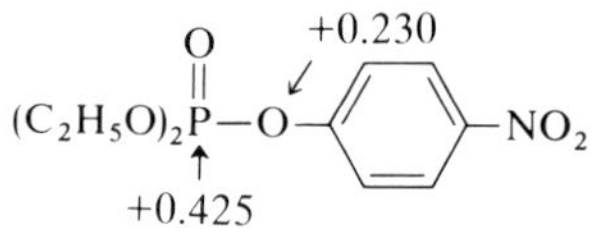

paraoxon

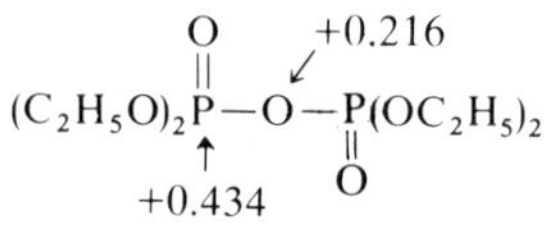

TEPP

A slightly larger positive charge was calculated for the phosphorus atom of TEPP (tetraethyl pyrophosphate), a compound of even greater inhibitory potency than paraoxon.

The parallel between charge density and anticholinesterase activity for these examples agrees with a mechanism of inhibition in which a nucleophile (serine hydroxyl moiety) in the active site attacks the electrophilic phosphorus atom and displaces an appropriate leaving group, resulting in phosphorylation of the enzyme. Esters with groups attached to the phosphorus atom which induce greater electrophilic character to it are expected to be more susceptible to attack by nucleophilic agents and hence should be more effective inhibitors of AChE. This is consistent with the relationship observed between anticholinesterase activity and σ constants for substituents in the diethyl-substituted phenyl phosphates. Calculation of superdelocalizability, $Sp^{(N)}$, and electron density of the phosphorus atom of a series of diethyl-substituted phenyl phosphates by application of frontier electron theory has revealed a parallel between $Sp^{(N)}$ and alkaline hydrolysis of the esters, anticholinesterase activity, and insecticidal activity (Fukui *et al.*, 1961). Esters with large values of $Sp^{(N)}$ were found to be the more effective anticholinesterases and insecticides.

3.1.2. Steric Effects in AChE Inhibition

It is apparent from structure–activity correlation studies that reactivity parameters alone often are inadequate in providing a satisfactory account of anticholinesterase activity, even within a limited series of organophosphorus esters. This is evident from data presented in Table I (Hansch and Deutsch, 1966). By means of regression analysis, equations (10) and (11) have been derived from the data:

	n	s	r	
$\log 1/I_{50} = 1.557\sigma + 5.806$	5	1.383	0.479	(10)
$\log 1/I_{50} = 3.451\sigma + 6.309$	6	0.507	0.954	(11)

Table I. Values for σ and Inhibition of Housefly Head AChE by meta- and para-Substituted Phenyl Diethyl Phosphates

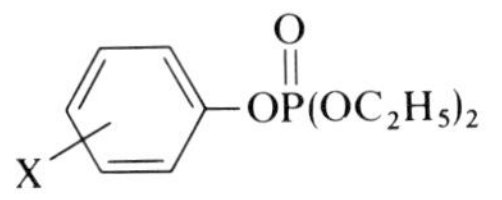

X	σ	Log $1/I_{50}$[a]
m-NO_2	0.71	7.30
m-OCH_3	0.12	3.89
m-t-C_4H_9	−0.12	6.05
m-$N(CH_3)_2$	−0.21	6.40
m-SF_5	0.61	7.12
p-NO_2	0.78	7.59
p-Cl	0.23	4.52
p-SO_2CH_3	0.73	6.60
p-SCH_3	−0.05	4.48
p-CN	0.63	6.89
p-t-C_4H_9	−0.20	4.00

[a] I_{50} is the molar concentration required to inhibit 50% of a fixed amount of enzyme after 15 min.

Equation (10) was obtained for the *meta* series and equation (11) for the *para* series (n is the number of compounds, s is the standard error, and r is the correlation coefficient). Inclusion of π in either of the regression equations did not improve the correlation, and therefore hydrophobic bonding as estimated by π appears to be of little significance in predicting the activity of these compounds. With the *meta* derivatives correlation of anticholinesterase activity with σ was quite poor, while excellent correlation was obtained with the *para* derivatives.

Subsequent results with a related series of substituted phenyl diethyl phosphates indicate that steric effects must also be considered in accounting for the activity of the *meta* derivatives (Hansch, 1970), as indicated by

$$\log 1/I_{50} = -1.06E_s^m + 2.19\sigma_p^- - 1.40X + 5.77 \qquad (12)$$

n	s	r
12	0.285	0.985

In this equation, E_s^m is the steric substituent constant for *meta* substituents and X is a position term ($X = 1$ for *meta* derivatives and $X = 0$ for *para* derivatives). Significantly better correlation was obtained with σ_p^- than with σ values for *para* substituents. This is expected since these esters are derivatives of substituted phenols. Thus by including a steric parameter in the analysis it was possible to combine *meta* and *para* isomers into a single equation with excellent results.

Analogous results have been obtained with other types of organophosphorus esters. Equations (13), (14), and (15) were obtained from data for the inhibition of HFAChE by a series of ethyl *p*-nitrophenyl alkylphosphonates, where R is alkyl or aralkyl (Fukuto and Metcalf, 1959*a*; Hansch, 1970):

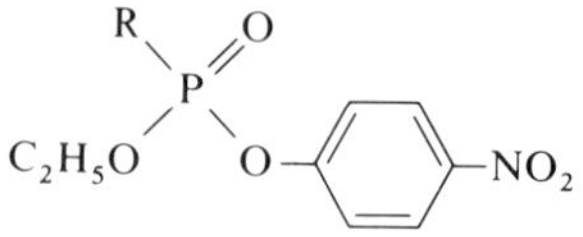

	n	s	r	
$\log k_i = 5.68\sigma^* + 6.31$	13	1.574	0.407	(13)
$\log k_i = 2.58E_s^c + 7.94$	13	0.648	0.927	(14)
$\log k_i = 2.89E_s^c + 2.72\sigma^* + 7.92$	13	0.616	0.940	(15)

The electronic parameter, Taft's polar substituent constant (σ^*), gave poor correlation, while significant correlation was obtained with E_s^c, Hancock and Falls' (1961) modified values for Taft's steric substituent constant (E_s). Combination of E_s^c and σ^* gave a slightly improved relationship. The equations reveal that steric effects attributable to R are by far the most important factor in the inhibition of AChE by these esters.

Another example that illustrates the importance of steric effects in AChE inhibition is found in results obtained for the inactivation of HFAChE by a series of *O*-methyl *O*-2,4,5-trichlorophenyl *N*-alkylphosphoramidates in which substitution on the amido nitrogen atom was varied (Fukuto *et al.*, 1963). Regression analysis of the data provided equations (16), (17), and (18) (Hansch and Deutsch, 1966):

	n	s	r	
$\log k_i = 2.709\sigma^* + 4.490$	8	0.816	0.712	(16)
$\log k_i = 1.119E_s + 4.541$	8	0.563	0.875	(17)
$\log k_i = 2.359E_s - 3.913\sigma^* + 4.948$	8	0.438	0.939	(18)

It is apparent from the values of s and r that the rate constant for AChE inhibition is dependent on both reactivity (σ^*) and steric (E_s) effects and a highly satisfactory correlation is obtained when both parameters are included in the analysis. The importance of steric effects with this series is demonstrated graphically in Figs. 1 and 2. The plot between $\log k_i$ and σ^* (Fig. 1) reveals a linear relationship except for the branched isopropyl and *tert*-butyl derivatives, which are substantially less effective as anticholinesterases than predicted from σ^* values. In comparison, Fig. 2, which shows a plot between experimental values of $\log k_i$ and values calculated from equation (18), gives an excellent linear relationship and provides support that the lower than expected anticholinesterase activity of the bulky isopropyl and *tert*-butyl derivatives is attributable to steric interference.

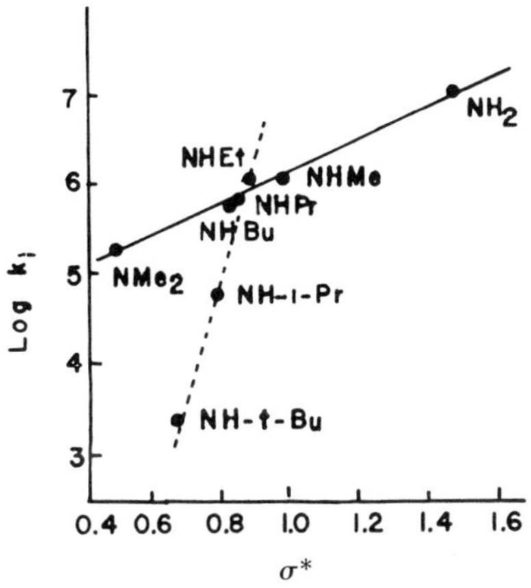

Fig. 1. Relation between log k_i for HFAChE and Taft's σ^* values for methyl 2,4,5-trichlorophenyl *N*-alkylphosphoramidates.

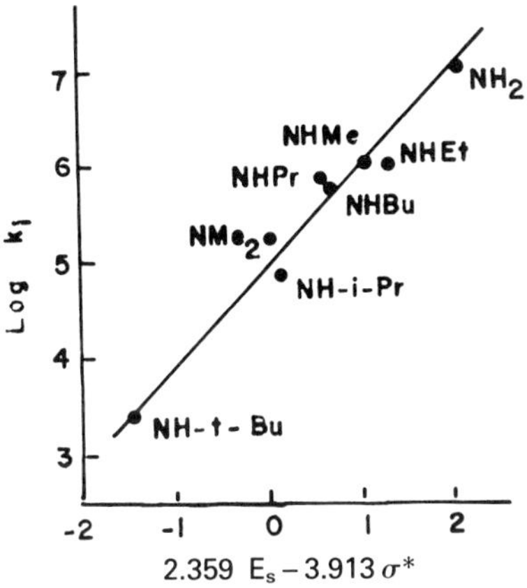

Fig. 2. Relation between log k_i for HFAChE and log k_i calculated from ($2.359E_s$–$3.913\sigma^*$) for methyl 2,4,5-trichlorophenyl *N*-alkylphosphoramidates.

3.1.3. Stereospecificity in AChE Inhibition

The cholinesterase enzymes frequently exhibit marked stereospecificity in their reactions with geometric and chiral isomers of organophosphorus esters. Effects attributable to chirality at the phosphorus atom often are quite profound and enormous differences in anticholinesterase activity between enantiomers of asymmetrical organophosphorus esters have been observed (De Jong and Van Dijk, 1972; Wustner and Fukuto, 1973; Keijer and Wolring, 1969), particularly with the acetylcholinesterases. For example, the rate constant for the inhibition of bovine erythrocyte acetylcholinesterase (BAChE) by (−)-*O*-isopropyl *S*-2-(trimethylammonium)ethyl methylphosphonothioate is approximately 3000-fold greater than that observed for the (+)-enantiomer (De Jong and Van Dijk, 1972). Differences in rate constants greater than 10,000-fold also have been observed between other chiral phosphorus esters (Wustner and Fukuto, 1973; Keijer and Wolring, 1969) (see Table II). In virtually all cases, the (−)-rotating isomer was more effective as an anticholinesterase than the (+)-isomer. The absolute configurations of the isomers of *O*-ethyl *S*-2-(ethylthio)ethyl ethylphosphonothioate and *O*-2-butyl *S*-2-(dimethylammonium)ethyl ethylphosphonothioate have been established by X-ray analysis of the corresponding *O*-alkyl ethylphosphonothioic acid. The configuration of the (−)-phosphorus esters has been established as *S* (Wustner and Fukuto, 1974; Lin *et al.*, 1974).

The differences in anticholinesterase activity between chiral phosphorus esters must be rationalized on steric grounds since on a purely chemical basis

Table II. Affinity (K_a), Phosphorylation (k_p), and Bimolecular Inhibition (k_i) Constants for BAChE and HFAChE by the Isomers of O-2-Butyl S-2-(Dimethylammonium)ethyl Ethylphosphonothioate

No.	Configuration	Enzyme	K_a ($M^{-1}\times 10^5$)	k_p (min^{-1})	k_i ($M^{-1}\,min^{-1}\times 10^5$)
1	S_CS_P	BAChE	0.944	87.4	92.6
2	S_CR_P	BAChE	234	15·5	0.0648
3	R_CS_P	BAChE	2.00	56.7	28.4
4	R_CR_P	BAChE	298	5.02	0.0174
1	S_CS_P	HFAChE	0.670	111	165
2	S_CR_P	HFAChE	86.1	6.08	0.0706
3	R_CS_P	HFAChE	0.548	80.3	146
4	R_CR_P	HFAChE	195	3.13	0.0160

reactivity arguments are not applicable to rate differences between chiral isomers. Support for this is found in the K_a and k_p data presented in Table II for the inhibition of BAChE and HFAChE by the four isomers of *O*-2-butyl *S*-2-(dimethylammonium)ethyl ethylphosphonothioate (Wustner and Fukuto, 1974) in which asymmetry resides at the 2-butyl carbon and phosphorus atom.

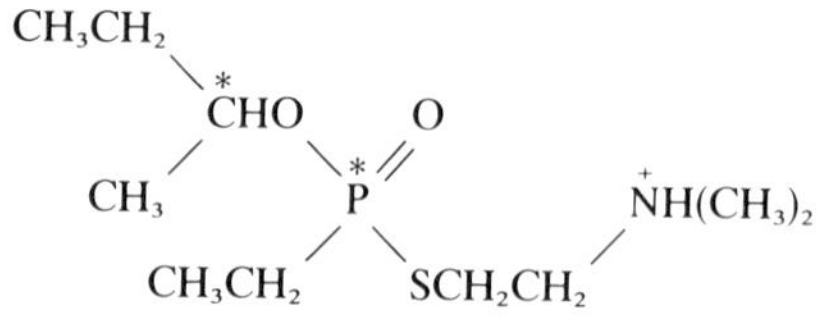

The symbols R_C or S_C and R_P or S_P refer to the absolute configuration of the chiral carbon and phosphorus centers. The data indicate that the large differences in k_i for the various isomers against both BAChE and HFAChE are attributable mainly to K_a and that differences in k_p contribute to a lesser degree. For example, comparison of isomers No. 1 and 4 for the inhibition of HFAChE reveals K_a values of 6.70×10^{-6} and 1.95×10^{-3} M, a difference of 290-fold, and k_P values of 111 and 3.13 min^{-1}, a difference of 35-fold. These differences in affinity and phosphorylation constants combine to give a difference in k_i of 10,340-fold. Similarly, the large difference (3071-fold) in the inhibitory activity of the enantiomers of *O*-isopropyl *S*-2-(trimethylammonium)ethyl methylphosphonothioate against BAChE has been attributed to a 540-fold difference in K_a and a 5.8-fold difference in k_p (De Jong and Van Dijk, 1972). Thus it appears that the stereochemical properties which are responsible for the high affinity (K_a) of the more active isomer also are responsible, although to a much smaller extent, for rapid phosphorylation (k_p).

Asymmetry in other parts of an organophosphorus anticholinesterase, i.e., centers other than phosphorus, contribute to differences in inhibition, but the effects are substantially smaller, e.g., compare isomer No. 1 with No. 3 and isomer No. 2 with No. 4 in Table II. A fourfold difference in the inhibition of BAChE by the enantiomers of malaoxon, a compound where chirality resides on the α-carbon of the diethyl mercaptosuccinate moiety, has been attributed to twofold differences in both K_a and k_p (Chiu and Dauterman, 1969).

Since toxicity to insects or to mammals parallels anticholinesterase activity—i.e., chiral isomers which are stronger inhibitors are more effective toxicants—it appears that the individual isomers are being metabolized by the test animal at more or less equal rates. Table III presents anticholinesterase and toxicity data for the various chiral isomers of *O*-2-butyl *S*-2-(ethylthio)ethyl ethylphosphonothioate (Wustner and Fukuto, 1973). The results show that toxicity to houseflies (*Musca domestica*), mosquito larvae (*Culex pipiens fatigans*), and white mice is consistent with anticholinesterase activity, and the isomers containing phosphorus in the *S* configuration show greater toxicity. The toxicological results are in agreement with those obtained for other chiral organophosphorus esters (Hilgetag and Lehmann, 1959; Fukuto and Metcalf, 1959*b*).

Stereospecificity in AChE inhibition also has been observed with geometric isomers of organophosphorus esters. The principal constituents of the insecticide mevinphos are the *cis*- and *trans*-crotonate forms

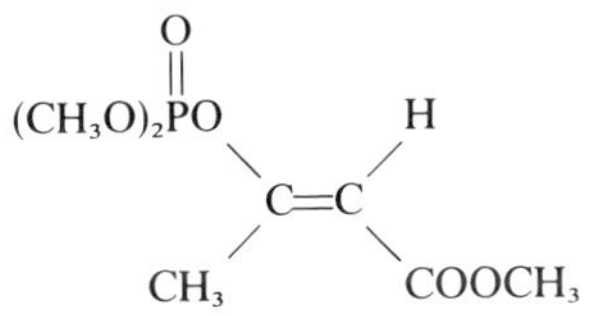

cis-crotonate

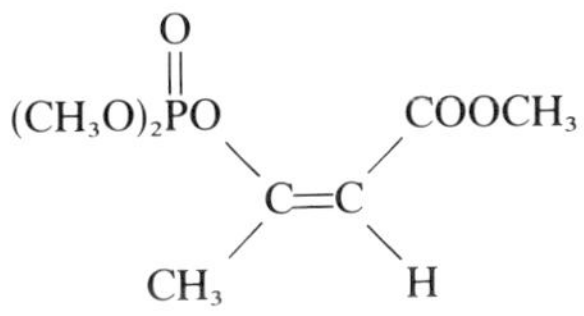

trans-crotonate

The *cis*-crotonate is about ten- to twentyfold more effective in inhibiting HFAChE and BAChE than the *trans*-crotonate (Fukuto *et al.*, 1961; Chiu and Dauterman, 1969). Corresponding differences in the toxicity of the isomers to insects also have been observed (Casida, 1955). From estimations of K_a and k_p for BAChE, it appears that the approximately twentyfold larger phosphorylation constant (k_p) for the *cis*-crotonate is primarily responsible for the difference in the inhibition constant k_i. K_a values for the two isomers are similar, indicating that anticholinesterase potency of these isomers depends more on reactivity than on steric effects.

3.2. Carbamate Esters

The toxicity of carbamate esters also is associated with the inhibition of the cholinesterase enzymes (Metcalf, 1971). The inhibition process is considered

Table III. Anticholinesterase and Toxicity Data for the Chiral Isomers of O-2-Butyl-S-2-(Ethylthio)ethyl Ethylphosphonothioate

No.	Configuration	k_i, BAChE (M^{-1} min^{-1})	k_i, HFAChE (M^{-1} min^{-1})	LD_{50} (mg/kg) Housefly	LD_{50} (mg/kg) White mouse	LC_{50} (ppm) *Culex* mosquito larvae
1	S_CS_P	6.53×10^4	1.71×10^6	6.7	3.2	0.25
2	S_CR_P	6.29×10^2	1.35×10^3	>500	110	>1
3	R_CS_P	5.45×10^4	1.69×10^6	6.9	2.8	0.15
4	R_CR_P	1.45×10^3	6.62×10^3	>500	125	>1

to be similar to that taking place with organophosphorus esters, the sequence of steps involving complex formation between the carbamate ester and enzyme, and eventual carbamylation of the enzyme. From kinetic analysis, it is evident that carbamylation is responsible for inhibition (Wilson *et al.*, 1960, 1961; O'Brien *et al.*, 1966). Although the mechanism of inhibition of AChE by carbamate esters is similar to that of organophosphorus esters, the chemical properties necessary for high anticholinesterase activity by carbamate esters are markedly different from those of organophosphorus esters. While chemical reactivity is of primary significance in the inhibition of AChE by organophosphorus esters, affinity or binding of the carbamate to the enzyme active site appears to be of greater importance in inhibition by carbamate esters. In general, those carbamate esters which structurally resemble the natural substrate acetylcholine, and which are therefore complementary to the enzyme active site, are found to be the most potent inhibitors (Metcalf, 1971).

3.2.1. Hydrophobic Effects in AChE Inhibition

By means of regression analysis of a large series of substituted phenyl methylcarbamate esters

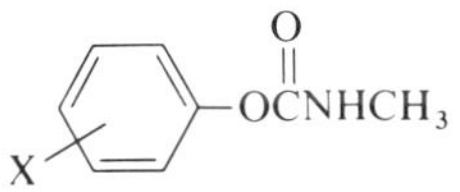

equations (19), (20), and (21) have been obtained, which relate anticholinesterase activity (HFAChE) with appropriate free energy parameters (Hansch and Deutsch, 1966). For the *para*-substituted derivatives equation (19) provided best fit to the data, equation (20) for the *meta* derivatives, and equation (21) for the *ortho* derivatives.

	n	s	r	
$\log 1/I_{50} = 0.714\pi - 0.868\sigma + 3.486$	23	0.399	0.839	(19)
$\log 1/I_{50} = 0.714\pi - 1.405\sigma + 4.618$	30	0.508	0.845	(20)
$\log 1/I_{50} = 3.895E_s + 2.799\pi + 4.246\sigma + 2.542$	7	0.494	0.962	(21)

The equations indicate that hydrophobic bonding (π) and chemical reactivity (σ) both play a role in AChE inhibition by carbamate esters, and, in the case of *ortho*-substituted derivatives, steric effects (E_s) also contribute to inhibition. In each of the series where separate analyses were made by using π or σ alone, π was substantially more significant than σ, suggesting that hydrophobic interaction between the carbamate and enzyme is more important than chemical reactivity in inhibition. Further, for the *meta* and *para* series the negative coefficient associated with σ indicates that ring substituents which release electrons to the carbamyl moiety, i.e., substituents which decrease the reactivity of the carbamate ester linkage, tend to increase anticholinesterase activity. The relationship obtained with the *ortho* derivatives appears to be a special case since it was necessary to introduce the steric substituent constant, E_s, into the equation and the coefficient associated with σ is positive. It should be pointed out, however, that relatively few *ortho* derivatives were examined.

A subsequent analysis (Hansch, 1970) in which monosubstituted *meta* and *para* derivatives were analyzed together provided equation (22), which gave best fit to the data.

$$\log 1/I_{50} = 0.69\pi - 0.95\sigma + 1.19X + 3.50 \qquad \begin{matrix} n & s & r \\ 53 & 0.415 & 0.913 \end{matrix} \qquad (22)$$

where X is a position term and a value of 1 is assigned to all *meta* isomers and 0 to *para* isomers. Again, when π or σ was used alone in separate analyses, π was more significant than σ. This coupled with the fact that the *meta* derivatives are approximately fifteenfold more active than the *para* derivatives indicates that hydrophobic bonding is most important for AChE inhibition and *meta* substituents are situated in positions which allow greater interaction with the enzyme active site. These results suggest that substituents located in strategic positions in the phenyl nucleus are capable of interacting hydrophobically with a lipophilic site situated approximately 5 Å away from the esteratic site where carbamylation actually takes place. This may be a separate lipophilic site (Bracha and O'Brien, 1968) or may be the anionic site (Cohen and Oosterbaan, 1963) in which resides a lipophilic pool.

The inhibition of AChE by carbamate esters is envisioned to take place by the following process (Metcalf, 1971):

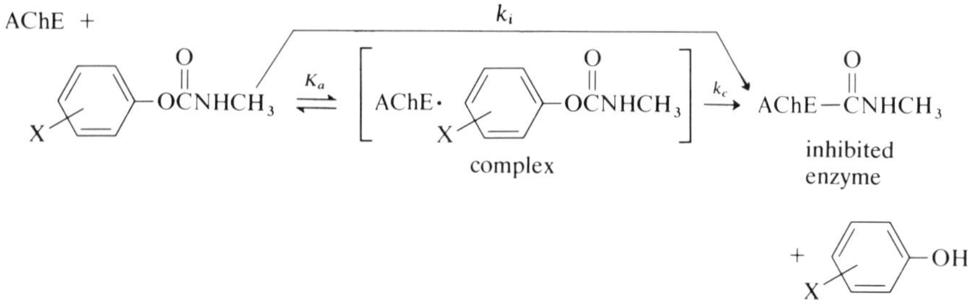

The constant for the dissociation of the enzyme–inhibitor complex is represented by K_a, the carbamylation rate constant by k_c, and the overall bimolecular inhibition constant by k_i. From the foregoing discussion on the relative roles of hydrophobic bonding and chemical reactivity on anticholinesterase activity, it is anticipated that carbamate esters with highest inhibitory activity will be those which readily combine with the enzyme through hydrophobic interaction to form the complex, i.e., compounds with low K_a values. Estimates of K_a and k_c for a variety of aryl methyl- and dimethylcarbamates reveal that affinity of the inhibitor for the enzyme (K_a) is the primary force which determines the inhibitory potency of carbamate esters (O'Brien *et al.*, 1966). Subsequent studies (Hastings *et al.*, 1970), however, reveal that the carbamylation step (k_c) is dependent on the magnitude of K_a and, in general, carbamylation occurs with a faster rate constant with compounds having high affinity (low K_a) for the enzyme.

3.2.2. Stereospecificity in AChE Inhibition

Stereochemical effects also have been observed in inhibition by carbamate esters (White and Stedman, 1937; Fukuto *et al.*, 1964). The L isomer of 2-*s*-butylphenyl methylcarbamate is sixfold more effective in inhibiting HFAChE than the D isomer and also is more toxic to insects. *l*-Miotine [2-(1-dimethylamino)ethylphenyl methylcarbamate] is two to ten times more toxic to a variety of test animals than the *d* isomer and is the more effective anticholinesterase. In both cases, the chiral center resides in the ring substituent and the stereochemical effects apparently are associated with hydrophobic binding of the substituent to the lipophilic site.

3.2.3. AChE Inhibition and Toxicity

The relationship between anticholinesterase activity and toxicity of carbamate esters to insects is complicated by the variability in which these compounds are metabolized in the insect. However, it is possible to minimize the rate of metabolic degradation of these esters by simultaneous treatment of the insect with a synergist such as piperonyl butoxide. When synergized toxicity (LD_{50}) is plotted against anticholinesterase activity, a highly satisfactory linear relationship is observed for a wide variety of substituted phenyl methylcarbamate esters (Metcalf, 1971).

3.3. DDT and Analogues

Despite the fact that it has been over three decades since the discovery of the insecticidal properties of DDT, relatively little information is available on the relationship between physicochemical properties of DDT and its analogues and insecticidal activity. While the toxicity of organophosphorus and carbamate esters has been attributed to the inhibition of the cholinesterase enzymes, an enzyme system has not been revealed which can be associated with the insecticidal activity of DDT. Although a number of suggestions have been made to account for the unique insecticidal activity of DDT, most of these have been unsatisfactory and have been discarded for various reasons (Brooks, 1973).

3.3.1. DDT Receptor Site

From the symptoms of intoxication developed in insects after exposure to DDT, it is clear that this insecticide acts as a neurotoxin (Chapter 9). The symptoms—hyperexcitability, ataxia, and convulsion—have been associated with the repetitive discharge observed in the sensory, central, and motor nervous systems in DDT-poisoned insects (Roeder and Weiant, 1948). Prolongation of the falling phase of the action potential in DDT-treated insect nerve also has been observed (Yamasaki and Narahashi, 1957) and this has been attributed to continued permeability of the nerve membrane to sodium ions (Narahashi and Hass, 1968; Hille, 1968). Evidently, the toxicity of DDT to insects is related to its effect on nerve membrane conductance and current belief is that DDT interacts directly with the nerve membrane through physicochemical forces (Mullins, 1956; Gunther *et al.*, 1954; Holan, 1969).

That the basis for the insecticidal activity of DDT is its ability to interact in some manner with a hypothetical receptor site located in a nerve membrane, thus affecting nerve conductance, was first suggested about two decades ago. Insecticidal activity of DDT and several of its carbon isosteres was correlated with the sum of the van der Waal's bonding energies for the various substituents in the molecule, suggesting that activity was related to van der Waal's forces of attraction between the molecule and a receptor site (Gunther *et al.*, 1954). Subsequently, based on the insecticidal activity of DDT and lindane (γ-HCH), a DDT receptor site visualized as a cavity in the interspace between three membrane macromolecules was proposed by Mullins (1956) (Fig. 3). According to this model, the phenyl rings of DDT must fit into the cavity in an "end-on" orientation for maximum interaction between the rings and other substituents of DDT and the lipoprotein of the receptor site. The presence of DDT in the membrane cavity is believed to distort the membrane and cause leakage of sodium ions, leading to abnormal transmission of nervous activity.

The Mullins' model for the receptor site was modified by Holan (1969) to explain the insecticidal activity of several DDT analogues which could not readily be explained by the original model. The model is illustrated in Fig. 4 by projections of DDT and a *p*-ethoxy analogue, 1,1-di-(*p*-ethoxyphenyl)-2,2-dichlorocyclopropane, in the receptor site. The two compounds are equitoxic to houseflies and therefore should fit in the receptor site with equal facility. The broken line in Fig. 4A shows the limit in size of the receptor site for alkyl group substituents on the ring and the solid line denotes van der Waal's limits of negative atom dipoles. Thus, according to Holan's model, both compounds can be accommodated and DDT or its equivalent is able to distribute itself at the

Fig. 3. Fit of DDT in receptor site. Phenyl rings of DDT in the cavity are observed as lying perpendicular to the page as illustrated by structure on right.

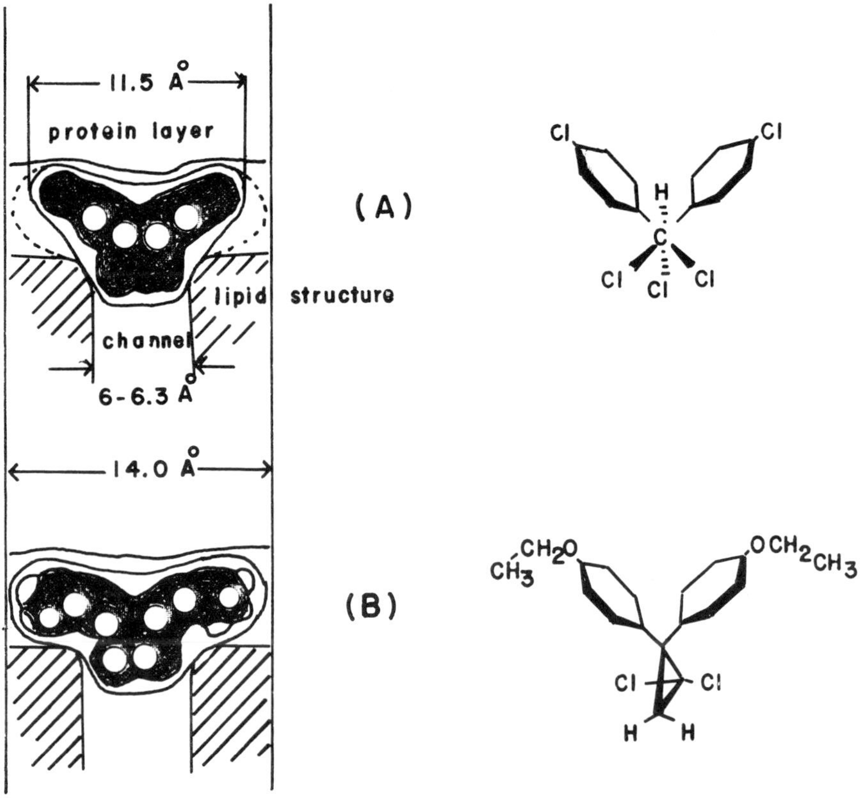

Fig. 4. Fit of DDT (A) and 1,1-di-(*p*-ethoxyphenyl)-2,2-dichlorocyclopropane (B) in the Holan receptor.

lipid–protein nerve membrane interface. In the interface, the substituted phenyl rings bind to the protein layer, perhaps through a charge transfer complex (O'Brien and Matsumura, 1960), and the side chain (—CCl_3 or dichlorocyclopropane) fits within the channel in the nerve membrane. In this position, DDT somehow is able to keep the channel open for sodium leakage through the membrane.

3.3.2. Substituent Effects on Toxicity

The Holan model represents a major advance toward a rational approach for the design of effective DDT-type insecticides and is directly responsible for the discovery of a number of DDT analogues with high insecticidal activity (Holan, 1971). Although the model is useful in predicting and explaining the activity of symmetrically substituted DDT analogues, it is not applicable to unsymmetrically substituted analogues, many of which have recently been observed to possess high insecticidal activity (Metcalf *et al.*, 1971). More recently, a modification of the Holan model has been proposed which takes into account the activity of unsymmetrically substituted DDT derivatives (Fahmy *et al.*, 1973). This revised model was suggested after analysis of the relationship between insecticidal activity and structure through a multiple-parameter free energy approach. The basic premise in this model is that DDT or its analogue fits into a receptor site in a nerve membrane similar to the one proposed by Holan (1969). The receptor site, however, was visualized as a cavity or pouch with a limited amount of flexibility and not as rigid as Holan's model. Maximum activity was anticipated where maximum interaction occurred between the receptor site and the four key substituents X, Y, L, and Z in the structure below:

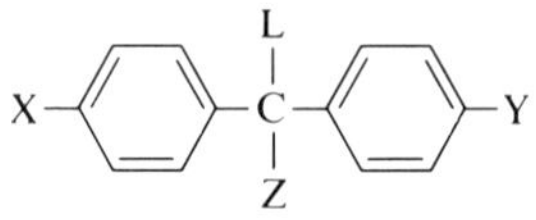

For maximum interaction, the overall size of the DDT molecule, i.e., summation of the size of X, Y, L, and Z, was considered to be critical and any deviation from this size was expected to result in reduced interaction, and therefore reduced toxicity. The concept of overall size and its relation to fit is illustrated in Fig. 5. M is a substituent approximately the size of chlorine, S is smaller than M, and L is larger than M. The point to be made is that good fit with the receptor site is not restricted to symmetrical molecules but also may be obtained with unsymmetrical analogues, as long as the overall size of the molecule remains within the flexible framework of the receptor site.

With this concept in mind, the relationship between the structure and insecticidal activity to houseflies and mosquito larvae of a large series of DDT analogues was examined by multiple regression analysis of activity and a variety of free energy constants. For a series of 25 compounds of the general structure in which X and Y are halogen, alkoxy, alkyl, and thioalkyl moieties,

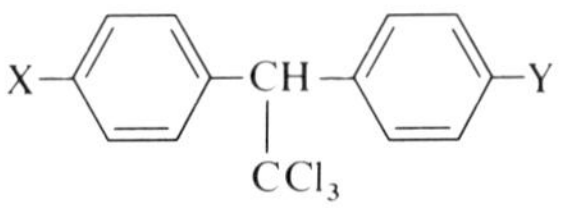

the following regression equations relating toxicity (LD_{50}) against houseflies, with (equation 23) and without (equation 24) the synergist piperonyl butoxide, were obtained:

	n	s	r	
$\log LD_{50} = 3.24 + 1.52 \sum E_s + 0.65 \sum E_s^2$	25	0.4	0.736	(23)
$\log LD_{50} = 2.69 + 1.85 \sum E_s + 0.85 \sum E_s^2$	25	0.31	0.874	(24)

Highest correlation was obtained with E_s and E_s^2, and introduction of other free energy parameters, e.g., σ and π, had no significant effect in improving the relationship. The use of piperonyl butoxide synergist resulted in improved correlation, and since this compound is an established inhibitor of the mi-

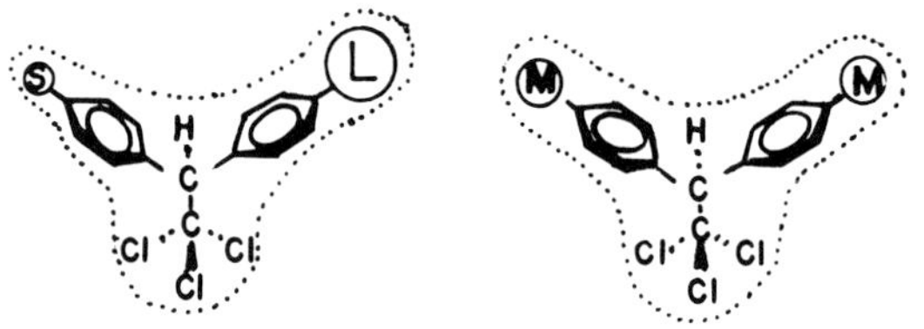

Fig. 5. Hypothetical models showing the fit of DDT analogues containing different-size ring substituents with the DDT receptor site.

crosomal mixed-function oxidase (Metcalf *et al.*, 1971) the improvement in correlation points out the importance of oxidative metabolism to the insecticidal activity of DDT analogues. In general, synergized toxicity may be regarded as a better measure of intrinsic toxicity owing to minimization of metabolic effects (Brooks, 1973). Against mosquito larvae equation (25) was obtained but in this case fewer compounds were available for analysis.

	n	s	r	
$LC_{50} = 1.63 \sum E_s + 0.93 \sum E_s^2$	12	0.21	0.933	(25)

The reasonably satisfactory correlation between insecticidal activity and E_s for the X and Y substituents provides support for the proposal that the sum of the size of these substituents, as estimated by E_s, is critical and that the receptor site is not as rigid as previously postulated (Holan, 1969). The squared term for E_s in the equations suggests that the relationship between insect toxicity and E_s is parabolic, i.e., analogues with substituents which are either too small or too large will be ineffective. This point is illustrated by the relationship given in Fig. 6, where log LD_{50} is plotted against E_s for a series of

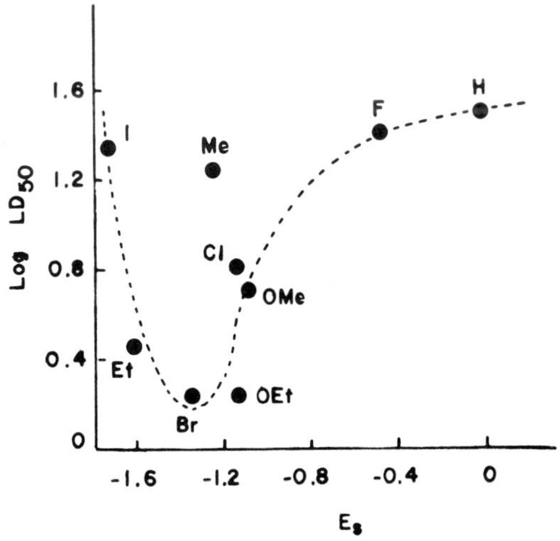

Fig. 6. Relation between observed synergized housefly toxicity and E_s for substituent X for 1,1,1-trichloro-*p*-methyl-*p'*-X-diphenylethanes.

1,1,1-trichloro-*p*-methyl-*p*′-X-diphenylethanes. In this case, one of the *para* substituents is restricted to the methyl group and the other is varied as indicated in the figure. With the exception of the point representing the methyl derivative (Me), the figure shows that there is a region where maximum insecticidal activity is obtained and activity decreases when E_s is above or below this region.

For compounds substituted in the L and Z positions of the structure

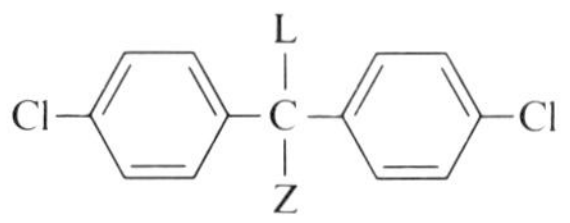

where L is either hydrogen or fluorine and Z is a halogenated methyl or ethyl moiety, the following regression equations were obtained. Analysis of unsynergized housefly toxicity data gave equation (26), and equation (27) and (28) were obtained for mosquito larvae.

	n	s	r	
$\log \mathrm{LD}_{50} = 14.14 + 9.36 \sum E_s + 1.12 \sum E_s^2 + 4.31 \sum \sigma^*$	9	0.31	0.856	(26)
$\log \mathrm{LD}_{50} = 5.29 + 4.90 \sum E_s + 0.76 \sum E_s^2 + 1.76 \sum \sigma^*$	14	0.29	0.889	(27)
$\log \mathrm{LD}_{50} = 7.27 + 6.03 \sum E_s + 0.76 \sum E_s^2 + 2.27 \sum \sigma^* + 0.37 \sum \pi^2$	14	0.16	0.970	(28)

Significant correlation was not obtained when each of the parameters was used alone, but the combination of E_s and σ^* was highly significant, indicating that steric and polar effects contribute to the interaction of L and Z with the receptor site in both houseflies and mosquito larvae. Against mosquito larvae, significant improvement in correlation was obtained by the introduction of a π^2 term (equation 28).

Overall, the steric substituent constant E_s appeared to be the single most important parameter for the correlation of housefly and mosquito larvae toxicity when substituents in the X, Y, L, and Z positions in DDT analogues were varied. The successful use of $\sum E_s$ and $\sum E_s^2$ in correlating insecticidal activity with structure suggests that there is a reasonable amount of flexibility in the DDT receptor site and that DDT analogues with substituents which provide best fit can be expected to show the highest insecticidal activity.

3.4. Nicotinoids

3.4.1. Mode of Action

Nicotinoids, a group of alkaloids isolated from the leaves of various species of *Nicotiana* and related plants, are cholinergic in their action; i.e., they mimic the action of acetylcholine at the synaptic junction in insects and mammals as well as the neuromuscular junction in mammals. At low concentration nicotine causes stimulation of the postganglionic fiber or muscle, and at high concentration blockage occurs. Because of the similarity in physiological action of nicotine and acetylcholine, it is believed that the nicotinoids exert their toxic effects by acting on the acetylcholine receptor sites located on the postganglionic membrane (Yamamoto, 1965) (Chapter 8). Nicotine, nornicotine, and anabasine represent a few of the insecticidally active constituents obtained from tobacco plants (Metcalf, 1955).

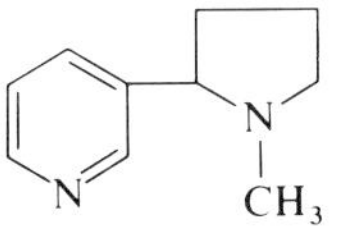

nicotine

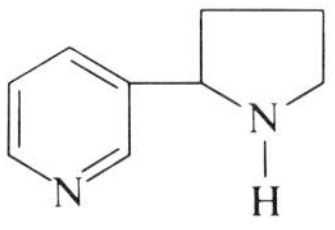

nornicotine

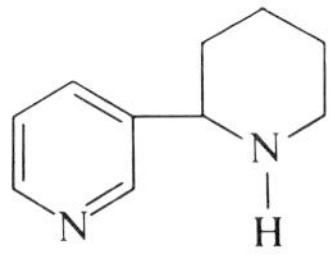

anabasine

3.4.2. *Structure Requirements for Toxicity*

Insecticidally active nicotinoids contain two basic centers in the molecule, a center of high basicity with pK_{a1} of the conjugate acid in the vicinity of 7.0–9.4 (pyrrolidine or other side chain nitrogen) and a center of low basicity with pK_{a2} of approximately 3.0 (pyridine nitrogen). Nicotine analogues with pK_{a1} values substantially lower than 7.0 are invariably ineffective insecticides. The pK_{a1} and pK_{a2} of nicotine itself are 7.9 and 3.1, respectively. At pH 7.0, i.e., near physiological pH, approximately 90% of nicotine in an aqueous environment exists in the protonated form, the remaining 10% as the free base.

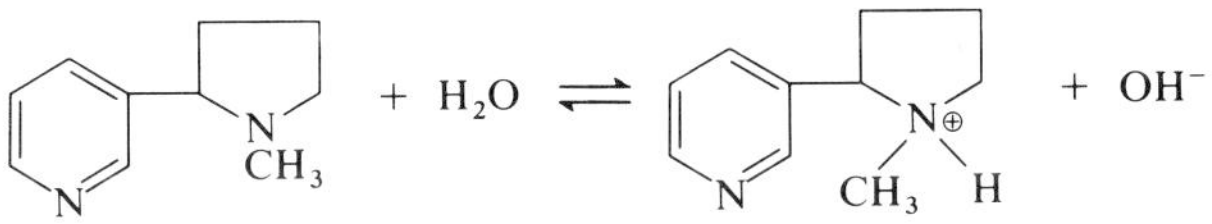

10% 90%

In the protonated form, nicotine bears a strong structural resemblance to acetylcholine as indicated below:

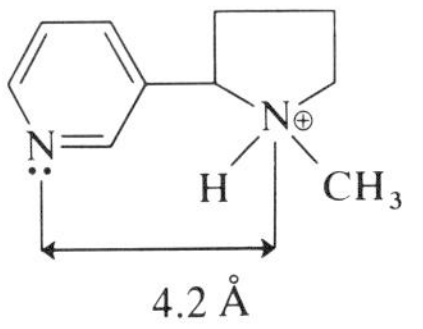

protonated nicotine

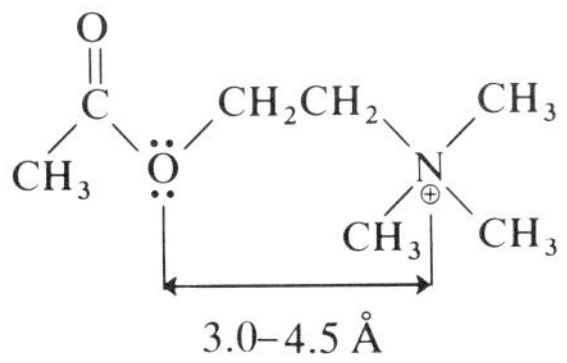

acetylcholine

The interspatial distance between the protonated pyrrolidine nitrogen and weakly basic pyridine nitrogen of nicotine is similar to the distance between the quaternary ammonium nitrogen and ester oxygen (ether) of acetylcholine. In view of the similarity in their physiological properties and structural dime sions, it is reasonable to believe that the site of nicotine action is the acety choline receptor site (Yamamoto, 1965). In mechanistic terms, the protonat pyrrolidine nitrogen is visualized as being attracted to the anionic site of t receptor by coulombic and hydrophobic forces, thus allowing the pyridi nitrogen to activate the membrane in a manner analogous to that effected the acetoxy moiety of acetylcholine.

In addition to the requirement for high pK_{a1} value for the side-cha nitrogen, the distance between the two nitrogen centers is critical for hig insecticidal activity. This is evident from the poor activity of nicotine analogu containing these centers closer together or farther apart than the 4.2 Å nicotine. For example, the anabasine analogues given below are practical ineffective as insecticides, even though *l*-anabasine is generally more effecti than nicotine:

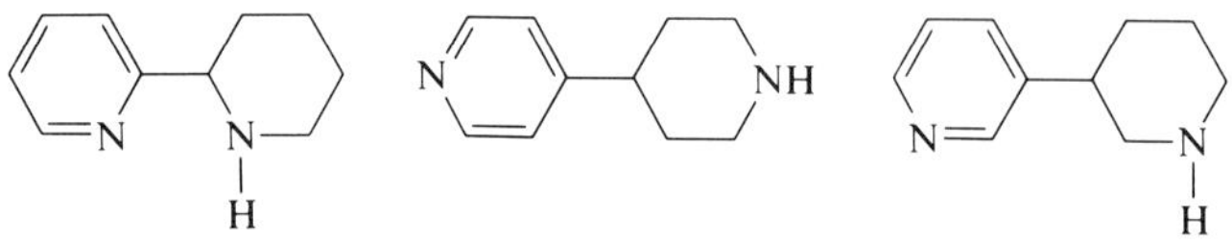

The inactivity of these compounds clearly indicates the rigid structural requir ments for target site interaction and shows that even small deviations fro 4.2 Å result in substantial loss of insecticidal activity.

3.4.3. Substituent Effects on Toxicity

Further insight into the mode of action of the nicotinoids is obtained fro regression analysis of housefly toxicities with physicochemical parameters for series of synthetic nicotine analogues of the general structure given belov where R_1 is methylene and R_2 and R_3 are hydrogen or alkyl (Fujita *et al.*, 1970)

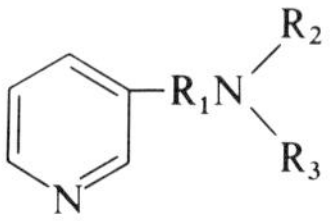

Nicotine, nornicotine, dihydronicotyrine, anabasine, and methylanabasine, a as the *dl* mixture, also were included in the analysis. For housefly toxicity th following equations were obtained:

	n	*s*	*r*	
$\log 1/\mathrm{LD}_{50} = 0.686 - 0.821 \sum \sigma^*$	19	0.309	0.720	(29
$\log 1/\mathrm{LD}_{50} = 0.525 + 0.126 \sum \pi$	19	0.429	0.270	(30
$\log 1/\mathrm{LD}_{50} = 0.877 - 0.247 \sum \pi - 1.258 \sum \sigma^*$	19	0.271	0.807	(31
$\log 1/\mathrm{LD}_{50} = 0.517 + 0.281 n_{\mathrm{H}} - 0.265 \sum \pi - 1.661 \sum \sigma^*$	19	0.259	0.838	(32

In equation (32), n_H is the number of hydrogen atoms on the protonated side-chain nitrogen, i.e., 3 for a primary nitrogen, 2 for a secondary nitrogen, and 1 for a tertiary nitrogen atom. The parameter n_H provides a measure of the extent of stabilization of the ammonium ion attributable to hydration energy and therefore would have an effect on K_{a1} and hydrogen-bonding properties of the conjugate acid.

Extension of the regression analysis to compounds with side chains longer than methylene, i.e., when R_1 is $—(CH_2)_2—$, $—(CH_2)_4—$, or $—CH{=}CHCH_2CH_2—$, resulted in

$$\log 1/LD_{50} = 0.527 + 0.274 n_H + 0.282 \sum \pi - 1.574 \sum \sigma^* - 0.689\, d_1 - 0.892\, d_2 - 0.613\, d_3 \quad n = 29 \quad s = 0.242 \quad r = 0.851 \qquad (33)$$

In this equation, the parameters d_1, d_2, and d_3 are used to account for the length of the side chain and d_1, d_2, and d_3 are, respectively, 1, 0, 0 for $—(CH_2)_2—$; 0, 1, 0 for $—(CH_2)_4—$; and 0, 0, 1 for $—CH{=}CHCH_2CH_2—$. The negative values for the coefficients reveal that toxicity decreases as chain length is increased beyond that of methylene.

It is apparent from equations (29)–(33) that the single most important physicochemical parameter which relates the toxicity of these nicotine analogues is $\sigma^*(\sum \sigma^* = \sigma^*_{R_1} + \sigma^*_{R_2} + \sigma^*_{R_3})$. The negative sign for the coefficient of $\sum \sigma^*$ shows that insecticidal activity increases with increase in the electron-donating properties of the substituents. This may be explained simply as a result of substituent effects on the basicity of the amino nitrogen atom. The authors (Fujita *et al.*, 1970), however, propose that electron-donating substituents promote greater insecticidal activity by virtue of their ability to reduce or delocalize the positive charge on the nitrogen atom of the protonated nicotinoid. This, in turn, is expected to cause a reduction in the iceberg structure of water molecules surrounding the ammonium nitrogen moiety, permitting stronger attraction of the cation to the oppositely charge anionic site of the receptor.

Of some interest is the absence of E_s in the regression equations, particularly since binding of the nicotinoids to the acetylcholine receptor site is believed to be responsible for the toxicity of these compounds. However, in light of the nature of the substituents R_1, R_2, and R_3, and the general parallel between E_s and π for alkyl groups, it is possible that π is serving the same purpose as E_s in the regression analysis. Comparison of equations (29) and (31) shows that significant improvement in correlation was obtained by the addition of $\sum \pi$ to $\sum \sigma^*$. Since E_s evidently was not introduced into the regression analysis, it is difficult at this point to decide whether steric effects play a role in the interaction of nicotinoids with the receptor site.

In vertebrates, the toxic action of nicotine and its analogues is attributed to the protonated form (Barlow and Hamilton, 1962) and this is probably also true in insects. It should be pointed out that penetration of nicotine into the insect and particularly through the lipoid ion barrier surrounding the nerve synapse most likely occurs in the form of the free base. Having once penetrated to the target site, the free base is converted to the conjugated acid, which interacts with the receptor site.

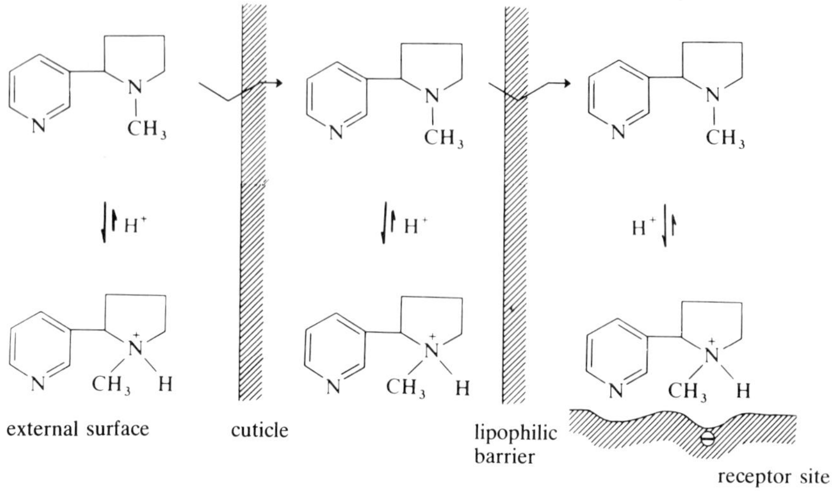

Because of the large pK_{a1} values of toxic nicotinoids, a minor portion of the material applied to insects exists in the free base form. This is consistent with the relatively low toxicity of nicotine to most insects and its high toxicity to soft-bodied insects, e.g., aphids, in which penetration of the ion may be less restricted (Yamamoto, 1965).

3.5. Insecticide Synergists

The use of adjuvants to enhance or synergize insecticidal activity has been practiced for many years. Compounds which are commonly used to synergize the toxicity of insecticidal carbamate esters or pyrethroids contain the methylenedioxyphenyl or 1,3-benzodioxole moiety, piperonyl butoxide and sesamex being typical examples. There is general agreement that the synergistic action of the 1,3-benzodioxole synergists is attributable to their ability to inactivate the mixed-function oxidases, the ubiquitous enzymes which are responsible for the oxidative metabolism of foreign compounds in animals (Wilkinson and Brattsten, 1972).

3.5.1. Substituent Effects in 1,3-Benzodioxole Synergists

The toxicity of the carbamate insecticide carbaryl (1-naphthyl methylcarbamate) to houseflies is synergized to varying degrees, depending on the nature of the 1,3-benzodioxole molecule (Wilkinson *et al.*, 1966; Wilkinson, 1967). 1,3-Benzodioxole synergists need not necessarily be complex molecules such as piperonyl butoxide and sesamex, and relatively simple compounds of the

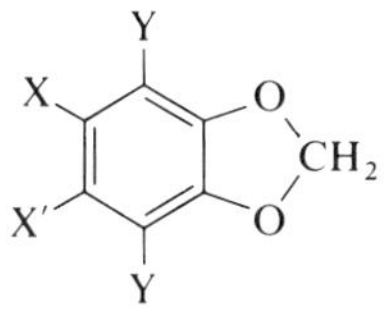

general structure given above, where X and X′ are halogens or nitro or methoxy moieties, and Y is hydrogen or halogen, have proved to be outstanding synergists (Wilkinson, 1967). For example, the toxicity of carbaryl to houseflies has been synergized approximately 500-fold by cotreatment with 4-methoxy-5-nitro-1,3-benzodioxole ($X = OCH_3$, $X' = NO_2$, $Y = H$).

Analysis of the synergistic activity of a series of these compounds (Hansch, 1968) revealed that synergism was related to the physicochemical parameters indicated in

	n	s	r	
$\log SR_5 = -0.195\,\pi^2 + 0.670\,\pi + 1.316\,\sigma\cdot + 1.612$	13	0.171	0.929	(34)

where SR_5 (synergistic ratio) is the ratio of LD_{50} values when carbaryl is used alone and in the presence of 5 parts of substituted 1,3-benzodioxole synergist, and $\sigma\cdot$ is a parameter based on radical phenylation of substituted benzenes (Williams, 1961). Use of $\sigma\cdot$, π and $\sigma\cdot$, or π^2 and π separately gave poor correlations, with $r \leqslant 0.638$. Substitution of σ_p, σ_I, or σ^+ for $\sigma\cdot$ in equation (34) also resulted in poor correlation.

In the analysis leading to equation (34), derivatives with bulky substituents in both the 4 and 5 positions, i.e., $X = NO_2$ and $X' = NO_2$, Br and Cl, were omitted since inclusion of data for these compounds resulted in poor correlation, attributable to steric interference of radical stabilization by large adjacent substituents. However, by introducing E_s into the analysis to account for *ortho* interaction, a highly satisfactory correlation was obtained:

	n	s	r	
$\log SR_5 = -0.206\,\pi^2 + 0.706\,\pi + 1.460\,\sigma\cdot + 0.875 E_s + 1.586$	16	0.164	0.943	(35)

Disregarding steric effects between adjacent substituents, equations (34) and (35) show that synergism of carbaryl by substituted 1,3-benzodioxole derivatives depends on both hydrophobic effects (π) and electronic effects ($\sigma\cdot$) related to radical reactions. The presence of $\sigma\cdot$ in the equations suggests that a radical intermediate is involved in the inactivation of microsomal mixed-function oxidase enzymes. For those derivatives having high synergistic activity, the radical formed after homolytic hydrogen atom abstraction is stabilized by resonance delocalization of the odd electron by the substituent. In turn, the relatively stable radical may inhibit the mixed-function oxidase by interfering with the free-radical-generating enzyme system responsible for oxidation of foreign molecules. Alternatively, the synergist may react directly with the radical process associated with the enzyme and cause inhibition by terminating the radical reaction. Either explanation is consistent with the analysis. This conclusion is contrary to the proposal that hydride ion removal is involved in the mode of action of 1,3-benzodioxole synergists (Hennesy, 1965). The inclusion of π and π^2 in the equation indicates that hydrophobic bonding also is important and this is most likely related to the effect of substituents on binding of the synergist to the oxidase enzymes.

3.5.2. Substituent Effects in 1,2,3-Benzothiadiazole Synergists

Observations similar to those described above for the substituted 1,3-benzodioxoles have been reported for a series of 1,2,3-benzothiadiazoles, substituted in the 5-, 6-, and 5,6-positions according to the general structure

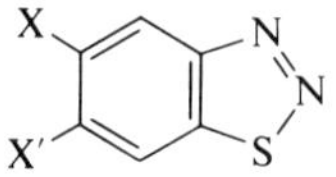

where X and X′ are either H, halogen, NO_2, OCH_3, or CH_3 (Gil, 1973). Several of these compounds were outstanding in synergizing the toxicity of carbaryl against houseflies; e.g., 5-methoxy-6-chloro-1,2,3-benzothiadiazole (X = OCH_3, X′ = Cl) increased toxicity 454-fold. As in the case of the 1,3-benzodioxoles, highest correlation of toxicity data was obtained when the free radical constant $\sigma\cdot$ and hydrophobic constant π were combined in the analysis:

	n	*s*	*r*	
$\log SR_5 = -0.138\,\pi^2 + 0.267\,\pi + 0.903\,\sigma\cdot + 1.816$	14	0.118	0.891	(36)

Substitution of $\sigma\cdot$ with σ in the regression analysis resulted in poor correlation ($r = 0.632$). The presence of $\sigma\cdot$ and the relative magnitude of its coefficient (0.903) suggest that radical reactions also are involved in the mode of action of the 1,2,3-benzothiadiazole synergists.

Concurrent *in vitro* studies on the inhibition of insect microsomal oxidase by the substituted 1,2,3-benzothiadiazoles, however, did not completely support this conclusion. While a reasonably satisfactory correlation was obtained among inhibition of aldrin epoxidation, π, and $\sigma\cdot$ (equation 37), a statistically more significant relationship was obtained when σ was substituted for $\sigma\cdot$ (equation 38).

	n	*s*	*r*	
$\log 1/I_{50} = -0.281\,\pi^2 + 0.561\,\pi + 0.0590\,\sigma\cdot + 3.863$	14	0.243	0.894	(37)
$\log 1/I_{50} = -0.347\,\pi^2 + 0.502\,\pi - 0.385\,\sigma + 3.960$	14	0.193	0.934	(38)

Since an ionic mechanism for the inhibition process is suggested by equation (38) and a radical process by equation (37), interpretation of these results in terms of carbaryl synergism is not immediately obvious. The rather small coefficient for $\sigma\cdot$ also reduces the importance of this parameter in the inhibition reaction. Whatever the case may be, it is clear, however, that hydrophobic and electronic effects are contributing factors in oxidase inhibition by the 1,2,3-benzothiadiazoles.

4. Conclusion

The foregoing discussion has focused on the use of physicochemical parameters in relating chemical structure to the properties associated with insecticidal action. While not all types of compounds are amenable to analysis using free energy related parameters, the multiple parameter approach for correlation of molecular structure with insecticidal action has been of immense value to our understanding of the toxicological action of several groups of compounds. The classes of chemicals discussed in this chapter—i.e., organophosphorus and carbamate esters, DDT analogues, nicotinoids, and insecticide synergists—cover a broad range of compounds, each representing special cases where different physicochemical parameters are employed to rationalize activity.

For the organophosphorus esters, reactivity (σ, σ^*) of the phosphorus atom is of primary significance for high anticholinesterase activity, although steric effects (E_s) often must be taken into account. On the other hand, reactivity appears to be of lesser importance in correlating anticholinesterase activity of carbamate esters, where hydrophobic effects (π) play a more significant role. With both types of compounds, chirality and geometric isomerism may have a profound effect on activity.

In constrast to the organophosphorus and carbamate esters, the insecticidal activity of DDT analogues and the nicotinoids is attributed to effects produced by the binding of these compounds to their respective receptor sites. In the case of DDT, appropriate molecular size, as estimated by $\sum E_s$ for the ring and side-chain substituents, is crucial for maximum insecticidal activity. It is possible that other physicochemical parameters to estimate size, e.g., molar volume, may serve equally well in place of E_s. Binding of the nicotinoids to the acetylcholine receptor site appears to be of a different nature since in this case polar (σ^*) and hydrophobic effects (π) both contribute to the forces involved in the interaction of these compounds with the receptor.

In each of the classes of insecticides discussed, the use of physicochemical parameters for analysis of biological activity has added notably to the elucidation of the mode of action of these compounds. Further, by relating biological activity to fundamental physicochemical concepts, it has been possible to identify the essential structural features necessary for insecticidal activity and use this information for an orderly approach to the design of new insect control agents.

5. References

Aldridge, W. N., and Davison, A. N., 1952, The inhibition of erythrocyte cholinesterase by tri-esters of phosphoric acid, *Biochem. J.* **51:**62.

Barlow, R. B., and Hamilton, J. T., 1962, Effects of pH on the activity of nicotine and nicotine mono-methiodide on the rat diaphragm preparation, *Br. J. Pharmacol.* **18:**543.

Bracha, P., and O'Brien, R. D., 1968, Trialkyl phosphate and phosphorothiolate anticholinesterases. I. Amiton analogs, *Biochemistry* **7:**1545.

Brooks, G. T., 1973, The design of insecticidal chlorohydrocarbon derivatives, *in: Drug Design*, Vol. 4 (S. Ariens, ed.), pp. 379–444, Academic Press, New York.

Brown, H. C., and Okamoto, Y., 1958, Electrophilic substituent constants, *J. Am. Chem. Soc.* **80:**4979.

Cammarata, A., and Martin, A. N., 1970, Physical properties and biological activity, *in: Medicinal Chemistry* (A. Burger, ed.), pp. 118–168, Wiley Interscience, New York.

Casida, J. E., 1955, Isomeric substituted-vinyl phosphates as systemic insecticides, *Science* **122:**597.

Chiu, Y.-C., and Dauterman, W. C., 1969, The affinity and phosphorylation constants of the optical isomers of *O,O*-diethyl malaoxon and the geometric isomers of phosdrin with acetylcholinesterase, *Biochem. Pharmacol.* **18:**359.

Cohen, J. A., and Oosterbaan, R. A., 1963, The active site of acetylcholinesterase and related esterases and its reactivity towards substrates and inhibitors, *in: Handbuch der Experimentellen Pharmakologie*, Vol. 15 (G. B. Koelle, ed.), pp. 299–373, Springer-Verlag, Berlin.

Cohen, L. A., and Jones, W. M., 1963, A study of free energy relationships in hindered phenols: Linear dependence for solvation effects in ionization, *J. Am. Chem. Soc.* **85:**3397.

Cram, D. J., 1965, *Fundamentals of Carbanion Chemistry*, p. 58, Academic Press, New York.

De Jong, L. P. A., and Van Dijk, C., 1972, Inhibition of acetylcholinesterase by the enantiomers of isopropyl *S*-2-trimethylammonioethyl methylphosphonothioate iodide: Affinity and phosphorylation constants, *Biochim. Biophys. Acta* **268:**680.

Fahmy, M. A. H., Fukuto, T. R., Metcalf, R. L., and Homstead, R. L., 1973, Structure–activity correlations in DDT analogs, *J. Agr. Food Chem.* **21:**585.

Ferguson, J., 1939, The use of chemical potentials as indices of toxicity, *Proc. Roy. Soc. London Ser. B* **127:**387.

Fujita, T., Iwasa, J., and Hansch, C., 1964, A new substituent constant, π, derived from partition coefficients, *J. Am. Chem. Soc.* **86:**5175.

Fujita, T., Yamamoto, I., and Nakajima, M., 1970, Analysis of the structure–activity relationship of nicotine-like insecticides using substituent constants, *in: Biochemical Toxicology of Insecticides* (R. D. O'Brien and I. Yamamoto, eds.), pp. 21–32, Academic Press, New York.

Fukui, K., Morokuma, K., Nagota, C., and Imamura, A., 1961, Electronic structure and biochemical activities in diethyl phenyl phosphates, *Bull. Chem. Soc. Jpn.* **34:**1224.

Fukuto, T. R., 1971, Relationship between the structure of organophosphorus compounds and their activity as acetylcholinesterase inhibitors, *Bull. WHO* **44:**31.

Fukuto, T. R., and Metcalf, R. L., 1956, Structure and insecticidal activity of some diethyl substituted phenyl phosphates, *J. Agr. Food. Chem.* **4:**930.

Fukuto, T. R., and Metcalf, R. L., 1959*a*, The effect of structure on the reactivity of alkylphosphonate esters, *J. Am. Chem. Soc.* **81:**372.

Fukuto, T. R., and Metcalf, R. L., 1959*b*, Insecticidal activity of the enantiomorphs of *O*-ethyl *S*-2-(ethylthio)ethyl ethylphosphonothiolate, *J. Econ. Entomol.* **52:**739.

Fukuto, T. R., Hornig, E. D., Metcalf, R. L., and Winton, M. Y., 1961, The configuration of the α and β isomers of methyl 3-dimethoxyphosphinyloxy) crotonate (phosdrin), *J. Org. Chem.* **26:**4620.

Fukuto, T. R., Metcalf, R. L., Winton, M. Y., and March, R. B., 1963, Structure and insecticidal activity of alkyl 2,4,5-trichlorophenyl *N*-alkylphosphoramidates, *J. Econ. Entomol.* **55:**889.

Fukuto, T. R., Metcalf, R. L., and Winton, M. Y., 1964, Carbamate insecticides: Insecticidal properties of some optically active substituted phenyl *N*-methylcarbamates, *J. Econ. Entomol* **57:**10.

Gil, D. L., 1973, Structure–activity relationships of 1,2,3-benzothiadiazole insecticide synergists Ph.D. dissertation, Cornell University, Ithaca, N.Y.

Gunther, F. A., Blinn, R. C., Carman, G. E., and Metcalf, R. L., 1954, Mechanisms of insecticidal action: The structure topography theory and DDT-type compounds, *Arch. Biochem. Biophys.* **50:**504.

Hammett, L. P., 1940, *Physical Organic Chemistry*, McGraw-Hill, New York.
Hancock, C. K., and Falls, C. P., 1961, Quantitative separation of hyperconjugation effects from steric substituent constants, *J. Am. Chem. Soc.* **83:**4211.
Hansch, C., 1968, The use of homolytic, steric, and hydrophobic constants in a structure–activity study of 1,3-benzodioxole synergists, *J. Med. Chem.* **11:**902.
Hansch, C., 1969, A quantitative approach to biochemical structure–activity relationships, *Accounts Chem. Res.* **2:**232.
Hansch, C., 1970, The use of physicochemical parameters and regression analysis in pesticide design, *in: Biochemical Toxicology of Insecticides* (R. D. O'Brien and I. Yamamoto, eds.), pp. 33–40, Academic Press, New York.
Hansch, C., and Deutsch, E. W., 1966, The use of substituent constants in the study of structure–activity relationships in cholinesterase inhibitors, *Biochim. Biophys. Acta* **126:** 117.
Hansch, C., and Dunn, W. J., III, 1972, Linear relationship between lipophilic character and biological activity of drugs, *J. Pharm. Sci.* **61:**1.
Hansch, C., and Fujita, T., 1964, ρ-σ-π analysis: A method for the correlation of biological activity and chemical structure, *J. Am. Chem. Soc.* **86:**1616.
Hastings, F. L., Main, A. R., and Iverson, F., 1970, Carbamylation and affinity constants of some carbamate inhibitors of acetylcholinesterase and their relation to analogous substrate constants, *J. Agr. Food Chem.* **18:**497.
Hennesy, D. J., 1965, Hydride-transferring ability of methylenedioxybenzenes as a basis of synergistic activity, *J. Agr. Food Chem.* **13:**218.
Helgetag, G., and Lehmann, G., 1959, Optically active thiophosphates, *J. Prakt. Chem.* **280:**224.
Hille, B., 1968, Pharmacological modifications of the sodium channels of frog nerve, *J. Gen. Physiol.* **5:**199.
Holan, G., 1969, New halocyclopropane insecticides and the mode of action of DDT, *Nature (London)* **221:**1025.
Holan, G., 1971, Rational design of degradable insecticides, *Nature (London)* **232:**644.
Hollingworth, R. M., Fukuto, T. R., and Metcalf, R. L., 1967, Selectivity of sumithion compared with methyl parathion: Influence of structure on anticholinesterase activity, *J. Agr. Food Chem.* **15:**235.
Jaffe, H. H., 1953, Reexamination of the Hammett equation, *Chem. Rev.* **53:**191.
Keijer, J. H., and Wolring, G. Z., 1969, Stereospecific aging of phosphonylated cholinesterases, *Biochim. Biophys. Acta* **185:**465.
Kier, L. B., 1971, *Molecular Orbital Theory in Drug Research*, Academic Press, New York.
Lin, G. H. Y., Wustner, D. A., Fukuto, T. R., and Wing, R. M., 1974, Absolute configuration of chiral *O*-2-butyl ethylphosphonothioic acid, *J. Agr. Food Chem.* **22:**1134.
Main, A. R., 1964, Affinity and phosphorylation constants for the inhibition of esterases by organophosphates, *Science* **144:**992.
Metcalf, R. L., 1955, *Organic Insecticides*, Interscience, New York.
Metcalf, R. L., 1971, Structure–activity relationships for insecticidal carbamates, *Bull. WHO* **44:**43.
Metcalf, R. L., Kapoor, I. P., and Hirwe, A. S., 1971, Biodegradable analogs of DDT, *Bull. WHO* **44:**363.
Meyer, H., 1899, The theory of alcohol narcosis, *Arch. Exp. Pathol. Pharmakol.* **42:**109.
Meyer, K. H., and Hemmi, H., 1935, The theory of narcosis, *Biochem. Z.* **277:**39.
Mullins, L. J., 1956, The structure of nerve cell membranes, *in: Molecular Structure and Functional Activity of Nerve Cells* (R. G. Grenell and L. J. Mullins, eds.), pp. 123–166, American Institute of Biological Sciences, Washington, D.C.
Narahashi, T., and Hass, H. G., 1968, The interaction of DDT and components of lobster nerve membrane conductance, *J. Gen. Physiol.* **51:**177.
Némethy, G., 1967, Hydrophobic interactions, *Angew. Chem. Int. Ed.* **6:**195.
O'Brien, R. D., and Matsumura, F., 1960, Absoption and binding of DDT by the central nervous system of the American cockroach, *J. Agr. Food Chem.* **14:**36.

O'Brien, R. D., Hilton, B. D., and Gilmour, L., 1966, The reaction of carbamates with cholinesterase, *Mol. Pharmacol.* **2:**593.

Overton, E., 1901, *Studien Uueber die Narkose,* Fischer, Jena, Germany.

Pullman, B., and Valdemoro, C., 1960, Electronic structure and activity of organophosphorus inhibitors of esterases, *Biochim. Biophys. Acta* **43:**548.

Roeder, K. D., and Weiant, E. A., 1948, The effect of DDT on sensory and motor structures in the cockroach leg, *J. Cell. Comp. Physiol.* **32:**175.

Shorter, J., 1970, The separation of polar, steric, and resonance effects in organic reactions by the use of linear free energy relationships, *Quart. Rev. Chem. Soc.* **24:**433.

Swain, C. G., and Lupton, E. C., 1968, Field and resonance components of substituent effects, *J. Am. Chem. Soc.* **90:**4328.

Taft, R. W., 1956, Separation of polar, steric, and resonance effects in reactivity, *in: Steric Effects in Organic Chemistry* (M. S. Newman, ed.), pp. 556–675, Wiley, New York.

White, A. C., and Stedman, E., 1937, The pharmacological and toxic actions of *d*- and *l*-miotine, *J. Pharmacol.* **60:**198.

Wilkinson, C. F., 1967, Penetration, metabolism, and synergistic activity with carbaryl of some simple derivatives of 1,3-benzodioxide in the housefly, *J. Agr. Food Chem.* **15:**139.

Wilkinson, C. F., 1973, Correlations of biological activity with chemical structure and physical properties, *in: Pesticide Formulations* (W. Van Valkenburg, ed.), pp. 1–64, Marcel Dekker, New York.

Wilkinson, C. F., and Brattsten, L. B., 1972, Microsomal drug metabolizing enzymes in insects, *Drug Metab. Rev.* **1:**153.

Wilkinson, C. F., Metcalf, R. L., and Fukuto, T. R., 1966, Some structural requirements of methylenedioxyphenyl derivatives as synergists of carbamate insecticides, *J. Agr. Food Chem.* **14:**73.

Williams, G. H., 1961, Some aspects of the organic chemistry of free radicals, *Chem. Ind. (London),* 1286.

Wilson, J. B., Hatch, M. A., and Ginsburg, S., 1960, Carbamylation of acetylcholinesterase, *J. Biol. Chem.* **235:**2312.

Wilson, J. B., Harrison, M. A., and Ginsburg, S., 1961, Carbamyl derivatives of acetylcholinesterase, *J. Biol. Chem.* **236:**1498.

Wustner, D. A., and Fukuto, T. R., 1973, Stereoselectivity in cholinesterase inhibition, toxicity, and plant systemic activity by the optical isomers of *O*-2-butyl *S*-2-(ethylthio)ethyl ethylphosphonothioate, *J. Agr. Food Chem.* **21:**756.

Wustner, D. A., and Fukuto, T. R., 1974, Affinity and phosphonylation constants for the inhibition of cholinesterases by the optical isomers of *O*-2-butyl *S*-2-(dimethylammonium)ethyl ethylphosphonothioate hydrogen oxalate, *Pestic. Biochem. Physiol.* **4:**365.

Yamamoto, I., 1965, Nicotinoids as insecticides, *in: Advances in Pest Control Research,* Vol. 6 (R. L. Metcalf, ed.), pp. 231–260, Interscience, New York.

Yamasaki, T., and Narahashi, T., 1957, Studies on the mechanism of action of insecticides. XIII. Increase in the negative afterpotential of insect nerve by DDT, *Botyu-Kagaku* **22:**296.

IV

Selectivity and Resistance

12

The Biochemical and Physiological Basis of Selective Toxicity

R. M. Hollingworth

1. Introduction

It may not be a travesty of the truth to suppose that as soon as the first two living organisms appeared in the primeval soup where life began, one tried to poison the other to its own advantage. Certainly the biological world is now full of poisoners and potential victims in a chemical warfare of a subtlety and virulence which man, fortunately, has not yet equaled. The potent natural neurotoxins described by Shankland in Chapter 6 are eloquent testimony to one route which has been exploited repeatedly by organisms in this struggle to survive and gain ascendency. Any organism thus finds itself continually challenged by the presence of potentially harmful chemicals from a variety of sources in the food, water, and air it takes in. In some cases these naturally occurring toxicants may be put to good use by man for insecticidal or medicinal purposes; for example, the root of false hellebore (*Veratrum* spp.) which contains the potent neurotoxic veratrine alkaloids has been used as an insect stomach poison (Crosby, 1971), and John Gerard (1633) in his classical herbal noted its vigorous therapeutic benefits:

> The root of white Hellebor procureth vomite mightily, wherein consisteth his chief vertue, and by that means voideth all superfluous slimes and naughtie humors.

Obviously, hellebore is not a selective stomach poison.

R. M. Hollingworth • Department of Entomology, Purdue University, West Lafayette, Indiana.

Naturally, these challenges to survival do not find a helpless victim. In response to the unfavorable aspects of their chemical environment, organisms have evolved a fascinating array of defensive mechanisms of varying generality. Thus Krieger *et al.* (1971) concluded that the level of detoxifying microsomal oxidase enzymes in the guts of various species of lepidopterous larvae was related to the range of plants on which the larvae fed, and thus these enzymes served as a general protection from food poisoning. Their conclusion echoed that of Gordon (1961), who considered that the generally high tolerance for contact insecticides among the polyphagous larvae of holometabolous insects was due to the built-in defenses which had evolved to allow these insects to withstand a variety of biochemical stresses.

Undoubtedly enhancement of detoxifying enzymes is a very common defensive strategy, but others exist; for example, consider the range of mechanisms by which certain insects can avoid poisoning by the natural insecticide nicotine while feeding on tobacco (*Nicotiana tabacum*). The green peach aphid (*Myzus persicae*) simply avoids exposure by confining its feeding to the phloem while the nicotine is in the xylem; thus no nicotine is taken in (Guthrie *et al.*, 1962). Interestingly, this same aphid rapidly succumbs on the related species *Nicotiana gossei*, which puts out a toxic exudate, perhaps nicotine, from the leaf hairs (Thurston and Webster, 1962); this the aphid cannot evade and the plant wins. A second strategy for survival is found in the tobacco hornworm (*Manduca sexta*) larva and possibly some other caterpillars which can rapidly and completely excrete large doses of nicotine unchanged (Self *et al.*, 1964*a*,*b*). Finally, there are a number of insect pests of tobacco, such as the tobacco wireworm (*Conoderus vespertinus*), differential grasshopper (*Melanoplus differentialis*), and cigarette beetle (*Lasioderma serricorne*), which have a high capacity to degrade nicotine to nontoxic metabolites, e.g., cotinine, as an effective means of defense against intoxication (Self *et al.*, 1964*a*). Along this line of evolutionary attack and defense between plants and insects, it is remarkable that even the lordly pyrethrum flower, the source of an extremely potent family of natural insecticides, is not immune to attack by insects, e.g., thrips, which somehow avoid self-destruction and raise the incongruous need to spray pyrethrum with an insecticide (Bullock, 1961).

Defensive strategies are particularly essential for many of the organisms which produce and store these potent toxins internally, in order to avoid self-poisoning. Thus millipedes which emit hydrogen cyanide as a defensive mechanism not only store the toxicant as mandelonitrile, an innocent cyanhydrin with benzaldehyde which is broken down to HCN only at the time of release, but also are themselves almost immune to cyanide poisoning. This fact is explained by the very low sensitivity of the normal biochemical target of cyanide, the mitochondrial respiratory chain, in these millipedes (Hall *et al.*, 1971). Equally fascinating is the fact that the completely disparate species which produce the powerful axonal blocking agent tetrodotoxin, i.e., puffer fishes and certain newts, have nervous systems which are highly insensitive to their own toxin, and the animals are themselves notably immune to poisoning by it (Kao and Fuhrman, 1967). In addition then to avoidance of exposure, rapid excretion, and metabolic degradation as routes to survival, as we see with nicotine, the above examples with cyanide and tetrodotoxin illustrate a fourth possibility, the development of a biochemical target which is insensitive to poisoning. It is significant that, on a greatly foreshortened time scale, the selection pressure imposed by insecticides has produced resistant strains of insects which utilize each of these mechanisms for survival (Chapter 13).

Upon this ancient toxicological chessboard, with evolutionary move and countermove, has appeared in very recent times a new and powerful force represented by the multitude of potentially toxic synthetic chemicals, including the modern organic insecticides, which man now releases into the environment. The old defense mechanisms, varying in nature and degree from species to species, may or may not prove equal to this unexpected challenge, and even quite closely related organisms may vary considerably in their ability to withstand such toxicants. Organisms separated by a wider evolutionary gulf, with major morphological differences and entirely divergent physiological systems in addition to their varying chemical defenses, may show an extreme range of susceptibility. Thus arises the subject of this chapter, selective toxicity, which deals with the occurrence, significance of, and reasons for the variations in toxicity of insecticides between different organisms.

1.1. Vertebrate Selectivity Ratios and Their Significance

The variations in susceptibility to many common insecticides, and the extreme differences in acute toxicity which are possible, are shown in Table I for a mammal, the rat, and an insect, the housefly. A particularly useful index of selectivity is the vertebrate selectivity ratio (VSR), which is LD_{50} vertebrate/LD_{50} insect. Obviously a high VSR indicates a compound much more toxic to insects than to vertebrates. From the VSR values in Table I, it is clear that at one extreme end of the range there are toxicants which are almost completely specific against insects. Naturally, little attention has been given to developing selective mammalicides at the other end, but the examples quoted in Table I are impressive enough, and no doubt these limits could be pushed much farther if it were useful. In fact, some cases of extraordinary specificity to vertebrates are known; for example, tetanus toxin, lethal to mammals at about 1 ng/kg (Chapter 6), is apparently completely innocuous to insects (Pappenheimer and Williams, 1952).

From Table I, it is obvious that, in comparing the rat and housefly, the great majority of commercial insecticides show at least some degree of selective insecticidal action (VSR 1–10); most are quite highly selective (VSR 10–100), a fair number have outstanding selectivity (VSR 100–1000), and a conspicuous few, particularly pyrethroids and juvenile hormone analogues (JHAs), show extreme specificity for insects (VSR > 1000).

However, before we place heavy reliance on such VSR values as they are used here and elsewhere, we should consider certain limitations on their

Table I. Comparative Acute Toxicities of a Number of Insecticides and Other Toxicants to the Rat and Housefly, and the Vertebrate Selectivity Ratio (VSR)

Compound	LD_{50} (mg/kg) Rat, oral (dermal)	Housefly, topical	VSR	Sources of data (rat, fly)
Selective mammalicides				
1. Quaternized tetram $[(EtO)_2P(O)SC_2H_4N(Et)_3{}^+I^-]$	0.17[a] (−)	>1,000[b]	<0.00017	*g,g*
2. Strychnine	1.4 (−)	>1,000[b]	<0.0014	*h,h*
3. Prostigmine	7.5[c] (−)	>5,000	<0.0015	*i,i*
4. Stauffer R-16,661 (keto derivative)	0.4–0.6 (−)	240	0.0017–0.0025	*j,j*
5. $[(CH_3)_2N]_2P(O)OC_6H_4NO_2(p)$	7.0[c] (−)	>500	<0.014	*i,i*
6. Schradan	42	1,932	0.022	*k,l*
7. Nicotine	20[a] (−)	500[b]	0.040	*h,h*
8. Aldicarb	0.6 (2.5)	5.5	0.11	*k,m*
9. Mexacarbate	25 (1,500–2,500)	65	0.38	*k,m*
10. Carbaryl	500 (>4,000)	>900	<0.56	*k,m*
11. Carbofuran	4.0 (−)	4.6	0.87	*m,m*
Nonselective				
12. Sulfotepp	5.0 (−)	5.0	1.0	*n,n*
Selective insecticides (VSR = 1–10)				
13. Endrin	7.5 (15)	3.15	2.4	*k,m*
14. Mevinphos	3.7 (4.2)	1.5	2.5	*k,m*
15. Demeton	2.5 (8.2)	0.75	3.3	*k,m*
16. Propoxur	86 (>2,400)	25.5	3.4	*k,m*
17. Ethyl parathion	3.6 (6.8)	0.9	4.0	*k,m*
18. Azinphosmethyl	11 (220)	2.7	4.1	*k,m*
19. Toxaphene	80 (780)	11.0	7.3	*k,m*
20. Chlorfenvinphos	13 (30)	1.4	9.3	*k,o*
Selective insecticides (VSR = 10–100)				
21. Methyl parathion	24 (67)	1.2	20	*k,m*
22. Aldrin	60 (98)	2.25	27	*k,m*
23. Malathion	1,000 (>4,444)	26.5	38	*k,m*
24. DDT	118 (2,510)	2	59	*k,m*
25. Allethrin	920 (11,300)[d]	15[e]	61	*n,n*
26. Abate	13,000 (>4,000)	205	63	*k,p*
27. Heptachlor	162 (250)	2.25	72	*k,m*
28. Diazinon	285 (455)	2.95	97	*k,m*
29. Aspon	1,450 (=)	15	97	*n,n*
Selective insecticides (VSR = 100–1000)				
30. Fenthion	245 (330)	2.3	107	*k,m*
31. Lindane	91 (900)	0.85	107	*k,m*
32. Fenitrothion	570 (300–400[!])	2.3	248	*k,m*

Table 1. continued

Compound	LD$_{50}$ (mg/kg) Rat, oral (dermal)	Housefly, topical	VSR	Sources of data (rat, fly)
Selective insecticides—*cont.*				
33. Dimethoate	245 (610)	0.55	445	*k,m*
34. Bromophos	1,730 (>5,000)	3.2	541	*k,m*
35. Stirofos	1,125 (>4,000)	1.6	703	*k,o*
36. Alodan	>15,000 (–)	15.5	>968	*q,q*
Selective insecticides (VSR > 1000)				
37. Dihydroheptachlor	5,000 (–)	3.75	1,333	*m,m*
38. Methoxychlor	6,000 (>6,000)[d]	3.4[e]	1,765	*n,n*
39. Permethrin	1,500 (–)	0.7[e]	2,143	*r,r*
40. Phoxim	8,500 (–)	2.3	3,696	*m,m*
41. Bioresmethrin	>8,000 (–)	0.4[e]	>20,000	*r,r*
42. Stauffer R-20458	>4,640 (–)	0.0125[f]	>371,000	*s,s*
43. Methoprene	>34,600 (–)	0.020[f]	>1,730,000	*t,u*

[a]Mouse, intraperitoneal.
[b]Injected.
[c]Mouse, oral.
[d]Rabbit, dermal.
[e]Original data in μg/insect.
[f]*Tenebrio molitor* pupae; original data in μg/insect.
[g]O'Brien (1959).
[h]O'Brien and Fisher (1958).
[i]Metcalf (1964).
[j]Fukuto *et al.* (1972).
[k]Gaines (1969), female rat.
[l]O'Brien (1967*a*).
[m]Metcalf (1972*b*).
[n]Negherbon (1959).
[o]Sun (1972).
[p]Leesch and Fukuto (1972).
[q]Winteringham (1969).
[r]Elliott *et al.* (1973).
[s]Pallos and Menn (1972).
[t]Kenaga and End (1974).
[u]Henrick *et al.* (1975).

validity, which also illustrate quite graphically the complexities of evaluating and exploring selective toxicity.

First, VSR values deal with acute toxicity, when in some cases the major hazard to nontarget organisms is from chronic effects. Limitations of space and the fact that the basis for selectivity of chronic effects is not well understood rule out any comprehensive discussion of this area here, but quite clearly selectivity in chronic and delayed effects occurs and may be of extreme practical significance (Chapter 14). In several studies dieldrin has been reported to be a carcinogen after prolonged exposure in mice, but such effects have not been seen in other mammals such as the dog, rat, or monkey, nor are all strains of mice necessarily of equal sensitivity (Walker *et al.*, 1972). In the field of teratology, carbaryl shows distinct selectivity, causing birth defects at low dietary levels in dogs and guinea pigs, but not in hamsters and rabbits even at much higher doses (Hathway and Amoroso, 1972). Furthermore, repeated administration of DDT rapidly increases the level of hepatic microsomal mixed-function oxidases (MFOs) in rats and most other mammals, but this has been difficult to show in mice and houseflies, and in quail and some other birds DDT causes a decrease in activity of these enzymes (Gillett, 1971). Dieldrin meanwhile is an almost universal inducing agent. On another front, the delayed neurotoxicity of some organophosphates has been particularly easy to induce in some species such as man and the hen, but is more difficult to observe in others such as the rat and the duck (Johnson, 1969). Finally, one can quote the example of those chlorinated hydrocarbons which have the ability to cause thinning of calcareous egg shells. Necessarily this type of reproductive toxicity is largely limited to birds, but even these may differ widely in their sensitivity, from the resistant gallinaceous species to the susceptible falcons (Cooke, 1973). Each of these instances illustrates selectivity in a nonacute toxic effect which has important and sometimes controversial implications for pesticide toxicology.

The example of the differential induction of MFO by chlorinated hydrocarbons introduces another aspect of the toxic effects of pesticides considered only in passing here, the sublethal actions such as effects on behavior, fertility, and growth rate. Nevertheless, these sublethal effects are not unimportant and are equally subject to selectivity. Really, all these examples go to prove that wherever there is toxicity there will be selectivity, whether that toxicity be chronic or acute, lethal or sublethal.

A further comment on the VSR values of Table I is that they represent relative values when absolute ones are also important. Thus all three hypothetical compounds below have the seemingly excellent VSR value of 100:

Compound	LD_{50}, vertebrate (mg/kg)	LD_{50}, insect (mg/kg)	VSR
A	2	0.02	100
B	200	2	100
C	20,000	200	100

However, compound A still has high hazard to the vertebrate which may prevent its use, and compound C is of such mediocre toxicity to insects that it is probably not useful either, while compound B may indeed be ideal. A practical instance of this is found in Table I in comparing compounds 21 and 23, methyl parathion and malathion. Although their VSR values are comparable, methyl parathion presents a relatively high mammalian hazard while malathion does not.

A further problem in using the VSR approach and in assessing selectivity in general is that the results of toxicity determinations represent only the response of a limited population, at a given time, and under restricted test conditions. It is not the purpose of this chapter to consider in detail the multitude of factors which can affect the outcome of toxicity tests, but it is universally accepted that the conditions of the test (e.g., route and form of dosage, time of day, and holding temperature after dosage) and the nature of the animals used (e.g., sex, age, nutrition, size, and strain) may greatly affect the outcome. Thus any LD_{50} (or related) value for any species is only one estimate from a potentially wide range of values. The VSR consists of the ratio of two such estimates and thus may vary widely even for the same compound in the same two species.

This fact deserves further illustration. Thus in Table I compounds 9, 10, and 16, all carbamates, appear to have quite unfavorable VSR values. This is in part due to their low toxicity to the housefly rather than their extraordinarily high mammalian toxicity, but the actual danger to mammals under conditions of use is much less than these VSR values would predict since their toxicity by the dermal route is much lower than by the oral one, and dermal exposure is probably the more likely hazard in practice.

The effect of dosage form as well as route is dramatically shown with the experimental chlorinated insecticide Allied GC-9160 (Kenaga and End, 1974), which is reported to have an acute oral LD_{50} to rats of 240–290 mg/kg when given in corn oil and 9000–15,000 mg/kg when given as an aqueous suspension (Allied Chemical Technical Bulletin). The anticuticular compound Thompson-Hayward TH-6040 (Dimilin), to be discussed later, is toxic only to immature insects, only by the oral route, and even then the potency is highly dependent on the fineness of the particle size administered (Mulder and Gijswijt, 1973). Holden (1973) quotes data which show that DDT is much more toxic to fish when used in an oil formulation than as an aqueous suspension, while the reverse is true for carbaryl, and Alabaster (1969) has also pointed out the overriding effects which differences in formulation may have on the toxicity of pesticides to fish.

Age may be a vital factor in the toxicity observed. Thus in Table I the toxicity data for the JHA compounds 42 and 43 are given for pupae rather than adults, since the latter are highly insensitive. Even in the immatures, sensitivity to JHA is present only for brief periods during development, perhaps only a few hours or 1 or 2 days. If this period is missed, even a 10^6-fold increase in dose may have no lethal effects (Slama *et al.*, 1974). Less dramatic but considerable effects of age, sex, and diet on toxicity of carbamates to houseflies were found by El-Aziz *et al.* (1969); for example, the topical LD_{50} for *m*-isopropylphenyl *N*-methylcarbamate to 1-day-old male flies was 104 mg/kg and showed a progressive drop to 7.8 mg/kg with flies which were 7 days old. In this case the variation in sensitivity is probably due to differences in metabolic capacity with age, since pretreatment with synergists which inhibit detoxication reduced the age-related differential toxicity greatly. On the other hand, not all species of Diptera show this age dependence of response, and age has only a minor effect on toxicity of the chlorinated hydrocarbons aldrin and dieldrin to houseflies (Brattsten and Metcalf, 1973*a*). Sensitivity also shows an extraordinary dependence on age with the formamidine insecticide, chlordimeform, which with lepidopterans is toxic only to eggs and the youngest larvae (Hollingworth, 1975*b*).

Finally, many examples of differences in sensitivity to toxicants between the sexes can be quoted. The rat is unusual and notorious among mammals in this respect largely because of the greater activity of mixed-function oxidases in the male; for example, the acute oral LD_{50} for parathion is 13 mg/kg for the male rat and 3.6 mg/kg for the female. On the other hand, the respective values for schradan are 9.1 and 42, with the male rat more susceptible (Gaines, 1969).

These rather random examples of the wide variability of toxicity of a single compound to individuals within the same species hardly scratch the surface of the subject, but they do reveal the considerable problems of using any single value (such as the VSR) to represent the degree of selectivity of any compound between two species and imply that selectivity in the laboratory and even more so in the field may be highly variable and unexpectedly complex.

1.2. Selectivity at the Species Level

To continue this rather pessimistic line of thought, what if we could with good precision define with any insecticide an absolute VSR for an insect and vertebrate pair? Would this be sufficient to usefully define its overall selectivity? Undoubtedly not, because one of the major sources of variability in toxicity tests arises in comparing different species. Thus the VSR for one insect–vertebrate combination may be absolutely different from that for another; for example, in Table I the VSR for Abate (compound 26) at 63 is quite favorable, but obviously we have here an unusually poor toxicant for the housefly. In fact, a VSR derived for toxicity to the mosquito larva rather than fly would place Abate among the most selective compounds since it is an excellent larvicide with an LC_{50} of 6.3 ppb to *Aedes aegypti* (Leesch and Fukuto, 1972). Also, continuing the comparison of malathion and methyl parathion, the former is often presented as a model for the desirable selective insecticide of low environmental hazard while the latter quite definitely is not. What then do we make of the relative toxicity of these two organophosphates to the bluegill, where the 96-h LC_{50} for malathion is 55 ppb and for methyl parathion is 5720 ppb (Holden, 1973)!

Since the species level is a most crucial one for comparisons of selectivity, three further, rather extreme situations are presented in Tables II, III, and IV. In the first instance, the VSR value for the organophosphate chlorfenvinphos varies from 7.4 in comparing the rat and housefly to more than 8824 in comparing the dog and housefly, a situation which will be explored in greater depth later. The variation between the bird species is less immense but still considerable.

Table II. Variations in the Vertebrate Selectivity Ratio (VSR) of Chlorfenvinphos with Different Species of Mammals and Birds[a]

	LD_{50} (mg/kg)	VSR (vertebrate/housefly)
Housefly (topical)	1.36	—
Rat	10	7.4
Mouse	100	74
Rabbit	500	368
Dog	>12,000	>8,824
Pigeon	16.4	12
Pheasant	107	79
Quail	148	109

[a]Sources of data: housefly, Sun (1972); mammals, Donninger (1971); birds, Bunyan *et al.* (1971).

Table III. Variations in the Vertebrate Selectivity Ratio (VSR) of Carbaryl Applied Topically to Different Insect Species[a]

	LD_{50} (mg/kg)	VSR (rat/insect)
Rat (acute oral)	540	—
Lygus lineolaris (tarnished plant bug)	0.19	2840
Photinus pyralis	0.6	900
Apis mellifera (honeybee)	2.3	235
Epilachna varivestis (Mexican bean beetle)	2.7	200
Ostrinia nubilalis (European corn borer)	12.3	44
Estigmene acrea (saltmarsh caterpillar, adult)	62	8.8
Periplaneta americana (American cockroach)	190	2.8
Musca domestica (housefly)	>900	<0.60
Dermestes ater (black larder beetle)	3500	0.15
Sarcophaga bullata	4000	0.14
Pogonomyrmex barbatus (red harvester ant)	>5800	<0.09

[a]Sources of data: rat, Metcalf (1972*b*); insects, Brattsten and Metcalf (1970, 1973*b*).

Table IV. Variations in the Vertebrate Selectivity Ratio (VSR) of Azinphosmethyl with Different Species of Fish[a]

	LC_{50} (96 h, ppb)	VSR (fish/mosquito)
Culex pipiens larvae	19	—
Brown trout	4	0.21
Perch	13	0.68
Bluegill	22	1.2
Sunfish	52	2.7
Minnow	235	12
Carp	695	37
Bullhead	3500	184
Goldfish	4270	225

[a]Sources of data: *Culex*, Mulla *et al.* (1961); fish, Macek and McAllister (1970).

In the second case (Table III), the extreme variability among insects in their susceptibility to the carbamate carbaryl is illustrated with the VSR values ranging over more than 30,000-fold using the rat as the vertebrate representative. Compare in particular two insects in the same order, the Hymenoptera, with the honeybee (VSR = 235) highly sensitive (a fact of very practical importance in the field use of carbaryl) while the red harvester ant (VSR < 0.09) is virtually immune. Just as great a variability is found in the Coleoptera; for example, compare *P. pyralis* (VSR = 900), the Mexican bean beetle (VSR = 200), and the black larder beetle (VSR = 0.015).

The last example (Table IV) is rather different in considering an aquatic system with fish as the nontarget species and mosquito larvae as the target. The great range in toxicity of azinphosmethyl to the different fish is revealed by the over 1000-fold variation of VSR from the brown trout (VSR = 0.21) to the goldfish (VSR = 225). Perhaps it is not surprising under these circumstances that azinphosmethyl has been studied for use as a selective piscicide in catfish culture (Meyer, 1965). In fact, many of these seemingly damaging deficiencies in vertebrate selectivity of insecticides have been put to good use; for example, high toxicity to the rat encourages thoughts of use as rodenticides [e.g., the organophosphate gophacide (Eto, 1974, p. 332) or sodium fluoroacetate as described later] and rotenone's high toxicity to fish makes it useful as a general piscicide. The otherwise excellently selective organophosphate fenthion (Table I, compound 30, VSR = 107) is of unusual toxicity to birds, a defect which has been turned to good account in its use as a selective avicide against the weaver bird, *Quelea quelea*, a significant agricultural pest in East Africa (Pope and Ward, 1972).

Other examples of wide species variation in toxicity of a single insecticide are quoted by O'Brien (1967*a*): dimethoate, which has a 13-fold range of toxicity to vertebrates and greater than a 1500-fold range to insects, and schradan, which has over a 100-fold range in LD_{50} to insects.

Such extreme variations in VSR values with the same compound in related species are not necessarily typical but they do make generalizations about selectivity between such broad categories as insects and vertebrates quite suspect in many instances, which is not to say that some useful generalizations cannot, with care, be made. Thus it is well accepted that the response of acarines and insects to poisons may be quite different, and there exists a group of quite specific acaricides with little toxicity to insects or mammals (Knowles, 1975). Certain groups of compounds are toxic largely to one order of insects; for example, the malonitriles have been described as rather specific lepidopterocides (Darlington *et al.*, 1972), and occasionally, such group specificity may extend nearly to the species level, e.g., with butacarb (3,5-di-*tert*-butylphenyl *N*-methylcarbamate) which has high specificity for sheep blowflies (Fraser *et al.*, 1967), or the now obsolete tetranitrocarbazole, Nirosan, with toxic action limited to grape berry moths (Unterstenhöfer, 1970).

1.3. Correlation of Toxicity Between Organisms

If it is usually misleading to think of selectivity on such a broad canvas as insect vs. vertebrate, and since even closely related species can vary in sensitivity to a toxicant quite widely, as the data in Tables II, III, and IV show, the question arises as to how meaningfully toxicity estimates for one species can be used to predict the toxicity of the same compound to a second species. This has particular significance in attempts to assess the potential toxicity of compounds to man, where, inevitably, the test data are obtained from a limited range of laboratory vertebrates. In fact in most toxicity tests one or a few species are chosen to represent the responsiveness of a general class of organisms.

In view of the taxonomic gulf which separates the insects and mammals, one would hardly expect any correlation of toxicities to exist over a broad range of insecticides. However, within a restricted group of materials, sometimes such correlation is possible; for example, Darlington *et al.* (1971) obtained a correlation coefficient of 0.97 in the regression of the log LD_{50} values for eight ethylphosphonate esters to the rat and fly, and Sun (1972) shows plots of log LD_{50} to the mouse and fly for 15 chlorinated hydrocarbons and for 14 analogues of stirofos (Gardona). Although no regression statistics are given, there is a clear linear relationship between the toxicities, particularly when the flies were treated with a synergist to block detoxication.

However, such elements of order are the exception rather than the rule and in many cases no obvious correlations exist in toxicity between species. For example, in a series of studies, Henderson and Pickering examined the toxicity of common organophosphate and chlorinated hydrocarbon insecticides to several species of fish. Their general conclusion was that there was no correlation between the relative toxicities of these compounds to fish, to the rat, and to insects (Pickering *et al.*, 1962). Similarly, Tucker and Haegele (1971) observed that the acute oral toxicity of 16 common insecticides to six species of birds varied randomly, with an average range of toxicities for any compound over the six species of about ten-fold. No relationship of toxicity to phylogeny was evident and no species could act as a "representative" for the group.

The linear correlations presented in Table V are one attempt to explore this substantial question further as it relates to commonly tested laboratory vertebrates. Acute oral LD_{50} values for the rat, mouse, dog, rabbit, guinea pig, and chicken were collected for as many insecticides as possible from the compendia of Negherbon (1959), Martin (1972), and Kenaga and End (1974). Only those compounds were included where toxicity data for three or more of the six species were available, and wherever possible the comparative data for any one compound were all determined in the same laboratory to minimize this source of variability. Wherever a range of toxicities for any species was given, the mean value was used. The data in Table V show regression statistics for log LD_{50} values between species pairs which were obtained for all the insecticides where more than ten data points were available (row A). A second series

Table V. Regression Parameters for the Comparison of the Acute Toxicities of Insecticides to Different Pairs of Vertebrate Species

Species compared		n[a]	r[b]	Slope	Intercept	Standard deviation
Rat–mouse	A[c]	54	0.869[e]	0.735	0.677	0.351
	B[d]	32	0.889[e]	0.721	0.722	0.342
Rat–guinea pig	A	28	0.777[e]	0.734	0.496	0.542
	B	17	0.881[e]	0.757	0.623	0.395
Rat–rabbit	A	42	0.840[e]	0.915	0.210	0.543
	B	26	0.819[e]	0.790	0.535	0.535
Rat–dog	A	20	0.751[e]	0.652	0.502	0.472
	B	9	0.847[f]	0.716	0.655	0.488
Rat–chicken	A	15	0.547[f]	0.438	1.049	0.622
	B	8	0.161[g]	0.123	1.786	0.637
Mouse–guinea pig	A	23	0.698[e]	0.820	0.112	0.630
	B	15	0.929[e]	1.045	−0.136	0.322
Mouse–rabbit	A	32	0.762[e]	1.026	−0.213	0.657
	B	22	0.843[e]	1.015	−0.155	0.507
Mouse–dog	A	16	0.560[f]	0.644	0.401	0.602
	B	8	0.785[f]	0.898	0.056	0.546
Rabbit–guinea pig	A	14	0.930[e]	1.023	−0.065	0.389
	B	9	0.918[e]	1.146	−0.254	0.416
Rabbit–dog	A	14	0.800[e]	0.694	0.546	0.501
	B	7	0.855[f]	0.924	0.344	0.550

[a]Number of compounds in regression.
[b]Correlation coefficient.
[c]A = All insecticides.
[d]B = Organophosphates only.
[e]Significant at $P = 0.01$ or below.
[f]Significant at $P = 0.05$.
[g]Not significant at $P = 0.10$.

of correlations was made considering only the organophosphorus insecticides (row B) since it seemed likely that better predictability would be found by restricting the correlation to compounds with a common mode of action. The values for the rat in Table V are derived from LD_{50} values for the female, wherever a choice was possible. However, the quality of the correlations was, in most cases, not greatly changed by using data for the male rat.

In some cases, e.g., rabbit vs. guinea pig ($r = 0.930$) or rat vs. mouse ($r = 0.869$), a quite reasonable correlation of toxicities was possible as judged by the correlation coefficient (r) and the standard deviation for the regression. This is particularly impressive in view of the heterogeneous sources for the data, and the variety of chemical structures involved. An example of such a regression is shown in Fig. 1 for the mouse vs. guinea pig data with the organophosphates only. In this case, an organophosphate with LD_{50} of 100 mg/kg to the mouse should have an LD_{50} of 90 mg/kg to the guinea pig with the 95% confidence limits of 60–136 mg/kg. In other cases, e.g., rat vs. chicken ($r = 0.547$) and mouse vs. dog ($r = 0.560$), much less satisfactory correlations were found, although the regressions are significant ($P = 0.05$).

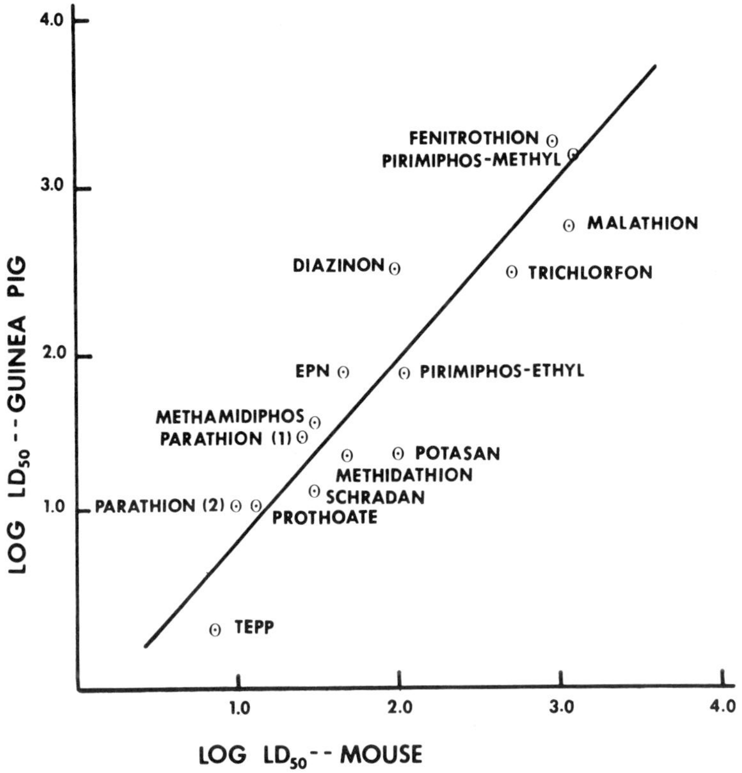

Fig. 1. Correlation of the acute oral toxicity of a range of organophosphorus insecticides to the mouse and to the guinea pig.

As expected, in most cases limitation of the comparison to the organophosphates significantly improves the correlation coefficient. A striking exception is in the rat–chicken comparison, where, with the limited data available, a clear decline in the reliability of the correlation occurs. However, this probably does not indicate any generally peculiar difference in the reaction of birds and the rat to organophosphates since similar correlations with the extensive toxicity data to birds provided by Schafer (1972) gave, in comparing log LD_{50} values for the rat and redwing blackbird, quite comparable results for a general series of toxicants and for organophosphates alone:

	n	r	Slope	Intercept	Standard deviation
A	20	0.746	0.565	0.394	0.354
B	44	0.748	0.670	−0.424	0.467

where case A represents a range of insecticides and drugs of mixed structure and case B is for organophosphates alone. It is, however, interesting to note that while the mean LD_{50} for the organophosphates to the rat was 378 mg/kg,

Table VI. Mean Correlation Coefficient (r) for Acute Toxicity of Insecticides to Pairs of Vertebrate Species Classed According to the Degree of Taxonomic Similarity Between Them[a]

Level of comparison	Number of comparisons	Mean r value	
		All compounds	O–P only
Genus	1	0.869	0.889
Family	5	0.820	0.885
Order	3	0.715	0.831
Class	1	0.547	0.161

[a]Data from Table V.

to the blackbird it was 23 mg/kg. Thus the blackbird, unlike the chicken, is peculiarly sensitive to these anticholinesterases.

The comparisons in Table V lie at several different taxonomic levels; for example, the rat and mouse are of the same family (Muridae), and they share the same order (Rodentia) with the rabbit (Leporidae) and the guinea pig (Caviidae). The dog is in a different order (Carnivora), but all these organisms are in the class Mammalia while the chicken is in the class Aves. The mean correlation coefficients for the comparisons in Table V at each of these taxonomic levels (obtained by conversion of the r values to Fisher's z before averaging) are presented in Table VI and reveal an interesting relationship between taxonomy and relative toxicity. For both the "all insecticide" and "organophosphate only" comparisons, it is clear that the closer the taxonomic similarity of the species compared the greater the reliance which can be placed on the extrapolation of toxicity from the one species to the other. Comparisons at the order level are of special interest since the extrapolation of data from such species as the rat, mouse, rabbit, guinea pig, and dog to man also lies at this level. Although the number of species comparisons here, each of which involves the dog as one member, is small, it is encouraging to find that a reasonably high degree of correlation is possible even at this level, particularly if the data are restricted to a single class of compounds, i.e., the organophosphates, where the mean r value is 0.831. Thus despite the examples cited previously, which show extreme species variability, there is a general similarity in relative sensitivity to most compounds between related species which on occasion may even extend to such broad comparisons as mammal vs. insect. Unfortunately such a conclusion deals with compounds *en masse* and we cannot predict with certainty the behavior of any single compound, which might be quite aberrant.

1.4. Scope of Selective Toxicity

Even with the material presented thus far, it will be clear that the scope and significance of selective toxicity as it relates to pesticides have broadened immensely in the last decade. Whereas in earlier reviews (O'Brien, 1961, 1967*a*; Metcalf, 1964, 1972*a*; Winteringham, 1969) the major, if not sole, concern could be the relationship of the acute toxicity of insecticides between insects and mammals, the vastly increased emphasis on environmental effects, chronic and sublethal actions, and compatibility with pest management concepts now demands a much broader consideration of insecticide selectivity as it relates to a wide range of nontarget vertebrates and invertebrates, with numerous aspects of toxicity beyond acute lethal actions. This is too large a burden for one chapter, and thus, although consideration will be given to selectivity over as broad a range of species as possible, chronic and sublethal effects will not be included, as already indicated. Limitation to lethal effects also precludes consideration of the selectivity of nonlethal means of pest control by chemicals such as chemosterilants, attractants, and feeding deterrants, although these methods may be exquisitely selective, at times acting on only a single species.

The selective toxicity of a compound may be expressed between any two levels of biological organization from single cells (e.g., the challenge in cancer chemotherapy) to a comparison of toxicities between the plant and animal kingdoms. We are here concerned primarily with selectivity at the level of the species and above, although selectivity between individuals, sexes, and strains of one species obviously exists and may have great importance. The most evident example of insecticidal selectivity below the species level is the development of resistance to insecticides by certain strains of organisms. This is discussed in the following chapter, but since resistance gives us some of our purest and best-understood examples of selective toxicity in action, from time to time an example from this field will be used here also.

In addition to the uniformly excellent reviews of selectivity already referred to, a number of other general references contain much material on selective toxicity among certain compounds or groups of organisms, e.g., selective toxicity of xenobiotics in general (Albert, 1973), anticholinesterases (Hollingworth, 1971; Corbett, 1974; Eto, 1974), and antifeeding agents (Hollingworth, 1975*a*). Selectivity to fish is outlined by Holden (1973), and in a pair of stimulating reviews Edwards and Thompson (1973) and Croft and Brown (1975) discuss insecticide toxicity and selectivity and their influence on population dynamics of the soil fauna and of natural control agents, respectively. Acute toxicity data for pesticides against many nontarget organisms have been compiled by Pimentel (1971).

2. The Twofold Way to Selective Toxicity

The theoretical grounds on which any type of selectivity may be explained have been established for some time. Ripper *et al.* (1951) distinguished between *ecological selectivity* and *physiological selectivity*. In the former case one organism succumbs to a poison, while another survives because, although it may be sensitive to the toxicant, it in some way avoids exposure. Thus an alternative name for this type of selectivity would be *extrinsic selectivity* or *avoidance*. In the second situation both organisms are exposed, but one, by some physiological or biochemical mechanism, can better tolerate the exposure and thus survives. This could also be termed *intrinsic selectivity* or *tolerance*.

2.1. Ecological Selectivity

Although our subject here relates strictly to physiological selectivity as defined above, it would be wrong to completely ignore the topic of ecological selectivity, since in practical terms it is of immense significance. In general, ecological selectivity represents the difference between toxicity and hazard; that is, the innate toxicity of a compound does not represent a practical hazard to nontarget organisms if their exposure can be avoided or minimized. A multitude of possibilities for minimizing exposure of nontarget organisms exist and many are in general use. The simplest must surely be the wearing of protective clothing when applying toxic materials. However, much more sophisticated examples based on a thorough knowledge of the ecology of the pest and the appropriate nontarget species are available; for example, the area over which DDT must be applied to control the tsetse fly (and thus its nonspecific ecological effects) can be drastically reduced by limitation of spraying to certain wooded areas and drainage courses where DDT is applied only on the underside of branches more than 2 inches wide which lie from the horizontal to 80° inclined to the horizontal and at 5–14 ft above the ground (MacLennan, 1967)! Other examples of selectivity through specific placement, through timing of applications (e.g., application of carbamates at dusk or night when the highly susceptible bees are not foraging), by the use of specific attractants or baits (e.g., in control of the fire-ant with Mirex baits), or by minimization of application rates or frequency through better understanding of pest population dynamics and their relation to economic injury are presented by Smith (1970), Metcalf (1975), and Croft and Brown (1975). Other means of minimizing exposure include the use of special formulations, e.g., oily formulations rather than dusts to protect bees (Johansen, 1972) and, in an area of great current interest, microencapsulation and other controlled release formulations which may allow less frequent or lower rates of application in the field, and which are safer to handle. An excellent example of decreased hazard is found with Penncap-M, a microencapsulated form of methyl parathion which is 4–5 times less toxic orally to the mouse than the pure compound and has a dermal toxicity (LD_{50} to the rabbit of over 2200 mg/kg) that is even more favorable (Pennwalt Corporation, Agchem-Decco Division, Technical Bulletin). A particularly satisfactory type of ecological selectivity is obtained by the use of plant systemic insecticides, which in general are toxic only to insects feeding directly on the treated plant and not to parasites and predators.

Ecological selectivity is also closely related to biodegradability. Whatever their innate toxicity, compounds which are ephemeral in the environment and do not accumulate in organisms, e.g., the pyrethroids, JHAs, and some organophosphates and carbamates, quickly dissipate after use and present only a brief threat to populations of nontarget organisms which may well be avoided completely or tolerated with rapid recovery. Highly persistent compounds such as some of the organochlorines, however, can present an enduring danger to nontarget organisms, and, because they may accumulate in tissues, also possess an enhanced risk of chronic toxicity and danger particularly to animals toward the top of food chains where accumulation is likely to be greatest (Matsumura, 1975). Biodegradable compounds thus will tend to be more selective in use.

Taken together, such strategies to enhance ecological selectivity have immense impact, enabling more satisfactory use of the many current insecticides which are deficient in intrinsic selectivity. As our knowledge of the ecology and behavior of pest and nontarget species increases, and new concepts in formulation are explored, the significance of ecological selectivity can be expected to increase yet further.

2.2. Physiological Selectivity

It is to O'Brien (1960) that we owe the generally accepted approach for analyzing physiological selectivity. As he points out, a finite number of processes and events are influential in the action of an insecticide from the time it first contacts the organism until the organism either dies or is safely beyond risk. These are shown diagrammatically in Fig. 2 by means of a multicompartment model. The initial process must be penetration through the integument or other outer barriers (e.g., of the gut or lung) in order for the toxicant to enter the general circulation of the organism. There it may be subject to more or less reversible binding to components in the circulatory fluid or within other tissues at sites which cause no detrimental effects, i.e., indifferent sites. Some portion will generally be converted in one or several tissues to metabolites which may be more toxic (activation) or less toxic (detoxication) than the parent compound. The metabolites reenter the general circulation and, along with the parent compound in a few cases, will be excreted from the body. The reaction which initiates the toxic response arises by the interaction of the parent compound or an active metabolite with some receptor system at the site of action. This "biochemical lesion" leads by a more or less complex series of disruptive effects on physiological and biochemical homeostasis to the eventual death of the organism. At several stages during the unfolding of these events, the compound must move through the structural barriers which invest tissues, cells, and organelles and which are generally of a lipoprotein nature. It is then only reasonable to suppose that a significant difference between two organisms in the rate or degree of any of these processes after exposure to a given toxicant could provide the basis for physiological selectivity.

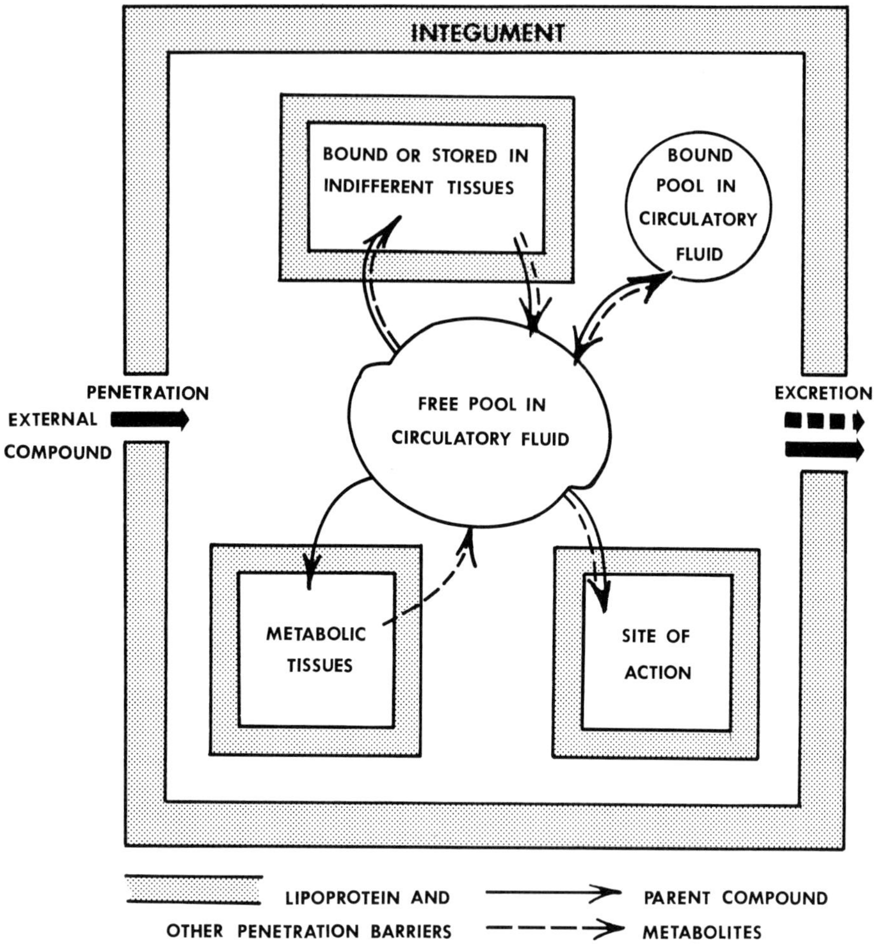

Fig. 2. Diagrammatic representation of the processes which govern the dynamics of action of an insecticide (*cf.* Fig. 1. Chapter 1).

3. Penetration

3.1. External Penetration Barriers

Although measuring the rate at which an insecticide enters an organism may seem to be an easy task, it has turned out to be both complex and controversial and, consequently, the relationship of penetration processes to selectivity is in many cases ill defined. This is particularly true in comparing insects and mammals.

O'Brien (1967*a*) emphasized some major differences between the integuments of insects (rigid, chitinized, usually greasy) and mammals (flexible, keratinized, and often damp). One could of course stress the similarities (both are heterogeneous lipophilic–hydrophilic barriers with outer lipid layers (wax or sebum) in which lipophilic compounds may initially be trapped), but the differences do seem sufficient to cause major variations in the rates of penetration of externally applied insecticides and thus act as a potentially significant source of selective toxicity between insects and mammals.

Also, the integuments of different insects, even if built on a common plan, vary widely in character from the rigid, thick armor of an iron-clad beetle to the delicate, flexible membrane of a young mosquito larva. Considerable differences in the amount and properties of the surface lipids occur, even in different life stages or ages of the same species (Hackman, 1974), giving rise to the thought that here too could lie a basis for selectivity between insect species, sexes, ages, and life stages.

However, it is clear that the study and interpretation of the rates of pesticide movement through the integument are strewn with difficulties. The continuing controversy over the method of entry of insecticides into insects (direct penetration vs. lateral diffusion as suggested by Gerolt) has been reviewed by Brooks in Chapter 1 and by Ebeling (1974) in his comprehensive survey of penetration through the insect cuticle. Two other complications in studying penetration are that (a) considerable metabolism of the compound may occur during penetration and (b) the compound may be firmly bound within the integument, thus disappearing from the surface, but not entering the body proper.

These factors explain in part why different techniques for assessing penetration may give incongruent results; for example, penetration is often assessed by rinsing the cuticle to remove the unpenetrated compound at different times after treatment. This of course does not indicate whether the unrecovered compound is held up in the cuticle, or is free to act internally. Excision of the treated area of the integument and subsequent digestion make it possible to estimate both bound material in the integument and free material in and on it, but may not allow adequately for lateral diffusion from the application site. Neither method indicates how much loss occurs by metabolic destruction in the integument. Even less direct are techniques which measure enzyme inhibition (e.g., plasma cholinesterase) or urinary excretion of metabolites after dermal application as an index of penetration rate.

The ambiguities of our current knowledge of the comparative penetration of insecticides are well illustrated in the case of DDT. Two lines of evidence have been put forward regarding the role of penetration in its selectivity.

3.1.1. Topical/Injected Toxicity Ratios

The basic argument behind and many examples of topical/injected toxicity ratios (TI ratios) have been presented in Chapter 1. In brief, it is presumed that if a compound penetrates slowly after topical application it will be relatively less toxic by this route than by direct injection and the ratio of its

Table VII. Ratios of Toxicities of DDT and Malathion When Applied Topically and by Intraperitoneal Injection to Various Species

	LD_{50} (mg/kg)		
	Topical	Injected	TI ratio
DDT[a]			
Rat	3000	100–200	15–30
American cockroach	10	5–8	1.3–2
Malathion[b]			
German cockroach	120	8.0	15.0
American cockroach	16	7.2	2.2

[a]Negherbon (1959).
[b]Krueger and O'Brien (1959).

LD_{50} values by the topical and injected routes will be high, while if it penetrates readily it will be more toxic and the ratio should be much lower, approaching a value of 1.0 for very rapid penetration.

Such data for DDT with the rat and American cockroach are shown in Table VII.

Obviously it matters little to the cockroach whether the dose is received externally or internally, but the relative safety of DDT by dermal application to the rat is very clear, and, of course, this safety is borne out by several decades of heavy use of DDT without appreciable acute toxicity to man. The conventional interpretation of these data (and related TI ratios for many other insecticides) is that DDT penetrates the insect cuticle readily but is seriously hindered in its passage through the mammalian skin (Winteringham, 1969; Metcalf, 1972*a*).

3.1.2. "Direct" Assay of Penetration

In contrast, more direct measurements of the penetration rates for DDT through rat skin (excision assay) and the cockroach integument (surface rinse method) show virtually no differences in rate between these organisms (Table VIII). Penetration in both cases proceeds slowly, with 50% penetration occurring in about 1550–1600 min. However, the difference in method with the two species should be noted since "penetration" does not mean the same thing in the two cases, and for the rat, unlike the cockroach, penetration is not a first-order process. Thus comparisons of penetration levels other than 50% would not necessarily be as similar. However, these results do stand in clear contradiction to those derived by the TI ratio method. Unfortunately, the conclusions of O'Brien and Dannelley (1965) using the excision method with shaved rat skin for DDT and several other insecticides are in fundamental disagreement also with those of Maibach and Feldman (1974), who applied the same compounds to the untreated human forearm and measured penetration

Table VIII. Comparative Rates of Penetration of Insect and Mammalian Integuments and Midguts by Common Insecticides and Their Relation to Partition Coefficients

Compound	Integument			Midgut		Partition coefficient (olive oil/water)			
	American cockroach[a] ($t_{0.5}$, min)	Rat, 50% penetration time[b] (min)	Human, % penetration[c] (5 days)	Tobacco hornworm[d,e] ($t_{0.5}$, min)	Mouse[d,e] ($t_{0.5}$, min)	Ref.[a]	Ref.[b]	Ref.[e]	Ref.[f]
Dimethoate	27	—	—	30	64	0.34	—	—	0.59
Nicotine	—	—	—	16	125	—	—	21	—
Carbaryl	—	870	73.9	36	49	—	64.5	38	—
Malathion	—	330	6.8	58	209	—	413	72	—
Dieldrin	320	210	7.7	1476	2585	64	1805	281	—
DDT	1584	1560	15.0	1785	1722	316	932	785	199

[a]Olson and O'Brien (1963) (surface rinse).
[b]O'Brien and Dannelley (1965) (disc excision).
[c]Maibach and Feldman (1974) (urinary clearance).
[d]Shah *et al.* (1972).
[e]Shah and Guthrie (1970).
[f]Buerger and O'Brien (1965).

rather indirectly by observing appearance of the radioactive label in the urine. Allowance was made for incomplete excretion of the dose. In this study (Table VIII) they found that DDT is quite comparable to most other pesticides in its rate of penetration, and is a considerably more rapid penetrant than dieldrin or malathion, a conclusion completely at odds with that of O'Brien and Dannelley. The comparative data for carbaryl in the two studies seem equally incompatible. Perhaps if one notes the methodological differences, assumes extensive binding of these compounds in the integument, and allows for the species difference these results can be reconciled, but it is hard to tell where the best evidence lies and whether one should believe that DDT is a selective insecticide because of permeability relationships or not. Certainly any role of penetration in DDT's selectivity remains to be proved despite the evidence from TI ratios.

A similar impasse was reached by Krueger and O'Brien (1959) in considering the role of integumental penetration in the differential toxicity of malathion to the German and American cockroaches. The TI ratio method (Table VII) again clearly may be interpreted to mean that malathion penetrates the cuticle of the German cockroach much less readily than that of the American. However, once more the direct assay of penetration (surface rinse method) found no such evidence for species difference. In this case, Krueger and O'Brien chose to accept the TI ratio as the valid indicator because the rinse method does not show how much material actually enters the insect's internal tissues. The implication of this view is that much more of the malathion which left the cuticular surface of the German cockroach was bound tightly within the cuticle than in the American cockroach, although to this author the difference in TI ratio between the species seems rather extreme for this interpretation.

Thus the whole question of how far differences in rate of entry of insecticides into the body govern toxicity is much more difficult to answer than one might expect, a difficulty due in large part to the fact that none of the methods used directly assays the amount of the parent compound which enters the general circulation of the animal and that several factors may interact to make such penetration a complex, nonlinear process.

Having criticized the methodology of "direct" measurement of penetration (and other criticisms are also possible, e.g., the role of the application solvent in disrupting surface lipid layers), it is equally possible to attack the meaningfulness of the conclusions based on TI ratios. In particular, the assumption that all other toxicological processes (Fig. 2) operate with identical efficiency whether the compound is applied topically or by injection may be seriously in error; for example, sites of loss of the compound, particularly metabolism, may be overwhelmed by the sudden flush of injected material, which does not occur in the slower presentation by the topical route. Thus toxicity by injection woud be enhanced and, by the TI ratio assumptions, it would appear that the topically applied compound has penetrated the integument more slowly than it really has. Furthermore, even the injection of small volumes of solvents may alter the toxicity of insecticides to insects. Thus, for xample, Krueger and O'Brien (1959) in their study with malathion, using ropylene glycol as solvent, injected 10 μl/American cockroach and .3 μl/German cockroach. This is about 10,000 and 25,000 mg/kg, respec-vely, hardly an inconsiderable dose, and one which might be capable of gnificantly affecting toxicity with consequent distortion of the TI ratio. If the olvent synergizes the insecticide, the enhanced toxicity by injection would gain be interpreted as indicating retarded penetration after topical applica-on. It is likely that this has occurred in many published studies where insects re injected with organic solvents as the carrier; for example, Gerolt (1972) ound that methomyl was 6 times more toxic to houseflies when injected in .1 μl acetone (approximately 5000 mg/kg of acetone) than when injected in μl water.

A third major point can be appreciated if we imagine the quite unlikely ase of a compound which once it enters the body is not subject to loss by any neans. It then does not matter at all how fast this compound penetrates after a ethal external dose, since eventually enough will enter to cause death. In this ase, whatever the rate of penetration, the TI ratio is 1.0. Obviously the effect f altered penetration on toxicity is significant only when there is a reasonable ate of removal of the penetrating compound, and we are concerned not with eciding rates of penetration alone but rather with assessing the balance etween entry and loss which governs the rate and extent of accumulation of he toxicant internally as the major criterion for the influence of penetration on electivity.

Thus, consider the simple situation where an externally applied compound enetrates and upon penetration is subject to loss, both processes being first rder:

$$\underset{\text{external compound}}{A} \xrightarrow{k_1} \underset{\text{internal compound}}{B} \xrightarrow{k_2} \underset{\text{compound lost}}{C}$$

Then the rate of change in internal concentration at any time is given as

$$\frac{d[B]}{dt} = k_1[A] - k_2[B]$$

hich can be restated as

$$[B]_t = \frac{k_1[A]_0}{k_1 + k_2}(e^{-k_1 t} - e^{-k_2 t})$$

here $[A]_0$ is the amount initially applied and $[B]_t$ the amount free internally t time t. Some accumulation curves based on this simple relationship with ifferent values of k_1 and k_2 are shown in Fig. 3.

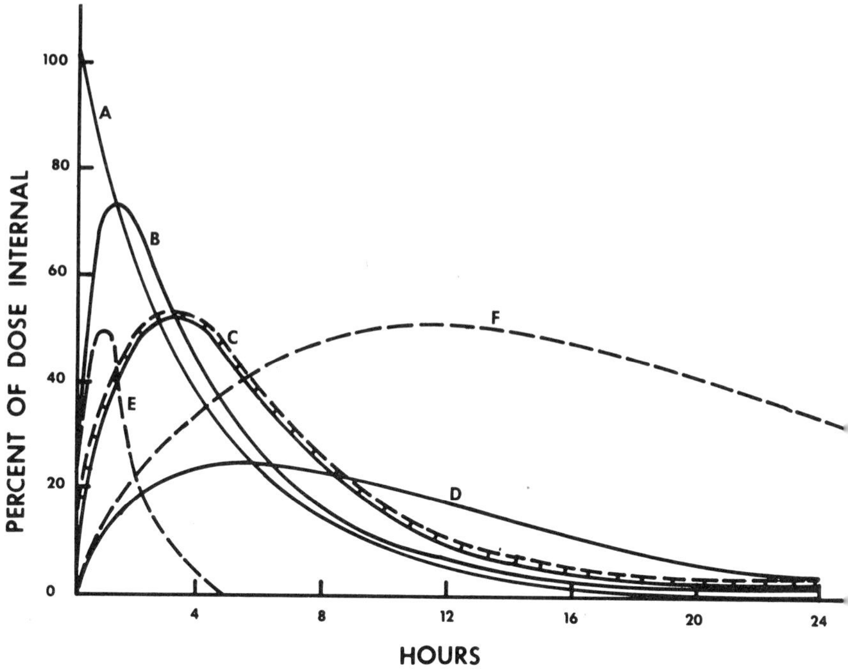

Fig. 3. Accumulation curves for an insecticide in a model system with varying rates of penetration and loss.

The rate constants used to derive these curves are

	Penetration (k_1, h^{-1})	Loss (k_2, h^{-1})	k_1/k_2
A	Injected	0.25	—
B	2.00	0.25	8
C	0.50	0.25	2
D	0.125	0.25	0.5
E	2.00	1.00	2
F	0.125	0.0625	2

In Fig. 3 there are two families of curves which describe the kinetics o accumulation of the toxicant. The solid lines (A, B, C, and D) show th situation where k_1 (penetration rate) varies from instantaneous after injectio (case A) to a rather slow 0.125 h^{-1} (case D), while k_2 (the rate of loss) i constant at 0.25 h^{-1}. In these four cases the peak internal concentration, whic is a function of k_1/k_2, decreases as k_1 decreases, but the area under the curve which represents the total exposure of the organism, is the same in each cas since it depends only on k_2 and is independent of the rate of penetration.

The second family of curves (E, C, and F) are shown by broken lines an represent the situation where k_1 and k_2 both vary but the ratio k_1/k_2 is fixed

Thus peak internal concentration is the same, but the area under the curve varies.

The crucial question now is whether toxicity depends on the maximum internal level (intensity of exposure), in which situation cases C, E, and F are equitoxic, or whether it depends on the area under the curve (total exposure), in which situation cases A, B, C, and D are equitoxic, and penetration rate has no influence on toxicity. The usual situation is probably an intermediate one where, varying with the compound and organism, both peak height and area are important, and thus penetration rate does have an influence, but a variable one. If this is so, the TI ratio has no direct relationship to the rate of penetration; for example, a TI value approaching 1 may indicate a compound which penetrates very rapidly, but it could just as well arise if penetration is slow but the organism responds equally whether the compound arrives as a large, short-term jolt (case A) or with a less intense but longer-lasting action (e.g., case C or D). Such a situation, for example, is envisioned for some rapidly degradable JHAs by Staal (1975), who states that "flash saturation of JH receptors is far less effective than a drawn-out exposure" and concludes that it is advantageous for high potency if the cuticle acts as a sink for the JHA with subsequent slow release into the system rather than penetration being very rapid accompanied by rapid loss. Also, Sun and Johnson (1971) observed that with several azodrin analogues the amount needed to kill houseflies was the same whether given by injection or by the slower exposure of infusion.

Whether such situations are common is not known, but even the rather simplistic analysis above shows that the dynamics of insecticide action are complex enough for elementary approaches such as the TI ratio method to be fallible. None of this of course proves that results based on TI ratios are necessarily wrong in all cases.

Turning now to a comparison of penetration rates between different insect species, there is indisputable evidence that the relative permeabilities of their cuticles differ; for example, Buerger and O'Brien (1965) found with a series of insecticides of varying polarity that the rate of cuticular penetration was proportional to polarity for the American cockroach and adult yellow mealworm, but not for the house cricket.

Numerous other authors have compared penetration rates between insect species, almost always by the surface rinse method, and considerable differences are not uncommon. Brady and Arthur (1962) examined the rate of loss of the phosphoramidate Ruelene from the cuticular surfaces of 19 insect species. After 4 h, the honeybee and Mexican bean beetle had the greatest surface loss (87–88%), while several Diptera, lepidopterous larvae, and cockroaches were in the 50–75% range, with no clear phylogenetic basis for the variation. However, ticks and two insect species (the boll weevil and bed bug) were clearly in a different group, with losses from the surface of only 14–25%. Camp and Arthur (1967) extended this study with the carbamate carbaryl on four insect species and found great differences between the internal levels of labeled compounds after 4 h in flies (housefly and stablefly; 39% of the recovered dose) and weevils (rice weevil and boll weevil; 5.4–7.5%). However, this did not correlate well with the relative toxicity of carbaryl to these species since the stablefly and boll weevil are relatively susceptible while the housefly and rice weevil are not. Thus it was concluded that metabolism of carbaryl was a major additional influence on toxicity, a conclusion amply confirmed by the work of Brattsten and Metcalf (1970, 1973*b*) with the synergist piperonyl butoxide.

Other comparative studies of penetration into insects include that of Uchida *et al.* (1965) with dimethoate (rate of penetration in the order housefly > milkweed bug > Colorado potato beetle > American cockroach), that of Eldefrawi and Hoskins (1961) with carbaryl (housefly > milkweed bug > German cockroach), and that of Tomlin and Forgash (1972) with both stirofos and DDT (housefly > gypsy moth larva). Although in these studies metabolism and other factors were often found to influence toxicity also, it is clear that differences in the rate of penetration of insecticides commonly occur and that they have toxicological significance and can be a considerable factor in selectivity among insects.

This view is strongly supported by observations of the reduced speed of action and toxicity of insecticides, especially of the more lipophilic and biodegradable type, to houseflies (and other insects) which have the *pen* gene for reduced penetration (Chapter 13). However, it is significant that in the studies outlined above the housefly is invariably among the species in which penetration is most rapid and yet this insect is not notably sensitive to many insecticides, which again underlines the fact that penetration rates alone do not determine susceptibility but act in concert with other processes, particularly mechanisms of loss, to influence the overall dynamics and internal accumulation of the toxicant.

Most studies of insecticide penetration have been directed toward the integument, but obviously other portals of entry—e.g., the gut or respiratory tract—may be very important in practice as routes of poisoning for both insects and vertebrates.

An excellent series of studies by A. H. Shah, P. V. Shah, and F. E. Guthrie throws considerable light on the comparative rates of penetration of a range of insecticides through the isolated midguts of the mouse and several insects. Typical results are shown in Table VIII and can be compared there to studies of integumental penetration with the same compounds. Whereas penetration of the mammalian skin showed no correlation with olive oil/water partition coefficients (immensely variable as these are), penetration through both the hornworm and mouse midguts paralleled penetration through the cockroach cuticle in generally being more rapid with more polar compounds, although the correlation is imperfect. In these studies, penetration of the gut of the cockroach *Blaberus cranifer* was also studied and found to be similarly related to polarity. Despite some differences in penetration rate between the species Shah and Guthrie (1970) concluded there was no consistent pattern of midgut permeability which related to selective toxicity. This was also the conclusion of Shah *et al.* (1972) after a related study with dimethoate analogues in which toxicity to mice and insects varied quite widely, but the rate of midgut penetration varied by less than twofold within the series for each species.

A rather special case where penetration can have great significance for selectivity is the entry of pesticides into aquatic organisms. Clearly here the whole body may be bathed in a dilute solution of the toxicant and can act as a "lipophilic sponge" to rapidly absorb and concentrate those compounds with a high lipid/water partition coefficient. Thus Derr and Zabik (1974) concluded that the uptake of DDE by chironomid larvae occurred mainly by passive entry through the cuticle, uptake being related to cuticular surface area, and the data of Wilkes and Weiss (1971) point to a similar conclusion with DDT in dragonfly nymphs, where, after 6 days of exposure, the level in the insects was 2000–2700 times greater than in the ambient water. The high efficiency of uptake of lipophilic pesticides has been suggested as a major factor in the markedly selective insecticidal action of several pesticides to mosquito larvae compared to terrestrial species, e.g., various derivatized carbamates (Black *et al.*, 1973) and the organophosphate Abate (Leesch and Fukuto, 1972). In the case of Abate, the larvae picked up 99% of the compound originally present in the water at 57 ppb within 1 h, and in a series of Abate analogues their toxicity to the larvae was proportional to their lipophilicity.

Fish too are well known to be outstandingly sensitive to many highly lipophilic compounds, particularly the chlorinated hydrocarbons and pyrethroids, with LC_{50} values below 1 ppb in some instances (Holden, 1973). Again they are capable of rapidly and efficiently removing and concentrating these compounds from aqueous solution, especially through the gills, which have a large area of lipoprotein membrane and are richly vascular (Fromm and Hunter, 1969). Thus Kapoor *et al.* (1970) found in their model ecosystem that with a water concentration of DDT of 0.22 ppb the final concentration of DDT in mosquito larvae was 1.8 ppm and in mosquito fish was 18.6 ppm, which represent concentration factors of about 8000 and 85,000 times, respectively, either by direct penetration or through the food chain. Rapid and effective uptake is similarly shown in the report of Gakstatter and Weiss (1967) that about 60% of the dieldrin and 70% of the DDT added to the water at 30 ppb was taken up by 60 bluegills (345 g) in 100 liters of water within 5 h.

This remarkable and, in the case of nontarget species, unfortunate tendency of aquatic organisms to concentrate lipophilic compounds both directly and through the food chain is further discussed by Matsumura (1975, p. 376) and Johnson (1973).

3.2. *Internal Penetration Barriers*

3.2.1. *Blood–Brain and Related Barriers Around the Central Nervous System*

The blood–brain barrier and related barriers around the central nervous system exclude, or more accurately slow penetration of, highly polar or charged compounds. Barriers of this type protect both the vertebrate and insect central nervous systems, although they differ in nature and may be lacking in some other organisms—e.g., crustaceans and leeches, as reviewed by Treherne and Pichon (1972). In Table I, compounds 1, 3, 5, 6, and 7 are obviously potent poisons but unfortunately to mammals rather than insects. They are highly polar or charged cholinergic neurotoxicants, or in the case of the phosphoramidates (compounds 5 and 6) are converted to polar activation products *in vivo*, and they represent only a small fraction of the compounds in this general class which are selective mammalicides. This selective mammalicidal action depends on the blood–brain barrier, which is quite effective in restricting access of such compounds to the CNS of both mammals and insects. The key to selectivity lies in the fact that mammals have numerous essential peripheral cholinergic sites which are unprotected, while insects do not. Thus the mammalian nervous system is disrupted while that of the insect is not affected. This work has been reviewed in detail by O'Brien (1967*a,b*) and Winteringham (1969), and in Chapters 1 and 6, and will not be addressed further here except for three further eclectic observations:

a. The species described by Treherne and Pichon (1972) as lacking a blood–brain barrier should be peculiarly sensitive to charged or polar neurotoxicants, a point worth further investigation.
b. Fascinating group variability occurs between invertebrates in their relative sensitivities to such polar neurotoxicants as nicotine, tetram, and the phosphoramidates such as schradan and mipafox. Some hemipterans (aphids, some plant bugs) and mites are frequently "mammalian" in their high susceptibility. The obvious explanation that in these species the blood–brain barrier is deficient unfortunately turns out to be wrong, at least in the case of the willow aphid (*Tuberolachnus salignus*) studied by Toppozada and O'Brien (1967). Beyond this nothing further is known about this interesting, and practically significant, selectivity phenomenon.
c. O'Brien (1967*a,b*) after extensive studies concluded that in the insect CNS the polarity barrier at best slowed penetration of ionic compounds by about 5- to 15-fold. Yet often effects on toxicity are rather dramatic; for example, Metcalf *et al.* (1968) found that the *m*-trimethylaminophenyl *N*-methylcarbamate ion, an extremely potent anticholinesterase (I_{50} for housefly cholinesterase $= 1.6 \times 10^{-8}$ M), had no effect on nervous transmission when perfused over the isolated cockroach ventral nerve cord at 10^{-4} M, and only moderate effect at the massive concentration of 10^{-3} M. Uncharged carbamates were effective at near their I_{50} values. This hardly seems compatible with a mere 5- to 15-fold slower penetration, and one wonders whether additional defensive processes in the nerve, e.g., metabolic degradation, play an ancillary role in protecting against some neurotoxicants. This is certainly so for acetylcholine itself (Treherne and Pichon, 1972).

3.2.2. General Lipoprotein Membranes as Barriers

Cells and organelles are invested with lipoprotein membranes which must be traversed repeatedly by a toxicant in its odyssey to the site of action. In the absence of assisted-diffusion mechanisms a compound must therefore have a sufficient balance in its lipophilic–hydrophilic properties to be able to move both in and out of lipid–water interfaces. This fact is discussed in detail by Hansch (1971), who showed with many series of compounds that biological activity is related parabolically to partitioning behavior. Thus within such series there will be an optimum partition coefficient for biological activity. An interesting question then is whether organisms differ in this optimum value, in which case a role in selectivity can be envisioned. Little is known regarding this with insecticides, but at least the possibility can be illustrated with equations taken from Hansch (1971) which relate toxicity (BR, biological response) of a series of *n*-alkylthiocyanates for the green peach aphid and thrips to their partition coefficients (P):

$$\text{Aphid:}\ \log \text{BR} = -0.11(\log P)^2 + 1.46 \log P - 0.47$$

$$\text{Thrips:}\ \log \text{BR} = -0.24(\log P)^2 + 2.45 \log P - 2.03$$

The optimum log P value for aphid toxicity is 6.6 and for the thrips is 5.2. The above equations can be combined as a selectivity ratio

$$\log \left(\frac{\text{BR aphid}}{\text{BR thrips}}\right) = 0.13(\log P)^2 - 0.99 \log P + 1.56$$

Solution of this equation shows that little selectivity occurs below a log P value of 6, but compounds with higher values are increasingly selective aphicides such that at a log P of 8 there is 100-fold selectivity with little loss of aphicidal effectiveness. Obviously this example is meant to illustrate the possibilities of the approach rather than to be a peculiar attempt to design an aphicide selective in the presence of thrips.

Before leaving the subject of penetration, it is worth recalling a familiar thought (Winteringham, 1969) that the insect (and even more so the mite) is at a severe disadvantage compared to a mammal in its exposure to contact insecticides because of its greater surface to volume ratio. Some quantitative data may help to bear this out. The surface area of the human skin is about 17,000 cm^2 or 0.24 cm^2/g, assuming 70 kg as the body weight (Maibach and Feldman, 1974). Measurements with third instar black cutworm larvae (weighing about 10 mg) and calculations of their surface area based on the assumption that they are closed cylinders (which considerably underestimates the complexity of their surface) yield a surface area to weight ratio of 29.6 cm^2/g. Thus this insect exposes well over 100 times more surface area than a human being for the same internal cellular volume, which must be a significant factor in increasing the risks from exposure to externally applied materials for the insect.

4. Binding and Loss at Indifferent Sites

The subject of binding and loss at indifferent sites has until very recently been extraordinarily neglected by students of insecticide toxicology, particularly as it pertains to insects themselves. The same cannot be said for workers in the field of drug action, where the crucial role of tissue binding in influencing the distribution and level of drugs in the body has received extensive study (e.g., see Gillette, 1973, and companion papers in that volume). Consequently, except for some fragmentary yet tantalizing glimpses of the possible role of binding in the selectivity of insecticides, little more can be said. However, it is quite evident that binding to tissue components, especially proteins, in some cases may be very extensive (more than 90% of the compound present internally may be thus sequestered), and this in turn could lead to several different effects on the dynamics of insecticide action depending on such factors as reversibility of binding and rate of metabolism.

a. Reversible binding temporarily removes the insecticide from circulation and the bound compound then acts as a reservoir, thus decreasing the peak concentration free in the body but prolonging its action in a way comparable to slowing the rate of penetration, as shown in Fig. 3. At the same time, the bound compound is protected from metabolic destruction. However, under some circumstances where such metabolism is very rapid, readily reversible binding in the plasma may actually increase delivery to the metabolic systems and thus speed elimination (Gillette, 1973). This introduces a second effect of reversible binding: as a means of increasing the efficiency of solubilization and transport of pesticides in the circulatory fluid, especially with compounds of limited aqueous solubility. To quote one instance from many, Fromm and Hunter (1969) found in perfusion studies with isolated trout gills that dieldrin was absorbed by the gills from the surrounding medium only if the perfusion fluid contained lipoprotein as a binding agent for the dieldrin. Additional examples are provided in Chapter 1. In some cases binding may be to specific carrier proteins, as with natural juvenile hormone which is transported and protected by a lipoprotein in insect hemolymph (Kramer *et al.*, 1974). A similar process might be important with synthetic JHAs (Hammock *et al.*, 1974*a*).
b. Irreversible binding is toxicologically equivalent to metabolic detoxication as a source of insecticide loss. One well studied example is the role of phosphorylation of nonspecific sites in the detoxication of organophosphates (Murphy, 1969). Removal of free insecticide from circulation by either reversible or irreversible means also steepens the concentration gradient at the site of application and in several studies has been implicated as a mechanism for enhancing penetration rates. At the same time, binding and penetration are also linked by the sometimes extensive sorption of toxicants which occurs to macromolecules within the integument (see Chapter 1 for further discussion in these areas).

Obviously there is no lack of complexity in the possible interpretations of binding phenomena, which provide a pharmacokineticist's paradise and a fertile field for the student of selective toxicity.

A few studies do exist which compare binding effects in insects and vertebrates, but it is hard to interpret their implications for selectivity. Boyer (1967) using stirofos and chlorfenvinphos compared binding to mouse and human plasmata and to fly head homogenates as it affected AChE inhibition studies. His conclusion was that a most significant binding action occurred in the plasmata but not in the insect preparation. A companion study of binding in the hemolymph of the American cockroach was recently presented (Boyer, 1975). A comparison of the two papers shows that with a free concentration of stirofos of 10^{-7} M (0.1 nmol/ml) at equilibrium, 13.5 nmol is bound per milliliter of mouse plasma, 7.7 nmol is bound per milliliter of human plasma, and 38 nmol is bound per milliliter of cockroach hemolymph (Boyer, 1975, Fig. 1). Thus in this case binding in insect blood is more effective than in that of mammals and must have a major influence on the toxicity of stirofos to the cockroach. The great significance of such sorption effects on the free level of diazoxon and pyrethrin in the insect is also graphically illustrated in the pioneering studies of Burt and his coworkers described in Chapter 1. Skalsky and Guthrie (1975) isolated several plasma proteins from the rat and American cockroach which bound DDT and dieldrin rather firmly but not parathion. Considerably less transport of these compounds occurred in cockroach than in mammalian blood cells, but the toxicological significance of these observations was not explored. In view of the great differences in protein composition of the blood of insects and vertebrates, one might well expect toxicologically significant differences in binding. On the other hand, when they compared the insect and mammalian midgut, Shah and Guthrie (1970) did not find any major differences in the binding of carbaryl, malathion, or nicotine during penetration, although somewhat higher degrees of binding of DDT and dieldrin were found with the tobacco hornworm midgut than with that of the cockroach *Blaberus*, or the mouse. In keeping with the data of Skalsky and Guthrie (1975), the degree of binding was inversely related to the polarity of the insecticide.

Mammalian species also show significant differences in their ability to bind xenobiotics in the serum; for example, Albert (1967) generalized that although man is a rather poor metabolizer of drugs he is a good binder of them by comparison with other mammals. In a comparative study of 12 vertebrates a wide variability (14.5–40.7% of the dose) in binding of amphetamine to plasma proteins was observed (Baggot *et al.*, 1972), although, in contrast to results with acidic materials, with this basic compound human serum was not highly active.

A second and rather different type of reversible removal from circulation occurs when a lipophilic compound is stored in fat depots. This has not been shown to have a major role in selectivity, but may on occasion have some effect; for example, the female American cockroach has a fat body threefold larger than that of the male. It has been suggested with schradan (Saito quoted in O'Brien, 1967*a*) and several chlorinated hydrocarbons (Brown, 1960) that this can explain why the female is severalfold more tolerant to these compounds than the male. In the same vein, Bacon *et al.* (1964) observed some quite large changes in the DDT sensitivity of the lygus bug with diet and season and correlated these with alterations in the lipid content of the body. Numerous other workers have attempted, often with limited success, to relate lipid content or type in insects to the development of resistance (Brown, 1960; Winteringham, 1969).

5. Excretion

The role of differences in excretory efficiency as a basis for selectivity can be dismissed very rapidly, partly because of our ignorance, especially with insects, but mainly because it is irrelevant for most insecticides.

Current insecticides are generally nonionic lipophilic compounds, a class of materials not readily excretable by insects or mammals without further metabolism (see Chapter 1). When such metabolism occurs, it almost invariably results either directly or indirectly in compounds of low toxicity. In this case, although efficient excretion is necessary to prevent the eventual accumulation of such metabolites, the immediate hazard has already abated before excretion occurs. Only with those insecticides which are ionizable at physiological pH (e.g., amines, formamidines, fluoroacetate, and phenols) is excretion of the parent compound in quantity at all likely, and instances of significant direct excretion even here seem rare (e.g., the rapid elimination of nicotine by the tobacco hornworm already mentioned).

However, as ever there are exceptions: Gupta *et al.* (1971) report that larvae of the khapra beetle, which are naturally tolerant to DDT, excrete large amounts of it apparently unchanged (37% of the topical dose in 12 h). Adult beetles, which are less tolerant, excrete less (11%). We cannot pretend to understand the excretion of xenobiotics in insects nearly to the degree that excretion in the mammalian system is understood.

The possibility for the removal of toxicants from aquatic organisms by direct dialysis into the surrounding medium should be kept in mind—and, as a reminder of the infinite resourcefulness of nature, one or two rather bizarre manifestations of resistance through enhanced excretion such as the case of resistant mosquito larvae which produce copious volumes of peritrophic membrane on which are adsorbed large amounts of the DDT taken in (Abedi and Brown, 1961).

6. Metabolic Alteration of Toxicants

6.1. Relationship of Metabolism and Toxicity

It is quite rare for any foreign compound (xenobiotic) to enter the body and not be converted, at least in part, to metabolites that are usually more polar than the parent compound and can be rapidly excreted from the body. A battery of more or less specialized enzymes provides this protective barrier against damage from foreign chemicals (Chapters 2, 3, 4, and 5). Metabolism to compounds less toxic than the parent xenobiotic constitutes *detoxication*, but not infrequently a metabolite may have significantly higher biological potency than its parent, an effect termed *activation*. Numerous examples of both types of metabolism are found with insecticides, and there is no doubt that differences in metabolic rate between organisms represent a (probably *the*) most important force behind selective toxicity.

Common sense and abundant evidence suggest that the more rapidly a compound is detoxified and eliminated from the body, the less toxic it will be, and that if two organisms differ widely in the rate at which this occurs the slower detoxifier of the two stands at greater risk. The reverse argument naturally applies to comparative rates of activative metabolism. This relationship between toxicity and metabolism is solidly supported by numerous studies of synergism, resistance, induction of detoxifying enzymes, and other conditions in which natural or induced changes in the ability of organisms to metabolize a toxicant can be directly related to changes in their sensitivity to it (e.g., see Hollingworth, 1971, and the several other chapters of this book which deal with the above topics). Some additional examples are cited below.

6.2. *Differing Capacity of Organisms to Metabolize Xenobiotics*

There is an abundant literature of pesticide and drug metabolism which clearly establishes that species may differ quite widely in the rate, and less often in the route, by which they metabolize xenobiotics. A few quite varied examples will illustrate this diversity in metabolism both *in vivo* and *in vitro*, which is so significant a basis for selectivity.

A very simple example is provided by the cat and the Gunn strain of rat. These organisms are deficient in the ability to form glucuronide conjugates with many (but not all) of the substrates such as phenols and carboxylic acids which other mammals readily detoxify and excrete in this form. Consequently phenols are uncommonly toxic to cats (Miller *et al.*, 1973) and there is a general correlation of toxicity with the difficulty of glucuronide conjugation in a series of xenobiotics in the Gunn rat (Yeary, 1970). In a related example, it has been suggested that the reason that 3-trifluoromethyl-4-nitrophenol can be used as a selective toxicant to remove parasitic sea lamprey larvae from the salmonid populations of the Great Lakes lies in the fact that the trout have a higher capacity than lampreys to glucuronidate this compound (Lech and Statham, 1975) and thus are less sensitive to it. Insects and plants incidentally do not form glucuronides but have a parallel mechanism for producing glucoside conjugates which may be toxicologically equivalent (Smith, 1968, and Chapter 5).

Additional evidence for species variation in metabolic capacity is provided by the *in vitro* data of Table IX. This table contains some wide differences (and many similarities) between certain species in several important enzyme systems for insecticide metabolism; for example, compare the horse with the rat or

Table IX. Comparative in Vitro Activity of Some Enzyme Systems Important in the Metabolism of Insecticides

	A-esterase				Epoxide hydrase		MFO (epoxidase)	Glutathione *S*-alkyltransferase
	Liver microsomes		Plasma					
	MP[a]	EP[b]	MP[a]	EP[b]	HEOM[c]	HCE[d]	Aldrin[e]	MP[a]
Rat	93	212	97	195	1.8	1	84(♂), 21(♀)	364
Rabbit	46	149	741	1893	14.5	10	344	40
Horse	2155	37	145	74	—	—	—	—
Pig	—	—	—	—	31.4	30	142	—
Chicken	60	0	3	0	—	—	—	30
Pigeon	24	0	5	1	0.0016	0	—	70
Quail	—	—	—	—	0.019	2	108(♂), 9(♀)	—
Green sunfish	7	5	0	0	—	—	6 (trout)	10
Catfish	7	3	—	—	—	—	—	15
Housefly	0	0	—	—	11.6	0	197	1.8–5.6

[a]Methyl paraoxon: data in nmol product/g liver/min (Hollingworth, unpublished data).
[b]Ethyl paraoxon: as above.
[c]1,2,3,4,9,9-Hexachloro-6,7-epoxy-1,4,4a,5,6,7,8,8a-octahydro-1,4-methanonaphthalene: data as μg diol/mg protein/min. Varying pH and temperature. From Brooks (1972).
[d]1,2,3,4,9,9-Hexachloro-*exo*-5,6-epoxy-1,4,4a,5,6,7,8,8a-octahydro-1,4-methanonaphthalene: data as % conversion in 30 min (37°C) except birds 90 min (42°C). From Brooks (1972).
[e]Data as pmol dieldrin produced/mg protein/min. Data from several sources (Hollingworth, 1971).

rabbit in the ability of hepatic microsomal A-esterases to hydrolyze methyl paraoxon (MP) to *p*-nitrophenol, yet this great difference does not appear with the closely related ethyl paraoxon (EP). Meanwhile houseflies, and several other insects tested, seem to be entirely devoid of this enzyme system. In the blood plasma, with either methyl or ethyl paraoxon as substrate, the rabbit has an extremely high A-esterase activity compared to other vertebrates. A rather graphic demonstration of the dominant role of A-esterase in a case of selective toxicity was uncovered by Lee (1964) when he found that individual sheep can be classified as either "sensitive" or "insensitive" to the delayed neurotoxicity of the organophosphate haloxon depending on whether they have a genetically based low or high level of an A-esterase in the plasma. This enzyme is capable of rapidly hydrolyzing haloxon but not the very closely related substrate coroxon, which is degraded by a second genetically independent A-esterase isozyme (see Table XIV for structures).

Although the data in Table IX for the activity of epoxide hydrase were gathered under various conditions, which makes numerical comparisons difficult, this enzyme, which converts epoxides to diols, is clearly very active in hepatic microsomes from the pig and rabbit, and in microsomes from the housefly, much less active in the rat, and almost absent in the two birds, the quail and pigeon. These values were obtained using HEOM, a dieldrin analogue lacking the 5,8-methylene bridge, as substrate. However, with a closely related epoxide (HCE), which differs from HEOM only in the position of the epoxy group, activity is completely lacking in the housefly, but the quail now has a capacity in the same range as the rat, while values for the rat, rabbit, pig, and pigeon are consonant with the HEOM data.

Undoubtedly the mixed-function oxidase system (MFO) is the most general and important metabolic force for insecticides. The data in Table IX show variation between species in the rate of a typical reaction, the epoxidation of aldrin. The considerable sex-related variation found with the rat and quail is notable. Houseflies show a quite appreciable activity in this and other MFO reactions. The aldrin epoxidase activity in tissues from numerous insect species has been determined, and although there is wide species variation it is clear that insects in general are not inferior to vertebrates in their oxidative capacity (Hollingworth, 1971; Gillett, 1971; Gilbert and Wilkinson, 1974). However, a note of caution is needed here since estimates of MFO activity in particular may vary quite widely in the literature even with the same species and substrate, as a comparison of the tables in the above references will show.

Finally, another rather general detoxifying system is the family of glutathione transferases. Activity in Table IX is related to the ability of liver-soluble fractions to *O*-demethylate methyl paraoxon, a reaction in which the rat excels. Housefly tissues in these assays show a relatively low activity which varies from strain to strain, as indicated by the range of values. It is noteworthy that in all these reactions the activity in fish is invariably toward the lower end of the scale, while birds have quite vigorous activity in most instances. Many other fascinating examples of species variation in the rates or paths of xenobiotic metabolism are described by Smith (1964, 1968), Walker (1974), and Williams (1967*a*, 1974).

In addition to illustrating the wide variability which may exist, the data of Table IX also illustrate some of the very real problems in trying to define the metabolic competence of a species (or a higher taxon) toward xenobiotics. Obviously, even for a single species, closely related compounds may vary widely and unexpectedly in their relative activities as substrates. Furthermore, such variables within the species as sex, age, nutrition, strain, and exposure to inducing agents may make a very great difference to the observed rate of metabolism. Finally, there are major problems in extrapolating metabolic rates *in vitro* (often obtained with a single tissue, under optimal conditions, and in the absence of competitive enzyme systems, inhibitors, and substrates) to conditions *in vivo* where these ideal conditions may be absent (Hollingworth, 1971). Perhaps the best example of this is provided by Williams (1967*b*), who quotes data to show that, judged by *in vitro* capacity, the dog has enough rhodanese activity in the liver to detoxify 4 kg of cyanide in 15 min with the additional capacity for 1.8 kg in the muscle tissue. Yet cyanide is highly toxic to dogs, with an oral LD_{50} of 1.6 mg/kg! Clearly the *in vitro* values are not predictive of *in vivo* capacity.

6.3. A Rational Basis for Variation in Metabolic Capacity?

Despite these problems, some courageous authors have attempted to wrestle order from the chaos of recorded species variability in xenobiotic metabolism. Where reasonably accurate, such generalizations would have great value for understanding and predicting selectivity. Unfortunately they are relatively few, and are subject to significant exceptions.

A familiar concept was put forward by Brodie and Maickel (1962), who claimed that aquatic vertebrates were incapable of carrying out many of the normal metabolic reactions with xenobiotics and explained this on the basis that these organisms can rid themselves of such compounds directly by outward dialysis into the surrounding water, while terrestrial species must convert them to the water-soluble forms appropriate for urinary excretion. This concept is further discussed by Smith (1968) and obviously does not apply when the ambient water itself is polluted with the xenobiotic. However, numerous subsequent studies have shown that aquatic vertebrates do have a reasonably normal armamentarium of metabolic defense reactions, although in many cases they are of reduced capacity, i.e., in the range of 5–15% of the rate found with rodents (Table IX; Adamson, 1967; Stanton and Khan, 1973; Pohl *et al.*, 1974). It seems likely that this lower metabolic capability may be related in part to the high sensitivity of fish to many xenobiotics (Holden, 1973). However, there is nothing to suggest that aquatic insects such as mosquito larvae are peculiarly defective in their general capacity to detoxify insecticides, and in several studies they show quite respectable metabolic activity (Hollingworth, 1971; Leesch and Fukuto, 1972; Krieger and Lee, 1973).

The theory of Krieger *et al.* (1971) which related the level of intestinal mixed-function oxidase to the breadth of host plant range in lepidopterous larvae has already been mentioned. Despite its generally reasonable (if oversimplified) basis and its correlative success with this restricted group of insects, other studies of the degree of synergism of carbaryl by MFO inhibitors as an index of MFO activity with a much broader range of insect orders (Brattsten and Metcalf, 1970, 1973*b*) have led to the conclusion that it is "difficult to discern any obvious correlation between the carbaryl tolerance and synergistic ratio and the phylogenetic position, food habits, and biological speciation" of the insects studied. Unfortunately, this attractively simple but rather indirect approach to defining MFO activity does not always agree with more direct measurements—e.g., in its prediction of low MFO activity in honeybees (*cf.* Gilbert and Wilkinson, 1974)—and probably suffers from the influence of unknown variables (Hollingworth, 1971). A parallel conclusion to that of Brattsten and Metcalf was reached by Pohl *et al.* (1974), who studied hepatic MFO in ten species of marine fish and concluded there was no correlation of enzyme activity with habitat or diet. However, as H. T. Gordon has pointed out (personal communication), the distribution of metabolic detoxication mechanisms must be the result of natural selection by endogenous and exogenous chemical compounds (including the kinds and quantities of xenobiotics in the natural diet or synthesized by gut microflora or parasites) to the extent that some biochemical target is vulnerable to the concentrations that actually occur, and for which no other defense mechanisms exist. Obviously it would be an immense task to define completely the chemical selective forces in each case, even though they provide an evolutionary rationale for the existence and specificity of metabolic detoxication systems.

6.4. Place of Man in the Metabolic Hierarchy

Man naturally has been concerned with his own capacity to metabolize toxicants in his environment, although this is a sadly neglected field of investigation as far as pesticides are concerned, and most of the available data are derived from studies of drug metabolism. A major question is that of extrapolation of metabolic studies from animals to man. This subject is reviewed by Williams (1974), who concludes that no species is an adequate surrogate for man, but that, as phylogeny would suggest, the Old World monkeys are among our closest metabolic cousins, with the New World monkeys next, but less closely related. Rodents, particularly the rat, may be used as a predictor for man only with an acute awareness of their many differences from him.

Man may not only sometimes differ from laboratory animals in the mechanisms by which he eliminates xenobiotics, but often in the rate, also. In some important reactions (e.g., with mixed-function oxidases and glutathione transferases, but not epoxide hydrase) there is evidence that man, at least in *in vitro* studies, has a rather low capacity to degrade foreign compounds compared to his laboratory animals (Hollingworth, 1971; Chasseaud, 1973), a conclusion reinforced by several comparative studies of the MFO system and its components in man (e.g., Nelson *et al.*, 1971). Such a conclusion must be tempered by consideration of the problems of procurement of satisfactory human tissue samples, the fact that the human population is genetically far more variable than inbred rodent strains, and the dependence of comparative metabolic rates on the substrate chosen for the assay. However, if this relatively poor degradative capacity is true, it may be no coincidence that in many cases man is also more easily poisoned by insecticides than are commonly tested mammals. Although this is hardly a subject which lends itself to controlled experimentation, there is reasonable evidence that such varied compounds as DDT, dieldrin, nicotine, endrin, diazinon, parathion, and malathion are considerably but variably more acutely toxic to man (especially children) than to rats (Hayes, 1967; Heyndrickx, 1969). The same generalization has also been made with respect to the relative toxicity of drugs to man and laboratory mammals (Lehman, 1959). This should be kept in mind in assessing the many studies of selectivity discussed here where rodents are the mammalian representative, since the degree of selectivity relative to man may often be overestimated.

6.5. Some Examples of Selectivity Based on Differential Detoxication

After this general background on the variation and significance of species differences in insecticide metabolism, it is logical to look at a number of specific cases where metabolism is a major determinant of selectivity. Most of those which have been investigated in any depth involve organophosphates.

The case of malathion, where selectivity between insects and mammals depends on the greater carboxylesterase activity of the latter, and the more complex situation with dimethoate, where selectivity between mammalian species depends in part on the variable activity of an amidase, have been reviewed in detail by O'Brien (1967*a*) and Murphy (1969), and need not be recapitulated here. However, from these studies have arisen, largely due to the efforts of O'Brien and his coworkers, two important general concepts. The first is that of a *selectophore*, by which is meant a chemical grouping which confers selective properties (generally but not necessarily between mammals and insects) on a toxicant. In the case of malathion the selectophore is the hydrolyzable carboxylester group, —C(O)OC_2H_5, while with dimethoate it is the carboxylamide group, —C(O)NHCH_3. O'Brien (1967*a*) describes how, pursuing this concept, these selectophores were used to devise a novel series of selective organophosphates, e.g., acethion, a derivative with an ethyl mercaptoacetate moiety in place of the diethyl mercaptosuccinate group of malathion. Fortune was smiling in this case because it was later found that acethion is probably not degraded by the same carboxylesterases as malathion in either mammals or insects (O'Brien, 1967*a*, p. 76). Once again the dangers of extrapolation are evident. The malathion–carboxylesterase theory is firmly

Table X. Liver Carboxylesterase and Its Relationship to the Toxicity of Malathion and TOTP Synergism in Several Vertebrates[a]

		Malathion[c] (ID_{50}, mg/kg)		
	Liver carboxylesterase[b]	Control	TOTP pretreated	Control/TOTP
Mouse	1.00	860	30	29
Quail	0.16	68	4	17
Bullhead	0.13	600	50	12
Frog	0.13	1300	13	100
Sunfish	0.07	43	4	11

[a]Data from Cohen and Murphy (1970).
[b]Relative activity in malathion hydrolysis.
[c]Dose of malathion causing 50% inhibition of AChE *in vivo*. TOTP pretreatment has no significant effect on AChE activity but lowers carboxylesterase activity by 80% or more in all species.

based on such evidence as comparative enzymology and metabolism, mechanistic studies of insect resistance, and the synergism of malathion by carboxylesterase inhibitors such as TOTP (tri-*o*-tolylphosphate) and the organophosphonate ester EPN. Despite this, the work of Cohen and Murphy (1970) reviewed in Table X shows that over a wider range of vertebrates there is no relationship between sensitivity to malathion and carboxylesterase titer in the liver, nor is the synergistic effect of TOTP related to carboxylesterase levels. Thus the frog, with little liver carboxylesterase activity, is highly insensitive to malathion, perhaps because of an insensitive AChE (see later). However, this cannot explain the very high degree of TOTP synergism observed with this animal. Additionally, Chen and Dauterman (1971) found that although amidase hydrolysis may be a factor in the low toxicity of dimethoate to mice, toxicity was not related to amidase activity (assessed with a sheep liver enzyme) in a series of dimethoate analogues with various *N*-alkyl and *O*-alkyl substituents.

The second central concept applies to compounds which are latent toxicants and depend on activation reactions for their toxicity (*proinsecticides* in current terminology). While these nontoxic precursors, such as phosphorothionates, await activation in the body they are susceptible to degradation, and if one organism (e.g., a mammal) is more efficient in this degradation than another (e.g., an insect) the compound will show a greater degree of selectivity. Thus in most cases phosphorothionates are more selective than their directly acting phosphate analogues. This effect has been termed the *opportunity factor*, and it has application far beyond the organophosphates. Its generality may be judged from the fact that of the 20 or so commercial organophosphates listed by Kenaga and End (1974) which have either carboxylester or carboxylamide groups, those which are direct inhibitors—e.g., monocrotophos and mevinphos—have acute oral LD_{50} values in the general range of 1–100 mg/kg to the rat, while the phosphorothionates have toxicities in the 100–1000 mg/kg (or higher) range. Apart from a fluoroethyl ester which may act through fluoroacetate, only one compound contravenes this rule, i.e., prothoate, the *O,O*-diethyl *N*-isopropyl analogue of dimethoate, which despite being a phosphorothionate has an LD_{50} of 8–25 mg/kg. However, this raises the interesting point that prothoate is the only diethyl ester in this series of phosphorothionates, all the others being dimethyl analogues, and the dimethyl analogue of prothoate fits our rule above perfectly (LD_{50} = 250 mg/kg; Eto, 1974, p. 281). Cyanthoate, the only diethyl ester among the phosphate analogues above, is also the most toxic, with an LD_{50} of 2–4 mg/kg. Additionally, the *O,O*-diethyl analogue of dimethoate is about twelvefold more toxic to mice than dimethoate itself (Chen and Dauterman, 1971). Similarly, the *O,O*-diethyl analogue of malathion is severalfold more toxic to mice than malathion itself, while the *O*-methyl ethylphosphonodithioate variant of malathion has an LD_{50} (acute, oral) to the rat of 45 mg/kg and a VSR value (rat/fly) of only about 2.0 (J. J. Menn, personal communication). Thus we may conclude that for good selectivity not only do we need the selectophore and opportunity factor, but also the compound should be an *O,O*-dimethyl ester. We will return to this point later.

Before leaving this area it should be noted that, in addition to the role of the opportunity factor in the low mammalian toxicity of the phosphorothionates, a second factor peculiar to substrates degraded by carboxylesterases and amidases may operate; that is, the phosphate analogues such as malaoxon and dimethoxon not only are substrates for these enzymes but also act as inhibitors by phosphorylation (Dauterman and Main, 1966; Chen and Dauterman, 1971). This complex self-synergism may be a contributing factor in the relatively high toxicity of the phosphate analogues in some cases but of course does not operate in the many compounds lacking such hydrolyzable groups which also demonstrate the opportunity factor.

Such useful rules of thumb as selectophores and opportunity factors are far from infallible, and in reality toxicity and selectivity depend more on the properties of the molecule as a whole than on those of any individual part. Rather clear proof of this is provided by the failure of O'Brien and Hilton (1965) to improve selectivity of highly toxic phosphates (e.g., DFP and mipafox) by making their phosphorothionate analogues, and the attempts of Sanborn *et al.* (1974) to apply the selectophore concept to the substituted phenyl *N*-methylcarbamates. In the latter study, such potential selectophores as carboxylester, carboxylamide, nitrile, and carbamoyloxime groups were introduced in the ring, but the most notably selective compound obtained was a selective mammalicide with a VSR of <0.03. The authors concluded that in addition to a selectophore these compounds also needed an opportunity factor, as in malathion and dimethoate, to be selective. Plausible as this may be, it was not shown in this study that the selectophoric groups were in fact performing their role of differentially enhancing biodegradability, so the reason for failure to obtain selectivity could as easily be attributed to this possibility.

In the above investigation, the nitrile group was chosen as a potential selectophore on the basis of a previous study by Vinopal and Fukuto (1971) of the reasons for the very high selectivity of phoxim (phenylglyoxylonitrile oxime *O,O*- diethyl phosphorothionate, VSR = 3696 in Table I). The selectivity is all the more remarkable since phoxim is a diethyl ester. It was concluded that two major factors operated in concert to favor survival of the mouse over the fly in this case. First, overall detoxication was more rapid in the mouse, so that the toxic oxygen analogue (phoximoxon) accumulated in the fly but not the mouse. Superior detoxication in the mouse involved both extensive cleavage of phoximoxon itself to diethyl phosphate, and hydrolysis of the nitrile group to a carboxylic acid, a reaction seen only in the mouse, which was particularly effective at high dosage levels. The second advantage of the mouse lay in the higher (270-fold) sensitivity of the fly acetylcholinesterase to inhibition by phoximoxon.

Diazinon is another reasonably selective organophosphate (VSR = 97, Table I) in which initial studies (Krueger *et al.*, 1960) suggested that metabolism was more rapid in the mammal (mouse) than insect (American cockroach), leading to a higher level of the activation product diazoxon in the insect in *vivo*. Several detailed studies have since been conducted on the comparative metabolism of diazinon *in vitro* which not only bear this conclusion out but also demonstrate the great enzymatic complexity which may underlie metabolism in the living organism. Diazinon, and the points of metabolic attack in both degradative and activative reactions, is shown in Fig. 4. At least three different

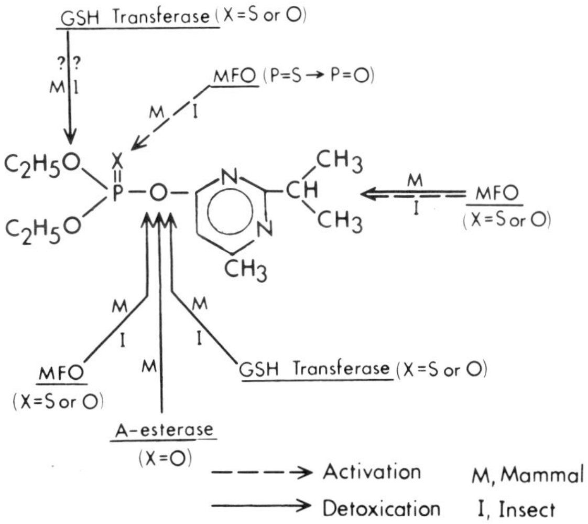

Fig. 4. Sites of enzymatic attack on diazinon (X = S) and diazoxon (X = O) *in vitro* by enzymes from rat liver and American cockroach fat body.

enzyme classes are important in its biotransformation—the MFO, glutathione transferases, and A-esterases—and the first two attack several different sites on the molecule. A simplified summary of the relative activity of these enzymes against diazinon and diazoxon in the mouse liver and cockroach fat body is shown below, based on the studies of Shishido *et al.* (1972*a*,*b*) and Shishido and Fukami (1972):

MFO:	liver $\gg$ fat body
A-esterase:	liver $\gg$ fat body
GSH transferase:	fat body $>$ liver

Since the activation of diazinon is an MFO function, the higher MFO activity in the mouse liver could be a toxicological disadvantage. However, it was concluded that in fact the liver acts as an overall detoxicative organ since the diazoxon formed is rapidly degraded, e.g., by the A-esterase which is highly active in the mammal but absent from the insect (*cf.* paraoxon data in Table IX). Overall the mouse clearly has a considerably greater detoxicative capacity than the cockroach against diazinon. However, the interesting question remains as to where diazinon is activated in the mouse if in fact diazoxon does not escape from the liver. A series of studies of the metabolism of diazinon by rat and housefly tissues gives rather a similar picture (Yang *et al.*, 1971*a*,*b*).

Although less fully studied, there is no doubt that metabolic differences are important in the toxicity and selectivity of most other groups of insecticides. Examples with the carbamates and the variable effectiveness of synergists in insects have already been mentioned. In Table I, three chlorinated hydrocarbons (Alodan, VSR $>$ 968, β-dihydroheptachlor, VSR = 1333, and methoxychlor, VSR = 1765) have unusually high selectivity and low mammalian toxicity. This is probably due to their relatively high degree of biodegradability in mammals, as discussed by Brooks (1969) and Metcalf (1972*a*), although uptake from the gut may also be inefficient in the cases of alodan and dihydroheptachlor. Unfortunately, neither of these safe compounds was developed commercially. The very favorable selectivity of the pyrethroids in general (e.g., compounds 25, 39, and 41 in Table I) has usually been attributed to their extremely rapid metabolism in mammals, although direct evidence from comparative studies of their toxicological action and fate in mammals and insects *in vivo* is not extensive. The role of metabolism in governing their toxicity is evident, however, from the fact that in addition to extensive metabolism by MFO, esterase hydrolysis may occur readily. However, esterases attack primarily the *trans*-chrysanthemic and related esters of primary alcohols, and it is often found that these hydrolyzable *trans* isomers are notably less toxic to mammals than the more resistant *cis* isomers, and also more selective. Thus (+)-*trans*-bioresmethrin has a VSR of 32,000 while its (+)-*cis* isomer has a VSR of only 140 (Abernathy *et al.*, 1973). Most of this difference is attributable to difference in toxicity to the rat rather than to the fly, which is relatively deficient in such esterase activity (Jao and Casida, 1974).

6.6. Activation as a Selective Force

It is curious that despite the existence of several oxidative mechanisms of insecticide activation, some of them crucial to the toxicity of the compound, examples where selectivity arises mainly through differential rates of MFO activation have rarely been described, and, furthermore, decreased MFO activation does not appear to be a very significant factor in the development of resistance to insecticides. A plausible reason for this is the fact that the MFO system is invariably involved also in detoxication reactions which proceed simultaneously with activation so that a generally higher level of MFO activity in an organism does not necessarily give rise to a higher level of the activation product. Of course, at the other end of the scale this balance of activation and detoxication reactions fails if there is a virtually complete absence of MFO activation. Such a situation was described in the helminth *Ascaris lumbricoides* by Knowles and Casida (1966). They could find no evidence for MFO activation of phosphorothionates in these animals, which in turn were very resistant to them, although they were quite susceptible to their phosphate analogues. Some examples mentioned elsewhere in this chapter for which it has been suggested that enhanced MFO activation is at least partly responsible for selectivity include isopropyl parathion (fly > honeybee), dimethoate (fly > cricket > milkweed bug and American cockroach) and Abate (mosquito larva > fly).

Several more definitive examples of activation as a source of selectivity are found with non-MFO metabolism, and particularly with *N*- and *O*-deacylation reactions. Acylation of the hydroxyethylphosphonate trichlorfon, especially with butyric acid to yield butonate, improves the VSR value about eightfold. Probably differential activation by hydrolysis at the butyric ester group is responsible (Eto, 1974, p. 199). A similar effect, although of unknown basis, is seen after *N*-acetylation of methamidophos (*O*,*S*-dimethyl phosphoramidate, rat oral $LD_{50} = 13\text{–}30$ mg/kg) to create acephate ($LD_{50} = 866\text{–}945$ mg/kg) without loss of insecticidal potency (Eto, 1974, p 200).

Another probable example of selective toxicity through differential activation in target and nontarget species is found with the fluoroacetate precursors.

Sodium fluoroacetate is a naturally occurring toxicant with interesting plant systemic properties which cannot be utilized directly in plant protection because of its high mammalian acute toxicity. Oral LD_{50} values generally are in the range of 0.1–10 mg/kg with toxicity to man estimated as 0.7–2.1 mg/kg, and that to the nutria, a rodent, as low as 0.056 mg/kg (Atzert, 1971). This toxicity has been put to sometimes controversial use in predator and rodent control using fluoroacetate as "compound 1080." Curiously, toxicity to amphibians ($LD_{50} = 50$ to >500 mg/kg; Atzert, 1971) and fish ($LC_{50} = 580$ ppm; Alabaster, 1969) is quite low. The uses, toxicology, and biochemistry of fluoroacetate have been reviewed by O'Brien (1967*a*), Atzert (1971), and Corbett (1974).

The desire to utilize such a unique and powerful toxicant for insect control has led to a number of attempts to find derivatives with satisfactory selectivity. In particular, esters or amides of fluoroacetic acid or fluoroethanol (which is oxidized to the acid *in vivo*) have been extensively examined. Enzymatic hydrolysis of the ester or amide is then necessary as an activation step to release the toxicant, and variations between species in the ability to carry out such hydrolyses should generate selectivity.

This theory seems to work smoothly in the case of fluoroacetamide, as shown in Table XI. Matsumura and O'Brien (1963) found that the amide is less toxic than the parent acid to mice and has improved potency to the housefly and American cockroach, a conclusion confirmed in general by the later results of Johannsen and Knowles (1974*a*). Matsumura and O'Brien additionally studied the rate of hydrolysis of fluoroacetamide by mouse and cockroach tissues and found that fluoroacetate was indeed released much more rapidly by the insect, in keeping with the pattern of selective toxicity.

However, improvement beyond fluoroacetamide is possible. Two more recent commercial compounds with significantly improved selectivity (Table XI) are fluenethyl or 2-fluoroethyl-(4-biphenylyl)acetate, a nonsystemic acaricide–insecticide, and Nissol or 2-fluoro-*N*-methyl-*N*-(1-naphthyl)-acetamide, a systemic acaricide. Clearly with these compounds the VSR is well improved over fluoroacetate, and acaricidal activity is retained. In studies of the metabolism and mode of action of these two compounds in mice, houseflies, and spider mites (Johannsen and Knowles, 1972, 1974*a*,*b*) it was concluded that the mode of action was the same in these organisms (i.e., release of the fluoroacetate or its precursor fluoroethanol by hydrolysis, and subsequent lethal accumulation of citrate), but no detailed study was made of the basis for their selectivity. However, Noguchi *et al.* (1968) compared the toxicities of Nissol to several mammals and found a close relationship between toxicity and the rate of hydrolysis to release fluoroacetate; for example, the guinea pig ($LD_{50} = 2$ mg/kg) is 60- to 130-fold more sensitive to Nissol than the rat or mouse by oral dosage, and the rate of activation is 30- to 45-fold higher in guinea pig liver homogenates than in those of rat or mouse. However, the guinea pig is 10- to 15-fold more sensitive to fluoroacetate itself, which also must play a role in determining the level of selectivity observed.

Although even the best VSR values of Table XI are meager compared to those of such highly selective compounds as the pyrethroids, they are sufficient to have allowed the harnessing of the power of fluoroacetate for agricultural purposes, and thus these derivatives must be regarded as a successful example of the rational attainment of selective toxicity.

Finally, a general class of selective *N*-derivatized analogues of familiar *N*-methylcarbamates have been developed, particularly by Fukuto and his coworkers at Riverside, based on the initial observations of Fraser *et al.* (1965) with *N*-acetyl *N*-methylcarbamates. Typically structures of such

Table XI. Improvements in the Vertebrate Selectivity Ratio (VSR) by Derivatization of Fluoroacetate

	LD_{50} (mg/kg)			LC_{50} (ppm)	VSR	
		Housefly				
	Mouse, intraperitoneal	Injected	Topical	Two-spotted spider mite	Mouse/Housefly	Mouse/Mite
Na fluoroacetate (FCH_2COONa)[a]	18	21	—	—	0.86	—
Fluoroacetic acid (FCH_2COOH)[b]	15	17	230	231	0.88	0.06
Fluoroethanol (FCH_2CH_2OH)[b]	20	17	350	163	1.18	0.12
Fluoroacetamide (FCH_2CONH_2)[a]	85	9.5	—	—	8.9	—
Fluoroacetamide (FCH_2CONH_2)[b]	110	27	235	1971	4.1	0.06
Fluenethyl ($FCH_2CH_2OC(O)CH_2$-ϕ-ϕ)[b]	42	13	300	284	3.2	0.15
Nissol ($FCH_2C(O)N(CH_3)(C_{10}H_7)$)[c]	200	14	525	250	14.3	0.80

[a]Matsumura and O'Brien (1963).
[b]Johannsen and Knowles (1974*a*).
[c]Johannsen and Knowles (1972).

compounds are

$$\mathrm{XC_6H_4OC(O)N}\begin{array}{l}\diagup \mathrm{CH_3}\\ \diagdown \mathrm{Z}\end{array}$$

where Z is $CH_3C(O)$—, $(RO)_2P$(O or S)—, $XC_6H_4OC(O)N(CH_3)$—S—, o $X'C_6H_4S$—.

A rather coherent toxicological picture, reviewed through 1973 by Hol lingworth (1975*a*), emerges with these derivatives, which are almost invariabl less toxic than their parent *N*-methylcarbamates to mammals and yet retai their toxicity to some insects such as houseflies and mosquito larvae. Since th derivatives themselves are invariably less effective anticholinesterases tha their parents, and in some cases may be completely inactive (Boulton *et al.* 1971), the parent carbamate is the active toxicant in all probability and one ca see two likely reasons for the improved selectivity:

a. The balance between activation to the *N*-methyl analogue and detoxi cation of this product is such that the activation product accumulates i insects but not in mammals.
b. The opportunity factor is operating here so that, while waiting fo activation, the parent compound is destroyed rapidly in mammals bu not insects.

These two possibilities are not mutually exclusive and both may operat together.

Although there is evidence to support the theory that more of th activation product accumulates in insects than mammals (e.g., studies o metabolism of *N*-acetyl Zectran in spruce budworms and mice by Miskus *et al.* 1969), definitive mechanistic studies have yet to be made of this type o selectivity. A third possible mechanism behind the selectivity was suggested b Cheng and Casida (1973), who postulated that during activation (assumed t be primarily an MFO reaction) the moiety removed ($Z = X'C_6H_4S$— in thi case) remains attached to the microsomal oxidase system, inhibiting it an preventing detoxication. Whether this would also occur with the other deriva tives, e.g., the *N*-acetyl analogues, is unknown. A further factor, particularl with the mosquito larvae, is the possibility of enhanced uptake of the mor lipid-soluble derivatized carbamate, leading to higher toxicity.

Taken together these results show there is an excellent record of success i improving selectivity by derivatization of known toxicants, even though th biochemical basis is not always known.

7. Site of Action

There is no doubt that the events occurring at the site of action are often major source of selectivity. An immediately useful classification of these sites i into those which are present and essential for life in both the target and nontarget organisms and those which are absent from or nonessential in the friend but present in the foe. Obviously the second case presents a much more favorable setting for selectivity, and, as we shall see, such opportunities are beginning to be exploited. But there is also ample evidence that enough differences exist in the molecular architecture of enzymes and other receptor sites which occur in both friend and foe to allow very significant selectivity. The target for the organophosphates and carbamates, acetylcholinesterase, is a case in point.

7.1. Sites of Action Common to Target and Nontarget Organisms

7.1.1. Acetylcholinesterase

Acetylcholinesterase (AChE) performs an analogous and vital function in the central nervous system of most animals that have one (Chapters 6 and 7) and yet inhibitors exhibiting high selectivity toward the AChEs from different organisms can be found. Thus in comparing, for example, "insect" and "vertebrate" AChEs one could take either of two positions: that they are remarkably similar in general properties and physiological function or that significant and useful differences do exist which may explain and be exploited for selectivity. This latter view will be stressed here, and some of the differences between "insect" and "vertebrate" AChEs most relevant to selectivity are outlined below. The categories "insect" and "vertebrate" are used here for convenience and it should be recognized that the enzymes from each group are far from homogeneous in their properties.

Most vertebrate AChEs are highly specific for ACh and have little activity against butyrylcholine (BuCh) as a substrate. Some insect AChEs (e.g., that of the housefly and other diptera, Brady, 1970) hydrolyze both substrates rapidly, while others (e.g., that of house cricket, honeybee, and German cockroach) are, like the vertebrate AChEs, poorly active against BuCh (Lee *et al.*, 1974). This suggests that the dipteran enzyme has an active site which will accept larger acyl groups than the vertebrate and many other insect enzymes, a situation which could be exploited for the design of selective inhibitors. However, the complexities of the AChE active site, as described by O'Brien in Chapter 7, are apparent here. Thus Yu *et al.* (1972*a*) studied the ability of several phenyl *N*-alkyl carbamates to bind to and carbamylate AChEs from bovine erythrocytes and from housefly, bee, and cricket heads. Increasing the length of the *N*-alkyl chain is structurally equivalent to lengthening the acyl chain in the choline esters, with the *N*-ethyl carbamate approximating BuCh. Such a compound might thus be a selective inhibitor for the fly AChE. In fact, no such pattern was seen, and binding to the active site of the fly AChE as estimated by Main's K_a value fell much more in passing from *N*-methyl to *N*-ethyl to *N*-propyl than it did to the bovine or bee AChE, which showed only a slight change in K_a over this series. From these results it seems likely that the choline esters and carbamates orient differently about the esteratic site of AChE.

Much other evidence exists for differences in the binding surfaces around the esteratic site among insect and vertebrate AChEs. For example, consider Fig. 5, which shows the effect of changing the *m*-alkyl substituent in a series of 3-alkyl-4-nitrophenyl dimethyl phosphates on comparative potency against AChEs from fly head and two mammalian sources (Hollingworth *et al.*, 1967*a*). Quite clearly, increasing the size of the *m*-alkyl group favors inhibition of the fly enzyme but decreases potency against the mouse and bovine AChEs. Based on previous observations which indicated differences in the spatial arrangement of binding sites on insect and vertebrate AChEs, it was hypothesized that bulky groups in the *m*-position of the ring enhance binding to the insect AChE, but obstruct it with the mammalian AChE. A similar conclusion for the fly enzyme was reached by Hansch (1970) in an analysis of the inhibition of fly head AChE by substituted-phenyl diethyl phosphates, who found that potency could be related to electronic effects only for *p*-substituents but that with *m*-substituents a further steric term had to be introduced, which

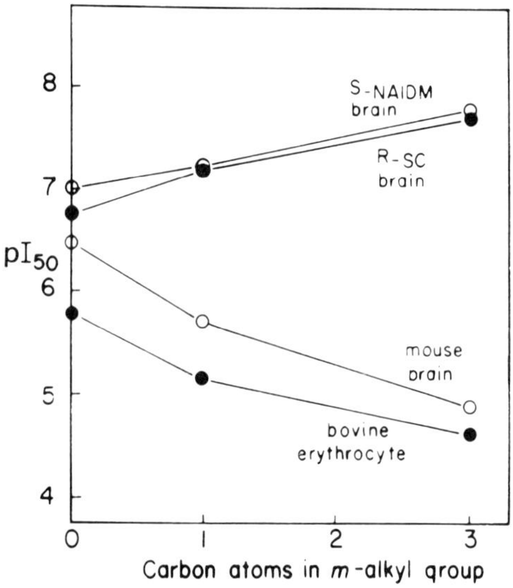

Fig. 5. Change in inhibitory potency ($pI_{50} = -\log I_{50}$) with alkyl chain size in dimethyl 3-alkyl-4-nitrophenyl phosphates using acetylcholinesterases from two housefly strains (R-SC and S-NAIDM) and two mammalian sources. Reprinted from Hollingworth *et al.* (1967*a*), with permission. Copyright by the American Chemical Society.

indicated that bulky *m*-substituents favored inhibition. Subsequently other authors have noted the same relationships with analogous inhibitors, as discussed by Hollingworth (1975*a*). Furthermore, the *m*-substituted analogues are often more selective and safer insecticides. This can be seen clearly in Table I by comparing methyl parathion (compound 21) and its *m*-methyl analogue, fenitrothion (compound 32), although such target differences are only one factor among several which govern these fascinating and important selectivity relationships among parathion analogues (Miyamoto, 1969; Metcalf, 1972*a*; Metcalf and Metcalf, 1973).

In an even broader generalization, Hellenbrand (1967) suggested that fly head AChE is more sensitive to carbamates than bovine erythrocyte AChE (the two enzymes most widely used in comparative studies). Several studies which strongly support this important conclusion are reviewed in Table XII. In this table the relative potency of the anticholinesterases to the fly and bovine enzymes is expressed as the selective inhibitory ratio (SIR), and data are presented to show the range of SIR values observed in each study, the mean value of SIR (data averaged as log SIR), and the total number of compounds studied, with the number of instances where the bovine enzyme was more sensitive given in parentheses. Perhaps the most striking fact is that, in five separate studies with 114 carbamates of quite varied structure, in not one case was the bovine enzyme shown to be more susceptible to inhibition than the fly AChE. Generally the difference in susceptibility was considerable, in the 10- to 100-fold range, with an occasional case of more extreme selectivity against the fly enzyme being over 1000-fold.

Such an observation with the carbamates at once raises the question of the relative sensitivity of the two enzymes to inhibition by organophosphates. This cannot be taken for granted, since O'Brien (Chapter 7) has shown that the nature of the binding of these two groups of anticholinesterases to AChE is probably quite different. However, the studies collected in Table XII demonstrate that the organophosphates also are generally better inhibitors of the fly AChE, although this varies rather widely with inhibitor structure. Of the 54 compounds studied, only two are preferential inhibitors of bovine AChE, and an SIR value as high as 2330 was found by Vinopal and Fukuto (1971) with a phosphinate analogue of phoxim.

Despite this surprisingly consistent record of selective target sensitivity in the fly, it would be a great oversimplification to extrapolate this to all insect–vertebrate comparisons. To quote only one instance of a reversal of this effect, Joiner *et al.* (1973) found that ethyl *bis*-*p*-nitrophenyl phosphate was over 12,000 times more potent against rat brain AChE and 13-fold more potent against a fish brain AChE than against the boll weevil enzyme.

Several other instances of selective inhibition (e.g., with DDVP and Ruelene) in which considerably greater sensitivity of the insect AChE can be correlated with the greater toxicity of these compounds to insects are presented by O'Brien (1967*a*). However, as O'Brien points out, these are extreme rather than typical examples.

Table XII. Selective Inhibition of Housefly Head Acetylcholinesterase by Carbamates and Organophosphates

Compounds	Selective inhibitory ratio (SIR)[a]			Reference
	SIR range	SIR mean	n[b]	
Carbamates				
1. Aryl *N*-methyl	4.7–701	34	23(0)	Sanborn *et al.* (1974)
2. Benzotriazolyl *N*-methyl and *N,N*-dimethyl	15–50	27	12(0)	Sacher *et al.* (1973)
3. *m*-Acylamido- and thioureidophenyl *N*-methyl	>0.7–1667	—	59(0)	Sacher and Olin (1972)
4. Carbofuran analogues	5.8–16.4	11	6(0)	Yu *et al.* (1972*b*)
5. Bicyclic *N*-methyl	6.2–32	13	14(0)	Yu *et al.* (1974)
Organophosphates				
6. Dimethoxon analogues	0.38-230	25	16(1)	Hastings and Dauterman (1971), Haung *et al.* (1974).
7. Aryl methyl ethylphosphonates	>0.7–44	—	7(0)	Darlington *et al.* (1971)
8. *O,O*-dimethyl *S*-aryl phosphorothioates	10.5–450	85	7(0)	Lee and Metcalf (1973)
9. Optical isomers of an ethylphosphonothiolate	0.92–5.5	2.0	4(1)	Wustner and Fukuto (1974)
10. Phoxim analogues	49–2330	450	5(0)	Vinopal and Fukuto (1971)
11. Methyl paraoxon derivatives	4.7–529	82	15(0)	Metcalf and Metcalf (1973)

[a]Ratio of k_i or I_{50} values showing relative sensitivity of fly head AChE/bovine erythrocyte AChE.
[b]Number of compounds examined; in parentheses, the number of compounds more active against the bovine enzyme.

In the studies described above where Main kinetic analysis of inhibition was conducted, variations in selective inhibition usually were much more dependent on changes in the relative binding strength of the inhibitor to AChE (K_a) than on variations in the rate of acylation of the enzymes.

So far we have concentrated mainly on insect–vertebrate comparisons. However, the properties of the AChEs in both these classes may vary quite widely. Thus in their study with *O,O*-dimethyl *S*-aryl phosphorothioates (Table XII) Lee and Metcalf (1973) also gathered data with cricket head AChE. Although the cricket resembled the fly enzyme in being uniformly more sensitive to inhibition than bovine AChE, the effect was less obvious; that is, the mean SIR for the cricket is 12 compared to 85 for the fly, with a maximum SIR of 35 for the cricket and 450 for the fly.

A more extensive comparison is possible with the housefly and honeybee esterases, which differ considerably in both substrate (Brestkin *et al.*, 1973; Lee *et al.*, 1974) and inhibitor specificity. In terms of inhibition, the study of Yu *et al.* (1972*b*) with carbofuran analogues showed that both the fly and bee AChEs were more sensitive than the bovine enzyme; however, the bee AChE (mean SIR of 26) was even more susceptible than the fly (mean SIR of 11).

Extensive studies have been made of the high (>200-fold) and desirable selective toxicity of isopropyl parathion to flies as compared to bees (and other beneficial Hymenoptera). This is particularly interesting because the closely related ethyl parathion is not very selective between these insects. Earlier studies in this area have been reviewed by O'Brien (1967*a*). More recent work by Camp *et al.* (1969*a*) has confirmed these earlier results in finding that at least a part of this selectivity can be related to the 40-fold greater potency of isopropyl paraoxon as an inhibitor of fly than bee head AChE. A major additional factor is probably metabolic since there is a greater rate of activation of isopropyl parathion in the fly than in the bee (Camp *et al.*, 1969*b*).

In another insect–insect comparison, Leesch and Fukuto (1972) determined that the AChE of the housefly was nearly 20-fold more sensitive to Abate-oxon than that of mosquito larvae (*Aedes aegypti*), although these enzymes were of similar sensitivity to the sulfoxide and sulfone analogues of this compound.

Considerable variability is also found in both the catalytic and inhibitory properties among vertebrate AChEs. Frogs, for example, show prodigious and general insensitivity to the toxic effects of anticholinesterases (see Table X and the data for the bullfrog in Pimentel, 1971), and it has long been known that frog AChE is unusually insensitive to the carbamate eserine. Potter and O'Brien (1963) studied this phenomenon further with paraoxon in the frog and mouse. Compared to the mouse, the 22-fold lower toxicity to the frog was paralleled by a 79-fold lower sensitivity of its AChE to inhibition by paraoxon. Since, surprisingly, these organisms did not differ very widely in the metabolism and distribution of injected paraoxon, target sensitivity seems to be the major determinant of selectivity here.

Further extensive comparisons of the substrate and inhibitor specificity of AChE from all corners of the animal kingdom are to be found in the review of Brestkin *et al.* (1973).

These examples also illustrate the relationship between selective inhibition and selective toxicity. The quantitative relationship between inhibitory potency and toxicity may be complex, but the qualitative basis is so obvious that it hardly needs stressing further except to note that extremely direct examples of selective toxicity based on variation in AChE sensitivity are to be found in the realm of insect resistance. Thus Tripathi and O'Brien (1973) found that a strain of houseflies has >1500-fold resistance to the organophosphate stirofos, which might be satisfactorily explained by the 206-fold lower sensitivity of its soluble AChE and 38-fold lower sensitivity of its particulate AChE to inhibition.

If species may differ so widely in the inhibitability of their AChEs, and this has the predictable toxicological consequences, it is interesting to ask how far we may expect differences in the AChEs of target and nontarget species to provide the basis for selective toxicity. One interesting attempt to answer this question with a series of substituted-phenyl methyl ethylphosphonates is presented by Darlington *et al.* (1971). In this study, satisfactory correlations of I_{50} values with LD_{50} values for both the housefly and rat (using bovine erythrocyte rather than rat AChE) were obtained, thus allowing prediction of the degree of enzyme selectivity necessary to provide a given level of selective toxicity. The results were rather depressing for anyone hoping to obtain selective toxicants solely on the basis of selective inhibition of AChE. Setting the desired toxicity levels at 1 mg/kg for the fly and 500 mg/kg for the mouse, their data predict that one would need a difference in sensitivity between the two AChEs of 34,000! However, if the opportunity factor applies here, the related phosphonothionates would have exhibited a more favorable selectivity pattern in this series.

A similar series of calculations is possible from the data of Yu *et al.* (1974) with bicyclic *N*-methylcarbamates and of Yu *et al.* (1972*b*) with carbofuran analogues using synergized toxicity to the housefly. In the former case a selectivity ratio of 500/1 would require a 5100-fold better inhibitor of the fly AChE. Although the data of Yu *et al.* (1972*b*) are more scattered, they do suggest a much more encouraging situation since only a 50- to 100-fold difference in anticholinesterase activity is needed for this degree of selective toxicity. Unfortunately, in neither of these series could the requisite degree of selective inhibition be achieved (Table XII).

In addition to major differences in the rate of inhibition, insect and vertebrate AChE show variations in their rates of reactivation after both carbamylation and phosphorylation (Table XIII). While vertebrate AChE recovers from phosphorylation more or less rapidly, depending on the nature of the alkyl groups attached to the phosphorus atom, insect AChE, if typified by the housefly, recovers not at all *in vitro*. It has been claimed, however, with good evidence, that rapid recovery does occur *in vivo* in insects (Mengle and

Table XIII. Spontaneous Reactivation of Several Acylated Acetylcholinesterases in Vitro and Its Variation with Species

Form of enzyme (E)	Half-life for reactivation (h)			
	Honeybee head	Housefly head	Cricket head	Bovine erythrocyte
$CH_3NHC(O)$-E[a]	1.33	1.30	0.90	0.62
$(CH_3)_2NC(O)$-E[a]	1.17	1.70	0.55	0.95
$(n\text{-}C_3H_7)NHC(O)$-E[a]	14.0	11.3	93.0	2.03
$(n\text{-}C_6H_{11})NHC(O)$-E[a]	565	No recovery	400	2.60
$(CH_3O)_2P(O)$-E	—	No recovery[b]	—	1.26[c]
$(C_2H_5O)_2P(O)$-E	—	No recovery[b]	—	58[d]

[a]Yu *et al.* (1972*a*), pH 7.8, 23°C.
[b]Mengle and O'Brien (1960), pH unstated, 38°C.
[c]Skrinjaric-Spoljar *et al.* (1973), pH 7.4, 37°C.
[d]From Reiner (1971), human erythrocytes, pH 7.4, 37°C.

O'Brien, 1960), a fact disputed by Brady and Sternburg (1967) in an equally careful and detailed study. For further consideration of this dilemma, the reader is referred to the discussion in their paper, and to that of Khoo and Sherman (1973), who also failed to find significant recovery of insect AChE *in vivo* after organophosphate poisoning. Carbamylated insect AChE does recover *in vitro*, as shown in Table XIII. However, as will be seen by comparing the rates of recovery for housefly and bovine erythrocyte AChEs, there are significant differences between them, particularly where longer *N*-alkyl chains are concerned, as well as differences between the insect species themselves.

A striking example of the role of dephosphorylation in selective toxicity is found with haloxon, which is unique among commercial organophosphates in being a di-2-chloroethyl ester. Extensive studies (Lee, 1964; Hart and Lee, 1966; Lee and Pickering, 1967; Pickering and Malone, 1967) reviewed in Table XIV have shown that with some but not all species recovery of AChE from inhibition by haloxon is extremely rapid and the rate of reactivation correlates directly with toxicity in several instances. Thus coroxon, the corresponding diethyl ester, is highly toxic to both rats and hens, while haloxon itself is extremely safe. In the rat, the AChE inhibited by haloxon recovers over 100-fold more rapidly than that inhibited by coroxon. Probably a similar situation holds with the hen, but data for spontaneous reactivation of diethylphosphorylated AChE are lacking for this species. A comparison of the susceptibilities of the hen, duck, and goose to haloxon is equally interesting. The range of oral toxicities is extremely wide and is paralleled by the relative stabilities of the phosphorylated brain AChE from these species. Finally, the usefulness of haloxon arises from its action as a selective antihelminthic agent, and here Hart and Lee (1966) concluded that a major factor in determining the relative susceptibility of these parasites was again the rate of AChE reactivation, which ranged from zero (e.g., with *Haemonchus contortus*) to very rapid in

Table XIV. Relationship of Reactivation Rate of Phosphorylated Acetylcholinesterase and Toxicity of Diethyl and Di-2-chloroethyl Phosphates in Several Species[a]

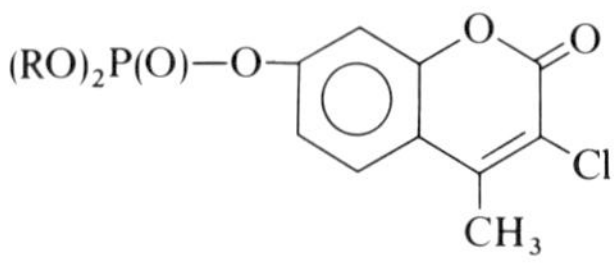

	Rat	Hen	Duck	Goose	Sheep	*Haemonchus contortus*
Coroxon (R = C_2H_5—)						
Toxicity (oral, mg/kg)	9.8[b]	2.2[b]	—	—	—	—
AChE (half-life, h.)	44	?	—	—	—	—
Haloxon						
(R = $ClCH_2CH_2$—)						
Toxicity (oral, mg/kg)	896[b]	5000[c]	500[c]	25[c]	"Low toxicity"	"High toxicity"
AChE (half-life, h.)	0.35	0.37	1.73	No recovery	0.39	No recovery

[a]Sources of AChE: Birds, brain; mammals, erythrocytes. Data from Lee (1964), Lee and Pickering (1967), Pickering and Malone (1967), and Reiner (1971).
[b]LD_{50}.
[c]Dose for "Severe toxic symptons."

the seven species studied. Reactivation in the host (sheep) occurs readily. Despite these correlations, it is probable that other factors enter into these cases of selectivity; for example, the relation of the A-esterase level in sheep and susceptibility to delayed neurotoxicity from haloxon has already been mentioned.

Thus we may conclude that although differences among AChEs in sensitivity to or recovery from anticholinesterase inhibition rarely account for selectivity in its entirety, there are numerous examples where such differences make a significant and perhaps major contribution. Several more examples of this are mentioned in Section 8 of this chapter.

7.1.2. Differences in Common Targets Other Than AChE

Naturally there exist many other targets for insecticides besides AChE, but since none has been as extensively studied, and some are completely obscure, their contribution to selectivity is uncertain. Where the direct interaction of insecticide and target in friend and foe cannot be assessed, only indirect and possibly misleading clues to this process as a source of selectivity are available.

Many pesticides exert their toxic action on the biochemical pathways associated with respiration and energy conservation (Chapter 10). This may be through uncoupling of mitochondrial oxidative phosphorylation, and Corbett (1974, p. 41) has presented comparative values for the uncoupling potency of several compounds against mammalian, invertebrate, and plant mitochondria.

There is very little difference between the two groups of animals, although plant mitochondria show larger variations. This lack of selectivity could arise if, as in the chemiosmotic theory, uncouplers do not bind to a receptor *per se* but act rather as physical shuttles of ions across the mitochondrial membrane and thereby discharge energy gradients.

The situation with inhibitors of the respiratory chain or of glycolysis is less coherent, and, not unexpectedly with such a range of enzymatic reactions, there is evidence for considerable variations in sensitivity between species, some instances of which are quoted by Corbett (1974). The example of the target insensitivity of the millipede to the respiratory inhibitor cyanide has already been given. In studying the mode of action of organotin acaricides in mites, houseflies, and mice, Ahmad and Knowles (1972) found that the presumed target, magnesium-stimulated ATPase, was of almost identical sensitivity to organotins in mice and flies, but ATPase from the spider mites was completely refractory to inhibition. Several different explanations for this are plausible, but there is a real possibility that a true species difference in sensitivity exists. Obviously this would mean that this ATPase is not the site of action in the mites at least.

Even less is known about target selectivity among the axonal poisons such as DDT or pyrethrins, or with other chlorinated hydrocarbon neurotoxicants such as the cyclodienes and lindane. But if no detailed account can be given of their role in selectivity, there are good indications that differences in sensitivity exist and may be significant. The case of tetrodotoxin insensitivity in the nerves of puffer fish and newts, attributed by Kao and Fuhrman (1967) to differences in the sensitivity of the axonal membrane itself, is an excellent illustration of this. Other examples can be taken from the field of insecticide resistance, where strains of flies (e.g., those with the *kdr* or *kdr-O* gene) show lowered sensitivity to axonal poisons by having developed "nerve insensitivity" of an as yet vaguely defined nature (Chapter 13). Other authors have concluded that the brains of the dogfish (Dvorchik and Maren, 1974) and rainbow trout (Mayer *et al.*, 1972) are severalfold more sensitive to DDT than those of mammals and birds as judged by the levels of DDT present there at doses necessary to cause death. But such observations may be only a pale reflection of comparative events at the DDT-receptor site within the axonal membrane. Snails also have been shown to withstand extremely high levels of chlorinated hydrocarbons in their tissues without ill effects (Hansen *et al.*, 1972).

The pyrethroids are among the most highly selective compounds in mammal–insect comparisons, although they are often highly toxic to fish (Pimentel, 1971). However, their low mammalian toxicity by the oral or dermal route is often accompanied by relatively high toxicity intravenously (Verschoyle and Barnes, 1972). This, and their rapid metabolism *in vivo*, has often been taken to indicate that little selectivity arises at the target level. Thus Verschoyle and Barnes (1972) state "it is clear that to the natural pyrethrins the nervous systems of insects and mammals are almost equally sensitive and respond in a similar manner." However, direct measurements of the relative sensitivity of crayfish, frog, and rat peripheral nerves to natural pyrethrins (Camougis, 1973) show considerable differences, with the crayfish, like the insect, sensitive, but the frog and rat nerves, even after desheathing, quite resistant to blockade. How this relates to the poisoning syndrome and high intravenous toxicity in mammals remains to be seen, but there is reason to be cautious in attributing the basis of pyrethroid selectivity solely to metabolic effects.

Finally, although a direct relationship to the action of most current insecticides may be remote, it is of interest to note the differences which, even at this early stage in our knowledge, can be discerned in some elements of the cholinergic synaptic mechanism other than AChE. Shankland in Chapter 6 pointed out the apparent similarities of the acetylcholine receptors (AChRs) from insect and vertebrate sources. However, it is probable that, as with AChE, significant differences do exist in the specificity of binding of cholinergic ligands to different AChRs (Chapter 8), and after extensive study of the nature and evolution of AChRs in vertebrates Michelson (1974) concluded that there may be greater variability in the AChR structures than among AChEs.

Strangely, few potent inhibitors of choline acetylase (ChA) are known and those that have been tested have not been useful as insecticides (Yu and Booth, 1971). However, the work of Dauterman and Mehrotra described by O'Brien (1967*a*, p. 286) shows some clear differences between the ChAs of mites, houseflies, and rats in their substrate specificity, and quite marked variations in sensitivity between insect and mouse ChAs to inhibitors such as the styrylpyridines (unfortunately with the mouse enzyme more sensitive) were recorded by Yu and Booth (1971). These observations suggest a potential for the production of selective toxicants which awaits exploitation.

7.2. Targets Unique to the Pest: Insect Growth Regulators

Among the most exciting recent advances in the development of highly selective insecticides is the discovery of a number of compounds which have been termed *insect growth regulators* (IGRs). These materials, whenever in its life cycle they may enter the insect, act only at specific critical stages of growth and development and are at other times virtually nontoxic. They are remarkably specific agents, disrupting insect morphogenesis with little toxicity beyond the class Insecta. The most extensive group of IGRs are the juvenile hormone analogues (JHAs). Others include the phenyl benzoylureas such as Thompson-Hayward TH-6040 (Dimilin) and a series of 2,6-di-*t*-butyl phenols such as Monsanto MON-0585. Although all fall under the broad heading of IGRs, this need not imply a common mode of action for these groups.

7.2.1. JHAs

The discovery, biology, chemistry, uses, and commercial prospects of the JHAs have been extensively reviewed (Menn and Beroza, 1972; Slama *et al.*

1974; Sorm, 1974; Menn and Pallos, 1975) and will not be presented here. Suffice it to say that the JHAs, by mimicking natural JHs, have a range of novel biological effects on insects, particularly through disruption of metamorphosis which results in the formation of supernumerary larvae or nonviable larval–pupal, larval–adult, or pupal–adult intermediates.

In certain cases the potency of JHAs is extreme; for example, one L-alanyl-*p*-aminobenzoic acid derivative was morphogenetically active on the European linden bug (*Pyrrhocoris apterus*) at 2 pg per insect (Sorm, 1974) and a terpenyloxypyridine compound was effective on *Tenebrio molitor* at 15 pg per pupa (Menn and Pallos, 1975). This represents about 75 ng/kg, or, more graphically, 75 μg per metric ton of insect, and such compounds fall within the range of even the most potent mammalian neurotoxins such as botulinus toxin and tetrodotoxin (Chapter 6).

The JHAs also exhibit favorable and often extreme selectivity at several important taxonomic levels. The extremely low acute mammalian toxicity of two JHAs is shown in Table I (compounds 42 and 43), and this is typical of several other JHAs (Bagley and Bauernfeind, 1972; Kenaga and End, 1974; Siddall and Slade, 1974). The acute toxicity to other vertebrates is also very favorable. Thus the LC_{50} for both methoprene and hydroprene to trout and bluegills exceeds 100 ppm (Siddall and Slade, 1974), and that for R-20458 to mosquito fish is over 10 ppm (Pallos and Menn, 1972).

Since the JHAs are such novel compounds, extensive study has been made of their toxicity to a very broad range of target and nontarget invertebrates. The results can be summarized briefly by noting that there is a high degree of selectivity toward noninsectan invertebrates (Miura and Takahashi, 1973) and this extends to the mites, which are uniformly insensitive (Staal, 1975), and to the crustaceans, as very close relatives of the insects. A characteristic of most JHAs is their selectivity between insect species, which in some cases may be very striking; for example, certain JHAs such as juvabione or peptidic JH analogues are active only on a single insect family, the pyrrhocorid bugs (Slama *et al.*, 1974, p. 214). Slama *et al.* state that selectivity is most generally shown at the family or higher taxonomic levels and that variation between species and genera is much less apparent. Consequently, toxicity to nontarget insects, particularly the beneficial parasites and predators, may present a problem, as with conventional insecticides. However, because toxicity varies between such species, particularly under conditions of field use, generalizations are impossible. More specific details of toxicity to nontarget insects are provided in the reviews of Bagley and Bauernfeind (1972) and Staal (1975).

Similarly, effectiveness against target insects varies quite widely, as illustrated in Table XV. Compounds A and B, an ester–amide pair, beautifully illustrate how extreme selectivity may be. Whereas the ester is reasonably toxic to the three heteropterans, it is relatively innocuous to the coleopteran *Tenebrio molitor.* The amide exactly reverses this situation, and the range of relative potencies exceeds 10^8. Less dramatic but more typical selectivity is found in comparing compounds C (methoprene) and D (hydroprene), both of

Table XV. Selective Toxicity of Four Juvenile Hormone Analogues to Several Insect Species[a]

	ID$_{50}$ (50% inhibition of metamorphosis)					
	Pyrrhocoris apterus[b]	*Dysdercus cingulatus*[b]	*Graphosoma italicum*[b]	*Tenebrio molitor*[c]		
A. 	0.01	0.005	0.1	50		
B. 	8	4	>250	0.0005		
Relative potency (A/B)	800	800	>2500	0.00001		
	Aedes aegypti[d]	*Musca domestica*[c]	*Galleria mellonella*[c]	*Heliothis virescens*[d]	*Tenebrio molitor*[c]	*Acyrtho-siphon pisum*[e]
C[f]. 	0.00025	0.0056	1.1	0.77	0.0040	0.0054
D[g]. 	0.0078	18	0.040	0.30	0.25	0.0039
Relative potency (C/D)	31	3200	0.036	0.39	63	0.72

[a]Sources of data: A,B, Slama *et al.* (1974) (compounds T-62 and T-63); C,D, Henrick *et al.* (1975).
[b]μg/larva, topical.
[c]μg/pupa or prepupa, topical.
[d]ppm in medium.
[e]% solution in spray.
[f]Methoprene (Altosid, ZR-515).
[g]Hydroprene (Altozar, ZR-512).

which are of some commercial interest. Whereas methoprene is markedly more active than hydroprene against Diptera (*Aedes* and *Musca*) and Coleoptera (*Tenebrio*), hydroprene has the clear advantage with Lepidoptera (*Galleria* and *Heliothis*). The compounds are both effective on the pea aphid (*Acyrthosiphon*). Obviously these patterns of selectivity presage quite different uses for the two compounds.

Much further information on variation in toxicity of JHAs between insect species is provided by Slama *et al.* (1974) and Staal (1975).

The reasons for these often extreme examples of selectivity in the JHAs have hardly been investigated, but there is good reason to suppose that more than one factor is involved. So far there seems to be no good evidence that any group outside the insects depends on the hormonal actions of a natural JH. Thus although such processes as metabolism and penetration of JHAs undoubtedly vary widely in noninsects, there is really no reason to look further than the lack of the normal JH receptors and their associated functions as the ultimate source of insensitivity in these organisms. The situation is more complex in dealing with selectivity between different insects. The crucial question of whether (or rather how much) insect JH receptors vary cannot yet be answered. Such factors as differential penetration, binding to sequestering proteins, and metabolism must also be considered as possible sources of selectivity. Several workers have found both quantitative and qualitative variations in the way different insect species degrade both natural JHs (Slade and Zibitt, 1972; Ajami and Riddiford, 1973) and synthetic JHAs (Hammock *et al.*, 1974*b*). A further dimension to this problem has been added by the discovery of Solomon and Metcalf (1974) that in the milkweed bug methoprene is activated *in vivo* by conversion to its more effective *O*-demethyl analogue. Evidence (unfortunately indirect) suggests that this metabolite is no more effective than methoprene in *Tenebrio molitor*, a finding which the authors attribute to possible differences in the JH receptors of the two species.

7.2.2. Other IGRs

Much of the extremely favorable selectivity of the JHAs is also found with other IGRs such as TH-6040, 1-(4-chlorophenyl)-3-(2,6-difluorobenzoyl)urea, and MON-0585, 2,6-di-*t*-butyl-4-(α,α-dimethylbenzyl)-phenol. These compounds and others which affect the insect cuticle have been surveyed by Hollingworth (1975*a*).

The toxic action of TH-6040 has been clearly defined by Mulder and Gijswijt (1973) as involving a lethal disruption of cuticular synthesis, particularly in the laying down of the chitinous endocuticle. This results in a thin, fragile integument at the next molt, which cannot support muscular activity, or which readily splits, leading in either case to death. Whether this arises by inhibition of chitin synthesis, as suggested by Marks and Sowa (1974) and Post *et al.* (1974), or by stimulation of chitinase and phenoloxidase (Ishaaya and Casida, 1974), or both, remains to be established.

The acute toxicity of TH-6040 to vertebrates is very low; for example, the oral LD_{50} to the male rat is 10,000 mg/kg, LC_{50} values to fish are in the range 135–255 ppm, and 8-day feeding studies with quail and ducks show LD_{50} values over 4640 ppm (TH-6040 Technical Information Bulletin, Thompson-Hayward Chemical Co., Kansas City, Ks.). No doubt this exemplary safety in acute studies depends on the absence of the cuticular target site in vertebrates. The potential for toxicity to other nontarget invertebrates seems to vary widely and unpredictably, some crustaceans and aquatic insects being sensitive while others are relatively resistant (TH-6040 Technical Information Bulletin).

Sacher (1971) described MON-0585 as being highly specific for mosquito larvae, blocking development and melanization early in pupation (in contrast to JHAs, which prevent emergence of the new adults from the pupal exuviae). The mode of action is not known but may be related to its properties as an antioxidant and free radical inhibitor (Desmarchelier and Fukuto, 1974). Toxicity to vertebrates is low, with an oral LD_{50} to the rat of 1890 mg/kg and an LC_{50} for fish of over 10 ppm (Sacher, 1971). However, selectivity at the insect level may be less than perfect. Steelman and Schillings (1972) report that MON-0585 reduced the number of dytiscid beetle larvae, and possibly hydrophilids, under field conditions.

These IGRs therefore have much in common in terms of selectivity. Acute toxicity to vertebrates is extremely low, and most nontarget invertebrates are relatively insensitive. Some shadows begin to appear with the crustaceans, and definite problems with toxicity to beneficial insects are possible with some IGRs in some uses. This can be determined only under appropriate field conditions since the JHAs at least are rapidly degraded in the environment. The IGRs do, however, fully bear out the idea that the best approach to high selectivity is by attacking biochemical systems peculiar to the pest.

7.3. Unexploited Targets for Insecticides

The relative success of the IGRs leads us to the question of other special targets in the biochemistry of the insect which might be utilized to develop new toxicants broadly selective between vertebrates and insects. This has been considered by Winteringham (1965), O'Brien (1967*a*), Bergmann (1972), and Hollingworth (1975*a*), and Shankland in Chapter 6 has outlined some major differences between the nervous systems of these two groups which might be exploitable. Here then will be mentioned only a few of the areas where much effort is being expended to understand and take advantage of such biochemical differences.

The insect neuromuscular junction is probably glutaminergic, in contrast to the vertebrate junction which is cholinergic. Rapid progress is being made in the study of the pharmacology of such junctions, including attempts to isolate the glutamate receptor and study the mechanism by which glutamate is removed to terminate junctional trasmission. Many of these results are reviewed by McDonald (1975). Unfortunately it seems that the glutamate receptor has much higher specificity than the AChR, making insecticidally effective agonists or antagonists harder to find (Clements and May, 1974), and none having commercial promise has appeared so far.

Insects are incapable of synthesizing the steroid nucleus and therefore must rely on dietary sources for their sterols. Possible targets are thus the uptake of sterols from the food, which can be blocked in some cases by polyene antibiotics such as filipin (Schroeder and Bieber, 1975), or the biochemical apparatus by which insects modify the dietary sterols to essential metabolites such as cholesterol and the molting hormone, ecdysone. Several types of compounds—e.g., triparanol and some azasteroids—have shown promising inhibitory effects on these conversions in a continuing program reviewed by Thompson *et al.* (1973).

Further exploitation of the unique aspects of the insect endocrine system is also receiving great attention (Ruscoe, 1974). One particularly interesting concept is that of developing "anti-JHs" and "antiecdysones." JHAs have a major flaw as agents to control agricultural pests since they do not affect the damaging larval forms. An anti-JH agent would prevent synthesis, release, or action, or would speed destruction of the natural JH and thus should lead to premature metamorphosis of the larvae and sterility in the female adults. On the other hand, an antiecdysone by the same general means should prevent further larval development. Although in one possible area, the development of JH receptor blocking agents, progress so far has not been encouraging (Slama *et al.*, 1974, p. 263), a recent report (Bowers, 1976) indicates that certain chromene derivatives from the plant *Ageratum houstonianum* are potent antagonists of the normal action of JH and treated insects have the appearance of being allatectomized. This is a most promising area for further study.

Finally, the complex biochemistry of the insect cuticle presents many opportunities for attack, some of which are outlined by Hollingworth (1975*a*). However, intriguing as these and other prospects are, two things must be kept in mind. First, some of these processes use biochemical pathways present also in vertebrates (e.g., glutamate and sterol conversions), and disrupting them may not be as selectively toxic as we intend. Additionally, compounds based on these biochemical peculiarities may not be selective among insect species. Second, it is far easier to conceive of such likely targets than to reduce them to practice for economical insect control. The easiest prospect, as with the JHAs, is to find lipophilic, highly potent natural products which are simple enough to be mimicked economically and thus produce contact insecticides. This restricts us largely to lipid biochemistry, and compounds of this kind are not readily found.

3. *Polyfactorial Nature of Selectivity*

The term *polyfactorial selectivity* is used to describe the situation where more than one of the processes shown in Fig. 2 contribute significantly to selectivity. From the previous discussion it will be evident that although differences between organisms in one of these processes may often predominate as a cause of selectivity, very frequently additional factors are important; that is, selectivity, especially in broad taxonomic comparisons such as insect–vertebrate, is generally polyfactorial. This is to be expected since such organisms as the housefly and the rat differ so widely in their anatomy, physiology, and the detailed nature of their biochemical defenses and target sites. In fact, selectivity is probably a good deal more complex in its causes than our discussion has indicated so far since it is extremely rare for all the pertinent influences and their interactions to be assessed in any study of selective toxicity. Thus the explanations offered are often partial and oversimplified.

One of the most detailed studies available is that of Hutson and Hathway (1967) concerned with the high differential toxicity of chlorfenvinphos to rats (sensitive) and dogs (insensitive) as shown in Table II. Instead of any extreme difference between these species in one of the events controlling toxicity, a series of smaller differences were found, each of which favored the dog and which interacted to create the high degree of selectivity observed. In the dog, uptake of chlorfenvinphos from the gut was less efficient, metabolic detoxication was probably more rapid, binding in the blood was tighter, penetration into the brain was slower, and the sensitivity of the brain AChE was lower—a fortuitous concurrence of advantages to one species over another.

The interaction of multiple factors is also indicated in the study of Morello *et al.* (1968) into the selective toxicity of the geometrical isomers of the thiono analogue of mevinphos between the mouse and housefly. VSR values (mouse intraperitoneal/fly topical) for the *cis* and *trans* isomers were 13 and 42, respectively. Here it was concluded that overall detoxication is much more effective in the mouse, mouse brain AChE is four- to five-fold less sensitive to inhibition by mevinphos than fly head AChE, and the dimethylphosphorylated mouse enzyme recovers much more rapidly from inhibition than the fly AChE (*cf.* Table XIII). Additionally, these compounds are phosphorothionates and the opportunity factor probably operates *in vivo*, since their activation products, the respective mevinphos isomers, are rather poorly selective (Table I, compound 14, VSR = 2.5).

A third and final example of polyfactorial selectivity has already been touched on in discussing the selectivity of carboxylesters and carboxylamides, i.e., that very generally dimethyl organophosphate esters are less toxic to vertebrates but not necessarily less potent as insecticides than their diethyl counterparts. The greater selectivity of dimethyl esters may be considerable (Eto, 1974, p. 239). It is surprising that this general and significant difference between the two most common types of organophosphates has received so little systematic investigation. However, enough is known of the biochemical toxicology of the organophosphates to predict that all the events shown in Fig. 2 will be different in our methyl–ethyl comparison, but of these the largest differences are likely to occur in metabolism and the inhibition of AChE:

a. Metabolism: *O*-Dealkylation by GSH transferases is highly specific for methyl esters (Chapter 4). A second detoxication reaction which varies in significance between methyl and ethyl phosphates is A-esterase hydrolysis, as shown in Table IX, although the relative rate of hydrolysis is species dependent and often favors ethyl esters.
b. Reaction with AChE: Some differences may be expected in the potency with which methyl and ethyl phosphates attack the target AChE. More significantly, however, a major difference exists in the rate of spotaneous reactivation of the dimethyl and diethyl phosphorylated esterase (Table XIII), which clearly favors selectivity of the dimethyl analogues.

Presumably the relative importance of differences in these processes varies, and the basis for methyl–ethyl selectivity need not be the same in every case.

In one study of particular relevance to this comparison (Benke *et al.*, 1974), the toxicology of methyl and ethyl parathion was compared in the sunfish and the mouse. The ethyl analogue was over 20-fold more toxic than its methyl counterpart to the sunfish. Probably this difference was not due to penetration effects, since it still existed after intraperitoneal injection. Fish AChE was about 100-fold less sensitive to inhibition by both oxygen analogues than that of the mouse, but no major differences were found in the relative potencies of the two oxons themselves. In a companion study (Benke and Murphy, 1974) it was shown that, by comparison with mice, inhibition developed *in vivo* rather slowly in the fish after intraperitoneal dosage, but persisted with only slow recovery of AChE activity for both compounds. Consequently, recovery from inhibition is not a factor here. In fish, the two parathions were similar in the rate of oxidative activation and in the oxidative release of *p*-nitrophenol. The two major differences between the analogues which might explain their differential toxicity to the fish lay in the high detoxicative activity of GSH-dependent *O*-dealkylation with both methyl parathion and methyl paraoxon and the fivefold greater hepatic A-esterase activity with methyl paraoxon than ethyl paraoxon. This must be coupled to the fact that the oxons themselves are relatively poor anticholinesterases in fish, and since inhibition builds up slowly there is a greater opportunity for the metabolic destruction of the methyl ester; that is, a version of the opportunity factor operates. It was in fact noted that the time to death at the LD_{50} for fish was considerably longer than for mice. However, it is worth noting that intraperitoneal injection as used here is not necessarily equivalent in fish and mammals since uptake of the compound in fish is probably not rapid (Adamson, 1967).

Adequate as this explanation may be for fish, it cannot be a general one for the methyl–ethyl selectivity situation since not all organophosphates are substrates for GSH-mediated *O*-dealkylation and A-esterase hydrolysis, nor are all A-esterases preferential in hydrolyzing methyl analogues.

Other examples of polyfactorial effects in selectivity are described by O'Brien (1967*a*), Winteringham (1969), and Hollingworth (1971).

Recognition of the polyfactorial nature of selectivity leads us to one of the least satisfactory aspects of selectivity research, namely the interpretation of the quantitative relationship and interactions among the several contributing processes. Thus in a typical (but imaginary) situation, a compound has an LD_{50} to species A of 500 mg/kg and to species B of 1 mg/kg. We find a 40-fold difference in the rate of detoxication and a six-fold difference in target sensitivity, both favoring A. Are these two effects together enough to account for the 500-fold selectivity, or should we look further?

Here we need to digress a moment to recall O'Brien's (1967*a*, p. 253) plea that in studies of selectivity events in the organisms concerned should be compared at numerically identical rather than equitoxic doses such as the LD_{50}. This is important but incomplete advice. An adequate investigation of selectivity would involve study at both numerically and toxicologically equal doses. This is because, to use our hypothetical example, we are really asking two questions, not one:

a. Why does species A survive at 1 mg/kg while B does not?
b. Why does species A survive a 500-fold further increase in dosage before succumbing?

The answers to these two questions are not necessarily the same, and the use of identical doses answers only the first question, although it is often taken to answer both. Thus after such an investigation the assumption may be made that since species A can metabolize our compound 40-fold faster than species B when given at the same dosage, it can therefore metabolize a 40-fold greater dose than B with equal efficiency. This ignores the fact, illustrated by O'Brien (1967*a*, p. 254), that biological events are linearly related to dosage only over a limited range. Thus metabolism in species A may well not respond linearly to a 40-fold higher dose (e.g., because of enzyme saturation or depletion of cofactors), and it may be able to support only a 20-fold increase in dose before its efficiency falls to that of species B at 1 mg/kg. Thus it would be incorrect to simply multiply the 6- and 40-fold advantages of species A to conclude that it should succumb at about 240 mg/kg. Consideration of the events in Fig. 2 shows that a less than proportional increase with dose in penetration, activation, or reaction at the target tends to negate the assumed increase in toxic effect with dosage, and, conversely, the same effect with detoxication, excretion, or loss by nonspecific binding will make the compound more toxic than we expect from a linear relationship. Such "saturation" effects were found to play an imporant role in determining the ultimate resistance levels to methyl parathion and fenitrothion in houseflies (Hollingworth *et al.*, 1967*b*), and several other examples exist in the literature.

The dynamics of insecticide action and thus of selectivity are complex and are not readily accessible to rapid study. This may be taken as a plea for future studies to be more detailed and complete, gathering enough data so that even simple pharmacokinetic models can be made of the type described in Chapter 1, by Hollingworth (1971), or in several papers in SCI Monograph No. 29 (1968) (Society of Chemical Industry, London). Such models can then be used to test quantitatively some of the hypotheses proposed to explain the selectivity observed and also to better define the presently obscure influence of the interactions between events in poisoning as well as the nonlinearities with dosage which are a fundamental part of selectivity.

9. Conclusion: Prospects for Selective Insecticides

Having developed at length the biochemical and physiological background for insecticide selectivity, and pondered the future possibilities of new insecticide targets, two important practical questions remain. First, do we want selective insecticides, and, second, how can we obtain them? Since the topic has been the subject of a more detailed discussion (Hollingworth, 1975*a*) and has also been discussed by others (Unterstenhöfer, 1970; Twinn, 1972), only an outline need be given here.

A simple answer to the first question is not possible; it depends what we mean by "selective." To understand this, we must consider events far beyond pesticide biochemistry, particularly the shifting economic, sociological, and political forces which govern pesticide development. The changed regulatory climate for insecticide development of the last decade now dictates that highly toxic or persistent compounds are unlikely to be marketed. Extensive tests for acute and chronic mammalian and environmental toxicity precede registration, and if these are usefully predictive it is unlikely that a new compound will present any extensive hazard to most nontarget species.

However, such constraints applied to the development of new compounds have obvious economic consequences in the lower success ratio for discovery of novel pesticides, and the growing time, cost, and risk involved in their development, which in turn may lead to a decrease in innovative research (Djerassi *et al.*, 1974). At this point, there is a balance of forces between what society demands in insecticide selectivity and what it is prepared to pay for. Clearly if because of such regulation a new insecticide now costs \$10 million or more to develop, it is economic only to consider those compounds with a large market potential. Thus, with the exception of activity against a few of the more important pests, it is increasingly unlikely that a compound having a narrow spectrum of activity will be economically viable. Therefore, attainment of the highest level of selectivity—i.e., to a single species (monotoxicity)—is not only technologically very difficult (Unterstenhöfer, 1970) but economically unwise. The fate of two compounds, Butacarb and Nirosan, described earlier as nearly monotoxic, is illustrative.

Butacarb is selectively toxic to the sheep blowfly and was introduced in Australia in 1966–1967 to control this insect. By 1969–1970, a high enough degree of resistance to this and other carbamates had developed to make it ineffective (Shanahan and Roxburgh, 1974), and since it has few other uses it is no longer produced. Much earlier Nirosan had been displaced from its small niche as a control agent for grape berry moths since despite its effectiveness it was too selective to control the other pests of this crop. This could be achieved with less selective agents having a broader market potential (Unterstenhöfer, 1970).

Two further examples of the problems of developing safer agents with limited insecticidal activity are worth brief mention. The simple and selective compound Aspon (tetra-*n*-propyl dithiopyrophosphate; compound 29 in Table I) has a VSR of 97. It is clearly much less hazardous than its close relatives TEPP and thiotepp (compound 12, VSR = 1.0), but despite its safety and a fair insecticidal effectiveness it has found only a limited market (e.g., in chinch bug control) and probably would not be considered worth development for these limited purposes under current economic conditions. A parallel case is discussed in a most interesting article by Kohn (1975), who describes Chevron Chemical Company's efforts to find an economic use for RE-11775, the *N*-benzenesulfenyl derivative of *m*-*sec*-butylphenyl *N*-methylcarbamate. In keeping with our previous discussion of *N*-derivatized carbamates, this compound is an excellent mosquito larvicide, particularly against resistant strains, and is of decreased mammalian toxicity. However, because of its greatly reduced spectrum of activity, and despite vigorous efforts, no way could be found to put this most useful compound on the market at a profit, or, in the last resort, to give it to the State of California.

Thus, despite their safety, we are unlikely to see narrow-spectrum compounds of these kinds appearing without a change in the incentives structure (Djerassi *et al.*, 1974). In turn, this implies that the broader-spectrum compounds which are marketed may not be highly selective toward beneficial insects either, and recourse must be made to the ecological selectivity concept to maintain compatibility with pest management ideals.

In response to the second question of strategies to obtain selective agents, Hollingworth (1975*a*) considered both "broad-scale design," i.e., the development of novel chemical classes based on unexploited targets of the kind already mentioned, and "fine-scale design," which involves the molecular manipulation of current compounds for improved selectivity. However, it would be quite wrong to leave the impression that there is any easy solution to this problem or that the custom synthesis of chemicals to control a given pest is anything but a pipe-dream. The complexities of insecticide toxicology are such that our capacity to rationally influence the toxicity of molecules is quite limited, although it grows encouragingly and deserves better support. However, for the near future at least, most selective insecticides will continue to come, only partly planned at best, from large-scale screening programs. This is not to denigrate less empirical approaches which can also bear fruit, as the JHAs attest, but advances in this field are hard-bought, unpredictable, and unlikely to be rapid.

10. References

Abedi, Z. H., and Brown, A. W. A., 1961, Peritrophic membrane as vehicle for DDT and DDE excretion in *Aedes aegypti* larvae, *Ann. Entomol. Soc. Am.* **54:**539.

Abernathy, C. O., Ueda, K., Engel, J. L., Gaughan, L. C., and Casida, J. E., 1973, Substrate-specificity and toxicological significance of pyrethroid-hydrolyzing esterases of mouse liver microsomes, *Pestic. Biochem, Physiol.* **3:**300.

Adamson, R. H., 1967, Drug metabolism in marine vertebrates, *Fed. Proc.* **26:**1047.

Ahmad, S., and Knowles, C. O., 1972, Biochemical mode of action of tricyclohexylhydroxytin, *Comp. Gen. Pharmacol.* **3:**125.

Ajami, A. M., and Riddiford, L. M., 1973, Comparative metabolism of the cecropia juvenile hormone, *J. Insect Physiol.* **19:**635.

Alabaster, J., 1969, Survival of fish in 164 herbicides, insecticides, fungicides, wetting agents and miscellaneous substances, *Int. Pest Control.* **11(2):**29.

Albert, A., 1967, Patterns of metabolic disposition of drugs in man and other species, in: *Drug Responses in Man* (G. Wolstenholme and R. Porter, eds.), pp. 55–63, Churchill, London.

Albert, A., 1973, *Selective Toxicity: The Physico-chemical Basis of Therapy*, 5th ed., Halsted Press, New York.

Atzert, S. P., 1971, A Review of Sodium Monofluoroacetate (Compound 1080): Its Properties, Toxicology, and Use in Predator and Rodent Control, Special Scientific Report—Wildlife No. 146, Bureau of Sport Fisheries and Wildlife, U.S. Department of the Interior, Washington, D.C.

Bacon, O. G., Riley, W. D., and Zweig, G., 1964, The influence of certain biological and environmental factors on insecticide tolerance of the lygus bug, *Lygus hesperus, J. Econ. Entomol.* **57:**225.

Baggot, J. D., Davis, L. E., and Neff, C. A., 1972, Extent of plasma protein binding of amphetamine in different species, *Biochem. Pharmacol.* **21:**1813.

Bagley, R. W., and J. C. Bauernfeind, 1972, Field experience with juvenile hormone mimics, in: *Insect Juvenile Hormones: Chemistry and Action* (J. J. Menn and M. Beroza, eds.), pp. 113–151, Academic Press, New York.

Benke, G. M., and Murphy, S. D., 1974, Anticholinesterase action of methyl parathion, parathion, and azinphosmethyl in mice and fish: Onset and recovery of inhibition, *Bull. Environ. Contam. Toxicol.* **12:**117.

Benke, G. M., Cheever, K. L., Mirer, F. E., and Murphy, S. D., 1974, Comparative toxicity, anticholinesterase action and metabolism of methyl parathion and parathion in sunfish and mice, *Toxicol. Appl. Pharmacol.* **28:**97.

Bergmann, E. D., 1972, The future of insecticides—A problem of human environment, *Proc. 2nd Int. IUPAC Congr. Pestic. Chem.* **1:**1.

Black, A. L., Chiu, Y., Fahmy, M. A. H., and Fukuto, T. R., 1973, Selective toxicity of *N*-sulfenylated derivatives of insecticidal methylcarbamate esters, *J. Agr. Food Chem.* **21:**747.

Boulton, J. J. K., Boyce, C. B. C., Jewess, P. J., and Jones, R. F., 1971, Comparative properties of *N*-acetyl derivatives of oxime *N*-methylcarbamates and aryl *N*-methylcarbamates as insecticides and acetylcholinesterase inhibitors, *Pestic. Sci.* **2:**10.

Bowers, W. S., 1976, Discovery of insect antiallatotropins, in: *The Juvenile Hormones* (L. I. Gilbert, ed.), Plenum Press, New York.

Boyer, A. C., 1967, Vinyl phosphate insecticide sorption to proteins and its effect on cholinesterase I_{50} values, *J. Agr. Food Chem.* **15:**282.

Boyer, A. C., 1975, Sorption of tetrachlorvinphos insecticide (Gardona) to the hemolymph of *Periplaneta americana, Pestic. Biochem. Physiol.* **5:**135.

Brady, U. E., 1970, Localization of cholinesterase activity in housefly thoraces: Inhibition of cholinesterase with organophosphate compounds, *Entomol. Exp. Appl* **13:**423.

Brady, U. E., Jr., and Arthur, B. W., 1962, Absorption and metabolism of Ruelene by arthropods, *J. Econ. Entomol.* **55:**833.

Brady, U. E., and Sternburg, J., 1967, Studies on *in vivo* cholinesterase inhibition and poisoning symptoms in houseflies, *J. Insect Physiol.* **13:**369.

Brattsten, L. B., and Metcalf, R. L., 1970, The synergistic ratio of carbaryl with piperonyl butoxide as an indicator of the distribution of multifunction oxidases in the Insecta, *J. Econ. Entomol.* **63:**101.

Brattsten, L. B., and Metcalf, R. L., 1973*a*, Age-dependent variations in the response of several species of Diptera to insecticidal chemicals, *Pestic. Biochem. Physiol.* **3:**189.

Brattsten, L. B., and Metcalf, R. L., 1973*b*, Synergism of carbaryl toxicity in natural insect populations, *J. Econ. Entomol.* **66:**1347.

Brestkin, A. P., Brick, I. L., and Grigor'eva, G. M., 1973, Comparative pharmacology of cholinesterases, in: *Comparative Pharmacology*, Vol. 1 (M. J. Michelson, ed.), pp. 241–344, Section 85: *International Encyclopedia of Pharmacology and Therapeutics*, Pergamon Press, New York.

Brodie, B. B., and Maickel, R. P., 1962, Comparative biochemistry of drug metabolism, in: *Metabolic Factors Controlling Duration of Drug Action*, Proceedings of the First International Pharmacology Meeting, Vol. 6 (B. B. Brodie and E. G. Erdös, eds), pp. 299–324, Macmillan, New York.

Brooks, G. T., 1969, The metabolism of diene-organochlorine (cyclodiene) insecticides, *Residue Rev.* **27:**81.

Brooks, G. T., 1972, Pathways of enzymatic degradation of pesticides, in: *Environmental Quality and Safety*, Vol. 1 (F. Coulston and F. Korte, eds.), pp. 106–164, G. Thieme, Stuttgart.

Brown, A. W. A., 1960, Mechanisms of resistance against insecticides, *Annu. Rev. Entomol.* **5:**301.

Buerger, A. A., and O'Brien, R. D., 1965, Penetration of non-electrolytes through animal integuments, *J. Cell. Comp. Physiol.* **66:**227.

Bullock, J. A., 1961, The pests of pyrethrum in Kenya, *Pyrethrum Post* **6:**22.

Bunyan, P. J., Jennings, D. M., and Jones, F. J. S., 1971, Organophosphorus poisoning: A comparative study of the toxicity of chlorfenvinphos (2-chloro-1-(2′,4′-dichlorophenyl)vinyl diethyl phosphate) to the pigeon, pheasant and the Japanese quail. *Pestic. Sci.* **2:**148.

Camougis, G., 1973, Mode of action of pyrethrum on arthropod nerves, in: *Pyrethrum: The Natural Insecticide* (J. E. Casida, ed.), pp. 211–225, Academic Press, New York.

Camp, H. B., and Arthur, B. W., 1967, Absorption and metabolism of carbaryl by several insect species, *J. Econ. Entomol.* **60:**803.

Camp, H. B., Fukuto, T. R., and Metcalf, R. L., 1969*a*, Selective toxicity of isopropyl parathion: Effect of structure on toxicity and anticholinesterase activity, *J. Agr. Food Chem.* **17:**243.

Camp, H. B., Fukuto, T. R., and Metcalf, R. L., 1969*b*, Selective toxicity of isopropyl parathion: Metabolism in the housefly, honey bee, and white mouse, *J. Agr. Food Chem.* **17:**249.

Chasseaud, L. F., 1973, The nature and distribution of enzymes catalyzing the conjugation of glutathione with foreign compounds, *Drug Metab. Rev.* **2:**185.

Chen, P. R. S., and Dauterman, W. C., 1971, Studies on the toxicity of dimethoate analogs and their hydrolysis by sheep liver amidase, *Pestic. Biochem. Physiol.* **1:**340.

Cheng, H., and Casida, J. E., 1973, Metabolites and photoproducts of 3-(2-butyl)phenyl *N*-methylcarbamate and *N*-benzenesulfenyl-*N*-methylcarbamate, *J. Agr. Food Chem.* **21:**1037.

Clements, A. N., and May, T. E., 1974, Pharmacological studies on a locust neuromuscular preparation, *J. Exp. Biol.* **61:**421.

Cohen, S. D., and Murphy, S. D., 1970, Comparative potentiation of malathion by triorthotolyl phosphate in four classes of vertebrates, *Toxicol. Appl. Pharmacol.* **16:**701.

Cooke, A.S., 1973, Shell thinning in avian eggs by environmental pollutants, *Environ. Pollut.* **4:**85.

Corbett, J. R., 1974, *The Biochemical Mode of Action of Pesticides*, Academic Press, New York.

Croft, B. A., and Brown, A. W. A., 1975, Responses of arthropod natural enemies to insecticides, *Annu. Rev. Entomol.* **20:**285.

Crosby, D. G., 1971, Minor insecticides of plant origin, in: *Naturally Occurring Insecticides* (M. Jacobson and D. G. Crosby, eds.), pp. 177–239, Dekker, New York.

Darlington, W. A., Partos, R. D., and Ratts, K. W., 1971, Correlation of cholinesterase inhibition and toxicity in insects and mammals. 1. Ethylphosphonates, *Toxicol. Appl. Pharmacol.* **18:**542.

Darlington, W. A., Ludvik, G. F., and Sacher, R. M., 1972, MON-0856: A promising new selective insecticide, *J. Econ. Entomol.* **65:**48.

Dauterman, W. C., and Main, A. R., 1966, Relationship between acute toxicity and *in vitro* inhibition and hydrolysis of a series of carbalkoxy homologs of malathion, *Toxicol. Appl. Pharmacol.* **9:**408.

Derr, S. K., and Zabik, M. J., 1974, Bioactive compounds in the aquatic environment: Studies on the mode of uptake of DDE by the aquatic midge, *Chironomus tentans* (Diptera: Chironomidae), *Arch, Environ. Contam. Toxicol.* **2:**152.

Desmarchelier, J. M., and Fukuto, T. R., 1974, Toxicological effects produced by some 1,3-benzodioxoles, catechols, and quinones in *Culex* mosquito larvae, *J. Econ. Entomol.* **67:**153.

Djerassi, C., Shih-Coleman, C., and Diekman, J. 1974, Insect control of the future: Operational and policy aspects, *Science* **186:**596.

Donninger, C., 1971, Species specificity of phosphate triester anticholinesterases, *Bull. WHO* **44:**265.

Dvorchik, B. H., and Maren, T. H., 1974, Distribution, metabolism, elimination and toxicity of ^{14}C-DDT in the dogfish, *Squalus acanthias, Comp. Gen. Pharmacol.* **5:**37.

Ebeling, W., 1974, Permeability of insect cuticle, in: *The Physiology of Insecta,* 2nd ed., Vol. VI (M. Rockstein, ed.), pp. 271–343, Academic Press, New York.

Edwards, C. A., and Thompson, A. R., 1973, Pesticides and the soil fauna, *Residue Rev.* **45:**1.

El-Aziz, S. A., Metcalf, R. L., and Fukuto, T. R., 1969, Physiological factors influencing the toxicity of carbamate insecticides to insects, *J. Econ. Entomol.* **62:**319.

Eldefrawi, M. E., and Hoskins, W. M., 1961, Relation of the rate of penetration and metabolism to the toxicity of Sevin to three insect species, *J. Econ. Entomol.* **54:**401.

Elliott, M., Farnham, A. W., Janes, N. F., Needham, P. H., Pulman, D. A., and Stevenson, J. H., 1973, A photostable pyrethroid, *Nature* (*London*) **246:**169.

Eto, M., 1974, *Organophosphorus Pesticides: Organic and Biological Chemistry,* CRC Press, Cleveland, Ohio.

Fraser, J., Clinch, P. G., and Reay, R. C., 1965, *N*-Acylation of *N*-methylcarbamate insecticides and its effect on biological activity, *J. Sci. Food Agr.* **16:**615.

Fraser, J., Greenwood, D., Harrison, I. R., and Wells, W. H., 1967, The search for a veterinary insecticide. II. Carbamates active against sheep blowfly, *J. Sci. Food Agr.* **18:**372.

Fromm, P. O., and Hunter, R. C., 1969, Uptake of dieldrin by isolated perfused gills of rainbow trout, *J. Fish. Res. Board Can.* **26:**1939.

Fukuto, T. R., Shrivastava, S. P., and Black, A. L., 1972, Metabolism of 2-(methoxy(methylthio)phosphinylimino)-3-ethyl-5-methyl-1,3-oxazolidone in the cotton plant and houseflies, *Pestic. Biochem. Physiol.* **2:**162.

Gaines, T. B., 1969, Acute toxicity of pesticides, *Toxicol. Appl. Pharmacol.* **14:**515.

Gakstatter, J. H., and Weiss, C. M., 1967, The elimination of DDT-^{14}C, dieldrin-^{14}C and lindane-^{14}C from fish following a single sublethal exposure in aquaria, *Trans. Am. Fish. Soc.* **96:**301.

Gerard, J., 1633, *The Herbal or General History of Plants.* Republished by Dover Publications Inc., New York (1975).

Gerolt, P., 1972, Mode of entry of oxime carbamates into insects, *Pestic. Sci.* **3:**43.

Gilbert, M. D., and Wilkinson, C. F., 1974, Microsomal oxidases in the honey bee, *Apis mellifera* (L.), *Pestic. Biochem. Physiol.* **4:**56.

Gillett, J. W., 1971, Induction in different species, *Proc. 2nd Int. IUPAC Congr. Pestic. Chem.* **2:**197.

Gillette, J. R., 1973, Review of drug-protein binding, *Ann. N.Y. Acad. Sci.* **226:**6.
Gordon, H. T., 1961, Nutritional factors in insect resistance to chemicals, *Annu. Rev. Entomol.* **6:**27.
Gupta, B., Agarwal, H. C., and Pillai, M. K. K., 1971, Distribution, excretion, and metabolism of ^{14}C-DDT in the larval and adult *Trogoderma granarium* in relation to toxicity, *Pestic. Biochem. Physiol.* **1:**180.
Guthrie, F. E., Campbell, W. V., and Baron, R. L., 1962, Feeding sites of the green peach aphid with respect to its adaptation to tobacco, *Ann. Entomol. Soc. Am.* **55:**42.
Hackman, R. H., 1974, Chemistry of the insect cuticle, in: *The Physiology of Insecta,* 2nd ed., Vol. VI (M. Rockstein, ed.), pp. 216–270, Academic Press, New York.
Hall, F. R., Hollingworth, R. M., and Shankland, D. L., 1971, Cyanide tolerance in millipedes: The biochemical basis, *Comp. Biochem. Physiol.* **38B:**723.
Hammock, B. D., Gill, S. S., and Casida, J. E., 1974*a*, Synthesis and morphogenetic activity of derivatives and analogs of aryl geranyl ether juvenoids, *J. Agr. Food Chem.* **22:**379.
Hammock, B. D., Gill, S. S., and Casida, J. E., 1974*b*, Insect metabolism of a phenyl epoxygeranyl ether juvenoid and related compounds, *Pestic. Biochem. Physiol.* **4:**393.
Hansch, C., 1970, The use of physicochemical parameters and regression analysis in pesticide design, in: *Biochemical Toxicology of Insecticides* (R. D. O'Brien and I. Yamamoto, eds.), pp. 33–40, Academic Press, New York.
Hansch, C., 1971, Quantitative structure–activity relationships in drug design, in: *Drug Design,* Vol. 1 (E. J. Ariens, ed.), pp. 271–342, Academic Press, New York.
Hansen, L. G., Kapoor, I. P., and Metcalf, R. L., 1972, Biochemistry of selective toxicity and biodegradability: Comparative *O*-dealkylation by aquatic organisms, *Comp. Gen. Pharmacol.* **3:**339.
Hart, R. J., and Lee, R. M., 1966, Cholinesterase activities of various nematode parasites and their inhibition by the organophosphate anthelmintic Haloxon, *Exp. Parasitol.* **18:**332.
Hastings, F. L., and Dauterman, W. C., 1971, Phosphorylation and affinity constants for the inhibition of acetylcholinesterase by dimethoxon analogs, *Pestic. Biochem. Physiol.* **1:**248.
Hathway, D. E., and Amoroso, E. C., 1972, The effects of pesticides on mammalian reproduction, in: *Biodegradation and Efficacy of Livestock Pesticides* (M. A. Khan and W. O. Haufe, eds.), pp. 218–251, Swets and Zeitlinger, Amsterdam.
Hayes, W. J., 1967, Toxicity of pesticides to man: Risks from present levels, *Proc. R. Soc.* (*London*) *Ser. B* **167:**101.
Hellenbrand, K., 1967, Inhibition of housefly acetylcholinesterase by carbamates, *J. Agr. Food Chem.* **15:**825.
Henrick, C. A., Willy, W. E., Garcia, B. A., and Staal, G. B., 1975, Insect juvenile hormone activity of the stereoisomers of ethyl 3,7,11-trimethyl-2,4-dodecadienoate, *J. Agr. Food Chem.* **23:**396.
Heyndrickx, A., 1969, Toxicology of insecticides, rodenticides, herbicides, and phytopharmaceutical compounds, *Progr. Chem. Toxicol.* **4:**179.
Holden, A. V., 1973, Effects of pesticides on fish, in: *Environmental Pollution by Pesticides* (C. A. Edwards, ed.), pp. 213–253, Plenum Press, New York.
Hollingworth, R. M., 1971, Comparative metabolism and selectivity of organophosphate and carbamate insecticides, *Bull. WHO* **44:**155.
Hollingworth, R. M., 1975*a*, Strategies in the design of selective insect toxicants, in: *Pesticide Selectivity* (J. C. Street, ed.), pp. 67–111, Dekker, New York.
Hollingworth, R. M., 1975*b*, Chemistry, biological activity, and uses of formamidine pesticides, *Environ. Health Perspect.* (in press).
Hollingworth, R. M., Fukuto, T. R., and Metcalf, R. L., 1967*a*, Selectivity of Sumithion compared with methyl parathion: Influence of structure on anticholinesterase activity, *J. Agr. Food Chem.* **15:**235.
Hollingworth, R. M., Metcalf, R. L., and Fukuto, T. R., 1967*b*, The selectivity of Sumithion compared with methyl parathion: Metabolism in susceptible and resistant houseflies, *J. Agr. Food Chem.* **15:**250.

Huang, C. T., Dauterman, W. C., and Hastings, F. L., 1974, Inhibition of flyhead acetylcholinesterase by dimethoxon analogs, *Pestic. Biochem. Physiol.* **4:**249.

Hutson, D. H., and Hathway, D. E., 1967, Toxic effects of chlorfenvinphos in dogs and rats, *Biochem. Pharmacol.* **16:**949.

Ishaaya, I., and Casida, J. E., 1974, Dietary TH 6040 alters composition and enzyme activity of housefly larval cuticle, *Pestic. Biochem. Physiol.* **4:**484.

Jao, L. T., and Casida, J. E., 1974, Insect pyrethroid-hydrolyzing esterases, *Pestic. Biochem. Physiol.* **4:**465–472.

Johannsen, F. R., and Knowles, C. O., 1972, Citrate accumulation in twospotted spider mites, houseflies, and mice following treatment with the acaricide 2-fluoro-*N*-methyl-*N*-(1-naphthyl)acetamide, *J. Econ. Entomol.* **65:**1754.

Johannsen, F. R., and Knowles, C. O., 1974*a*, Toxicity and action of fluenethyl acaricide and related compounds in the mouse, housefly and twospotted spider mite, *Comp. Gen. Pharmacol.* **5:**101.

Johannsen, F. R., and Knowles, C. O., 1974*b*, Metabolism of fluenethyl acaricide in the mouse, housefly, and twospotted spider mite, *J. Econ. Entomol.* **67:**5.

Johansen, C., 1972, Spray additives for insecticidal selectivity to injurious vs. beneficial insects, *Environ. Entomol.* **1:**51.

Johnson, D. W., 1973, Pesticide residues in fish, in: *Environmental Pollution by Pesticides* (C. A. Edwards, ed.), pp. 181–212, Plenum Press, New York.

Johnson, M. K., 1969, Delayed neurotoxic action of some organophosphorus compounds, *Br. Med. Bull.* **25:**231.

Joiner, R. L., Chambers, H. W., and Baetcke, K. P., 1973, Comparative inhibition of boll weevil, golden shiner, and white rat cholinesterases by selected photoalteration products of parathion, *Pestic. Biochem. Physiol.* **2:**371.

Kao, C. Y., and Fuhrman, F. A., 1967, Differentiation of the actions of tetrodotoxin and saxitoxin, *Toxicon* **5:**25.

Kapoor, I. P., Metcalf, R. L., Mystrom, R. F., and Sangha, G. K., 1970, Comparative metabolism of methoxychlor, methiochlor, and DDT in mouse, insects, and in a model ecosystem, *J. Agr. Food Chem.* **18:**1145.

Kenaga, E. E., and End, C. S., 1974, Commercial and Experimental Organic Insecticides (1974 revision), Entomological Society of America, Special Publication 74-1.

Khoo, B. K., and Sherman, M., 1973, Toxicity and anticholinesterase activity of halogenated organophosphates to *Boettcherisca peregrina, J. Econ. Entomol.* **66:**595.

Knowles, C. O., 1975, Basis for selectivity of acaricides, in: *Pesticide Selectivity* (J. C. Street, ed.), pp. 155–176, Dekker, New York.

Knowles, C. O., and Casida, J. E., 1966, Mode of action of organophosphate anthelmintics: Cholinesterase inhibition in *Ascaris lumbricoides, J. Agr. Food Chem.* **14:**566.

Kohn, G. K., 1975, Target-specific pesticides: An industrial case history, in: *Pesticide Selectivity* (J. C. Street, ed.), pp. 113–133, Dekker, New York.

Kramer, K. J., Sanburg, L. L., Kezdy, F. J., and Law, J. H., 1974, The juvenile hormone binding protein in the haemolymph of *Manduca sexta* Johannson (Lepidoptera: Sphingidae), *Proc. Natl. Acad. Sci. U.S.A.* **71:**493.

Krieger, R. I., and Lee, P. W., 1973, Properties of the aldrin epoxidase system in the gut and fat body of a caddisfly larva, *J. Econ. Entomol.* **66:**1.

Krieger, R. I., Feeny, P. P., and Wilkinson, C. F., 1971, Detoxication enzymes in the guts of caterpillars: An evolutionary answer to plant defenses? *Science* **172:**579.

Krueger, H. R., and O'Brien, R. D., 1959, Relationship between metabolism and differential toxicity of malathion in insects and mice, *J. Econ. Entomol.* **52:**1063.

Krueger, H. R., O'Brien, R. D., and Dauterman, W. C., 1960, Relationship between metabolism and differential toxicity in insects and mice of diazinon, dimethoate, parathion, and acethion, *J. Econ. Entomol.* **53:**25.

Lech, J. J., and Statham, C. N., 1975, Role of glucuronide formation in the selective toxicity of 3-trifluoromethyl-4-nitrophenol (TFM) for the sea lamprey: Comparative aspects of TFM uptake and conjugation in sea lamprey and rainbow trout, *Toxicol. Appl. Pharmacol.* **31:**150.

Lee, A., and Metcalf, R. L., 1973, *In vitro* inhibition of acetycholinesterase by *O,O*-dimethyl *S*-aryl phosphorothioates, *Pestic. Biochem. Physiol.* **2:**408.

Lee, A., Metcalf, R. L., and Kearns, C. W., 1974, Purification and some properties of house cricket (*Acheta domesticus*) acetylcholinesterase, *Insect Biochem.* **4:**267.

Lee, R. M., 1964, Di-(2-chloroethyl) aryl phosphates: A study of their reaction with B-esterases, and of the genetic control of their hydrolysis in sheep, *Biochem. Pharmacol.* **13:**1551.

Lee, R. M., and Pickering, W. R., 1967, The toxicity of haloxon to geese, ducks and hens, and its relationship to the stability of the di-(2-chloroethyl)phosphoryl cholinesterase derivatives, *Biochem. Pharmacol.* **16:**941.

Leesch, J. G., and Fukuto, T. R., 1972, The metabolism of Abate in mosquito larvae and houseflies, *Pestic. Biochem. Physiol.* **2:**223.

Lehman, A. J., 1959, Introduction, in: *Appraisal of the Safety of Chemicals in Foods, Drugs, and Cosmetics,* Association of Food and Drug Officials of the U.S., Texas State Department of Health, Austin, Tex.

Macek, K. J., and McAllister, W. A., 1970, Insecticide susceptibility of some common fish family representatives, *Trans. Am. Fish. Soc.* **99:**20.

MacLennan, K. J. R., 1967, Recent advances in techniques for tsetse-fly control, *Bull. WHO* **37:**615.

Maibach, H. I., and Feldman, R., 1974, Systemic absorption of pesticides through the skin of man, in: *Occupational Exposure to Pesticides,* pp. 120–127, Report to the Federal Working Group on Pest Management, January 1974, Washington, D.C.

Markes, E. P., and Sowa, B. A., 1974, An *in vitro* model system for the production of insect cuticle, in: *Mechanism of Pesticide Action* (G. K. Kohn, ed.), pp. 144–155, Symposium Series No. 2, American Chemical Society, Washington, D.C.

Martin, H., 1972, *Pesticide Manual,* 3rd ed., British Crop Protection Council.

Matsumura, F., 1975, *Toxicology of Insecticides,* Plenum Press, New York.

Matsumura, F., and O'Brien, R.D., 1963, A comparative study of the modes of action of fluoroacetamide and fluoroacetate in the mouse and American cockroach, *Biochem. Pharmacol.* **12:**1201.

Mayer, F. L., Jr., Street, J. C., and Neuhold, J. M., 1972, DDT intoxication in rainbow trout as affected by dieldrin, *Toxicol. Appl. Pharmacol.* **22:**347.

McDonald, T. J., 1975, Neuromuscular pharmacology of insects, *Annu. Rev. Entomol.* **20:**151.

Mengle, D. C., and O'Brien, R. D., 1960, The spontaneous and induced recovery of fly-brain cholinesterase after inhibition by organophosphates, *Biochem. J.* **75:**201.

Menn, J. J., and Beroza, M., 1972, *Insect Juvenile Hormones: Chemistry and Action,* Academic Press, New York.

Menn, J. J., and Pallos, F. M., 1975, Development of morphogenetic agents in insect control, *Environ. Lett.* **8:**71.

Metcalf, R. A., and Metcalf, R. L., 1973, Selective toxicity of analogs of methyl parathion, *Pestic. Biochem. Physiol.* **3:**149.

Metcalf, R. L., 1964, Selective toxicity of insecticides, *World Rev. Pest Control* **3:**28.

Metcalf, R. L., 1972*a*, Selective toxicity of insecticides, in: *Toxicology, Biodegradation and Efficacy of Livestock Pesticides* (M. A. Kahn and W. O. Haufe, eds.), pp. 350–378, Swets and Zeitlinger, Amsterdam.

Metcalf, R. L., 1972*b*, Development of selective and biodegradable pesticides, in: *Pest Control Strategies for the Future,* pp. 137–156, National Academy of Sciences, Washington, D.C.

Metcalf, R. L., 1975, Insecticides in pest management, in: *Introduction to Insect Pest Management* (R. L. Metcalf and W. H. Luckmann, eds.), pp. 235–273, Wiley, New York.

Metcalf, R. L., Gruhn, W. B., and Fukuto, T. R., 1968, Electrophysiological action of carbamate insecticides in the central nervous system of the American cockroach, *Ann. Entomol. Soc. Am.* **61:**618.

Meyer, F. P., 1965, The experimental use of Guthion as a selective fish eradicator, *Trans. Am. Fish. Soc.* **94:**203.

Michelson, M. J., 1974, Some aspects of evolutionary pharmacology, *Biochem. Pharmacol.* **23:**2211.

Miller, J. J., Powell, G. M., Olavesen, A. H., and Curtis, C. G., 1973, The metabolism and toxicity of phenols in cats, *Biochem. Soc. Trans.* **1:**1163.

Miskus, R. P., Andrews, T. L., and Look, M., 1969, Metabolic pathways affecting toxicity of *N*-acetyl Zectran, *J. Agr. Food. Chem.* **17:**842.

Miura, T., and Takahashi, R. M., 1973, Insect developmental inhibitors. 3. Effects on nontarget aquatic organisms, *J. Econ. Entomol.* **66:**917.

Miyamoto, J., 1969, Mechanism of low toxicity of Sumithion towards mammals, *Residue Rev.* **25:**251.

Morello, A., Vardanis, A., and Spencer, E. Y., 1968, Comparative metabolism of two vinyl phosphorothionate isomers (thiono Phosdrin) by the mouse and the fly, *Biochem. Pharmacol.* **17:**1795.

Mulder, R., and Gijswijt, M. J., 1973, The laboratory evaluation of two promising new insecticides which interfere with cuticle deposition, *Pestic. Sci.* **4:**737.

Mulla, M. S., Axelrod, H., and Isaak, L. W., 1961, Effectiveness of new insecticides against mosquito larvae, *Mosquito News* **21:**216.

Murphy, S. D., 1969, Mechanisms of pesticide interactions in vertebrates, *Residue Rev.* **25:**201.

Negherbon, W. O., 1959, *Handbook of Toxicology,* Vol. 3: *Insecticides: A Compendium,* Saunders, Philadelphia.

Nelson, E. B., Raj, P. P., Belfi, K. J., and Master, B. S. S., 1971, Oxidative drug metabolism in human liver microsomes, *J. Pharmacol. Exp. Ther.* **178:**580.

Noguchi, T., Yoshinobu, H., and Miyata, H., 1968, Studies of the biochemical lesions caused by a new fluorine pesticide, *N*-methyl-*N*-(1-naphthyl)monofluoroacetamide, *Toxicol. Appl. Pharmacol.* **13:**189.

O'Brien, R. D., 1959, Effect of ionization upon penetration of organophosphates to the nerve cord of the cockroach, *J. Econ. Entomol.* **52:**812.

O'Brien, R. D., 1960, *Toxic Phophorus Esters,* Academic Press, New York.

O'Brien, R. D., 1961, Selective toxicity of insecticides, *Advan. Pest. Control. Res.* **4:**75.

O'Brien, R. D., 1967*a*, *Insecticides: Action and Metabolism,* Academic Press, New York.

O'Brien, R. D., 1967*b*, Barrier systems in insect ganglia and their implications for toxicology, *Fed. Proc.* **26:**1056.

O'Brien, R. D., and Dannelley, C. E., 1965, Penetration of insecticides through rat skin, *J. Agr. Food Chem.* **13:**245.

O'Brien, R. D., and Fisher, R. W., 1958, The relation between ionization and toxicity to insects for some neuropharmacological compounds, *J. Econ. Entomol.* **51:**169.

O'Brien, R. D., and Hilton, B. D., 1965, Effect of a thiono substituent on toxicity of fluorophosphates to insects and mice, *J. Agr. Food Chem.* **13:**381.

Olson, W. P., and O'Brien, R. D., 1963, The relation between physical properties and penetration of solutes into the cockroach cuticle, *J. Insect Physiol.* **9:**777.

Pallos, F. M., and Menn, J. J., 1972, Synthesis and activity of juvenile hormone analogs, in: *Insect Juvenile Hormones: Chemistry and Actions* (J. J. Menn and M. Beroza, eds.), pp. 303–316, Academic Press, New York.

Pappenheimer, A. M., Jr., and Williams, C. M., 1952, The effects of diphtheria toxin on the cecropia silkworm, *J. Gen. Physiol.* **35:**727.

Pickering, Q. H., Henderson, C., and Lemke, A. E., 1962, The toxicity of organic phosphorus insecticides to different species of warm-water fishes, *Trans. Am. Fish. Soc.* **91:**175.

Pickering, W. R., and Malone, J. C., 1967, The acute toxicity of dichloroalkyl aryl phosphates in relation to chemical structure, *Biochem. Pharmacol.* **16:**1183.

Pimentel, D., 1971, *Ecological Effects of Pesticides on Non-target Species,* Office of Science and Technology, Executive Office of the President, Washington, D.C.

Pohl, R. J., Bend, J. R., Guarino, A. M., and Fouts, J. R., 1974, Hepatic microsomal mixed-function oxidase activity of several marine species from coastal Maine, *Drug Metab. Dispos.* **2:**545.

Pope, G. G., and Ward, P., 1972, The effects of small applications of an organophosphorus poison, fenthion on the weaver-bird *Quelea quelea, Pestic. Sci.* **3:**197.

Post, L. C., de Jong, B. J., and Vincent, W. R., 1974, 1-(2,6-Disubstituted benzoyl)-3-phenylurea insecticides: Inhibitors of chitin synthesis, *Pestic. Biochem. Physiol.* **4:**473.

Potter, J. L., and O'Brien, R. D., 1963, The relation between toxicity and metabolism of paraoxon in the frog, mouse, and cockroach, *Entomol. Exp. Appl.* **6:**319.

Reiner, E., 1971, Spontaneous reactivation of phosphorylated and carbamylated cholinesterases, *Bull. WHO* **44:**109.

Ripper, W. E., Greenslade, R. M., and Hartley, G. S., 1951, Selective insecticides and biological control, *J. Econ. Entomol.* **44:**448.

Ruscoe, C. N. E., 1974, The exploitation of insect endocrine systems, *Chem. Ind.* (*London*), Aug. 17, p. 648.

Sacher, R. M., 1971, A mosquito larvicide with favorable environmental properties, *Mosquito News* **31:**513.

Sacher, R. M., and Olin, J. F., 1972, The insecticidal and the anticholinesterase activity of *meta*-acylamidophenyl and *meta*-thioureidophenyl *N*-methylcarbamates, *J. Agr. Food Chem.* **20:**354.

Sacher, R. M., Alt, G. H., and Darlington, W. A., 1973, Insecticidal and anticholinesterase activity of benzotriazolyl methyl and dimethylcarbamates, *J. Agr. Food Chem.* **21:**132.

Sanborn, J. R., Lee, A., and Metcalf, R. L., 1974, Investigations into carbamate insecticide selectivity. 1. Evaluation of potential selectophores, *Pestic. Biochem. Physiol.* **4:**67.

Schafer, E. W., 1972, The acute oral toxicity of 369 pesticidal, pharmaceutical and other chemicals to wild birds, *Toxicol. Appl. Pharmacol.* **21:**315.

Schroeder, F., and Bieber, L. L., 1975, Studies on hypo- and hypercholesterolemia induced in insects by filipin, *Insect Biochem.* **5:**201.

Self, L. S., Guthrie, F. E., and Hodgson, E., 1964*a*, Metabolism of nicotine by tobacco-feeding insects, *Nature* (*London*), **204:**300.

Self, L. S., Guthrie, F. E., and Hodgson, E., 1964*b*, Adaptation of tobacco hornworns to the ingestion of nicotine, *J. Insect Physiol.* **10:**907.

Shah, A. H., and Guthrie, F. E., 1970, Penetration of insecticides through the isolated midgut of insects and mammals, *Comp. Gen. Pharmacol.* **1:**391.

Shah, P. V., Dauterman, W. C., and Guthrie, F. E., 1972, Penetration of a series of dialkoxy analogs of dimethoate through the isolated gut of insects and mammals, *Pestic. Biochem. Physiol.* **2:**324.

Shanahan, G. J., and Roxburgh, N. A., 1974, The sequential development of insecticide resistance problems in *Lucilia cuprina* Wied. in Australia, *Pestic. Artic. News Summ. (PANS)* **20:**190.

Shishido, T., and Fukami, J., 1972, Enzymatic hydrolysis of diazoxon by rat tissue homogenates, *Pestic. Biochem. Physiol.* **2:**39.

Shishido, T., Usui, K., and Fukami, J., 1972*a*, Oxidative metabolism of diazinon by microsomes from rat liver and cockroach fat body, *Pestic. Biochem. Physiol.* **2:**27.

Shishido, T., Usui, K., Sato, M., and Fukami, J., 1972*b*, Enzymatic conjugation of diazinon with glutathione in rat and American cockroach, *Pestic. Biochem. Physiol.* **2:**51.

Siddall, J. B., and Slade, M., 1974, Tests for toxicity of juvenile hormone and analogs in mammalian systems, in: *Invertebrate Endocrinology and Hormonal Heterophylly* (W. J. Burdette, ed.), pp. 345–347, Springer-Verlag, New York.

Skalsky, H. L., and Guthrie, F. E., 1975, Binding of insecticides to macromolecules in the blood of the rat and American cockroach, *Pestic. Biochem. Physiol.* **5:**27.

Skrinjaric-Spoljar, M., Simeon, V., and Reiner, E., 1973, Spontaneous reactivation and aging of dimethylphosphorylated acetylcholinesterase and cholinesterase, *Biochim. Biophys. Acta* **315:**363.

Slade, M., and Zibitt, C. H., 1972, Metabolism of cecropia juvenile hormone in insects and mammals, in: *Insect Juvenile Hormones: Chemistry and Action* (J. J. Menn and M. Beroza, eds.), pp. 155–176, Academic Press, New York.
Slama, K., Romanuk, M., and Sorm, F., 1974, *Insect Hormones and Bioanalogues*, Springer-Verlag, New York.
Smith, J. N., 1964, Comparative biochemistry of detoxification, in: *Comparative Biochemistry* (M. Florkin and H. S. Mason, eds.), pp. 403-457, Academic Press, New York.
Smith, J. N., 1968, The comparative metabolism of xenobiotics, in: *Advances in Comparative Physiology and Biochemistry*, Vol. 3 (O. Lowenstein, ed.), pp. 173–232, Academic Press, New York.
Smith, R. F., 1970, Pesticides: Their uses and limitations in pest management, in: *Concepts of Pest Management* (R. L. Rabb and F. E. Guthrie, eds.), pp. 103–113, North Carolina State University Press, Raleigh, N.C.
Solomon, K. R., and Metcalf, R. L., 1974, The effect of piperonyl butoxide and triorthocresyl phosphate on the activity and metabolism of Altosid (isopropyl 11-methoxy-3,7,11-trimethyldodeca-2,4-dienoate) in *Tenebrio molitor* L. and *Oncopeltus fasciatus* (Dallas), *Pestic. Biochem. Physiol.* **4:**127.
Sorm, F., 1974, Insect hormones and their bioanalogues as potential insecticides, *FEBS Lett. Suppl.* **40:**S128.
Staal, G. B., 1975, Insect growth regulators with juvenile hormone activity, *Annu. Rev. Entomol.* **20:**417.
Stanton, R. H., and Khan, M. A. Q., 1973, Mixed-function oxidase activity towards cyclodiene insecticides in bass and bluegill sunfish, *Pestic. Biochem. Physiol.* **3:**351.
Steelman, C. D., and Schillings, P. E., 1972, Effect of a juvenile hormone mimic on *Psorophora confinnis* (Lynch-Arribalzaga) and non-target aquatic insects, *Mosquito News* **32:**350.
Sun, Y. P., 1972, Correlation of toxicity of insecticides to the housefly and to the mouse, *J. Econ. Entomol.* **65:**632.
Sun, Y. P., and Johnson, E. R., 1971, A new technique for studying the toxicology of insecticides with house flies by the infusion method, with comparable topical application and injection results, *J. Econ. Entomol.* **64:**75.
Thompson, M. J., Kaplanis, J. N., Robbins, W. E., and Svoboda, J. A., 1973, Metabolism of steroids in insects, *Advan. Lipid Res.* **11:**219.
Thurston, R., and Webster, J. A., 1962, Toxicity of *Nicotiana gossei* Domin to *Myzus persicae* (Sulzer), *Entomol. Exp. Appl.* **5:**233.
Tomlin, A. D., and Forgash, A. J., 1972, Penetration of Gardona and DDT in gypsy moth larvae and house flies, *J. Econ. Entomol.* **65:**942.
Toppozada, A., and O'Brien, R. D., 1967, Permeability of the ganglia of the willow aphid, *Tuberolachnus salignus*, to organic ions, *J. Insect Physiol.* **13:**941.
Treherne, J. E., and Pichon, Y., 1972, The insect blood–brain barrier, *Advan. Insect Physiol.* **9:**257.
Tripathi, R. K., and O'Brien, R. D., 1973, Insensitivity of acetylcholinesterase as a factor in resistance of houseflies to the organophosphate Rabon, *Pestic. Biochem. Physiol.* **3:**495.
Tucker, R. K., and Haegele, M. A., 1971, Comparative acute oral toxicity of pesticides to six species of birds, *Toxicol. Appl. Pharmacol.* **20:**57.
Twinn, D. C., 1972, Research on selectivity and structure–activity relationships as a basis for future agricultural insecticides, *Proc. 2nd Int. IUPAC Congr. Pestic. Chem.* **1:**353.
Uchida, T., Rahmati, H. S., and O'Brien, R. D., 1965, The penetration and metabolism of ^{3}H-dimethoate in insects, *J. Econ. Entomol.* **58:**831.
Unterstenhöfer, G., 1970, Integrated pest control from the aspect of industrial research on crop protection chemicals, *Pflanzenschutz–Nachr.* **23:**264.
Verschoyle, R. D., and Barnes, J. M., 1972, Toxicity of natural and synthetic pyrethrins to rats, *Pestic. Biochem. Physiol.* **2:**308.
Vinopal, J. H., and Fukuto, T. R., 1971, Selective toxicity of phoxim (phenylglyoxylonitrile oxime *O,O*-diethyl phosphorothioate), *Pestic. Biochem. Physiol.* **1:**44.

Walker, A. I. T., Thorpe, E., and Stevenson, D. E., 1972, The toxicology of dieldrin (HEOD). I. Long-term oral toxicity studies in mice, *Food Cosmet. Toxicol.* **11:**415.

Walker, C. H., 1974, Comparative aspects of the metabolism of pesticides, in: *Environmental Quality and Safety* (F. Coulston and F. Korte, eds.), Vol. 3, pp. 113–153, Thieme, Stuttgart.

Wilkes, F. G., and Weiss, C. M., 1971, The accumulation of DDT by the dragonfly nymph, *Tetragoneuria, Trans. Am. Fish. Soc.* **100:**222.

Williams, R. T., 1967*a*, Comparative patterns of drug metabolism, *Fed. Proc.* **26:**1029.

Williams, R. T., 1967*b*, The biogenesis of conjugation and detoxication products, in: *Biogenesis of Natural Compounds*, 2nd ed. (P. Bernfeld, ed.), pp. 589–639, Pergamon Press, New York.

Williams, R. T., 1974, Inter-species variations in the metabolism of xenobiotics, *Biochem. Soc. Trans.* **2:**359.

Winteringham, F. P. W., 1965, Some distinctive features of insect metabolism, in: *Aspects of Insect Biochemistry* (T. W. Goodwin, ed.), pp. 29–37, Academic Press, New York.

Winteringham, F. P. W., 1969, Mechanisms of selective insecticidal action, *Annu. Rev. Entomol.* **14:**409.

Wustner, D. A., and Fukuto, T. R., 1974, Affinity and phosphonylation constants for the inhibition of cholinesterases by the optical isomers of *O*-2-butyl *S*-2-(dimethylammonium)ethyl ethylphosphonothioate hydrogen oxalate, *Pestic. Biochem. Physiol.* **4:**365.

Yang, R. S. H., Hodgson, E., and Dauterman, W. C., 1971*a*, Metabolism *in vitro* of diazinon and diazoxon in rat liver, *J. Agr. Food. Chem.* **19:**10.

Yang, R. S. H., Hodgson, E., and Dauterman, W. C., 1971*b*, Metabolism *in vitro* of diazinon and diazoxon in susceptible and resistant houseflies, *J. Agr. Food Chem.* **19:**14.

Yeary, R. A., 1970, Comparative toxicity studies on glucuronide-forming compounds in icteric and nonicteric newborn Gunn rats, *J. Pediat.* **77:**139.

Yu, C., and Booth, G. M., 1971, Inhibition of choline acetylase from the house fly (*Musca domestica* L.) and mouse, *Life Sci.* (*II*) **10:**337.

Yu, C., Kearns, C. W., and Metcalf, R. L., 1972*a*, Acetylcholinesterase inhibition by substituted phenyl *N*-alkyl carbamates, *J. Agr. Food Chem.* **20:**537.

Yu, C., Metcalf, R. L., and Booth, G. M., 1972*b*, Inhibition of acetylcholinesterase from mammals and insects by carbofuran and its related compounds and their toxicities toward these animals, *J. Agr. Food Chem.* **20:**923.

Yu, C., Park, K. S., and Metcalf, R. L., 1974, Correlation of toxicity and acetylcholinesterase (AChE) inhibition in 2-alkyl substituted 1,3-benzodioxolyl-4*N*-methylcarbamates and related compounds, *Pestic. Biochem. Physiol.* **4:**178.

13

Biochemistry and Physiology of Resistance

F. J. Oppenoorth and W. Welling

1. Introduction

The development of insect resistance to modern organic insecticides has attracted so much attention, and its various aspects have been studied so intensively, that it is hardly possible to do justice to the subject in a single chapter. It has therefore been necessary to restrict this chapter to the mechanisms underlying the better-known cases of resistance. Even with this restriction, only examples have been given in many cases and the treatment can in no way be considered complete. By emphasis on certain aspects of methodology, an attempt has been made to illustrate the possibilities and limitations of the different techniques that contribute to our knowledge of the problem, in the hope that this will be helpful for a critical evaluation of the available information.

2. Resistance as an Acquired Character

Resistance has been defined as "the developed ability in a strain of insects to tolerate doses of toxicants which would prove lethal to the majority of individuals in a normal population of the same species" (Anonymous, 1957). Although the acquired character is important, it should be stressed that this

F. J. Oppenoorth and W. Welling • Laboratory for Research on Insecticides, Marijkeweg, Wageningen, The Netherlands.

definition indicates resistance to be a property of the population and not the result of alterations within individual insects. Individual insects inherit resistance, because they are offspring of insects that initially survived treatment through an existing genetically determined property. This implies that normal (unexposed) populations of insects exhibit polymorphism with respect to their degree of susceptibility, and it should be pointed out that in all likelihood a continuum exists from species that lack more resistant variants, and consequently do not become resistant, to species in which all individuals are already insensitive to the insecticide. It is only when the majority of the population was initially susceptible and when this disappears on selection with an insecticide that the term *resistance* is applied. For an extensive description of the development and mechanisms of resistance in various insect vectors of disease, the reader is referred to Brown (1958) and Brown and Pal (1971). In this book a list of about 40 reviews on the subject is provided. A similar book for agricultural pests is being prepared by Georghiou and Reynolds.

The above concept of resistance as a *preadaptive phenomenon* (in contrast to the possible direct effect of the insecticide in changing the insects) rests on several studies in which it was invariably found that resistance was entirely dependent on selection and could not be developed by treatments with sublethal insecticide doses (Crow, 1957). In some instances, selection can act on reproductive capacity rather than on survival. For example, oogenesis in the housefly can be adversely affected by DDT, and selection in this case will favor those individuals with normal oogenesis. Since there seems to be a correlation between reproductive capacity and resistance, it is possible (in this case) to obtain resistance without killing individuals in the population (Beard, 1965).

As discussed elsewhere, treatment of individuals with insecticides can sometimes lead to a decreased susceptibility due to *induction* of a greater detoxication capacity. It has never been shown that such induced alterations are inherited.

3. Dependence of Degree of Resistance on Method of Determination

Insects can have various degrees of resistance and these are usually expressed as the ratio of the mean susceptibility (LD_{50}) of the R-population to that of the S-population. This value can depend greatly on the materials and methods used in gathering the data for estimating the LD_{50} and it should therefore be emphasized that there is no such thing as an absolute resistance factor. If different treatment methods are employed, the resistance mechanism will have a different opportunity to affect the final toxicity, which is determined by the interaction of factors such as penetration, transport, and detoxication and intoxication reactions. A good example of this was provided by Busvine (1951), who showed that a certain strain of houseflies was 300-fold resistant to DDT applied topically in acetone, but only 16-fold when mineral oil was used as the solvent. Higher resistance factors are often obtained when dosing is slow, such as in a contact method where the toxicant is picked up gradually. Although we cannot speak of an absolute resistance factor, this does not imply that different methods provide results that are all equally meaningful. The penetration rate is generally only proportional to dose until a certain upper limit is reached. Above this limit, the effect of increasing the dose gradually becomes smaller. Therefore, R-strains can appear to be so resistant that they cannot be killed at all or only with very high doses. In such cases, a different method of treatment (larger amount of solvent, use of certain adjuvants) may give more satisfactory results.

4. Importance of Genetics for an Understanding of Resistance Mechanisms

4.1. Altered Enzymes vs. Altered Amounts of Enzyme

Selection with an insecticide will tend to favor alleles which cause greater resistance, but at the same time natural selection will operate to restrict the genetic variability that is available. The so-called structural genes determine the composition of the amino acid chains. These, alone or in combination, form the proteins, which as structural elements and as the catalytic machinery of the body are of immense importance for life and for the processes involved in intoxication. The amount of protein formed by a structural gene, although not altogether independent of its structure, is mainly determined by a more or less complex regulatory system. There are two ways in which insecticide selection can affect the activity of an enzyme that can lead to resistance: by selecting for an aberrant structural gene producing an enzyme with different properties due to an alteration of its amino acid sequence or by selecting regulatory factors that determine the amount of the normal enzyme produced.

These two possible mechanisms have rather different consequences. In the case of aberrant structural genes, all allelomorphs can, in principle, produce products (enzymes) with different properties. As we will see, there is evidence in the literature that differences in resistance are of this type. Examples are DDTases, cholinesterases, and hydrolytic detoxication enzymes. If, on the other hand, regulatory factors are involved, only the quantity of the enzyme and not its properties will be different. Whether this occurs has not been established, although the altered oxidation in the cabbage looper, as discussed under the mixed-function oxidases (see Section 6.2.4), may be an example (Kuhr, 1971).

If a mutant gene resulting in a certain level of resistance has spread in a population which continues under the same selection pressure, there is, of course, a possibility that secondary mutations might modify the gene to give higher resistance levels. Some evidence for the existence of an R-allele causing low parathion resistance which developed into one for higher resistance has been given by Kikkawa (1964) in irradiated *Drosophila*. Also, for DDTases more than one mutational step may be involved. DDTases intermediate in activity between R- and S-strains have been found and may represent primary mutations which may be modified by a second mutation to produce the more active enzyme (Oppenoorth, 1965*b*). In normal genetic analysis, such multistep alterations would still show up as monogenic cases of resistance.

4.2. Monogenic vs. Polygenic Resistance

Many examples of resistance dependent on a single gene have been found, and in these cases the level of resistance can be quite high. Well-known examples are the 2000-fold resistance to organophosphates in spider mites (e.g., Ballantyne and Harrison, 1967), DDT resistance in houseflies (e.g., Lichtwardt, 1964), and dieldrin resistance in several Diptera (e.g., Busvine *et al.*, 1963). In other cases, genetic analysis has clearly shown polygenic inheritance (Sawicki and Farnham, 1967; Georghiou, 1971). Whether one or the other type of inheritance will occur depends both on the nature of the factors available (if one factor provides immunity it is not likely that a polygenetic system will occur) and on the time period of selection (over long periods there is a greater chance for producing combinations of R-factors). In most cases of resistance, the number of genes involved is relatively small, and many of the genes when isolated in separate strains have effects that can be clearly demonstrated.

4.3. Contribution of Genetics to Resistance Studies

Genetics offers a valuable tool in analysis of resistance. Biochemical studies are greatly facilitated if they can be performed on insect strains in which only one R-gene, and therefore only one biochemical mechanism, is operating. Even if this ideal situation cannot be easily attained, genetics can at least be helpful in analyzing the strains studied. Genetic studies have been helpful in those cases where more than one mechanism can lead to resistance to the same insecticide. Several examples will be discussed in this chapter. They can further be of great help in establishing a relation between a resistance factor and a biochemical or physiological character in an R-strain that deviates from that in S-strains. If genetic analysis shows that they are associated, it becomes very likely that we are dealing with the cause of resistance, although definite proof cannot be derived in this way. The best techniques that are available to show the influence of single chromosomes on resistance are factorial analyses of backcross material (Tsukamoto, 1964) and isolation of strains with single chromosomes followed by reconstruction of the original strain (Georghiou, 1971; Sawicki, 1973*a*).

A more detailed analysis of factors on one chromosome is sometimes required. It should be pointed out that it is very difficult to distinguish resistance due to a single factor from that due to closely linked genes. It is also difficult to establish whether genes are alleles, and since definite conclusions cannot be obtained genetically other techniques are required.

5. Cross-Resistance vs. Multiple Resistance

The term *cross-resistance* denotes the resistance of a strain of insects to compounds other than the selective agent, due to the same mechanism. In contrast, *multiple resistance* is the resistance of a single strain to several different compounds but resulting from different mechanisms. The latter often results from the simultaneous or consecutive use of several insecticides under field conditions. It is important to distinguish between these two phenomena, because a specific cross-resistance pattern can often provide information on the resistance mechanism involved and is therefore important in determining which alternative insecticides are most suitable for the control of R-populations.

It is sometimes difficult to discriminate between the two, however, since genetic linkage may result in apparent cross-resistance, particularly during selection in the laboratory. If selection with an insecticide favors a rare gene *A*, it will simultaneously select the gene-alleles that are linked with it. If another resistance gene *B* already occurs in the strain, either this gene or its S-allele may be linked with gene *A*. Selection for gene *A* can thus increase or decrease gene *B*, depending on the accidental concurrence of *A* with *B* or its S-allele. If selection favors gene *B*, this would suggest cross-resistance whereas in actuality multiple resistance is present.

Cross-resistance patterns show characteristics of the resistance gene involved, provided that other mechanisms are absent. A well-known example is cyclodiene resistance in houseflies and several species of mosquitos, where characteristic ratios of resistance factors are found for the compounds of the group (Busvine, 1954). In other cases, the patterns show up only after genetic isolation of single factors, and again these may serve to characterize the gene-alleles involved (Farnham, 1973).

6. Causes of Resistance

Theoretically, alteration in any of the processes that determine the penetration, distribution, or target site interaction of an insecticide can contribute to resistance (Chadwick, 1955; Winteringham, 1969). Of major importance are altered site of action, increased detoxication, and reduced penetration.

6.1. Altered Site of Action

6.1.1. Altered Cholinesterase

Any study of alterations occurring at the site of insecticide action requires, of course, that the normal interactions at this site be fully understood and that techniques be available for studying the nature and properties of the site. There is at present little doubt that the organophosphates (OP) and carbamates exert their action mainly by inhibition of acetylcholinesterase (AChE) (Chapter 7), and alterations at this site are the best understood examples of this kind of resistance.

The first indication that resistance might be associated with changes in AChE was the observation of Smissaert (1964) that in OP-resistant spider mites, *Tetranychus urticae*, the activity of AChE toward acetylcholine (ACh) and acetylthiocholine in an R-strain was only one-third that in an S-strain. Since the R-strain had been genetically "purified" by backcrossing the main R-gene into the S-genome, it did not appear likely that the difference between the esterases was due simply to chance. Subsequent studies on the rate of inhibition of the enzyme by several OPs confirmed this. Bimolecular rate constants for paraoxon were 10^5 and 10^2 liters mole^{-1} min^{-1}, and for diazoxon 3×10^6 and 2×10^4 in S- and R-strains, respectively. The larger difference in the rate constants for paraoxon is in agreement with the higher resistance of the R-strain to this compound. Genetic studies revealed that in hybrids both enzymes were present, since the enzyme controlled by the S-allele, representing 75% of the activity (the S-enzyme was more active), was rapidly inhibited, whereas that from the R-allele was inhibited only slowly. This shows that the differences in the rate constants between the S- and R-strains are due to differences in the enzymes themselves and are not caused by other substances present in the homogenates. Further supportive evidence was obtained with other substrates and inhibitors (Smissaert *et al.*, 1970), and with the latter the ratio of the bimolecular constants in S- and R-strains varied from 8 to 10^4. The work was extended to other R- and S-strains, and as can be seen in Fig. 1 a good correlation was obtained between resistance to malathion and the "insensitivity index" of AChE, which is the ratio of the dose required for 50% inhibition in the R-strain divided by that of the S-strain (Zahavi and Tahori, 1970). It is of

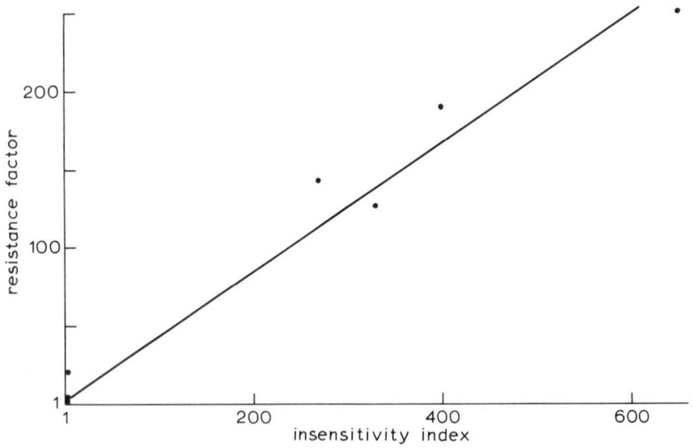

Fig. 1. Relation between resistance to malathion and "insensitivity index" of acetylcholinesterase for malaoxon of seven strains of spider mites. Data from Zahavi and Tahori (1970).

interest that although the AChE of the R-strain with the highest level of resistance had the lowest rate of inhibition by OP compounds its activity per milligram of protein toward ACh was about the same as that from the S-strain. This clearly shows that alteration of the enzyme is not necessarily connected with a reduced capacity to perform its normal physiological function. The existence of strains with different degrees of resistance and possessing AChEs with different levels of OP insensitivity suggests that the alterations in the enzyme are caused by alleles. Allelism was indeed found in two of the R-strains (Schulten, 1968).

An altered acetylcholinesterase is also a major resistance factor in the cattle tick, *Boophilus microplus* (see Wharton and Roulston, 1970, for a review). As in the spider mite strain studied by Smissaert, the AChE of several R-strains of the cattle tick exhibit reduced substrate activity and a reduction in their rate of inhibition by OP insecticides (Table I). Resistance is due to a single gene and there is some evidence that the genes are allelic (Stone, 1972). As shown in Table I, the Ridgelands strain suffered the largest reduction in enzyme activity, but 80–90% of this can still be irreversibly inhibited in surviving ticks (Schuntner and Smallman, 1972). A decreased rate of AChE inhibition occurs in both the Ridgelands and Biarra strains, although the magnitude of this factor for the various inhibitors is quite different in each strain, and they have different cross-resistance patterns (Wharton and Roulston, 1970; Stone, 1972). A curious situation is found in the Mackay strain. The AChE activity of this strain was only 27% that of the reference strain, although when first examined its rate of inhibition appeared practically identical with that of the S-strain. This initially indicated that the reduced AChE activity was unrelated to the resistance, which was in accord with monogenic resistance being due to increased detoxication. Later a much lower inhibition rate constant was reported for coroxon in this strain, suggesting the existence of an altered AChE probably due to a second R-gene.

In two strains, Yeerongpilly and Biarra, a study was made of the AChE activity in different subcellular fractions and in protein fractions separated by electrophoresis (Nolan *et al.*, 1972). In all subcellular fractions, strain differences in the rate of inhibition of AChE by coroxon were observed and were the same as those in unfractionated homogenates. Column chromatography allowed separation of five fractions possessing AChE activity. Two of these, which together accounted for 10% of the activity in the Yeerongpilly and 30% of that of the Biarra strain, were identical in the strains and showed a very high rate of inhibition. The remaining three fractions showed a 500-fold difference with regard to their rate of inhibition by coroxon. It would be interesting to know more about the role and localization of the enzymes that do not differ between the strains, since to date no evidence has been provided to show that they represent AChE from the nervous system.

The first insect in which an altered AChE was suggested as a possible cause of resistance was the sheep blowfly, *Lucilia cuprina* (Schuntner and Roulston, 1967). The only evidence to suggest any difference in the AChE of the S- and R-strains was that in thorax homogenates of one of two R-strains a longer

Table I. Resistance Factors and Properties of AChE in Four Strains of the Cattle Tick[a]

Strain of *B. microplus*	AChE activity as percentage of that of Yeerongpilly strain	Insensitivity index[b] of AChE		Resistance factor			
		Coroxon	Diazoxon	Dioxathion	Coumaphos	Diazinon	Dimethoate
Yeerongpilly	100	1	1	1	1	1	1
Ridgelands	14	2.9	21	7	2	10	400
Biarra	30	380	110	13	25	28	110
Mackay	27	1.1(8.3)[c]	1.3	25	9	8	65

[a]Data from Wharton and Roulston (1970) and Schuntner and Smallman (1972).
[b]Insensitivity index is the reciprocal of the bimolecular rate constant of inhibition divided by that for the Yeerongpilly strain.
[c]Data obtained later, indicating a change in the strain.

incubation time with the inhibitor was required to achieve 50% inhibition. No such differences were observed in head homogenates. The significance of this difference is open to question since it could result from several changes in the homogenates other than those relating directly to AChE.

A resistant strain of the green rice leafhopper, *Nephotettix cincticeps*, was shown to have an altered AChE which is less susceptible to inhibition by carbamates (Hama and Iwata, 1971; Iwata and Hama, 1972). I_{50} values were obtained by first preincubating homogenates with the carbamates and subsequently incubating with acetylcholine. The values obtained in this way showed R- and S-strain differences of 115-fold for propoxur, 49-fold for mipcin, and 47-fold for carbaryl. The values obtained for propoxur with head and whole body homogenates were almost identical, so that it seems very unlikely that they are due to *in vitro* artifacts associated with other components in the homogenates. A rather good correlation was found between the resistance factors against seven *N*-methylcarbamates and the ratio of the I_{50} values (R/S) obtained *in vitro*. No important strain difference was found with regard to the activity of AChE towards acetylcholine. Whether the differences in I_{50} values reflect a difference in affinity, carbamylation, or decarbamylation has not yet been determined.

A strain of houseflies has been found that appears to be resistant to Rabon due to the presence of an altered cholinesterase (Tripathi and O'Brien, 1973). Since this is the insect with the best-studied AChE, the availability of mutants provides for important opportunities for the study of the details of the reaction with the inhibitors. The enzyme showed a considerably decreased rate of inhibition, which was found to be mainly due to a greatly reduced affinity (increased dissociation constant) of the enzyme–insecticide complex. This was partly counteracted by a small increase in phosphorylation constant. The V_{max} for acetylcholine of the R-strain appears to be 4.5 times increased and the K_m 4.3 times "worse," i.e. larger. The authors conclude from this that apparently a change in enzyme structure affects binding sites for Rabon that are not involved in acetylcholine binding. A second strain, resistant to dimethoate, has been found by Devonshire and Sawicki (1974). A tenfold reduction in bimolecular velocity constant of inhibition by omethoate for AChE seems responsible for part of the resistance. This difference is controlled by a gene on chromosome 2.

Recently a mutant AChE strain of *Anopheles albinanus* has been found, which is very resistant to parathion and propoxur, but not to fenthion (Ayad and Georghiou, 1975). Correspondingly, a very large difference in inhibition rate compared to normal strains was found with propoxur but none with fenoxon.

In summary, the most important observations on AChE mutants are:

a. Mutants are found in which the AChE exhibits an enormously reduced susceptibility to organophosphate and carbamate inhibitors.
b. Several different alterations are possible leading to enzymes which cause different cross-resistance patterns.

c. Some alterations impair the natural function of the enzyme (i.e., activity toward ACh), but this is not always so.

It should be mentioned that the decreased rate of inhibition by itself would only serve to postpone death in the absence of other processes to reduce the amount of inhibitor present. Some level of detoxication capability exists in most normal insects, and it is likely that as a result of the much longer survival time caused by the decreased rate of AChE inhibition in R-strains the detoxication rate is sufficient to allow removal of the toxicant.

6.1.2. Resistance Due to Gene kdr

Early observations indicated that there were different kinds of DDT resistance in houseflies (Busvine, 1951). One Italian strain showed resistance which could not be related to DDTase activity (see Section 6.2.2.) (Winteringham *et al.*, 1951). Of particular interest was the fact that this strain survived without first being knocked down, whereas other strains were usually first knocked down and subsequently recovered. Busvine noted that the Italian strain showed cross-resistance to pyrethrins.

This type of resistance was the first that was thoroughly studied genetically. It was found to be due to a recessive gene, called *kdr* for "knockdown resistance" (Milani and Travaglino, 1957). It should be pointed out that although this name stresses only the effect of the gene on knockdown it has an effect on mortality as well. The gene was located on what is now called chromosome 3, and crossover distances with two marker genes were established. Later a similar gene on the same chromosome was found in an American strain, and since it appeared to be not allelic with *kdr* it was called *kdr-0* (Milani and Franco, 1959).

Only negative information is available on the action of these genes. Since their effect extends to DDT analogues that cannot be dehydrochlorinated, they are clearly not associated with DDTase activity (Plapp and Hoyer, 1968*a*). Further evidence that there is no influence on either DDTase or on the enzymes associated with the oxidative detoxication of DDT comes from studies with synergists that block these reactions. Such compounds often have strong effects on DDT toxicity to other DDT-R-strains, but have no effect on *kdr* strains (Grigolo and Oppenoorth, 1966). Also, as long as 18 hr after application large amounts of intact DDT can be extracted from *kdr* flies which are completely unaffected. That *kdr* causes low nerve sensitivity was shown by application of DDT to the thoracic ganglion and measurement of the electrical activity in the femur muscles (Tsukamoto *et al.*, 1965). Recent work in our laboratory employing the same techniques used by Barton Browne and Kerr (1967) also indicated a reduced sensitivity of sensory cells. In this technique, DDT is applied to labellar hairs and the electrical response in single sensory cells is recorded. As was the case in a strain with high DDTase activity, production of the typical DDT symptoms in our *kdr* strain (with low DDTase activity) required much higher dosages of DDT than in an S-strain.

In agreement with the early observation of Busvine (1951) on the cross-resistance of the original Italian strain to the pyrethrins, Farnham (1971, 1973) isolated several genes for pyrethrin resistance genetically, and concluded that a gene *kdr-NPR* was of major importance in determining resistance to natural as well as synthetic pyrethroids. He obtained results indicating that the three genes *kdr-NPR*, *kdr*, and *kdr-0* are allelic (Farnham, 1972), a conclusion at variance with earlier observations that *kdr* and *kdr-0* were not allelic (Milani and Franco, 1959).

The absence of any information concerning the physiological function of the *kdr* gene together with a cross-resistance pattern indicating an effect of insecticides which attack nerve and sensory cells invites speculation. Two hypotheses come to mind: *kdr* might alter the access of the insecticide to the site of action, or it might alter the site of action itself. Any hypothesis on the resistance mechanism controlled by the *kdr* gene has to take into account that (a) the alteration affects DDT and its analogues as well as the pyrethroids and (b) its action is quantitative, since at higher dosages a normal response is observed.

Holan (1971), on the basis of his hypothesis concerning the importance of molecular dimensions in determining the insecticidal action of DDT analogues, synthesized several compounds which were remarkably toxic. One of these compounds was tested on *kdr* flies and a high degree of resistance was observed (Oppenoorth, 1971*b*). It would be interesting to evaluate the toxicity of other compounds with slightly different dimensions on *kdr* strains since this might provide a clue to the alteration caused by the *kdr* gene.

The effect of *kdr* is greatly affected by certain modifier genes (Milani and Travaglino, 1957), and one of these, causing a low level of DDTase activity, was found to render the gene more dominant (Grigolo and Oppenoorth, 1966).

No comparative study has yet been made to determine the frequency of the gene *kdr* relative to other types of genes causing DDT resistance in different parts of the world. Although its action is probably frequently masked by the DDTase gene, it appears to be predominant in certain areas and may well be responsible for the reported failure of applications of DDT in combination with synergists (Brown and Rogers, 1950). *Kdr* also confers resistance to methoxychlor, whereas DDTase does not, and the greater abundance of the *kdr* gene in Europe (Keiding, 1963) may be one cause of the lack of success with this compound (Gysin, 1971). Whether this type of resistance is of general importance in other insect species is difficult to assess, since its characterization is mainly negative. A similar type of resistance to that controlled by the *kdr* gene in houseflies has been reported in *Culex tarsalis* (Plapp and Hoyer, 1968*a*) and resistance in the stablefly, which is also independent of DDT detoxication, may belong in this class (Stenersen, 1965). It would be interesting to know more about cross-resistance to pyrethrins. DDT-resistant cattle ticks were shown to be cross-resistant to pyrethrins and there is no evidence to suggest that the small amount of detoxication to DDE is a cause of this resistance (see Wharton and Roulston, 1970, for references).

6.1.3. Cyclodiene and Lindane Resistance

As with DDT resistance due to gene *kdr*, evidence pointing to the importance of an altered site of action in cyclodiene and lindane resistance is mainly negative. There is ample evidence that in many Diptera one major gene is responsible for almost all resistance to the cyclodienes and for at least part of the resistance to lindane (see Oppenoorth, 1965*c*, for a review). Although factors which determine the amount of toxicants in the body contribute to lindane resistance, this does not appear to be true for cyclodiene resistance (Winteringham and Harrison, 1959; Brooks, 1960).

Since the mode of action of dieldrin remains unknown, it is not possible to test the hypothesis that resistance is associated with a changed target site. The suggestion of Wang *et al.* (1971) that localized formation of aldrin *trans*-diol is involved in the action of dieldrin has not yet been tested in its consequences for resistance, and it would be interesting to know whether this compound is less toxic to R-nerves or is formed more slowly in R-flies.

6.2. Increased Metabolism

6.2.1. Methodology

To determine the possible role of detoxication in resistance, the rate of detoxication can be measured directly either *in vivo* or *in vitro* or the effect on toxicity of combining the insecticide with compounds (synergists) which block specific metabolic reactions can be assessed. Each of these approaches has its own merits and limitations which will be discussed briefly.

a. In Vivo Evidence. In vivo metabolism studies constitute one method for evaluating the role of detoxication in resistance. If in R-strains a more rapid decrease of the amount of internal insecticide is found than in S-strains, or if there is a more rapid formation of some degradation product, an increased detoxication capacity is indicated. However, since the R-flies usually offer better conditions for detoxication than S-flies (they live longer) it is often difficult to determine whether an increased rate of metabolism is a direct result of resistance rather than its cause. Furthermore, if the enhanced detoxication in the R-flies is indeed a cause of resistance it is difficult to determine its quantitative importance.

Only rarely can *in vivo* studies alone provide solutions to these problems. Sometimes the question of cause or effect can be studied with the help of nontoxic analogues of the insecticide in question. If these, too, are metabolized more rapidly in the R-strain, this is indicative of metabolism as a cause of resistance. This technique has been applied in the case of hexachlorocyclohexane resistance in houseflies, where the toxic γ-isomer as well as the nontoxic α- and δ-isomers were found to be degraded more rapidly in resistant strains (Oppenoorth, 1956). Another example is the use of nontoxic DDT analogues in studies to assess the significance of the formation of kelthanelike substances (Tsukamoto, 1961).

In some cases, the rate of disappearance of the parent insecticide can be used as a measure of the *in vivo* rate of metabolism, but this provides little or no information concerning the actual route by which metabolism occurs. If there are several routes by which detoxication can occur, the product of any particular reaction can be measured. However, the value of such studies is limited because many primary metabolic products undergo secondary reactions. Many of the more polar metabolites are conjugated prior to excretion (see Chapter 5). Also, nonpolar metabolites such as DDE can be degraded further in some cases (Oppenoorth, 1967).

As discussed below, the rate of penetration of an insecticide can have a profound effect on its rate of degradation, and thus on resistance; this is also true with respect to passage through internal membrane barriers. If there are strain differences in any of these processes, the interpretation of *in vivo* results is further complicated.

Despite these limitations, metabolic data obtained *in vivo* are often indispensable in studies of resistance and can provide a useful starting point in studies to determine the relative importance of this factor. In many cases, methods are not readily available for a study of the same reactions *in vitro*. For example, no DDT-dehydrochlorinase activity could be demonstrated in preparations from mosquitos (*Aedes aegypti*) *in vitro*, in spite of the fact that DDE was a major metabolite *in vivo* (Brown, 1956; Chattoraj and Brown, 1960). Similarly, Tsukamoto (1961) was unable to demonstrate kelthane production in *Drosophila melanogaster* preparations *in vitro*, whereas its *in vivo* production was evident in this species.

b. In Vitro Evidence. If differences in the detoxication potential of R- and S-strains can be shown *in vitro*, this is a great step forward. Preferably both V_{max} (maximum velocity of the enzyme reaction) and K_m (as a measure of the dependence of the reaction rate on substrate concentration) should be measured, since both of these parameters are of importance in determining *in vivo* enzyme activity. Differences in *in vitro* activity in various insect preparations can also be caused by the presence of a variety of endogenous activators or inhibitors acting during either preparation or incubation. This is particularly true with respect to the *in vitro* measurement of mixed-function oxidase activity, where the presence of endogenous inhibitors is often a severe problem. Thus the technique which allows work with housefly microsomes was found to be unsuitable for aphids, and aphid homogenates when added to housefly microsomes abolished all activity (Devonshire, 1973).

If the *in vitro* work requires the use of specific subcellular fractions (mitochondria, microsomes, etc.), as if often the case, further complications arise from the loss of activity in other fractions. Thus a portion of the microsomal fraction is always lost in the mitochondrial and nuclear fractions and this often makes it difficult to calculate the total enzyme activity present in whole insects.

Consequently, the major problem in all *in vitro* studies is whether the results obtained are a true reflection of the *in vivo* situation or whether they are artifactual due to the conditions employed. It should be recalled that, *in vitro*, DDT was equally well degraded by preparations from both S- and R-strains of body lice but only the R-strain metabolized DDT *in vivo* (Perry and Buckner, 1958). The cause of this discrepancy has never been established.

For these reasons, interpretations based solely on *in vitro* data can often lead to erroneous conclusions. This is strikingly illustrated in comparisons between oxidase and DDTase activities in the Fc strain DDT-R houseflies (Oppenoorth, 1967; Oppenoorth and Houx, 1968). The effect of synergists on DDT toxicity, as well as genetic analysis, clearly showed that oxidation and not DDTase was the important resistance mechanism in the Fc strain. In spite of this, measurements of *in vitro* oxidase activity and DDTase activity were 0.5 and 9 μg per fly per hour, respectively! The precise reasons for this discrepancy are not fully understood, but the results clearly emphasize the need for caution in interpreting *in vitro* data.

c. Use of Synergists. Since it is often difficult to obtain a clear-cut indication of the importance of a certain detoxication pathway from *in vivo* and *in vitro* metabolism studies, the availability of more or less specific enzyme inhibitors (synergists) can provide additional information on the relative role of various metabolic routes *in vivo* (Casida, 1970; Wilkinson, 1968, 1971).

If, for example, resistance can be overcome by combining the insecticide with a synergist known to inhibit mixed-function oxidation it is assumed that the resistance mechanism is associated with this oxidase system. A major problem, however, is that the action of the synergist is not necessarily specific for one enzyme system. Also, if activation and inactivation, which can both be oxidative in nature, are blocked by the same inhibitor, the effects on toxicity are generally far too complex for meaningful interpretation (Chapter 2). Since a maximum effect is desired, rather high dosages of the synergist are required and this increases the danger of other nonspecific effects such as those related to penetration. Possible effects of the synergist on insecticide penetration can be minimized, but perhaps not entirely eliminated, by application of synergist and insecticide on different parts of the insect.

Another problem which has to date received little attention is whether the action of various synergists on different species can be considered the same. Even between different strains of the housefly, quantitative differences in *in vitro* inhibition of oxidase activity have been found with sesamex and piperonyl butoxide (Schonbrod *et al.*, 1973). Sawicki (1973*b*) has described a gene in a Dimethoate-R housefly strain which is responsible for an enzyme that probably causes oxidation but which is not affected by sesamex. If the enzyme is indeed a mixed-function oxidase, this provides further warning that correlations between the action of synergists and the presence of certain types of detoxication mechanisms need not be absolute. Differences in other processes such as the rate of metabolism and penetration of the synergist undoubtedly can also come into play. Consequently, the failure of a particular synergist to enhance insecticidal action in a certain species cannot be taken as an absolute indication of the absence in that species of the enzymatic reaction which it usually inhibits.

Where synergists are effective with both the S- and R-strains, the resistance of the synergized R-strain can be expressed in relation to either the synergized or the nonsynergized S-strain. The former is probably more correct, but if the synergistic effect is larger in the S- than in the R-strain this leads to so-called increased resistance (Sawicki, 1973*b*), although the addition of the synergist does in fact decrease the LD_{50} of the R-strain (Sawicki and Farnham, 1968). A synergistic ratio larger numerically in the S-strain than in the R-strain need not mean that the detoxication capacity of the S-strain is larger, since other factors may be involved. This is apparent, for example, from the much larger difference generally found between S and F_1 (with one dose of detoxication enzyme) than between F_1 and R (with two doses, since generally the amount of enzyme in the homozygote is double that of the heterozygote). This is the well-known semidominance of resistance due to detoxication.

In summary, a careful combination of *in vivo, in vitro,* and synergist studies can often provide important information on the role of various detoxication mechanisms in resistance. For a complete understanding of more complex cases, the technique of genetic analysis and reconstitution is required (Sawicki and Farnham, 1968; Georghiou, 1971; Sawicki, 1973*a*).

6.2.2. DDT-dehydrochlorinase

DDT-dehydrochlorinase, which is now generally called DDTase, and which converts DDT to the relatively nontoxic DDE, has been extensively studied, mainly in the housefly. It now seems certain that this enzyme is the principal cause of resistance to DDT, although other mechanisms are by no means rare (Sections 6.1.2 and 6.2.4). Since the older literature has already been covered in several excellent reviews (Lipke and Kearns, 1960; Perry, 1964; O'Brien, 1967), emphasis here will be placed on some of the more recent information.

Elucidation of the role of this enzyme in DDT resistance is the classical example illustrating the combined use of *in vivo* and *in vitro* data together with information obtained from the use of synergists and genetic analysis. The *in vivo* formation of DDE from DDT was first demonstrated in DDT-R flies (Sternburg *et al.*, 1950; Perry and Hoskins, 1950) and was followed by the demonstration of enzymatic activity in homogenates and acetone powders of many R- but not S-strains (Sternburg *et al.*, 1954). A rough correlation between resistance and *in vitro* DDTase was shown in ten housefly strains by Lipke and Kearns (1960) and represented graphically by O'Brien (1967).

DDT resistance was considerably reduced when the insecticide was applied in combination with several nontoxic DDT analogues (analogue synergists) such as DMC (Perry *et al.*, 1953), and these were found to inhibit DDTase activity *in vitro* (Moorefield and Kearns, 1955). DMC, however, was found to be metabolized rapidly *in vivo*, and another more stable compound, FDMC (Cohen and Tahori, 1957), is a more suitable synergist to indicate DDTase action.

In highly resistant strains, DDTase activity was found to be due to a single gene on chromosome 2. However, if only a single gene is responsible, this would be expected to result in an all-or-none effect, which does not explain the correlation between resistance and DDTase activity mentioned above. The most probable answer to this apparent anomaly seems to be that there are several alleles of the gene that produce different DDTase activities.

The advent of new analytical techniques and instrumentation, particularly gas–liquid chromatography, has provided ultrasensitive methods for DDT assay. This now permits measurement of as little as 0.1% of the DDTase activity of a single R-fly and therefore obviates the necessity of using large numbers of mixed flies with high and low DDTase activities. Using single flies, several levels of DDTase activity have been demonstrated in different strains (Oppenoorth, 1965*a*). If the high DDTase activity of a DDT-DMC selected strain is set at 100% (180 μg DDT per fly per hour), two other levels were found: 5% in two strains with factors for DDT resistance other than DDTase and 0.3% in a S-marker strain. No measurements were made in other strains exhibiting various levels of resistance, and this would be needed for a full understanding of the correlation shown by Lipke and Kearns. However, the occurrence of different DDTase levels suggests the presence of different alleles, and the "5%" activity was shown to be due to an allele of the "100%" gene, or to be closely linked with it (no crossover was shown in 100 flies).

The enzymes, although dependent on allelic genes, are not identical. Berger and Young (1962), studying activity toward several bromo analogues of DDT, suggested that there might be two different DDTases, one in S- another in R-strains. This was further elaborated by Oppenoorth (1965*b*) on the basis that the two enzymes with low activity appeared to have a different substrate specificity and susceptibility to inhibitors from that of the "100%" enzyme (Table II). This strongly suggests that there is a qualitative difference between the enzymes, but whether this alone is responsible for the increased activity in R-strains or whether there is also a quantitative difference involved cannot be decided at the present time.

The enzyme from R-strains has been purified by Lipke and Kearns (1960) and more recently by Dinamarca *et al.* (1969, 1971). It appears to be a lipoprotein with a molecular weight of about 120,000 consisting of four subunits. The monomer aggregates to form the tetramer in the presence of DDT and only the trimer and tetramer exhibit DDTase activity. Glutathione does not cause aggregation, but prevents the dissociation of the aggregates. It is required for activity, but its role is still unexplained. It does not take part in the enzymatic reaction, and its action is not related to the maintenance of sulfhydryl groups. Goodchild and Smith (1970), working with strains with low DDTase activity, have also found multiple forms of DDTase in electrophoretic studies. Unfortunately, these workers used different methods from those of Dinamarca *et al.*, so that the DDTases of the two strains cannot be compared. This should now be possible, however, particularly if the substrates of Berger and Young, which are split relatively rapidly by S-strains, are employed. A

Table II. Susceptibility for DDT and Properties of DDTase in Different Strains of Housefly

		Strains					
		CSMA[a]	CS[a]	DDT-45[a]	*acr*[b]	Fc[b]	L[b]
LD_{50} (μg DDT/fly)		0.08	2.6	>200	0.15	>10[c]	>10
Activity (moles × 10^8/fly/hr)							
DDT		0	1.8	7.2	0.17	2.5	51
TDE		0	4.7	21.7	0.07	0.5	217
Br-TDE		9.5	25.3	110.0	5.9	144.0	355
Inhibition[b] (%)	DDT–DMC 50–50[d]				—	37	99.4
	DDT–DMC 50–5				—	3	95
	Br-TDE–DMC 50–50				22	30	97
	DDT–FDMC 50–2.5				—	60	98

[a]Data of Berger and Young (1962) with supernatants of homogenates.
[b]Data of Oppenoorth (1965*a*,*b*) with acetone powders.
[c]Resistance in Fc due not to DDTase but to oxidative degradation (see text).
[d]Inhibitor and substrate mixtures as indicated (M × 10^6). Inhibitors in solution, substrates as colloidal suspension.

thorough comparative study of various "allelic" DDTases might shed considerable light on details of the development of the DDTase genes and might confirm whether, as has been suggested (Oppenoorth, 1965*b*), the DDTase with the very high activity might have arisen from the enzyme with the lower activity by more than one mutational step. So far, nothing is known on the natural function of the DDTase in S- or R-strains. DDTase activity has been shown to be inducible by cyclodienes, phenobarbital, and hormone analogues (Terriere and Yu, 1973; Yu and Terriere, 1973), but the significance of this observation cannot yet be assessed.

DDTase is present in many different insect organs, and it is important to note that it probably exists at the main sites of action, the peripheral sense organs. Thus, in a study of DDT action on labellar sense cells, Barton Browne and Kerr (1967) found that the R-strain carrying the DDTase gene is much less affected than S-strains. As shown in Fig. 2, the repetition of action potentials caused by DDT was less pronounced and occurred only at high DDT levels in R-strains.

DDTase has received only little attention in insects other than the housefly, and since the enzyme can differ even between strains of the fly it would not be surprising to find considerable differences in other species. In *Aedes aegypti*, where *in vivo* DDE formation has been found to correlate with DDT resistance in American but not in Asiatic strains, a DDTase is present that can degrade *o*-chloro-DDT but not deutero-DDT (deuterium on the tertiary carbon atom); this is the reverse of the situation in the housefly (Pillai *et al.*, 1963). A DDTase in *Culex* species seems to have still another pattern of substrate specificity (Kimura *et al.*, 1965).

6.2.3. Hydrolases and Glutathione-Dependent Transferases

Although many insecticides are esters (OP compounds, carbamates, pyrethrins), resistance due to hydrolases or glutathione-dependent alkyl- or aryltransferases (GSH-*S*-transferases) has been found to be of importance only with OP esters. The reactions and properties of these enzymes have already been discussed in some detail (Chapters 4 and 5) and will not be repeated here. However, in view of the importance of these enzymes to this discussion the following examples will summarize theoretically possible primary sites of attack on two insecticides (an arrow does not necessarily mean that the mechanism has been found). For review, see Motoyama and Dauterman (1974).

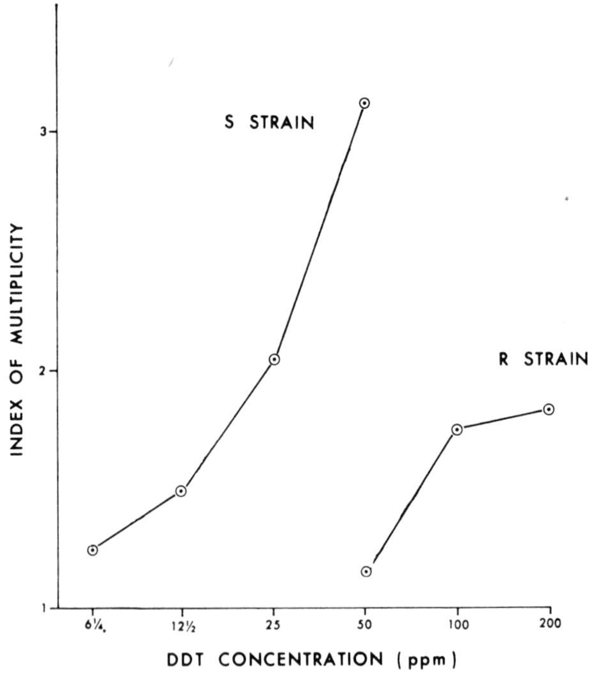

Fig. 2. Nerve impulse multiplicity of NaCl-stimulated neurons in labellar chemoreceptor hairs in two strains of the housefly upon DDT treatment. From Barton Browne and Kerr (1967).

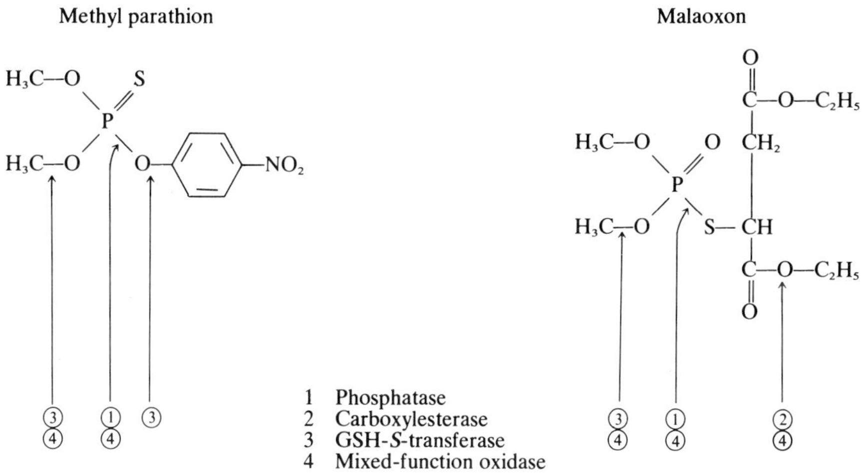

Since the activity of the hydrolases, GSH-*S*-transferases, and mixed-function oxidases (Section 6.2.4.) on a certain insecticide often generates the same metabolic products, it is not possible to infer the nature of detoxication mechanism involved solely from the structure of the products found.

a. Hydrolases. Two kinds of enzymes have been found to cause resistance in certain insects: phosphatases and carboxylesterases (1 and 2, respectively, in the scheme). The latter can be of importance only in OP compounds with a carboxylic ester such as malathion. The importance of the two enzymes will first be discussed, after which their mutual relationship and affiliation with carboxylesterases attacking other noninsecticidal esters will be dealt with.

Phosphatases. It is generally accepted that the properties of an OP compound that determine its susceptibility to phosphatase cleavage are the same as those that enable it to inhibit AChE, i.e., an electrophilic phosphorus atom. Consequently, it is likely that phosphates are much more readily attacked than phosphorothioates. Whereas phosphatase activity could not be detected in S-strains of houseflies, a low level of activity was found in homogenates of Parathion-R and Diazinon-R strains (Oppenoorth and van Asperen, 1960, 1961; Welling *et al.*, 1971; Lewis and Sawicki, 1971). In some strains, paraoxon and diazoxon were degraded at a rate of only 25 pmoles per fly per hour, in several other strains at 150 pmoles per fly per hour. The product formed has been shown to be diethyl phosphate (from paraoxon, Welling *et al.*, 1971; from diazoxon, Lewis and Sawicki, 1971), and dimethyl phosphate would be the expected product of dimethyl OP compounds, e.g., malaoxon. According to Matsumura and Hogendijk (1964*a*), this compound is degraded in a Malathion-R strain to what are apparently phosphatase products (not identified definitely). In *Myzus persicae* resistant to many OP compounds, a phosphatase action was recently found degrading paraoxon to diethyl phosphate and malaoxon to dimethyl phosphate (Oppenoorth and Voerman, 1975).

The phosphatases present in different resistant strains of houseflies are under the control of alleles of a gene on chromosome 2. The enzyme studied by Welling *et al.* (1971) has a K_m for paraoxon of 4×10^{-9} M. It was readily inhibited by *n*-propyl paraoxon and calculations showed it to be present at a concentration of about 12 pmoles per fly; its turnover number was calculated to be 0.2 per minute.

Since in several strains with phosphatase activity other resistance factors are present (e.g., in strain Fc a factor for oxidative attack has been found on chromosome 5 and a penetration factor on chromosome 3; see Sections 6.2.4 and 6.3), the question arises of how much of the resistance is due to the hydrolytic enzyme. A degradation of 150 pmoles per fly per hour represents only 0.04 μg paraoxon, and O'Brien (1966, 1967) pointed out that such degradation rates are too low to explain the level of resistance observed (e.g., the LD_{50} of the strain studied by Welling was 0.5 μg paraoxon per male fly without an inhibitor of oxidative detoxication but 0.1 μg in the presence of such an inhibitor; Oppenoorth, 1972). The importance of the hydrolysis factor has been shown in two ways: First, two strains differing only in one aspect, the presence or absence of the hydrolyzing enzyme, were found to differ considerably in susceptibility (about fifteenfold for parathion by a contact method) (van Asperen *et al.*, 1965). Second, if the oxidative degradation of paraoxon and diazoxon is blocked with sesamex in two R-strains, one with the gene for phosphatase activity and one without, the resistance is suppressed only in the latter, but a seventeenfold resistance remains in the former (Oppenoorth, 1972). The extremely low K_m value of the phosphatase makes it particularly suited to remove the last traces of inhibitor, and helps explain its relatively large contribution to resistance, in spite of its low maximal detoxication capacity.

Collins and Forgash (1970), studying diazinon resistance in houseflies, presented evidence that at moderate resistance levels the decrease in susceptibility correlated well with the observed increase in the hydrolytic degradation of diazoxon. With further selection, however, resistance was further increased with no concomitant increase in diazoxon hydrolysis capacity. This clearly shows that above a certain resistance level other R-factors were involved in the selection process.

As has been mentioned, it seems unlikely that phosphatases can attack phosphorothioates to any extent, and since the product which would be formed from this reaction, dialkyl phosphorothioic acid, can be produced *in vivo* either oxidatively or by a GSH-*S*-transferase the question as to whether phosphatases contribute to phosphorothioate resistance can be solved only by *in vitro* investigations. Several authors claim to have demonstrated the direct esteratic cleavage of phosphorothioates *in vitro* (Matsumura and Hogendijk, 1964*b*; Jarczyk, 1966, 1967). Jarczyk reported the presence of two hydrolytic enzymes, one attacking parathion and its analogues and another acting on paraoxon. The activities of these enzymes in R-flies were less than twofold greater than those in a S-strain. In contrast, no direct hydrolytic degradation of parathion could be demonstrated *in vitro* by Nakatsugawa *et al.* (1969) in either R- or S-houseflies, and Welling *et al.* (1971) were unable to confirm the results of Matsumura and Hogendijk (1964*b*).

Carboxylesterases. The presence of carboxylester groups in malathion makes this compound and its oxygen analogue malaoxon vulnerable to attack by carboxylesterases. Only recently was it shown that these groups can also be degraded oxidatively (Welling *et al.*, 1974) and since, as previously mentioned, degradation by several mechanisms can produce the same products, great caution is required in the interpretation of data in the older literature. Increased degradation by carboxylesterase has been established in Malathion-R strains of houseflies. Welling and Blaakmeer (1971) found a carboxylesterase in both S- and R-strains which attacks malathion and is present in the soluble cell fraction. In the R-strain an additional particulate carboxylesterase was found and had a fivefold greater activity than the soluble enzyme. This enzyme is readily inhibited by *n*-propyl paraoxon and produces both the α- and β-monoacids of malathion in a ratio of about 4 : 1. It occurs at a level of about 12 pmoles per fly and its turnover number is 55 per minute.

Carboxylesterase inhibitors such as *n*-propyl paraoxon, EPNO, and TPP are excellent synergists for malathion against this R-strain, which would indicate that malathion breakdown to the monocarboxylic acid is of importance for resistance. However, it cannot be concluded that this is the only resistance mechanism involved since *in vivo* the monoacid represents a relatively small fraction of the total metabolites produced (Matsumura and Hogendijk, 1964*a*). Furthermore, low concentrations of malaoxon inhibit malathion carboxylic ester cleavage *in vitro* (Welling, unpublished) and the synergists mentioned could therefore block malathion degradation either directly or indirectly by inhibiting the detoxication of malaoxon. Main and Dauterman (1967) showed that a partially purified carboxylesterase from rat liver catalyzed the hydrolysis of malaoxon at a carboxylester bond, and was simultaneously irreversibly inhibited by phosphorylation. In other words, malaoxon is both a substrate and an inhibitor of carboxylesterase. A similar situation may occur *in vivo* in resistant houseflies and, if so, the precise resistance mechanism will be extremely difficult to unravel.

Malathion resistance in the blowfly *Chrysomya putoria* appears to be quite similar to that in the housefly. R-flies degrade malathion faster than S-flies, the breakdown product being exclusively malathion monocarboxylic acid (Townsend and Busvine, 1969). In accordance with this, a variety of esterase inhibitors (TPP, TBTP, TOCP, EPN) synergize malathion to R-flies, whereas inhibitors of oxidation (sesamex, SKF-525A) do not.

In the mosquito *Culex tarsalis*, malathion resistance also results from metabolism of malathion to its monocarboxylic acid, but rather surprisingly there is no cross-resistance to methyl or ethyl acethion (Matsumura and Brown, 1961, 1963; Darrow and Plapp, 1960; Bigley and Plapp, 1962). The literature on malathion resistance in *Culex tarsalis* has been reviewed and should be consulted for a further critical evaluation (Oppenoorth, 1965*c*).

Resistance to malathion in some strains of spider mites (*Tetranychus urtica*) has also been found to be due to increased degradation (Matsumura and Voss, 1964, 1965). In the resistant Blauvelt strain, there are two enzymes degrading malathion to carboxylesterase products. Only one of these enzymes is specific for the R-strain and it is characterized by a low K_m value and a susceptibility to inhibition by materials such as DFP and malaoxon. A somewhat similar situation has been described for the housefly by Welling and Blaakmeer (1971). Resistance levels to a large number of malathion analogues have been determined in the housefly (*Musca domestica*), the blowfly (*Chrysomya putoria*), the mosquito (*Culex tarsalis*), and the spider mite (*Tetranychus urticae*). The results are shown in Table III, and from it the following general conclusions may be drawn:

a. For spider mites, there appears to be an inverse correlation between the susceptibility of the S-strain and the R-factor observed. Thus the lower the LD_{50} of the S-strain, the higher the resistance of the R-strain. Voss *et al.* (1964) have given a graphical representation of this correlation, including a larger number of compounds than shown in Table III.
b. Resistance to malathion analogues in each of the three insect species studied is not determined by the nature of the carboxylester group but is highly dependent on the nature of the alkoxy substituents at the phosphorus atom. With ethoxy and larger groups, there is only a very small degree of resistance. It would appear, therefore, that the specificity of the carboxylesterase in the R-strain is related more to the size of the alkoxy groups than to the nature of the carboxylester group to be hydrolyzed. An alternative conclusion might be that the degradation by carboxylesterases contributes less to resistance than might be expected on the basis of *in vitro* measurements.

In *Tribolium castaneum* malathion resistance without cross-resistance to other OP compounds has been studied by Dyte and Rowlands (1968). The only difference between S- and R-strains was an increased rate of formation *in vivo* of carboxylesterase products (both malathion mono- and diacid) in the R-strain. Phosphatase metabolites (dimethyl phosphorothioate, dimethyl phosphorodithioate, and dimethyl phosphate) were produced at equal rates in both strains, and in about the same amount as carboxylesterase products in the S-strain. TPP strongly synergized malathion in the R-strain and suppressed the production of carboxylesterase metabolites without affecting that of the phosphatase products.

Malathion resistance in the bedbug *Cimex lectularius* has been studied by Feroz (1971). Metabolic breakdown of malathion *in vivo* was higher in the R-strain but relative amounts of the metabolites produced depended strongly on the concentration of malathion used. It seems that several degradation pathways involving a desmethylase, a carboxylesterase, and possibly a

Table III. Resistance to Malathion and Malathion Analogues in Four Species of Arthropods[a]

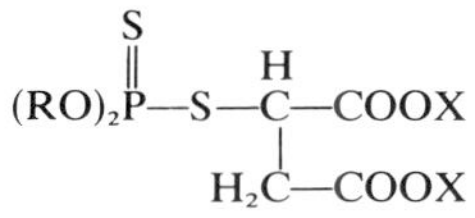

Number of carbon atoms		*Chrysomya putoria*		*Musca domestica*				*Culex tarsalis* larvae				*Tetranychus urticae*	
				LD_{50} S-strain		Resistance factor							
R	X	LD_{50} S-strain	Resistance factor	Application	Contact	Application	Contact	LC_{50} S-strains		Resistance factors		LC_{50} S-strain	Resistance factor
1	1	5.9	144	21	8	146	>375	0.1	0.1	10	1	0.15	2
1	2	9.6	260	27	17	157	106	0.03	0.025	100	60	0.005	60
1	n3	13	>182	19	15	250	>200	0.08	0.09	>125	34	0.004	75
1	i3	5.9	>800	19	8	110	>250	0.08	0.19	19	4.5	0.004	75
1	4	24	>200	21	45	182	66	0.2	0.4	50	28	0.015	20
2	1	4.8	16	5.7	—	5	—	—	—	—	—	—	—
2	2	3.8	10	6.5	7	4	1.7	0.08	—	19	—	0.0008	250
n3	2	12	3	6.5	70	6	4.3	1.0	—	2.5	—	>0.5	(1)
i3	2	14	5	27	13	4	2	1.5	—	3	—	>0.5	(1)
4	2	81	3	400	—	8	—	—	—	—	—	>0.5	(1)

[a]Data from Dauterman and Matsumura (1962), Voss *et al.* (1964), Plapp *et al.* (1965), and Townsend and Busvine (1969). The various analogues are designated by the number of carbon atoms of the alkyl groups; *n* = normal, *i* = *iso*. LD_{50} values are given in $\mu g/g$ (application) or in μg/jar (contact), LC_{50} values in ppm. With *Musca domestica* data for application and contact methods are given. With *Culex tarsalis* the two columns in the upper half of the table are results from two different groups of investigators.

phosphatase cleaving malathion at the P—S bond contribute to this increased metabolic rate.

Origin of the Hydrolytic Detoxication Enzymes of the R-Strains. There is abundant literature describing correlations between OP resistance and altered carboxylesterase (aliesterase) activity toward substrates such as methyl butyrate, α- or β-naphthyl acetate, and other esters. Sometimes resistance appears to be associated with decreased carboxylesterase activity, sometimes with an increased activity. In studying such correlations, one must keep in mind that carboxylesterase activity in insect tissues is brought about by a more or less complex mixture of enzymes, in which individual members cannot easily be distinguished because of broad and overlapping substrate specificities. In cases where higher or lower carboxylesterase activity is found in R-strains, it is commonly found that one esterase is strongly different in activity.

Lower carboxylesterase activity is observed in many field strains of houseflies resistant to different OP insecticides, although the difference in activity found is dependent to a large extent on the substrate used for assay. Thus decreased carboxylesterase activity was found with respect to the substrates α-naphthyl acetate, methyl butyrate, and some other esters, but not with β-naphthyl acetate, triacetin, and others, which probably are degraded by other esterases. These resistant strains invariably had high hydrolytic detoxication activity (see Oppenoorth, 1965*c*, for a review). This observation led to the so-called mutant aliesterase theory, which suggested that in these strains the OP resistance was at least partly due to phosphatases or (in the case of malathion resistance) carboxylesterases, which were presumed to be "mutants" of a carboxylesterase (aliesterase) present in S-strains (Oppenoorth and van Asperen, 1960).

The alteration in the enzyme brought about by the mutation is considered to cause a change in its reaction with OP compounds so that in contrast to the original carboxylesterase, which is irreversibly phosphorylated, the mutant enzymes are capable of dephosphorylation. As a result, they can degrade the OP compounds (phosphatase action) with turnover numbers of 0.05–0.4 per minute, or cleave the carboxylester group of malathion with a turnover number of about 55. At the same time, this change in the enzyme apparently causes a modification in its original substrate specificity, with the result that activity toward esters such as methyl butyrate is reduced by a much as 98%. Unequivocal proof of the relationship between the carboxylesterase in S-strains and the hydrolytic enzymes degrading OP compounds in the R-strains will require purification of the enzymes and a detailed analysis of their structure. Although such evidence is presently lacking, there is both genetic and biochemical evidence to support the proposed relationship.

Resistance due to the hydrolytic enzymes cannot be separated from low carboxylesterase activity in backcrosses (Oppenoorth, 1959). It has been pointed out by Lichtwardt (1964) that this could be due to genetic linkage, since insufficient opportunity was provided for crossing over of unrelated factors. This is true, and consequently linkage of two independent factors cannot be excluded (one for low esterase activity, the other for the detoxifying enzyme). Why such mutants would be present together only in R-strains, however, remains a mystery.

The different hydrolytic enzymes attacking OP esters and causing resistance to parathion and malathion in various R-strains are each due to single genes on chromosome 2, which appear to be allelic. Two strains with low carboxylesterase activity but without hydrolytic OP degradation action (and without resistance) have also been found (Franco and Oppenoorth, 1962; van Asperen *et al.*, 1965). The most likely genetic explanation for this seems to be that in most S-strains a gene a^+ on chromosome 2 is responsible for high carboxylesterase activity. This is replaced in R-strains by alleles of this gene producing the hydrolytic degradation enzymes, or, in the S-strains mentioned above, by an allele which does not produce any demonstrable activity.

The normal carboxylesterase and the hydrolytic degradation enzymes (phosphatase or carboxylesterase) have many properties in common. They are both particle bound, rather labile, and present in approximately the same amount (12 pmoles per fly). Moreover, they react in the same way with some OP compounds such as *iso*- and *n*-propyl paraoxon. Combined with the genetic evidence, these similarities support the theory, an attractive feature of which is that it explains the origin of the detoxication enzymes in the R-strains, through slight alterations in enzymes already present.

The theory is further supported by the fact that a relationship between carboxylesterase activity and OP resistance has been found in many insect species. Thus, in addition to the housefly, it has been established in malathion-resistant *Chrysomya putoria* (Busvine *et al.*, 1963; Townsend and Busvine, 1969), *Laodelphax striatellus* (Ozaki and Kasai, 1970), *Nephotettix cincticeps*, and OP resistant *Myzus persicae* (Needham and Sawicki, 1971; Sudderuddin, 1973*a,b*; Beranek, 1974). In the first of these species, the situation is very similar to that in malathion-R houseflies. In Malathion-R *Laodelphax* and *Nephotettix*, a carboxylesterase with *increased* (relative to S-strains) activity toward β-naphthyl acetate is present, and in the former species gel electrophoresis has shown that this is due to a single carboxylesterase. In *Nephotettix*, the relationship of the high β-naphthyl acetate hydrolysis activity to malathion and OP hydrolysis has not yet been established (Kojima *et al.*, 1963; Ozaki *et al.*, 1966; Ozaki, 1969). In *Myzus*, high activity toward α-naphthyl acetate was found in nine Dimethoate-R strains and low activity in 21 S-strains (Beranek, 1974). The cause of this resistance is not known.

No relationship between esterase activity and OP resistance was found in the malathion resistance already described for *Culex tarsalis* and *Cimex lectularius*.

In summary, hydrolytic enzymes have been found to be important causes of resistance to OP compounds and in several cases the formation of these enzymes has been shown to involve an alteration in existing esterases with concomitant changes in their substrate specificity.

Glutathione-S-transferase. The importance of glutathione-*S*-transferase in OP resistance has been recognized only during the last few years. Although the enzyme has been known to be involved in insecticide metabolism for some time (Fukami and Shishido, 1963, 1966), its importance as a possible resistance mechanism seems to have been neglected due to the fact that the products of its action on OP compounds are identical to those formed by other enzymes (Chapter 5).

The GSH-*S*-alkyltransferase is probably of most importance in OP metabolism in the desalkylation of dimethyl phosphates and phosphorothioates. There is also evidence for the existence of a GSH-*S*-aryltransferase which removes the acidic or "leaving" group of the OP (Shishido *et al.*, 1972). Although many papers indicate that the *in vivo* formation of metabolites resulting from these reactions is often enhanced in R-insects, it has been assumed that they were formed by either oxidative or hydrolytic reactions. Consequently, the role of the GSH-*S*-transferases in OP resistance is generally difficult to evaluate.

Even when it can be shown that the glutathione-*S*-transferase activity in *in vitro* preparations of R-insects is higher than in S-strains, it remains difficult to estimate to what extent this contributes to the observed resistance in the presence of other resistance factors. Comparisons among several resistant strains show that, *in vitro*, glutathione-*S*-transferase activity does not always parallel resistance (Oppenoorth *et al.*, 1972), one strain with a very high transferase activity having only four times increased resistance to parathion. There are no suitable synergists available that can specifically block the transferase in insects and the only means to accurately assess its contribution is genetic isolation of the gene involved; this method has not been employed to date.

In the housefly, the gene regulating glutathione-*S*-transferase has been localized on the second chromosome (Lewis, 1969; Lewis and Sawicki, 1971; Oppenoorth *et al.*, 1972). Although Lewis claimed that the transferase was due to the presence of gene *a* for low carboxylesterase activity, it was later realized that there must be a separate factor, close to gene *a* (Lewis and Sawicki, 1971; Oppenoorth *et al.*, 1972). Oppenoorth *et al.* have also found the transferase in a strain without gene *a*.

Yang *et al.* (1971) found that in R-houseflies (strain Rutger-R) the GSH-dependent cleavage of diazinon and diazoxon to the corresponding diethylphosphorothioic (DEPTA) and phosphoric (DEPA) acids was five and 15 times greater, respectively, than in an S-strain. However, these authors did not conclusively establish that the reaction was indeed catalyzed by a GSH-*S*-aryltransferase, and no desethylated products were observed. Nolan and O'Brien (1970), on the other hand, showed that these R-flies could desethylate paraoxon after injection. Lewis and Sawicki (1971) have shown that in the multiresistant housefly strain SKA the GSH-*S*-transferase activity was considerably increased and desethylated diazinon, parathion, and diazoxon. The malathion-resistant strain G studied by Oppenoorth *et al.* (1972) was found to metabolize parathion to both desethyl parathion and DEPTA. The latter product also seems to be formed by the SKA strain (Devonshire, 1973).

That GSH-*S*-transferase also seems to play a role in resistance to azinphosmethyl and its oxon analogue was shown by Motoyama and Dauterman (1972) for houseflies and by Motoyama *et al.* (1971) for the predaceous mite *Neoseiulus (T.) fallacis*. In houseflies there was enzyme activity toward both methyl iodide (*S*-alkyltransferase) and 3,4-dichloronitrobenzene (*S*-aryltransferase; Johnson, 1966), but the insecticidal compounds were only desmethylated and not desarylated. In *Neoseiulus*, with a high level of resistance, the glutathione-dependent desmethylation reaction seems to be the only resistance mechanism.

6.2.4. Mixed-Function Oxidases

a. Contributions of Oxidases to Resistance. As discussed in Chapter 2, mixed-function oxidases (mfo) are now known to be of great importance in the metabolism of almost all groups of insecticides in both mammals and in insects.

The first real evidence for their role in insect resistance to insecticides was obtained *in vivo* from the action of various synergists known to inhibit these enzymes (see Wilkinson, 1968, and Casida, 1970, for reviews). That certain synergists, particularly the methylenedioxy compounds such as piperonyl butoxide and sesamex, acted specifically on the enzymes involved in oxidation was first suggested by *in vitro* work on mammalian liver preparations, where they were shown to inhibit the oxidation of carbamates (Hodgson and Casida, 1960). A similar action was indicated in insects since oxidation reactions such as the epoxidation of aldrin to dieldrin and the activation of phosphorothionates to phosphates were also inhibited by these compounds (Sun and Johnson, 1960). Subsequently, it was found that the tenfold resistance to carbaryl in a DDT-R strain of houseflies could be overcome if the carbaryl was applied in combination with sesamex (Eldefrawi *et al.*, 1960). Piperonyl butoxide was found to decrease the metabolism of 3-isopropylphenyl-*N*-methylcarbamate in a resistant strain of houseflies and at the same time reduced the resistance of this strain (Georghiou and Metcalf, 1961). Studies with these synergists have now become an established and widely used technique for indicating the *in vivo* importance of oxidative degradation, although as previously discussed it is possible that they may not be equally dependable as indicators of oxidation in all insect species.

With the advent of *in vitro* methods for the study of mfo activity in insects, the role of this enzyme system in resistance has become more evident. Most of the available information to date is derived from studies in houseflies, but there is now increasing information available on other insects. Increased naphthalene oxidation was shown to be associated with resistance to this compound in houseflies (Schonbrod *et al.*, 1965), and Tsukamoto and Casida (1967*a,b*) demonstrated an increased *in vitro* oxidation of methylcarbamates in R-strains. Subsequently, a large number of other insecticides have been studied, mostly in the housefly, and it is now quite clear that insect resistance to some chlorinated hydrocarbons, many OPs and carbamates, and some pyrethroids (to mention only the most important groups) can be due at least partly to increased oxidase activity.

An important problem is how many resistance factors are involved in these various types of resistance and whether there could be specific factors for certain types of oxidative reactions. Casida (1969) found that selection with a carbamate increased the rate of many types of oxidations, including epoxidation, hydroxylation, and *O*- and *N*-demethylation. The selection may therefore theoretically cause cross-resistance to a large number of unrelated compounds. Further analysis is required to establish whether one or more factors are involved.

Another method which has been used in an attempt to establish which compounds are degraded by the same enzyme system has been to study the inhibition of a certain type of oxidase reaction in the presence of other potential substrates (Khan *et al.*, 1970). This type of inhibition is referred to as "alternative substrate" inhibition and depends on the fact that if two compounds are indeed metabolized by the same enzyme they should each inhibit the metabolism of the other. Inhibition of aldrin epoxidation was obtained in the presence of *p*-nitroanisole, hexobarbital, aniline, parathion, and methoxychlor, and was interpreted as evidence for a common conversion mechanism for all these compounds. This may be true but should be accepted with caution since some compounds could inhibit one oxidase system and be converted by another.

Genetic analysis in the housefly has provided evidence that microsomal oxidation is controlled by genes on chromosomes 2 and 5 (Oppenoorth, 1967; Tsukamoto *et al.*, 1968; Schonbrod *et al.*, 1968; Khan *et al.*, 1973). Georghiou (1971) has called attention to the fact that the total oxidation capacity (as measured by the model epoxidation system, aldrin to dieldrin) in an R-strain was larger than the activity in any substrain carrying only one chromosome derived from it. This points to an interaction of genetic factors not only on the level of resistance but also on the level of degradation capacity. Khan *et al.* (1973) found that the effects of the two chromosomes are simply additive when combined in one strain. The fact that so far only two of the five chromosomes have been found to carry important factors for the control of oxidation may indicate that the number of genes involved is small. Only little information is available on the location of these genes, and nothing is known concerning the number of alleles in which each of the genes may occur. The gene on chromosome 5 showed 46% crossover with the marker gene *ocra* (Oppenoorth, 1967). This gene is responsible for the oxidation of DDT and diazoxon, but contributes little if anything to resistance to carbaryl (Plapp, 1970), epoxidation of aldrin, and resistance to propoxur (Plapp and Casida, 1969; Schonbrod *et al.*, 1973). A factor for low resistance to pyrethroids has also been reported to be associated with this chromosome (Farnham, 1973). One or more genes on chromosome 2 seem to be more commonly involved in increased mfo activity, with the notable exception of that to DDT. Thus increased oxidation of carbamates (Shrivastava *et al.*, 1969; Plapp and Casida, 1969), aldrin (Khan, 1969; Georghiou, 1971), naphthalene (Schafer and Terriere, 1970), organophosphates (Plapp and Casida, 1969; Oppenoorth, 1972), and pyrethrins (Plapp and Casida, 1969) can all be due to factors on this chromosome. There is evidence for more than one gene. A factor for dithion and diazinon resistance showed 17% crossover with the marker gene *car* (Oppenoorth, 1967). Khan (1969) reported 39% crossover between the aldrin epoxidation factor and the gene *Deh* for high DDTase activity, and a crossover of 34% with the marker gene *stw.* We found no crossover between a gene causing oxidation of malaoxon (described by Welling *et al.*, 1974) and the marker gene *cm.* Since *car, Deh,* and *cm* are clustered together, the three genes seem to be localized far apart.

b. DDT. The role of oxidases in DDT resistance first became apparent when a DDT-R housefly (strain Fc) was found to have relatively low DDTase activity and a DDT resistance which could be almost completely abolished by cotreatment with sesamex (Oppenoorth, 1965*a*, 1967). Similar observations were made in the related SKA strain (Fig. 3), where sesamex but not the DDTase inhibitor (Warf antiresistant) had a strong synergistic effect (Sawicki and Farnham, 1967). A mixed-function oxidase was active in microsomes of this strain, which were prepared and incubated by the method of Tsukamoto and Casida (1967*a*). In the presence of NADPH, DDT was rapidly metabolized to a mixture of at least five products, none of which was kelthane or benzophenone (Khan and Terriere, 1968; Oppenoorth and Houx, 1968).

The rate of oxidative degradation of DDT in this strain was only 0.5 μg DDT per fly per hour compared with DDTase levels of 180 μg per fly per hour in acetone powders of other R-strains. However, in view of difficulties in *in vitro* measurements of oxidase activity in insects this may not truly reflect the *in vivo* activity of this system. Oxidase activities in preparations of an S-strain and in a DDT-R strain with high DDTase activity were negligible and differed only slightly in the presence or absence of NADPH.

The DDT-R gene on chromosome 5, which was called *ses* by Sawicki and Farnham (1967) and *DDT-md* by Oppenoorth and Houx (1968), is probably identical to that controlling the oxidation of diazoxon.

By use of another technique, DDT oxidation has been demonstrated in two S-strains, Fc, and P2/sel, a strain selected with pyrethrum and cross-resistant to DDT (Gil *et al.*, 1968). Oxidase activities in these strains, as calculated from the data in this paper, are 43 and 50 (S-strains), 56 (Fc), and 79 (P2/sel) ng/fly/hr. The activity of the Fc strain is therefore much lower than reported by Oppenoorth and Houx, and the activity measured in each of the S-strains is only slightly lower than that in the R-strains. Furthermore, one of the metabolites found by Gil *et al.* cochromatographed with kelthane. It is possible that some of the quantitative differences in activity can be explained by differences in the techniques employed, but since Fc oxidizes DDE rapidly

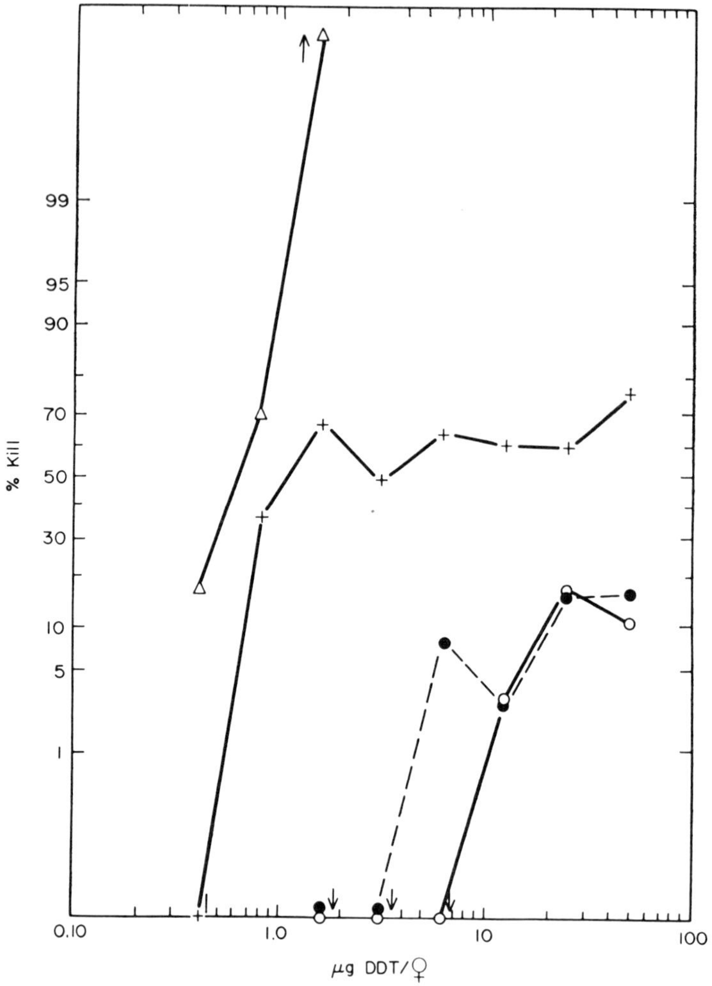

Fig. 3. Effect of pretreatment with FDMC and/or sesamex on response to DDT in housefly strain SKA. All individuals have an oxidation mechanism, 30% of them also DDTase. ○, DDT alone; ●, DDT plus FDMC; +, DDT plus sesamex; △, DDT plus FDMC plus sesamex. From Sawicki (1973*b*).

(Oppenoorth, 1967) and P2/sel does not there must be a difference in the oxidative mechanism of the two strains.

The degree of resistance caused by gene *DDT-md* seems to be dependent on the presence of other genes. According to Sawicki and Farnham (1968), it is tenfold for DDT and twentyfold for methoxychlor applied topically but becomes much larger (eight times larger in the case of DDT) when combined with the factor *Pen*, which reduces penetration (Sawicki, 1973*a*). It is also dependent on the method of application employed and is larger when a contact method is used (Schonbrod *et al.*, 1973).

It was found by Gil *et al.* and later discussed by Agosin (1971) that DDT hydroxylase activity is inducible. Thus a 25% increase in activity was found in Fc and 60% in P2/sel 3 hr after topical application of flies with DDT. This raises the interesting question of whether the induced hydroxylation capacity might contribute to the ability of the flies to survive DDT treatment. It is not warranted to state, as Agosin (1971) did, that this is "against the classical view of preadaptation." Even if the R-genes were effective only on induction (which is certainly not so in the strains studied, since they show *in vitro* activity without having been treated with DDT), their presence would still be a case of preadaptation. Only if the induced state were inheritable would this be different from the preadaptation concept.

In this connection Perry and Agosin (1974) have stated:

> The origin and development of resistance is biphasic: phase I is due to selection of variants in the population which carry preadaptive genes for resistance, ultimately attaining a resistance level commensurate with the gene pool originally present; in phase II, enhancement of resistance takes place by induction of preexisting detoxifying enzymes toward higher activity, thus resulting in a faster breakdown of the chemical. The inducer is the insecticide itself which used to control the insect.

It should be emphasized that there is no evidence that this postulated phase II plays any significant role in resistance. High *in vitro* detoxication can easily be demonstrated in R-strains that have not been treated with insecticides for many generations, e.g., DDTase activity or mixed-function oxidation of DDT. Although these activities can be further increased experimentally by induction, the dose levels of the insecticides required and the time course of the induction process are such that a role in resistance seems unlikely (Terriere and Yu, 1974), and has never been demonstrated.

c. *Organophosphates.* As has been discussed, resistance to organophosphates can be due to altered cholinesterase, hydrolytic breakdown, and detoxication by glutathione-*S*-transferase; it can also be caused by enhanced oxidative degradation.

Increased levels of oxidation in several R-housefly strains have been reported; e.g., six- to sixteenfold for paraoxon and diazoxon (ElBashir and Oppenoorth, 1969), four- and sixfold for diazinon and diazoxon (Yang *et al.*, 1971), and about ninefold for azinphosmethyl (Motoyama and Dauterman, 1972). These factors depend to a large extent on the strains and techniques employed and the rates of degradation differ markedly with the insecticide under investigation. For paraoxon and diazoxon, respectively, *in vitro* degradation rates of about 0.2 and 4 μg per abdomen per hour have been reported.

Since OP insecticides may be activated (phosphorothioates) or degraded by the mixed-function oxidases, a very complex situation exists. A reduced rate of activation is one rather obvious theoretical mechanism for phosphorothioate resistance. It appeared, however, that in strains with an increased oxidative detoxication mechanism both activation and detoxication are usually considerably increased (ElBashir and Oppenoorth, 1969). The rate of activation of parathion in two R-strains was 16 and 6 times greater than that of the S-strains, and for diazinon it was 7 and 6 times greater. In Malathion-R strains, the rate of activation was two- to threefold greater than in the S-strain (Welling, unpublished). The greater activation rate could explain why after injection of parathion the R-strain was knocked down much faster than the S-strain. That the reverse is true after topical application, when the insecticide has first to penetrate the insect cuticle, indicates that the increased capacity for activation has much less influence than that for detoxication.

To make the whole situation even more complex, it appears that small amounts of the thiono compounds strongly inhibit the oxidative degradation of the oxon analogues in the same strains that were resistant because of increased oxidative detoxication (Oppenoorth, 1971*a*; Oppenoorth *et al.*, 1971). This inhibition does not seem to be of the competitive "alternative substrate" type since it is found even after the inhibitor has been completely degraded in the incubation mixture (Oppenoorth, unpublished). This leads to such interesting questions as why parathion does not act as its own synergist and how paraoxon degradation can escape being blocked by parathion *in vivo*. When larger doses of a nontoxic P=S compound, SV_1, were applied in combination with either parathion or paraoxon, a strong synergism was indeed found. One has to imagine, therefore, that parathion under the usual conditions does not reach the sites that are converting the paraoxon derived from it.

In the strains of houseflies discussed so far, resistance to thiono compounds differs only slightly from that to the oxon analogues in spite of the inhibitory effect just discussed. There are, however, several cases where this is not so, as in the case of the housefly strain E_1, which is 66-fold resistant to malaoxon, but only fourfold resistant to malathion (Welling *et al.*, 1974). It was shown that the oxidative degradation of malaoxon in *in vitro* preparations of this strain was inhibited by very small amounts of malathion. Similar observations indicating the inhibition of oxidation by thiono compounds have been made by others (Morello *et al.*, 1968; Yang *et al.*, 1971).

In the flour beetle, *Tribolium castaneum*, a type of resistance to malathion was found that was also characterized by a much higher resistance to malaoxon than to malathion and by a very broad cross-resistance, particularly to oxon analogues of dimethoxy- and diethoxyphosphorothionates (Dyte *et al.*, 1970). It appeared that the R-strain metabolized malathion at about the same rate as did S-beetles but that only in the R-strain was malaoxon converted to desmethyl malaoxon. Neither the carboxylesterase inhibitor TPP nor piperonyl butoxide synergizes malathion in this strain. However, the latter compound was found to be oxidized very rapidly (Dyte, personal communication). This suggests that resistance might be due to increased oxidation.

d. Carbamates. There is now ample evidence that carbamate resistance is mainly due to oxidative degradation and that mechanisms involving esterase action are of little importance. The evidence for the role of oxidation in propoxur resistance in houseflies and in the mosquito *Culex pipiens fatigans* will serve as a useful example. An extensive genetic and biochemical study on propoxur resistance has been made in houseflies (Tsukamoto *et al.*, 1968; Shrivastava *et al.*, 1969). The influence of different chromosomes was studied on both *in vivo* and *in vitro* oxidative metabolism and on resistance levels in the presence and absence of piperonyl butoxide. Resistance is dependent on factors on chromosomes 2, 5, and 3 in decreasing order of importance. Chromosomes 2 and 5 are responsible for *in vivo* and *in vitro* enzymatic activity. The effect of the factor on chromosome 2 (one or more genes may be involved) is by far the largest and influences the *in vivo* and *in vitro* formation of three major oxidative metabolites 5-hydroxy propoxur, 2-hydroxyphenyl methylcarbamate, and *N*-hydroxymethyl propoxur. *In vivo* several other products are formed, probably all by secondary reactions. Piperonyl butoxide inhibits the oxidative degradation of propoxur and reduces but does not completely eliminate the resistance to this insecticide in the R-strain. Penetration differences were not found and there was no evidence of any hydrolytic activity. In this particular type of resistance, there was little interaction between the effects contributed by the different chromosomes, which in some cases such as carbaryl resistance in the housefly may be significant (Fig. 4). The nature of the resistance due to chromosomes 5 and 3 is not known, although in another strain chromosome 5 has an influence on oxidative degradation (Plapp and Casida, 1969).

In *Culex pipiens fatigans* larvae, the mfo system also constitutes an important factor in propoxur resistance (Shrivastava *et al.*, 1970). Major *in vivo* and *in vitro* metabolites were the same as those formed in the housefly, and

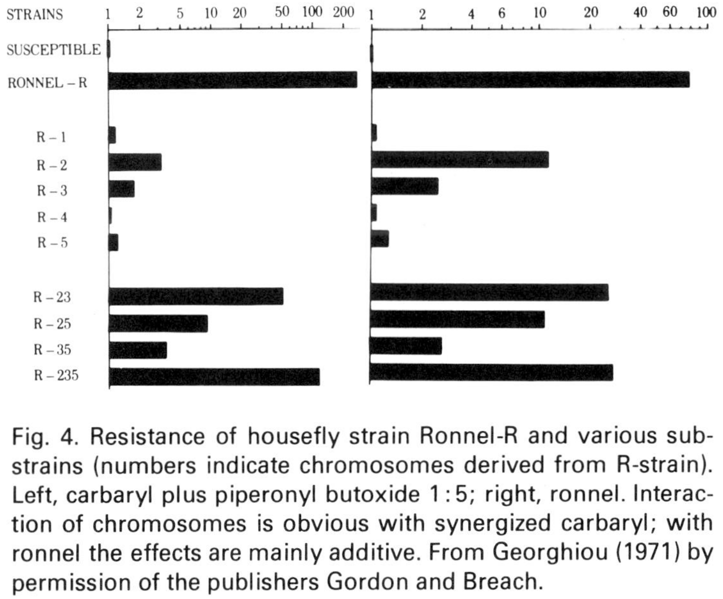

Fig. 4. Resistance of housefly strain Ronnel-R and various substrains (numbers indicate chromosomes derived from R-strain). Left, carbaryl plus piperonyl butoxide 1 : 5; right, ronnel. Interaction of chromosomes is obvious with synergized carbaryl; with ronnel the effects are mainly additive. From Georghiou (1971) by permission of the publishers Gordon and Breach.

a fourth was identified as *N*-demethyl propoxur. *In vitro* the difference in oxidative capacity between the two strains was much clearer than *in vivo.* The effects of the synergist were much smaller than in houseflies and a reduced penetration in the R-strain probably enhances the effects resulting from increased metabolism.

e. Pyrethrins. There is little doubt that in pyrethrin resistance in the housefly a factor associated with increased mfo activity is of importance in addition to the *kdr*-type mechanism. This was shown in houseflies by Yamamoto and Casida (1966), who found a complex mixture of metabolic products of which the major one was that formed by oxidation at one of the methyl groups in the acid moiety. Plapp and Casida (1969) found that allethrin oxidase was controlled by a factor on chromosome 2. Different resistance genes were isolated from a Pyrethrin-R strain by Farnham (1973), and their relative importance was studied in the presence and absence of sesamex. A factor *py-ses* on chromosome 5 caused slight resistance to natural pyrethrins, but not to some synthetic pyrethroids. There is evidence that this factor is concerned with the alcohol moiety of the pyrethrin, a fact which combined with the metabolic data of Yamamoto and Casida suggests that there is more than one factor for mfo degradation.

f. Cross-Resistance to Hormones. The apparent lack of substrate specificity of the mfo system enables it to metabolize insecticides of many different groups and causes a broad spectrum of cross-resistance. This already has been shown to extend to juvenile hormones and hormone-mimics and represents a real threat to the future effectiveness of this promising group of materials. Dyte (1972) compared the susceptibility of S- and R-strains of *Tribolium castaneum* to a synthetic *cis/trans* mixture of methyl-10,11-epoxy-7-ethyl-3,11-dimethyl-2,6-tridecadienoate, one of the isomers of which is a natural juvenile hormone of *Hyalophora cecropia.* The R-strain was resistant to DDT, lindane, bromodan, all four carbamates, and 22 organophosphates tested, and showed an approximately threefold resistance to the hormone.

Cross-resistance in certain R-strains of houseflies has also been found to extend to several juvenile hormone analogues (Cerf and Georghiou, 1972; Plapp and Vinson, 1973; Terriere and Yu, 1973; Vinson and Plapp, 1974). Cerf and Georghiou applied the analogue ZR-515 (isopropyl-11-methoxy-3,7,11-trimethyldodeca-2,4-dienoate) to pupae of one S-strain and several R-strains. A DDT-Lindane-R strain was only 2.7 times more resistant, whereas strains selected with fenthion, chlorthion, or parathion were about 4.5 times more resistant. Strains selected with *m*-isopropylphenyl methylcarbamate, *O*-ethyl-*O*-(2,4-dichlorophenyl)phosphoramidothioate, and dimethoate showed the highest levels of cross-resistance (40 times for the latter at LD_{50}, 150 times at LD_{95}). The factor(s) responsible for the cross-resistance have not yet been elucidated, but mixed-function oxidases are among the most likely candidates.

Vinson and Plapp exposed third instar housefly larvae to films of juvenile hormone analogue Ent. 70460 and observed that several strains showed some cross-resistance. The strain with the highest cross-resistance, the Baygon-R strain, had 31 times increased resistance.

g. Nature of the Difference. A characteristic capacity of the mfo system is that it can be induced with certain compounds to higher activity by activation of the genes synthesizing the system. Resistance might theoretically represent a situation in which the genes were already active without an inducer, and thus would not be further inducible. Terriere and Yu (1974) showed, however, that induction is possible in both S- and R-strains of houseflies.

As discussed in Section 4.1, resistance can theoretically be due to the production of an altered enzyme or to an increased amount of an enzyme. In the case of the mfo system, there are indications that both situations may occur.

In the cabbage looper, *Trichoplusia ni*, a parallel increase in metabolism of carbaryl and amount of cytochrome P450 was found in fat bodies and guts of three strains (Kuhr, 1971) (Fig. 5). The increase in this cytochrome may be the only difference, but it is also possible that all components of the mfo system are produced in greater quantities. In houseflies, the amount of cytochrome P450 is generally only twofold greater in R-strains than in S-strains, a difference that is much smaller than that in the oxidation capacity (Perry *et al.*, 1971; Tate *et al.*, 1973).

The preceding observation points to qualitative differences besides this quantitative one. The nature of these is not yet clear, but there is some evidence that a difference is present between the cytochromes P450 of S- and R-strains (Perry *et al.*, 1971; Tate *et al.*, 1973, 1974; Hodgson *et al.*, 1974) (Chapter 3).

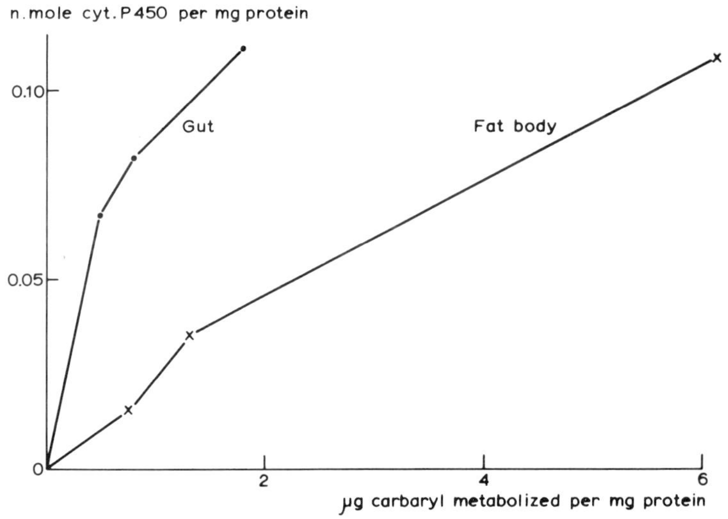

Fig. 5. Relation between oxidative metabolism of carbaryl and amount of cytochrome P450 in gut and fat body of three strains of cabbage looper. Data from Kuhr (1971).

6.3. Reduced Penetration

Reduced penetration has been reported as a possible resistance mechanism for a number of species against several insecticides (see Plapp and Hoyer, 1968*b*, for references). In one case which has been studied in some detail, the effects of reduced penetration of many insecticides were found to be caused by a gene on the third chromosome of the housefly. By employing genetic analysis to combine the gene with other factors, much insight into its effect has been obtained (Sawicki and Farnham, 1968; Hoyer and Plapp, 1968; Plapp and Hoyer, 1968*b*; Sawicki, 1970; Georghiou, 1971). The expression of the gene is greatly dependent on the way in which the insecticide is administered and contact methods usually give much larger differences than topical application. The reduced rate of penetration caused by this gene can sometimes cause resistance on its own, as is the case with the organotin compound tributyltin chloride. Because of this, the gene has been called *organotin-R* by Plapp and Hoyer (1968*b*), but because of its more general importance the name *pen* given by Sawicki and Farnham (1968) seems more appropriate. Sometimes the gene only delays knockdown, without having any effect on the final mortality, but when it is combined with other resistance factors it often acts to enhance resistance. In particular, it can enhance the level of resistance due to a moderate rate of detoxication, as is the case with the hydrolytic and glutathione-dependent degradation in Diazinon-R strain SKA (enhanced up to five to ten times), or the oxidative DDT degradation in strains Fc and SKA. It seems likely that its action is always dependent on the presence of a slow rate of detoxication and that in those cases where it confers resistance on its own some level of detoxication was already present in the S-strains.

The effectiveness of the *pen* gene is also dependent on the nature of the insecticide. Only a small effect is apparent on topical application of dieldrin, a larger one with DDT (Plapp and Hoyer, 1968*b*), and a much larger one with carbaryl (Georghiou, 1971).

It is rather surprising that the nature of the difference caused by the *pen* gene has not been further investigated. Since the effect is caused by a single gene, a simple difference between R- and S-strains is indicated, and elucidation of the nature of the difference could provide insight into the rather neglected area of the factors controlling insecticide penetration. Two studies on the nature of the reduced penetration factor have been made, but unfortunately both of these have been in strains where the genetic background was lacking. One was a comparison of an R- and an S-strain in *Heliothis virescens*, where the R-strain had a slow rate of DDT penetration. The cuticle contained more protein and lipid than the S-strain and showed a greater degree of sclerotization (Vinson and Law, 1971). The other study was made with a strain of houseflies showing an eighteenfold resistance to malathion and an approximately threefold reduction in the rate of penetration of this insecticide. The difference in penetration disappeared following treatment of the flies with activated silicic acids, although no quantitative or qualitative differences in hydrocarbons were found (Benezet and Forgash, 1972).

7. Concluding Remarks

It should be stressed that although we have discussed several mechanisms by which insect resistance to insecticides can occur there may be several others which are not yet recognized. The longer an insect is treated with insecticides the more complex its adaptation is likely to become. An example of this, particularly with respect to the effect of consecutive use of different compounds, has been provided by Georghiou and Hawley (1971), who observed that the resistance spectrum expanded progressively and that resistance levels rose gradually due to the accumulation of several different factors. Also, from the work of Sawicki (1974) and Farnham (1973) it is clear that high resistance levels to "hardy" insecticides develop through a combination of factors and that the mechanism by which some of these contribute to the overall resistance is still unknown.

A satisfactory answer to individual R-factors can usually be found, either by shifting to alternative insecticides or by the use of synergists to block the specific R-mechanism involved. A further possibility of coping with individual R-mechanisms consists of using new compounds that are toxic only to the R-insects. This principle of "negatively correlated cross-resistance" (see Brown, 1958, for a review) could theoretically lead to a back-selection to the S-state. However, the versatility and multiplicity of the insecticide resistance mechanisms constitute the major problem, and alternative resistance mechanisms can develop which invalidate each of the solutions outlined above. With four known mechanisms for DDT resistance and four or more for organophosphates, we may not yet have reached the limit, and as long as selection continues we can expect a continuous development of additional or more efficient mechanisms. In this respect, development of insecticide resistance is similar to the adaptation of insects to the defense mechanisms of plants (Dyte, 1969).

It is not possible to predict whether resistance will develop in any particular situation, since this will depend on the insecticide selection pressure applied, as well as on unknown characteristics of the insect population. Whether rare variants giving rise to resistance are already present in populations or whether they will develop over a period of years cannot be predicted. Many insects have already developed resistance to one or more insecticides, and yet insecticides still constitute an important means of protecting our crops and ourselves from insect attack. It is hoped that the development of new insecticides, based on our knowledge of their toxicology, will enable us to keep up with loss of effective compounds through insect resistance and other problems. Other methods of control that release selection pressure must obviously be exploited wherever practicable.

8. References

Agosin, M., 1971, Microsomal mixed function oxidases and insecticide resistance, *in: Proceedings of the 2nd International IUPAC Congress of Pesticide Chemistry*, Vol. II (A. S. Tahori, ed.), pp. 29–59, Gordon and Breach, New York.

Anonymous, 1957, *World Health Organisation Expert Committee on Insecticides, 7th Report*, WHO Technical Report Series No. 125.

Ballantyne, G. H., and Harrison, R. A., 1967, Genetic and biochemical comparisons of organophosphate resistance between strains of spider mites (*Tetranychus* species: Acari), *Entomol. Exp. Appl.* **10:**231.

Barton Browne, L., and Kerr, R. W., 1967, The response of the labellar taste receptor of DDT-resistant and non-resistant houseflies (*Musca domestica*) to DDT, *Entomol. Exp. Appl.* **10:**337.

Beard, R. L., 1965, Ovarian suppression by DDT and resistance in the housefly (*Musca domestica* L.), *Entomol. Exp. Appl.* **8:**193.

Benezet, H. J., and Forgash, A. J., 1972, Reduction of malathion penetration in houseflies pretreated with silicic acid, *J. Econ. Entomol.* **65:**895.

Beranek, A. P., 1974, Esterase variation and organophosphate resistance in populations of *Aphis fabae* and *Myzus persicae, Entomol. Exp. Appl.* **17:**129.

Berger, R. S., and Young, R. G., 1962, Specificity of diarylhaloethane dehydrohalogenase of susceptible and DDT-resistant houseflies, *J. Econ. Entomol.* **55:**533.

Bigley, W. S., and Plapp, F. W., 1962, Metabolism of malathion and malaoxon by the mosquito *Culex tarsalis, J. Insect Physiol.* **8:**545.

Brooks, G. T., 1960, Mechanisms of resistance of the adult housefly (*Musca domestica*) to "cyclodiene" insecticides, *Nature (London)* **186:**96.

Brown, A. W. A., 1956, DDT-dehydrochlorinase activity in resistant houseflies and mosquitoes, *Bull. WHO* **14:**807.

Brown, A. W. A., 1958, *Insecticide Resistance in Arthropods*, WHO Monograph Series 38, Geneva.

Brown, A. W. A., and Pal, R., 1971, *Insecticide Resistance in Arthropods*, WHO Monograph Series 38, Geneva.

Brown, H. D., and Rogers, E. F., 1950, The insecticidal activity of 1,1-dianisyl neopentane, *J. Am. Chem. Soc.* **72:**1864.

Busvine, J. R., 1951, Mechanism of resistance to insecticide in houseflies, *Nature (London)* **168:**193.

Busvine, J. R., 1954, Houseflies resistant to a group of chlorinated hydrocarbon insecticides, *Nature (London)* **174:**783.

Busvine, J. R., Bell, J. D., and Guneidy, A. M., 1963, Toxicology and genetics of two types of insecticide resistance in *Chrysomyia putoria* (Wied.), *Bull. Entomol. Res.* **54:**589.

Casida, J. E., 1969, Insect microsomes and insecticide chemical oxidations, *in: Microsomes and Drug Oxidations* (J. R. Gillette *et al.*, eds.), pp. 517–531, Academic Press, New York.

Casida, J. E., 1970, Mixed-function oxidase involvement in the biochemistry of insecticide synergists, *J. Agr. Food Chem.* **18:**753.

Cerf, D. C., and Georghiou, G. P., 1972, Evidence of cross-resistance to a juvenile hormone analogue in some insecticide-resistant houseflies, *Nature (London)* **239:**401.

Chadwick, L. E., 1955, Physiological aspects of insect resistance to insecticides, *in: Origins of Resistance to Toxic Agents* (M. G. Sevag and R. D. Reinolds, eds.), pp. 133–147, Academic Press, New York.

Chattoraj, A. N., and Brown, A. W. A., 1960, Internal DDE production by normal and DDT-resistant larvae of *Aedes aegypti, J. Econ. Entomol.* **53:**1049.

Cohen, S., and Tahori, A. S., 1957, Mode of action of di(*p*-chlorophenyl)trifluoromethylcarbino as a synergist to DDT against DDT-resistant houseflies, *J. Agr. Food Chem.* **5:**519.

Collins, W. J., and Forgash, A. J., 1970, Mechanisms of insecticide resistance in *Musca domestica* Carboxylesterase and degradative enzymes, *J. Econ. Entomol.* **63:**394.

Crow, J. F., 1957, Genetics of insect resistance to chemicals, *Ann. Rev. Entomol.* **2:**227.

Darrow, D. I., and Plapp, F. W., 1960, Studies on resistance to malathion in the mosquito *Culex tarsalis, J. Econ. Entomol.* **53:**777.

Dauterman, W. C., and Matsumura, F., 1962, Effect of malathion analogs upon resistant and susceptible *Culex tarsalis* mosquitoes, *Science* **138:**694.

Devonshire, A. L., 1973, The biological mechanisms of resistance to insecticides with especial reference to the housefly, *Musca domestica* and aphid, *Myzus persicae, Pestic. Sci.* **4:**521.

Devonshire, A. L., and Sawicki, R. M., 1974, The importance of the decreased susceptibility of acetylcholinesterase in the resistance of house flies to organophosphorus insecticides, *in: Third International Congress of Pesticide Chemistry,* Helsinki, Abst. No. 177.

Dinamarca, M. L., Saavedra, I., and Valdés, E., 1969, DDT-Dehydrochlorinase. I. Purification and characterization, *Comp. Biochem. Physiol.* **31:**269.

Dinamarca, M. L., Levenbook, L., and Valdés, E., 1971, DDT-dehydrochlorinase. II. Subunits, Sulfhydryl groups, and chemical composition, *Arch. Biochem. Biophys.* **147**:374.

Dyte, C. E., 1969. Evolutionary aspects of insecticide selectivity, *in: Proceedings of the 5th British Insecticide and Fungicide Conference,* 393.

Dyte, C. E., 1972, Resistance to synthetic juvenile hormone in a strain of the flour beetle *Tribolium castaneum, Nature (London)* **238:**48.

Dyte, C. E., and Rowlands, D. G., 1968, The metabolism and synergism of malathion in resistant and susceptible strains of *Tribolium castaneum, J. Stored Prod. Res.* **4:**157.

Dyte, C. E., Rowlands, D. G., Daly, J. A., and Blackman, D. G., 1970, *in: Pest Infestation Research 1969,* pp. 41–42, Agriculture Research Council, Report of the Pest Infestation Laboratory, Slough, England,

ElBashir, S., and Oppenoorth, F. J., 1969, Microsomal oxidations of organophosphate insecticides in some resistant strains of houseflies, *Nature (London)* **223:**210.

Eldefrawi, M. E., Miskus, R., and Sutcher, V., 1960, Methylenedioxyphenyl derivatives as synergists for carbamate insecticides on susceptible, DDT-, and parathion-resistant houseflies, *J. Econ. Entomol.* **53:**231.

Farnham, A. W., 1971, Changes in cross-resistance patterns of houseflies selected with natural pyrethrins or resmethrin (5-benzyl-3-furylmethyl (±)-*cis-trans*-chrysanthemate), *Pestic. Sci.* **2:**138.

Farnham, A. W., 1972, *Genetics of resistance to Pyrethroids in Houseflies,* Report Rothamsted Experimental Station, p. 179.

Farnham, A. W., 1973, Genetics of resistance of pyrethroid-selected houseflies, *Musca domestica* L., *Pestic. Sci.* **4:**513.

Feroz, M., 1971, Biochemistry of malathion resistance in a strain of *Cimex lectularius* resistant to organophosphorus compounds, *Bull. WHO* **45:**795.

Franco, M. G., and Oppenoorth, F. J., 1962, Genetical experiments on the gene for low aliesterase activity and organophosphate resistance in *Musca domestica* L., *Entomol. Exp. Appl.* **5:**119.

Fukami, J., and Shishido, T., 1963, Studies on the selective toxicities of organic phosphorus insecticides. III. The characters of enzyme system in cleavage of methyl parathion to desmethyl parathion in the supernatant of several species of homogenates, *Botyu-Kagaku* **28:**77.

Fukami, J., and Shishido, T., 1966, Nature of a soluble, glutathione-dependent enzyme system active in cleavage of methyl parathion to desmethyl parathion, *J. Econ. Entomol.* **59:**1338.

Georghiou, G. P., 1971, Isolation, characterization and resynthesis of insecticide resistance factors in the housefly, *Musca domestica, in: Proceedings of the 2nd International IUPAC Congress of Pesticide Chemistry,* Vol. II (A. S. Tahori, ed.), pp. 77–94, Gordon and Breach, New York.

Georghiou, G. P., and Hawley, M. K., 1971, Insecticide resistance resulting from sequential selection of houseflies in the field by organophosphorus compounds, *Bull. WHO* **45:**43.

Georghiou, G. P., and Metcalf, R. L., 1961, The absorption and metabolism of 3-isopropylphenyl *N*-methyl carbamate by susceptible and carbamate-selected strains of houseflies, *J. Econ. Entomol.* **54:**231.

Gil, L., Fine, B. C., Dinamarca, M. L., Balazs, I., Busvine, J. R., and Agosin, M., 1968, Biochemical studies on insecticide resistance in *Musca domestica, Entomol. Exp. Appl.* **11:**15.

Goodchild, B., and Smith, J. N., 1970, The separation of multiple forms of housefly 1,1,1-trichloro-2,2-bis-(*p*-chlorophenyl)ethane (DDT) dehydrochlorinase from glutathione *S*-aryltransferase by electrofocusing and electrophoresis, *Biochem. J.* **117:**1005.

Grigolo, A., and Oppenoorth, F. J., 1966, The importance of DDT-dehydrochlorinase for the effect of the resistance gene *kdr* in the housefly *Musca domestica* L., *Genetica* **37:**159.

Gysin, H., 1971, *Bull. WHO* **44:**374.

Hama, H., and Iwata, T., 1971, Insensitive cholinesterase in the Nakagawara strain of the green rice leafhopper, *Nephotettix cincticeps* Uhler (Hemiptera: Cicadellidae), as a cause of resistance to carbamate insecticides, *Appl. Entomol. Zool.* **6:**183.

Hodgson, E., and Casida, J. E., 1960, Biological oxidation of *N,N*-dialkyl carbamates, *Biochim. Biophys. Acta* **42:**184.

Hodgson, E., Tate, L. G., Kulkarni, A. P., and Plapp, F. W., 1974, Microsomal cytochrome P450: Characterization and possible role in insecticide resistance in *Musca domestica, J. Agr. Food Chem.* 22:360.

Holan, G., 1971, Rational design of insecticides, *Bull. WHO* **44:**355.

Hoyer, R. F., and Plapp, F. W., 1968, Insecticide resistance in the housefly: Identification of a gene that confers resistance to organotin insecticides and acts as an intensifier of parathion resistance, *J. Econ. Entomol.* **61:**1269.

Iwata, T., and Hama, H., 1972, Insensitivity of cholinesterase in *Nephotettix cincticeps* resistant to carbamate and organophosphorus insecticides, *J. Econ. Entomol.* **65:**643.

Jarczyk, H. J., 1966, Der Einfluss von Esterasen in Insekten auf den Abbau von Phosphorsäure-estern der E 605-Reihe, *Pflanzenschutz Nachr. (Bayer)* **19:**1.

Jarczyk, H. J., 1967, Influence on the degradation of parathion-compounds by insect and spider mites hydrolases *in vitro, Abstracts 6th International Congress of Plant Protection, Vienna, pp.* 294–295.

Johnson, M. K., 1966, Studies on glutathione *S*-alkyltransferase of the rat, *Biochem. J.* **98:**44.

Keiding, J., 1963, Possible reversal of resistance, *Bull. WHO Suppl.* **29:**51.

Khan, M. A. Q., 1969, Some biochemical characteristics of the microsomal cyclodiene epoxidase system and its inheritance in the housefly, *J. Econ. Entomol.* **62:**388.

Khan, M. A. Q., and Terriere, L. C., 1968, DDT-dehydrochlorinase activity in housefly strains resistant to various groups of insecticides, *J. Econ. Entomol.* **61:**732.

Khan, M. A. Q., Chang, J. L., Sutherland, D. J., Rosen, J. D., and Kamal, A., 1970, Housefly microsomal oxidation of some foreign compounds, *J. Econ. Entomol.* **63:**1807.

Khan, M. A. Q., Morimoto, R. I. Bederka, J. T., and Runnels, J. M., 1973, Control of the microsomal mixed-function oxidase by Ox^2 and Ox^5 genes in houseflies, *Biochem. Genet.* **10:**243.

Kikkawa, H., 1964, Genetic studies on the resistance to parathion in *Drosophila melanogaster.* II. Induction of a resistance gene from its susceptible allele, *Botyu-Kagaku* **29:**37.

Kimura, T., Duffy, J. R., and Brown, A. W. A., 1965, Dehydrochlorination and DDT-resistance in *Culex* mosquitoes, *Bull. WHO* **32:**557.

Kojima, K., Ishizuka, T., and Kitakata, S., 1963, Mechanism of resistance to malathion in the green rice leaf hopper, *Botyu-Kagaku* **28:**17.

Kuhr, R. J., 1971, Comparative metabolism of carbaryl by resistant and susceptible strains of the cabbage looper, *J. Econ. Entomol.* **64:**1373.

Lewis, J. B., 1969, Detoxication of diazinon by subcellular fractions of diazinon resistant and susceptible houseflies, *Nature (London)* **224:**917.

Lewis, J. B., and Sawicki, R. M., 1971, Characterization of the resistance mechanism to diazinon, parathion and diazoxon in the organophophorus-resistant SKA strain of houseflies (*Musca domestica* L.), *Pestic. Biochem. Physiol.* **1:**275.

Lichtwardt, E. T., 1964, A mutant linked to the DDT-resistance of an Illinois strain of houseflies, *Entomol. Exp. Appl.* **7:**296.

Lipke, H., and Kearns, C. W., 1960, DDT-dehydrochlorinase, *Advan. Pest Control Res.* **3:**253.

Main, A. R., and Dauterman, W. C., 1967, Kinetics for the inhibition of carboxylesterase by malaoxon, *Can. J. Biochem.* **45:**757.

Matsumura, F., and Brown, A. W. A., 1961, Biochemistry of malathion resistance in *Culex tarsalis*, *J. Econ, Entomol.* **54:**1176.

Matsumura, F., and Brown, A. W. A., 1963, Studies on carboxyesterase in malathion resistant *Culex tarsalis*, *J. Econ. Entomol.* **56:**381.

Matsumura, F., and Hogendijk, C. J., 1964*a*, The enzymatic degradation of malathion in organophosphate resistant and susceptible strains of *Musca domestica*, *Entomol. Exp. Appl.* **7:**179.

Matsumura, F., and Hogendijk, C. J., 1964*b*, The enzymatic degradation of parathion in organophosphate-susceptible and -resistant houseflies, *J. Agr. Food Chem.* **12:**447.

Matsumura, F., and Voss, G., 1964, Mechanism of malathion and parathion resistance in the two-spotted spider mite *Tetranychus urticae*, *J. Econ. Entomol.* **57:**911.

Matsumura, F., and Voss, G., 1965, Properties of partially purified malathion carboxyesterase of the two-spotted spider mite, *J. Insect Physiol.* **11:**147.

Milani, R., and Franco, M. G., 1959, Comportamento ereditario della resistenza al DDT in incroci tra il ceppo Orlano-R e ceppi *kdr* e kdr^+ di *Musca domestica* L., *Symp. Genet.* **6:**269.

Milani, R., and Travaglino, A., 1957, Ricerche genetiche sulla resistenza al DDT in *Musca domestica* concatenazione del gene *kdr* (knockdown-resistance) con due mutanti morfologigi, *Riv. Parasitol.* **18:**199.

Moorefield, H. H., and Kearns, C. W., 1955, Mechanism of action of certain synergists for DDT against resistant houseflies, *J. Econ. Entomol.* **48:**403.

Morello, A., Vardanis, A., and Spencer, E. Y., 1968, Comparative metabolism of two vinylphosphorothionate isomers (thiono phosdrin) by the mouse and the fly, *Biochem. Pharmacol.* **17:**1795.

Motoyama, N., and Dauterman, W. C., 1972, *In vitro* metabolism of azinphosmethyl in susceptible and resistant houseflies, *Pestic. Biochem. Physiol.* **2:**113.

Motoyama, N., and Dauterman, W. C., 1974, The role of nonoxidative metabolism in organophosphorus resistance *J. Agr. Food. Chem.* **22:**350.

Motoyama, N., Rock, G. C., and Dauterman, W. C., 1971, Studies on the mechanism of azinphosmethyl resistance in predaceous mite, *Neoseiulus* (T.) *fallacis*, *Pestic. Biochem. Physiol.* **1:**205.

Nakatsugawa, T., Tolman, N. M., and Dahm, P. A., 1969, Metabolism of S^{35}-parathion in the housefly, *J. Econ. Entomol.* **62:**408.

Needham, P. H., and Sawicki, R. M., 1971, Diagnosis of resistance to organophosphorus insecticides in *Myzus persicae*, *Nature (London)* **230:**125.

Nolan, J., and O'Brien, R. D., 1970, Biochemistry of resistance to paraoxon in strains of houseflies, *J. Agr. Food. Chem.* **18:**802.

Nolan, J., Schnitzerling, H. J., and Schuntner, C. A., 1972, Multiple forms of acetylcholinesterase from resistant and susceptible strains of the cattle tick, *Boophilus microplus* (Can.), *Pestic. Biochem. Physiol.* **2:**85.

O'Brien, R. D., 1966, Mode of action of insecticides, *Ann. Rev. Entomol.* **11:**369.

O'Brien, R. D., 1967, *Insecticides, Action and Metabolism*, Academic Press, New York.

Oppenoorth, F. J., 1956, Resistance to gamma-hexachlorocyclohexane in *Musca domestica* L., *Arch. Neerl. Zool.* **12:**1.

Oppenoorth, F. J., 1959, Genetics of resistance to organophosphorus compounds and low aliesterase activity in the housefly, *Entomol. Exp. Appl.* **2:**304.

Oppenoorth, F. J., 1965*a*, DDT-resistance in the housefly dependent on different mechanisms and the action of synergists, *Mededeel. Landbouwhogeschool Opzoekingsstat. Gent.* **30:**1390.

Oppenoorth, F. J., 1965*b*, Some cases of resistance caused by the alteration of enzymes, *in: Proceedings of the 12th International Congress of Entomology*, London, pp. 240–242.

Oppenoorth, F. J., 1965*c*, Biochemical genetics of insecticide resistance, *Ann. Rev. Entomol.* **10:**185.

Oppenoorth, F. J., 1967, Two types of sesamex-suppressible resistance in the housefly, *Entomol. Exp. Appl.* **10:**75.

Oppenoorth, F. J., 1971*a*, Resistance in insects: The role of metabolism and the possible use of synergists, *Bull. WHO* **44:**195.

Oppenoorth, F. J., 1971*b*, *Bull. WHO* **44:**377.

Oppenoorth, F. J., 1972, Degradation and activation of organophosphorus insecticides and resistance in insects, *in: Toxicology, Biodegradation and Efficacy of Lifestock Pesticides* (M. A. Khan and W. O. Haufe, eds.), pp. 73–92, Swets and Zeitlinger, Amsterdam.

Oppenoorth, F. J., and Houx, N. W. H., 1968, DDT resistance in the housefly caused by microsomal degradation , *Entomol. Exp. Appl.* **11:**81.

Oppenoorth, F. J., and van Asperen, K., 1960, Allelic genes in the housefly producing modified enzymes that cause organophosphate resistance, *Science* **132:**298.

Oppenoorth, F. J., and van Asperen, K., 1961, The detoxication enzymes causing organophosphate resistance in the housefly; properties, inhibition and the action of inhibitors as synergists, *Entomol. Exp. Appl.* **4:**311.

Oppenoorth, F. J., and Voerman, S., 1975, Hydrolysis of paraoxon and malaoxon in three strains of *Myzus parsicae* with different degrees of parathion resistance, *Pestic. Biochem. Physiol.* **5**:431.

Oppenoorth, F. J., Voerman, S., Welling, W., Houx, N. W. H., and Wouters van den Oudenweyer, J., 1971, Synergism of insecticidal action by inhibition of microsomal oxidation with phosphorothionates, *Nature New Biol.* **233:**187.

Oppenoorth, F. J., Rupes, V., ElBashir, S., Houx, N. W. H., and Voerman, S., 1972, Glutatione-dependent degradation of parathion and its significance for resistance in the housefly, *Pestic. Biochem. Physiol.* **2:**262.

Ozaki, K., 1969, The resistance to organophosphorus insecticides of the green rice leafhopper *Nephotettix cincticeps* Uhler and the smaller brown plant hopper *Laodelphax striatellus* Fallén, *Rev. Plant Protect. Res.* **2:**1.

Ozaki. K., and Kasai, T., 1970, Biochemical genetics of malathion resistance in the smaller brown plant hopper (*Laodelphax striatellus*), *Entomol. Exp. Appl.* **13:**162.

Ozaki, K., Kurosu, Y., and Koike, H., 1966, The relation between malathion resistance and esterase activity in the green rice leafhopper, *Nephotettix cincticeps* Uhler, *SABCO J.* **2:**98.

Perry, A. S., 1964, *In: The Physiology of Insecta*, Vol. 3 (M. Rockstein, ed.), pp. 285–378, Academic Press, New York.

Perry, A. S., and Agosin, M., 1974, *in: The Physiology of Insecta*, Vol. 6 (M. Rockstein, ed.), pp. 3–121, Academic Press, New York.

Perry, A. S., and Buckner, A. J., 1958, Biochemical investigations on DDT-resistance in the human body louse, *Pediculus humanus humanus, Ann. J. Trop. Med. Hyg.* **7:**620.

Perry, A. S., and Hoskins, W. M., 1950, The detoxification of DDT by resistant houseflies and inhibition of this process by piperonyl cyclonene, *Science* **111:**600.

Perry, A. S., Mattson, A. M., and Buckner, A. J., 1953, The mechanism of synergistic action of DMC with DDT against resistant houseflies, *Biol. Bull.* **104:**426.

Perry, A. S., Dale, W. E., and Buckner, A. J., 1971, Induction and repression of microsomal mixed function oxidases and cytochrome P-450 in resistant and susceptible houseflies, *Pestic. Biochem. Physiol.* **1:**131.

Pillai, M. K. K., Hennessy, D. J., and Brown, A. W. A., 1963, Deuterated analogues as remedial insecticides against DDT-resistant *Aedes aegypti, Mosquito News* **23:**118.

Plapp, F. W., 1970, Inheritance of dominant factors for resistance to carbamate insecticides in the housefly, *J. Econ. Entomol.* **63:**138.

Plapp, F. W., and Bigley, W. S., 1961, Carbamate insecticides and ali-esterase activity in insects, *J. Econ. Entomol.* **54:**793.

Plapp, F. W., and Casida, J. E., 1969, Genetic control of housefly NADPH-dependent oxidases: Relation to insecticide chemical metabolism and resistance, *J. Econ. Entomol.* **62:**1174.

Plapp, F. W., and Hoyer, R. F., 1968*a*, Possible pleiotropism of a gene conferring resistance to DDT, DDT analogs and pyrethrins in the housefly and *Culex tarsalis, J. Econ. Entomol.* **61:**761.

Plapp, F. W., and Hoyer, R. F., 1968*b*, Insecticide resistance in the housefly: Decreased rate of absoption as the mechanism of action of a gene that acts as an intensifier of resistance, *J. Econ. Entomal.* **61:**1298.

Plapp, F. W., and Vinson, S. B., 1973, Juvenile hormone analogs: Toxicity and cross-resistance in the housefly, *Pestic. Biochem. Physiol.* **3:**131.

Plapp, F. W., Orchard, R. D., and Morgan, J. W., 1965, Analogs of parathion and malathion as substitute insecticides for the control of resistant houseflies and the mosquito *Culex tarsalis. J. Econ. Entomol.* **58:**953.

Sawicki, R. M., 1970, Interaction between the factor delaying penetration of insecticides and the desethylation mechanism of resistance in organophosphorus-resistant houseflies, *Pestic. Sci.* **1:**84.

Sawicki, R. M., 1973*a*, Resynthesis of multiple resistance to organophophorus insecticides from strains with factors of resistance isolated from the SKA strain of houseflies, *Pestic. Sci* **4**:171.

Sawicki, R. M., 1973*b*, Recent advances in the study of the genetics of resistance in the housefly, *Musca domestica, Pestic. Sci.* **4:**501.

Sawicki, R. M., 1974, Genetics of resistance of a dimethoate-selected strain of houseflies (*Musca domestica* L.) to several insecticides and methylene dioxyphenyl synergists, *J. Agr. Food. Chem.* **22:**344.

Sawicki, R. M., and Farnham, A. W., 1967, Genetics of resistance to insecticides of the SKA strain of *Musca domestica.* I. Location of the main factors responsible for the maintenance of high DDT-resistance in diazinon-selected SKA flies, *Entomol. Exp. Appl.* **10:**253.

Sawicki, R. M., and Farnham, A. W., 1968, Examination of the isolated autosomes of the SKA strain of houseflies (*Musca domestica* L.) for resistance to several insecticides with and without pretreatment with sesamex and TBTP, *Bull. Entomol. Res.* **59:**409.

Schafer, J. A., and Terriere, L. C., 1970, Enzymatic and physical factors in housefly resistance to naphthalene, *J. Econ. Entomol.* **63:**787.

Schonbrod, R. D., Philleo, W. W., and Terriere, L. C., 1965, Hydroxylation as a factor in resistance in houseflies and blowflies, *J. Econ. Entomol.* **58:**74.

Schonbrod, R. D., Khan, M. A. Q., Terriere, L. C., and Plapp, F. W., 1968, Microsomal oxidases in the housefly: A survey of fourteen strains, *Life Sci.* **7:**681 (Part I).

Schonbrod, R. D., Hoyer, R. F., Yu, S. J., and Terriere, L. C., 1973, The epoxidation of aldrin by microsomes from the Fc strain of housefly, *Pestic. Biochem. Physiol.* **3:**259.

Schulten, G. G. M., 1968, Genetics of organophosphate resistance in the two-spotted spider mite (*Tetranuchus urticae* Koch), *Commun. Royal Trop. Inst. Amsterdam* **57:**1.

Schuntner, C. A., and Roulston, W. J., 1967, A resistance mechanism in organophosphorus-resistant strains of sheep blowfly (*Lucilia cuprina*), *Aust. J. Biol. Sci.* **21:**173.

Schuntner, C. A., and Smallman, B. N., 1972, Cholinergic systems in organophosphorus-resistant and -susceptible larvae of the cattle tick, *Pestic. Biochem. Physiol.* **2:**78.

Shishido, T., Usui, K., Sato, M., and Fukami, J., 1972, Enzymatic conjugation of diazinon with glutathione in rat and American cockroach, *Pestic. Biochem. Physiol.* **2:**51.

Shrivastava, S. P., Tsukamoto, M., and Casida, J. E., 1969, Oxidative metabolism of C^{14}-labeled Baygon by living houseflies and by housefly enzyme preparations, *J. Econ. Entomol.* **62**:483.

Shrivastava, S. P., Georghiou, G. P., Metcalf, R. L., and Fukuto, T. R., 1970, Carbamate resistance in mosquitoes: The metabolism of propoxur by susceptible and resistant larvae of *Culex pipiens fatigans, Bull. WHO* **42**:931.

Smissaert, H. R., 1964, Cholinesterase inhibition in spider mites susceptible and resistant to organophosphate, *Science* **143:**129.

Smissaert, H. R., Voerman, S., Oostenbrugge, L., and Renooy, N., 1970, Acetylcholinesterases of organophosphate-susceptible and -resistant spider mites *J. Agr. Food Chem.* **18:**66.

Stenersen, J. H. V., 1965, DDT-metabolism in resistant and susceptible stable flies and in bacteria, *Nature (London)* **207:**660.

Sternburg, J., Kearns, C., and Bruce, W., 1950, Absorption and metabolism of DDT by resistant and susceptible houseflies, *J. Econ. Entomol.* **43:**214.

Sternburg, J. G., Kearns, C. W., and Moorefleld, H., 1954, DDT-dehydrochlorinase, an enzyme found in DDT-resistant flies, *J. Agr. Food Chem.* **2:**1125.

Stone, B. F., 1972, The genetics of resistance by ticks to acaricides, *Aust. Vet. J.* **48:**345.

Sudderuddin, K. I., 1973*a*, An electrophoretic study of some hydrolases from an OP-susceptible and an OP-resistant strain of the green peach aphid, *Myzus persicae, Comp. Biochem. Physiol.* **44B:**923.

Sudderuddin, K. I., 1973*b*, An *in vitro* study of esterases hydrolysing non-specific substrates of an OP-resistant strain of the green peach aphid, *Myzus persicae, Comp. Biochem. Physiol.* **44B:**1067.

Sun, Y. P., and Johnson, E. R., 1960, Synergistic and antagonistic actions of insecticide–synergist combinations and their mode of action, *J. Agr. Food Chem.* **8:**261.

Tate, L. G., Plapp, F. W., and Hodgson, E., 1973, Cytochrome P-450 difference spectra of microsomes from several insecticide-resistant and -susceptible strains of the housefly, *Musca domestica* L., *Chem. Biol. Interact.* **6:**237.

Tate, L. G., Plapp, F. W., and Hodgson, E., 1974, Genetics of cytochrome P450 in two insecticide-resistant strains of the housefly, *Musca domestica* L., *Chem. Biol. Interact.* **11:**49.

Terriere, L. C., and Yu, S. J., 1973, Insect juvenile hormones: Induction of detoxifying enzymes in the housefly and detoxication by housefly enzymes, *Pestic. Biochem. Physiol.* **3:**96.

Terriere, L. C., and Yu, S. J., 1974, The induction of detoxifying enzymes in insects, *J. Agr. Food Chem.* **22:**366.

Townsend, M. G., and Busvine, J. R., 1969, The mechanism of malathion resistance in the blowfly *Chrysomya putoria, Entomol. Exp. Appl.* **12**:243.

Tripathi, R. K., and O'Brien, R. D., 1973, Insensitivity of acethylcholinesterase as a factor in resistance of houseflies to the organophosphate Rabon, *Pestic. Biochem. Physiol.* **3:**495.

Tsukamoto, M., 1961, Metabolic fate of DDT in *Drosophila melanogaster.* III. Comparative studies, *Botyu-Kagaku* **26:**74.

Tsukamoto, M., 1964, Methods for the linkage-group determination of insecticide-resistance factors in the housefly, *Botyu-Kagaku* **29:**51.

Tsukamoto, M., and Casida, J. E., 1967*a*, Metabolism of methylcarbamate insecticides by the $NADPH_2$ requiring enzyme system from houseflies, *Nature (London)* **213:**49.

Tsukamoto, M., and Casida, J. E., 1967*b*, Albumin enhancement of oxidative metabolism of methylcarbamate insecticide chemicals by the housefly microsome–$NADPH_2$ system, *J. Econ. Entomol.* **60:**617.

Tsukamoto, M., Narahashi, T., and Yamasaki, T., 1965, Genetic control of low nerve sensitivity to DDT in insecticide-resistant houseflies, *Botyu-Kagaku* **30:**128.

Tsukamoto, M., Shrivastava, S. P., and Casida, J. E., 1968, Biochemical genetics of housefly resistance to carbamate insecticide chemicals, *J. Econ. Entomol.* **61:**50.

van Asperen, K., van Mazijk, M., Oppenoorth, F. J., 1965, Relation between electrophoretic esterase patterns and organophosphate resistance in *Musca domestica, Entomol. Exp. Appl.* **8:**163.

Vinson, S. B., and Law, P. K., 1971, Cuticular composition and DDT resistance in the tobacco budworm, *J. Econ. Entomol.* **64:**1387.

Vinson, S. B., and Plapp, F. W., 1974, Third generation pesticides: The potential for the development of resistance by insects, *J. Agr. Food Chem.* **2:**356.

Voss, G., Dauterman, W. C., and Matsumura, F., 1964, Relation between toxicity of malathion analogs and OP-resistance in the two-spotted spider mite, *J. Econ. Entomol.* **57:**808.

Wang, C. M., Narahashi, T., and Yamada, M., 1971, The neurotoxic action of dieldrin and its derivatives in the cockroach, *Pestic. Biochem. Physiol.* **1:**84.

Welling, W., and Blaakmeer, P. T., 1971, Metabolism of malathion in a resistant and a susceptible strain of houseflies, *in: Proceedings of the 2nd International IUPAC Congress of Pesticide Chemistry,* Vol. II (A. S. Tahori, ed.), pp. 61–75, Gordon and Breach, New York.

Welling, W., Blaakmeer, P. T., Vink, G. J., and Voerman, S., 1971, *In vitro* hydrolysis of paraoxon by parathion resistant houseflies, *Pestic. Biochem. Physiol.* **1:**61.

Welling, W., de Vries, A. W., and Voerman, S., 1974, Oxidative cleavage of a carboxyester bond as a mechanism of resistance to malaoxon in houseflies, *Pestic. Biochem. Physiol.* **4:**31.

Wharton, R. H., and Roulston, W. J., 1970, Resistance of ticks to chemicals, *Ann. Rev. Entomol* **15:**381.

Wilkinson, C. F., 1968, The role of insecticide synergists in resistance problems, *World Rev. Pest Control* **7:**155.

Wilkinson, C. F., 1971, Effects of synergists in the metabolism and toxicity of anticholinesterases, *Bull. WHO* **44:**171.

Winteringham, F. P. W., 1969, Mechanisms of selective insecticidal action, *Ann. Rev. Entomol.* **14:**409.

Winteringham, F. P. W., and Harrison, A., 1959, Mechanisms of resistance of adult houseflies to the insecticide dieldrin, *Nature (London)* **184:**608.

Winteringham, F. P. W., Loveday, P. M., and Harrison, A., 1951, Resistance of houseflies to DDT, *Nature (London)* **167:**106.

Yamamoto, I., and Casida, J. E., 1966, *O*-Demethyl pyrethrin II analogs from oxidation of pyrethrin I, allethrin, dimethrin and phthalthrin by a housefly enzyme system, *J. Econ. Entomol.* **59:**1542.

Yang, R. S. H., Hodgson, E., and Dauterman, W. C., 1971, Metabolism *in vitro* of diazinon and diazoxon in susceptible and resistant houseflies, *J. Agr. Food Chem.* **19:**14.

Yu, S. J., and Terriere, L. C., 1973, Phenobarbital induction of detoxifying enzymes in resistant and susceptible houseflies, *Pestic. Biochem. Physiol.* **3:**259.

Zahavi, M., and Tahori, A. S., 1970, Sensitivity of acetylcholinesterase in spider mites to organo-phosphorus compounds, *Biochem. Pharmacol.* **19:**219.

V

Insecticide Toxicology

14

Teratogenic, Mutagenic, and Carcinogenic Effects of Insecticides

Lawrence Fishbein

1. Introduction

There is increasing concern over the possible toxicological hazards posed by a spectrum of synthetic organic chemicals including industrial pollutants, pesticides, food additives, and drugs. Because of their ubiquity, persistence, presence and/or concentration in food chains, and toxicological properties, insecticides constitute a major group of potential environmental hazards to mammalian species, including man.

However, there exists considerable disagreement concerning the methodology available for assessment of the teratogenic, mutagenic, and carcinogenic effects of these agents. Because of the variety of testing procedures employed, comparisons in any one area are difficult to effect and the unambiguous recognition and interpretation of the toxic event and the extrapolation of animal data to man are extremely complex.

To permit a more satisfactory assessment of toxic hazard, it is first necessary to define what is meant by teratogenicity, mutagenicity, and carcinogenicity, to examine whether or not any interrelationship exists among them, and to briefly outline and evaluate the various testing procedures employed for their detection.

Lawrence Fishbein • National Center for Toxicological Research, Jefferson, Arkansas.

2. Definition of Terms

Any chemical which can cause alterations in hereditary material potentially can produce carcinogenic, mutagenic, or teratogenic responses in mammalian systems. The ultimate consequence of exposure to DNA-altering substances depends on the developmental state of the organism, the type of cells affected, and the type of genetic alteration produced.

Teratology is generally defined as the study of the effects of intrinsic or extrinsic factors as they relate to permanent structural or functional deviations arising during embryogenesis. Hence a teratogenic agent is any agent that can induce or increase the incidence of congenital malformations. Teratogenicity is the result of damage or death of certain cells of a developing organism. In determining a teratogenic effect, the time and route of administration, the dose level, the number of doses, and the presence of additional stresses, e.g., nutritional deficiency, are all vital considerations. Teratogenic assessments in most studies to date have been limited to external, gross visceral, and skeletal examination.

Mutation in its broadest sense denotes any heritable change in the genetic material. The change can be a chemical transformation of an individual gene that causes it to have an altered function, or it can involve a rearrangement or a gain or loss of parts of a chromosome. If the change affects the genetic functioning of the cell while still permitting it to divide, this change can be transmitted to descendant cells, causing less localized damage. The effect may be cancerous or teratogenic, the latter being of major importance if the change takes place during embryonic development. A mutation or chromosome change transmitted via the sperm of egg to the next generation can affect every cell of the descendant individual with the possibility of disastrous consequences.

Mutations may occur in any somatic or germ cell. Mutation occurring in a somatic cell in a mature organism when the cells are duplicating can result in a neoplasm. The term *mutagenic agent* is usually restricted to an agent capable of producing a mutation in germ cells. Thus the major focus to date is undoubtedly on germinal mutation, although it is considered likely that somatic mutation can be used as a sensitive indicator of the germinal mutation rate (Bridges, 1971).

Mutations are known to cause a variety of effects, as genes influence every part of the body and control every metabolic process. They can induce lethal effects, physical abnormalities, and mental deficiencies and diseases. Inborn errors of metabolism, congenital malformations, and a spectrum of other physical and mental disabilities affecting future populations can be attributed to deleterious mutations. Mutations that are mainly recessive will be expressed only in the homozygous state, while dominant mutations can be expressed in either the heterozygous or the homozygous condition.

Chemical carcinogenesis (carcinogenicity) in its broadest, most widely accepted sense is considered to mean the induction or enhancement of neoplasia (cancer) by chemicals. Common usage, however, recognizes the induction of various types of neoplasms, with the result that terms such as "oncogen," "blastomogen," and "tumorigen" have been used synonymously with "carcinogen." Occasionally "tumorigen" is used specifically to denote the induction of benign tumors (International Agency for Research on Cancer, 1971). Many instances of carcinogenesis involve the induction of both benign and malignant tumors, and a new and apparently precedent-setting criterion for determining the carcinogenicity of a substance—that "there is no valid distinction between the induction of benign or malignant tumors in determining the carcinogenicity of a compound"—was used in the Environmental Protection Agency (1974) decision to suspend all major uses of Shell Chemicals' aldrin and dieldrin pesticides.

Carcinogenesis is defined by Miller (1970) as a heritable and at least quasipermanent loss of control of cell multiplication. There are at least two general basic hypotheses that might account directly for such a change in cell behavior. These are (a) interaction of carcinogens with DNA resulting in alterations in the information contained in the macromolecule and (b) alterations in specific proteins or RNAs which result in relatively stable and heritable changes in genome expressions. Two indirect mechanisms, the activation of a latent carcinogenic virus genome and the selection of preneoplastic cells by conditions that favor the multiplication of these cells, have also been proposed and are likely to operate through one of the previous mechanisms. Combinations of these direct and indirect mechanisms can conceivably also occur. However, none of these or any other hypotheses have yet been unequivocally demonstrated with any carcinogenic agent.

The categorical designation *cancer* encompasses a group of diseases caused by multiple factors acting either singly or in combination. These factors are operative both in the external environment and in the host. Age, sex, and physiological state of the target tissue also modify cancer incidence (Kotin, 1972).

3. Interrelationships of Teratogenicity, Mutagenicity, and Carcinogenicity

Mutagenesis and carcinogenesis are complex cellular processes which are grossly similar in that each produces heritable changes in phenotype. The concept that neoplasms arise from mutations in somatic cells has long been advocated as a theory of cancer causation by Boveri (1929) and Bauer (1928, 1963). The modern understanding of the chemical basis of mutation has enhanced its acceptance for the mechanism of action of chemical carcinogens (Clayson, 1962) since these or their metabolites have frequently been found to possess mutagenic properties.

Opposing the somatic mutation concept is the theory that neoplasia results from quasipermanent and heritable changes in genome expression without changes in the genome itself (Pitot and Heidelberger, 1963; Gelboin, 1967; Weinstein, 1969). This is the process by which cellular differentiation is generally considered to occur. Another theory suggests that chemical carcinogens may alter selection pressures in the cellular environment (e.g., immunological factors) and hence permit latent tumor cells to produce gross neoplasms.

Viral activity as the basis for the origin of chemically induced neoplasia is another popular concept since many carcinogenic viruses are now known. According to this concept, the chemical carcinogen activates a latent carcinogenic virus or causes the expression of part or all of an integrated viral genome which induces neoplastic changes.

There is no experimental basis at present for favoring any one of the above mechanisms, and it may be that each is a valid explanation for the induction of tumors by chemicals in specific instances (Miller and Miller, 1971).

It has been shown that most if not all chemical carcinogens either are themselves strong electrophils (containing relatively electrondeficient atoms) or are converted to potent electrophilic agents *in vivo*, and there is increasing evidence suggesting that this may be associated with carcinogenic activity (Miller and Miller, 1971).

Attempts to correlate the mutagenic and carcinogenic activity of chemicals are limited by the types and amounts of data available. Many of the chemical carcinogens have received only limited tests for mutagenic activity, and some of the better-known mutagenic agents have not been adequately tested for carcinogenic activity. It is also important to note that while a variety of assays are useful in the determination of each type of activity the results of different tests cannot always be readily combined or compared. For example, mutagens capable of interacting with DNA *in vitro* may have acted in the form administered or may have been converted nonenzymatically to an active product. Similarly those which are mutagenic in cellular systems may have acted directly or through the mediation of enzymatic or nonenzymatic reactions. Consequently, complications due to lack of penetration to critical target sites, lack of activation, or rapid enzymatic or nonenzymatic decomposition must all be considered in assessing the results of various tests.

Carcinogens which are active only at sites distant from the port of entry may be assumed to require activation to an ultimate carcinogenic form. However, the converse is not true, since compounds which are carcinogenic at the site of administration may be in the active form as administered, or may be activated in the cells at the site of treatment.

There appears to exist a general qualitative correlation among electrophilic reactivity, carcinogenicity, and mutagenicity wherever enough information is available on the nature of the probable carcinogenic metabolites. Thus many alkylating agents possess all three of these properties. The numerous examples of carcinogenic compounds which have not shown mutagenic activity have probably resulted in most cases from the failure of these compounds to be metabolized to reactive electrophils in the mutagenicity testing systems employed.

It appears that many, and perhaps all, chemical carcinogens are *potential* mutagens. Similarly, many, possibly not all, mutagens are *potential* carcinogens.

Two basic reasons have been suggested to account for the lack of a complete association between carcinogenesis and mutagenesis. These are (a) the lack of knowledge of *all* of the steps in carcinogenesis and (b) an uncertainty that the bioassay system employed will metabolize the carcinogen along the same pathways as all the tissues of animals or man.

Chemical carcinogenesis has always been considered to be a two- or multistep process, and since the first step involves a permanent change in the hereditary material (mutation) it can be detected by mutagenicity test systems. Consequently, test systems will usually detect initiators and complete carcinogens but not necessarily promotion.

Although mutagenesis appears to be reasonably well correlated with carcinogenesis in the compounds tested to date, it is likely that better correlations will be obtained if a more sensitive method is found to link biochemical drug-activating systems with mutagen-sensitive systems.

The interrelationships of carcinogenicity, mutagenicity, and teratogenicity have been reviewed by Wilson (1972). Points of similarity have been previously summarized in the Corvallis Task Force Report (1970) as (a) insidiousness of nature, (b) relatively long time lag between exposure and overt effect, (c) irreversibility of the diseases, (d) relatively greater susceptibility of immature or developing tissues, and (e) some similarity of etiological factors. Some of these, however, are more apparent than real. Thus in regard to the time lag between exposure and manifestation of an effect there may be orders of magnitude of difference. The interval for carcinogenesis can range from a few months to many years. Mutagenesis may be manifested in the succeeding generation (if dominant or sex-linked or involving chromosomal aberration), two or more generations later (if involving a common recessive), or perhaps never, if a rare recessive. Teratogenesis can be apparent within hours of exposure but is more typically evident at or soon after birth.

Immature tissues are inherently more susceptible to the action of teratogenic agents, and certain cancers have also been shown to be more easily induced in newborn or infant animals than in adults. However, this is not always true (Toth, 1968), and it should be noted that cancers in older individuals most often occur in tissues that retain a generative function (bone marrow, basal layer of skin, and crypt cells of intestine). Gene mutations appear to occur equally well in immature individuals (or germ cells) and in mature ones.

Etiological similarities of carcinogenesis, mutagenesis, and teratogenesis can be substantiated only to a limited extent. Except for a few chemical agents such as nitrogen mustard, urethan, aflatoxin, and benzo[*a*]pyrene, which are known to be active in mammals in all three areas, the case for common causation cannot easily be supported. Kalter (1971) reviewed 86 compounds believed to be either mutagenic or teratogenic in mammals and found some evidence for mutagenicity with 36 and teratogenicity with 40; only 15 of the 86 were shown to possess both properties and of these only three (aflatoxin, benzo[*a*]pyrene, and nitrogen mustard) were shown to be carcinogenic. Wilson (1972) also reviewed the converse point of view: i.e., how carcinogenesis, mutagenesis, and teratogenesis differ. Table I illustrates the complex sequence of events associated with teratogenesis. Although some similarities with carcinogenesis (e.g., certain causes) and mutagenesis (e.g., somatic mutations, chromosomal aberrations) are obvious, many of the factors involved are unique to teratogenesis.

The most important difference between the three disease processes is in their basic toxicological characteristics. Carcinogenesis is *usually* regarded as one of the standard forms of chronic toxicity, whereas teratogenesis is an acute toxicological phenomenon, requiring only short periods of exposure and usually manifesting itself within a relatively short time thereafter. Although mutations can be instantaneously induced with X-rays, they more usually appear to be the consequence of continuous or repeated exposure, and the final expression of mutagenesis can be delayed in the extreme. As summarized in Table II, the conditions of exposure often determine whether or to what extent a given tissue responds to carcinogens, mutagens, and teratogens. For these reasons, it is difficult to evaluate carcinogenesis, mutagenesis, and teratogenesis simultaneously in a single test without sacrifice of reliability and sensitivity in one or all of the areas concerned.

Table I. Summary of Teratogenesis

CAUSES →	MECHANISMS →	MANIFESTATIONS
Action of an agent from the environment on the embryo or the germ cells, e.g., Radiations Chemicals Dietary deficiency Infection Hypoxia, etc. Temperature Endocrine imbalance Physical trauma Placental failure	→ *Reaction* within the embryo or germ cells, such as one or more of the following: Mutation Chromosomal nondisjunction Mitotic interference Altered nucleic acid integrity or function Lack of precursors, substrates, etc. Altered energy sources Changed membrane characteristics Water–electrolyte imblance Enzyme inhibition	→ *Pathogenesis*, initiated by one or more of following: Cell death Mitotic rate change Reduced biosynthesis Altered differentiation schedules Impeded morphogenetic movement Etc. —and leading to abnormal tissue and organ development which determine the— NATURE AND INCIDENCE OF FINAL DEFECT

Table II. Conditions of Exposure That Determine the Response in Carcinogenesis, Mutagenesis, and Teratogenesis

	Susceptible tissues	Optimal time of exposure	Duration and level of dosage
Carcinogen	Proliferating tissues	Uncertain, probably all stages capable of mitosis	Usually chronic, possibly all doses
Mutagen	Germinal tissues	All stages of gametogenesis	Either acute or chronic, possibly all doses
Teratogen	Possibly all immature tissues	Highest during early differentiation	Acute only, above usual no-effect level

4. Testing Procedures: Principles, Problems, and Interpretations

A complete description of the current testing procedures for teratogenesis, mutagenesis, and carcinogenesis is beyond the scope of this chapter. However, a qualitative description of the more basic test systems is considered germane for the understanding of the potential risk of the classes of environmental agents in question.

4.1. Teratogenicity Tests

In teratogenesis, we are dealing with two biological systems—the pregnant female and the embryo—whose specific reactions can be completely different. A drug that is nontoxic to the female may still be able to kill the embryo or produce congenital malformations. The two factors in teratogenesis which have embryological bearing are the genotype and the stage of development at the time of treatment. The genotype determines whether any tissue or organ will be susceptible to a given teratogen, while the developmental stage determines which tissues or organs of a susceptible organism will be affected at a given time.

At present, three basic treatment schedules are normally utilized for teratogenic evaluations: (a) administration (single or multiple) on certain days of gestation, (b) administration throughout gestation, and (c) administration prior to and during gestation.

4.1.1. Chick Embryo and Yolk–Chemical Mixture

Although the chick embryo test for detecting teratogenic effects by injection of the chemical into the yolk is an easy, rapid, and inexpensive procedure that has been extensively utilized (McLaughlin *et al.*, 1963, 1964; Walker, 1968), it can only be considered as a *screening* system for assessing *potential* human hazards. Despite the fact that the age, diet, strain of maternal hen, storage temperature, and interval between laying and incubation can affect the results of tests on eggs, the chick embryo test allows studies on morphological and physiological disturbances during development.

4.1.2. Mammalian Testing

The response of animal species to teratogenic agents is known to vary widely. The choice of a suitable mammal may be based on the similarity of the metabolism of the test material in man and the proposed test animal. In addition to metabolic transformation, the concentration, distribution, and rate of excretion of the test material in man and test species are germane considerations (Tuchmann-Duplessis, 1972). Ideally, the species of choice should have an easily determinable time of fertilization, a relatively short gestation period, and high fertility under animal-house conditions, and the embryology, genetic background, incidence of spontaneous, and inducible malformations should be well defined (Clegg, 1971).

Factors influencing the choice of an appropriate testing procedure include human exposure pattern (route, levels, duration, and intensity), physical and chemical characteristics of test substance, pharmacological activity, and acute, subacute, and chronic toxicity data.

Studies involving short-term administration generally include exposing the animal to the test material during the period of organogenesis, which is thought to be the period of maximum sensitivity to teratogenic agents. This period is considered covered if the material is administered on days 6–15 in the mouse or the rat (Tuchmann-Duplessis, 1965), 6–18 in the rabbit, 4–14 in the hamster (Shenefeldt, 1972), 6–20 in the guinea pig (Robens, 1969), 7–35 in the pig, and 9–40 in the monkey (Witschi, 1955).

The significance of any teratogenic activity that may be observed can be assessed by taking into account the percentages of malformation obtained, the dose, the number of species at which the teratogenic effect was manifest, and the constancy of the results in several replicate experiments.

4.2. Mutagenicity Tests

A crucial point concerning the evaluation of genetic risks by chemicals is the choice of test systems on which to base practical decisions and permit proper and valid assessments. Tests in microorganisms, cell cultures, insects, and fertile avian eggs are acknowledged to be useful *screening* and ancillary procedures. In view of a spectrum of factors, including cell uptake, metabolism, detoxication, dosage, and method of administration, the human relevance of data obtained from ancillary test systems is uncertain; mammalian test systems appear to entail fewer of these limitations. It is recognized that no single method can detect all possible types of mutation, and consequently a combination of methods must be employed. A positive result in any mammalian system is suggestive evidence of a *potential* mutagenic hazard.

A number of procedures for mammals are presently available—the majority of recent origin—that can be used to determine the mutagenic activity of chemicals.

The basis for choosing standard, most relevant test systems for evaluating environmental mutagens has focused on three major aspects: (1) the tests employed must cover as far as possible the types of genetic damage which are generally considered causes of human disorders; (2) the tests should take into account the features of mammalian metabolism which can result in activation or inactivation of the tested chemical; (3) the cost and time required for the testing should not be prohibitive. The following mammalian test systems as a consequence are generally acknowledged to best fit the above criteria (a) dominant lethal tests, (b) host-mediated assay, (c) *in vivo* cytogenetic tests, and (d) the specific locus test (Table III).

Legator and Flamm (1973) recently reviewed aspects of environmental mutagenesis and repair and assessed tests (a), (b), and (c) above in regard to their respective advantages, disadvantages, present status, and future developments.

4.2.1. Dominant Lethal Test

The induction of dominant lethal mutations in animals can be assayed with a high degree of sensitivity following acute, subacute, or chronic administration of test materials, either orally or by any parenteral route, including the respiratory. Following administration to male rodents, the animals are mated sequentially with groups of untreated females over the duration of the spermatogenic cycle (Bateman, 1958). The recommended scheme for dominant lethal testing in male mice has been outlined by Epstein and Röhrborn (1971). The genetic basis for dominant lethality is mainly the induction of structural and numerical chromosomal anomalies, such as translocations and aneuploidies; sequentially, these may induce preimplantation losses of nonviable zygotes, early fetal deaths, and sterility or semisterility in F_1 progeny (Epstein, 1973).

4.2.2. Host-Mediated Assay

The host-mediated assay is the only technique involving microorganisms, and possibly cultured cells, that can detect metabolic products that induce point mutations, irrespective of how these products are formed, e.g., activation by intestinal flora or via enzymatic conversion by liver enzymes.

In the host-mediated assay, the mammal (during treatment with a potential chemical mutagen) is injected with an indicator organism, e.g., the histidine auxotrophs of *Salmonella typhimurium*, in which mutation frequency can be measured. After a sufficient time period (approximately 3 hr), the indicator microorganism is withdrawn from the animal and the induction of mutation is thence determined. The effect in the animal is compared with the effect of the chemical in an *in vitro* plate assay. Comparison between the mutagenic action

Table III. Mammalian Test Systems

Test system	Advantages	Disadvantages	Present status	Future development
Host-mediated assay	Only feasible screening method that detects point mutation, can detect transient metabolites as well as directly acting chemical Indirect detoxication as well as potentiation by comparison with direct test on indicators Simple, economical test Moderately skilled investigator required Used to correlate carcinogenicity with mutagenicity	Indirect indication of active metabolite as to organ or tissue may be difficult	Practical screening procedure, over 100 compounds	Indicators, such as mammalian cells, that can divide in host and mutate at same rate as *in vitro* (i.e., mouse lymphoma system) Localization of genetic effect in various organs In bacteria, detection of multiple genetic events with a single indicator

Cytogenic, direct analysis (mitotic)	Applicable to a variety of hosts Direct observation of chromosomal abnormalities Somatic and germ cells can be analyzed at comparatively low cost Moderately time consuming Can be adapted to standard toxicological protocol	Needs highly trained investigator Quantitative correlation to point mutations not known Usually detects a high percentage of nonviable cells Chromatid aberrations mainly detected	Practical screening procedure	Anaphase determination Micronuclei determination Application of newer staining techniques Automation of chromosome analysis
Dominant lethal	Genetic test conducted in variety of hosts Easy to carry out by trained technician Fits into ongoing toxicity protocol Moderate cost and time Determines stages of spermatogenesis affected Can be carried out on either chronic, subacute, or acute basis	Test for lethality Indirect test for cytogenetic abnormalities Sensitivity presumably less than viable inherited abnormalities	Practical screening procedure	Cytogenetic analysis of live embryos Development of improved statistical procedures A higher precision in timing of affected stages of spermatogenesis

of the compound on the microorganism directly and in the host-mediated assay indicates whether the host can detoxify the compound or whether mutagens can be formed as a result of host metabolism (Gabridge and Legator, 1969; Legator and Malling, 1971; Legator, 1970; Malling and Frantz, 1973).

The host-mediated assay in no way indicates the effect of DNA repair mechanisms of the host in response to specific chemicals and hence is only an indirect measure of mutagenicity in terms of the mammalian host.

4.2.3. In Vivo Cytogenetic Studies

In vivo cytogenetic test methods for mutagenicity offer the distinct advantages that they are essentially the only current systems that are directly applicable to humans and the only procedures that permit visualization of the entire genome in the light microscope. While there may be basic differences in the mechanism of production, a high correlation has been observed between the induction of chromosomal abnormalities and the induction of gene mutation produced by irradiation and chemicals. This means that cytogenetic methods can be used as an indicator of gene mutations as well as a direct system for assessing chromosomal mutations. In animals, somatic as well as germ cells can be analyzed, and when performed *in vivo* the assay is acknowledged to be meaningful and selective (Nichols, 1973; Brewen and Preston, 1973).

4.3. Carcinogenicity Tests

The numbers and types of tumors induced by a carcinogen are greatly influenced by the species and strain of the test animals as well as other factors including route of administration, dosage, frequency of exposure, purity of test substance, sex of test animal, age of the animal at the start of the test, basic diet and possible diet contaminants (pesticides, heavy metals, PCBs, estrogens, nitrosamines, mycotoxins, synergists, enzyme inducers), hormonal and immunological status, and duration of the experiment. Environmental factors (besides impurities in the diet, water, or air) that could influence the outcome of cancer experiments include bedding, housing conditions, crowding stress, intercurrent disease, and lung therapy instituted for its control. It should be noted that selection of doses, frequency of exposure, duration of test, and species and strain, as well as the factors concerning the final assessment of carcinogenicity, are still matters of controversy. There is by no means common agreement with respect to the assessment of either spontaneous or induced tumor incidence in laboratory animals or the diagnostic criteria applied to classify lesions including the differentiation of benign and malignant tumors, hyperplasia, metaplasia, and neoplasia.

The study of the carcinogenic potential of a chemical requires that the study be performed over the animals' average life span since tumors in experimental animals *usually* develop only after a long latent period. Hence the rat, mouse, and hamster with relatively short life spans are used for routine carcinogenic testing, with the dog and primate reserved for special cases.

Weanling or immediately post-weanling animals are used in most routine screening studies for carcinogenesis as this permits the greater part of their life span for tumor development to occur coincident to exposure to the test agent.

The selection of strain of animal is dictated by the nature of the test. Use of highly inbred strains is generally avoided for the determination of the potential carcinogenicity of an untested compound since the possibility exists that while one inbred strain may be very responsive to an agent another inbred strain may fail to develop tumors with the same carcinogen. Use of animals with a more heterogeneous genetic constitution can to some extent overcome this problem.

Treatment of the F_0 generation prior to mating and during gestation and treatment of the F_1 generation following parturition have been suggested to provide a more thorough assessment of the potential carcinogenicity of test compounds.

In regard to routes of administration, it is generally acknowledged that the test regimen should attempt to produce the qualitative conditions of possible human exposure and to exaggerate them quantitatively. Many studies have employed at least three dose levels, which is compatible with WHO (1969) recommendations, and it is generally intended that the lowest dose selected will not produce a real increase in tumor incidence over control animals.

5. *Teratogenic, Mutagenic, and Carcinogenic Agents and Their Effects*

5.1. *Chlorinated Hydrocarbons*

5.1.1. *DDT and Metabolites*

The principal reasons for the concern regarding the potential hazard to human health of DDT and its metabolites are (a) their ubiquity and persistence in the environment, (b) their capacity to accumulate in living organisms (including man), in the fetus and in mother's milk, (c) evidence indicating that DDT causes an increase in tumor incidence in laboratory animals, (d) evidence of the mutagenic potential of DDT and its metabolites in mammals and insects, and (e) implication of DDT in the declining reproduction of several species of birds (Chapter 17).

DDT has been shown to cross the placenta in the mouse (Bäckström *et al.*, 1965), dog (Finnegan *et al.*, 1949), and rabbit (Hart *et al.*, 1972), and since DDT residues have been detected in the aborted human fetus, placental transfer is likely in man (Wassermann *et al.*, 1967).

The long-term effects of DDT on mammalian reproduction are unclear, although DDT exerts a number of reproduction-related effects in mammalian species. These include an increased incidence of stillbirths in cows (Laben *et al.*, 1965), higher blood levels of DDT in female sealions that abort compared to females with normal gestation (DeLong *et al.*, 1972), and higher blood levels of *p,p'*-DDE, the principal stored metabolite of *p,p'*-DDT, in premature infants than in normal infants at term (O'Leary *et al.*, 1970). Treatment of pregnant New Zealand white rabbits at doses of 10 and 50 mg/kg of *p,p'*-DDT caused an increased incidence of prematurity, an increase in the number of fetal resorptions and resorption sites, and a decrease in fetal weight (Hart *et al.*, 1972). It is not considered teratogenic at this dose level, however, since only one malformed fetus was observed in both control and pesticide-treated groups. Furthermore, no increase in the incidence of chromosomal aberrations in the maternal or fetal tissue was observed.

Ingestion of technical DDT (80% *p,p'*-DDT, 2% *p,p'*-DDE) at levels of 0.20 and 200 ppm by Sprague-Dawley derived rats produced no significant effects on the fertility or fecundity of dams or the viability of young (Ottoboni, 1969.) Indeed, the reproductive performance of 52-week-old rats indicated that DDT may exert a protective effect against age decrement of the reproductive process. Similarly negative results were obtained with respect to the viability and fertility of mice fed DDT either alone or in combination with other organochlorine insecticides (Deichmann and Keplinger, 1966; Ware and Good, 1967). Beagle dogs receiving daily doses of either DDT (12.0 mg/kg), aldrin (0.15 and 0.3 mg/kg), or DDT plus aldrin (6.0 mg/kg DDT, 0.15 mg/kg aldrin) over a period of 5 years exhibited delayed estrus, reduction in libido, and lack of mammary development (reduced milk production), which led to a high mortality rate among the offspring (Deichmann, 1972*b*; Deichmann *et al.*, 1971).

A number of reports have cited the genetic effects of DDT, although several of these are implied rather than proven. Thus, based on chromosomal changes occurring in the fly *Drosophila pseudoobscura* over the past 24 years in the western United States (Dobzhansky *et al.*, 1964, 1966) and the distribution patterns of DDT residues in this same area (Cory *et al.*, 1971), it was suggested that DDT was responsible for alterations in the genetic content of the populations. Similarly, DDT was suspected as the causal agent in the unprecedented number of genetic and chromosomal mutants occurring in a natural population of mice in the foothills of the Andes (Wallace, 1971).

Johnson and Jalal (1973) reported more direct evidence for chromosomal damage in BALB/c mice treated with DDT at doses ranging from 100 to 400 mg/kg. Significantly higher proportions of chromosomal abnormalities in the form of deletions, stickiness, and, rarely, ring and metacentric chromosomes were observed in mice receiving dosages of 150 ppm or higher. Since the induction of chromosomal damage is closely associated with mutagenic events (point mutations) in mammals (Epstein and Legator, 1971), DDT would appear to be a *potential* mutagen. Markaryan (1966) had previously reported significantly higher proportions of stickiness and chromosomal damage in mice on treatment with a single dose of DDT (100 mg/kg).

Kelly-Garvert and Legator (1973) described cytogenetic and mutagenic effects of DDT and DDE in a Chinese hamster cell line. In all experiments DDE consistently produced a significant increase in the mutation frequency and the number of polyploid cells compared with the controls, while DDT proved inactive. DDE but not DDT also led to a significant increase in chromosome aberrations, with exchange figures and chromatid breaks being particularly evident.

Palmer *et al.* (1972) reported that chromosome abnormalities were produced by DDT and its metabolites in a cultured cell line from kangaroo rat. Mitotic inhibition was moderate in cultures treated with *p,p'*- or *o,p'*-DDT and slight in those treated with *p,p'*- and *o,p'*-DDD or DDE, and none was observed with *p,p'*-DDA. The *p,p'*-isomers were approximately twofold more active than the corresponding *o,p'*-isomers. At a concentration of 10 μg/ml, the *p,p'*-isomers of DDT, DDD, and DDE caused chromosome damage in 22.4, 15.5, and 13.7% of the cells, respectively, approximately 12% of the damage caused by *p,p'*-DDT and *p,p'*-DDE being in the form of exchange figures.

DDT and various DDT metabolites (DDE, DDD, DDOM, and DDA) were tested for genetic activity in male germ cells of *Drosophila melanogaster* (Vogel, 1972) using the Basc technique (Lindsley and Grell, 1968) to score for sex-linked recessive lethals. Only DDA, the principal urinary excretion product of DDT in mammals, was found to induce recessive lethals, mature sperm being more sensitive than either spermatids or spermatocytes. Some evidence was obtained suggesting that DDT is a very weak mutagen in *Drosophila*, although because of its high toxicity, doses above 0.14 mmole/liter could not be tested. Grosch (1967) failed to demonstrate DDT-induced mutations in brine shrimp and subsequently obtained similarly negative data with 99.9% *p,p'*-DDT and 95% dieldrin in the wasp *Bracon hebetor* (Grosch and Valcovic, 1967), which is uniquely suited for the detection of induced recessive lethals (Grosch and Valcovic, 1967; Atwood *et al.*, 1956).

The results of mutagenicity tests with DDT and its analogues are somewhat equivocal. DDT was reported to be nonmutagenic in dominant lethal tests with mice (Epstein and Shafner, 1968; Epstein *et al.*, 1972) but was found to be active in similar tests with rats (Legator, 1973). As a result, Epstein *et al.* (1972) stressed the importance of conducting tests with both rats and mice in acute and chronic testing procedures. Palmer *et al.* (1973) have recently reported that DDT is only marginally positive with respect to the dominant lethal test in rats. In a study employing the host-mediated assay with *Salmonella typhimurium* and *Serratia marcescens*, DDD produced an elevated mutation rate and DDT and DDA were adjudged possibly mutagenic (Buselmaier *et al.*, 1972).

In one of the first studies associating chromosomal damage with DDT, a near lethal dosage of DDT fed to dogs caused a complete breakdown of mitotic activity in bone marrow and resulted in tripolar mitosis and chromosomal damage (Gerelzoff *et al.*, 1950). However, subsequent studies to determine the cytotoxic effects of a variety of pesticides (including DDT and HCH) *in vitro* in cultured cells and in rat bone marrow after oral dosing showed that although DDT caused mitotic arrest at 10^{-5} M in culture it produced no significant increase in the rate of mutation (Styles, 1972).

Exposure of cultured human embryonic fibroblasts to DDT and several other pesticides at concentrations occurring in the environment had little effect on the number of degenerating cells up to 192 hr, although simultaneous exposure of the cells to pesticide combinations or to a DDT–hexachloran mixture considerably increased the number of degenerated cells (Perelygin *et al.*, 1973).

Legator *et al.* (1973) described a collaborative study of the *in vivo* cytogenetic analysis of bone marrow cells from rats treated orally or intraperitoneally with DDT or with trimethyl phosphate (positive control) or corn oil (solvent control) as a single treatment or daily for 5 days. The animals were sacrificed at 18, 24, or 48 hr after the single treatment, or 6 hr after the last of the multiple treatments, and bone marrow cells from the rats were examined independently by four laboratories to evaluate the *in vivo* cytogenicity of DDT. The results from each laboratory were in close agreement and indicated that *p,p′*-DDT exhibited no dose–response relationship and produced no increase in chromosomal aberrations from that of the solvent control; the response to trimethyl phosphate was positive.

It is in the area of the reported carcinogenicity of DDT that most confusion and conflicting literature exist. The experimental evidence for the carcinogenicity is based on a number of published reports. Fitzhugh and Nelson (1947) reported the occurrence of hepatic cell tumors in four of 156 Osborne-Mendel rats receiving technical DDT (81.8% *p,p′*-isomer and 18.2% *o,p′*-isomer) in the diet at levels of 200–800 ppm. Eleven other rats were described as showing adenomatoid hyperplasia of the liver. All 15 rats affected survived over 18 months and their age distribution corresponded to that in untreated controls, 1% of which developed hepatomas. In discussing this study, Barnes (1966) noted that although a very detailed account was given of the hepatocellular changes no specific mention was made of inequality of cell size, of variability in size and character of cell nuclei, or of other changes considered to be precancerous in animals receiving liver carcinogens. The complete absence of any reference to the frequency of tumors in the different groups was also noted by Barnes (1966) with the observation that "with an 8-fold difference in dose rates there should have been a striking difference in both the frequency and time of onset of the tumors in the groups on the highest and lowest doses, if DDT had behaved like a true liver carcinogen of which many are known."

In a test of 130 pesticides and related compounds, DDT significantly increased incidences of hepatomas in mice of two strains following its continuous administration for 80 weeks (Innes *et al.*, 1969). *p,p′*-DDT was given by gavage at a dose of 46.4 mg/kg in 0.5% gelatin from day 7 to day 28 and later on at the level of 140 ppm in the diet. The incidence of hepatomas in 36 treated males was 50% and in 36 treated females was 14%, while it was 7% in 169 untreated males and 0.5% in 169 untreated females. An increased incidence of lymphomas was observed among females of the (C57BI/6 X AKR)F_1 strain but not among females of the (C57BI/6 X C3H/Amf)F_1 strain.

Kemeny and Tarjan (1966) described a DDT study in BALB/c mice extending over five generations. DDT-fed mice received 2.8–3.0 ppm (corresponding to a daily intake of 0.4–0.7 mg/kg) for 6 months and the mice were observed for an additional 20 months. The basic diet of the control animals contained 0.2–0.4 ppm. Accumulated DDT storage level in the body fat of the treated mice was about 7–11 ppm for the F_3, F_4, and F_5 generations, a value close to that of the urban human population (Denes and Tarjan, 1964). It was estimated that blastemogenic levels of DDT might approach 0.3–0.6 ppm. A progressive increase in tumor incidence was observed from the second generation onward. In the experimental group of 684 mice, 24 animals (3.51%) developed leukemia (three types) and 37 (5.41%) developed tumors of extremely variable localization and structure. Malignant tumors were observed in only five (1.22%) of 406 control animals. In a subsequent report on these data, Tarjan and Kemeny (1969) indicated that tumor incidence of all five generations was higher than that of the controls. The most frequent tumor types were leukemia, reticulum cell sarcoma, carcinoma of the lungs, and hemangioendothelioma. A transgeneration cumulative effect, resulting in a progressive increase in tumor incidence from the second generation onward, was not substantiated (Terracini and Tarjan, 1970).

Tomatis *et al.* (1972) in a definitive study described the long-term exposure to DDT of CF-1 mice. CF-1 minimal inbred mice were fed technical DDT (73–78% *p,p*′-DDT, 20% *o,p*′-DDT, 1% *m,p*′-DDT, 0.5–1.5% *p,p*′-DDD, and 0.5% *p,p*′-DDE) in the diet at the dose levels of 2, 10, 50, and 250 ppm for the entire life span for two consecutive generations. (The DDT content of the control diet varied between 0.02 and 0.09 ppm; γ-HCH was found once at 0.17 ppm and once at 0.04 ppm, while no other organochlorine pesticides were detected at the 0.01 ppm level.) Exposure to all four dosage levels of DDT resulted in a significant increase in liver tumors in males, and was most evident at the highest level used. In females, the incidence of liver tumors was slightly increased following the exposure to 10 and 50 ppm, while a marked increase was observed following exposure to 250 ppm. In DDT-treated animals the liver tumors were observed at an earlier age than in untreated controls. The age at death with liver tumors and the incidence of liver tumors appeared to be directly related to the dose of DDT to which the mice were exposed. Four liver tumors (all occurring in DDT-treated mice) gave metastases. Histologically, liver tumors were either well-differentiated nodular growths pressing but not infiltrating the surrounding parenchyma or nodular growths in which the architecture of the liver was obliterated, showing glandular or trabecular patterns. These results basically support those described by Innes *et al.* (1969) and Fitzhugh and Nelson (1947) as well as the reported occurrence of hepatomas in rainbow trout fed technical DDT (Halver, 1967). They do not, however, confirm the DDT-induced increase in the incidence of a variety of tumors at extrahepatic sites as reported by Kemeny and Tarjan (1966) and Tarjan and Kemeny (1969).

Another study on the possible carcinogenic action of DDT was carried out over five consecutive generations with organ cultures of embryonal lungs from the fetuses of mice treated with 10 or 50 mg/kg DDT (Shabad, 1972). Although a variety of changes were observed in the organ cultures (growth stimulation, increase in diffuse and local hyperplasia of the epithelium, dystrophic and adenomatous changes), it was concluded that the blastemogenic effects of DDT are weak.

The predictive value of mouse liver tumor induction in carcinogenicity testing was discussed by Tomatis *et al.* (1973). A survey of literature data on 58 chemicals (including DDT, dieldrin, aldrin, HCH, mirex, and strobane) revealed a positive correlation between the capacity of a chemical to induce parenchymal liver tumors in the mouse and its ability to induce tumors at any site in the rat or the hamster. The strongest correlation is found when the chemical induces both hepatic and extrahepatic tumors in both sexes of adult mice, although the induction of liver tumors in the mouse does not signify that the liver will be a target organ in the rat or the hamster. In contrast, it has recently been concluded (Grasso and Crampton, 1972) that the induction of mouse hepatomas cannot be considered a valid demonstration of carcinogenicity.

A multigenerational study on the long-term effects of three dose levels of technical DDT (2, 20, and 250 ppm in the diet) in two separate colonies of BALB/c mice has been described by Terracini *et al.* (1973); *N*-nitrosodimethylamine (DMN) (0.003%) administered in the drinking water served as a positive control and treatment of parent mice with DDT started at 4–5 weeks of age. The salient results observed in the first two generations were as follows: (a) in females, the highest dose level of DDT produced liver-cell tumors in 44% of the parent mice and 74% of the F_1 mice, whereas no liver-cell tumors were found in mice given the control diet or lower dietary concentrations of DDT; (b) liver-cell tumors did not give metastases but grew after transplantation to syngeneic mice; (c) the ability of DDT to induce liver-cell tumors was identical in both colonies of mice given 250 ppm DDT; (d) malignant lymphomas occurred in about 50% of the mice in all colonies given 0, 2, or 20 ppm DDT; (e) the incidence or lung adenomas was not affected by DDT treatment; (f) tumors at sites other than liver, lymphatic system, and lungs were most numerous in control mice and least numerous in mice given 250 ppm DDT; (g) no linear dose response was found; (h) no particular trend was observed in the distribution of liver-cell tumors; (i) DMN-positive control mice consistently produced lung adenomas, liver blood cysts, and hemangioendotheliomas.

Liver tumors have been reported to occur in mice exposed to a number of other organochlorine insecticides including mirex, strobane (Innes *et al.*, 1969), aldrin and dieldrin (Davis and Fitzhugh, 1962), and HCH (Nagasaki *et al.*, 1971). Lehman (1952) reported that at 2000 ppm in the diet (but not at 500 ppm) the DDT analogue methoxychlor also produced liver tumors in rats. Of 20 survivors of a group of rats fed methoxychlor at a level of 2000 ppm for

18 months, three had benign liver tumors and a fourth had a malignant liver tumor (Food and Drug Administration, 1958).

Although the results of many of these studies are qualitatively similar, there are considerable variations in the dose level of DDT required to elicit an effect. This is clearly dependent on the species of animal employed in the test as well as on the strain of any particular species. Various strains of mice exhibit large variations in their susceptibility to tumor development. Thus CF-1 mice have an incidence of spontaneous liver-cell tumors of the order of 22% in males and 4% in females (Tomatis *et al.*, 1972), whereas in the BALB/c strain hepatoma incidence in control mice is very rare (Terracini *et al.*, 1973; Andervont and Dunn, 1948; Deringer, 1965; Madison *et al.*, 1968; Tarjan and Kemeny, 1969; Smith and Pilgrim, 1971). In addition to the inherent degree of susceptibility to tumor formation, it is likely that some of the species and/or strain variations observed reflect metabolic or other differences.

The considerable species variation observed in studies on DDT-induced tumor formation is of extreme importance in attempts to extrapolate the results of animal studies to man. The relatively high susceptibility of mice to DDT-induced hepatoma appears to be well documented (Innes *et al.*, 1969; Tomatis *et al.*, 1972; Terracini *et al.*, 1973), although only one study (Tomatis *et al.*, 1972) provided dose–response data. Rats generally appear less susceptible (Fitzhugh and Nelson, 1947) and Syrian golden hamsters much less susceptible (Agthe *et al.*, 1970) to the effects of DDT. It is also interesting to note that in a study of workers exposed in their occupation to high levels of DDT for 10–20 years (Laws *et al.*, 1967) no cases of cancer were reported in some 1300 man-years of exposure.

It is important to point out that several studies have failed to show any tumorigenic effects of DDT, and in some cases antitumorigenic effects have been reported. Neither benign nor malignant tumors were observed in mice painted weekly for 52 weeks with 5% DDT in kerosene (Bennison and Mostofi, 1950), although the treated mice did show acute and chronic inflammatory changes in the painted skin and very slight liver damage. The subcutaneous administration of 15,000 mg/kg body weight of DDT to newborn Swiss mice (which were sacrificed at the age of 6 months) failed to provide evidence of carcinogenicity after 6 months (Gargus *et al.*, 1969), and no evidence of carcinogenicity was obtained from one experiment with monkeys (Durham *et al.*, 1963).

In studies with various pesticides and pesticide mixtures (Deichmann *et al.*, 1967; Radomski *et al.*, 1965), little or no evidence of synergistic or additive carcinogenic effects was obtained. Although hypertrophy and hyperplasia of the hepatic cells were noted in rats fed 150 ppm DDT for 2 years (Deichmann *et al.*, 1967), only mild hepatic lesions were observed with dietary levels of 1–50 ppm DDT given for up to 27 weeks (Laug *et al.*, 1950). Leukemia was reported following long-term administration of 3.5 mg DDT per rat per day, but it was concluded by Kimbrough *et al.* (1964) that this resulted from the use of a purified diet rather than from DDT ingestion.

The suggestion of anticarcinogenic potential of DDT has been advanced (Jukes, 1970; Walker *et al.*, 1970*a,b*). Laws (1971) reported evidence of an antitumorigenic effect of technical DDT (5.5 mg/kg/day) toward transplants of an experimental ependymoma in C57B1 mice exposed to a chronic dosage. Compared to controls, DDT-treated animals showed a decrease in the "take" of tumor transplants and a significant increase in longevity. Autopsies revealed no evidence of liver tumors or other hepatic pathological findings. The study of Terracini *et al.* (1973) also suggested a decreased incidence of tumors at sites other than the liver, lung, and lymphatic system in female mice given 250 ppm DDT.

5.1.2. Cyclodienes

The results of studies on the effects of cyclodiene insecticides on reproduction and neonatal mortality are equivocal and vary with the compound and test species employed. The transport of dieldrin from mother to blastocyst and fetus in pregnant rabbits has been reported by Hathaway *et al.* (1967).

The incidence of pregnancy in rats receiving continuous exposure to 2.5 or 12.5 ppm dieldrin initially decreased, but in subsequent matings no effect on pregnancy was observed (Treon and Cleveland, 1955). Good and Ware (1969) reported the effects of dieldrin and endrin on reproduction in mice following the feeding of the insecticide at 5 ppm for 120 days beginning 30 days before mating. Dieldrin produced no parent mortality but significantly reduced the size of all litters, while endrin produced significant parent mortality and significantly smaller litters. Neither affected fertility, fecundity, or the number of young produced.

Smith *et al.* (1970) reported that injection of 5 to 10 mg of dieldrin into fertile hen eggs prior to incubation caused decreases in hatchability of 30% and 40%, respectively. In addition, chicks from the 10 mg treatment exhibited tremors and died within 24 hr, while the chicks given 5 mg all died within 1 week.

The effects of combinations of organochlorine insecticides on the reproduction of Swiss white mice through five or six generations were studied by Keplinger *et al.* (1968*a,b*). The insecticides fed individually and in combination were aldrin (3, 5, 10, and 25 ppm), dieldrin (3, 10, and 25 ppm), chlordane (25, 50, and 100 ppm), DDT (25, 100, and 250 ppm), and toxaphene (25 ppm). The experiment was designed to determine effects in the offspring caused by absorption of the insecticide or combination of insecticides, via the placenta, ingestion of mother's milk, and ingestion of contaminated food. Little or no adverse effect was noted through five or six generations in mice fed 25 ppm toxaphene, 25 ppm chlordane, or 25 ppm DDT. However, marked effects on fertility, gestation, viability, lactation, or survival indices were noted in the parent and second generations and in their offspring fed aldrin (25 ppm), dieldrin (25 ppm), and various combinations of these with the other insecticides. Histological examination of the organs and tissues revealed changes in the livers of all groups, and in the kidneys, lungs, and brains of most groups. The combination of aldrin and chlordane caused more severe changes in the liver than the other pesticides or combinations fed. In other studies, the maternal ingestion of high levels of aldrin caused an increased neonatal toxicity in dogs (Kitselman, 1953), and Boucard *et al.* (1970) reported that 2.5 mg/kg or 85% dieldrin administered to Wistar rats and Swiss mice produced fetal malformations.

Noda *et al.* (1972) administered an endrin emulsion (at 0.58 mg/kg) orally once a week for 4 weeks to virgin female mice and rats. The animals were mated once a week after the last administration and sacrificed on day 18 (mice) or day 20 (rats) of pregnancy. Fetal mortality rates were 13% for treated mice and 9% for controls, 34% for treated rats and 17% for controls. There were 15 malformed or immature fetuses in the treated mice and eight in the controls. The most frequently occurring malformation in mice was clubfoot, and the incidence of skeletal anomalies (primarily delayed and deviant ossification) was significantly elevated in treated rats.

Postmortem examination of sows which had received repetitive oral doses of up to 15 mg dieldrin/kg during the last 30 days of gestation revealed that although dieldrin concentrations in fetal liver and brain were significantly lower than those in maternal tissues, placental transfer had obviously occurred. Mild degeneration of the renal tubules and slight hepatic lipidosis were observed in the sows, but no lesions were detected in the fetuses and no fetal deaths or abortions occurred in the treated sows.

Ottolenghi *et al.* (1973) found that single oral doses of aldrin, dieldrin, and endrin, $\frac{1}{2}$ LD_{50} in corn oil, administered to pregnant golden hamsters on day 7, 8, or 9 of gestation caused a high incidence of fetal death, congenital anomalies, and growth retardation. The most frequent defects were cleft palate, open eye, and webbed foot, often occurring in combination. Pregnant CD-1 mice given equivalent oral doses of each pesticide on day 9 of gestation showed similar anomalies without concurrent increase in fetal mortality or growth impairment.

Cytogenetic effects of dieldrin, aldrin, hexachlorane, heptachlor, and DDT (at doses of 4.0–6.7% the LD_{50} values) on mouse bone marrow cell nuclei were reported by Markaryan (1966). Cytogenetic analysis carried out 21 hr after intraperitoneal administration of the test compounds in oil solution showed a significant increase in the frequency of nuclear disturbance. In addition to increased chromosome rearrangement and adhesion, other nuclear changes including vacuolation, karyopycnosis, karyorrhexis, and karyolysis were observed. In particular, pycnosis and vacuolation appeared in the nuclei during the telophase, with rearrangement and multiple attachment of the chromosomes. Markaryan (1966) concluded that "the high mutagenic activity of the chloro-organic insecticides demonstrated in this work on bone marrow cells makes urgent the further study of their mutagenic properties in different tissues and organs."

Bunch and Low (1973) described the effects on juvenile mallard ducks of parental exposure to dieldrin (0, 4, 10, and 30 ppm) for approximately 60 days. The level and duration of treatment for each juvenile matched those for its parent, and at the end of this period femoral bone marrow cultures were examined for chromosomal aberrations. No significant incidence of abnormalities was found in ducks at any of the four levels of treatment. The mitotic index varied little among the juveniles maintained on either the 0, 4, or 10 ppm dieldrin treatments but was reduced more than five times in ducks exposed to 30 ppm ($P < 0.05$). Lymphocytes treated with 100 ppm dieldrin exhibited an approximate twofold increase in incidence of chromosome structural alterations over controls and the mitotic index was significantly reduced in all treated cultures. These results suggest that the levels of dieldrin commonly found in free-living waterbirds (less than 4 ppm) are probably too low to elicit chromosome aberration.

Cytological changes have recently been reported to occur in male albino rats by Dikshith and Datta (1973) following the intratesticular injection of 0.25 mg of endrin. Effects observed included chromosome fragmentation and abnormal restitution, and the formation of single and double bridges with acentric fragments resulted in the transformation of the chromatin mass into an amorphous lump.

The possible mutagenic activity of 34 pesticides including aldrin, dieldrin, endrin, heptachlor, heptachlor epoxide, lindane, and methoxychlor to *Drosophila melanogaster* was evaluated by Benes and Sram (1969). None of these organochlorine insecticides showed any significant mutagenic activity.

Cerey *et al.* (1973) conducted a comprehensive study in which rats were continuously exposed to diets containing 1 and 5 ppm heptachlor over three generations and subjected to a variety of tests to determine reproductive performance and possible mutagenic effects. A significant increase in the number of resorbed fetuses was observed in the second and third generations, and cytogenetic studies revealed that this was accompanied by an increase in the incidence of abnormal mitoses, abnormalities of chromatids, and pulverization and translocation in the bone marrow cells of the treated animals. Dominant lethal changes were observed in male rats by Cerey *et al.* (1973), although Epstein *et al.* (1972) had previously reported both heptachlor and heptachlor epoxide to be nonmutagenic by the dominant lethal assay in ICR/HA Swiss mice.

As with DDT, evidence for the tumorigenicity or carcinogenicity of the cyclodiene insecticides appears to be conflicting. Davis and Fitzhugh (1962) concluded that since the long-term feeding of both aldrin and dieldrin to mice resulted in statistically significant increases in the incidence of hepatic tumors (no lung metastases were found) the compounds were tumorigens. However, since the tumors were morphologically benign and since no metastases occurred, they were not considered to be carcinogenic. Fitzhugh *et al.* (1964 later noted an increase of tumors in rats and dogs chronically fed aldrin or dieldrin, although the effects were not clearly dose related.

Studies in dogs and rats (Walker *et al.*, 1969), monkeys (Wright, personal communication, 1971), and man (Jager, 1970) have revealed no evidence of tumorigenic or carcinogenic activity by aldrin and dieldrin.

Cabral (1972) reported the absence of long-term effects in suckling Wistar rats (95) given five administrations of 10 mg/kg heptachlor in corn oil by stomach tube every second day starting at 10 days of age. Nine of 28 treated females developed a total of 12 tumors in various organs (including five mammary tumors and two renal lipomatous tumors), whereas four of 27 control females developed a total of four tumors (two of which were located in the breast). In view of the different locations of the tumors and the lack of reproducibility in results with males, the data were not considered to provide evidence for the carcinogenicity of heptachlor.

Walker *et al.* (1972) reported on two studies comprising five long-term oral studies in which dieldrin (purity greater than 99% HEOD) and *p,p'*-DDT (purity greater than 99.5%) both alone and in combination were fed to CF-1 mice at various dietary concentrations. No signs of ill health were associated with the feeding of dieldrin at levels of 10 ppm or less, or of DDT at levels at or below 100 ppm, although after long periods of exposure (depending on dose) liver enlargement was detectable and was associated with increased morbidity. Liver lesions observed in all the tested groups, including controls, were described as either (a) a simple nodular growth of liver parenchymal cells or (b) areas of papilliform and adenoid growth of tumor cells (sometimes accompanied in both control and treated animals by metastases to the lungs). These lesions are quite similar to those reported in rats by Fitzhugh and Nelson (1947), particularly the increased size of hepatocytes with cytoplasmic change in the centrilobular region. The morphological changes regress only slowly and to a variable extent after cessation of dieldrin exposure, but since the concentration of dieldrin in liver also decreases slowly after exposure is terminated, a rapid return to a normal histological appearance would not be expected.

Rosival and Kapeller (1970) described slight changes in the hepatic endoplasmic reticulum (irregularly shaped vacuoles) in rats receiving doses of heptachlor of 2 mg/kg for 7 days and 6 mg/kg for 28 days, but no lesions could be established on the basis of several enzymatic assays.

Lifetime tests with aldrin and dieldrin in albino rats were reported (Deichmann *et al.*, 1970). Both male and female Osborne-Mendel strain rats, fed either aldrin or dieldrin at a level of 20, 30, or 50 ppm, showed a *decrease* in the incidence of all tumors, particularly tumors of the mammary and lymphatic tissues, in relation to controls. It was suggested that these results might reflect the induction of microsomal enzyme activity by aldrin and dieldrin, which in turn could influence the incidence of hormone-dependent mammary tumors; also considered was the possibility of an action on the adrenal cortex suppressing lymphopoiesis.

5.1.3. Hexachlorocyclohexane (HCH)

Daily subcutaneous administration of β-HCH (100 mg/kg) to pregnant ICR-JCL mice from day 2 to day 6 of gestation caused various abnormalities such as retardation of growth and early death of the fetuses. A significant retardation of fetal growth was also observed following administration of β-HCH at 5 mg/kg in the diet for a period starting 1 month prior to mating and continuing to day 18 of gestation or oral administration of 100 mg/kg on days 6 and 7 of gestation (Yamagishi *et al.*, 1972).

Few investigations have been reported on the hepatocarcinogenic effects of HCH isomers. Nagasaki *et al.* (1971) found a 100% incidence of liver tumors in male dd mice fed 660 ppm technical HCH, but no effect was observed at 66 ppm. In a subsequent study in which several other isomers of HCH were fed to male dd mice at various concentrations, only the α-isomer at 500 or 250 ppm caused the appearance of hepatomas similar to those induced by crude HCH (Nagasaki *et al.*, 1972); the β-, γ-, and δ-isomers of HCH exhibited no carcinogenic activity.

Thorpe and Walker (1973) described a 2-year oral toxicity study in which highly purified samples of the β- and γ-isomers of HCH were fed to CF-1 mice at various dietary concentrations. Liver enlargement was detected in both sexes after 50–60 weeks with both γ- (400 ppm) and β-isomers (200 ppm). Liver lesions observed were classified as hyperplastic foci and consisted of either simple nodular growths of parenchymal cells or papilliform and adenoid growths of tumor cells sometimes associated with lung metastases.

Despite the massive size and multiplicity of liver lesions in the treated mice, only a minimum shortening of the life span occurred during the 2-year experiment. No increase in neoplasms in nonhepatic tissues was found in any of the treated groups, and in fact reduced numbers of tumors were reported in male mice exposed to 200 ppm β-HCH.

As pointed out by Thorpe and Walker (1973), the biological significance of the altered incidence of liver tumors in mice is difficult to evaluate. Roe and Grant (1970) as well as Grasso and Crampton (1972) have suggested that the increased incidence of liver tumors in mice constitutes insufficient evidence *per se* to classify a compound as carcinogenic. This view is supported by reports that the development of liver tumors in mice may be influenced by the protein or calorific value of the diet (Tannenbaum and Silverstone, 1949*a*,*b*), the source of the diet (Heston *et al.*, 1960), the strain of mouse (Bonser *et al.*, 1952; Keplinger *et al.*, 1968*a*,*b*; Tomatis *et al.*, 1972), the microbial flora of the animals (Roe and Grant, 1970), and alterations in endocrine status (Warwick, 1971).

Thorpe and Walker (1973) also discussed the diagnostic problems posed by the nodular liver lesions in mice and suggested that the term "hepatoma" was inappropriate to describe these lesions because of different connotations in mouse and human pathology (International Agency for Research on Cancer 1971). Although proof is apparently lacking, parenchymal cell hyperplasia benign neoplasms, and primary parenchymal cell carcinoma in rodent liver may be associated with a progressive spectrum of change (Popper and Schaffner, 1957; Anonymous, 1972). Warwick (1971) has shown that spontaneous liver nodules are not uncommon in adult males of certain mouse strains and the incidence of lesions can be increased by exposure of neonates or adults to various chemicals. Further, morphology alone is suggested to be inadequate to identify hyperplastic lesions from some liver cell neoplasms in mice (Thorpe and Walker, 1973).

5.1.4. Mirex and Kepone

The insecticide kepone has been shown to decrease the reproductive success of mice when administered in the diet at 5 ppm (Good *et al.*, 1965). In part, this appears to be due to the fact that test females are in constant estrus (Huber, 1965). Bioassay revealed a decreased activity of luteinizing hormone in the pituitary of the females, and although the level was probably adequate to act with the follicle-stimulating hormone and result in estrogen secretion it was insufficient for ovulation.

Naber and Ware (1965) reported that reduced hatchability in chickens was observed only when laying hens were maintained on diets containing at least 150 ppm kepone or 600 ppm mirex, levels approximately fifteenfold greater than those required to produce changes in mouse fertility and fecundity.

Large-scale feeding studies with the BALB/c strain of Swiss mice indicated that at a dietary level of 5 ppm mirex caused parent mortality and a reduction in litter size (Ware and Good, 1967). The latter was also observed with the CFW strain under a similar treatment schedule, although in this strain little or no parent mortality occurred (Ware and Good, 1967).

Mirex has been reported to be a hepatocarcinogen when administered daily to (C57BL/6 X C3H/Amf)F_1 and (C57BL/6 X AKR)F_1 mice at a dosage of 10 mg/kg in 0.5% gelatin during days 7–28 of age or when provided *ad libitum* in the diet at 26 ppm after 28 days of age (Innes *et al.*, 1969).

5.2. Organophosphorus Insecticides

5.2.1. Malathion, Parathion, and Methyl Parathion

With the apparent diminishing use of the chlorinated hydrocarbon insecticides in the United States and many Western European countries, the organophosphorus insecticides are assuming an increasingly important replacement role. The organophosphorus compounds currently in use are generally considered nonpersistent and biodegradable, but many have a high acute toxicity to mammals. Three of the most widely used organophosphorus compounds are malathion, parathion, and methyl parathion.

Interest in the possible teratogenic potential of organophosphorus insecticides in mammals was instigated by reports that this important class of insecticides caused nervous system lesions (Baron and Johnson, 1964) as well as skeletal abnormalities (Khera, 1966; Khera *et al.*, 1965, 1966; Khera and Bedok, 1967; Greenberg and La Ham, 1969; Meiniel *et al.*, 1970) in various species of birds. Although several organophosphorus pesticides are now recognized as being teratogenic (especially to avian embryos), the mechanisms responsible for their action have not yet been unequivocally elaborated. Ghadiri *et al.* (1967) reported a general correlation between the maximum quantity of an organophosphate which permitted some chicks to hatch and its acute oral toxicity to mice. Marliac *et al.* (1965) confirmed this with several insecticides, although diazinon was about 300-fold more toxic to the chicken embryo than the rat. Eggs from hens fed for 3 weeks on basal diets containing malathion and carbaryl (both singly and in combination) at levels of up to 600 ppm showed a concentration-dependent decrease in hatchability (Ghadiri *et al.*, 1967).

Malathion, one of the least acutely toxic organophosphates to mammals, is teratogenic to the chick embryo *in ovo* (McLaughlin *et al.*, 1963; Walker, 1967; Greenberg and La Ham, 1969). McLaughlin *et al.* (1963) reported that malathion lowered the percent hatch and produced abnormalities including short legs and bleaching effect on down in the chick embryo. Injections of 3.99 mg malathion into the yolk sac of 4- to 5-day-old incubated eggs consistently produced deformed chicks with a combination of sparse plumage, micromelia, overall growth retardation, and beak defects (Greenberg and La Ham, 1969).

Walker (1967), while confirming the results of McLaughlin *et al.* (1963), pointed out that the effects observed were to some extent dependent on the physical characteristics of the vehicle in which the malathion was injected. In view of this factor, which is presumably associated with the distribution of the injected material within the egg, Walker (1967) indicated that the effects of various chemicals following injection are not valid criteria for evaluating or comparing their activities. As a result of this finding, Walker (1968) developed a new administration technique whereby the material to be tested was first incorporated into egg yolk and this was subsequently transferred to the yolk sac cavity of a fertile egg. Employing this technique, he found that malaoxon was more teratogenic than malathion at a low level, but less so at a higher level (Walker, 1971). Abnormalities such as shortened tibiofibulae and toes, sparse or clubbed down, and hooked beaks were observed in embryos from malathion-treated eggs (Walker, 1968). As will be discussed later in relation to the teratogenic action of DDVP, it is possible that the effects of malaoxon and/or malathion result from inhibition of an unidentified esterase.

Wilson and Walker (1966) found that concentrations of malathion above 1 μg/ml (3.0×10^{-6} M) inhibited the growth rate and peak cell production in cultured 14-day-old chick fibroblasts and cell populations rapidly decreased in cultures containing 50 μg/ml. In a more recent study, Wilson *et al.* (1973) have demonstrated that malathion, malaoxon, parathion, and paraoxon all inhibit the growth of cultured chick embryo pectoral muscle cells. It is possible that the inhibition of cell growth by malathion can be attributed to its ability to inhibit protein synthesis (Gabliks and Friedman, 1965*a,b*), a mechanism already proposed to account for its teratogenic action (Greenberg and La Ham, 1969).

Despite the well-established teratogenic activity of malathion and several other organophosphates toward chick embryos, few investigations have been reported with mammals. Studies with malathion and diazinon in the Wistar rat (Dobbins, 1967) revealed that malathion (205.4 mg/kg) administered on day 10 of the gestation period and diazinon (63.8 mg/kg) on day 9 both produced morphological malformations manifested primarily as hydronephrosis and hydroureter. However, the malformations were mild, inconsistent, and apparently not dose related, and it was suggested they may have been related to factors other than the insecticides. In another study, the intraperitoneal administration of toxic doses of malathion (600–900 mg/kg) to female Sherman rats neither affected the weight of the fetus nor produced malformations (Kimbrough and Gaines, 1968). Breeding experiments in which Wistar rats were exposed continuously to 240 ppm malathion showed a decrease in litter size and weight and an increase in the incidence of ring-tails (Kalow and Martin, 1961).

Czeizel *et al.* (1973) studied human chromosome anomalies in 31 acute organophosphorus intoxication cases from either suicide attempts or accidental exposure to malathion, dimethoate, mevinphos, dichlorvos, methyl parathion, trichlorfon, and diazinon. The frequency of stable chromosome-type aberrations (chromosomal breaks) temporarily increased significantly in acutely intoxicated patients. It was suggested that these temporary somatic chromosome mutations might have significant long-term consequences for the acutely intoxicated individual, and questions were raised concerning the possible effects of chronic exposure to low doses of the pesticides. Of the insecticides shown to cause human chromosome anomalies, malathion was assessed as being the most dangerous.

Parathion (0.1 mg/egg) is reported to be teratogenic toward developing chick embryos (Marliac, 1964). Khera (1966) found that injection of parathion (as well as diazinon, trithion, and ruelene) at any stage of embryonic development caused the formation of cartilaginous and osseous skeleton, eye cataracts, ascites, and hepatic degeneration in both ducks and chicks. Skeletal defects observed were dwarfism, micromelia, ectrosyndactyly, stunted growth of cervical vertebrae with or without fusion, and irregular beak growth.

The reduction in fetal cerebral cholinesterase activity following intraperitoneal injection into pregnant Holzmann rats of parathion and methyl parathion (Fish, 1966) clearly indicated transplacental passage of these lipid-soluble phosphorothionates, as shown by others (Ackermann and Engst, 1970). Although large subcutaneous hepatomas were observed in the treated animals, no significant increases in resorption or developmental defects were noted. High stillbirth and neonatal death rates occurred in the offspring, and the weight gain in surviving neonates was substantially lower than that of the controls.

Single intraperitoneal injections of methyl parathion in Wistar rats (15 mg/kg) on day 12 of gestation and ICR-JCL mice (60 mg/kg) on day 10 (doses equivalent to respective LD_{50} values) caused severe systemic symptoms and a transient decrease in food and water intake (Tanimura *et al.*, 1967). In rats there were no embryotoxic effects other than suppression of fetal growth and ossification, while in mice high fetal mortality and incidence of cleft palate also occurred. Tanimura *et al.* (1967) suggested that these effects may have resulted from a combination of the transient depression of maternal food intake and the chemical actions of methyl parathion *per se* or its metabolites on the embryo. It has also been suggested that the reduced weight of the fetus and placenta and the high incidence of resorptions following treatment of Sherman rat dams with parathion (3.5 mg/kg intraperitoneally on day 11 of gestation) might result from the action of *p*-nitrophenol metabolically released from parathion (Kimbrough and Gaines, 1968). *p*-Nitrophenol is known to be a weak uncoupler of mitochondrial oxidative phosphorylation and as such could increase metabolic activity with consequent weight loss. The potent uncoupler 2,4-dinitrophenol has been associated with an increase in stillbirths in rats (Wulff *et al.*, 1953).

5.2.2. DDVP (Dichlorvos) and Trichlorfon (Dipterex)

Few pesticides have engendered as much recent concern with respect to potential health hazards as the organophosphate DDVP, which is the major insecticidal component of the widely used and controversial "No Pest" resin strips. These are commonly used in enclosed domestic and public situations (stores, restaurants, etc.) where humans may be chronically exposed through either direct inhalation of the released vapor or ingestion of contaminated food. It is in the area of its reported alkylating properties (Löfroth *et al.*, 1969; Löfroth, 1970; Bedford, 1972) and potential mutagenicity that most concern and controversy exist.

Much of the evidence for the ability of DDVP to alkylate DNA and to cause mutations has been obtained by direct addition of the insecticide to a variety of bacterial test systems in culture. Mutagenicity has been demonstrated in various strains of *Escherichia coli* (Löfroth *et al.*, 1969; Ashwood-Smith *et al.*, 1972; Mohn, 1973; Bridges *et al.*, 1973; Voogd *et al.*, 1972), *Serratia marescens* (Dean, 1972*a*), *Saccharomyces cerevisiae* (Fahrig, 1973), and others including *Klebsiella pneumoniae* and *Salmonella typhimurium* (Voogd *et al.*, 1972). Some of the data, however, are conflicting. Voogd *et al.* concluded that under the conditions employed in their study DDVP was about twice as mutagenic as caffeine, which is known to have weak mutagenic properties (Fishbein *et al.*, 1970; Voogd and Van der Vet, 1969).

Dean (1972*b*) stressed the fact that the conditions in the bacterial test systems, where the DDVP is in intimate contact with the cell and hence more readily available for DNA interaction, are vastly different from those existing *in vivo*, where the DDVP is confronted by a variety of physicochemical barriers and degradative enzymes (Hutson and Hoadley, 1972*a,b*). Thus in spite of the mutagenic action of DDVP when added directly to cultures of *S. typhimurium* (Voogd *et al.*, 1972) no such effect was found when the same organism was used in the host-mediated assay with Swiss mice (Cabridge and Legator, 1969). Similarly, Dean *et al.* (1972) were unable to demonstrate any enhanced mitotic gene conversion with *S. cerevisiae* in the host-mediated assay with CF-1 mice (dosed orally with up to 100 mg/kg DDVP or exposed for 5 hr to atmospheres containing up to 99 μg DDVP vapor/liter) despite the mutagenic activity of DDVP in stationary phase cultures of this microorganism (4 mg/ml),

Dean and Thorpe (1972*a*) demonstrated the absence of dominant lethal mutations in CF-1 mice following inhalation exposure to DDVP at concentrations of 30 and 55 μg/liter for 16 hr a day for 4 weeks. There were no preimplantation losses or early fetal deaths in subsequent matings and no impairment of male fertility. Dean and Thorpe (1972*a*) pointed out that the concentration of DDVP used in the repeated exposure study (5.8 μg/liter) of air is more than 100 times the air concentrations (0.4 μg/liter) reported for normal domestic use of "No-Pest" strips (Elgar and Steer, 1972). Dean and Thorpe (1972*b*) also reported that chromosome preparations from bone marrow and spermatocytes of Chinese hamsters following high oral and inhalation exposure to DDVP did not differ from those of controls. Although Epstein *et al.* (1972) found that administration of DDVP orally (up to 27.3 mg/kg) and intraperitoneally (up to 16.5 mg/kg) to ICR/Ha Swiss mice caused some early fetal deaths and preimplantation losses, the data were not significantly greater than those for the controls. The oral administration of DDVP (40 mg/kg) to mice produced vacuolization of germinal epithelium, increased the number of spermatids, decreased the number of spermatocytes, and caused degenerating meiosis in metaphase (Krause and Homola, 1972). This dose is far above the estimated maximum human exposure (100–200 μg/day) from ingestion of food and beverages exposed to DDVP strips used in the recommended manner (Gillett *et al.*, 1973).

At concentrations from 5 to 40 μg/ml, DDVP has been reported to be cytotoxic to cultures of human lymphocytes, but no evidence of chromosome aberrations was obtained (Dean, 1972*b*). It was noted that if indeed the cytotoxic action of DDVP results directly from the alkylation of DNA, dose-related chromatid-type aberrations should be produced. An increase in chromatid gaps was observed in one pair of test cultures, but these were not dose related and in general such breaks were of low incidence and evenly distributed between control and test cultures. DDVP also caused a 100% mortality within 24 hr of its addition to HEB cell cultures at concentrations up to 90 μg/ml (Kugaczewska *et al.*, 1972). Since similar cell mortality was observed when the DDVP had been preincubated with the culture medium 24 hr prior to addition of the cells, it is possible that some of the hydrolytic degradation products of DDVP are also active.

A number of conflicting reports have appeared concerning the teratogenicity of DDVP. Kimbrough and Gaines (1968) found three affected fetuses among 41 offspring from four female Sherman rats injected intraperitoneally with 15 mg/kg DDVP on day 11 of gestation, whereas Vogin *et al.* (1971) observed no teratogenic effects in the offspring of female rabbits given oral doses of 12 mg/kg DDVP during the period of major organogenesis. Upshall *et al.* (1968) reported that the inoculation of 1 mg DDVP into the chick egg was only mildly teratogenic. Negative results have also been obtained in studies to determine the effects of DDVP on reproduction in rats (Witherup *et al.*, 1971) and pigs (Collins *et al.*, 1971*b*), and Hendriksson *et al.* (1971) were unable to demonstrate any significant toxic effects with six calves, one heifer, six guinea pigs, seven sheep, 20 mice, or the personnel caring for them after exposure to various numbers of Vapona strips in an experimental stable.

Thorpe *et al.* (1972) studied the teratogenic potential of inhaled DDVP vapor in Dutch rabbits and Carworth Farm E strain rats exposed throughout pregnancy to concentrations up to 6.25 μg DDVP/liter. No indication was obtained that DDVP vapor is teratogenic in rabbits or rats even at exposure concentrations resulting in maternal deaths in rabbits, and causing depression of plasma, erythrocyte, and brain cholinesterase. Thorpe *et al.* (1972) in summarizing these results indicate that a man exposed for 24 hr to 0.04 μg of DDVP/liter (the concentration achieved in the first few weeks of normal use of a DDVP-impregnated resin strip) (Elgar and Steer, 1972) might inhale approximately 6.0 μg/kg. Gillett *et al.* (1973) have estimated a nominal human inhalation exposure of about 1–5 mg/day. At the lowest concentration of DDVP (0.25 μg/liter) used in the experiments of Thorpe *et al.* (1972), a rabbit probably inhales about 110 μg/kg and a rat 300 μg/kg in 24 hr. These doses are about 18 and 50 times, respectively, the theoretical dose inhaled by man under normal-use conditions.

In considering the possible mutagenicity of DDVP, it is important to note that trimethyl phosphate, which occurs as a minor component (0.3–0.8%) in Vapona strips (Slomka, 1970), has been shown to produce mutagenic effects in the dominant lethal assay when administered orally or parenterally to mice (Epstein *et al.*, 1970*a*). The effects, manifested by early fetal deaths and losses before implantation, were restricted to matings during the postmeiotic stages of spermatogenesis and were more marked than indicated by a previous report (Jackson and Jones, 1968). Trimethyl phosphate has also been reported to induce reverse mutations in *Neurospora* (Kolmark, 1956). The potential mutagenic activity of some minor component in an insecticidal formulation emphasizes the importance of establishing the purity of the test compound and illustrates the difficulties likely to be encountered in comparing data from different studies.

Trichlorfon (dipterex, chlorophos) is also present as an impurity (1.5–3%) in Vapona "No Pest" strips, and is used with polyvinyl alcohol for the preparation of "Insektopolimer" (Bromberg *et al.*, 1972) for analogous use. The insecticidal activity of trichlorfon is attributed to its metabolic conversion to DDVP under physiological conditions since trichlorfon itself is devoid of anticholinesterase activity.

Continous inhalation of trichlorfon at concentrations of 1.0, 0.2, and 0.02 mg/m^3 for 90 days produced some morphological alterations in albino male rats (Bonashevskaya and Tabakova, 1972). Exposure to 0.2 mg/m^3 trichlorfon produced a widening of the capillaries in the interalveolar membranes of the lung, slight Disse space enlargements, the occurrence of blood cells in the perivascular spaces of the liver, and moderate hypertrophy of the bile duct epithelium. Severe alterations of the medullar and cortical canaliculi, including a densification of cortical epithelium, were found in the kidney, while alterations of the thyroid were observed at the parenchymal level. More obvious changes were produced after exposure to 1 mg/m^3 trichlorfon and alterations in vascular permeability were considered the basic pathogenetic factor.

Gofmekler and Tabakova (1970) reported the development of embryonic abnormalities and placental changes following the administration of trichlorfon to mice via inhalation.

Trichlorfon (30 mg/kg) administered orally and subcutaneously to cats induced exophytic papillomas in the stomach, and liver damage was observed when it was applied topically for periods of 17 months. Subcutaneous administration of the insecticide to mice induced localized tumors, and as a result it was described by Preussman (1968) as a weak carcinogen.

5.2.3. Bidrin and Azodrin

Embryonic abnormalities have been induced by bidrin and several of its analogues, when injected into the chick egg during the first 9 days of development (Roger *et al.*, 1964). Teratogenic effects were occasionally observed using 100 μg/egg of the *cis*-crotonamide isomer of bidrin and always when 300 μg or more of bidrin was injected per egg, the severity of effect increasing progressively with higher doses. The marked teratogenic effects of bidrin could be alleviated by pre-, post-, or simultaneous injection of nicotinamide and certain of its analogues (Roger *et al.*, 1964).

Upshall *et al.* (1968) examined the effect of bidrin and other teratogenic and nonteratogenic neurotoxicants on esterase activity, on acetylcholine (ACh) levels, and on the distribution and metabolism of labeled ACh within the egg. Embryonic levels of ACh did not appear to be related to the degree of teratogenesis, thus indicating that ACh and the cholinergic system do not play a major role in the differentiating processes involved in organophosphate-induced teratogenesis.

In a classic paper, Flockhart and Casida (1972) postulated a relationship of the acylation of membrane esterases and proteins to the teratogenic action of organophosphorus insecticides and eserine in developing hen eggs. Among the organophosphorus compounds examined in this study were bidrin (technical), azodrin, bidrin amide, paraoxon, and EPN. The yolk sac membrane from hen eggs was found to contain esterases which hydrolyze phenyl phenylacetate (PPA) and which are sensitive to inhibition (both *in vitro* and *in vivo*) by many organophosphorus compounds and by eserine. These esterases are the principal proteins in the membrane which are phosphorylated. Inhibition of the mixture of PPA-hydrolyzing esterases *in vitro* did not correlate with teratogenesis since it was observed with both teratogenic and nonteratogenic compounds. However, electrophoretic separation of the yolk sac membrane esterases following *in ovo* treatment with various organophosphates clearly established the inhibition of specific esterases comprising only a very small proportion of the total mixture. This inhibition was observed only with compounds showing teratogenic effects. Thus it was postulated by Flockhart and Casida (1972) that phosphorylation or carbamoylation of this reactive membrane component(s) *in vivo* may initiate a sequence of events leading to embryonic abnormalities. Inhibition of activity of the reactive membrane esterases was exhibited by all compounds that were teratogenic at the administered dose and included *cis*-bidrin, technical bidrin, *cis*-azodrin, bidrin amide, chloro-bidrin, mevinphos (high dose), dichlorvos, paraoxon, and eserine, but not the compounds that were not teratogenic (EPN, *trans*-bidrin, and a low dose of mevinphos). It is of interest that this postulated mode of action for teratogenesis (Flockhart and Casida, 1972) is very similar to that proposed earlier (Johnson, 1969) for the production of delayed neurotoxicity by some organophosphorus compounds.

5.2.4. Diazinon

Diazinon is teratogenic in the chick embryo test at levels which are 300 less than the rat oral LD_{50} value (Khera, 1966; Marliac *et al.*, 1965).

When administered intraperitoneally to Sherman rats (200 mg/kg on day 11 of gestation; Kimbrough and Gaines, 1968) or by stomach tube to Wistar rats (63 mg/kg on day 9; Dobbins, 1967), diazinon caused a high incidence of fetal resorptions, a reduction in weight of the fetus and placenta, and fetal malformations. In contrast, Robens (1969) found diazinon to be non-teratogenic to New Zealand White rabbits (30 mg/kg) or Syrian golden hamsters (0.25 mg/kg), and in tests with several strains of *Salmonella* it was judged to be nonmutagenic (Seiler, 1973).

5.2.5. Fenthion, Demeton, and EPN

The rapid transplacental passage of demeton in CF3-1 mice has been demonstrated by Budreau and Singh (1973*a*) and the teratogenic and embryotoxic effects of this and fenthion have been studied in CF-1 mouse embryos (Budreau and Singh, 1973*b*). Demeton administered to CF-1 mice between days 7 and 12 of gestation either as a single intraperitoneal dose of 7 or 10 mg/kg or as three consecutive doses of 5 mg/kg was found to be embryotoxic, with only mild teratogenic activity; decreased fetal weight and slightly higher mortality of the young were the main embryotoxic effects. Fenthion administered as a single intraperitoneal dose of 40 or 80 mg/kg showed similar effects. Consecutive treatments caused a lower embryotoxic effect than an equivalent single dose, and in general the decrease in fetal weight (but not the number of abnormalities) was dose dependent. High dosages of fenthion (120–160 mg/kg) or demeton (10 mg/kg) administered on day 8, 9, or 10 of gestation had no effect on litter size at birth or on the survival rate of the young. High-dosage treatment with fenthion or demeton reduced birth weight, but only demeton reduced the growth rate of neonates. These results are very similar to those reported to occur with parathion and methyl parathion (Fish, 1966).

Toxic effects in ducklings hatched from eggs inoculated with systox (demeton) or EPN have also been reported (Khera *et al.*, 1965). Anticholinesterase activity was first observed 8 days after inoculation (21-day embryo), and although it was not associated with a lethal effect on the embryo it correlated with toxic symptoms observed in the hatched ducklings (Khera and La Ham, 1965). Ducklings hatched from eggs incubated under optimal conditions and inoculated via the yolk sac on day 13 with 100 μg of EPN manifested asthenia and lethargy. In 27% of the ducklings, the metatarsophalangeal and first and second phalangeal joints were affected, with a resultant permanent malformation. Ducklings hatched from eggs treated with demeton were hyperexcitable and in some cases exhibited leg paralysis and body tremors with intermittent convulsions. Ducklings from eggs treated with EPN during embryogenesis often suffer from a permanent foot deformity known as talipes varus. Khera *et al.* (1966) studied the incidence of this condition by inoculating eggs on day 13 of incubation with 100 μg EPN. During embryogenesis, they observed a progressive inhibition of cholinesterase at the motor end plates of the thigh skeletal muscles and associated this with dystrophic changes leading to the foot deformity.

5.3. Carbaryl

Although carbaryl possesses advantages in an environmental sense (it is biodegradable, its acute toxicity to birds and mammals is quite low, and there is no evidence of its accumulation in food chains), there is increasing concern over its reported teratogenic effects in mammals.

Carbaryl administered during organogenesis (for 10 consecutive days of gestation, from day 11 to day 20 at a level of 300 mg/kg), was teratogenic in the guinea pig (Robens, 1969); no teratisms were produced in either hamsters or guinea pigs at lower levels of exposure. Smalley *et al.* (1969) reported that after feeding carbaryl up to 30 mg/kg/day to miniature sows during their entire pregnancies there was no indication of any influence on reproduction.

In contrast, carbaryl fed to beagle dogs throughout the gestation period at levels from 3 to 50 mg/kg/day caused dystocia (difficult births) due to atonic uterine musculature, an apparent contraceptive effect at the highest dose level, and a teratogenic effect observed at all but the lowest dose levels (Smalley *et al.*,

1968). Fetal abnormalities produced by technical grade carbaryl (99.9% purity) were seen in 20 of a total of 181 pups born (11.6%) and included abdominal-thoracic fissures with varying degrees of brachygnathia, ecaudate pups, failure of skeletal formation, superfluous phalanges, and other defects that were difficult to characterize.

The effects of extremely high dietary levels (up to 10,000 ppm) of carbaryl on the reproduction of Osborne–Mendel rats and Mongolian gerbils over three generations were elaborated by Collins *et al.* (1971*a*). Dose-related decreases were observed in the averages of litter size, number of liveborn progeny, number of survivors to day 4, and number weaned in both rats and gerbils, and at the highest levels of exposure no litters were produced in either species. The no-effect level was estimated to be below 2000 ppm, although gerbils appeared more sensitive than rats (Collins *et al.*, 1971*b*). Similar results were obtained in a three-generation rat study reported by Weil *et al.* (1972) using dietary levels of carbaryl up to 200 mg/kg/day.

Orlova and Zhalbe (1968) reported that the reduced rat fertility caused by carbaryl was associated with decreased sperm motility and changes in the enzymatic activity of both the testes and ovaries. Spermatogenic damage and changes in the duration of the phases of the estrus cycle in rats were also reported in rats exposed to carbaryl (Vashakidze *et al.*, 1966; Vashakidze 1967), and Shternberg and Rybakova (1968) suggested that the effects on the reproductive organs are exerted indirectly through an interaction with hypothalamohypophyseal control functions.

Congenital malformations in chicken and duck embryos have been reported when carbaryl was applied directly to the embryo or injected into the yolk sac (Marliac, 1964; Marliac *et al.*, 1965; Ghadiri and Greenwood, 1966; Khera, 1966), and in a feeding study with laying hens and roosters decreased hatchability of eggs and an increase in the number of teratisms in the young were observed and found to be dose related (Ghadiri *et al.*, 1967).

No teratogenic effects were noted by Guthrie *et al.* (1971) when 12–14 generations of white mice were injected intraperitoneally with an LD_{50} of carbaryl (396–1000 mg/kg) at 6–8 weeks of age, and Benson *et al.* (1967) reported similarly negative data with mice and rabbits exposed to carbaryl for various periods of time during pregnancy at a level of 30 mg/kg/day.

Dougherty *et al.* (1971) were unable to find any teratisms in rhesus monkeys exposed to carbaryl at up to 20 mg/kg/day but did report the occurrence of several abortions; however, the data do not appear to be dose related or statistically significant (Weil *et al.*, 1972).

Defects in the cardiac septa of lambs from two of 23 ewes fed diets containing 250 ppm of carbaryl have been cited by Panciera (1967). This was compared with an incidence of ten cardiac anomalies among 3020 sheep of all ages used as a control group, and it was concluded that a teratogenic effect has been established. However, Weil *et al.* (1972) have pointed out that the control group employed was unsatisfactory for observations confined to embryos and neonatal lambs since many lambs born with anomalies probably did not survive to be included in the autopsy calculations.

D-32 and Meller-5 strain *Drosophila melanogaster* exposed to preparations containing 1% carbaryl by weight (approximately LD_{50}) for 24 hr were examined for deletion of X chromosomes, recessive sex-linked lethal and sublethal mutations in the F_2 generations, and male fertility (Brzheskiy, 1972). While carbaryl produced no deletion or change in fertility, it produced partial and full lethal mutations in all stages of spermatogenesis with a cumulative frequency of $0.20 \pm 0.07\%$. Spermatocytes were most sensitive to carbaryl, and partial mutations accounted for more than 50% of all mutations. Brzheskiy (1972) concluded that carbaryl had a slight mutagenic tendency and that its mutagenic characteristics were different from those of ionizing radiation and other chemical mutagens.

Carbaryl has also been reported to have cytogenetic effects on meiosis and on pollen viability of *Vicia faba* (Amer and Farah, 1968); stickiness, lagging chromosomes, and other irregularities were described.

Two reports (Andrianova and Alekseev, 1970; Zabezhinsky, 1970) indicate that carbaryl produces malignant tumors when administered in various ways to rats and mice.

5.4. Chemosterilants

5.4.1. Tepa, Thiotepa, Tem, and Apholate

Considerable interest in alkylating compounds (e.g., tepa and related compounds) has developed in recent years following recognition of their potential use as insect chemosterilants (antifertility agents).

The use of these compounds to date has been largely restricted to laboratory testing since the techniques of field application have not been fully developed. However, a more important drawback to their practical development is the potential health and genetic risk inherent in their biological alkylating properties and their potential mutagenicity and/or carcinogenicity.

The biological alkylating agents belong to very diverse chemical groups and are generally classified together only because they can all impart an alkyl group to some cellular receptor under *in vivo* conditions. It is convenient to consider an alkylating molecule as consisting of two parts, a reactive center and a carrier group. The reactive center initiates the alkylation process through formation of an electrophilic moiety (there may be one or several reactive centers per molecule) which can attack a variety of nucleophilic groups associated with cell macromolecules. The carrier group of the molecule may be one or several aromatic rings, a saturated or an unsaturated aliphatic chain, or a functional unit such as an amino acid, a polypeptide, a DNA base, or a polynucleotide. The presumed mode of action of aziridines such as tepa, tem, and apholate is through opening up of the aziridine ring and subsequent interaction with a target macromolecule such as a nucleoprotein (Jackson and Craig, 1969*a*).

Effects of some aziridine alkylating agents on domestic livestock (Khan, 1963; Younger and Young, 1963; Younger and Radeleff, 1964; Younger,

1965*a,b,c*) and on avian species (Davis, 1962; Herrick and Sherman, 1964; Herrick *et al.*, 1967; Sherman and Herrick, 1966; Shellenberger *et al.*, 1967) have been described.

Younger (1973) demonstrated that the gross teratogenic effects of tepa, metepa, and apholate in developing chick embryos after yolk sac injection were similar to those observed in mammalian fetuses after maternal treatment with aziridines (Murphy *et al.*, 1958; Gaines and Kimbrough, 1966; Kimbrough and Gaines, 1968). In contrast, thiotepa produced about 94% malformation in rat fetuses but did not produce a teratogenic response in 4-day chick embryos at the LD_{50} dose (Murphy *et al.*, 1958). Apholate administered continuously in the diet at various levels for various periods of time to Leghorn hens (Herrick and Sherman, 1966; Herrick *et al.*, 1967) and Japanese quail (Shellenberger *et al.*, 1967) failed to induce malformation of the embryo but did affect reproductive performance; the probable teratogenic effect of apholate in the sheep fetus has also been described (Younger, 1965*a*).

Metepa, tepa, and apholate affected reproduction in male rats and decreased both fetal and placental weights (Gaines and Kimbrough, 1964, 1966; Kimbrough and Gaines, 1968). Metepa produced a 100% and 30% incidence of ectrodactyly in the front and hind paws, respectively, and induced hydrocephalus, webbed toes, meningoceles, and kinky tails.

Tepa, metepa, thiotepa, and tem have been found to be mutagenic in the dominant lethal test in mice (Epstein *et al.*, 1970*b*, 1972) and the mutagenicity of tepa has been demonstrated in *Habrobracon* (Palmquist and La Chance, 1966), *Neurospora* (Kaney and Atwood, 1964), *Escherichia coli* (Szybalski, 1958), *Drosophila* (Benes and Sram, 1969; Obe *et al.*, 1971), and bacteriophage T_4 (Drake, 1963). The induction of aberrations in human chromosomes *in vitro* by tepa, thiotepa, tem, and apholate has also been reported (Chang and Klassen, 1968; Hampel *et al.*, 1966; Pankova, 1967). Chinese hamster cells are reported to be 23 times more sensitive than those of *Vicia faba* root tips to the chromosome-breaking activity of tepa (Sturelid, 1971).

Tem has been shown to be mutagenic in *Neurospora* (Westergaard, 1957), *Drosophila* (recessive lethal mutations) (Fahmy and Bird, 1952), and mice (Röhrborn, 1966; Bateman, 1960) and to induce chromosome aberrations in mice (Moutschen, 1961; Cattanach, 1967), *Drosophila* (Schalet, 1955; Ratnayake *et al.*, 1967), cultured human leukocytes (Hampel and Gerhartz, 1965), and *Escherichia coli* (Iyer and Szybalski, 1958). Apholate has been shown to be mutagenic in *Neurospora* (Kaney and Atwood, 1964).

Cattenach (1959) reported that a special risk from substerilizing doses is the transfer of heritable damage to offspring via damaged sperm. Thus exposure of male mice to a small dose (0.2 mg/kg) of tem resulted in a high yield of sterile and partially sterile F_1 male progeny.

The effects of alkylating chemicals in mammalian reproductive cells have been reviewed by Jackson and Craig (1969*a*). In general, polyfunctional aziridines interfere with the developing reproductive cells without affecting sexual activity, which implies a direct action on the seminiferous epithelium and epididymal sperm. In the experimental rodent, postmeiotic cells (spermatids and spermatozoa) are particularly sensitive to damage, although spermatocytes and spermatogonia (premeiotic cells) can also be affected. Cumulative antifertility potency shown by tem in the rat and rabbit (Bock and Jackson, 1957; Fox *et al.*, 1963) and by metepa in the rat (Gaines and Kimbrough, 1964, 1966) has been reported.

5.4.2. Hempa and Hemel

The discovery that structural analogues of tepa and tem such as hempa and hemel are also effective chemosterilants against male flies (Chang *et al.*, 1964) is of considerable interest since unlike the aziridines these are considered to be stable compounds and are apparently nonalkylating. Although the activity of hempa toward male flies and other organisms is characteristically much lower than that of tepa, both compounds produce strikingly similar physiological and cytological effects. Both hempa and tepa are mutagenic in the wasp *Bracon hebetor* (Chang *et al.*, 1964) and induce testicular atrophy in rats (Kimbrough and Gaines, 1966), and hempa induces a marked antispermatogenic effect in rats and mice (Kimbrough and Gaines, 1966). The latter effect appears to occur at a different locus from that of the corresponding aziridines. Hempa produced dramatic and prolonged antifertility effects in male rats and reversible aspermia in rabbits (Jackson and Craig, 1969*b*). A study of the mode of action of hemel in rats suggested an action on premeiotic cells (Jackson and Craig, 1969*a*).

Hempa and hemel have been shown to be mutagenic in *Drosophila* (Benes and Sram, 1969; Ninan and Wilson, 1969) and to induce aberrations in human chromosomes *in vitro* (Chang and Klassen, 1968).

5.5. Rotenone

Rotenone is a selective nonsystemic insecticidal compound of certain *Derris* and *Lonchocarpus* roots that has been used extensively during the last 40 years (e.g., 10,000–20,000 kg/year, world-wide consumption). Gosalvez and Merchan (1973) reported the induction of mammary adenomas in albino rats given intraperitoneal injections of rotenone (1.7 μg/g rat weight, dissolved in 0.1 ml sunflower oil) daily for 42 days; the total dose administered was 9.1 ± 1.6 mg of rotenone per rat. The batch of rotenone used contained less than 1% impurities (composed mostly of the rotenoids epirotenone, rotenolones, and dehydrorotenone as well as traces of a spiro compound). One-hundred percent incidence of mammary tumors was observed in the first series of ten rotenone-treated rats, 6–11 months after the end of treatment. In the other three series of ten rotenone-treated rats, an incidence of 60% mammary tumors was found, while none of the controls showed any tumor. The first series of rotenone tumors seemed to be characterized by mammary localization, histological benignancy, a structural mitochondrial deletion accompanied by poor mitochondrial functions, and a glycolysis rate similar to that found in malignant mammary tumors. In the study of Gosalvez and Merchan (1973), it was found that the age of the animal was not critical for tumorigenicity and that no histological change was observed in the liver or endocrine organs of the animals bearing tumors.

6. References

Ackermann, H., and Engst, R., 1970, Presence of organophosphate insecticides in the fetus, *Arch. Toxikol.* **26:**17–22.

Agthe, C., Garcia, H., Shubik, L., Tomatis, L., and Wenyon, E., 1970, Study of the potential carcinogenicity of DDT in the Syrian golden hamster, *Proc. Soc. Exp. Biol. Med.* **134:**113–116.

Amer, S. M., and Farah, O. R., 1968, Cytological effects of pesticides. III. Meiotic effects of *N*-methyl-1-naphthyl carbamates, *Cytologia* **33:**337–342.

Andervont, H. B., and Dunn, T. B., 1948, Efforts to detect a mammary-tumor agent in strain C mice, *J. Natl. Cancer Inst.* **8:**235–240.

Andrianova, M. M., and Alekseev, I. V., 1970, Carcinogenic properties of sevin, maneb, ziram and zineb, *Vopr. Pitan.* **29:**71–74.

Anonymous, 1972, Aetiology of liver cancer, *Br. Med. J.* **1:**261–263.

Ashwood-Smith, M. J., Trevino, J., and Ring, R., 1972, Mutagenicity of dichlorovos, *Nature (London)* **240:**418–420.

Atwood, K. C., Von borstel, R. C., and Whiting, A. R., 1956, An influence of ploidy on the time of expression of dominant lethal mutations in *Habrobracon, Genetics* **41:**804–813.

Bäckström, J., Hansson, E., and Ullbergh, J., 1965, Distribution of C^{14}-DDT and C^{14}-dieldrin in pregnant mice determined by whole body autoradiography, *Toxicol. Appl. Pharmacol.* **7:**90–96.

Barnes, J. M., 1966, Carcinogenic hazards from pesticide residues, *Residue Rev.* **13:**69–82.

Baron, R. L., and Johnson, H., 1964, Neurological disruption prolonged in hens by two organophosphate esters, *Br. J. Pharmacol.* **23:**295–304.

Bateman, A. J., 1958, Mutagenic sensitivity of maturing germ cells in the male mouse, *Heredity* **12:**213–232.

Bateman, A. J., 1960, The induction of dominant lethal mutations in rats and mice with triethylene melamine (tem), *Genet. Res.* **1:**381–382.

Bauer, K. H., 1928, *Mutations Theorie der Geschwulst-Enstehlung: Uebergang von Korperzellen in Geschwulstzellen durch Gen-Anderung*, Springer-Verlag, Berlin.

Bauer, K. H., 1963, *Das Krebs Problem*, Springer-Verlag, Berlin.

Bedford, C. T., and Robinson, J., 1972, The alkylating properties of organophosphates, *Xenobiotica* **2:**307–337.

Benes, U., and Sram, R., 1969, Mutagenic activity of some pesticides in *Drosophila melanogaster, Ind. Med. Surg.* **38:**50–52.

Bennison, B. E., and Mostofi, F. K., 1950, Observations on inbred mice exposed to DDT, *J. Natl. Cancer Inst.* **10:**989–992.

Benson, B. W., Scott, W. J., and Beliles, R. P., 1967, Sevin safety evaluation by teratological study in the mouse, unpublished Woodward Research Corporation report to Union Carbide *in:* S. W. Weil *et al.*, 1972, *Toxicol. Appl. Pharmacol.* **21:**390–404.

Bock, M., and Jackson, H., 1957, The action of triethylene melamine on the fertility of male rats, *Br. J. Pharmacol.* **12:**1–7.

Bonashevskaya, T. I., and Tabakova, S. A., 1972, Morphological alterations occurring in albino rat organs following inhalation of chlorofos, *Farmakol. Toksikol.* **35:**240–241.

Bonser, G. M., Clayson, D. B., Jull, J. W., and Pyrah, L. N., 1952, The carcinogenic properties of 2-amino-1-naphthol hydrochloride and its parent amine, 2-naphthylamine, *Br. J. Cancer* **6:**412–424.

Boucard, M., Beaulaton, I. S., Mestres, R., Allieu, M., and Cabane, S., 1970, Teratogenesis: Effect of the period and duration of treatment, *Therapie* **25:**907–913.

Boveri, T., 1929, *The Origin of Malignant Tumors*, Williams and Wilkins, Baltimore.

Brewen, J. G., and Preston, R. J., 1973, Chromosome aberrations as a measure of mutagenesis: Comparisons *in vivo* and *in vitro* and in somatic and germ cells, *Environ. Health Persp.* **6:**157–166.

Bridges, B., 1971, Environmental genetic hazards: The impossible problem, *EMS Newsletter* **5:**13–15.

Bridges, B. A., Mottershead, R. P., Green, M. H. L., and Gay, W. J. H., 1973, Mutagenicity of Dichlorvos and methyl methane sulphonate for *E. coli* WP_2 and some derivatives deficient in DNA repair, *Mutation res.* **19:**295–303.

Bromberg, A. I., Potsheba, T. L., Pozin, Z. S., Brikma, L. I., Volkova, A. P., Voronkina, T. M., and Kamenskiy, 1972, New applications form of trichlorfon—The preparation "insekto polimer," *Zh. Mikrobiol. Immunobiol.* **49:**133–138.

Brzheskiy, V. V., 1972, Study of the mutagenic properties of sevin, *Genetika* **8:**151–153.

Budreau, C. H., and Singh, R. P., 1973*a*, Transplacental passage of demeton in CF-1 mice, *Arch Environ. Health* **26:**161–163.

Budreau, C. H., and Singh, R. P., 1973*b*, Teratogenicity and embryotoxicity of demeton and fenthion in CF-1 mouse embryos, *Toxicol. Appl. Pharmacol.* **24:**322–326.

Bunch, T. D., and Low, J. B., 1973, Effects of dieldrin on chromosomes of semi-domestic mallard ducks, *J. Wildlife Management* **37:**51–57.

Buselmaier, W., Röhrborn, G., and Propping, P., 1972, Mutagenitaets Untersungen mit Pestiziden im host-mediated assay und mit dem dominanten letal Test an der Maus, *Biol. Zentral bl.* **91(3):**311–325.

Cabral, J. R., 1972, Lack of long-term effects of the administration of heptachlor to suckling rats, *Tumori* **58:**49–53.

Cattanach, B. M., 1959, The sensitivity of the mouse testes to the mutagenic action of triethylene melamine, *Z. Vererbungsl.* **90:**1–6.

Cattanach, B. M., 1967, Induction of paternal sex chromosome losses and deletions and of autosomal gene mutations by treatment of mouse postmeiotic germ cells with triethylene-melamine, *Mutation Res.* **4:**73–78.

Cerey, K., Izakovic, V., and Ruttkay-Nedecka, J., 1973, Effect of heptachlor on dominant lethality and bone marrow in rats, *Mutation Res.* **21:**26.

Chang, S. C., Terry, P. A., and Borkovec, A. B., 1964, Insect chemosterilants with low toxicity for mammals, *Science* **144:**57–58.

Chang, T. H., and Klassen, W., 1968, Comparative effects of tretamine, tepa, apholate and their structural analogs on human chromosomes *in vitro*, *Chromosoma* **24:**314–343.

Clayson, D. B., 1962, *Chemical Carcinogenesis*, Little, Brown, Boston.

Clegg, D. G., 1964, The hen egg in toxicity and teratogenicity studies, *Food Cosmet. Toxicol.* **2:**717–727.

Clegg, D. G., 1971, Teratology, *Ann. rev. Pharmacol.* **11:**409–424.

Collins, T. F. X., Hansen, W. H., and Keeler, H. V., 1971*a*, The effect of carbaryl (sevin) on reproduction of the rat and the gerbil, *Toxicol. Appl. Pharmacol.* **19:**202–216.

Collins, J. A., Schooley, M. A., and Singh, V. K., 1971*b*, The effect of the dietary dichlorvos on swine reproduction and viability of their offspring: Abstracts of the 10th Annual Meeting of the Society of Toxicology, *Toxicol. Appl. Pharmacol.* **19(2):** Abst. No. 41.

Corvallis Task Force, 1970, *Man's Health and the Environment—Some Research Needs*, Government Printing Office, Washington, D.C.

Cory, L., Fjeld, P., and Serat, W., 1971, Environmental DDT and the genetics of natural populations, *Nature (London)* **229:**128–130.

Czeizel, A., Trinh, V. B., Szabo, I., and Ruzicska, P., 1973, Human chromosome aberrations in acute organic phosphorus acid ester (pesticide) intoxication, Environmental Mutagen Society, European Branch, 3rd Annual Meeting, Uppsala, Sweden, June 4–7.

Davis, D. E., 1962, Gross effects of triethylene melamine on gonads of starlings, *Anat. Rec.* **142:**353–357.

Davis, K. J., and Fitzhugh, O. G., 1962, Tumorigenic potential of aldrin dieldrin for mice, *Toxicol. Appl. Pharmacol.* **4:**187–189.

Dean, B. J., 1972*a*, The mutagenic effects of organophosphorus pesticides on micro-organisms, *Arch. Toxikol.* **30:**67–74.

Dean, B. J., 1972*b*, The effects of dichlorvos on cultured human lymphocytes, *Arch. Toxikol.* **30:**75–85.

Dean, B. J., and Thorpe, E., 1972*a*, Studies with dichlorvos vapor in dominant lethal mutation tests on mice, *Arch. Toxikol.* **30:**51–59.

Dean, B. J., and Thorpe, E., 1972*b*, Cytogenic studies of dichlorvos in mice and chinese hamsters, *Arch. Toxikol.* **30:**39–49.

Dean, B. J., Doak, S. M. A., and Funnel, J., 1972, Genetic studies with dichlorvos in the host-mediated assay in liquid medium using *S. cerevisiae, Arch. Toxikol.* **30:**61–66.

Deichmann, W. B., 1972*a*, Toxicology of DDT and related chlorinated hydrocarbon pesticides, *J. Occup. Med.* **14:**285–292.

Deichmann, W. B., 1972*b*, The debate on DDT, *Arch. Toxicol.* **29:**1–27.

Deichmann, W. B., and Keplinger, M. L., 1966, Effect of combinations of pesticides on reproduction in mice, *Toxicol. Appl. Pharmacol.* **8:**337–338 (Abst.).

Deichmann, W. B., Keplinger, M., Sala, F., and Glass, E., 1967, Synergism among oral carcinogens. Part IV, *Toxicol. Appl. Pharmacol.* **11:**88–93.

Deichmann, W. B., MacDonald, W. E., Blum, E., Bevilacqua, M., Radomski, J., Keplinger, M., and Balkus, M., 1970, The tumorigenicity of aldrin, dieldrin and endrin in the albino rat, *Ind. Med. Surg.* **39:**426–434.

Deichmann, W. B., MacDonald, W. E., Beasley, A. G., and Cubit, D. A., 1971, Subnormal reproduction in beagle dogs induced by DDT and aldrin, *Ind. Med. Surg.* **40:**10–18.

DeLong, R., Gilmartin, W. G., and Simpson, J. G., 1973, Premature births in California sea lions: Association with high organochlorine pollutant residue levels, *Science* **181:**1168–1170.

Denes, A., and Tarjan, R., 1964, The accumulation of DDT in food in human fatty tissues, Conference of Nutritional Research, Budapest, April 9–13.

Deringer, M. K., 1965, Occurrence of mammary tumors, reticular neoplasms and pulmonary tumors in strain BALB/c and breeding female mice, *J. Natl. Cancer Inst.* **35:**1047–1052.

Dikshith, T. S. S., and Datta, K. K., 1973, Endrin induced cytological changes in albino rats, *Bull. Environ. Contam. Toxicol.* **9:**65–69.

Dobbins, P. K., 1967, Organic phosphate insecticides as teratogens in the rat, *J. Fla. Med. Assoc.* **54:**452–456.

Dobzhansky, T., Anderson, W. W., Pavlovsky, O., Spassy, B., and Wills, C. J., 1964, Genetics of natural populations: *Drosophila pseudoobscura* in the American Southwest, *Evolution* **18:**164–176.

Dobzhansky, T., Anderson, W. W., and Pavlovsky, O., 1966, Genetics of natural populations. XXXVIII. Continuity and change in populations of *Drosophila pseudoobscura* in western United States, *Evolution* **20:**418–427.

Dougherty, W. J., Golberg, L., and Coulston, F., 1971, The effect of carbaryl on reproduction in the monkey (*Macacca mulatta), Toxicol. Appl. Pharmacol.* **19:**365.

Drake, J. W., 1963, Polyimines mutagenic for bacteriophage T4B, *Nature (London)* **197:**1028.

Durham, W. F., Ortega, P., and Hayes, W. J., Jr., 1963, The effect of various dietary levels of DDT on liver function, cell morphology and DDT storage in the rhesus monkey, *Arch. Int. Pharmacodyn.* **141:**111–129.

Elgar, K. E., and Steer, B. D., 1972, Dichlorvos concentrations in the air of houses arising from the use of vapona strips, *Pestic. Sci.* **3:**59.

Environmental Protection Agency, 1974, Shell Chemical Co., Consolidated Aldrin/Dieldrin Hearing, *Federal Register* **39(203):**37246–37272.

Epstein, S. S., 1973, Use of the dominant-lethal test to detect genetic activity of environmental chemicals, *Environ. Health Persp.* **6:**23–26.

Epstein, S. S., and Legator, M. S., 1971, *The Mutagenicity of Pesticides*, pp. 3–8, MIT Press, Cambridge, Mass.

Epstein, S. S., and Röhrborn, G., 1971, Recommended procedures for testing genetic hazards from chemicals based on the induction of dominant lethal mutation in mammals, *Nature (London)* **230:**495.

Epstein, S. S., and Shafner, H., 1968, Chemical mutagens in the human environment, *Nature (London)* **219:**385–387.

Epstein, S. S., Bass, W., Arnold, E., and Bishop, Y., 1970*a*, Mutagenicity of trimethyl phosphate in mice, *Science* **168:**584–586.

Epstein, S. S., Arnold, E., Steinberg, K., Mackintosh, D., Shafner, H., and Bishop, Y., 1970*b*, Mutagenic and antifertility effects of tepa and metepa in mice, *Toxicol. Appl. Pharmacol.* **17:**23–40.

Epstein, S. S., Arnold, E., Andrea, J., Bass, W., and Bishop, Y., 1972, Detection of chemical mutagens by the dominant lethal assay in the mouse, *Toxicol. Appl. Pharmacol.* **23:**288–325.

Fahmy, O. G., and Bird, M. J., 1952, Chromosome breaks among recessive lethal induced by chemical mutagens in *Drosophila*, *Heredity Suppl.* **6:**149–159.

Fahrig, R., 1973, Genetic effects of organophosphorus insecticides, *Naturwissenschaften* **60:**50–51.

Finnegan, J. K., Haag, H. B., and Larson, P. S., 1949, Tissue distribution and elimination of DDD and DDT following oral administration, *Proc. Soc. Exp. Biol. Med.* **72:**357–360.

Fish, S. A., 1966, Organophosphorous cholinesterase inhibitors and fetal development, *Am. J. Obstet. Gynecol.* **96:**1148–1154.

Fishbein, L., Flamm, W. G., and Falk, H. L., 1970, *Chemical Mutagens*, p. 117, Academic Press, New York.

Fitzhugh, O. G., and Nelson, A. A., 1947, Chronic oral toxicity of DDT, *J. Pharmacol. Exp. Ther.* **89:**18–30.

Fitzhugh, O. G., Nelson, A. A., and Quaife, M. L., 1964, Chronic oral toxicity of aldrin and dieldrin in rats and dogs, *Food Cosmet. Toxicol.* **2:**551.

Flockhart, I. R., and Casida, J. E., 1972, Relationship of the alkylation of membrane esterases and proteins to the teratogenic action of organophosphorus insecticides and eserine in developing hen eggs, *Biochem. Pharmacol.* **21:**2591–2603.

Food and Drug Administration, 1958, Methoxychlor: carcinogenic, *Food Drug Cosmet. Law. J.* **13:**4–6.

Fox, B. W., Jackson, H., Craig, A. W., and Glover, T. D., 1963, effect of alkylating agents on spermatogenesi in the rabbit, *J. Reprod. Fertil.* **5:**13–22.

Gabliks, J., and Friedman, L., 1965*a*, responses of cell cultures to insecticides. I. Acute toxicity to human cells, *Proc. Soc. Exp. Biol. Med.* **120:**163–168.

Gabliks, J., and Friedman, L., 1965*b*, Responses of cell cultures to insecticides. II. Chronic toxicity and induced resistance, *Proc. Soc. Exp. Biol. Med.* **120:**168–171.

Gabridge, M. G., and Legator, M. S., 1969, A host-mediated microbial assay for the detection of mutagenic compounds, *Proc. Soc. Exp. Biol.* **130:**831–834.

Gains, T. B., and Kimbrough, R. D., 1964, Toxicity of metepa to rats with notes on two other chemosterilants, *Bull. WHO* **31:**737–745.

Gains, T. B., and Kimbrough, R. D., 1966, The sterilizing, carcinogenic and teratogenic effects of metepa in rats, *Bull. WHO* **34:**317–320.

Gargus, J. L., Paynter, C. E., and Reese, W. H., Jr., 1969, Utilization of newborn mice in the bioassay of chemical carcinogens, *Toxicol. Appl. Pharmacol.* **15:**552–559.

Gelboin, H. V., 1967, Carcinogens, enzyme induction and gene action, *Adv. Cancer Res.* **10:**1–75.

Gerelzoff, M. A., Dallemagne, M. J., and Philippot, E., 1950, Plasmocytomes renaux multiples et plasmocytose a la suite d'injections repetees de DDT en chein, *Soc. Biol. Compt. Rend.* **144:**1135–1137.

Ghadiri, N., and Greenwood, D. A., 1966, Toxicity and biological effects of malathion, phosdrin and sevin in the chick embryo, *Toxicol. Appl. Pharmacol.* **8:**342 (abst.).

Ghadiri, N., Greenwood, D. A., and Binns, W., 1967, feeding of malathion and carbaryl to laying hens and roosters, *Toxicol. Appl. Pharmacol.* **10:**392.

Gillett, J. W., Harr, J. R., Lindstrom, F. T., Mount, D. A., St. Clair, A. D., and Weber, L. J., 1973, *Residue Rev.* **44:**115–184.

Gofmekler, V. A., and Tabakova, S. A., 1970, Chlorphol action on the embryogenesis of rats, *Farmakol. Toksikol.* **33:**737.

Good, E. E., and Ware, G. W., 1969, Effects of insecticides on reproduction in the laboratory mouse. IV. Endrin and dieldrin, *Toxicol. Appl. Pharmacol.* **14:**201–203.

Good, E. E., Ware, G. W., and Miller, D. F., 1965, effects of insecticides on reproduction in the laboratory mouse. I. Kepone, *J. Econ. Entomol.* **58:**754–757.

Gosalvez, M., and Merchan, J., 1973, Induction of rat mammary adenomas with the respiratory inhibitory rotenone, *Cancer Res.* **33:**3047–3050.

Grasso, P., and Crampton, R. F., 1972, The value of the mouse in carcinogenicity testing, *Food Cosmet. Toxicol.* **10:**418–426.

Greenberg, J., and La Ham, Q. N., 1969, Malathion induced teratism in the developing chick, *Can. J. Zool.* **47:**539–542.

Grosch, D. S., 1967, Poisoning with DDT: Effect on the reproductive performance of *Artemia, Science* **155:**592–593.

Grosch, D. S., and Valcovic, L. R., 1967, Chlorinated hydrocarbon insecticides are not mutagenic in *Bracon hebetor* tests, *J. Econ. Entomol.* **60:**1177–1178.

Guthrie, F. E., Monroe, R. J., and Abernathy, C. O., 1971, response of the laboratory mouse to selection for resistance to insecticides, *Toxicol. Appl. Pharmacol.* **18:**92–101.

Halver, C. E., 1967, *Crystalline Aflatoxin and Other Vectors for Trout Hepatoma: Trout Hepatoma Research Conference Papers*; Bureau Sport Fisheries and Wildlife research Report, No. 70, pp.78–102.

Hampel, K. E., and Gerhartz, H. H., 1965, Strukturanomalien der Chromosomen menschlicher Leukozyten *in Vitro* durch Triaethylenmelamin, *Exp. Cell Res.* **37:**251–258.

Hampel, K. E., Kober, B., Rosch, D., Gerhartz, H., and Meinig, K. H., 1966, The action of cytostatic agents on the chromosomes of human leukocytes *in vitro, Blood* **27:**816–823.

Hart, M. H., Whang-Peng, J., Sieber, S. M., fabro, S., and Admanson, R. H., 1972, Distribution and effects of DDT in the pregnant rabbit, *Xenobiotica* **2:**567–574.

Hathaway, D. E., Moss, J. A., Rose, J. A., and Williams, D. J. M., 1967, transport of dieldrin from mother to blastocyst and from mother to foetus in pregnant rabbits, *Eur. J. Pharmacol.* **1:**167–175.

Hendriksson, K., Kalleca, K., Virjamo, M., and Pfaffli, P., 1971, The toxicity of DDVP evaporated from vapona strip, *Maataloustieteellinen Aikak.* **43:**187–200.

Herrick, R. B., and Sherman, M., 1964, Effect of an alkylating agent apholate on the chicken, *Poultry Sci.* **43:**1327–1328.

Herrick, R. B., Sherman, M., and Batra, T. R., 1967, The effect of insect chemosterilant apholate on the reproductive performance of Leghorn hens, *Poultry Sci.* **46:**1045–1050.

Henson, W. E., Vlahakis, G., and Deringer, M. K., 1960, High incidence of spontaneous hepatomas and the increase of this incidence with urethan in $C3H_1$, $C3H_f$ and $C3H_e$ male mice, *J. Natl. Cancer Inst.* **24:**425–432.

Huber, J. J., 1965, Some physiological effects of the insecticide Kepone in the laboratory mouse, *Toxicol. Appl. Pharmacol.* **7:**516–524.

Hutson, D. H., and Hoadley, E. C., 1972*a*, The metabolism of (^{14}C-methyl)dichlorovos in the rat and the mouse, *Xenobiotica* **2:**107–116.

Hutson, D. H., and Hoadley, E. C., 1972*b*, The comparative metabolism of (^{14}C-vinyl)dichlorvos in animals and man, *Arch. Toxikol.* **30:**9–18.

Innes, J. R. M., Ulland, B. M., Valerio, M. G., Petrucelli, L., Fishbein, L., Hart, R., Pallotta, A. J., Bates, R. R., Falk, H. L., Gart, J. J., Klein, M., Mitchell, I., and Peters, J., 1969, Bioassay of pesticides and industrial chemicals for tumorigenicity in mice: A preliminary note, *J. Natl. Cancer Inst.* **42:**1101–1114.

International Agency for Research on Cancer, 1971, *Report and Recommendations of Subcommittee on Morphology, Epidemiology and Pathology*, IARC Scientific Publications, No. 1, International Agency for Research on Cancer, Lyon, p. 173.

Iyer, V. N., and Szybalski, W., 1958, The mechanism of chemical mutagenwesis. I. Kinetic studies on the action of tem and azaserine, *Proc. Natl. Sci. U.S.A.* **44:**446–456.

Jackson, H., and Jones, A. R., 1968, Antifertility action and metabolism of trimethyl phosphate in rodents, *Nature (London)* **220:**591–592.

Jackson, H., and Craig, A. W., 1969*a*, Effects of alkylating chemicals on reproductive cells, *Ann. N.Y. acad. Sci.* **160:**215–217.

Jackson, H., and Craig, A. W., 1969*b*, Antifertility action and metabolism of hexamethyl phosphoramide, *Nature (London)* **212:**86–87.

Jager, K. W., 1970, *Aldrin, Dieldrin, Endrin, and Telodrin: An Epidemiological and Toxicological Study of Long-Term Occupational Exposure*, Elsevier, Amsterdam.

Johnson, G. A., and Jalal, S. M., 1973, DDT-induced chromosomal damage in mice, *J. Hered.* **64:**7–8.

Johnson, M. K., 1969, Phosphorylation site in brain and the delayed neurotoxic effect of some organophosphorus compounds, *Biochem. J.* **111:**487–495.

Jukes, T. H., 1970, DDT and tumors in experimental animals, *Int. J. Environ. Stud.* **1:**43–46.

Kalow, W., and Marton, A., 1961, Second-generation toxicity of malathion in rats, *Nature (London)* **192:**464–465.

Kalter, H., 1971, Correlation between teratogenic and mutagenic effects of chemicals in mammals, *in*: *Chemical Mutagens*, Vol. 1 (A. Hollaender, ed.), pp. 57–82, Plenum Press, New York.

Kaney, A. R., and Atwood, K. C., 1964, Radiomimetic action of polyimine chemisterilants in *Neurospora, Nature (London)* **201:**1006–1008.

Kelly-Garvert, F., and Legator, M. S., 1973, Cytogenetic and mutagenic effects of DDT and DDE in a Chinese hamster cell line, *Mutation Res.* **17:**223–229.

Kemeny, T., and Tarjan, R., 1966, Investigations on the effects of chronically administered small amounts of DDT in mice, *Experientia* **22:**748–749.

Keplinger, M. L., Deichmann, W. B., and Sala, F., 1968*a*, effects of combinations of pesticides on reproduction in mice, *Ind. Med. Surg.* **37:**525–531.

Keplinger, M. L., Deichmann, W. B., and Sala, F., 1968*b*, Effects of Combinations of Pesticides Symposia, 6th Inter-American Conference on Toxicology and Occupational Medicine, Miami, Fla., pp. 125–138.

Khan, M. A., 1963, toxicity of apholate to cattle, *Can. J. Comp. Med. Vet. Sci.* **27:**233–236.

Khera, K. S., 1966, Toxic and teratogenic effects of insecticides in duck and chick embryos, *Toxicol. Appl. Pharmacol.* **8:**345 (abst.).

Khera, K. S., and Bedok, S., 1967, Effects of thiol phosphates on mitochondrial and vertebral morphogenesis in chick and duck embryos, *Food Cosmet. Toxicol.* **5:**359–365.

Khera, K. S., and LaHam, Q. N., 1965, Cholinesterases and motor-end-plates in developing duck skeletal muscle, *J. Histochem. Cytochem.* **13:**559–562.

Khera, K. S., LaHam, Q. N., and Grice, H. C., 1965, toxic effects in ducklings hatched from enbryos inoculated with EPN or systox, *Food Cosmet. Toxicol.* **3:**581–586.

Khera, K. S., LaHam, Q. N., Ellis, C. F. G., Zawidzka, Z. Z., and Grice, H. C., 1966, Food deformity in ducks from injection of EPN during embryogenesis, *Toxicol. Appl. Pharmacol.* **8:**540–549.

Kimbrough, R. M., and Gains, T. B., 1966, Toxicity of hexamethyl phosphoramide in rats, *Nature (London)* **211:**146–147.

Kimbrough, R. D., and Gains, T. B., 1968, Effect of organic phosphorus compounds and alkylating agents on the rat fetus, *Arch. Environ. Health* **16:**805–808.

Kimbrough, R., Gaines, T. B., and Sherman, J. D., 1964, nutritional factors, Long-term DDT intake and chloroleukemia in rats, *J. Natl. Cancer Inst.* **33:**215–220.

Kitselman, C. H., 1953, Long term studies on dogs fed aldrin and dieldrin in sub-lethal dosages, with reference to the histopathological findings and reproduction, *J. Am. Vet. Med. Assoc.* **123:**28–36.

Kolmark, G., 1956, Mutagenic properties of certain esters of inorganic acids investigated by the *Neurospora* back-mutation test, *Compt. Rend. Trav. Lab. Carlsbad Ser. Physiol.* **26:**205–220.

Kotin, P., 1972, Cancer: A disease involving multiple factors, *in: Multiple Factors in the Causation of Environmentally Induced Disease* (D. H. K. Lee ed.), pp. 16–27, Academic Press, New York.

Krause, W., and Homola, S., 1972, Influence on spermiogenesis by DDVP (dichlorvos), *Arch. Dermatol. Forsch.* **244:**439–441.

Kugaczewska, M., Piekarski, L., Szutowski, M., Trzaskowski, J., and Ziemski, R., 1972, Cytotoxicity of dipterex and DDVP and their decomposition products, *Bromatol. Chem. Toksykol.* **5(4):**473–479. *Chem. Abst.* **78:**106752X (1973).

Laben, R. C., Archer, T. E., Crosby, D. G., and Peoples, S. A., 1965, Lactational output of DDT fed prepartum to dairy cattle, *J. Dairy Sci.* **48:**701–708.

Laug, E. P., Nelson, A. A., Fitzhugh, O. G., and Kunze, F. M., 1950, Liver cell alteration and DDT storage in the fat of the rat induced by dietary levels of 1 to 50 ppm DDT, *J. Pharmacol. Exp. Ther.* **98:**268–273.

Laws, E. R., Jr., 1971, Evidence of anti-tumorigenic effects of DDT, *Arch. Environ. Health* **23:**181–186.

Laws, E. R., Curley, A., and Biros, F. J., 1967, Men with intensive occupational exposure to DDT: A clinical and chemical study, *Arch. Environ. Health* **15:**766–775.

Legator, M. S., 1970, The host-mediated assay: A practical procedure for evaluating potential mutagenic agents, *in: Chemical Mutagenesis in Mammals and Man* (F. Vogel and G. Röhrborn, eds.), pp. 260–270, Springer-Verlag, New York.

Legator, M. S., 1973, Unpublished data, *in*: Epstein, S. S., 1973, Use of the dominant-lethal test to detect genetic activity of environmental chemicals, *Environ. Health Persp.* **6:**23–26.

Legator, M. S., and Flamm, W. G., 1973, Environmental mutagenesis and repair, *Ann Rev. Biochem.* **42:**683–708.

Legator, M. S., and Malling, H. V., 1971, The host-mediated assay: A practical procedure for evaluating potential mutagenic agents, *in: Chemical Mutagens*, Vol. 2 (A. Hollaender, ed.), pp. 569–589, Plenum Press, New York.

Legator, M. S., Palmer, K. A., and Adler, I. D., 1973, Collaborative study of *in vivo* cytogenetic analysis. I. Interpretation of slide preparations, *Toxicol. Appl. Pharmacol.* **24:**337–350.

Lehman, A. J., 1952, Chemicals in foods: A report to the association of food and drug officials on current developments. Part II. Pesticides, *Assoc. Food Drug Officials U.S. Quart. Bull.* **16:**126–132.

Lindsley, D. L., and Grell, E. H., 1968, genetic variations of *Drosophila melanogaster, Carnegie Inst. Wash. Publ.* **627:**472.

Löfroth, G., 1970, Alkylation of DNA by dichlorvos, *Naturwissenschaften* **57:**393–394.

Löfroth, G., Kim, C., and Hussain, S., 1969, alkylating property of 2,2-dichlorovinyl dimethyl phosphate: A disregarded hazard, *EMS Newsletter* **2:**21–27.

Madison, R. M., Rabstein, L. S., and Bryan, W., 1968, Mortality rate and spontaneous lesions found in 2,928 untreated BALB/c mice, *J. Natl. Cancer Inst.* **40:**683–685.

Malling, H. V., and Frantz, C. N., 1973, *In vitro* versus *in vivo* metabolic activation of mutagens, *Environ. Health Persp.* **6:**71–82.

Markaryan, D. W., 1966, Cytogenetic effect of some chlorinated insecticides on mouse bone marrow, cell nuclei, *Genetika* **2:**132–137.

Marliac, J. P., 1964, Toxicity and teratogenic effects of 12 pesticides in the chick embryo, *Fed. Proc.* **23:**105 (abst.).

Marliac, J. P., Verrett, M. J., McLaughlin, J., Jr., and Fitzhugh, O. G., 1965, A comparison of toxicity data obtained from 21 pesticides by the chicken embryo technique for acute, oral LD_{50}'s in rats, *Toxicol. Appl. Pharmacol.* **7:**490.

McLaughlin, J., Jr., Marliac, J. P., Verrett, M. J., Mutchler, M. K., and Fitzhugh, O. G., 1963, The injection of chemicals into the yolk sac of fertile eggs prior to incubation as a toxicity test, *Toxicol. Appl. Pharmacol.* **5:**760–767.

McLaughlin, J., Marliac, J. P., Verrett, J., Mutcher, M. C., and Fitzhugh, O. G., 1964, Toxicity of fourteen volatile chemicals as measured by the chick embryo method, *Am. ind. Hyg. Assoc. J.* **25:**282–284.

Meiniel, R., Lutz-Ostertag, Y., and Lutz, H., 1970, Effects tératogènes du parathion sur le sequelette embryonnaire de la cailler japonaise *(Coturnix japonica), Arch. Anat. Microsc. Morphol. Exp.* **59:**167–183.

Miller, E. C., and Miller, J. A., 1971, The mutagenicity of chemical carcinogens: Correlations, problems, and interpretations, *in: Chemical Mutagens,* Vol. 1 (A. Hollaender, ed.), pp. 83–119, Plenum Press, New York.

Miller, J. A., 1970, Carcinogenesis by chemicals: An overview, G. H. A. Clowes Memorial Lecture, *Cancer res.* **30:**559–576.

Mohn, G., 1973, Comparison of the mutagenic activity of eight organophosphorus insecticides in *E. coli,* Environmental Mutagen Society, European Branch, 3rd Annual Meeting, Uppsala, Sweden, June 4–7.

Moutschen, J., 1961, Differential sensitivity of mouse spermatogenesis to alkylating agents, *Genetics* **46:**291–299.

Murphy, M. L., Del Moro, A., and Lalon, C., 1958, The comparative effects of five polyfunctional alkylating agents on the rat fetus with additional notes on the chick embryo, *Ann. N. Y. Acad. Sci.* **68:**762–781.

Naber, E. C., and Ware, G. W., 1965, Effect of kepone and mirex on reproductive performance in the laying hen, *Poultry Sci.* **44:**875–880.

Nagasaki, H., Tomii, S., Mega, T., Marugami, M., and Ito, N., 1971, development of hepatomas in mice treated with benzene hexachloride, *Gann* **62:**431.

Nagasaki, H., Tomii, S., Mega, T., Marugami, M., and Ito, N., 1972, Hepatocarcinogenic Effects of α, β, γ, and δ-isomers of benzene hexachloride in mice, *Gann* **63:**393.

Nichols, W. W., 1973, Cytogenetic techniques in mutagencity testing, *Agents Actions* **3/2:** 86–92.

Ninan, T., and Wilson, G. B., 1969, Chromosome breakage by ethylenimines and related compounds, *Genetica* **40:**103–119.

Noda, K., Hirabayashi, M., Yonemura, I., Maruyama, M., Endo, I., 1972, Influence of pesticides on embryos II, *Oyo Yakuri* **6(4):**673–679.

Obe, G., Sperling, K., and Belitz, H. J., 1971, Some aspects of chemical mutagenesis in man and in *Drosophila, Angew. Chem. Int. Ed.* **10:**302–314.

O'Leary, J. A., Davis, J. E., Edmundson, W. F., and Feldman, M., 1970, Correlation of prematurity and DDE levels in fetal whole blood, *Am. J. Obstet. Gynecol.* **106:**939.

Orlova, N. V., and Zhalbe, E. P., 1968, Maximum permissible amounts of sevin in food products, *Vopr. Pitan.* **27:**49–55.

Ottoboni, A., 1969, Effect of DDT on reproduction in the rat, *Toxicol. Appl. Pharmacol.* **14:**74–81.

Ottolenghi, A. D., Haseman, J. K., and Suggs, F., 1973, Teratogenic effects of aldrin, dieldrin, and endrin in hamsters and mice, *Teratology* **9:**11–16.

Palmer, K. A., Green, S., and Legator, M. S., 1972, Cytogenetic effects of DDT and derivatives of DDT in a cultured mammalian cell line, *Toxicol. Appl. Pharmacol.* **22:**355–364.

Palmer, K. A., Green, S., and Legator, M. S., 1973, Dominant lethal study of *p,p'*-DDT in rats, *Food Cosmet. Toxicol.* **11:**53–62.

Palmquist, J., and La Chance, L. E., 1966, Comparative mutagenicity of two chemosterilants, tepa and hempa, in sperm of *Bracon hebetor, Science* **154:**915–917.

Panciera, R. J., 1967, Determination of teratogenic properties of oral administered sevin in sheep, Unpublished Report from the Department Veterinary Pathology, College of Veterinary Medicine, Oklahoma State University, to Union Carbide Corp. *in:* Weil, C. S., Woodside, M. D., Carpenter, C. P., and Smyth, H. F., Jr., 1972, Current tests of carbaryl for reproductive and teratogenic effect, *Toxicol. Appl. Pharmacol.* **21:**390–404.

Pankova, N. V., 1967, Mechanism of action of thio-tepa on human chromosomes, *Genetika* **1967:**62–67.

Perelygin, V. M., Shpirt, M. B., and Genis, V. I., 1973, Cytotoxicity of pesticides during their simultaneous action, *Vopr. Pitan.* **30:**44–47.

Pitot, H. C., and Heidelberger, C., 1963, Metabolic regulatory circuits and carcinogenesis, *Cancer Res.* **23:**1694–1704.

Popper, H., and Schaffner, F., 1957, *Liver: Structure and Function,* p. 593, McGraw-Hill, New York.

Preussman, R., 1968, Direct alkylating agents as carcinogens, *Food Cosmet. Toxicol.* **6:**576–577.

Radomski, J. L., Dęichmann, W. B., MacDonald, W. E., and Glass, E. M., 1965, Synergism among oral carcinogens. I. Results of the simultaneous feeding of four tumorigens to rats, *Toxicol. Appl. Pharmacol.* **7:**652–656.

Ratnayake, W., Strachen, C., and Auerbach, C., 1967, Genetical analysis of the storage effect of triethylene melamine (TEM) on chromosome breakage in *Drosophila, Mutation Res.* **4:**380–381.

Robens, J. F., 1969, Teratologic studies of carbaryl, diazinon, norea, disulfiram and thiram in small laboratory animals, *Toxicol. Appl. Pharmacol.* **15:**152–163.

Roe, F. J. C., and Grant, G. A., 1970, Inhibition by germ-free status of development of liver and lung tumors in mice exposed neonatally to 7,12-dimethyl-benz(*a*)-anthracene: Implications in relation to tests for carcinogenicity, *Int. J. Cancer* **6:**133–144.

Roger, J. C., Chambers, H., and Casida, J. E., 1964, Nicotinic Acid analogs: Effects on response of chick embryos and hens to organophosphate toxicants, *Science* **144:**539–540.

Röhrborn, G., 1966, Ueber einen Geschlechtsunterschied in der Mutagenen Wirkung von Trenimon bei der Maus, *Humangenetik* **2:**81–82.

Rosival, L., and Kapeller, K., 1970, Heptachlor induced morphological changes, *in: Proceedings of the Fourth International Congress of Rural Medicine—Whither Rural Medicine?* (H. Kurolwa, ed.), pp. 29–32, Japanese Association of Rural Medicine, Tokyo.

Schalet, A., 1955, The relationship between the frequency of nitrogen mustard induced translocations in mature sperm of *Drosophila* and utilization of sperm by females, *Genetics* **40:**594 (abst.).

Seiler, J. P., 1973, A survey on the mutagenicity of various pesticides, *Experientia* **29:**622–623.

Shabad, L. M., 1972, On a possible blastomogenicity of DDT, *Vopr. Pitan.* **31(1):**63–66.

Shellenberger, T. E., Skinner, W. A., and Lee, J. M., 1967, Effect of organic compounds on reproductive processes. IV. Response of Japanese quail to alkylating agents, *Toxicol. Appl. Pharmacol.* **10:**69–78.

Shenefeldt, R., 1972, Morphogenesis of malformations in hamsters caused by retionoic acid, relation of dose and stage of treatment, *Teratology* **5:**103–118.

Sherman, M., and Herrick, R. B., 1966, Acute and sub-acute toxicity of apholate to the chick and Japanese quail. *Toxicol. Appl. Pharmacol.* **9:**279–292.

Shternberg, A. I., and Rybakova, M. N., 1968, Effects of carbaryl on the neuro endocrine system of rats, *Food Cosmet. Toxicol.* **6:**461–467.

Slomka, M. B., 1970, *Facts about No-Pest DDVP Strips,* Shell Chemical Co., 18 pp.

Smalley, H. E., Curtis, J. M., and Earl, F. L., 1968, teratogenic Action of Carbaryl in beagle dogs, *Toxicol. Appl. pharmacol.* **13:**392–403.

Smalley, H. E., O'Hara, P. J., Bridges, C. H., and Radeleff, R. D., 1969, The effects of chronic carbaryl administration on the neuromuscular system of swine, *Toxicol. Appl. Pharmacol.* **14:**409–419.

Smith, C. S., and Pilgrim, H. I., 1971, Spontaneous neoplasms in germ-free BALB/cPi mice, *Proc. Soc. Exp. Biol. Med.* **138:**542–544.

Smith, S. I., Weber, C. W., and Reid, B. L., 1970, The effect of injection of chlorinated hydrocarbon pesticides on hatchability of eggs, *Toxicol. Appl. Pharmacol.* **16:**179–185.

Sturelid, S., 1971, Chromosome-breaking capacity of tepa and analogs in *Vicia faba* and Chinese hamster cells, *Hereditas* **68:**255–276.

Styles, J. A., 1972, Cytotoxic effects of various pesticides *in vivo*, and *in vitro*, 2nd Annual Meeting of European Environmental Mutagen Society, Zinkouycastle, Czechoslovakia, May 10–12.

Szybalski, W., 1958, Special micro-biological systems. II. Observations on chemical mutagenesis in microorganisms, *Ann. N. Y. Acad. Sci.* **76:**475–489.

Tanimura, T., Katsuya, T., and Nishimura, H., 1967, Embryotoxicity of acute exposure to methyl parathion in rats and mice, *Arch. Environ. Health* **15:**613–619.

Tannenbaum, A., and Silverstone, H., 1949*a*, The genesis and growth of tumors. IV. Effects of varying the proportion of protein (casein) in the diet, *Cancer Res.* **9:**162–167.

Tannenbaum, A., and Silverstone, H., 1949*b*, The influence of the degree of calorie restriction in the formation of skin tumors and hepatomas in mice, *Cancer Res.* **9:**724–730.

Tarjan, R., and Kemeny, T., 1969, Multigeneration studies on DDT in mice, *Food Cosmet. Toxicol.* **7:**215–222.

Terracini, B., and Tarjan, R., 1970, Multigeneration studies on DDT: Letters to the editor, *Food Cosmet. Toxicol.* **8:**478–481.

Terracini, B., Testa, M. C., Cabral, J. R., and Dau, N., 1973, The effects of long-term feeding of DDT to BALB/c mice, *Int. J. Cancer* **11:**747–764.

Thorpe, E., and Walker, A. I. T., 1973, The toxicology of dieldrin (heod) II. Comparative long-term oral toxicity studies in mice with dieldrin, DDT, phenobarbitone, β-BHC and γ-BHC, *Food Cosmet. Toxicol.* **11:**433–442.

Thorpe, E., Wilson, A. B., Dix, K. M., and Blair, D., 1972, teratological studies with dichlorvos vapor in rabbits and rats, *Arch. Toxikol.* **30:**29–38.

Tomatis, L., Turusov, V., Day, N., and Charles, R. J., 1972, The effect of long-term exposure to DDT of CF-1 mice, *Int. J. Cancer* **10:**489–506.

Tomatis, L., Partensky, C., and Montesamo, R., 1973, The predictive value of mouse liver tumor induction in carcinogenicity testing—A literature survey, *Int. J. Cancer* **12:**1–20.

Toth, B., 1968, A critical review of experiments in chemical carcinogenesis using newborn animals, *Cancer Res.* **28:**727–738.

Treon, J. F., and Cleveland, F. D., 1955, Toxicity of certain chlorinated hydrocarbon insecticides for laboratory animals, with special reference to aldrin and dieldrin, *J. Agr. Food Chem.* **3:**402–408.

Tuchmann-Duplessis, H., 1965, Design and interpretation of teratogenic tests, in: *Symposium on Embryopathic Activity of Drugs* (J. M. Robertson, M. F. Sullivan, and R. L. Smith, eds.), pp. 56–87, Churchill, London.

Tuchmann-Duplessis, H., 1972, Teratogenic drug screening. Present procedures and requirements, *Teratology* **5:**271.

Upshall, D. G., Roger, J. C., and Casida, J. E., 1968, Biochemical Studies on the teratogenic action of bidrin and other neuroactive agents in developing hen eggs, *Biochem. Pharmacol.* **17:**1529–1542.

Vashakidze, V. I., 1967, Mechanism of action of pesticides (granosan, sevin, dinoc) on the reproductive cycle of experimental animals, *Soobshch. Akad. Nauk. Gruz. SSSR* **48:**219–224. *Chem. Abst.* **68:**28750X (1968).

Vashakidze, V. I., Shavladze, N. S., and Guineriya, I. S., 1966, Harmful effects of nitrogen-containing pesticides (dinoc and sevin) on the sexual function of male rats, *Sb. Tr. Nauch. Issled. Inst. Gig. Tr. Prof. Zabol. Tiflis* **10:**205–208. *Chem. Abst.* **68:**113580M (1968).

Vogel, E., 1972, Mutageitats Untersuchungen mit DDT und den DDT-Metaboliten DDE, DDD, DDOM, und DDA an *Drosophila melanogaster*, *Mutation Res.* **16:**157–164.

Vogin, E. E., Carson, S., and Slomka, M. B., 1971, teratology studies with dichlorvos in rabbits, Abstracts of the 10th Annual Meeting of the Society of Toxicology, *Toxicol. Appl. Pharmacol.* **19(2):** Abst. no. 42.

Voogd, C. E., and Van der Vet, P., 1969, Mutagenic action of ethylene halohydrins, *Experientia* **25:**85–86.

Voogd, C. E., Jacobs, J. J. J. A. A., and Van der Stel, J. J., 1972, On the mutagenic action of dichlorvos, *Mutation Res.* **16:**413–416.

Walker, A. I. T., Stevenson, D. E., Robinson, J., Thorpe, E., and Roberts, M., 1969, the toxicology and pharmaco-dynamics of dieldrin (heod): two year oral exposures of rats and dogs, *Toxicol. Appl. Pharmacol.* **15:**345–352.

Walker, A. I. T., Thorpe, E., and Stevenson, D. E., 1972, The toxicology of dieldrin (heod). I. Long-term oral toxicity studies in mice, *Food cosmet. Toxicol.* **11:**415–432.

Walker, E. M., Jr., Gadsden, R. H., and Atkins, L. M., 1970*a*, Effects of *p,p'*-DDT on the Ehrlich ascites carcinoma, *Pestic. Symp.* **1:**295–301.

Walker, E. M., Jr., Gadsden, R. H., Atkins, L. M., and Gale, G. R., 1970*b*, Some effects of dietary commercial DDT on Ehrlich ascites tumor cells, *Ind. Med. Surg.* **39:**461–464.

Walker, N. E., 1967, Distribution of chemicals injected into fertile eggs and its effect upon apparent toxicity, *Toxicol. Appl. Pharmacol.* **10:**290–299.

Walker, N. E., 1968, use of yolk–chemical mixtures to replace hen egg yolk in toxicity and teratogenicity studies, *Toxicol. Appl. Pharmacol.* **12:**94–104.

Walker, N. E., 1971, The effect of malathion and malaoxon on esterases and gross development of the chick embryo, *Toxicol. Appl. Pharmacol.* **19:**590–601.

Wallace, M. E., 1971, An unprecedented number of mutants in a colony of wild mice, *Environ. Pollut.* **1:**175–184.

Ware, G. W., and Good, E. E., 1967, Effects of insecticides on reproduction in the laboratory mouse. II. Mirex, telodrin, DDT, *Toxicol. Appl. Pharmacol.* **10:**54–61.

Warwick, G. P., 1971, Metabolism of liver carcinogens and other factors influencing liver cancer induction, *in: Liver Cancer*, Proceedings of a Working Conference Held at Chester Beatty Research Institute, London, June 30–July 4, 1969, IARC Scientific Publication No. 1, International Agency for Research on Cancer, Lyon, p. 121.

Wassermann, M., Wassermann, D., Zellermayer, L., and Gon, M., 1967, Pesticides in people: Storage of DDT in the people of Israel, *Pestic. Monit. J.* **1:**15–20.

Weil, C. S., Woodside, M. D., Carpenter, C. P., and Smyth, H. F., Jr., 1972, Current status of tests of carbaryl for reproductive and teratogenic effect, *Toxicol. Appl. Pharmacol.* **21:**390–404.

Weinstein, I. B., 1969, Modifications in transfer RNA during chemical carcinogenesis, *in: Symposium Fundamental Cancer Research*, pp. 380–408, Williams and Wilkins, Baltimore.

Westergaard, M., 1957, Chemical mutagenesis in relation to the concept of the gene, *Experientia* **13:**224–234.

Wilson, B. W., and Walker, N. E., 1966, Toxicity of malathion and mercaptosuccinate to growth of chick embryo cells *in vitro*, *Proc. Soc. Exp. Biol. Med.* **121:**1260–1264.

Wilson, B. W., Stinnett, H. O., Fry, D. M., and Nieberg, P. S., 1973, Growth and metabolism of chick embryo cultures: Inhibition with malathion and other organophosphorus compounds, *Arch. Environ. Health* **26:**93–99.

Wilson, J. G., 1972, Interrelations between carcinogenicity, mutagenicity, and teratogenicity, *in: Mutagenic Effects of Environmental Contaminants* (H. E. Sutton and M. I. Harris, eds.), pp. 185–195, Academic Press, New York.

Witherup, S., Jolley, W. J., Stemmer, K., and Pfitzer, E. A., 1971, Chronic toxicity studies with DDVP in dogs and rats including observations on rat reproduction, Abstracts of the 10th Annual Meeting of the Society of Toxicology, *Toxicol. Appl. Pharmacol.* **19(2):** Abst. No. 40.

Witschi, E., 1955, Vertebrate gonadotropins, *in: The Comparative Endocrinology of Vertebrates.* I. (J. Chester and P. Eckstein, eds.), pp. 149–165, *Mem. Soc. Endocrinol. No. 4, Symp. Soc. for Endocrinol.*

World Health Organization, 1969, *Principles of the Testing and Evaluation of Drugs for Carcinogenicity*, WHO Technical Report Series No. 426.

Wright, A. S., 1971, Personal communication, *in:* Walker, A. I. T., Thorpe, E., and Stevenson, D. E., 1972, The toxicology of dieldrin (heod). I. Long term oral studies in mice, *Food Cosmet. Toxicol.* **11:**415–432.

Wulff, L. M. R., Emge, L. A., and Bravo, F., 1935, Some effects of alpha-dinitrophenol on pregnancy in the white rat, *Proc. Soc. Exp. Biol. Med.* **32:**678–680.

Yamagishi, T., Takeba, K., Fujimoto, C., Murimoto, K., and Haruta, M., 1972, On the effects of beta-BHC on the fetus of mouse, *Rinsho Elyo* **41:**599–604.

Younger, R. L., 1965*a*, Probable induction of congenital anomalies in a lamb by apholate, *Am. J. Vet. res.* **26:**291–295.

Younger, R. L., 1965*b*, Low level feeding of a polyfunctional alkylating agent to sheep, *Am. J. Vet. res.* **26:**1075–1078.

Younger, R. L., 1965*c*, Long term toxicity studies in sheep with apholate, an alkylating agent, *Am. J. Vet. Res.* **26:**1218–1220.

Younger, R. L., 1973, Effects of three aziridine alkylating compounds on the gross development of chicken embryos, *Toxicol. Appl. Pharmacol.* **24:**423–433.

Younger, R. L., and Radeleff, R. D., 1964, The toxicologic and pathologic effects of three insect chemosterilants in sheep. *Ann. N.Y. Acad. Sci.* **111:**715–728.

Younger, R. L., and Young, J. E., 1963, Toxicologic studies and associated clinical and hematologic effects of apholate in sheep—A preliminary report, *Am. J. Vet. Res.* **24:**659–669.

Zabezhinsky, M. A., 1970, The study of a possible carcinogenic effect of sevin, *Vopr. Onkol.* **16:**106–107.

15

Insecticide Interactions

C. F. Wilkinson

1. Introduction

In assessing the biological activity (toxicity, pharmacological action, etc.) of a chemical, it is usual to carry out the appropriate tests under carefully controlled conditions so that the only experimental variable is the amount of chemical employed. Since biological activity follows a normal dose–response relationship, a genetically homogeneous population of test organisms standardized with respect to age, sex, and rearing conditions will respond in a relatively fixed and predictable manner to a chemical administered by a particular route (topical, intraperitoneal, etc.) under specified physical conditions (temperature, humidity, etc.). Conversely, even small variations in any one of these parameters can lead to marked changes in biological response through effects on penetration, distribution, metabolism, or target site interaction of the chemical concerned. The discrepancies sometimes found in the results of similar tests conducted in different laboratories are undoubtedly often the result of minor variations in testing and emphasize the tenuous nature of the dose–response relationship.

Another and potentially more important way in which the biological activity of a chemical may be modified is through the prior or simultaneous exposure of the test organism to another chemical agent. The effects resulting from various combinations of chemicals are often as dramatic as they are unpredictable and are commonly referred to as *interactions*. If the net result of the interaction is to enhance biological activity the effect is said to be *synergistic*, whereas if activity is decreased the effect is described as *antagonistic*. Purely

C. F. Wilkinson • Department of Entomology, Cornell University, Ithaca, New York.

additive toxic effects resulting from combinations of chemicals having the same or opposite pharmacological actions are not usually considered as interactions.

It is probable that a variety of both synergistic and antagonistic interactions have long been known and that the beneficial aspects of the latter were first recognized in the course of early attempts to develop antidotes for various toxic agents. There is currently growing concern over the potential hazard to man of toxic interactions resulting from chronic exposure to low levels of the multitude of drugs, pesticides, and other chemical agents present in our environment. Indeed, in view of the enormous number and diversity of such chemicals the theoretical potential for some type of unforeseen interaction appears to be considerable.

The increasing use of multidrug therapy in human clinical pharmacology has in recent years provided many examples of both beneficial and deleterious interactions, and recognition of the potential toxicological significance of such interactions has done much to stimulate research on the mechanisms by which they occur. The literature in this area is now vast, but since it is outside the scope of this chapter the reader should consult the following reference sources for further information: Brown (1972), Garb (1971), Hansten (1971), Hartshorn (1971), and Sher (1971).

This chapter will be concerned primarily with the interactions of insecticides with each other and with a variety of other chemical agents. Even with this limitation, the subject is extremely broad and diffuse and no attempt will be made to review it in its entirety. Instead, the basic principles of interactions will be outlined and illustrated by reference to selected examples which are best understood. It should be emphasized from the outset that many of the interactions to be discussed are presently limited to laboratory observations and in most cases their toxicological significance remains obscure. Previous reviews on various aspects of insecticide interactions have been provided by Durham (1967) and Murphy (1969*a*) and the subject is covered in a section of a U.S. Government Report (Mrak, 1969).

2. Basic Principles and Sites of Interactions

2.1. Introduction

Since the concentration of any chemical (A) reaching a particular target site is governed by a complex series of interrelated equilibria associated with absorption, distribution (including tissue binding and storage), biotransformation reactions, and excretion, there are many potential sites at which interactions may occur (Fig. 1). The ability of another chemical (B) to cause a shift in the rate constants or equilibria by which one or more of these events are controlled is likely to be reflected by some change in the biological activity of A when the two materials are combined. In addition, there is always the possibility of direct interaction occurring at the target site itself. Although examples of interactions occurring at each of these sites have been encountered with

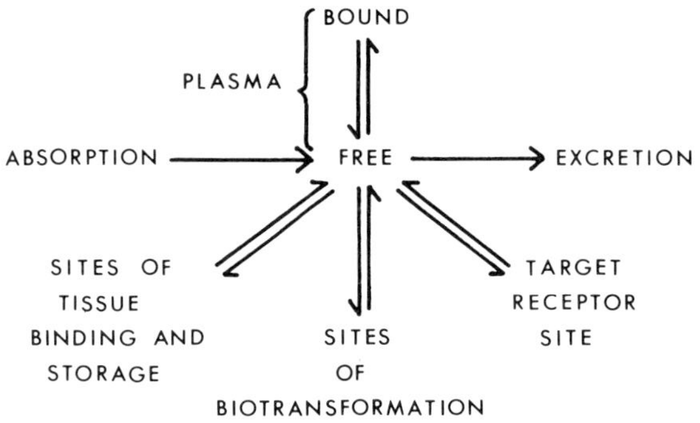

Fig. 1. Sites of interactions.

various combinations of drugs (Stockley, 1971, 1972), those involving insecticides have to date been subject to less intensive investigation and in most cases are not yet clearly understood.

One of the fundamental difficulties in all interaction studies is that of relating the observable biological effect of an interaction to the primary mechanism(s) by which it occurs. Figure 1 clearly shows that the several sites at which interactions may occur are closely linked to one another through a series of dynamic equilibria so that a change in the equilibrium at one site may cause compensatory changes elsewhere. It is probable, therefore, that with most interactions the overall observed effect on biological activity results not from a single primary mechanism (although initially one is usually dominant) but from a combination of this with several secondary events occurring in direct response to the primary interaction. Consequently, the results of many interaction studies are difficult to interpret and the primary mechanism(s) remain ill-defined.

In addition to such obvious factors as concentration, efficacy, and route of administration, the extent to which one chemical (B) can modify the biological activity of another (A) depends on the site at which the interaction occurs; more correctly, it depends on the relative importance of that site in limiting the activity of A applied alone. Consequently, if enzymatic degradation normally represents the major factor limiting the biological activity of A, a suitable metabolic inhibitor can be expected to exert a greater effect on its biological activity than, for example, a compound which interacts to enhance its rate of absorption. That the major factors limiting the activity of any given chemical are species dependent and in any one species often change with age, sex, and physiological condition adds an additional dimension of complexity to interaction studies.

2.2. *Interactions Affecting Absorption*

The process by which a chemical enters the circulation of an animal from its site of application outside the body is referred to as *absorption* and involves the penetration of the outer protective barriers represented by the skin and the epithelia of the alimentary and respiratory tracts (technically outside the body). Since the rate at which this occurs can clearly affect the accumulation of a chemical at its target site, interactions which affect absorption have potentially important consequences on biological activity.

The factors which determine the ability of compounds to penetrate the skin and the epithelia of the gut and the lungs are largely those associated with movement through lipidlike membranes. These have been reviewed in some detail elsewhere (Schanker, 1971; Wilkinson, 1973; see also Chapter 1) and will not be further discussed here. In general, absorption occurs by a process of simple diffusion and the rate of penetration increases with the lipid/water partition coefficient. Ionized materials have very low partition coefficients, and consequently the absorption efficiency of orally administered ionizable materials rests primarily on their degree of ionization (pK_a) in the acidic contents (pH 1) of the gastric lumen.

Little information is available on interactions affecting the absorption of insecticides administered either dermally or by inhalation. Absorption through the skin can be influenced by the vehicle in which a material is applied and is generally enhanced by the use of materials such as oils, ointment bases, and organic solvents (Brown, 1967; McDermot *et al.*, 1967). Detailed studies to determine the dermal absorption of insecticides from formulations containing various spray additives such as emulsifiers, surface-active materials, and organic solvents appear to have been largely neglected.

Considerably more information is available on the types of interactions which can influence the absorption of compounds from the gastrointestinal tract. Interactions which reduce absorption generally do so either by decreasing the amount of material available for uptake or by changing the physical conditions in the gut in a manner which decreases absorption efficiency.

Although of little-known significance with insecticides, some compounds (e.g., tetracycline) can react with various metal ions (Ca^{2+}, Mg^{2+}, etc.) present in milk or antacids to form chelate complexes or insoluble precipitates which are poorly absorbed from the gut. A similar reduction in absorption is often observed in the presence of chemically inert adsorbents such as kaopectate, attapulgite, or activated charcoal and is the basic principle behind the use of the last in acute antidotal therapy following ingestion of many drugs and insecticides (Chin *et al.*, 1970).

Materials causing changes in the peristaltic movement of the gut can also cause a decrease in absorption efficiency. Laxatives which cause a dilution of the gut contents as well as an increase in peristaltic movement and content flow rate decrease the time of contact of a material with the intestinal wall and therefore tend to reduce absorption. It is possible that anticholinesterases (the organophosphates and carbamates) could also interact to cause modifications in the gastric absorption of various materials through their stimulatory effects on the parasympathetic system and consequent action on peristaltic motility.

Interactions to enhance gastric absorption may result from the presence of detergents and other surfactants or the lipophilic oils commonly used as vehicles for oral administration. Antacids and other materials which change the pH of the gut contents could conceivably enhance the absorption of ionizable bases such as nicotine by increasing the amount of the nonionized form available for penetration.

2.3. Interactions Affecting Distribution

Following their absorption, drugs and insecticides are distributed to the tissues by the circulation. Since most have only limited aqueous solubility, their transport in the blood occurs not in solution in the plasma water but primarily through binding to plasma proteins, particularly albumin (Davison, 1971; Thorp, 1964). Indeed, almost all types of chemicals undergo measurable protein binding through ionic, hydrophobic, or van der Waals forces or dipole–dipole interactions such as those responsible for hydrogen bonding (Wilkinson, 1973). Protein binding is a reversible process so that the free and bound forms of a material are in equilibrium at all times, their ratio being dependent on the association constant of the reaction. Since the bound drug is in the form of a high molecular weight complex and is therefore unavailable for membrane transport, protein binding can have a significant effect on the tissue distribution of a given material by limiting its availability for enzymatic modification, glomerular filtration, or tubular secretion. Clearly, therefore, protein binding often regulates the action of a drug or insecticide and any modifications in the binding equilibria may significantly change the degree or duration of its biological effect.

It is now known that important interactions can occur through the displacement of one chemical by another from the protein binding sites in the plasma or tissues (Brodie, 1965; Davison, 1971) (Fig. 2). Drugs and other foreign compounds of very different structure may compete for the same protein binding sites and the degree of competition to be expected can be estimated from the relative association constants of the individual compounds concerned. With compounds that are extensively bound to the plasma proteins,

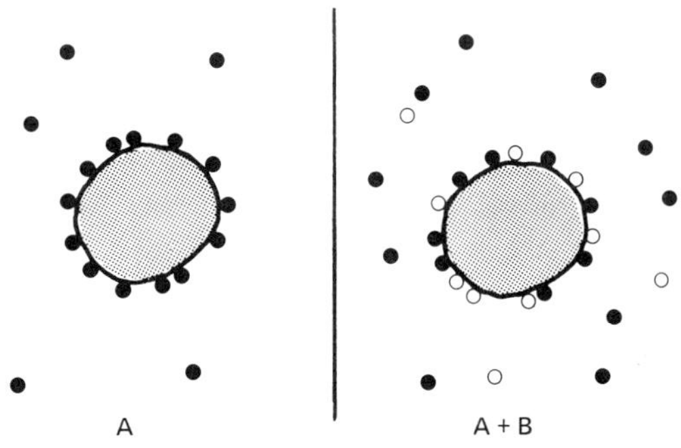

Fig. 2. Protein displacement interactions.

even very small percentage displacements can have profound effects on the amount of the free form available for biological action. Thus if a drug B competes to reduce the plasma binding of drug A from 99% (1% free) to 95% (5% free), this has effectively caused a fivefold increase in the free plasma concentration of drug A and might well cause a significant increase in its biological activity. On the other hand, a similar percentage decrease in binding with a drug normally bound to the extent of only 50% is likely to result in little or no increase in biological activity.

2.4. Interactions Involving Biotransformation

Almost all drugs, insecticides, and other foreign compound are metabolized by living organisms, and the extent to which this occurs is often rate limiting with respect to their biological activity. Accordingly, interactions which modify the metabolic capacity of an animal usually have profound effects on the degree or duration of action of a given chemical.

Of the several types of interactions recognized, those involving biotransformation are undoubtedly the most important and most frequently encountered. They occur through the ability of one chemical (A) to inhibit or stimulate the activity of the enzyme(s) responsible for the metabolism of another (B). The result of the interaction may be to either synergize (potentiate) or antagonize biological activity depending on whether the enzyme concerned activates or inactivates (detoxifies) the chemical in question.

Because of their overwhelming importance in foreign compound metabolism (Chapter 2), the enzymes comprising the hepatic microsomal oxidase system are primary sites at which interactions involving biotransformation can occur. The microsomal enzymes are noted for their extremely low degree of substrate specificity and their ability to catalyze several types of oxidation involving numerous functional groups. When two oxidizable substrates are presented simultaneously to the microsomal enzymes, one can competitively inhibit the metabolism of the other by a process termed "alternative substrate inhibition" (Mannering, 1968). This usually short-lasting inhibitory effect could account for many interactions observed under acute conditions where the concentrations of the two compounds are high enough to temporarily overload the system and they compete with one another for the available microsomal binding sites. The extent to which this kind of inhibition is likely to occur will depend on the relative affinities of the two compounds for the binding sites, and presumably compounds with high affinities and low rates of metabolism will be the most effective alternative substrate inhibitors.

In addition to this rather general phenomenon, several groups of compounds are known particularly for their acute inhibitory potency toward microsomal enzyme activity *in vitro* and their ability to modify the action of many drugs and insecticides *in vivo*. Possibly the best known of these is the drug potentiator SKF 525-A (Mannering, 1968; Anders, 1971), although in recent years interest in the commercial development of new types of insecticide synergists has led to the discovery of several other groups of compounds with similar activity (Casida, 1970; Wilkinson, 1971*a*,*b*, 1975; Hodgson and Philpot, 1974) (see Section 3.3). It is probable that the inhibitory action of several of these materials occurs initially through the alternative substrate mechanism (most are microsomal enzyme substrates), but in most cases the effect is subsequently intensified by additional noncompetitive inhibitory interactions of metabolic products or intermediates (Chapter 3).

Other materials such as carbon tetrachloride (reviewed by Brown, 1972) and a variety of sulfur-containing compounds such as carbon disulfide (Bond and De Matteis, 1969; Bond *et al.*, 1969; Magos and Butler, 1972) and the phosphorothionate insecticides (Norman *et al.*, 1974; Conney *et al.*, 1967) inhibit microsomal enzyme activity when administered chronically to mammals at quite low concentrations. Both carbon tetrachloride and carbon disulfide are recognized as potent hepatotoxins. Their effects on drug metabolism are associated with a decrease in the level of cytochrome P450 which is considered to result from the covalent binding to the microsomes of active radical species (e.g., $\cdot CCl_3$) or atomic sulfur (in the case of the sulfur-containing compounds) released during metabolism (Diaz Gomez *et al.*, 1973; Norman *et al.*, 1974; Dalvi *et al.*, 1974). That the effects of these agents are directly related to their own metabolism is strongly suggested by the fact that their action is enhanced by prior treatment of animals with inducing agents such as phenobarbital (De Matteis and Seawright, 1973; Norman *et al.*, 1974) and reduced following treatment with SKF 525-A.

More commonly the chronic administration of drugs and insecticides to animals results in a marked increase in hepatic drug-metabolizing activity through the process of enzyme induction. Consequently, a compound which when administered as a single dose can inhibit the metabolism of a particular drug 1–8 hr after treatment may stimulate the metabolism of that same drug 24 hr later or after chronic administration (Kato *et al.*, 1964; Kamienski and Murphy, 1971; see also Section 3.3.3). Such biphasic interactions are commonly encountered with many combinations of drugs and insecticides and can lead to extremely complex effects on biological activity.

Interactions resulting from induction of the hepatic drug-metabolizing enzymes are equally as dramatic as those arising from their inhibition. The phenomenon of enzyme induction has received a great deal of attention in recent years, and it is now clearly established that it may occur following treatment of animals with any of a large number of structurally diverse drugs, insecticides, and other chemical agents (Conney, 1967, 1971; Remmer, 1972). Since the subject of induction is discussed elsewhere in this book, it will not be given further consideration here.

2.5. Target Site Interactions

Interactions occurring directly at the target site are encountered only infrequently, and those which are documented are usually antagonistic in nature. The ability of a variety of oximes such as 2-PAM to reactivate organophosphate-inhibited (phosphorylated) cholinesterase is one example of an antagonistic interaction which is used in the antidotal treatment of insecticide poisoning (Namba, 1971), and another is the ability of certain carbamates and sulfur-containing acids to bind to the "neurotoxic esterase" and protect animals from the delayed neurotoxic effects of some organophosphorus insecticides (Aldridge and Johnson, 1971).

As pointed out elsewhere (Mrak, 1969), it is difficult to envisage two compounds interacting with the same target receptor to produce a synergistic effect unless they exert their action on the target by two different mechanisms. Such a situation might be feasible in the case of a combination of, for example, an organochlorine insecticide (causing hyperexcitability and release of acetylcholine at the synapse) and an organophosphorus or carbamate insecticide which prevents the breakdown of the acetylcholine by inhibiting cholinesterase.

3. Insecticide Interactions

3.1. Organochlorine Insecticides

3.1.1. Introduction

During the last three decades, organochlorine insecticides such as DDT and its analogues, hexachlorocyclohexane, and a variety of chlorinated cyclodienes have been used extensively in controlling pests of agricultural and public health importance around the world. In general, these materials have low acute toxicities to mammals and other warmblooded species. They are, however, extremely widely dispersed as persistent residues in the environment. Furthermore, because of their environmental ubiquity and highly lipophilic character they have accumulated in the tissues of many living organisms and through a process termed "biomagnification" have attained quite high levels, in some species occupying the topmost positions in ecological food chains (Chapter 17).

Small residues of organochlorine insecticides also occur in human tissues as a result of ingestion of a variety of consumer products containing traces of these materials. Studies indicate that for the general population of the United States during the years 1965–1968 the average daily intake of DDT and its metabolites was approximately 0.03 mg, which for an average 70 kg man is equivalent to a dosage of about 0.0004 mg/kg/day. This level of intake results in a storage in adipose tissue of about 4.0 ppm (Duggan and Lipscomb, 1969). Similar studies with dieldrin indicate adipose storage levels of 0.20–0.25 ppm. Considerably higher levels have been measured in the general populations of other countries and in those individuals whose occupations bring them into regular and sustained contact with relatively high concentrations of pesticides (spray formulators, applicators, and employees in chemical manufacturing plants).

There is evidence that human tissue residues of persistent organochlorine insecticides are presently lower than those which existed during the 1950s and that these will decrease still further as commercial use of the compounds continues to be phased out. It is probable, however, that human tissue residues will persist for some time in view of the continued presence of traces of the chlorinated hydrocarbons in the environment. Concern has been voiced over the potential human hazard posed by these residues and their possible interactions with drugs, insecticides, and other environmental agents.

Because of their generally low mammalian toxicity, there is only a remote possibility of interactions occurring to enhance the acute toxicity of the chlorinated hydrocarbons *per se*. Consequently, most studies have been concerned with interactions affecting tissue storage levels and with the possible consequences of these residues on the biological activity of other types of chemicals.

3.1.2. Interactions Affecting Tissue Storage

The concentration of an organochlorine insecticide in animal tissue results from a dynamic equilibrium involving on one hand the rate of intake and on the other the rate of metabolism and/or elimination from the body (Chapter 1). Although residues occur in the tissues of many organs, by far the major storage reservoir is the neutral fat comprising the adipose tissue. Since the steady-state concentrations observed in different tissues appear to be functionally related (Robinson, 1969), the distribution patterns probably reflect the physiochemical partitioning of the organochlorine insecticide between the blood and the tissue. Compounds influencing the rates of accumulation, metabolism, or excretion of organochlorine insecticides can interact to cause changes in their steady-state storage levels in the tissues.

The phenomenon was first reported by Street (1964), who observed that the simultaneous feeding of rats with dieldrin and DDT caused a marked reduction of dieldrin storage in the adipose tissues. Subsequent studies revealed that DDT administered in the diet of rats at 5 and 50 ppm over a 10-day period similarly reduced the tissue storage of several cyclodiene insecticides administered simultaneously at 1 and 10 ppm (Street and Blau, 1966; Street, 1968). When fed in combination with 50 ppm DDT, the retention of dieldrin and heptachlor (1 ppm in the diet) in rat adipose tissue was reduced from those in the controls by over 90%. A significant reduction in dieldrin storage was also observed with DDT at a dietary level of 5 ppm, and by extrapolation it was suggested that some effect could be expected to occur at concentrations as low as 0.5–1.0 ppm (Street, 1968). DDT was active when administered either prior to or simultaneously with dieldrin (Fig. 3), and it effectively accelerated the depeletion of preexisting dieldrin stores. The effect persisted for periods of more than 6 weeks after cessation of DDT exposure

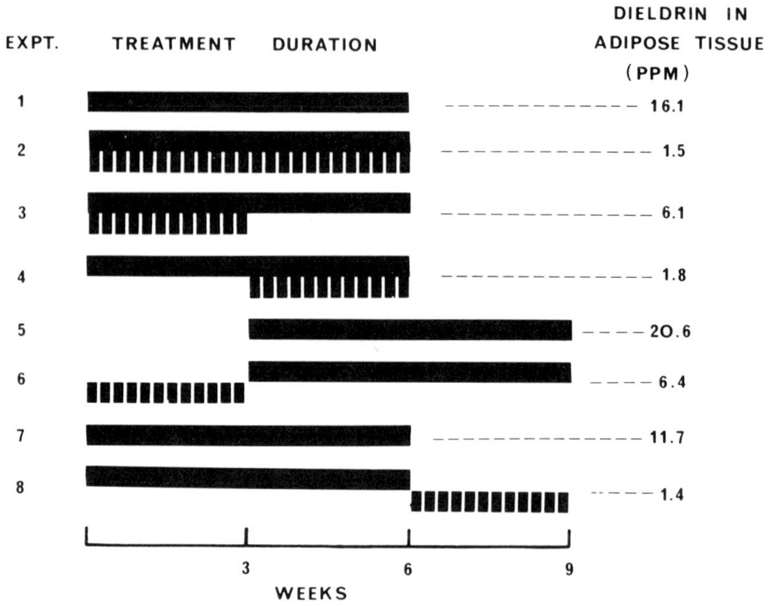

Fig. 3. Effects of DDT administration on the storage of dieldrin in rat adipose tissue. Solid bars indicate periods that rats were fed 1 ppm dieldrin and striped bars periods that they received 50 ppm DDT. The rats in experiments 1–4 and 6–8 were sacrificed at 6 and 9 weeks, respectively. Redrawn from Street *et al.* (1966*a*).

(Street, 1968; Street *et al.*, 1966*a*). Studies with other mammalian species have demonstrated less dieldrin storage in mice (Keplinger *et al.*, 1966), pigs, and to some extent sheep (Street *et al.*, 1966*a*) fed combinations of DDT and dieldrin. In contrast, Deichmann *et al.* (1969) found that dieldrin retention in dogs fed for 10 months on diets containing either 0.6 mg/kg aldrin or a combination of 0.3 mg/kg aldrin and 12 mg/kg DDT remained essentially the same. No effect on dieldrin storage was observed in either hens or trout fed combinations of DDT and dieldrin (Street, 1968).

Storage interactions with dieldrin are not limited to those involving DDT. Several DDT analogues and metabolites caused a similar although less marked reduction of dieldrin storage in rats (DDT > DDE > DDD > DDMU) (Street, 1968), and several barbiturates (phenobarbital, barbital, heptabarbital, seconal, etc.) were effective at dietary levels of 1200–1800 ppm (Street *et al.*, 1966*b*; Street, 1968). Other drugs including aminopyrine, phenylbutazone, tolbutamide, and chlorpromazine were less effective. Cueto and Hayes (1965) and Engebretson and Davison (1971) also reported reduced dieldrin storage in phenobarbital-treated rats, and Koransky *et al.* (1964) found a similar effect on the storage of α-HCH.

It has been suggested that the major mechanism through which these storage interactions occur involves microsomal enzyme induction (Street,

1968). Each of the drugs and insecticides shown to have an effect on the tissue storage levels of the organochlorine insecticides is known to induce the hepatic drug-metabolizing enzymes (Conney, 1967), and the chlorinated hydrocarbons are particularly well recognized for their potent, long-lasting effects (Hart and Fouts, 1965; Schwabe and Wendling, 1967). The probable involvement of induction is supported by the fact that depletion of dieldrin storage following exposure of DDE is associated with increases in the conversion of dieldrin to hydrophilic products and in the excretion of polar metabolites in the feces and urine, the latter being enhanced seventeenfold (Street, 1968; Street and Chadwick, 1967).

Dieldrin is susceptible to enzymatic hydroxylation *in vivo* (Brooks, 1972), and it is probable that this pathway and that involving epoxide hydration to the diol (not a microsomal oxidase) are susceptible to induction by DDT. Subsequent studies in which the effects on dieldrin metabolism of DDT, heptabarbital, α-chlordane, and 3-methylcholanthrene were compared indicated interesting qualitative differences in the metabolites excreted in the urine and feces. Chlordane (250 ppm) substantially enhanced the excretion of urinary metabolites but was not as active as DDT (50 ppm) in reducing tissue storage of dieldrin. Methylcholanthrene (500 ppm) had only a slight effect on dieldrin metabolism and mobilization (Street, 1968).

Enhanced metabolism through induction of the hepatic drug-metabolizing enzymes could certainly provide a satisfactory explanation of tissue storage interactions since an increased rate of removal of material circulating in the blood would establish a concentration gradient from the tissues and eventually lead to the attainment of a new steady-state tissue level. However, the repeated failure of inhibitors of protein synthesis to block the process (Street, 1968) challenges the suggestion that microsomal enzyme induction constitutes the sole mechanism by which DDT accelerates dieldrin depletion. Neither actinomycin D nor ethionine, both of which have been shown to inhibit the inducing action of DDT (Lange, 1967), had any significant effect on the ability of DDT to mobilize tissue residues of dieldrin even when administered at relatively high concentrations.

It has been proposed (Menzer and Rose, 1971) that in addition to their enzyme-inducing properties DDT and various drugs might enhance dieldrin metabolism and excretion by displacing it from the proteins to which it is normally bound in the plasma and liver. Since dieldrin is a highly lipophilic compound and is extensively bound to the plasma proteins (Moss and Hathway, 1964; Eliason and Posner, 1971), even a small degree of competitive displacement could substantially increase its concentration in the free form available for metabolism and excretion. The extent to which displacement can be expected to occur will depend on the relative concentrations of DDT and dieldrin in the plasma and on their relative affinities for the common binding sites. It is of interest that the greatest effects on dieldrin mobilization were observed at low dieldrin dosages and high levels of DDT, conditions which would presumably favor binding displacement. The plasma binding affinities of

DDT and various drugs relative to dieldrin have not yet been studied, but differences in this parameter could account for the fact that despite their almost equal effectiveness as enzyme inducers DDT is 30–40 times more active than the barbiturates in reducing dieldrin storage (Street, 1968; Street *et al.*, 1966*b*). Certainly there appears to be no direct relationship between induction *per se* and reduction in dieldrin storage levels, and the ability of DDT and to a lesser extent phenobarbital to stimulate the release of [^{14}C]dieldrin bound to liver slices *in vitro* (Matsumura and Wang, 1968) supports the possible involvement of protein displacement reactions.

Another factor which might play a role in the elimination of dieldrin tissue residues following treatment with enzyme-inducing agents is biliary excretion. The significance of this mechanism has received little attention to date despite its suggested importance in dieldrin excretion (Heath and Vandekar, 1964) and studies demonstrating enhanced biliary flow and accelerated plasma disappearance of some drugs following treatment of rats with phenobarbital (Klaasen, 1969; Klaasen and Plaa, 1968). However, the increase in biliary flow is apparently not related to microsomal induction *per se* since it is not observed to any great extent with several other known inducing agents (Klaasen, 1969); reports indicate that the effect may be associated with bile salt production (Eakins and Slater, 1973).

Although biliary excretion constitutes one route by which dieldrin could enter the alimentary tract, it is probable that the recently established ability of dieldrin to recycle from the blood to the gut via the salivary and gastric secretions (Cook, 1970) plays a more important role in determining the level of dieldrin excreted in the feces. This recycling process is the basis for the suggested use of activated charcoal as an antidote for acute dieldrin poisoning in ruminants (Wilson and Cook, 1970). If, in addition to their enzyme-inducing properties, DDT, the barbiturates, and various other drugs are able to displace dieldrin from the plasma proteins as previously discussed, they will presumably increase the amounts of dieldrin available for recycling and fecal excretion. It is entirely possible, therefore, that this mechanism could play an important role in the increased fecal excretion of dieldrin and its metabolites and the decreased levels of these materials in the tissues of rats treated with DDT and the barbiturates. The use of combined phenobarbital–charcoal treatments has been successfully applied in the removal of dieldrin residues from dairy cattle (Reeder, 1969) and swine (Dobson *et al.*, 1971), although neither phenobarbital nor charcoal alone had any effect on dieldrin levels in the eggs or livers of hens fed this insecticide (Mick and Long, 1973). Davies *et al.* (1969) have also reported that human patients receiving the anticonvulsant drug phenytoin (diphenylhydantoin) had significantly lower blood levels of DDE than similar patients not on drug therapy.

3.1.3. Inductive Effects

Induction of the hepatic microsomal drug-metabolizing enzymes is a well-established mechanism by which a large number of unrelated drugs and insecticides can interact to modify the biological action of other chemicals (Conney, 1967, 1971; Remmer, 1972; Sher, 1971). The inducing capacity of the organochlorine insecticides was first discovered accidentally following the spraying of laboratory animal rooms with chlordane and DDT (Hart *et al.*, 1963), and since then almost all insecticides of this class have been shown to induce a variety of drug oxidations *in vitro* and to exert effects on the *in vivo* metabolism and biological activity of numerous chemicals. They are now widely recognized as being among the most potent, long-lasting inducing agents known. Since their activity has been discussed elsewhere in this book and since specific interactions of organochlorine and organophosphorus insecticides (many of which involve induction) are covered in Section 3.2 of this chapter, only a few additional examples will be included here.

The stimulatory action of the organochlorine insecticides on hepatic drug metabolism can markedly influence the acute pharmacological action and/or toxicity of various drugs. Thus pretreatment of rats with chlordane or DDT substantially decreases barbiturate sleeping time (Conney *et al.*, 1967) and affords protection from the gastric lesions caused by phenylbutazone (Welch and Harrison, 1966) and the toxic action of the anticoagulant warfarin (Ikeda *et al.*, 1968). In view of the ubiquitous presence of residues of these and other chlorinated compounds in the environment, it has been suggested that this last effect might explain, at least in part, the resistance of rats to warfarin used as a rodenticide (Burns, 1969). Treatment of dogs with chlordane (5 mg/kg orally, three times a week for 5–6 weeks) enhanced the metabolism of antipyrine and phenylbutazone as measured by plasma levels of these materials, some effect being observed as long as 4 months after termination of treatments (Burns *et al.*, 1965; Burns, 1969) (Fig. 4).

Although there is overwhelming evidence to support the inducing action of relatively large doses of organochlorine insecticides in laboratory animals, the practical significance of these observations with respect to alterations in drug or pesticide toxicity in man is extremely difficult to evaluate. It has been established that the serum half-life of phenylbutazone in people occupationally exposed to DDT (factory workers) is about 20% lower than that in an appropriate control group (Conney *et al.*, 1971), supporting the findings of an earlier study with aminopyrine (Kolmodin *et al.*, 1969). It would appear, therefore, that under conditions of relatively high exposure, inductive interactions can occur in man, but it is not clear whether environmental levels are sufficiently high to cause any effect in members of the general population. Pentobarbital hypnosis in rats is significantly decreased following a single intraperitoneal injection of only 1 mg/kg DDT (Conney *et al.*, 1967), a dose which results in fat concentrations of DDT of about 10 ppm. This is only slightly more than the levels found in the adipose tissues of the general population of the United States and is certainly within the range of values reported from other countries and from those individuals who are occupationally exposed (Robinson, 1969). Few attempts have been made to establish the minimum doses of organochlorine insecticides required to elicit an inductive

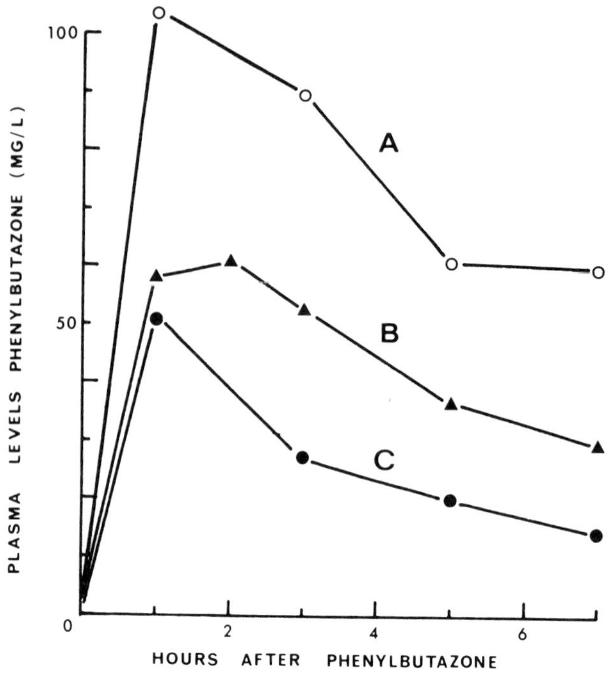

Fig. 4. Stimulatory effect of chlordane (5 mg/kg administered orally three times a week) on phenylbutazone (25 mg/kg, intraperitoneally) metabolism in the dog. (A) Before chlordane treatment, (B) 16 weeks after cessation of chlordane treatment, (C) 7 weeks on chlordane.

response. Kinoshita *et al.* (1966) reported enzyme induction in rats fed DDT and toxaphene at dietary levels of 1 and 5 ppm, respectively, but even these low doses are considerably in excess of the average daily intake of these materials in the U.S. total diet. On this basis, it appears unlikely that any significant induction occurs in man as a result of exposure to current environmental residues of organochlorine insecticides, although this is probably not true for some other species, particularly those birds and fish occupying the topmost positions of the ecological food chains. Another factor which may be a justifiable cause for concern is the possibility of additive induction. Thus although a large number of individual compounds may occur in the environment at concentrations which alone show no inductive effect, it is possible that they could exert a total cumulative effect of some significance.

The ability of the organochlorine insecticides to stimulate the hepatic microsomal hydroxylation of steroid hormones is another interaction which has caused some concern in recent years because of its potential to alter normal body metabolism. As a result of the similarities between the hepatic enzymes which hydroxylate foreign compounds and steroids (Kuntzman *et al.*, 1964), exposure to organochlorine insecticides usually leads to a marked increase in the metabolism and excretion of a variety of steroid hormones in animals (Conney, 1972; Conney *et al.*, 1967, 1971) and in man (Conney *et al.*, 1971). Welch *et al.* (1967) have reported that as in the case of drug metabolism the organochlorine insecticides can either stimulate or inhibit microsomal testosterone hydroxylation depending on whether they are administered chronically or acutely.

The toxicological significance of the increased metabolism and excretion of steroid hormones in man is not clear. It probably occurs only in response to relatively high inducer concentrations and is undoubtedly balanced by an increase in the rate of biosynthesis in the appropriate endocrine gland.

It is of interest that the potent and long-lasting inductive effects of organochlorine insecticides have led to their recognition as potential therapeutic agents. Thus the DDT analogue *o,p′*-DDD has been shown to ameliorate the ill-effects of Cushing's syndrome, which is characterized by an adrenal overproduction of cortisol (Bledsoe *et al.*, 1964), and DDT itself has been successfully used in the treatment of newborn infants suffering from hyperbilirubinemia (Thompson *et al.*, 1969). The latter effect appears to result from the induction by DDT of hepatic glucuronyltransferase activity. There is also one somewhat sketchy report of the apparently successful use of massive single oral doses of DDT (5 g) to treat three human patients suffering from acute barbiturate intoxication (Rappolt, 1970).

3.2. Organophosphorus Insecticides

3.2.1. Introduction

The organophosphorus insecticides are among the most toxic compounds employed for insect control, and since their use has recently increased considerably with the phasing out of the organochlorine compounds the need for a complete assessment of their potential toxicological hazards to man has assumed greater importance. Unlike the chlorinated hydrocarbons with their well-known propensity for persistence and accumulation in the tissues of living organisms, organophosphorus insecticides are usually rapidly degraded and eliminated. Consequently, the major effects of interactions involving these compounds are usually acute and are likely to occur within a short period of time following exposure.

The increase or decrease in organophosphate toxicity resulting from interactions with other chemicals or drugs has been recognized for many years. The possible combinations of mechanisms by which they can occur are many and complex, and few if any are yet completely understood. Useful reviews in this area have been provided by Murphy (1969*a*) and Durham (1967).

In order to fully understand the potential complexities of interactions involving organophosphorus insecticides, it is necessary to briefly review some

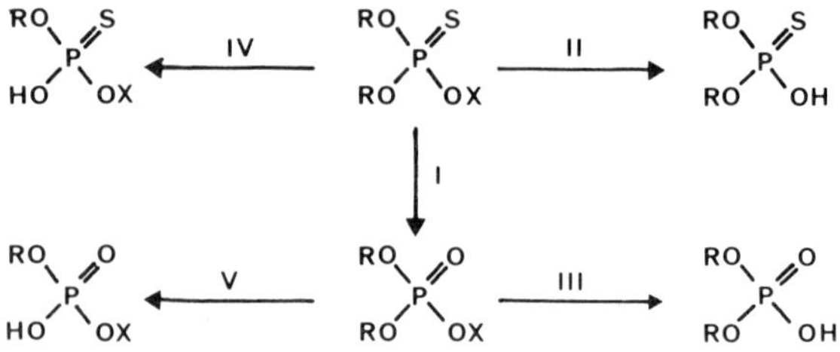

Fig. 5. Metabolic pathways of phosphorothionate insecticides.

of the factors which determine the biological activity of these compounds. Biotransformation reactions are of critical importance in this respect, and since many of the insecticides in commercial use are phosphorothionates a typical compound of this class will be used as an example. The principal metabolic pathways followed by a phosphorothionate are shown in Fig. 5, and although these have been discussed in detail in earlier chapters (Chapters 2, 4, and 5) some brief repetition is necessary for the purposes of the present discussion.

Based on the nature of the products formed, three major reactions may occur. These are conversions of the phosphorothionate (P═S) to the corresponding phosphate (P═O) (reaction I), cleavage of the P—O—X linkage with subsequent liberation of the acidic or leaving group X (reactions II and III), and cleavage of the P—O—R group or dealkylation (reactions IV and V). Reactions II, III, IV, and V all result in the formation of insecticidally inactive phosphodiesters and are therefore detoxication pathways. In contrast, reaction I leads to the formation of a highly potent anticholinesterase (phosphorothionates are inactive in this respect) and is an activation pathway.

Reaction I is catalyzed by a typical, rather nonspecific microsomal oxidase (Chapter 2) which is widely distributed in many species of insects and mammals. Initially both reactions II and III were thought to be catalyzed by phosphatases (Heath, 1961; O'Brien, 1967), but more recent studies have established that with the phosphorothionates (reaction II) the reaction is effected mainly by the microsomal oxidase system requiring NADPH and O_2 (Chapter 2). Consequently, microsomal oxidation of phosphorothionates can lead to products with more or less insecticidal potency. Whether esterases capable of hydrolyzing phosphorothionates directly are of any metabolic importance has not been fully established, and available evidence indicates that the phosphates (P═O) are the preferred (perhaps the only) substrates for phosphatase action. The phosphatases are typified by "paraoxonase," which was initially classified as an A-esterase by Aldridge (1953*b*). The range of substrates attacked by the A-esterases is not known, although it appears that the structural features on which this depends are similar to those which determine the ability of the phosphate to inhibit cholinesterase. Accordingly, good anticholinesterases are usually rapidly cleaved by A-esterases (Heath, 1961) (Chapter 4).

Dealkylation (reactions IV and V) can be effected by at least two types of enzymes, the relative importance of each depending on the nature of the alkyl group (R). In the case of dimethyl phosphates, the reaction is catalyzed primarily by a glutathione-*S*-alkyltransferase (GSAT) present in mammalian tissues (Fukami and Shishido, 1966; Hollingworth, 1969, 1970) and insect tissues (Fukunaga *et al.*, 1969) (Chapter 5). Compounds containing ethyl substituents or longer alkyl chains in this part of the molecule may also be attacked by GSATs but are more susceptible to oxidative dealkylation by the microsomal enzymes (Donninger *et al.*, 1972).

In addition to the general pathways outlined in Fig. 5, important biotransformation reactions often result from direct enzymatic modification of the leaving group (X) without ester cleavage. Some of these reactions are extremely important in organophosphate interactions but will be discussed later with reference to specific insecticides.

Although biotransformation reactions are undoubtedly of major importance in many interactions involving the organophosphate insecticides, it should be emphasized that any of the other factors previously discussed may also play an important role.

The toxicity of organophosphorus insecticides to mammals may be antagonized or potentiated when they are administered in combination with various drugs, insecticides, or other chemicals. Although, for convenience, the following discussion is divided into antagonistic and synergistic interactions, the distinction is seldom clear-cut and there are many exceptions and apparent data contradictions which make this area difficult to review. The results vary with the combinations of chemicals and animal species employed, and it should be remembered that since many interactions are biphasic in nature (Section 3.3.3) either antagonism or synergism may be observed depending on the time interval between administration of the two agents.

3.2.2. Antagonistic Interactions

Ball *et al.* (1954) were the first to observe that 3–4 days after receiving a single oral dose of aldrin, chlordane, or lindane female rats were significantly protected against the toxicity of orally administered parathion. Main (1956) confirmed the protective action of aldrin against orally or intravenously administered parathion in rats and found a similar effect with orally (but not intravenously) administered paraoxon and TEPP. Antagonism of the toxicities of parathion, paraoxon, malaoxon, guthion, EPN, TEPP, and DFP, but not OMPA, was subsequently observed (Table I) in mice exposed to a single oral dose of aldrin (16 mg/kg) (Triolo and Coon, 1966*a,b*; Cohen and Murphy, 1974), an effect against parathion being significant at an aldrin concentration as low as 1 mg/kg. Protection appeared maximal when exposure to parathion occurred about 4 days after the animals had received a single dose of either aldrin or dieldrin (16 mg/kg), and, indeed, 1 hr after dosage, parathion toxicity was greater than in the controls (Triolo and Coon, 1966*a,b*). Mice given a

Table I. Effect of a Single Oral Dose of Aldrin (16 mg/kg) on the Acute Toxicities of Several Organophosphorus Insecticides[a]

Insecticide (mg/kg)	Percent mortality	
	Control	Aldrin pretreated
Parathion (22)	35	0
Paraoxon (40)	100	44.4
TEPP (10)	95	0
DFP (50)	66.6	10
EPN (75)	50	0
Guthion (15)	84.6	15.4
TOCP (2000)	60	20
OMPA (25)	60	70

[a] Data from Triolo and Coon (1966*b*).

single oral dose of DDT or DDE (75 mg/kg) were similarly protected against parathion (although not paraoxon) (Bass *et al.*, 1972), but in another study no effect on parathion toxicity was observed in mice pretreated with DDT (Chapman and Leibman, 1971); however, an effect was observed following similar pretreatment with chlordane. The miticide ovex has also been shown to protect rats against parathion toxicity (Black *et al.*, 1973), and Neubert and Schaefer (1958) reported that pretreatment of rats with α-hexachlorocyclohexane (not the insecticidal isomer) decreased their susceptibility to OMPA but not to DFP.

Similar protective effects have been observed with several drugs. Welch and Coon (1964) reported that mice treated for 4 days with chlorcyclizine, cyclizine, SKF 525-A, and phenobarbital were less susceptible to parathion, malathion, and EPN, and chlorcyclizine (McPhillips, 1965) and tolbutamide (Takabatake *et al.*, 1968) were effective against OMPA and parathion poisoning, respectively. The toxicities of orally administered parathion and paraoxon were increased 1–6 hr after intraperitoneal treatment of mice with chlorpromazine but were reduced when determined 1 day after treatment (Vukovich *et al.*, 1971). The protective action of phenobarbital has been confirmed with malathion and EPN (Brodeur, 1967), parathion and paraoxon (Alary and Brodeur, 1969; Murphy *et al.*, 1959), and several other commonly used organophosphates (DuBois, 1971; Menzer, 1971). DuBois and Kinoshita (1968) showed that phenobarbital treatment decreased the toxicity to rats of all compounds tested except OMPA (Table II) and observed similar, although less well-defined effects with mice. Mice pretreated with phenobarbital were similarly protected against the toxicity of dicrotophos, phosphamidon, and their respective dealkylated derivatives but not against malathion (Menzer and Best, 1968; Menzer, 1971), and in agreement with an earlier report (O'Brien,

Table II. Acute Toxicities of Organophosphorus Insecticides to Normal Rats and Rats Pretreated with Phenobarbital (50 mg/kg) for 5 days[a]

Insecticide	Rat LD_{50} values (mg/kg) — Rats: Control	PB treated	Ratio of LD_{50} values (PB treated/control)
Parathion	2.5	7.3	2.9
Methyl parathion	7.0	8.0	1.1
EPN	7.3	75.0	10.3
Systox	1.4	5.8	4.1
Di-Syston	2.1	17.0	8.1
Guthion	8.7	11.4	1.3
Delnav	17.2	118.7	6.9
Phosdrin	1.2	2.4	2.0
Ethion	25.9	302.6	11.7
Trithion	10.1	66.5	6.6
OMPA	28.7	14.5	0.5
Co-Ral	7.5	13.8	1.8
Malathion	619.4	949.9	1.5
Ronnel	2822.6	3034.8	1.1
Folex	124.0	171.9	1.4

[a]Data from DuBois and Kinoshita (1968).

1967) the toxicity of dimethoate was enhanced three- to fourfold. Similar, although less pronounced effects were observed with respect to the toxicities of dicrotophos, phosphamidon, dimethoate, and malathion to mice maintained on diets containing various levels of DDT and dieldrin (Menzer, 1970, 1971).

It is clear from the foregoing that pretreatment of laboratory animals with a variety of chlorinated hydrocarbon insecticides and drugs known to induce the hepatic microsomal enzymes (Conney, 1967) often affords some degree of protection against organophosphate toxicity. The precise mechanism by which this occurs, however, is still not fully understood, despite a great deal of work in this area in recent years.

With the phosphorothionate insecticides, induction of mixed-function oxidase activity would be expected to enhance the formation of the toxic oxon analogue and lead to an increase in toxicity. However, it should be remembered that the same enzyme system can also result in the direct cleavage of the phosphorothionate, a detoxication pathway (Fig. 5). The overall effect of oxidase induction *per se* will therefore depend on the balance between these two opposing reactions and whether one of the pathways is more responsive to enzyme-inducing agents than the other. The evidence to date is somewhat equivocal. Neal (1967*a*) found that pretreatment of rats with phenobarbital or 3,4-benzo[*a*]pyrene caused increases of a similar magnitude in both reactions as measured by the *in vitro* metabolism of parathion in liver microsomes, although it was suggested that the reactions were probably catalyzed by separate enzymes (Neal, 1967*b*). Subsequent studies by Alary and Brodeur (1969) on the *in vitro* metabolism of parthion in rat liver preparations also indicated that both activation and detoxication pathways were stimulated following the phenobarbital treatment. The results indicated that phenobarbital stimulated the direct degradation of parathion to a larger extent than its activation, and this conclusion was supported by the enhanced urinary output of diethylphosphorothioic acid from parathion in corresponding *in vivo* studies. Unfortunately, the validity of the data in this study is somewhat questionable since no direct measurements of paraoxon formation were made. Whether or not one pathway is stimulated significantly more than the other is still open to question and appears to depend on the inducing agent (possibly also its route of administration) and the animal species employed. DDT and chlordane caused an equal stimulation of both pathways of parathion metabolism in rat liver preparations, whereas 3-methylcholanthrene preferentially enhanced that leading to paraoxon (Chapman and Leibman, 1971). In mice, however, DDT appeared to cause a preferential stimulation of parathion degradation while 3-methylcholanthrene and chlordane reduced and stimulated both pathways, respectively. No relationship was found between these *in vitro* data and the *in vivo* effect of the inducers on parathion toxicity in rats and mice (Chapman and Leibman, 1971).

Even allowing for the expected errors in extrapolating from *in vitro* and *in vivo* metabolism data to toxicity, it seems unlikely that the protective action of various drugs and insecticides toward organophosphate toxicity can be satisfactorily accounted for simply on the basis of a difference in inducer response between the microsomal activation and detoxication pathways. Furthermore, it should be emphasized that the protective action is not limited to phosphorothionates but extends to a variety of phosphates (Main, 1956; Triolo and Coon, 1966*a,b*; Alary and Brodeur, 1969; Welch and Coon, 1964; Menzer, 1971), which in general are not detoxified primarily by microsomal oxidation. Clearly in the case of the phosphates other protective mechanisms have to be considered.

Since pretreatment of mice with aldrin significantly decreases the subsequent inhibition of brain cholinesterase by paraoxon and malaoxon (Triolo and Coon, 1966; Cohen and Murphy, 1974) (Fig. 6), the protective action apparently results from some mechanism causing a decrease in the level of the insecticide available for target site interaction. One or more of several events may occur to cause this. The paraoxon may of course be hydrolyzed by phosphatases (A-esterases, reaction III in Fig. 5) in the blood or tissues, or alternatively it may bind to and inhibit aliesterases (B-esterases) without itself being hydrolyzed; it may also bind to a variety of other nontarget sites in the tissues.

Main (1956) attributed the protective effect of aldrin against paraoxon to an increase in liver A-esterase activity, although at the same time the A-esterase activity in the plasma was found to decrease. These aldrin-induced

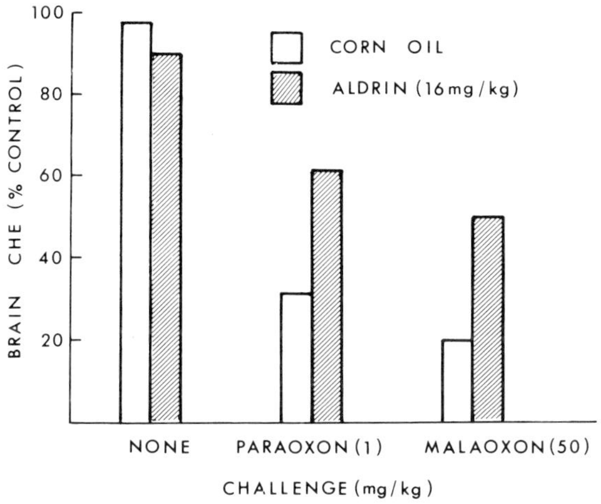

Fig. 6. Effect of aldrin on the *in vivo* inhibition of mouse brain cholinesterase activity (CHE) by paraoxon and malaoxon. The mice were challenged 4 days after a single oral dose of aldrin (16 mg/kg) and sacrificed 30 min and 10 min after intraperitoneal treatment with malaoxon and paraoxon, respectively. Redrawn from Cohen and Murphy (1974).

changes in rat liver and plasma A-esterase levels were confirmed by Ghazal *et al.* (1964) and Triolo and Coon (1966*b*), and Welch and Coon (1964) found a similar effect in mice following treatment with chlorcyclizine. Main (1956) suggested that the reason aldrin effected protection only against orally and not intravenously administered paraoxon was that with the oral route the paraoxon went first to the liver, where it could be detoxified by A-esterase action. However, Triolo and Coon (1966*b*) showed that despite its ability to increase liver A-esterase activity aldrin did not increase the *in vivo* enzymatic degradation of paraoxon and subsequently argued (Triolo *et al.*, 1970) that because of the apparently high K_m of the liver A-esterase for paraoxon (10^{-3} to 10^{-4} M, *in vitro*) it was unlikely to represent a significant factor in detoxication. More recent studies were not able to confirm an increase in liver A-esterase activity following treatment of mice with aldrin (Cohen and Murphy, 1974).

In addition to reportedly modifying liver and plasma A-esterase titers, treatment of rats with a variety of drugs and insecticides causes an increase in liver and serum aliesterase (B-esterase) activity. Crevier *et al.* (1954) reported that single oral doses of aldrin, dieldrin, chlordane, and DDT produced this effect, and Ball *et al.* (1954) suggested this as an explanation for the protective action of aldrin against parathion (after oxidation to paraoxon) poisoning. Increases in serum aliesterase have also been observed in mice following treatment with aldrin (Triolo and Coon, 1966; Cohen and Murphy, 1974) and chlorcyclizine (Welch and Coon, 1964), and aldrin is reported to cause a similar increase in liver aliesterase activity (Cohen and Murphy, 1974).

Since many organophosphates are effectively bound to aliesterases, an increase in aliesterase titer might constitute a general mechanism for decreasing the levels of organophosphates and phosphorothionates (after activation to the oxons) available for cholinesterase inhibition. Of interest and possible importance in this respect is the fact that the protein synthesis inhibitor ethionine blocked the aldrin-induced increases in serum B-esterase (as well as liver A-esterase) and at the same time eliminated the protective action of aldrin against orally administered paraoxon (Triolo and Coon, 1966*b*). Lauwerys and Murphy (1969) working with rats clearly demonstrated that low concentrations of paraoxon were "detoxified" by a nonenzymatic binding process to noncritical sites in the liver and plasma. Pretreatment of rats with a concentration of tri-*o*-cresyl phosphate (TOCP) (125 mg/kg) which itself caused no inhibition of either phosphatase (A-esterase, paraoxonase) or brain cholinesterase activities but which completely blocked these binding sites *in vitro* caused a rapid increase in the rate of inhibition of brain cholinesterase (Fig. 7) and an approximately twofold increase in paraoxon toxicity (oral, intraperitoneal, or subcutaneous). Pretreatment of mice with TOCP (125 mg/kg) similarly potentiated brain anticholinesterase action of paraoxon *in vivo* (Cohen and Murphy, 1974).

Triolo and Coon (1969) and Triolo *et al.* (1970) established that the *in vitro* protein binding of paraoxon was increased in the plasma of mice treated with aldrin and that this resulted in a marked decrease in the level of both "free paraoxon" and cholinesterase inhibition (Table III). A good correlation was subsequently observed between paraoxon toxicity to rats and the "free paraoxon" levels in the plasma of animals treated with DDT, chlordane, and dieldrin (Triolo *et al.*, 1970) (Table IV). Since the time at which the maximum

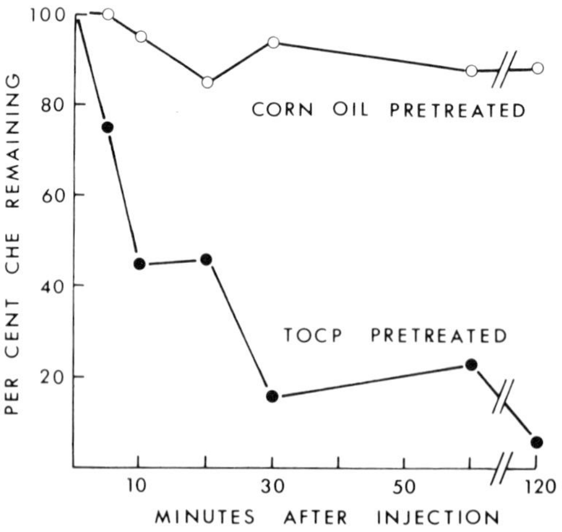

Fig. 7. *In vivo* inhibition of brain cholinesterase activity after treatment with paraoxon (0.25 mg/kg, intraperitoneally) in rats pretreated with corn oil (control) or TOCP (125 mg/kg, intraperitoneally) 16 hr before paraoxon. Redrawn from Lauwerys and Murphy (1969).

Table III. Effect of Aldrin Treatment on the in Vitro Plasma Binding and Detoxication of Paraoxon[a]

Plasma	Paraoxon added to plasma (μg/0.5 ml)	Percent paraoxon "bound"	Percent paraoxon free	Percent cholinesterase inhibition[b]
Control mice	1.6	86.2±4.3	13.9±4.3	84.2±5.8
Aldrin-treated mice[c]	1.6	99.4±0.1	0.6±1.0	4.2±3.4

[a]Data from Triolo *et al.* (1970).
[b]Acetylcholinesterase added to plasma.
[c]Single oral dose of aldrin (16 mg/kg) 4 days before assay.

protective action of aldrin against paraoxon occurred (4 days) coincided with that at which maximum stimulation of plasma aliesterase was observed, and since paraoxon completely inhibited this enzyme at about 10^{-6} M, Triolo *et al.* (1970) suggested that aliesterase was in fact the major paraoxon-binding protein in the plasma. A similar conclusion has been reached by Cohen and Murphy (1974). The possible importance of other binding sites should not be discounted, however (Cohen and Murphy, 1972), and it is of interest that Welch and Coon (1964) reported an increase in serum tosylarginine methyl ester (TAME) hydrolysis (considered an indicator of proteolytic enzyme activity) following treatment of mice with chlorcyclizine.

It therefore appears that several possible mechanisms exist which can alone or in combination account for interactions which antagonize the toxicity of the organophosphate insecticides in mammals. The relative importance of these in any particular situation will depend on numerous factors including the organophosphate in question, the nature of the interacting chemical, and the animal species concerned.

Table IV. Relationship Between in Vitro Plasma Binding and Toxicity of Paraoxon in Mice Pretreated with Organochlorine Insecticides[a]

Insecticide[b] (mg/kg)	Percent mortality paraoxon (2 mg/kg)	Percent free paraoxon in plasma
Control	60	17.3±1.9
DDT (75)	40	7.4±4.9
Dieldrin (16)	20	0.7±0.1
Chlordane (150)	15	0.4±0.1

[a]Data from Triolo *et al.* (1970).
[b]Single oral dose 2 days before assay.

3.2.3. Synergistic Interactions

The best-known and most thoroughly studied examples of organophosphate synergism are those involving combinations of various compounds with malathion. This insecticide usually has low mammalian toxicity due to its rapid detoxication by tissue carboxylesterases (aliesterase), which hydrolyze an α-carboxylester group of malathion and malaoxon (the active oxon analogue) to yield products with little or no anticholinesterase activity (Chapter 4).

$$\underset{\text{malathion}}{\begin{array}{l}(CH_3O)_2P(S)SCHCOOC_2H_5\\ \qquad\qquad\quad |\\ \qquad\qquad CH_2COOC_2H_5\end{array}} \xrightarrow{\text{carboxylesterase}} \underset{\alpha\text{-monoacid}}{\begin{array}{l}(CH_3O)_2P(S)SCHCOOH\\ \qquad\qquad\quad |\\ \qquad\qquad CH_2COOC_2H_5\end{array}}$$

In 1957, Frawley *et al.* reported the ability of EPN to dramatically potentiate the acute toxicity of malathion to rats and dogs, and as a direct result of this unexpected discovery the U.S. Food and Drug Administration instituted new regulations requiring that all new anticholinesterase insecticides be tested in combination with other insecticides of the same class.

Subsequent studies on the mechanism of the EPN–malathion interaction revealed that single doses of EPN causing no cholinesterase inhibition *per se* significantly impaired the ability of rat liver to detoxify malaoxon (Murphy and DuBois, 1957). *In vitro* studies further established that the oxygen analogue of EPN and several other organophosphates inhibited the enzymatic hydrolysis of malathion and other carboxylesters in rat liver preparations (Cook *et al.*, 1958; Murphy and DuBois, 1957, 1958). It is now generally accepted that in agreement with its classification as a B-esterase (Aldridge, 1953*b*; Main and Braid, 1962) the enzyme hydrolyzing malathion (carboxylesterase, aliesterase) is susceptible to inhibition by a large number of organophosphates and consequently many of these are capable of synergizing the toxicity of malathion (Casida, 1961; DuBois, 1961; Mei-Quey *et al.*, 1971). The fact that malaoxon can be either a substrate or an inhibitor of carboxylesterase (Main and Dauterman, 1967) probably explains, at least in part, the interesting observation of Murphy (1967) that treatment of guinea pigs and rats with doses of malathion below those required to inhibit acetylcholinesterase increased their subsequent susceptibility to the same insecticide. The degree of malathion potentiation observed varies considerably with different compounds and is particularly high with EPN and the noninsecticidal aliesterase inhibitor TOCP (Murphy *et al.*, 1959). Since there appeared to be a general correlation between the degree of potentiation and the inhibition of liver and plasma carboxylesterase activity (Casida, 1961; Murphy and Cheever, 1968), it was suggested that measurement of the latter might provide a useful *in vitro* indicator of compounds likely to potentiate the toxicity of malathion (DuBois, 1969; Murphy, 1969*a*). Following pretreatment of mice (16–18 hr) with low doses of TOCP, a close relationship was observed between the degree of potentiation of malathi-

Table V. Inhibition of Liver Carboxylesterase Activity and Potentiation of Brain Cholinesterase Inhibition by Malathion in Mice Pretreated (16-18 hr) with TOCP[a]

Pretreatment dose of TOCP (mg/kg)	Dose of malathion (mg/kg) for 50% inhibition of brain ChE	Degree of potentiation of ChE inhibition	Percent inhibition liver carboxylesterase	
			Triacetin	Malathion
Control	1250	—	—	—
5	660	1.9	21	22
10	230	5.4	45	47
15	145	8.6	60	59
20	76	16.4	75	67

[a] Data from Cohen and Murphy (1971*a*).

on's anticholinesterase action *in vivo* and the inhibition of liver carboxylesterase activity toward malathion and other carboxylester substrates such as triacetin (Cohen and Murphy, 1971*a*) (Table V). Additional studies, however, with both malathion (Cohen and Murphy, 1971*b*) and malaoxon (Cohen *et al.*, 1972; Cohen and Murphy, 1972) revealed that potentiation by TOCP continued to increase with doses in excess of that required to cause maximum inhibition of carboxylesterase activity. Evidence was obtained of a rapid nonenzymatic (temperature-independent) loss of malaoxon in liver homogenates which could account for about 50% of the total amount "detoxified" (Cohen and Murphy, 1972), and it was suggested that this occurred through the binding of malaoxon to both carboxylesterase and other "noncritical" sites. The fact that at concentrations of TOCP above that required for complete carboxylesterase inhibition there were both a decrease in malaoxon binding by liver and an increase in the inhibition of liver pseudocholinesterase activity suggests that the latter may constitute an important binding site. Consequently, potentiation of the toxicity of malathion and malaoxon by a variety of organophosphates probably occurs through a combination of carboxylesterase inhibition and a blocking of several malaoxon binding sites in the liver and plasma.

In addition to potentiating the toxicities of malathion, malaoxon, and other carboxylester-containing insecticides, EPN and TOTP are able to synergize the toxicity of dimethoate (containing a carboxylamide group) to mice (Seume and O'Brien, 1960). Subsequent studies have established that the carboxylamidase which metabolizes dimethoate to the corresponding acid is also susceptible to inhibition by several organophosphates (Chapter 4).

$$(CH_3O)_2P(S)SCH_2C(O)NHCH_3 \xrightarrow[\text{carboxylamidase}]{} (CH_3O)_2P(S)SCH_2COOH$$

dimethoate — dimethoate acid

Its hydrolytic activity seems to be limited to carboxylamide phosphorothionates since several amide-containing phosphates (azodrin, bidrin, dimethoxon) are not substrates and like other phosphates inhibit the enzyme (Chen and Dauterman, 1971). Consequently, the carboxylamidase appears to be another B-esterase and some degree of potentiation can be expected to result from combinations of a variety of organophosphates with carboxylamide-containing phosphorothionates. The degree of potentiation will, of course, depend on the carboxylamidase titer of the species concerned, and Uchida *et al.* (1966) demonstrated that in guinea pigs (with low carboxylamidase activity) the potentiation of dimethoate toxicity by EPN was much lower than that observed in mice. Although the mechanism responsible for the potentiation of dimethoate toxicity in mice pretreated with enzyme inducers such as phenobarbital (O'Brien, 1967; Menzer and Best, 1968) and the organochlorine insecticides (Menzer, 1970, 1971) is not immediately obvious, it is possible that this, too, is associated with changes in tissue carboxylamidase activity. Potentiation of dimethoate toxicity has also been reported in mice pretreated for 48 hr with the insecticide synergist piperonyl butoxide (Kamienski and Murphy, 1971).

Several other cases of potentiation have been reported with various combinations of organophosphates and phosphorothionates which contain neither carboxylester nor carboxylamide groups (DuBois, 1961; Seume and O'Brien, 1960; McCollister *et al.*, 1959; Rosenberg and Coon, 1958*a*). For the most part, the mechanisms involved in these interactions are not clearly understood. In most cases, the potentiation observed probably depends on the net result of one or more of a series of competitive reactions involving on one hand the relative ability of the two compounds (or their primary metabolites) to bind to, be metabolized by, or inhibit the various tissue esterases with which they can interact and on the other hand their relative capacity to inhibit brain acetylcholinesterase. The end result of any particular interaction will therefore depend on the enzymatic makeup of the species concerned and on subtle differences in the affinity of the organophosphates for the sites, which are usually rate limiting with respect to the availability of one of the pair at the target site. Thus 16 hr after treatment of rats with TOCP (125 mg/kg, intraperitoneally) the paraoxon-binding capacity of their liver and plasma was completely abolished and the animals were approximately twice as susceptible to poisoning by paraoxon administered orally, intraperitoneally, or subcutaneously (Lauwerys and Murphy, 1969). Results obtained in a similar study by Lynch and Coon (1972) indicated that TOCP could either antagonize or potentiate the toxicity of several organophosphates to mice depending on the route by which the compounds were administered. Presumably this reflects effects of the route of administration on the tissue distribution of the interacting compounds and the relative distribution of the potential sites of loss in these tissues. Fleischer *et al.* (1963) showed that potentiation of the toxicity of the nerve gas, sarin, following treatment with EPN was due to a reduction of binding sites in the lung and a consequent shift in the distribution of sarin to other tissues including the brain.

3.2.4. Effects on Hepatic Drug Metabolism

In contrast to the organochlorine insecticides, few studies have been made to determine the effects of organophosphates on microsomal drug metabolism; in those which have been reported there is a general consensus that the effects are inhibitory rather than inductive. Direct addition of chlorthion, parathion, malathion, and to a lesser extent paraoxon (all at 10^{-5} or 10^{-4} M) to rat liver microsomes inhibited the hydroxylation of testosterone, an effect reflected by the *in vivo* inhibition of metabolism of the latter and other steroids in rats chronically exposed to chlorthion (Conney *et al.*, 1967). These data are in agreement with the results of other studies which have demonstrated that several organophosphorus insecticides inhibited hepatic drug metabolism and prolonged hexobarbital hypnosis (Rosenberg and Coon, 1958*b*; Stevens *et al.*, 1972*a*). The overall inhibitory action of the phosphorothionates probably results from the combined effects of alternative substrate inhibition and that resulting from the oxidative release and binding of reactive sulfur (Norman *et al.*, 1974; see Section 2.4). However, their action on drug metabolism appears to be complex since different kinetic characteristics were observed with different compounds (Stevens *et al.*, 1972*a*). Thus although paraoxon prolonged hexobarbital sleeping time in mice and inhibited the *in vitro* metabolism of both hexobarbital and ethyl morphine, it acted consistently as an uncompetitive activator of aniline hydroxylation (Stevens *et al.*, 1972*a*). In contrast to these results, Stevens *et al.* (1972*b*, 1973) have shown that repeated subacute administration (3–10 days) of parathion, paraoxon, disulfoton, and the carbamate carbaryl produces an increase in the rate of hexobarbital and aniline metabolism in mice and a concomitant decrease in hexobarbital sleeping time. However, the doses required to produce these effects were equivalent to half the LD_{50} values.

In studies to determine changes in hepatic drug-metabolizing activity in female rats treated simultaneously with parathion (1.5 mg/kg) and several inducing agents (organochlorine insecticides, drugs, etc.), a bewildering variety of effects were noted depending on the combination of compounds employed and the microsomal reaction measured (MacDonald *et al.*, 1970). In some cases parathion decreased the action of the inducing agent and in others there appeared to be a distinct synergistic effect. There are clearly many factors which remain to be elucidated before the mechanisms underlying these complex interactions are fully understood.

3.3. Insecticide Synergists

3.3.1. Introduction

Insecticide synergists represent a unique class of compounds in the context of this discussion since the entire basis of their usefulness depends on their ability to increase the toxicity of an insecticide with which they are combined. Consequently, in contrast to the largely inadvertent and unpredictable nature of most interactions, the synergism of insecticides to insects is intentional and synergists are designed specifically for this purpose.

The best-known, most thoroughly studied, and most important class of compounds of this type are derivatives of methylene dioxybenzene (1,3-benzodioxole). The activity of these compounds was first recognized in the early 1940s, when it was observed that sesame oil synergized the insecticidal action of the pyrethrins. Subsequent evaluation of a large number of synthetic and naturally occurring 1,3-benzodioxoles has established the general ability of these materials to enhance the toxicity of a variety of insecticides with which they are combined, and four compounds, piperonyl butoxide, sulfoxide, propyl isome, and tropital, have been used commercially in aerosol formulations of the pyrethrin insecticides. Piperonyl butoxide, prepared from the natural product safrole, is the most important commercial synergist with an annual U.S. production of about 800,000 lb.

In recent years, the search for new commercial synergists has led directly to the discovery of several other groups of compounds with activities similar to that of the 1,3-benzodioxoles. These include several series of aryl propynyl ethers, pyopynyl oxime ethers, and propynyl phosphonate esters as well as a variety of other materials such as the benzylthiocyanates. *N*-Alkyl compounds such as SKF 525-A and Lilly 18947 (long known as drug potentiators in mammals) can also act as insecticide synergists, and more recently there has been considerable interest in the activity of various nitrogen-containing compounds such as the 1,2,3-benzothiadiazoles and the substituted imidazoles. Since the structure–activity relationships, spectrum of activity, and mode of action of these compounds have been discussed in some detail in several comprehensive reviews (Casida, 1970; Wilkinson, 1971*a*,*b*, 1975; Hodgson and Philpot, 1974) as well as in an earlier chapter in this book (Chapter 3), they will not be given further consideration here.

The common feature of each of these structurally diverse groups of compounds is their ability to inhibit the microsomal mixed-function oxidases responsible for the metabolism of drugs, insecticides, and other foreign compounds. In theory, therefore, they are capable of synergizing the action of any of a large variety of biologically active foreign compounds thus degraded and antagonizing those which require oxidative activation. The degree to which this occurs is often directly related to the rate at which the insecticide or other foreign compound is metabolized, and this depends on, among other factors, the structure of the compound in question and the metabolic capacity of the animal concerned. The observed effects are often modified considerably by variations in the metabolic stability of the synergist itself in various species.

3.3.2. Synergist Interactions in Mammals

Most insecticide synergists have quite low acute and subacute toxicities to mammalian species (Casida, 1970; Hodgson and Philpot, 1974). However, in view of the fundamental similarities between the insect and mammalian microsomal oxidases (Wilkinson and Brattsten, 1972) and the facility with which most synergists inhibit the enzymes from both sources, some concern has been voiced regarding possible hazards to man of interactions arising from the use of synergized insecticide formulations or from exposure to other sources of naturally occurring or synthetic compounds of this type.

At relatively high doses, most insecticide synergists have been shown to prolong hexobarbital sleeping time in mammals. The effect was first demonstrated by Fine and Molloy (1964) following intraperitoneal injection of mice with either piperonyl butoxide or sesamex (both at 50 mg/kg) and was subsequently confirmed in rats by Anders (1968), who also showed that the *in vivo* metabolism of hexobarbital was decreased in the presence of these compounds. Piperonyl butoxide and tropital were reported to be slightly more active than sesamex and slightly less potent than the drug potentiator SKF 525-A in prolonging hexobarbital sleeping time (Anders, 1968; Graham *et al.*, 1970). In another study to compare the activities of several synergists as potentiators of hexobarbital sleeping time in mice, the order of effectiveness was found to be piperonyl butoxide > sulfoxide > sesamex > tropital > propyl isome (Skrinjaric-Spoljar *et al.*, 1971). The ability of piperonyl butoxide and a large number of other methylene dioxyphenyl compounds to prolong barbiturate sleeping time and zoxazolamine paralysis time has now been amply confirmed by numerous investigators, the activity varying with both the compound and the test animal employed (Fujii *et al.*, 1968, 1970; Jaffe and Neumeyer, 1970; Kamienski and Murphy, 1971; Conney *et al.*, 1972). Similar effects on hexobarbital sleeping time have been reported for *O*-2-methylpropyl-*O*-2-propynyl phenylphosphonate (Skrinjaric-Spoljar *et al.*, 1971), 5,6-dichloro-1,2,3-benzothiadiazole (Gil, 1973; Skrinjaric-Spoljar *et al.*, 1971), and several 1- and 4(5)-substituted imidazoles (Wilkinson *et al.*, 1972, 1974). The last compounds are among the most potent microsomal enzyme inhibitors yet reported and in mice are active *in vivo* at concentrations as low as 1 mg/kg. Despite the established inhibitory activity of insecticide synergists toward microsomal oxidation, surprisingly few studies have been made to determine their effects on the metabolism and/or biological activity of other chemical agents in mammals. Piperonyl butoxide increases the acute toxicity to infant mice of freons 112 and 113, griseofulvin, and benzo[*a*]pyrene (Epstein *et al.*, 1967*a*) and enhances the incidence of tumor formation (cocarcinogenicity) with the two freons (Epstein *et al.*, 1967*b*). The cocarcinogenic activity of several naturally occurring methylene dioxyphenyl compounds with benzo[*a*]pyrene was first revealed in studies with the components of sesame oil (Morton and Mider, 1939) and is probably a general property of compounds of this type when administered at high dosages. Piperonyl butoxide decreases the rate of biliary excretion of benzo[*a*]pyrene metabolites when injected intravenously into rats at 262 mg/kg (Falk *et al.*, 1965; Falk and Kotin, 1969), but the effect is considerably decreased following oral or intraperitoneal administration; similar results were obtained with other natural and synthetic methylene dioxphenyl compounds. Conney *et al.* (1972) also reported a decreased elimination of benzo[*a*]pyrene metabolites in animals treated with 1000–2000 mg/kg of piperonyl butoxide.

In another study, Conney *et al.* (1971) showed that while antipyrine metabolism in rats was inhibited only by dosage levels of piperonyl butoxide in excess of about 100 mg/kg (intraperitoneal), a significant effect occurred in mice exposed to a single dose of 1 mg/kg (oral or intraperitoneal). In one of the few reports on the action of piperonyl butoxide in man, these workers found that oral exposure of human volunteers to a dose of 0.71 mg/kg had no influence on antipyrine metabolism.

Few studies have been made to determine the ability of insecticide synergists to potentiate the acute toxicity of insecticides to species other than insects. Robbins *et al.* (1959) reported that the toxicity to mice of coumaphos [*O*-(3-chloro - 4 - methylumbelliferone) - *O*,*O*-dimethylphosphorothionate] was increased following joint oral administration of piperonyl butoxide, and a similar synergism of the toxicities of coumaphos, dimetilan, and rotenone to mice has been reported with piperonyl butoxide and compounds representing several other groups of synergists (Skrinjaric-Spoljar *et al.*, 1971). In a subsequent study, however (Kamienski and Murphy, 1971), piperonyl butoxide was found to synergize or antagonize a variety of phosphates and phosphorothionates depending on the time interval between treatment with the synergist and the insecticide. Piperonyl butoxide has also been reported to increase the pathological changes occurring in rat liver as a result of exposure to pyrethrins (Kimbrough *et al.*, 1968), and these were additive to similar changes resulting from treatment with DDT at 50 mg/kg.

3.3.3. Biphasic Interactions of Synergists

One problem in evaluating the interactions of synergists such as piperonyl butoxide is that the observed effect is dependent on the time at which it is measured after synergist treatment (Fujii *et al.*, 1968; Falk and Kotin, 1969). It is now clearly established that the effects of piperonyl butoxide and other insecticide synergists are biphasic and that their acute inhibitory effects on microsomal metabolism are usually followed by a stimulation of enzyme activity (Kamienski and Murphy, 1971; Skrinjaric-Spoljar *et al.*, 1971). Hexobarbital sleeping times were markedly prolonged when measured in mice 0.5–12 hr after exposure to piperonyl butoxide or tropital but were shortened when the barbiturate was given 24–72 hr after the synergist (Fig. 8). The biphasic effects of piperonyl butoxide were reflected in the susceptibility of the mice to the toxic action of several organophosphates. Thus 1 hr after treatment with 400 mg/kg piperonyl butoxide mice were protected against the acute toxicities of methyl parathion, guthion, and dimethoate but were more susceptible to parathion and ethyl guthion. After 48 hr, they were protected against all compounds except dimethoate (Kamienski and Murphy, 1971). Subsequent studies have clearly shown that prolonged exposure to relatively high concentrations of piperonyl butoxide and related methylene dioxyphenyl compounds

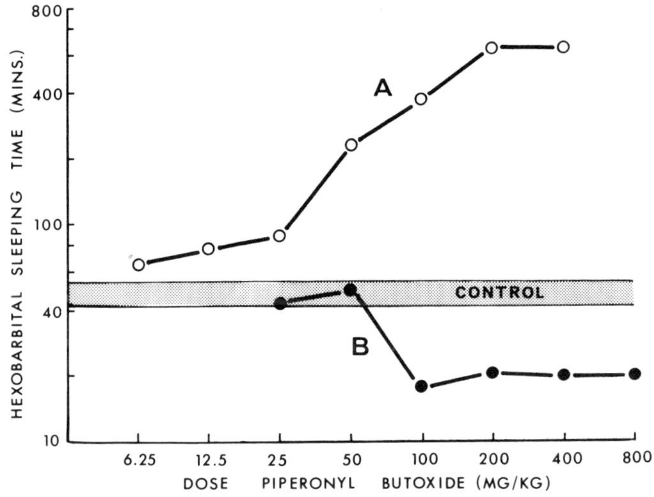

Fig. 8. Hexobarbital (125 mg/kg, intraperitoneally) sleeping times in mice 30 min (A) and 48 hr (B) after treatment with single doses of piperonyl butoxide.

causes an induction of the microsomal enzymes in rat liver (Wagstaff and Short, 1971; Goldstein *et al.*, 1973) and in insects (Thongsinthusak and Krieger, 1974). The potential interactions of insecticides synergists are consequently dependent on numerous factors and are extremely difficult to predict. It is also of interest that when administered acutely piperonyl butoxide (Falk and Kotin, 1969) and sesamex (Wilkinson, 1968) can block the microsomal enzyme induction by other foreign compounds, a fact which may account for the ability of piperonyl butoxide to reverse the protective effect of aldrin or chlordane on the toxicity of some carbamates in rats (Williams and Casterline, 1970).

4. Interactions Involving Exogenous Factors

4.1. Introduction

There is growing recognition that, in addition to the "classical" interactions in which one chemical modifies the biological activity of another, a variety of nutritional, physical, and social factors can markedly influence drug and insecticidal action (Mrak, 1969). Such interactions have received little serious attention to date, although under certain conditions they could be of considerable toxicological significance.

4.2. Nutritional Factors

Malnutrition and undernutrition are the most common diseases in the world, and approximately 60–70% of the world's population currently exists on a suboptimal daily caloric intake of less than 2200. Since the social and economic conditions under which nutritional problems exist are often those associated with the incidence of other diseases requiring drug therapy and since pesticide usage in many underdeveloped areas of the world is substantial, there appears to be considerable potential for nutritional interactions with drugs and insecticides. Both quantitative and qualitative nutritional factors can effect the absorption, distribution, and storage of insecticides and can determine the ability of the animal to synthesize and maintain its detoxication enzymes. In addition, it is possible that effects could result from steroid-mediated mechanisms resulting from the general stress imposed on the animal by nutritional imbalance.

As has previously been discussed, the adipose tissues of animals tend to accumulate organochlorine insecticides such as DDT and the cyclodienes and serve as an inert storage compartment to prevent such materials reaching the sensitive nervous tissues. Mammals (Spicer *et al.*, 1947) are reported to be generally more resistant to DDT if they are fat, and, in theory at least, the depletion of adipose tissues through starvation, illness, dieting, or increased catabolism in response to injury could mobilize the organochlorine residues they contain.

Rats maintained for some time on diets containing 600 ppm DDT were observed to develop characteristic DDT tremors during subsequent starvation (Fitzhugh and Nelson, 1947), and in another study (Dale *et al.*, 1962) an increased excretion and general mobilization of DDT and its metabolites from the fat to other tissues (plasma, brain, liver, and kidney) were observed. In contrast to these reports, neither surgical stress nor complete starvation caused any significant changes in the levels of dieldrin in human blood in spite of losses of body depot fat (Robinson, 1969; Hunter, 1968; Durham *et al.*, 1965). The nature of the diet can also influence the accumulation and biological action of some insecticides. Thus the enhanced tissue levels or the β-isomer of HCH in rats fed diets containing high levels of fat or the nonionic emulsifier Tween 20 were reportedly due to enhanced gastric absorption (Frawley and Fitzhugh, 1950; Fitzhugh, 1966) and confirmed earlier observations indicating that rats and mice on high fat diets exhibited increased acute toxic effects from DDT (Sauberlich and Baumann, 1947). Conversely, dieldrin and DDT accentuate the effects of diets deficient in essential fatty acids in the rat (Tinsley, 1966; Tinsley and Lowry, 1969).

Rats fed on protein-deficient diets show an increased susceptibility to γ-HCH, dieldrin, and chlordane (Boyd and Chen, 1968; Lee *et al.*, 1964; Stoewsand and Bourke, 1968) and to the anticholinesterases parathion and banol (Casterline and Williams, 1969). These effects could in part result from a decrease in the activity of the hepatic drug-metabolizing enzymes (Kato, 1967; Kato and Takanaka, 1967). Hexobarbital sleeping time is reportedly prolonged in malnourished rats (Catz *et al.*, 1971), and this, combined with the finding that the levels of several components of the microsomal oxidase system are decreased in protein-deficient monkeys, led Rumack *et al.* (1973) to suggest a "need to reduce drug dosage in the presence of protein malnutrition."

Possible interactions associated with other nutritional factors have received only scant attention. In ascorbic acid deficient guinea pigs, the activity of the microsomal enzymes is decreased (Kato *et al.*, 1969) and their susceptibility to induction by dieldrin is also impaired (Wagstaff and Street, 1971). The increased storage of dieldrin in the tissues of vitamin A deficient rats has been suggested to be due to impaired microsomal enzyme activity (Street, 1968), although the livers of rats exposed to dietary levels of 10–100 ppm DDT had decreased levels of vitamin A (Phillips, 1963) as did those fed low levels of dieldrin in a protein-deficient diet (Lee *et al.*, 1964).

4.3. Physical Factors

Temperature appears to be the major physical factor which has been studied in relation to insecticide toxicity, and since the effects of this parameter may be mediated through several different mechanisms (absorption, distribution, and metabolism) they are often difficult to interpret. In general, animals seem to be more susceptible to a variety of insecticides at higher temperatures. This is true for parathion in mice (Baetjer and Smith, 1956), sarin in monkeys (Craig *et al.*, 1959), and DDT and dinitrophenols in rats (Keplinger *et al.*, 1959). Rats fed malathion are reportedly less able to withstand exposure to cold (Marton *et al.*, 1962), an effect possibly associated with the hypothermia observed in animals treated with anticholinesterases (Meeter and Wolthuis, 1968).

The effects of cold and other types of stress on microsomal drug metabolism have been studied (Furner and Stitzel, 1969), and Murphy (1969*b*) has reported that organophosphorus insecticides, irritants, and cold stress all increase the levels of plasma glucocorticosteroids. The possible interactions between stress and insecticide toxicity are not yet understood.

4.4. Other Factors

There are innumerable additional factors (chemical, physical, and social) which could in theory interact to modify the toxicity of insecticides or drugs to which man is currently exposed. Man's current predilections for nicotine, caffeine, alcohol, and across-the-counter drugs and the widespread use of oral contraceptives are factors which could play a role. Add to these the possible effects of the large number of chemical agents commonly used around the home and those of the air pollutants and other materials to which man is currently exposed on the city streets and it becomes clear that there are real possibilities for interactions to occur.

5. Toxicological Significance of Interactions

It should be clear from the foregoing discussion that under certain conditions a large number of interactions can occur to modify the pharmacological and/or toxic actions of drugs and insecticides. Most of the studies described, however, have been conducted with relatively high concentrations of materials (often several orders of magnitude greater than those to which man is normally exposed) administered by numerous "unphysiological" routes such as intraperitoneal or intravenous injection. Few attempts have yet been made to assess whether indeed the interactions observed can realistically occur under current environmental conditions and more commonly encountered types of exposure (oral, dermal, and inhalation).

At the present time, determination of the human safety or hazard of a new drug or pesticide depends, at least initially, on the successful extrapolation to man of data obtained in carefully controlled laboratory studies with animals exposed to pure single compounds. Even assuming that the difficult task of extrapolation can be successfully accomplished, the "real-life" validity of such data is questionable in view of the potential for some kind of interaction with one or more of the vast number of other chemicals to which man is daily exposed. Since tests to establish the potential interactions of various combinations of two or more chemicals are in general far too complex to undertake on a routine basis, and since in any case the results of such tests would be almost impossible to interpret, there is always the possibility that unforeseen interactions will occur. The use of all chemical agents is fundamentally inseparable from some risk, and absolute safety can never be guaranteed even following the most rigorous laboratory evaluation. Consequently, there will undoubtedly be new and startling revelations in store for us from time to time. All we can hope for is to continue to increase our basic understanding of the mechanisms by which major interactions can occur and to use this knowledge to recognize and avoid situations of obvious potential hazard to man.

6. References

Alary, J. G., and Brodeur, J., 1969, Studies on the mechanism of phenobarbital induced protection against parathion in adult female rats, *J. Pharmacol. Exp. Ther.* **169:**159.

Aldridge, W. N., 1953*a*, Serum esterases. 1. Two types of esterases (A and B) hydrolysing *p*-nitrophenul acetate, propionate and butyrate, and a method for their determination, *Biochem. J.* **53:**110.

Aldridge, W. N., 1953*b*, Serum esterases. 2. An enzyme hydrolysing diethyl *p*-nitrophenyl phosphate (E600) and its identity with the A-esterase of mammalian sera, *Biochem. J.* **53:**117.

Aldridge, W. N., and Johnson, M. K., 1971, Side effects of organophosphorus compounds: Delayed neurotoxicity, *Bull. WHO* **44:**259.

Anders, M. W., 1968, Inhibition of microsomal drug metabolism by methylenedioxybenzenes, *Biochem. Pharmacol.* **17:**2367.

Anders, M. W., 1971, Enhancement and inhibition of drug metabolism, *Ann. Rev. Pharmacol.* **11:**37.

Baetjer, A. M., and Smith, R., 1956, Effect of environmental temperature on reaction of mice to parathion an anticholinesterase agent, *Am. J. Physiol.* **186:**39.

Ball, W. L., Sinclair, J. W., Crevier, M., and Kay, K., 1954, Modification of parathion's Toxicity for rats by pretreatment with chlorinated hydrocarbon insecticides, *Can. J. Biochem. Physiol.* **32:**440.

Bass, S. W., Triolo, A. J., and Coon, J. M., 1972, Effect of DDT on the toxicity and metabolism of parathion in mice, *Toxicol. Appl. Pharmacol.* **22:**684.

Black, W. D., Wade, A. E., and Talbot, R. B., 1973, A study of the effect of ovex on parathion toxicity in rats, *Can. J. Biochem. Physiol.* **51:**682.

Bledsoe, T., Island, D. P., Ney, R. L., and Liddle, G. W., 1964, An effect of *o,p'*-DDD on the extra-adrenal metabolism of cortisol in man, *J. Clin. Endocrinol. Metab.* **24:**1303.

Bond, E. J., and De Matteis, F., 1969, Biochemical changes in rat liver after administration of carbon disulphide with particular reference to microsomal changes, *Biochem. Pharmacol.* **18:**2531.

Bond, E. J., Butler, W. H., De Matteis, F., and Barnes, J. M., 1969, Effects of carbon disulphide on the liver of rats, *Br. J. Ind. Med.* **26:**335.

Boyd, E. M., and Chen, C. P., 1968, Lindane toxicity and protein deficient diet, *Arch. Environ. Health* **17:**156.

Brodeur, J., 1967, Studies on the mechanism of phenobarbital-induced protection against malathion and EPN, *Can. J. Biochem. Physiol.* **45:**1061.

Brodie, B. B., 1965, Displacement of one drug by another from carrier or receptor sites, *Proc. Roy. Soc. Med.* **58:**946.

Brooks, G. T., 1972, Pathways of enzymatic degradation of pesticides, *in: Environment, Quality and Safety*, Vol. I, (F. Coulston and F. Korte, eds.), pp. 106–164. Academic Press, New York.

Brown, S. S., 1972, Interactions of drugs and foreign compounds, *in: Foreign Compound Metabolism in Mammals*, Vol. 2 (D. E. Hathway, senior reporter), pp. 456–474, The Chemical Society, London.

Brown, V. K., 1967, Solubility and solvent effects as rate determining factors in the acute percutaneous toxicology of pesticides, *Soc. Chem. Ind. (London) Monogr.* No. 29.

Burns, J. J., 1969, Interaction of environmental agents and drugs, *Environ. Res.* **2:**352.

Burns, J. J., Cucinell, S. A., Koster, R., and Conney, A. H., 1965, Application of drug metabolism to drug toxicity studies, *Ann. N.Y. Acad. Sci.* **123:**273.

Casida, J. E., 1961, Specificity of substituted phenyl phosphorus compounds for esterase inhibition in mice, *Biochem. Pharmacol.* **5:**332.

Casida, J. E., 1970, Mixed-function oxidase involvement in the biochemistry of insecticide synergists, *J. Agr. Food Chem.* **18:**753.

Casterline, J. L., and Williams, C. H., 1969, Effect of pesticide administration upon esterase activities in serum and tissues of rats fed variable casein diets, *Toxicol. Appl. Pharmacol.* **14:**266.

Catz, C. S., Brasel, J. A., Winick, M., and Yaffe, S. J., 1971, Influence of early malnutrition on drug metabolism and effect, *Pediat. Res.* **5:**420.

Chapman, S. K., and Leibman, K. C., 1971, The effects of chlordane, DDT, and 3-methylcholanthrene upon the metabolism and toxicity of diethyl-4-nitrophenyl phosphorothionate (parathion), *Toxicol. Appl. Pharmacol.* **18:**977.

Chem, P. R. S., and Dauterman, W. C., 1971, Studies on the toxicity of dimethoate analogs and their hydrolysis by sheep liver amidase, *Pestic. Biochem. Physiol.* **1:**340.

Chin, L., Picchioni, A. L. and Duplisse, B. R., 1970, The action of activated charcoal on poisons in the digestive tract, *Toxicol. Appl. Pharmacol.* **16:**786.

Cohen, S. D., and Murphy, S. D., 1971*a*, Malathion potentiation and inhibition of hydrolysis of various carboxylic esters by triorthotolyl phosphate (TOTP) in mice, *Biochem. Pharmacol.* **20:**575.

Cohen, S. D., and Murphy, S. D., 1971*b*, Carboxylesterase inhibition as an indicator of malathion potentiation in mice, *J. Pharmacol. Exp. Ther.* **176:**733.

Cohen, S. D., and Murphy, S. D., 1972, Inactivation of malaoxon by mouse liver, *Proc. Soc. Exp. Biol. Med.* **139:**1385.

Cohen, S. D., and Murphy, S. D., 1974, A simplified bioassay for organophosphate detoxication and interactions, *Toxicol. Appl. Pharmacol.* **27:**537.

Cohen, S. D., Callaghan, J. E., and Murphy, S. D., 1972, Investigation of multiple mechanisms for potentiation of malaoxon's anticholinesterase action by triorthotolyl phosphate, *Proc. Soc. Exp. Biol. Med.* **141:**906.

Conney, A. H., 1967, Pharmacological implications of microsomal enzyme induction, *Pharmacol. Rev.* **19:**317.

Conney, A. H., 1971, Environmental factors influencing drug metabolism, *in: Fundamentals of Drug Metabolism and Disposition* (B. N. LaDu, H. G. Mandel, and E. L. Way, eds.), pp. 253–278, Williams and Wilkins, Baltimore.

Conney, A. H., Welch, R. M., Kuntzman, R., and Burns, J. H., 1967, Effects of pesticides on drug and steroid metabolism, *Clin. Pharmacol. Ther.* **8:**2.

Conney, A. H., Welch, R., Kuntzman, R., Chang, R., Jacobson, M., Munro-Faure, A. D., Peck, A. W., Bye, A., Poland, A., Poppers, P. J., Finster, M., and Wolff, J. A., 1971, Effects of environmental chemicals on the metabolism of drugs, carcinogens and normal body constituents in man, *Ann. N.Y. Acad. Sci.* **179:**155.

Conney, A. H., Chang, R., Levin, W. M., Garbut, A., Munro-Faure, A. D., Peck, A. W., and Bye, A., 1972, Effects of piperonyl butoxide on drug metabolism in rodents and man, *Arch. Environ. Health* **24:**97.

Cook, J. W., Blake, J. R., Yip, G., and Williams, M., 1958, Malathionase. I. Activity and inhibition, *J. Assoc. Offic. Agr. Chemists* **41:**399.

Cook, R. M., 1970, Dieldrin recycling from the blood to the gastrointestinal tract, *J. Agr. Food Chem.* **18:**434.

Craig, F. N., Rales, P. D., and Frankel, H. M., 1959, Lethality of sarin in a warm environment, *J. Pharmacol. Exp. Ther.* **127:**35.

Crevier, M., Ball, W. L., and Kay, K., 1954, Observations on toxicity of aldrin. II. Serum esterase changes in rats following administration of aldrin and other chlorinated hydrocarbon insecticides, *Arch. Ind. Hyg.* **9:**306.

Cueto, C., and Hayes, W. J., 1965, Effect of phenobarbital on the metabolism of dieldrin, *Toxicol. Appl. Pharmacol.* **7:**481.

Dale, W. E., Gaines, T. B., and Hayes, W. J., Jr., 1962, Storage and excretion of DDT in starved rats, *Toxicol. Appl. Pharmacol.* **4:**89.

Dalvi, R. R., Poore, R. E., and Neal R. A., 1974, Studies on the metabolism of carbon disulfide by rat liver microsomes, *Life Sci.* **14:**1785.

Davies, J. E., Edmundson, W. F., Carter, C. H., and Barquet, A., 1969, Effect of anticonvulsant drugs on dicophane (DDT) residues in man, *Lancet* **1969:**7.

Davison, C., 1971, Protein binding, *in*: *Fundamentals of Drug Metabolism and Disposition* (B. N. LaDu, H. G. Mandel, and E. L. Way, eds.), pp. 63–75, Williams and Wilkins, Baltimore.

Deichmann, W. B., Keplinger, M., Dressler, I., and Sala, F., 1969, Retention of dieldrin and DDT in the tissues of dogs fed aldrin and DDT individually and as a mixture, *Toxicol. Appl. Pharmacol.* **14:**205.

De Matteis, R., and Seawright, A. A., 1973, Oxidative metabolism of carbon disulphide by the rat: Effect of treatments which modify the liver toxicity of carbon disulphide, *Chem. Biol. Interact.* **7:**375.

Diaz Gomez, M. I., Castro, J. A., de Ferreyra, E. C., D'Acosta, N., and de Castro, C. R., 1973, Irreversible binding of ^{14}C from $^{14}CCl_4$ to liver microsomal lipids and proteins: From rats pretreated with compounds altering microsomal mixed-function oxidase activity, *Toxicol. Appl. Pharmacol.* **25:**534.

Dobson, R. C., Fehey, J. E., Ballee, D. L., and Baugh, E. R., 1971, Reduction of chlorinated hydrocarbon residues in swine, *Bull. Environ. Contam. Toxicol.* **6:**189.

Donninger, C., Hutson, D. H., and Pickering, B. A., 1972, The oxidative dealkylation of insecticidal phosphoric acid triesters by mammalian liver enzymes, *Biochem. J.* **126:**701.

DuBois, K. P., 1961, Potentiation of the toxicity of organophosphorus compounds, *Advan, Pest Control Res.* **4:**117.

DuBois, K. P., 1969, Combined effects of pesticides, *Can. Med. Assoc. J.* **100:**173.

DuBois, K. P., 1971, The toxicity of organophosphorus compounds to mammals, *Bull. WHO* **44:**233.

DuBois, K. P., and Kinoshita, F., 1968, Influence of induction of hepatic microsomal enzymes by phenobarbital on toxicity of organic phosphate insecticides, *Proc. Soc. Exp. Biol. Med.* **129:**699.

Duggan, R. E., and Lipscomb, G. Q., 1969, Dietary intake of pesticide chemicals in the United States (II), June 1966–April 1968, *Pestic. Monit. J.* **2:**153.

Durham, W. F., 1967, The interaction of pesticides with other factors, *Residue Rev.* **18:**21.

Durham, W. F., Armstrong, J. F., and Quinby, G. E., 1965, DDT and DDE content of complete prepared meals, *Arch. Environ. Health* **11:**641.

Eakins, M. N., and Slater, T. F., 1973, Investigations on the possible relationship between hepatic endoplasmic reticulum enzymes and bile secretion in the rat, *Trans. Biochem. Soc. (London)* **1:**931.

Eliason, B. C., and Posner, H. S., 1971, Placental passage of [^{14}C]-Dieldrin altered by gestational age and plasma proteins, *Am. J. Obstet. Gynecol.* **111:**925.

Engebretson, K. A., and Davison, K. L., 1971, Dieldrin accumulation and excretion by rats fed phenobarbital and carbon, *Bull. Environ. Contam. Toxicol.* **6:**391.

Epstein, S. S., Clapp, J. A. P., and Mackintosh, D., 1967*a*, Enhancement by piperonyl butoxide of acute toxicity due to freons, benzo[α]pyrene and griseofulvin in infant mice, *Toxicol. Appl. Pharmacol.* **11:**442.

Epstein, S. S., Joshi, S., Andrea, J., Claph, P., Falk, H., and Mantel, N., 1967*b*, Synergistic toxicity and carcinogenicity of freons and piperonyl butoxide, *Nature (London)* **214:**526.

Falk, H. L., and Kotin, P., 1969, Pesticide synergists and their metabolites: Potential hazards, *Ann. N.Y. Acad. Sci.* **160:**299.

Falk, H. L., Thompson, S. J., and Kotin, P., 1965, Carcinogenic potential of pesticides, *Arch. Environ. Health* **10:**847.

Fine, B. C., and Molloy, J. O., 1964, Effects of insecticide synergists on duration of sleep induced in mice by barbiturates, *Nature (London)* **204:**789.

Fitzhugh, O. G., 1966, Problems related to the use of pesticides, *Can. Med. Assoc. J.* **94:**598.

Fitzhugh, O. G., and Nelson, A. A., 1947, Chronic oral toxicity of DDT [2,2-bis-(*p*-chlorophenyl)-1,1,1-trichloroethane], *J. Pharmacol. Exp. Ther.* **89:**18.

Fleisher, J. H., Harris, L. W., Prudhomme, C., and Bursel, J., 1963, Effects of ethyl *p*-nitrophenyl thiobenzene phosphonate (EPN) on the toxicity of isopropyl methyl phosphonofluoridate (G.B.), *J. Pharmacol. Exp. Ther.* **139:**390.

Frawley, J. P., and Fitzhugh, O. G., 1950, Factors effecting tissue distribution following oral ingestion of lipid soluble substances, *Fed. Proc.* **9:**273.

Frawley, J. P., Fuyat, H. N., Hagàn, E. C., Blake, J. R., and Fitzhugh, O. G., 1957, Marked potentiation in mammalian toxicity from simultaneous administration of two anticholinesterase compounds, *J. Pharmacol. Exp. Ther.* **121:**96.

Fujii, K., Jaffe, H., and Epstein, S. S., 1968, Factors influencing the hexobarbital sleeping time and zoxazolamine paralysis time in mice, *Toxicol. Appl. Pharmacol.* **13:**431.

Fujii, K., Jaffe, H., Bishop, Y., Arnold, E., Mackintosh, D., and Epstein, S. S., 1970, Structure–activity relations for methylenedioxyphenyl and related compounds on hepatic microsomal enzyme function as measured by prolongation of hexobarbital narcosis and zoxazolamine paralysis in mice, *Toxicol. Appl. Pharmacol.* **16:**482.

Fukami, J., and Shishido, T., 1966, Nature of a soluble glutathione-dependent enzyme system active in cleavage of methyl parathion to desmethyl parathion, *J. Econ. Entomol.* **59:**1338.

Fukunaga, K., Fukami, J., and Shishido, T., 1969, The *in vitro* metabolism of organophosphorus insecticides by tissue homogenates from mammal and insect, *Residue Rev.* **25:**223.

Furner, R. L., and Stitzel, R. E., 1969, Stress-induced alterations in microsomal drug metabolism in the adrenalectomized rat, *Biochem. Pharmacol.* **17:**121.

Garb, S., 1971, *Clinical Guide to Undesirable Drug Interactions and Interferences*, Springer, New York.

Ghazal, A., Koransky, W., Portig, J., Vohland, H. W., and Klempau, I., 1964, Beschleunigung von Entgiftungsreaktionen durch verschiedene Insekticide, *Naunyn-Schmiedebergs Arch. Exp. Pathol. Pharmakol.* **249:**1.

Gil, L. D., 1973, Structure–activity relationships of -1,2,3-benzothiadiazole insecticide synergist: Ph.D. dissertation, Cornell University.

Goldstein, J. A., Hickman, P., and Kimbrough, R. D., 1973, Effects of purified and technic piperonyl butoxide on drug metabolizing enzymes and ultrastructure of rat liver, *Toxico Appl. Pharmacol.* **26:**444.

Graham, P. S., Hellyer, R. O., and Ryan, A. J., 1970, The inhibition of drug metabolism enzyme by some naturally occurring compounds, *Biochem. Pharmacol.* **19:**759.

Hansten, P. D., 1971, *Drug Interactions,* Lea and Febiger, Philadelphia.

Hart, L. G., and Fouts, J. R., 1965, Further studies on the stimulation of hepatic microsomal dru metabolizing enzymes by DDT and its analogs, *Naunyn-Schmiedebergs Arch. Exp. Patho Pharmakol.* **249:**486.

Hart, L. G., Shultice, R. W., and Fouts, J. R., 1963, Stimulatory effects of chlordane on hepati microsomal drug metabolism in the rat, *Toxicol. Appl. Pharmacol.* **5:**371.

Hartshorn, E. A., 1971, *Handbook of Drug Interactions,* Hamilton Press, Hamilton, Ill.

Heath, D. F., 1961, *Organophosphorus Poisons,* Pergamon Press, New York.

Heath, D. F., and Vandekar, M., 1964, Toxicity and metabolism of dieldrin in rats, *Br. J. Ind. Me* **21:**269.

Hodgson, E., and Philpot, R. M., 1974, Interaction of methylenedioxyphenyl (1,3-benzo dioxole) compounds with enzymes and their effect *in vivo* on animals, *Drug Metab. Rev.* 3 231.

Hollingworth, R. M., 1969, Dealkylation of organophosphorus esters by mouse liver enzymes *i vitro* and *in vivo, J. Agr. Food Chem.* **17:**987.

Hollingworth, R. M., 1970, The dealkylation of organophosphorus triesters by liver enzymes, *i Biochemical Toxicology of Insecticides* (R. D. O'Brien and I. Yamamoto, eds.), pp. 75–9 Academic Press, New York.

Hunter, C. G., 1968, Human exposures to aldrin and dieldrin, *in: Symposium on Science an Technology of Residual Insecticides in Food Production with Special Reference to Aldrin an Dieldrin,* pp. 118–129, Shell Chemical Co.

Ikeda, M., Conney, A. H., and Burns, J. J., 1968, Stimulatory effect of phenobarbital an insecticides on warfarin metabolism in the rat, *J. Pharmacol. Exp. Ther.* **162:**338.

Jaffe, H., and Neumeyer, J. L., 1970, Comparative effects of piperonyl butoxide and *N*-(4 pentynyl) phthalimide on mammalian microsomal enzyme functions, *J. Med. Chem.* **13:**901

Kamienski, F. X., and Murphy, S. D., 1971, Biphasic effects of methylenedioxyphenyl synergis on the action of hexobarbital and organophosphate insecticides in mice, *Toxicol. App Pharmacol.* **18:**883.

Kato, R., 1967, Effects of starvation and refeeding on the oxidation of drugs by liver microsome *Biochem. Pharmacol.* **16:**871.

Kato, R., and Takanaka, A., 1967, Effect of starvation on the *in vivo* metabolism and effect of drug in female and male rats, *Jpn. J. Pharmacol.* **17:**208.

Kato, R., Chiesara, E., and Vassanelli, P., 1964, Further studies on the inhibition and stimulatio of microsomal drug metabolizing enzymes of rat liver by various compounds, *Biochen Pharmacol.* **13:**69.

Kato, R., Takanaka, A., and Oshima, T., 1969, Effect of vitamin C deficiency on the metabolism c drugs and NADPH-linked electron transport system in liver microsomes, *Jpn. J. Pharmaco* **19:**25.

Keplinger, M. L., Lanier, G. E., and Deichman, W. B., 1959, Effects of environmental temperatur on the acute toxicity of a number of compounds in rats, *Toxicol. Appl. Pharmacol.* **1:**156.

Keplinger, M. L., Dressler, I. A., and Deichmann, W. B., 1966, Tissue storage in mice fe combination of pesticides, *Toxicol. Appl. Pharmacol.* **8:**344.

Kimbrough, R. D., Gaines, T. B., and Hayes, W. J., 1968, Combined effect of DDT, pyrethrur and piperonyl butoxide on rat liver, *Arch. Environ. Health* **16:**333.

Kinoshita, F. K., Frawley, J. P., and DuBois, K. P., 1966, Quantitative measurement of inductio of hepatic microsomal enzymes by various dietary levels of DDT and toxaphene in rat *Toxicol. Appl. Pharmacol.* **9:**505.

Klaasen, C. D., 1969, Biliary flow after microsomal enzyme induction, *J. Pharmacol. Exp. Ther.* **168:**218.

Klaasen, C. D., and Plaa, G. L., 1968, Studies on the mechanism of phenobarbital-enhanced sulfobromophthalein disappearance, *J. Pharmacol. Exp. Ther.* **161:**361.

Kolmodin, B., Azarnoff, D. L., and Sjöqvist, 1969, Effect of environmental factors on drug metabolism: decreased plasma half-life on antipyrine in workers exposed to chlorinated hydrocarbon insecticides, *Clin. Pharmacol. Ther.* **10:**638.

Koransky, W., Portig, J., Vohland, H. W., and Klempau, I., 1964, Elimination of α- and γ-hexachlorocyclohexane and the effects of liver microsomal enzymes, *Naunyn-Schmiedebergs Arch. Exp. Pathol. Pharmakol.* **247:**49.

Kuntzman, R., Jacobson, M., Scheidman, K., and Conney, A. H., 1964, Similarities between oxidative drug metabolizing enzymes and steroid hydroxylases in liver microsomes, *J. Pharmacol. Exp. Ther.* **146:**280.

Lange, G., 1967, Different stimulation of the microsomal *N*- and *p*-hydroxylation of aniline and *N*-ethylaniline in rabbits, *Naunyn-Schmiedebergs Arch. Exp. Pathol. Pharmakol.* **257:**230.

Lauwerys, R. R., and Murphy, S. D., 1969, Interaction between paraoxon and tri-*o*-tolyl phosphate in rats, *Toxicol. Appl. Pharmacol.* **14:**348.

Lee, M., Harris, J., and Trowbridge, H. J., 1964, Effect of the level of dietary protein on the toxicity of dieldrin for the laboratory rat, *J. Nutr.* **84:**136.

Lynch, W. T., and Coon, J. M., 1972, Effect of tri-*o*-tolyl phosphate pretreatment on the toxicity and metabolism of parathion and paraoxon in mice, *Toxicol. Appl. Pharmacol.* **21:**153.

MacDonald, W. E., MacQueen, J., Diechmann, W. B., Hamill, T., and Copsey, K., 1970, Effect of parathion on liver microsomal enzyme activities induced by organochlorine pesticides and drugs in female rats, *Int. Arch. Arbeitsmed.* **26:**31.

Magos, L., and Butler, W. H., 1972, Effect of phenobarbitone and starvation on hepatotoxicity in rats exposed to carbon disulphide vapour, *Br. J. Ind. Med.* **29:**95.

Main, A. R., 1956, The role of A-esterase in the acute toxicity of paraoxon, TEPP, and parathion, *Can. J. Biochem. Physiol.* **34:**197.

Main, A. R., and Braid, P. E., 1962, Hydrolysis of malathion by aliesterases *in vitro* and *in vivo*, *Biochem. J.* **84:**255.

Main, A. R., and Dauterman, W. C., 1967, Kinetics for the inhibition of carboxylesterase by malaoxon, *Can. J. Biochem.* **45:**757.

Mannering, G. J., 1968, Significance of stimulation and inhibition of drug metabolism, *in: Selected Pharmacological Testing Methods* (A. Burger, ed.), pp. 51–119, Marcel Dekker, New York.

Marton, A. V., Sellers, E. A., and Kalow, W., 1962, Effect of cold on rats chronically exposed to malathion, *Can. J. Biochem. Physiol.* **40:**1671.

Matsumura, F., and Wang, C. M., 1968, Reduction of dieldrin storage in rat liver: Factors affecting *in situ*, *Bull. Environ. Contam. Toxicol.* **3:**203.

McCollister, D. D., Oyen, F., and Rowe, V. K., 1959, Toxicological studies of *O,O*-dimethyl-*O*-(2,4,5-trichlorophenyl)phosphorothioate (ronnel) in laboratory animals, *J. Agr. Food Chem.* **7:**689.

McDermot, H. L., Finkbeiner, A. J., Wills, W. J., and Heggie, R. M., 1967, The enhancement of penetration of an organophosphorus anticholinesterase through guinea pig skin by dimethyl sulfoxide, *Can. J. Physiol. Pharmacol.* **45:**229.

McPhillips, J. J., 1965, Effect of chlorcyclizine on the toxicity and metabolism of octamethyl pyrophosphoramide, *Toxicol. Appl. Pharmacol.* **7:**64.

Meeter, E., and Wolthius, O. L., 1968, The effect of cholinesterase inhibitors on the body temperature of the rat, *Eur. J. Pharmacol.* **4:**18.

Mei-Quey, S., Kinoshita, F. K., Frawley, J. P., and DuBois, K. P., 1971, Comparative inhibition of aliesterases and cholinesterases in rats fed eighteen organophosphorus insecticides, *Toxicol. Appl. Pharmacol.* **20:**241.

Menzer, R. E., 1970, Effect of chlorinated hydrocarbons in the diet on the toxicity of several organophosphorus insecticides, *Toxicol. Appl. Pharmacol.* **16:**446.

Menzer, R. E., 1971, Effect of enzyme induction on the metabolism and selectivity of organophosphorus insecticides, *in: Proceedings of the 2nd International Congress of Pesticide Chemistry*, Tel-Aviv, p. 51.

Menzer, R. E., and Best, N. H., 1968, Effect of phenobarbital on the toxicity of several organophosphorus insecticides, *Toxicol. Appl. Pharmacol.* **13:**37.

Menzer, R. E., and Rose, J. A., 1971, Effect of enzyme-inducing agents on fat storage and toxicity of insecticides, *in: Pesticide Chemistry*, Vol. II (A. S. Tahori, ed.), pp. 257–265, Gordon and Breach, New York.

Mick, D. L., and Long, K. R., 1973, The effects of dietary dieldrin on residues in eggs and tissues of laying hens and the effects of phenobarbital and charcoal on these residues, *Bull. Environ. Contam. Toxicol.* **9:**197.

Morton, J. J., and Mider, G. B., 1939, Effect of petroleum ether extract of mouse carcasses as solvent in production of sarcoma, *Proc. Soc. Exp. Biol. Med.* **41:**357.

Moss, J. A., and Hathway, D. E., 1964, Transport of organic compounds in the mammal: Partition of dieldrin and telodrin between the cellular components and soluble proteins of the blood, *Biochem. J.* **91:**384.

Mrak, E. M. (Chairman), 1969, *Report of the Secretary's Commission on Pesticides and Their Relationship to Environmental Health*, Government Printing Office, Washington, D.C.

Murphy, S. D., 1967, Malathion inhibition of esterases as a determinant of malathion toxicity, *J. Pharmacol. Exp. Ther.* **156:**352.

Murphy, S. D., 1969*a*, Mechanisms of pesticide interactions in vertebrates, *Residue Rev.* **25:** 201.

Murphy, S. D., 1969*b*, Some relationships between effects of insecticides and other stress conditions, *Ann. N.Y. Acad. Sci.* **160:**366.

Murphy, S. D., and Cheever, K. L., 1968, Effect of feeding insecticides: Inhibition of carboxylesterase and cholinesterase activities in rats, *Arch. Environ. Health* **17:**749.

Murphy, S. D., and DuBois, K. P., 1957, Quantitative measurement of inhibition of the enzymatic detoxification of malathion by EPN (ethyl *p*-nitrophenyl thionobenzene phosphonate), *Proc. Soc. Exp. Biol. Med.* **96:**813.

Murphy, S. D., and DuBois, K. P., 1958, Inhibitory effect of dipterex and other organic phosphates on detoxification of malathion, *Fed. Proc.* **17:**396.

Murphy, S. D., Anderson, R. L., and DuBois, K. P., 1959, Potentiation of the toxicity of malathion by triorthotolyl phosphate, *Proc. Soc. Exp. Biol. Med.* **100:**483.

Namba, T., 1971, Cholinesterase inhibition by organophosphorus compounds and its clinical effects, *Bull. WHO.* **44:**289.

Neal, R. A., 1967*a*, Studies on the metabolism of diethyl 4-nitrophenyl phosphorothionate (parathion) *in vitro*, *Biochem. J.* **103:**183.

Neal, R. A., 1967*b*, Studies on the enzymic mechanism of the metabolism of diethyl 4-nitrophenyl phosphorothionate (parathion) by rat liver microsomes, *Biochem. J.* **105:**289.

Neubert, D., and Schaefer, J., 1958, Wirkungsverlust des Diäthyl-*p*-nitrophenylphosphats und Octamethyl-pyrophosphoramids nach Vorbehandlung mit α-Hexachlorocyclohexan, *Naunyn-Schmiedebergs Arch. Exp. Pathol. Pharmakol.* **233:**151.

Norman, B. J., Poore, R. E., and Neal, R. A., 1974. Studies of the binding of sulfur released in the mixed-function oxidase-catalyzed metabolism of diethyl *p*-nitrophenylphosphorothionate (parathion) to diethyl *p*-nitrophenyl phosphate (paraoxon), *Biochem. Pharmacol.* **23:**1733.

O'Brien, R. D., 1967, Effects of induction by phenobarbital upon the susceptibility of mice to insecticides, *Bull. Environ. Contam. Toxicol.* **2:**163.

Phillips, W. E. J., 1963, DDT and the metabolism of vitamin A and carotene in the rat, *Can. J. Biochem. Physiol.* **41:**1793.

Rappolt, R. T., Sr., 1970, Use of oral DDT in three human barbitutate intoxications: CNS arousal and/or hepatic enzyme induction by reciprocal detoxicants, *in: Pesticides Symposia* (W. M. Deichmann, J. L. Radomski, and R. A. Penalver, eds.), pp. 269–271, Halos and Associates, Miami.

Reeder, N., 1969, Fast way to clean pesticides out of cows, *Farm J.*, Aug. 25.

Remmer, H., 1972, Induction of drug metabolizing enzyme system in the liver, *Eur. J. Clin. Pharmacol.* **5:**116.

Robbins, W. E., Hopkins, T. L., and Darrow, D. I., 1959, Synergistic action of piperonyl butoxide with Bayer 21/199 and its corresponding phosphate in mice, *J. Econ. Entomol.* **52:**660.

Robinson, J., 1969, The burden of chlorinated hydrocarbon pesticides in man, *Can. Med. Assoc. J.* **100:**180.

Rosenburg, P., and Coon, J. M., 1958*a*, potentiation between cholinesterase inhibitors, *Proc. Soc. Exp. Biol. Med.* **97:**836.

Rosenberg, P., and Coon, J. M., 1958*b*, Increase of hexobarbital sleeping time by certain anticholinesterases, *Proc. Soc. Exp. Biol. Med.* **98:**650.

Rumack, B. H., Holtzman, J., and Chase, H. P., 1973, Hepatic drug metabolism and protein malnutrition, *J. Pharmacol. Exp. Ther.* **186:**441.

Sauberlich, H. E., and Baumann, C. A., 1947, Effect of dietary variations on the toxicity of DDT to rats and mice, *Proc. Soc. Exp. Biol. Med.* **66:**642.

Schanker, L. S., 1971, Drug adsorption, *in: Fundamentals of Drug Metabolism and Disposition* (B. N. LaDu, H. G. Mandel, and E. L. Way, eds.), pp. 22–43, Williams and Wilkins, Baltimore.

Schwabe, U., and Wendling, I., 1967, Beschleunigung des Arzneimittel-Abbaus durch kleine Dosen von DDT und anderen Chlorkohlenwasserstoff-Insekticiden, *Arzneimittel-Forschung* **17:**614.

Seume, F. W., and O'Brien, R. D., 1960, Metabolism of malathion by rat tissue preparations and its modification by EPN, *J. Agr. Food Chem.* **8:**36.

Sher, S. P., 1971, Drug enzyme induction and drug intereactions: Literature tabulation, *Toxicol. Appl. Pharmacol.* **18:**780.

Skrinjaric-Spoljar, M., Matthews, H. B., Engel, J. L., and Casida, J. E., 1971, Response of hepatic microsomal mixed-function oxidases to various types of insecticide synergists administered to mice, *Biochem. Pharmacol.* **20:**1607.

Spicer, S. S., Sweeney, T. R., Von Oettingen, W. F., Lillie, R. D., and Neal, P. A., 1947, Toxicological observations on goats fed large doses of DDT, *Vet. Med.* **42:**289.

Stevens, J. T., Stitzel, R. E., and McPhillips, J. J., 1972*a*, Effects of anticholinesterase insecticides on hepatic microsomal metabolism, *J. Pharmacol. Exp. Ther.* **181:**576.

Stevens, J. T., Stitzel, R. E., and McPhillips, J. J., 1972*b*, The effects of subacute administration of anticholinesterase insecticides on hepatic microsomal metabolism, *Life Sci.* **11:**423.

Stevens, J. T., Green, F. E., Stitzel, R. E., and McPhillips, J. T., 1973, Effects of anticholinesterase insecticides on mouse and rat liver microsomal mixed-function oxidases, *in: Pesticides in the Environment* (W. B. Deichmann, ed.), pp. 498–501, Intercontinental Medical Book Corp., New York.

Stockley, I. H., 1971, The practical significance of drug interactions, *Pharm. J.* **207:**351.

Stockley, I. H., 1972, Basic principles of drug interaction, *Chem. Br.* **8:**114.

Stoewsand, G. S., and Bourke, J. B., 1968, The influence of dietary protein on the resistance of dieldrin toxicity in the rat, *Ind. Med. Surg.* **37:**526.

Street, J. C., 1964, DDT antagonism to dieldrin storage in adipose tissue of rats, *Science* **146**:1580.

Street, J. C., 1968, Modifications of animal responses to toxicants, *in: Enzymatic Oxidations of Toxicants* (E. Hodgson, ed.), pp. 197–226, North Carolina University, Raleigh.

Street, J. C., and Blau, A. D., 1966, Insecticide interactions affecting residue accumulation in animal tissues, *Toxicol. Appl. Pharmacol.* **8:**497.

Street, J. C., and Chadwick, R. W., 1967, Stimulation of dieldrin metabolism by DDT, *Toxicol. Appl. Pharmacol.* **11:**68.

Street, J. C., Chadwick, R. W., Wang, M., and Phillips, R. L., 1966*a*, Insecticide interactions affecting residue storage in animal tissues, *J. Agr. Food Chem.* **14:**545.

Street, J. C., Wang, M., and Blau, A. D., 1966*b*, Drug effects on dieldrin storage in rat tissue, *Bull. Environ. Contam. Toxicol.* **1:**6.

Takabatake, E., Ariyoshi, T., and Schimizu, K., 1968, Effect of tolbutamide on the metabolism of parathion, *Chem. Pharm. Bull.* **10:**1065.

Thompson, R. P. H., Pilchner, C. W. T., Robinson, J., Strathers, C. M., McLean, A. E. M., and Williams, R., 1969, Treatment of unconjugated jaundice with dicophane, *Lancet* **2:**4.

Thongsinthusak, T., and Krieger, R. I., 1974, Inhibitory and inductive effects of piperonyl butoxide on dihydroisodrin hydroxylation *in vivo* and *in vitro* in black cutworm (*Agrotis ypsilon*) larvae, *Life Sci.* **14:**2131.

Thorp, J. M., 1964, Influence of plasma proteins on the action of drugs, *in: Absorption and Distribution of Drugs* (T. B. Binns, ed.), pp. 64–76, Livingston, Edinburgh.

Tinsley, I. J., 1966, Nutritional interactions in dieldrin toxicity, *J. Agr. Food Chem.* **14:**563.

Tinsley, I. J., and Lowry, R. R., 1969, Nutritional interactions and organochlorine insecticide activity, *Ann. N.Y. Acad. Sci.* **160:**291.

Triolo, A. J., and Coon, J. M., 1966*a*, Toxicologic interactions of chlorinated hydrocarbon and organophosphate insecticides, *J. Agr. Food Chem.* **14:**549.

Triolo, A. J., and Coon, J. M., 1966*b*, The protective action of aldrin against the toxicity of organophosphate anticholinesterases, *J. Pharmacol. Exp. Ther.* **154:**613.

Triolo, A. J., and Coon, J. M., 1969, Binding of paraoxon by plasma of aldrin-treated mice, *Toxicol. Appl. Pharmacol.* **14:**622.

Triolo, A. J., Mata, E., and Coon, J. M., 1970, Effects of organochlorine insecticides on the toxicity and *in vitro* plasma detoxication of paraoxon, *Toxicol. Appl. Pharmacol.* **17:**174.

Uchida, T., Zschintzsch, J., and O'Brien, R. D., 1966, Relation between synergism and metabolism of dimethoate in mammals and insects, *Toxicol. Appl. Pharmacol.* **8:**259.

Vukovich, R. A., Triolo, A. J., and Coon, J. M., 1971, The effect of chlorpromazine on the toxicity and biotransformation of parathion in mice, *J. Pharmacol. Exp. Ther.* **178:**395.

Wagstaff, D. J., and Short, C. R., 1971, Induction of hepatic microsomal hydroxylating enzymes by technical piperonyl butoxide and some of its analogs, *Toxicol. Appl. Pharmacol.* **19:**54.

Wagstaff, D. J., and Street, J. C., 1971, Ascorbic acid deficiency and induction of hypatic microsomal hydroxylative enzymes by organochlorine pesticides, *Toxicol. Appl. Pharmacol.* **19:**10.

Welch, R. M., and Coon, J. M., 1964, Studies on the effect of chlorcyclizine and other drugs on the toxicity of several organophosphate anticholinesterases, *J. Pharmacol. Exp. Ther.* **143**:192.

Welch, R. M., and Harrison, Y., 1966, Reduced drug toxicity following insecticide treatment, *Pharmacologist* **8:**217.

Welch, R. M., Levin, W., and Conney, A. H., 1967, Insecticide inhibition and stimulation of steroid hydroxylases in rat liver, *J. Pharmacol. Exp. Ther.* **155:**167.

Wilkinson, C. F., 1968, Detoxication of pesticides and the mechanism of synergism, *in: Enzymatic Oxidations of Toxicants* (E. Hodgson, ed.), pp. 113–149, North Carolina State University, Raleigh.

Wilkinson, C. F., 1971*a*, Insecticide synergists and their mode of action, *in: Pesticide Chemistry*, Vol. II, (A. S. Tahori, ed.), pp. 117–159, Gordon and Breach, New York.

Wilkinson, C. F., 1971*b*, Effects of synergists on the metabolism and toxicity of anticholinesterases, *Bull. WHO*, **44:**171.

Wilkinson, C. F., 1973, Correlation of biological activity with chemical structure and physical properties, *in: Pesticide Formulations* (W. Van Valkenberg, ed.), pp. 1–65, Dekker, New York.

Wilkinson, C. F., 1975, Insecticide synergists, *in: Insecticides for the Future: Needs and Prospects* (R. L. Metcalf and J. McKelvey, eds.), Wiley, New York.

Wilkinson, C. F., and Brattsten, L. B., 1972, Microsomal drug metabolizing enzymes in insects, *Drug Metab. Rev.* **1:**153.

Wilkinson, C. F., Hetnarski, K., and Yellin, T. O., 1972, Imidazole derivatives—A new class of microsomal enzyme inhibitors, *Biochem. Pharmacol.* **21:**3187.

Wilkinson, C. F., Hetnarski, K., Cantwell, G. P., and Di Carlo, F. J., 1974, Structure-activity relationships in the effects of 1-alkylimidazoles on microsomal oxidation *in vitro* and *in vivo*, *Biochem. Pharmacol.* **23:**2377.

Williams, C. H., and Casterline, J. L., Jr., 1970, Effects on toxicity and on enzyme activity of the interactions tions between aldrin, chlordane, piperonyl, *Proc. Soc. Exp. Biol. Med.* **135:**46.

Wilson, K. A., and Cook, R. M., 1970, Use of activated charcoal as on antidote for pesticide poisoning in ruminants, *J. Agr. Food Chem.* **18:**437.

16

The Treatment of Insecticide Poisoning

John Doull

1. Introduction

Poisoning by pesticides is a serious and growing problem both in this country and throughout the world. This increase is mainly the result of the greater availability and utilization of pesticides and other agricultural chemicals. However, it results also in part from the banning of relatively nontoxic pesticides such as DDT and the phenoxy herbicides and the subsequent increase in the use of replacement pesticides which are more acutely toxic such as the organophosphorus insecticides and paraquat. Pesticides are like antibiotics in the sense that the best agents are those which exhibit a highly selective toxicity to the invading parasite or pest yet are relatively nontoxic to the host or nontarget species. As with antibiotics such as penicillin, when a first-line pesticide such as DDT is abandoned, the available replacements almost invariably exhibit less selectivity and are more likely, therefore, to create poisoning problems in the host. Some of the reported increase in pesticide poisoning may also be more apparent than real, since with the increased awareness and concern about the ability of pesticides to produce both acute and chronic poisoning there tends to be better reporting of these cases. Thus pesticide poisonings which were formerly attributed to other causes are more likely today to be properly diagnosed and tabulated.

John Doull • Professor of Pharmacology and Toxicology, The University of Kansas Medical Center, College of Health Sciences and Hospital, Kansas City, Kansas.

Prior to the establishment of the National Clearing House for Poison Control Centers and the development of a reporting system for poisoning, a reliable estimate of the magnitude of the total problem and particularly the relative importance of the various causative agents was unavailable. Even with the Clearing House reporting system, the actual number of poisoning cases which occur each year is not known since the reporting system does not include cases treated in a doctor's office or elsewhere and not reported to a Poison Control Center. Regardless of whether one accepts the lower estimate of 1 million cases per year or the higher estimate of 4 million cases, it is clear that in comparison with the incidence of other disease states poisoning represents a major medical problem in this country. Current statistics from the National Clearing House for Poison Control Centers indicate that pesticides account for about 6% of all reported cases of acute poisoning, and it has been estimated that the mortality rate associated with pesticide poisoning in the United States is about 0.65 per 1 million population (Hayes, 1969). Most cases of acute pesticide poisoning result from accidental ingestion, attempts at suicide, or occupational exposure; as is the case with other types of poisoning, pesticide poisoning occurs most frequently in young children. Several episodes of mass poisoning have occurred in recent years as a result of pesticide contamination of food, but these have been more common in foreign countries than in the United States (Namba *et al.*, 1971).

What can be done to reduce the incidence of acute pesticide poisoning? One approach would be to use only those pesticides which exhibit the highest degree of selective toxicity between the target and nontarget species. Unfortunately, such agents do not always meet the pest control requirements, are extremely difficult to discover, and are not usually economically feasible to develop. Another approach would be to encourage the use of the least toxic members of each class of pesticides regardless of their selective toxicity. In the case of the organophosphorus compounds, for example, one might consider replacing the highly toxic parathion and methyl parathion (which are responsible for most cases of OP poisoning) with less toxic agents such as sumithion. It has also been suggested that the use of the more hazardous pesticides be restricted to trained and licensed pest control operators who would be required to handle these agents under the same type of restrictions which exist for prescription drugs. Other approaches already in use include the requirement for warning labels on pesticide products and efforts to educate pesticide users. Most individuals who are occupationally exposed to pesticides and even the general public are aware of the rules for preventing pesticide poisoning, such as (a) read, understand, and follow the recommendations on the label, (b) use protective clothing and equipment where indicated, (c) store pesticides under lock and key and in the original container, (d) buy only the amount of pesticide needed for the specific job, and, most important, (e) when exposure occurs, get help (family physician, Poison Control Center, hospital emergency room, etc.). It is evident that, in order for these rules to be effective, health professionals and particularly emergency room personnel not only must be capable of the diagnosis and treatment of the various types of pesticide poisoning but also must have accurate information about pesticides in general.

Although the pesticides are certainly an important class of agents in toxicology, they have in the past received and continue to receive more than their share of publicity. Most of this is adverse publicity and it tends to focus on two generalizations. The first is that the pesticides are widely distributed and persistent and that they get into our food chain, and the second is that they are extremely toxic. Like most generalizations, these are only partly true. DDT is widely distributed, it persists in the environment and gets into the food chain, but it is relatively nontoxic and acute poisoning is extremely rare in man. On the other hand, some of the organophosphorus insecticides are quite toxic, but most of these are rapidly hydrolyzed in the environment and therefore do not persist for any appreciable time after their application. What is usually lacking in this kind of publicity is the recognition that the term "pesticide" includes other classes of compounds, such as the herbicides, defoliants, fungicides, nematocides, desiccants, rodenticides, and molluscicides, and that generalizations about insecticides such as DDT or the organic phosphates do not apply to all of these groups. Another common misconception is that pesticides all act in the same way. The only feature which pesticides have in common is their deleterious effects on an organism or life form deemed to be undesirable by the person or society that applies these agents. There is as much diversity in the mechanisms by which pesticides produce their effects in biological systems as there is diversity in their chemical and physical chemical properties. Even within a group of compounds which have a similar chemical structure and mechanisms of action such as the organic phosphates, the individual compounds may range in toxicity from extremely toxic, such as tetraethylpyrophosphate (TEPP), to relatively nontoxic, such as dipterex. It is obvious, therefore, that one cannot generalize either qualitatively or quantitatively about the toxicity of pesticides. The failure to distinguish between different classes of pesticides also complicates the evaluation of risk–benefit ratios for this group of chemicals. The benefits of pesticides in reducing disease, for example, are due mainly to the use of insecticides, whereas the benefits of increased food production depend not only on the insecticides but also on herbicides and other classes of pesticides.

Since the management of acute pesticide poisoning is basically no different from the management of other types of poisoning, it may be useful to consider some of the more general aspects of poisoning before discussing the individual classes of insecticide poisoning. There are three basic rules for the clinical treatment of all types of acute poisoning. These are (a) *support vital functions* (check respiration and insert artificial airway if necessary, treat hypotension with fluids or plasma, avoid vasopressors if possible, etc.), (b) *minimize further exposure* (induce emesis or lavage, remove contaminated clothing and wash exposed skin areas, increase excretion, insert Foley catheter, administer charcoal, etc.), and (c) *treat clinical symptoms* (provide good nursing care, give antidote where indicated and available, check for trauma, obtain a detailed history, continue to monitor vital functions, etc.). It is important to keep in mind the fact that all ingestions do not result in poisoning, and that in many cases there may be no medical reason to see the patient unless he has a history of repeated previous exposures or severe anxiety.

Although the inhalation and dermal routes are fairly common in industrial exposure, ingestion of the poison is the most common route encountered in both accidental and intentional cases of poisoning seen in the emergency room. Since the poison is functionally outside of the patient for as long as it remains in the gastrointestinal tract, one would like to remove the poison from the gut through either the mouth or the anus before it ever gets into the circulation. Removal of the poison through the mouth can be accomplished either by inducing emesis or through gastric lavage. However, neither of these procedures should be considered in individuals who have ingested caustics (either acids or bases) or who have taken a convulsant such as strychnine or who are in coma. Actually, coma is only a relative contraindication since such a patient can be lavaged once he has had a cuffed endotracheal tube inserted and properly inflated to protect the lungs. Because of the danger of aspiration in a comatose patient who vomits, it is recommended that such patients be transported on the left side with the head and shoulders down. This is also a more desirable position for lavage than with the patient flat on his back. In lavage therapy, it is essential to use a large-bore tube and to avoid water loading of the patient. The easiest way to induce vomiting is to use syrup of ipecac, which can be obtained without prescription and kept in the home so that treatment can be initiated even before the patient is transported to the emergency room. When ipecac is used, the patient should be given additional fluids since adding ipecac to an empty stomach is like squeezing an empty balloon. Many pesticide formulations contain kerosene or a similar hydrocarbon as the vehicle and physicians have sometimes been reluctant to utilize emesis or lavage in cases of pesticide poisoning because of the dangers of pneumonitis with hydrocarbon aspiration. Although hydrocarbon ingestion is commonly listed as a contraindication to emesis or lavage in many of the poisoning textbooks, most major poison treatment centers today routinely induce emesis in patients who are mentally alert and who have ingested appreciable quantities of the hydrocarbon or in those cases where the hydrocarbon is a vehicle for a highly toxic agent such as a pesticide.

Purgation or the use of a laxative is an effective way to accelerate the passage of the poison through the gut. Saline cathartics such as magnesium citrate or sulfate are more commonly used for this purpose than oils such as mineral oil or castor oil. Another related procedure is to give activated charcoal or some other agent to bind the poison and prevent its absorption. Activated charcoal can also be combined with lavage to remove additional poison from the stomach.

Once the poison has been absorbed into the circulation, the next stage of treatment is directed toward facilitating its excretion into the urine either through forced diuresis or by altering the urinary pH. Although diuresis is effective for many poisons that are filtered in the glomerulus, it is ineffective for agents that are tightly protein bound. Another treatment procedure which may be considered when these measures are ineffective is the use of some type of dialysis, although most poison treatment centers do not utilize dialysis as a part of the initial therapy except perhaps in poisonings with *Amanita phalloides*, ethylene glycol, or methanol. A recent development in this area is the use of coated charcoal or resin hemoperfusion as a mechanism for enhancing the therapeutic efficacy of dialysis.

It is estimated that there are currently about 250,000 products available on the market which may cause poisoning. With this large number of potentially toxic products, one would anticipate that every emergency room would be stocked with an equally large number of antidotes. Since the antidote tray of a good emergency room is likely to contain fewer than two dozen agents, it is obvious that there are relatively few poisons for which specific antidotes are available. In the case of the pesticides, the situation is somewhat better in that we do have antidotes for several of the more important types of pesticides, such as the organic phosphates, the carbamates, several of the rodenticides, and the metal-containing pesticides. However, even in those situations where specific antidote therapy is available, good supportive care is essential and must be performed whether there is an antidote or not. In general, the good supportive care of a poisoned individual does not include the use of analeptics or respiratory stimulants, the use of antipyretics, or the use of vasopressors.

Despite the fact that the initial history obtained in a poisoning situation is often incorrect, and may in fact represent a deliberate attempt by the patient or his relatives to conceal the true nature of the cause of his illness, every attempt should be made to obtain as complete a history as possible. This should include not only the history of the patient's illness but also details of the circumstances surrounding his exposure. The container from which the poison came should be brought in with the patient as well as any bottles of medication in the area either empty or full. Since toxicological analysis may be necessary in order to establish the diagnosis or for subsequent medical-legal purposes, it is also essential to obtain and save appropriate biological samples such as blood, urine, gastric washings, and vomitus. There are two types of information which the attending physician would like to obtain from the poison container or in some cases from the laboratory toxicological analysis: (a) What is the poison? (b) How much of it is contained in the product or in the blood or urine of the patient? This type of information is particularly helpful in cases of pesticide poisoning because of the number and variety of agents involved in such products, the use of multiple trade names for a single agent, and the likelihood that the attending physician will be less familiar with the various classes of pesticides than he would be with drugs.

In most types of poisoning, the attending physician does not utilize the clinical laboratory for toxicological analysis, at least during the early treatment phase, since the answers come too slowly or in most cases are not available on a routine basis from his laboratory. Thus the emergency room physician is more likely to rely on his clinical diagnostic skills or perhaps on some simple bedside tests (salicylate, paraquat, etc.) and to use the laboratory only for blood gas and electrolyte determinations. This situation is changing, and clinical laboratories capable of providing 24-hr availability of sophisticated, rapid toxicological analysis are proving the potential benefits of such an arrangement to busy poison treatment centers. There is no question about the importance of this type of laboratory investigation in centers which have research programs under way, and the pesticide area would appear to represent a good place for the clinical chemistry laboratory to demonstrate its ability to assist the emergency room physician in the management of poisoning. There are now rapid methods for detecting cholinesterase inhibition in serum, metals in biological fluids by atomic absorption, hydrocarbons by gas chromatograph, and other agents where the results could be of real diagnostic value. Some of the problems and benefits of this kind of team approach have been described in a recent Ciba Foundation Symposium dealing with the role of the laboratory in the poisoned patient (Curry, 1974).

Toxicology information sources are also of particular value in the management of pesticide poisonings since some of these agents have antidotes, and the early identification of the specific chemical involved improves the probability of a good antidotal response. The major toxicology information sources in current use include handbooks of product and ingredient listings, the card file provided by the National Clearing House for Poison Control Centers, on-line computer terminals, microfiche systems, and a variety of other reference sources (books on plants, pharmacology and toxicology texts, drug lists such as PDR, pharmacopeias, etc.). Pesticides are listed by trade name, chemical name, class of agent, and various other categories in all of the general handbooks (e.g., Gleason *et al.*, 1969; Deichmann and Gerarde, 1969; Dreisbach, 1974; Arena, 1973; Moeschlin, 1965; Sax, 1975) and in several handbooks devoted specifically to pesticides (e.g., Brown, 1966; Spencer, 1973; Frear, 1961). They are also tabulated in various governmental publications such as the Toxic Substances List from NIOSH (Christensen *et al.*, 1974). In addition, all of the major pesticide manufacturing companies issue manuals and collections of toxicological data on their products. A major problem with these sources is that frequent republication is needed to keep the information up to date. The card file from the National Clearing House for Poison Control Centers avoids this problem by issuing new cards as new products appear on the market or product changes occur. Our experience with this card file system is that it provides an adequate vehicle for responding to about 80% of our poison control center telephone inquiries. However, the treatment recommendations provided on these cards appear to be designed for minimal exposure situations, and they are therefore inappropriate for most suicidal or massive overexposure situations. Another minor problem with our use of this system is the need for continual replacement of lost and misplaced cards which our house staff either fails to replace or sends along with patients when they are transferred from the emergency room to the intensive care unit.

The on-line computer terminal provides, at least in theory, the best way to maintain a current toxicology information source in the emergency room, and systems of this type have been established by Dr. Vernon Green at The Children's Mercy Hospital in Kansas City, at the National Clearing House for Poison Control Centers in Bethesda, and at other locations. The main problem with these systems is the expense of maintaining the terminals (computer hardware, staff, etc.) and the difficulty of ensuring that the system will be available on a 24 hr a day basis. Although not intended to serve as a source of poison control information, the Toxline system of the National Library of Medicine is an example of a toxicology information service which provides on-line bibliographic searching capability to the terminal user in addition to a variety of other useful services (Chemline for CA numbers and related information, off-line literature searches, etc.). Since this system places most of the toxicology-related resources of the National Library of Medicine in the hands of the user, it is ideally suited to support the library needs of a major medical center even though its cost, restricted access, and other factors make it impractical for emergency room use.

The use of a microfiche system to provide poison control information is a relatively new development in the toxicology information source area, and there are at least two such systems currently available (Toxifile, Chicago Micro Corporation, Chicago, Illinois, and Poisindex, Micromedex, Inc., Denver, Colorado). Since the Toxifile system is based primarily on the National Clearing House card file system, it has many of the disadvantages associated with that system, although it is clearly an advance compared to previously available systems. The microfiche system is economical, compact, easy to use, always available, and convenient. Since the microfiche cards cannot be read without a viewer, they do not disappear like books, and updated cards can be inexpensively generated as frequently as needed. The Poisindex system, which was developed by the Director of the Rocky Mountain Poison Center, Dr. Barry Rumack, uses a computer-generated microfiche to provide greater legibility. This system also includes drug imprint codes and photographs of plants and mushrooms in addition to the product and ingredient information on over 100,000 entries. Since the treatment recommendations in the Poisindex system were prepared by a group of physicians representing several major poison treatment centers, they provide poisoning management recommendations which are clinically sound and empirically tested in the emergency room environment.

To conclude this section, it may be helpful to reemphasize some of the previously covered points. From what has been said thus far, it is clear that in treating all types of poisoning the focus must be on the patient and not on the poison. The old adage that the physician should treat the patient and not the poison is as true for pesticide poisoning as it is for poisoning by other agents. The second general conclusion is that the diagnosis of poisoning requires a high index of suspicion. The possibility of poisoning should be considered in any case of acute illness for which another cause is not firmly established. Poisoning should be considered in the differential diagnosis of any patient who exhibits vomiting, convulsions, collapse, coma, and other types of symptoms, particularly where the onset is sudden and severe. A word of caution should be added here: the physician should exercise care in attaching a firm diagnosis of pesticide poisoning or for that matter of any type of poisoning to a patient until he has adequate evidence to support this conclusion. This course of action not only helps to protect the patient from inappropriate treatment during the acute phase of his illness but also may avoid future embarrassment for both the patient and the physician when the true facts regarding the poisoning become known. Finally, it is evident that the treatment of acute pesticide poisoning is the same as that with other agents and that it involves mainly symptomatic and supportive care. Panic, the use of heroic measures, or overtreatment with stimulants, sedatives, or other therapeutic agents is likely to result in more damage than that produced by the poison.

2. Treatment of Organophosphorus Insecticide Poisoning

Organophosphorus insecticide poisoning can and does occur throughout the year, but the incidence is typically increased during periods of peak agricultural use. Even during these periods, however, most cases of severe OP poisoning result from accidental exposure rather than from occupational exposure and occur most frequently in children. OP insecticide poisoning exhibits geographic patterns in both morbidity and mortality, and these geographic patterns are dependent not only on whether the community is located in an agricultural area but also on the type of OP insectides and the purpose for which they are being used in that area. In a survey on the incidence of OP insecticide poisoning in Dade County, Florida, it was observed (Reich *et al.*, 1968) that during the period 1956–1965 pesticides were exceeded only by drugs and alcohol as a cause of fatal poisoning and that in children under 5 years old pesticides were the leading cause of poisoning deaths. Parathion was the chief offender among the OP insecticides in this study, and there was a progressive increase in the number of fatal poisonings with this agent during the period examined. In evaluating these studies, the medical examiner suggested that the occurrence of OP insecticide poisoning is probably being seriously underestimated in this country since the clinical signs of intoxication can be confused with other disease entities with which the physician or nurse may be more familiar. Irrespective of the validity of this evaluation, poisoning by the OP insecticides provides the physician with a situation in which a high index of suspicion, background knowledge of the community, and familiarity with the symptoms and treatment of exposure can be expected to have a greater influence on the mortality statistics than is the case with most other types of poisoning.

2.1. Clinical Symptoms

At the present time, there are over 100 different OP insecticide formulations on the market, and although they all produce their toxic effects by the same mechanism of action they differ greatly in their inherent toxicity. The acute lethal dose of parathion, for example, is only one-hundredth that of malathion, but it is over ten times that of tetraethylpyrophosphate (TEPP). Further, the toxicity of the OP insecticides is influenced by the type of formulation, the vehicle, and the route of absorption. In addition, some of the OP insecticides produce their effects directly (TEPP, phosdrin, etc.), whereas others are not active until they have been altered by chemical or enzymatic change (parathion, malathion, guthion, etc.). Their toxicity can be influenced by other insecticides, such as the chlorinated hydrocarbons, and agents which by influencing the metabolism of the OP insecticides may either increase or decrease their toxicity (Chapter 15). Prior exposure to phenothiazines, theophyllines, succinylcholine, or other cholinesterase inhibitors or parasympathomimetics may also increase the toxicity of OP insecticides. Exposure to certain combinations of OP insecticides, such as malathion plus EPN, produces in animals a greater than additive effect (potentiation).

The clinical manifestations of OP insecticide poisoning are due to the inhibition of cholinesterase, which leads to an accumulation of acetylcholine at cholinergic nerve synapses and at neuroeffector junctions including the myoneural junction. The symptoms of poisoning can be grouped in relation to whether the effects are primarily muscarinic (stimulation of the parasympathetic nervous system), nicotinic (myoneural junction stimulation), or central nervous system effects. The muscarinic effects are usually the first to appear and include anorexia, nausea, sweating, epigastric and substernal tightness, and heartburn. More severe exposure may produce abdominal cramps, vomiting, diarrhea, salivation, lacrimation, profuse sweating, pallor, dyspnea, and perhaps audible wheezing. Myosis may be present but is not a constant feature, and occasionally mydriases may occur. Involuntary micturition and defecation, excessive bronchial secretions, and pulmonary edema are also seen in severely poisoned individuals. The sequence of symptoms and their severity and rapidity of onset vary with the type of agent to which the individual is exposed and with the route and severity of exposure. As one might anticipate, gastrointestinal effects are usually among the first symptoms seen after ingestion. Sweating and sometimes muscle fasciculation may appear as the first symptoms after dermal exposure, and respiratory effects follow inhalation exposure.

The nicotinic effects of OP insecticides usually appear after the muscarinic effects have reached moderate severity and include muscular twitching, fasciculation and cramps, and mild generalized weakness which is increased with exertion. Severe exposure produces a pronounced weakness of all muscles including those of respiration. Among the central nervous system effects of the OP insecticides are such complaints as apprehension, restlessness, giddiness, emotional lability, and headache (which may be due to the mercaptan odor associated with some of the thiophosphates). These symptoms may be the only complaints in minimal exposure and with continued mild exposure may progress to insomnia, excessive dreaming, difficulties in concentration, poor memory, drowsiness, and confusion. With severe exposure, tremors, ataxia, slurring of words, convulsions, and coma may occur. Death from OP poisoning usually results from respiratory failure.

The respiratory failure associated with OP insecticide exposure is caused by a combination of their effects on the lungs (bronchoconstriction, excessive pulmonary secretions and edema, and paralysis of the respiratory musculature), the respiratory center, and the heart (brachycardia and various degrees of AV block). Some of the OP insecticides produce a delayed paralysis due to demyelination of the spinal cord similar to that seen in Ginger–Jake paralysis, but chronic poisoning, as such, does not occur with OP insecticide exposure. Since OP insecticide poisoning may be confused with encephalitis, brain hemorrhage, and other types of injury, laboratory tests may be necessary to establish or to exclude the diagnosis and to evaluate the severity of exposure. The single most useful laboratory test for OP insecticide exposure is measurement of the blood cholinesterase activity. Although lowered blood cholinesterase activity levels can occur as a result of disease or from exposure to other toxic agents or drugs such as the phenothiazines or as a result of genetic factors, a careful consideration not only of the degree of inhibition but also of the rate at which it changes will generally provide a useful and satisfactory diagnostic standard. In evaluating such laboratory results it is important to keep in mind the fact that irreversible inhibition of the red cell cholinesterase activity will persist until these cells are regenerated and that in such an individual the measurement of red cell cholinesterase activity may give a false indication of exposure during the regeneration period.

2.2. Treatment

As is the case with many other types of poisoning, the removal of the poison and the symptomatic care of the patient are of primary importance in the management of OP insecticide poisoning. In fact, severely poisoned patients have recovered with no additional treatment. Fortunately, there are two antidotes useful in the treatment of OP insecticide poisoning. The first of these is atropine, which reverses the muscarinic and some of the central nervous system effects but does not block the nicotinic effects of the cholinesterase inhibitors. Atropine has no effect on the inhibited blood cholinesterase, but it does block the action of acetylcholine on the parasympathetic receptors; thus it alleviates the bronchial spasm, reduces respiratory secretions, and reduces miosis at least temporarily. Individuals who are severely poisoned require large doses of atropine and the usual procedure is to give 2–4 mg intravenously as soon as cyanosis is overcome and to repeat the dose at

5–10 min intervals until signs of early atropine toxicity occur (dry mouth, dry hot skin, mydriasis, and tachycardia). The ability of the OP insecticide poisoned individual to tolerate large doses of atropine may be of diagnostic help in establishing the cause of his illness.

The other antidote for organic phosphate poisoning is 2-PAM or pralidoxime chloride. 2-PAM is a specific antidote in that it will reactivate the reversibily inhibited (phosphorylated) cholinesterase (Chapter 7). The oximes are only moderately effective when given alone, but since they act by a mechanism different from that of atropine the two antidotes do not interfere with each other's activity when given together in severe poisoning cases. OP insecticide poisoned individuals require close surveillance during the first 24–48 hr after exposure since a favorable initial response may be followed by a sudden and fatal relapse when the antidotal effects of atropine "wear off" before those of the OP insecticide. The use of oxime therapy in the treatment of OP poisoning is more effective if treatment is started early, since there is a secondary irreversible reaction between the enzyme and the OP insecticide (aging) which reduces the ability of the oxime to reactivate the enzyme. Normally, the symptoms of OP poisoning occur fairly rapidly following ingestion, although they may be delayed for several hours following occupational exposure or with the ingestion of certain types of insecticidal formulations. In individuals where the symptoms do not appear for 12–24 hr after exposure, one should consider the possibility that the poisoning is the result of an agent other than an OP insecticide. With massive exposure, symptoms and even death may occur within a few minutes, but even in such severely poisoned individuals prompt and energetic treatment of the patient can produce a dramatic clinical recovery.

Occasionally in massively overexposed individuals the symptoms will continue to progress even though the patient has received adequate therapeutic doses of atropine and 2-PAM. In this situation, it may be worthwhile to consider the use of additional therapeutic measures to reduce the body burden of the poison. In one such case recently treated at the Denver General Hospital by Dr. B. Rumack, the use of exchange transfusion resulted in a distinct clinical improvement in a 4-year-old child poisoned with parathion. It may be that some of the exchange resins being investigated for use in treating severe poisoning by other agents may also prove to be beneficial in the treatment of such nonresponsive cases of OP insecticide poisoning.

With highly toxic OP compounds such as parathion, it is also of critical importance that the patient be thoroughly decontaminated. In a case recently treated in our emergency room, we were unable to explain the unusual persistence of symptoms in a patient exposed to an OP insecticide until it was discovered that small amounts of the insecticide were present under his fingernails. When this material was removed, the patient promptly recovered. Similar reports have appeared in which patients have exhibited a dramatic recovery following a vigorous shampoo of the hair. A final word of caution is indicated in regard to the administration of atropine to a person who is suspected of an OP exposure. If the diagnosis in such an individual is incorrect, the atropine treatment may itself produce poisoning.

3. Treatment of Carbamate Insecticide Poisoning

Poisoning resulting from exposure to the carbamate insecticides occurs less frequently than poisoning by OP insecticides and is usually easier to treat. Although the carbamates produce their toxic effects by the same mechanism as the OP insecticides (cholinesterase inhibition), the carbamylated enzyme recovers more rapidly than that which has been phosphorylated, and this, together with the rapid metabolism of the carbamates, results in cholinergic symptoms which disappear quickly following exposure. There are two other major differences between the effects of carbamate and OP insecticide poisoning. The first difference is in the antidotal management. Atropine is an effective antidote for both types of poisoning (or for any other situation where cholinergic symptoms predominate), but 2-PAM is ineffective and is in fact contraindicated in the management of at least some types of carbamate insecticide poisoning. The other difference between OP and carbamate poisoning is related to the value of the laboratory determination of blood cholinesterase. With the conventional methods used to measure serum or erythrocyte cholinesterase activity, blood samples from carbamate-poisoned individuals may appear to have normal or near normal cholinesterase activity values. Although there are special radioactive methods by which enzyme inhibition can be demonstrated in the blood of carbamate insecticide exposed patients, these are usually available only in pesticide research laboratories.

At the present time, there are about two dozen different carbamate insecticides in commercial usage, but each of these may be encountered under a variety of trade names. They vary in toxicity from agents such as temik (aldicarb), furadan (carbofuran), and tirpate which are relatively toxic (oral LD_{50} of about 10 mg/kg or less), to relatively nontoxic agents such as sevin (carbaryl) and fenethcarb (LD_{50} over 500 mg/kg). The carbamate insecticides are used mainly as contact and systemic insecticides, nematocides, miticides, and aphicides, and they may be formulated with synergists (piperonyl butoxide, tropital, etc.) to enhance their insecticidal effect. Most of these agents are methylcarbamate or methylcarbamoyl oxime derivatives and they should not be confused with the thiocarbamate herbicides and fungicides such as vernolate and maneb, which produce different symptoms and clinical signs in poisoning.

3.1. Clinical Symptoms

The general signs and symptoms of carbamate insecticide poisoning are secondary to cholinesterase inhibition and include miosis, blurred vision, weakness, epigastric pain, abdominal distress, nausea and vomiting, sweating, salivation, lacrimation, urination, diarrhea, muscle fasciculation, pulmonary edema, areflexia, and convulsions, etc. As is the situation in OP insecticide poisoning, the time of onset of these symptoms, their severity and rate of progression, and even the type of symptom may depend on the agent involved, the formulation, the route of exposure, and various other environmental factors (see Section 2).

3.2. Treatment

Support of the vital functions and removal of the poison constitute the primary management recommendations for the initial treatment of carbamate insecticide poisoning. In the absence of cyanosis and where life-threatening cholinergic symptoms exist, atropine is the drug of choice. A test dose of 2 mg intravenously will produce no signs of atropinism in a poisoned patient and the atropine may then be repeated at 5–10 min intervals as needed to produce cessation of secretions and dilatation of the pupils. A mild degree of atropinization should be maintained for 24 hr or more if cholinergic symptoms reappear when the atropine is discontinued. Decontamination of the skin (including the hair) can be accomplished by using an initial soap washing followed by alcohol washing and a final soap washing. Emesis and/or lavage are indicated for orally ingested agents unless the patient is comatose, is convulsing, or has lost the gag reflex. As is the case for other types of poisoning, endotracheal intubation should precede gastric lavage if these contraindications are present. Other contraindications include the use of morphine, other cholinergic drugs, 2-PAM, and other oxime reactivators of cholinesterase. Since most carbamate insecticides are formulated in a hydrocarbon vehicle which sensitizes the myocardium to the effects of epinephrine and related drugs, such agents should be used with caution or avoided in cases of carbamate and other types of pesticide poisoning. Even in patients who are severely poisoned with a carbamate insecticide, rapid recovery and the absence of any chronic sequelae are the typical response when proper atropinization is combined with careful supportive treatment.

4. Treatment of Chlorinated Hydrocarbon Insecticide Poisoning

The chlorinated hydrocarbon or organochlorine insecticides include the chlorinated ethane derivatives such as DDT, the cyclodienes such as dieldrin, and the hexachlorocyclohexanes such as lindane. These agents vary widely in toxicity when compared between classes or even within a single class such as the cyclodienes. DDT or chlorphenothane has only rarely caused acute clinical poisoning, and individuals have survived the single ingestion of 10 or more of this agent whereas dieldrin and endrin have both produced fatalities with ingestions of less than 2 g.

4.1. Clinical Symptoms

The ingestion of DDT or other organochlorine insecticides may be expected to produce general malaise, loss of appetite, and nausea as early symptoms, and with sufficient exposure these symptoms may progress to include muscle fibrillation, gross tremor, and vomiting. Severe exposure to the more toxic organochlorine insecticides may produce convulsions and other central nervous system effects. Paresthesia of the lips, face, and extremities, a clouded sensorium, and liver, renal, and myocardial toxicity have also been reported for organochlorine insecticides as well as for other types of chlorinated hydrocarbons. Since all of these symptoms can be produced by other classes of pesticides as well as by other types of agents, it may be necessary to utilize laboratory analysis to demonstrate poisoning by DDT or other organochlorine insecticides. Unlike with many of the other types of pesticides, dermal poisoning with the organochlorine insecticides is rare since the combination of low solubility and application in the form of a powder decreases markedly the possibility of absorption.

4.2. Treatment

Since severe poisoning with DDT is rarely encountered in the emergency room, optimal treatment recommendations for this type of acute poisoning are not well established. Poisoning with other more toxic organochloride insecticides may require sedation (intravenous diazepam). Decontamination of an exposed individual is more difficult with the organochlorine insecticides than with water-soluble insecticides but should be a required and early part of the management regime. Oil-based cathartics should be avoided in such individuals since they may increase absorption, and the hepatic, renal, and myocardial functions should be monitored in severely poisoned individuals. Once the initial symptoms have been treated, the physician may wish to consider prolonged anticonvulsant therapy since epileptiform convulsions have been reported to occur for days and even weeks after the initial exposure to such organochlorine insecticides. Hypothermia has been reported in cases of severe poisoning with dieldrin, and there is some evidence which suggests that this effect results from a direct action of the insecticide on the temperature-regulating center in the hypothalamus. It has also been reported that lindane (the γ-isomer of hexachlorocyclohexane) is capable of producing hematopoietic damage in addition to the usual organochlorine effects. In animals, severe exposure to hexachlorocyclohexane produces CNS effects similar to those seen with other organochlorine insecticides, but the mechanism for this appears to involve CNS stimulation with the γ and α isomers and CNS depression with the β and δ isomers. Calcium gluconate has been recommended in the past as an adjuvant to barbiturate therapy in the treatment of organochlorine insecticide exposure, but it does not appear to be widely used for this purpose in current therapy. The major cause of death in organochlorine insecticide poisonings has been reported to be respiratory failure and/or ventricular fibrillation. Since hydrocarbons sensitize the myocardium to the effects of epinephrine and other vasopressors, such agents should be used with caution, if at all, in treating organochlorine insecticide poisoning. Pneumonitis due to aspiration of the hydrocarbon vehicle is a fairly common complication of organochlorine insecticide poisoning cases and is considered by some emergency room physicians to be a potentially more serious problem than organochlorine insecticide poisoning.

5. Treatment of Botanical Insecticide Poisoning

The three major groups of botanical insecticides are the pyrethroids, rotenoids, and nicotine. Although most users regard these agents as nontoxic because of their plant origin, nicotine is highly toxic, and the oral LD_{50} for pyrethrum and rotenone would place them in the moderately toxic range. In addition, there are a number of new synthetic pyrethroid derivatives which may be used commercially in the future that are considerably more toxic than the agents now in use. There are also a number of alkaloids in addition to nicotine which have been used in the past or are still used as natural insecticides. These include such agents as anabasine, hellebore, sabadilla, quassia, and ryania. The toxic effects of ryania are due mainly to the alkaloid ryanodine, and the symptoms and treatment recommendations are similar to those for rotenone. Sabadilla contains alkaloids (cevadine) which produce cardiac effects similar to those seen in veratrum poisoning, but the initial effects following ingestion are related to severe gastrointestinal irritation. Hellebore poisoning has occurred mainly in veterinary medicine (false or green hellebore), and the toxic effects are also due to veratrumlike alkaloids. Anabasine is a nicotinoid, and the symptoms and treatment for this agent (and also nornicotine) are similar to those described in Section 5.3.

5.1. Pyrethrins

Pyrethrum is one of the oldest insecticides known to man and is obtained from the flowers of the plant *Chrysanthemum cinerariaefolium.* It is still widely used in home insecticide products because of its fast-acting insecticidal properties (quick knockdown). The major toxic principles are the esters pyrethrin I and pyrethrin II and cinerin I and cinerin II, and commercial formulations usually contain piperonyl butoxide, sesamin, or other synergists; the vehicle may be deodorized kerosene or other hydrocarbons. Pressurized or aerosol formulations often contain methyl chloride or dichlorofluoromethane. Allethrin is a synthetic pyrethrin derivative, and in recent years significant progress has been made in the development of numerous other synthetic pyrethroids with enhanced insecticidal activity. Severe poisoning with the pyrethrins has not been common but fatalities have occurred. Contact dermatitis is the most common clinical manifestation of exposure, and sensitization is reported to occur more frequently in individuals who are sensitive to ragweed pollen. Following massive ingestion or inhalation, clinical symptoms are most likely to involve the central nervous system and may include hyperexcitability, incoordination, tremors, muscular paralysis, numbness of the lips and tongue, sneezing, nausea, and vomiting. In those cases where fatalities have occurred, death was due primarily to respiratory failure. Severe anaphylactic reactions with peripheral vascular collapse have occurred in individuals who develop sensitivity to the pyrethrins.

The treatment of pyrethrin poisoning is mainly symptomatic and should include support of the vital functions, removal of the poison by emesis or lavage, and good supportive care. As is the situation with other pesticides, the treatment of pyrethrum poisoning may be complicated by pneumonitis resulting from aspiration of hydrocarbons into the lungs. Proper decontamination following exposure is a particularly important step in the management of exposure to the pyrethrins because of their irritant properties and propensity to produce dermatitis.

5.2. Rotenone

Rotenone is a selective contact insecticide with some ascaricidal activity. It was first used by primitive populations to paralyze fish and continues to enjoy some popularity for this purpose. Rotenone is obtained by extraction from the root of *Derris elliptica* and *Lonchocarpus* species. The toxic principles of derris include some 13 compounds in addition to rotenone. Acute poisoning with rotenone is rare in man, and rotenone preparations have been used to treat scabies, head lice, and other ectoparasites because of their relative nontoxicity. Rotenone is a powerful irritant and produces local effects of conjunctivitis, dermatitis, pharyngitis, rhinitis, and severe pulmonary irritation when inhaled. Following ingestion, the major clinical symptom is gastrointestinal irritation with epigastric pain, abdominal cramps, nausea, vomiting, etc. The treatment is largely symptomatic and consists of good support of the vital functions and removal of the poison by emesis or lavage. Rotenone has a strong inherent emetic action which usually produces vomiting when toxic doses are taken internally. With massive overexposure, spasmolytic effects similar to those produced by papaverine overdosage have been reported. Rotenone inhibits the oxidation of reduced NAD, and this effect is thought to be partly responsible for the toxicity of rotenone in both insects and mammals.

5.3. Nicotine

Nicotine poisoning occurs not only as a result of the ingestion of nicotine-containing insecticide formulations such as Black Leaf 40 but also from the ingestion of various plants containing nicotinelike alkaloids. These include conine poisoning from poison hemlock, lobeline poisoning from Indian tobacco, cytisine poisoning from *Laburnum* species, and nicotine poisoning from tobacco plants. The pharmacological effect of these alkaloids is to stimulate nicotinic receptors in the autonomic ganglia, at the neuromuscular junction, and in some pathways of the central nervous system. Since nicotine is an extremely toxic alkaloid (60 mg may be fatal), nicotine poisoning has resulted from the ingestion of relatively small amounts of tobacco products (swallowing tobacco quids or pipe residues). Confirmed smokers, however, develop tolerance to nicotine, and survival has followed exposure to over 2 g of nicotine in such individuals. The clinical effects of nicotine poisoning generally occur rapidly after a short lag period for absorption, and they may include ataxia, bloody vomiting, coma, convulsion, diaphoresis, salivation, hallucinations, headache, hyperthermia, hyperventilation, irritation of the mucous membranes, mydriasis, tachycardia, nausea, and weakness. With lethal exposures, renal failure, hypotension, paralysis, and coma may precede death, which usually results from respiratory failure. Nicotine can be absorbed through the skin as well as from the gastrointestinal tract and the lungs, and the type of symptoms, severity, and rate of onset will depend on the route of exposure.

The treatment of poisoning by nicotine and related alkaloids consists of vital function support and early removal of the poison. Since activated charcoal will absorb these alkaloids, it is a valuable adjunct to lavage therapy. Potassium permanganate can also be used in the lavage fluid (diluted 1 : 10,000) to inactivate the alkaloids, and autonomic blocking drugs may be of some value in antagonizing some of the visceral symptoms of poisoning by nicotine and the related alkaloids. Patients exhibiting convulsions may require sedation (e.g., intravenous diazepam), and such patients may require vasopressor drugs if their hypotension does not respond to the usual therapeutic regime. Patients who survive for more than a few hours are likely to recover since nicotine is metabolized fairly rapidly to less toxic metabolites.

6. Treatment of Pentachlorophenol Poisoning

Pentachlorophenol is used fairly widely as an insecticide for termite control and also as a herbicide, molluscicide, fungicide, and bactericide. The toxic effects of pentachlorophenol (PCP) are similar to those produced by dinitrophenol derivatives (some of which are also used as insecticides) and they result from the ability of these agents to uncouple oxidative phosphorylation. Thus poisoning by these agents is characterized by increased oxygen consumption, hyperthermia, and an increase in respiration and heart rate. Two fatalities and several nonfatal poisonings occurred several years ago in a St. Louis hospital nursery where PCP was used in the laundry. Subsequent investigation demonstrated that the poisoned infants were exposed to the small amounts of PCP remaining in diaper following the laundry treatment. PCP and other phenol derivatives are readily absorbed through the skin as well as from the gastrointestinal tract and by inhalation.

6.1. Clinical Symptoms

The early symptoms of PCP poisoning include restlessness, anxiety, excitement, and occasionally early convulsions. These may progress to hyperthermia, dehydration, increased thirst, profuse sweating, flusing, tachycardia, hyperpnea, dyspnea, cyanosis, coma, and death due to myocardial or respiratory failure.

6.2. Treatment

Since the symptoms of acute PCP poisoning may develop rapidly, the support of vital functions is critical in the management of this type of poisoning. Cold packs, tepid water baths, and other therapeutic measures to reduce body temperature have been reported to be beneficial in PCP and similar types of poisoning, but the possible adverse effects of these measures on subsequent temperature regulation mechanisms should be considered. Dialysis was reported to be beneficial in the treatment of the PCP poisonings in the St. Louis nursery.

7. Conclusion

A major class of insecticides not discussed in this chapter are the metal-containing agents such as paris green and lead arsenate. The metals usually encountered in this type of poisoning are lead and arsenic, but thallium and mercury have also been used for insecticidal purposes. The clinical symptoms and treatment recommendations for metal-containing insecticides are similar to those for other types of products containing these metals and can be found in most textbooks of pharmacology or toxicology (Casarett and Doull, 1975).

Additional classes of insecticides not included in this chapter because of their infrequent involvement in poisoning or because they are commonly encountered in non-pesticide-related exposures are the organic thiocyanates such as lethane and thanite, the fluorides such as cryolite, sulfur-containing agents such as thiovit and lime-sulfur mixtures, and the borates. There are also a variety of adjuvants and auxiliary agents which are used in insecticide formulations such as emulsifiers, wetting agents, diluents, spreaders, attractants, repellents, perfumes, and others which may produce toxic effects independently or contribute to the toxic effects of the insecticides. Also, many of the vehicles used in insecticide formulations such as kerosene, turpentine, and sage oil may modify toxicity or are capable of producing independent toxic effects.

To conclude this chapter it is appropriate to consider briefly how some of the newer approaches to insect pest control may affect the future insecticide poisoning situation. There is currently a great deal of interest in the use of pheromones and juvenile hormones for insect control, and it does not appear likely that the use of these agents will create serious acute poisoning problems. Another new approach involves the use of chemosterilants and formamidines, and since some of these produce serious acute toxic effects consideration should be given to the management of poisoning by those agents most likely to be used commercially. Another type of agents which have already produced acute poisoning are the organotin compounds, and the poisoning potential of any new derivatives of this type needs to be carefully weighed against their insecticidal importance. An enlightened approach to the development of new insecticides and pest control techniques should include the recognition that both accidental and intentional exposure to these agents will occur and that it is much easier and more effective to deal with a poisoned individual when the management is preplanned than when it is dictated by a crisis situation involving an unknown agent.

8. References

Arena, J. M., 1974, *Poisoning*, Charles C Thomas, Springfield, Ill.

Brown, R. L., 1966, *Pesticides in Clinical Practice*, Charles C Thomas, Springfield, Ill.

Casarett, L. J., and Doull, J., 1975, *Toxicology, the Basic Science of Poisons*, Macmillan, New York.

Christensen, H. E., Luginbyhl, T. T., and Carroll, B. S., 1974, *The Toxic Substances List 1974 Edition*, Department of Health, Education, and Welfare, NIOSH, Rockville, Md.

Curry, A. S., 1974, *Symposium on the Poisoned Patient: Role of the Laboratory*, Ciba Foundation Symposia 26, Elsevier, Amsterdam.

Deichmann, W. B., and Gerarde, H. W., 1969, *Toxicology of Drugs and Chemicals*, Academic Press, New York.

Dreisbach, R. H., 1974, *Handbook of Poisoning*, Lange Medical Publications, Los Altos, Calif.

Frear, D. E. H., 1961, *Pesticide Index*, College Science Publishers, State College, Pa.

Gleason, M. N., Gosselin, R. E., Hodge, H. C., and Smith, R. P., 1969, *Clinical Toxicology of Commercial Products*, Williams and Wilkins, Baltimore.

Hayes, W. J., 1969, Pesticides and human toxicity, *Ann. N.Y. Acad. Sci.* **160:**40–54.

Moeschlin, S., 1965, *Poisoning, Diagnosis and Treatment*, Grune and Stratton, New York.

Namba, T., Nolte, E. T., Jackrel, G., and Grob, D., 1971, Poisoning due to organophosphate insecticides, *Am. J. Med.* **50:**475–492.

Reich, G. A., Davis, J. H., and Davies, J. E., 1968, Pesticide poisoning in South Florida, *Arch. Environ. Health* **17:**768–772.

Sax, I. N., 1975, *Dangerous Properties of Industrial Materials*, Van Nostrand Reinhold, New York.

Spencer, E. Y., 1973, *Guide to the Chemicals Used in Crop Protection*, Information Canada, Ottawa.

17

Environmental Toxicology

Wendell W. Kilgore and Ming-yu Li

1. Introduction

Synthetic organic pesticides when used properly have been of tremendous benefit to man and his environment, but when misused or used carelessly they have caused considerable harm. Fortunately, the adverse effects have been relatively minor in comparison to the great benefits from pest control. There is little doubt that pesticides have played, and most likely will continue to play, an important role in the production of food as the world's supply of raw agricultural products continues to decline in proportion to the increase in population.

During recent years, the overall use patterns of pesticides have changed considerably. The risks or hazards of using chemical pesticides have increased in recent years with the sharp rise in their consumption by agriculture, industry, householders, and government. In 1972, the U.S. production of pesticides and related products amounted to 1158 million pounds—1.9% more than the 1136 million pounds reported for 1971. Sales in 1972 were 1022 million pounds, valued at $1092 million, compared with 946 million pounds, valued at $979 million, in 1971. Some 800 million pounds of pesticides are used each year in the United States, and about 40% is applied by agriculture. Today, more than 34,000 products made from one or more of 900 chemical compounds are registered by the U.S. Environmental Protection Agency (EPA).

Wendell W. Kilgore and Ming-yu Li • Department of Environmental Toxicology, University of California, Davis, California.

Pest control chemicals are poisons; they may present immediate danger to the user if applied improperly or without sufficient knowledge of their toxic effects. Some are highly toxic and may cause serious illness and even death if spilled on the skin, inhaled, or otherwise used carelessly. In addition, potential future hazards to human health and wildlife can be created by residues from some long-lived pesticides that may build up in the food chain and cause widespread contamination of the environment.

Pesticides are of great benefit to man. They have saved millions of lives through control of disease-carrying insects. They have minimized catastrophic crop damage by insects, weeds, plant diseases, rodents, and other pests, preserved valuable forests and parklands from insect destruction, and protected households against damaging beetles, moths, and other bugs. Generally, they have provided a higher quality of life for man. We cannot afford to lose the advantages gained through pesticides, but neither can we ignore the potential dangers. Obviously, we must derive the maximum benefits by safe pesticide use. At the same time, we must find ways to minimize or eliminate the hazards that may accompany the application of these chemicals.

2. Insecticide Residues

2.1. Environmental Contamination

Insecticides can be widely dispersed in the environment, mainly by the action of wind and water. The most significant concentrations are found in and near the areas of intensive use, but traces have been found in the Antarctic and other areas far from the area of application.

In recent years, there have been many reports of residues of persistent pesticides in air, rainwater, dust, rivers, and the sea, and in the bodies of aquatic and terrestrial invertebrates, fish, birds, mammals, and man. However, the importance of these residues in the environment has not yet been fully assessed. The present status of persistent pesticide residues in the environment has been discussed in two comprehensive monographs (Edwards, 1973*a*,*b*).

Contamination of the environment by pesticides has been a subject of public concern for the last 25 years. The pesticides most frequently involved are the organochlorine insecticides DDT, TDE, endrin, heptachlor, aldrin, dieldrin, chlordane, toxaphene, strobane, and BHC or its γ isomer, lindane (Nicholson, 1969). Among these, there is now little doubt that DDT and, to a lesser extent, dieldrin are major long-term contaminants of the total environment. Small traces of these chemicals can be found in almost all compartments of our ecosystems. In upstate New York, five commercial apple orchards were surveyed in 1972 for residues of DDT and its metabolites (Kuhr *et al.*, 1974). Despite the fact that these orchards have not received regular DDT treatments for about 13 years, their soil is still contaminated with DDT and some of its metabolites. It was estimated that, in one particular orchard, approximately half the DDT applied from 1947 to 1960 continues to persist in the soil a

DDT, DDD, and DDE. According to this survey, within one heavily residued orchard, DDT and analogues were detected in all animal species captured and in a small sample of growing vegetation. Samples of stream water and bottom mud contained minute quantities of DDT residues, implying that movement of the insecticide from the soil into the waterway was not excessive. It was reported that "the low levels present in the stream environment coupled with the high soil load suggest that the DDT residues are present in a relatively static situation, moved about primarily by animal life and redistributed principally as part of a mobile food chain." Another survey which included orchards across the United States revealed a DDT soil load ranging from 0.07 to 245.4 ppm (Stevens *et al.*, 1970).

Persistent insecticides in the soil may create a variety of hazards for the living organisms. According to Edwards (1973*a,b*): "(1) they may be directly toxic to elements of the soil fauna or microflora; (2) they may affect these organisms genetically so that resistant populations develop; (3) they may have a range of sublethal effects on activity, behavior, reproduction, or metabolism of soil organisms; and (4) they may be taken up into the body tissue of the soil fauna and flora."

Pesticides can enter the atmosphere by a variety of routes, particularly from spray drift or volatilization from soil or water. Akesson and Yates (1964) presented a summary and critique of pertinent pesticide drift literature. They showed the importance of particle size to the drift potential of pesticides. Akesson *et al.* (1970) reported that pesticide movement as spray drift was much more widespread and that contamination of a whole airshed could result when large-scale treatments are made. Many workers have shown that only a portion of pesticides sprayed onto crops reach their target; the rest fall to the ground or are taken up into the atmosphere by air currents or turbulence. Spencer *et al.* (1973) reviewed the subject of pesticide volatilization. They indicated that volatilization is obviously a major pathway for loss of applied pesticides from plant, water, and soil surfaces. The vapor pressure of the pesticide is the major factor influencing volatilization. Other routes of entry include wind erosion and agricultural burning. Widespread transport of pesticides by air and rain was reported by Cohen and Pinkerton (1966). Large amounts of DDT were found in Texas, 9 miles distant from and 1600 ft higher than the place of application (Lasher and Applegate, 1966), and pesticides found in the Sierra Nevada at altitudes of up to 12,000 ft were supposed to be windborne drifts of aerosol DDT released from the Central Valley of California (Cory *et al.*, 1970). Risebrough *et al.* (1968) regarded the European–African land areas as the source of chlorinated hydrocarbon insecticides found in airborne dust at Barbados. The insecticides (adsorbed on the dust) were carried some 3727 miles by the transatlantic movement of the northeast trade winds. Those workers claimed that the amounts of pesticides contributed to the tropical Atlantic Ocean by the trade winds are comparable to those carried to the ocean by the major river systems emptying into that portion of the ocean. The question of how much global transport of pesticides occurs is still unresolved because the evidence available is rather circumstantial. In a review entitled "Pesticides in the Atmosphere," Wheatley (1973) indicates that "the difficulty of predicting pesticide redistribution and behavior in the atmosphere lies primarily in our limited understanding of the complex interactions of the processes involved." Current evidence shows that hazards from pollution of the atmosphere by pesticides are still small in comparison with those of other major air pollutants.

Rivers, streams, lakes, ponds, oceans, and bottom mud are major reservoirs for residues of persistent pesticides. There are many routes in which pesticides can reach the aquatic environment (Edwards, 1973*a,b*; Nicholson, 1970; Nicholson and Hill, 1970; U.S. Department of Health, Education, and Welfare, 1969; Foy and Bingham, 1969; Westlake and Gunther, 1966; West, 1966). These routes are (a) surface runoff and sediment transport from treated soil, (b) industrial wastes discharged into factory effluent, (c) direct application as aerial sprays or granules to control water-inhabiting pests, (d) spray drift from normal agricultural operations, (e) atmospheric transport, (f) municipal wastes discharged into sewage effluent, (g) agricultural wastes and (h) accidents and spills.

Runoff is generally considered to be the major movement into the water environment. Edwards *et al.* (1970) reported that runoff water and surface runoff are more important routes than leaching in transport of dieldrin from soil into pond. Nicholson and coworkers have conducted intensive studies on the routes of pesticide movement into the water environment, and in their view (Nicholson and Hill, 1970) runoff is probably the single most widespread and significant source of low-level contamination (less than 1 ppb) of surface water by pesticides. During runoff, the pesticide may be adsorbed on eroding soil particles suspended in the runoff water. Chlorinated hydrocarbon pesticides, because of their low solubilities in water, are probably transported with the soil particles in the adsorbed state rather than in solution. Heavy rainfall immediately after application of pesticides will have a higher potential for pesticide transport into the water environment. For dieldrin-incorporated soils, losses were appreciable when erosion occurred and reached 2.2% of the amount applied (Caro and Taylor, 1971).

Industrial waste constitutes perhaps the second most significant source of pesticides in water. The wastes from manufacturing and formulating plants, unless very closely controlled, contain pesticides. Additionally, the effluent from plants that use pesticides in their manufacturing processes may contain various amounts of pesticides. According to a report issued by the Alabama Water Improvement Commission (1961), a parathion and methyl parathion manufacturing plant dumped its effluent untreated into a creek when its treatment plant failed in 1961. Fish, turtles, and snakes died along 28 miles of the creek. Traces of parathion residues were recovered 90 miles down the Coosa River, into which the creek emptied.

Burnett (1971) studied the distribution of DDT residues in *Emerita analoga* (Stimpson) along coastal California and concluded that animals near the Los Angeles County sewer outfall contain over 45 times as much DDT as animals near major agricultural drainage areas. The probable source of this high concentration of DDT in the sewer outfall was thought to be a plant that manufactures DDT. Burnett pointed out that this observation suggested that historically the buildup of residues in California coastal marine organisms could be attributed, to a significant degree, to industrial waste discharge rather than merely to extensive agricultural usage.

Pesticides may also enter water along with municipal wastes such as sewage effluents. A review of pesticide monitoring programs in California prepared by an Ad Hoc Working Group of the Pesticide Advisory Committee to the State Department of Agriculture (California Water Resources Control Board, 1971) reported that substantial quantities of pesticides, mainly chlorinated hydrocarbons, have been discharged to surface waters through municipal and industrial waste discharges. The quantities of waste water discharged are so large that even low concentrations of pesticides in the water result in a large emission. The group pointed out that the discharge of the Los Angeles County Sanitation Districts contained high concentrations of pesticides at least from December 1969 through May 1970. Los Angeles County Sanitation emissions of total identified chlorinated hydrocarbons during that period exceeded all of the other known discharges of pesticides to the ocean by a wide margin. The group also pointed out that if such emissions from the Los Angeles County Sanitation Districts' outfalls have been occurring for a number of years, this one source may overshadow all other sources of DDT in Southern California marine waters.

The pesticide disposal problem has been reviewed by Stojanovic *et al.* (1972) and Kennedy *et al.* (1969). This problem can be classified into three general categories—disposal by land burial, disposal by chemical and thermal methods, and recycling of waste and containers.

The burial of pesticides presents problems beyond the contamination of water. Biodegradation in soil may result in suppression of bacterial population, thus endangering important processes such as nitrification, nitrogen fixation, and sulfur transformation, among others. It has been shown that incineration is superior to chemical methods as a means of pesticide disposal. However, the process in itself is not entirely satisfactory, and incineration without the entrapment of pesticides in the resulting gases represents an air pollution hazard. If disposal of pesticide residues is by burning, it is possible that further chemical treatment will be needed. The disposal of pesticide containers presents a particular problem. These containers usually retain substantial amounts of residue (Hsieh *et al.*, 1972). If the container, such as a metal drum, can be safely recycled, this problem is much lessened. However, if these containers should be disposed of by dumping, the buildup of toxic material could be very significant, and the material could subsequently be transported to other areas by water movement.

Disposal of pesticide wastes has become a major problem because these wastes (surplus, discharge, and used containers) may contaminate surface and ground waters. Mitchell *et al.* (1970) did an exploratory study of pesticide migration from waste disposal pits in 1967 and found concentrations of DDT, toxaphene, and methyl parathion of up to 7.40, 129.50, and 99.20 ppm, respectively. Concentrations were highest in the vicinity of the pit bottom and at the water table (about 9 or 10 ft deep).

The movement of pesticides from waste disposal sites is a potential hazard both to the environment and to human health. Old and rusting containers represent some of the worst hazards. Small quantities of the chemical can drip out of rusty containers, imperiling the water supply and human health. The problem of disposal of pesticide wastes is further complicated by the lack of data on pesticide persistence beyond 6 inches of soil depth under different conditions of temperature, pressure, oxygen tension, and other environmental conditions. The California Department of Agriculture requires that all pesticide containers be discarded at class I dump sites (capable of handling hazardous material by being so situated that no liquid drainage can later reach groundwaters). Used pesticide containers, mostly 1- or 5-gal cans, have piled up in many locations, largely because of the difficulty of finding an approved location for dumping them. Dump sites approved for disposal of agricultural chemicals are few and far between. A special disposal campaign, conducted jointly in March 1971 by the Department of Agriculture and Public Health and the Water Resources Control Board, attempted to alleviate the problem temporarily by identifying existing dump sites that might be temporarily approved for use in the campaign. A 1970 survey of 75 counties in Tennessee indicated that 3332 empty pesticide containers were discarded as trash (Thorton and Walker, 1970).

Many organic pesticides are applied to water to control aquatic weeds, rough fish, and aquatic insect pests. Most of these applications are made for a specific purpose and are managed by professionals, and the amount of pesticides used is usually closely controlled. However, instances of damage to the aquatic environment from direct application of pesticides to water have been reported. For example, dieldrin was applied at the rate of 1 lb/acre in 1955 to 2000 acres of salt marsh in St. Lucie county, Florida, during a sand fly eradication program. An estimated 1,117,000 fish, representing some 30 species, were killed. Reproduction was not observed during the following 4 weeks (Nicholson, 1959). Toxaphene was first used for control of rough fish in lakes in the early 1950s and has sometimes caused problems. Although a toxaphene-treated lake may generally be restocked within a month to a year later, occasionally a lake may remain toxic to restocked fish for 5 years, according to Kallman *et al.* (1962) and Terriere *et al.* (1966).

2.2. Current Problems

Pesticides are introduced into the environment via various routes of transport. Figure 1 attempts to portray the various pathways by which pesticides cycle through the environment (Li and Fleck, 1972). The complexity of

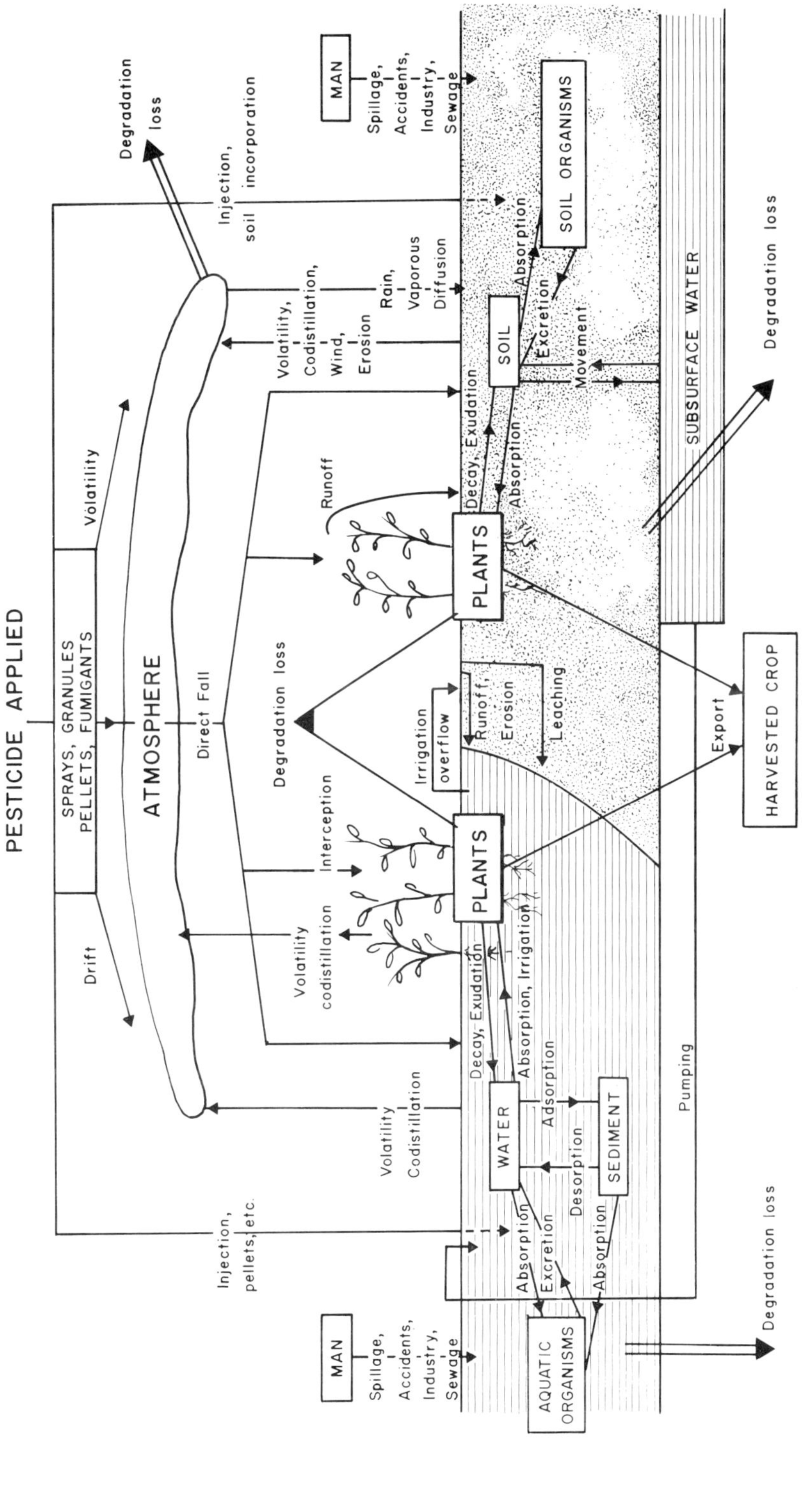

Fig. 1. Pesticide cycling in the environment.

pathways shown in this figure emphasizes the need to approach the problem at its origin—the initial pesticide application. It further demonstrates that the task of performing a material balance on pesticides is indeed not a simple matter.

The problem of pesticide contamination is a unique environmental problem in that pesticides are deliberately introduced into the environment for beneficial purposes and/or monetary gains. Modern agriculture requires large inputs of energy to maintain its stable state. Pesticides, together with fertilizers, fuel for tractors, and electric power for irrigation, represent the major energy inputs required for this stability. The energy applied as pesticides is cheaper and far more efficient that that of a man with a hoe. Unfortunately, in our drive to increase agricultural production, little attention has been paid to the broader aspects of the agriecosystem and the sound principles of crop rotation and other cultural practices which benefit many pest control efforts. We have now come to realize that future trends and developments must take into consideration their effects on the quality of the total environment. Clearly, from now on all pest-management programs employing pesticides will be assessed by the public on the basis of three important criteria: economics, public health, and environmental pollution. All three are important, and a serious deficiency in any one of these could easily prevent a pest-control technique from being used.

To gain a better perspective about the potential for pesticide contamination in the environment, one must appreciate the amount of pesticides used now and in the past. Table I shows pesticide sales in the United States for the period 1962–1972. The rate of increase in total pesticide use has averaged more than 7% a year. According to U.S. Department of Agriculture (USDA) reports, 85% of all farmers used pesticides in 1966, and 90% of the amount was applied to crops. The remaining 10% was shared about equally between

Table I. Pesticides Used in the United States (Sales in Million Pounds)[a]

Year	Fungicides	Herbicides	Insecticides	Total
1962	97	95	442	634
1963	93	123	435	651
1964	95	152	445	692
1965	106	183	475	764
1966	118	222	482	822
1967	120	288	489	897
1968	130	319	511	960
1969	124	311	493	928
1970	129	308	444	881
1971	132	316	497	945
1972	129	354	540	1023
Total	1273	2671	5253	9197

[a]Source: U.S. Tariff Commission, *Synethetic Organic Chemicals*, Government Printing Office, Washington, D.C.

Table II. Pesticides Used in California (Pounds)[a]

Pesticide	1970	1971	1972	1973
Toxaphene	2,659,274	2,779,189	1,783,722	2,903,885
Parathion	1,172,492	1,069,107	1,097,091	1,951,982
DDT	1,164,700	111,058	80,800	1,296
Malathion	1,100,241	1,461,752	922,034	1,021,715
Carbaryl	936,059	1,077,712	845,586	1,335,397
Methyl parathion	930,242	1,011,368	768,603	1,204,023
Chlordane	661,318	445,099	592,447	1,030,001
Guthion	467,989	523,998	436,424	384,746

[a]Source: California Department of Food and Agriculture, *Pesticide Use Report*, Sacramento, Calif.

livestock application and various uses around the farm. In terms of crop application, USDA estimated that, of 250 million acres of cultivated land in 1966, herbicides were applied to 27% insecticides to 12%, and fungicides to 2.6%. Table II lists the application of some of the common insecticides reported to the California Department of Food and Agriculture from 1970 through 1973. These figures represent approximately 70% of all the actual pesticide use in California. In 1973, California's use of pesticides totaled 183,656,527 lb, representing an increase of approximately 43 million pounds from 1970, the year preceding the restricted use of several of the pesticides.

Even though pesticide chemicals are used extensively today, the loss to agriculture caused by pests is enormous. In agriculture, even after effective use of pesticides, pests still cause an annual loss of about $4 billion in the United States and $21 billion on a worldwide scale (Edwards, 1973*a*). These losses would have been astronomical if use of persistent chemicals had not been permitted.

Pesticide residues in foods are often a matter of major concern to the public. However, in a study conducted by Corneliussen (1972) it was found that the pesticide residue levels detected in ready-to-eat foods remained at relatively low levels during the sixth year of the Total Diet Study carried out by the Food and Drug Administration (FDA) in its present form. The residues of most pesticide chemicals present in a high-consumption well-balanced diet have been below and in most cases substantially below the limits established in 1967 for acceptable daily intakes by the United Nations Committees (Food and Agricultural Organization and World Health Organization). The daily intake of total chlorinated organics dropped 22% from the previous reporting period (June 1968 through April 1969) (Corneliussen, 1970), while daily intake of organic phosphorus residues rose 14% (Duggan and Corneliussen, 1972). It appears that the small amounts of pesticide residues consumed by man in his daily diet do not constitute an immediate threat to his health. However, extensive studies have revealed that certain pesticides, particularly chlorinated hydrocarbons, do accumulate in selected body tissues. These studies have been focused primarily on DDT and related chemicals because they are regarded as persistent and because they accumulate in the food chain. The data have been used to establish the average amount of DDT and DDE present in the adipose tissue of the average U.S. citizen.

It is extremely difficult to assess the long-term effects of low-level pesticides which accumulate in the tissues of man. Data presently available do not suggest that man is being harmed by the small quantities of pesticides in his tissues.

Of particular concern is the fact that chlorinated hydrocarbon pesticides are present in the aquatic environment; they concentrate in sediment and aquatic organisms. DDT in Lake Michigan provides a remarkable example of the extent to which persistent substances may concentrate in food chains. DDT is present in the water of Lake Michigan at approximately 2 ppm, yet bottoms contain an average of 0.014 ppm, amphipods 0.41 ppm, fish such as coho and lake trout, 3–6 ppm, and herring gulls, at the top of the food chain, as much as 99 ppm. The long-term effects of this process (biomagnification) on the biota of the lake are difficult to assess, but they are likely to result in disturbing changes in species composition (Harrison *et al.*, 1970).

Estuarine waters, especially those containing oysters, shrimp, and other seafood particularly vulnerable to pesticides, are of great concern because of agricultural drainage and dispersions. In nine California estuaries, the shellfish are exposed to pesticide runoff containing as much as 11,000 ppb of DDT, dieldrin, and endrin, while those geographically isolated from agriculture generally contain less than 100 ppb.

The recent shift in pesticide usage from the chlorinated hydrocarbons to the so-called substitutes or nonpersistent pesticides is an encouraging trend. However, the less persistent carbamates and organophosphates, which are gaining in popularity, are also more toxic. Although there is a considerable knowledge about the residues of these other pesticides and about their metabolism in target organisms, we know little about their overall effects on our environment.

The term "nonpersistent" is misleading and tends to provide a false sense of security. Simply because under a certain set of conditions a chemical is unstable does not necessarily mean that it will always be unstable. Until recently, it was thought that parathion, when applied in the field, degraded within a few days to nontoxic substances. Studies conducted by Kilgore *et al.* (1972) indicate that parathion is quite stable and does not always degrade to nontoxic substances as previously believed. After application, about 15% of the parathion which reaches the plant is absorbed and translocated to most, if not all, internal components of the plants. The larger portion (85%) is volatilized as parathion, and no metabolic or degradation products have been detected. In addition, parathion residues have been detected in grape vines 15 months after application.

The increasing use of parathion and other organophosphorus pesticides replacing DDT and other chlorinated hydrocarbons has presented a new

Table III. Parathion and Methyl Parathion Used in California

Year	Total pounds parathion used		Total pounds methyl parathion used	
1970	1,172,492	(1,775,656)[a]	930,242	(1,445,427)
1971	1,069,107	(1,496,312)	1,011,368	(963,097)
1972	1,097,091	(1,271,111)	768,603	(975,917)
1973	1,951,983	(1,373,878)	1,204,023	(1,088,038)

[a]Figures in parentheses indicate acres treated.

dimension of occupational hazards to the agricultural industry. Table III shows the usage of parathion and methyl parathion in California for 1970–1973. The California Department of Health has been summarizing the acute effects of pesticides and other agricultural chemicals on California workers as reported by physicians under the State's Workman's Compensation System since 1950. From these data, the Department indicates that "occupational disease caused by pesticides and other agricultural chemicals is one of the most important occupational health problems in the State."

It is obvious from Table III that the usage of parathion and methyl parathion in California has been enormous. It is therefore not surprising that an increasing frequency of parathion poisoning being reported in California and elsewhere in the nation. According to a report entitled "Occupational Disease in California Attributed to Pesticides and Other Agricultural Chemicals, 1970" issued by the California Department of Public Health in May 1972, "organophosphorus pesticides were implicated in 22 percent of the 1,493 reports of occupational illness attributed to agricultural chemicals in 1970 . . . . The increased number of reports reflects in part a series of pesticide poisoning outbreaks in 1970 among citrus pickers and other farm laborers in the San Joaquin Valley. Residues of several organophosphorus compounds on foliage were responsible."

Determining the effect of pesticides on environmental quality is an extremely complex problem. Considerable information is presently available on the metabolism, degradation, and mechanism of action of pesticides in target organisms, but unfortunately little is known about the fate and distribution of pesticides in the total environment. When one considers that these pesticides represent some 45,000 different formulations of over 900 compounds, problems of monitoring, identifying, and generally studying their fate in the environment assume tremendous proportions. When they are used with reasonable caution, serious problems generally do not occur. More often, it is misuse that creates problems. Accidental spillage on the farm, industrial wastes from pesticide manufacturers, residues on agricultural by-products and wastes, drift in air, and losses from fields in runoff or adsorbed on sediments following heavy rains are all contributory.

At present, scientists are much more competent in confirming the presence of pesticides in the environment than in assessing the significance of such residues. They have not yet established the seriousness of the problem. There is little doubt that extensive monitoring programs for pesticide residues in the environment are urgently needed. In the United States, the National Pesticide Monitoring Program has completed its first 10 years of operation. Nevertheless, the monitoring of residues in various parts of the physical environment is still inadequate. Major rivers in the United States are now being monitored annually for pesticide residues at large numbers of stations. The monitoring of soils for pesticides is much more sporadic and so far has been confined mainly to agricultural soils. There is very little information on the amount of pesticides that occur in the large areas of untreated land. Such information is essential for adequate assessment of the hazards caused by persistent pesticides.

At present, there is considerable information on amounts of pesticide residues in the biota, but this is scattered and very uneven. As a matter of fact, there have been many more investigations into residues in fish, birds, and birds' eggs than in other vertebrates and invertebrates. A considerable amount of data on residues in the human diet and in bodies of human beings in many parts of the world has been collected, and it is now much more feasible to assess possible hazards to man.

There is now little doubt that pesticides will be used in the foreseeable future. This is the collective judgment of the National Academy of Sciences Committee on Persistent Pesticides (1969):

> The nonchemical methods may be used more widely in the future, but speed of adoption and ultimate usefulness are restricted by cost and by lack of applicability to many problems. Often there is an unavoidable delay before control is attainable by biological means. For most purposes, nonchemical methods of control are not expected to supplant the use of chemicals in the foreseeable future.

The Mrak Commission (U.S. Department of Health, Education, and Welfare, 1969) reached a similar conclusion:

> Our society has gained tremendous benefits from the usage of pesticides to prevent disease and to increase the production of foods and fibers. Our needs to use pesticides and other pest control chemicals will continue to increase for the foreseeable future. However, recent evidence indicates our need to be concerned about the unintentional effects of pesticides on various life forms within the environment and on human health. It is becoming increasingly apparent that the benefits of using pesticides must be considered in the context of the present and potential risks of pesticide usage. Sound judgments must be made.

However, insofar as environmental quality is concerned, this considered indispensability cannot justify the use of pesticides on an infinitely increasing scale or irresponsible use. On the contrary, it seems quite clear that changes must be made in pest control practices and, in some cases, in the nature of pesticide chemicals themselves if the agricultural industry is to maintain public confidence in its practices and to observe an appropriate and responsible regard for public health and environmental quality.

2.3. Laws and Regulations

On October 21, 1972, the President signed into law H.R. 10729, the Federal Environmental Pesticide Control Act (FEPCA) of 1972 (Public Law 92-516), amending the Federal Insecticide, Fungicide, and Rodenticide Act (FIFRA) of 1947. Some sections of the new act became effective immediately, while others have deadlines for later enforcement, pending the establishment of regulations and development of federal standards to guide states in implementing the legislation. All of the provisions of the new act must be in effect by Ocober 1976.

The new act extends federal registration and regulation to all pesticides, including those distributed or used within a single state. It requires the proper application of pesticides to ensure greater protection of man and the environment. Penalties may be levied against a purchaser who misuses pesticides.

Perhaps most important of all, the Federal Environmental Pesticide Control Act, by helping the farmer and others to apply pesticide products properly and safely, can help ensure the continued use of these products with their wide range of benefits for the American people.

The FIFRA was administered by the U.S. Department of Agriculture (USDA) until the authority was transferred to the Environmental Protection Agency (EPA) when it was established in December 1970. FIFRA requires that all pesticide products shipped in interstate commerce be registered with EPA. Before registration may be granted for a pesticide product, the manufacturer is required to provide scientific evidence that the product, when used as directed, will (a) effectively control the pest(s) listed on the label, (b) not injure humans, crops, livestock, and wildlife or damage the total environment, and (c) not result in illegal residues in food or feed.

EPA has the authority to cancel a pesticide registration if it can be determined that the directed use of the pesticide poses a serious hazard to man or the environment. A registrant can appeal an EPA cancelation notice through a complicated process that includes public hearings and review by a scientific advisory group. Suspension of a pesticide registration, unlike cancelation, stops interstate shipments immediately, but can be initiated only when the product presents an "imminent hazard."

The Pesticide Amendment to the Federal Food, Drug, and Cosmetic Act is a law closely related to the FIFRA and its 1972 amendments. It provides protection to consumers against harmful residues in food. It requires that where necessary to protect the public health a tolerance or legal limit be established for any residues that might remain in or on a harvested food or feed crop as a result of the application of a chemical for pest control. The authority to establish tolerance levels was transferred from the Food and Drug Administration of the U.S. Department of Health, Education, and Welfare to EPA in December 1970. Enforcement of tolerance levels established by EPA is the responsibility of two other federal agencies. The Food and Drug Administration (FDA) continually inspects agricultural commodities to ensure that residues in foods offered for sale in the marketplace do not exceed the established tolerance levels. In addition, the U.S. Department of Agriculture (USDA) inspects meat and poultry for such residues. Any food found to have residues in excess of the legal tolerance level is subject to seizure and destruction. Thus EPA in cooperation with FDA and USDA acts to protect the public welfare through the establishment and strict enforcement of tolerances.

Major provisions of the 1972 act are as follows:

- The use of any registered pesticide in a manner inconsistent with labeling instructions is prohibited, effective immediately. Label directions and precautions on all pesticide products registered for sale are designed to prevent injury to man and to the environment.
- Pesticides must be classified for "general" use or "restricted" use within 2 years after October 1974. Those placed in the restricted category may be used only by, or under the supervision of, certified applicators or under such other conditions as the EPA administrator may require to protect man and the environment.
- The states will certify pesticide applicators for use of restricted pesticides; the law allows 4 years for development of certification programs. Federal standards for certification must be set forth by EPA by October 1973, and the states must submit their certification programs based on these standards to the agency by 1975. Such plans are to be approved by the EPA administrator within 1 year of submission.
- Farmers and other private applicators can be fined up to $1000 or given 30 days in prison or both upon criminal conviction for a knowing violation of the law. The same users also are liable to civil penalties of up to $1000 on their second and subsequent offenses.
- Any registrant, commercial applicator, wholesaler, dealer, retailer, or other distributor who knowingly violates the law is liable to a $5000 civil or $25,000 criminal fine or 1 year in prison or both. The sharply increased criminal penalties are now in effect.
- The EPA administrator may issue a "stop sale, use, and removal" order when it appears that a pesticide violates the law or its registration has been suspended or finally canceled. Pesticides may also be seized if they violate the law. These provisions were effective immediately.
- Pesticide manufacturing plants must be registered with the federal government (EPA) 1 year after enactment of the new law. Information on the types and amounts of pesticides produced, distributed, and sold must be submitted upon registration of the firm and annually thereafter. An EPA agent may enter and inspect such an establishment and take samples.
- Federal assistance to the states to enforce provisions of the law and help develop and administer applicator certification programs is authorized.

- The EPA administrator is required to develop procedures and regulations for storage or disposal of pesticide containers. He must accept at convenient locations for disposal a pesticide registration which is suspended, then canceled.
- The agency may issue experimental use permits, conduct research on pesticides and alternatives, and monitor pesticide use and presence in the environment.

Other provisions of the 1972 law include:

- Indemnification is authorized for certain owners of pesticide registrations which are suspended, then canceled. This provision is effective immediately.
- Applicants for registration of a pesticide product retain proprietary rights to their test data submitted in support of the application. However, the law authorizes a system whereby such data may be used by a second applicant upon payment of reasonable compensation to the original applicant.
- States can be authorized to issue limited registrations for pesticides intended for special local needs.
- States may impose more stringent regulations on pesticides than the federal government, except for packaging and labeling.
- The views of the Secretary of Agriculture are required to be solicited before the publishing of regulations under the act.
- Federal registration of *all* pesticide products, whether they are shipped in interstate or intrastate commerce, is required under the new act.

Within the last few years, EPA has taken several control actions against a number of pesticides. In 1971, EPA initiated cancelation proceedings under FIFRA against DDT, mirex, aldrin, and dieldrin. After extensive hearings, the agency announced cancelation of nearly all remaining uses of DDT in June 1972, based on potential future hazard to man and his environment. The agency has also limited the use of mirex against the imported fire ant in the southeastern United States, primarily because of hazard to aquatic life.

In June 1972, cancelation of most major registered uses of aldrin and dieldrin on corn and fruit and for seed treatments was extended until the conclusion of a public hearing and announcement of a final decision by the agency on possible use restrictions. On August 2, 1974, EPA administrator Russell E. Train issued suspension notices for aldrin and dieldrin, citing evidence of "imminent hazard" from five areas:

a. Medical data indicate that two strains of mice and rats fed a diet containing 0.1 ppm dieldrin (a metabolite of aldrin) for several weeks showed increasing signs of liver tumor, "many malignant."

b. A 1973 Food and Drug Administration market basket survey found "measurable amounts" of dieldrin in composite samples of foods, such as dairy products.

c. A 1971 EPA sampling of human fat tissues revealed that 99.5% of them showed detectable residues of dieldrin.

d. Based on National Cancer Institute methods for estimating human cancer risk, the present average human daily dietary intake of dieldrin subjects the human population to an unacceptably high cancer risk.

e. Children, particularly infants from birth to 1 year of age, because of their high dairy product diet, consume considerably more dieldrin on a body weight basis than any other age segment of our population.

Mr. Train proposed prohibiting the production and further use of aldrin and dieldrin for agricultural purposes, primarily for insect control in corn and citrus and for seed treatments, after August 2, 1974. However, the suspension order placed no restriction on the sales of existing supplies. The manufacturer, Shell Chemical Company, immediately requested an "expedited suspension hearing." EPA administrative law judge, Herbert L. Perlman, who had presided over the hearing, concluded that the two chemicals, which have been linked to cancer, constitute an "imminent hazard" to humans and that their federal registrations should be suspended pending the outcome of separate cancelation hearings still in progress. He further indicated that the suspension of the pesticides will cause no "significant macroeconomic or microeconomic consequences resulting from the suspension of the pesticides for use on corn in 1975." On October 1, 1974, EPA administrator Train ordered an immediate suspension of further production of aldrin and dieldrin, because of evidence they may cause cancer. Mr. Train's decision allows the continued sale and use of existing stocks of the pesticides. But it prohibits further production starting immediately, until proceedings began in 1971 are completed and a decision is reached whether or not to ban the sale and use of aldrin and dieldrin permanently.

In 1971 and 1972, EPA issued suspension and cancelation notices for mercury pesticides. Used heavily by industry, mercury builds up in the food chain and persists in the environment.

3. Biomagnification of Insecticide Residues

The problem of biological concentration and magnification of insecticides in plants, animals, and the environment has been reviewed by Hunt (1966), Macek (1970), Moriarty (1972), and Edwards (1973*a*,*b*). In general, magnification is more pronounced in animals than in plants, water, soil, and other components of the environment.

3.1. Movement Through Food Chains

Small amounts of chemicals absorbed by plankton and insects are transferred in increasing concentrations to fish, birds, animals, and eventually man, and there is evidence that pesticide residues act adversely on the reproduction and behavior of certain birds and may threaten the survival of some wildlife species. In certain instances, pesticide residues accumulated through food chains have been directly implicated in the death of certain species. Thus the loss of fish-eating waterbirds at Tule Lake and Clear Lake, California, was attributed to pesticides which had traveled through biological networks and accumulated in the tissues of these birds (Hunt, 1966).

According to Macek (1970), "there are two basic processes intrinsic to the phenomenon of biological magnification of pesticide residues in the food chain: biomagnification of the pesticide at some trophic level and biotransfer from one trophic level to the next. In order to be biomagnified, a compound must be persistent in the physical environment, available to the organisms, and persistent once assimilated into the biological system." Among the major classes of pesticides, only the organochlorine insecticides can fulfill these requirements. It is now evident from laboratory and field studies that the biological magnification of DDT and its metabolites in the food chain is occurring in nature and is a major source of contamination for higher trophic levels in terrestrial as well as aquatic communities. Furthermore, there are reports suggesting that other organochlorine insecticides such as dieldrin (Chadwick and Brocksen, 1969) may also undergo biological magnification in the food chain.

However, Moriarty (1972) questioned the belief that organochlorine insecticides concentrate along the food chains. He felt that "the confusion comes possibly from the unappreciated difference between persistence in the physical environment and persistence within organisms. The amounts of such chemicals retained by different organisms probably depend more on different rates of metabolism and excretion."

3.2. Accumulation and Magnification in Aquatic Animals

The distribution of pesticides in water influences the pathway of biological uptake. The quantity accumulated by each biological entity is dependent on the chemical properties of the pesticide, the physiology and behavior of the organism, and the seasonal variation of the quantities of pesticides available within a given aquatic habitat.

One of the first studies of the buildup of organochlorine insecticides in an aquatic ecosystem was conducted in California in 1958 by Hunt and Bischoff (1960). It uncovered a striking example of the extent of pesticide accumulation and concentration, as well as the routes taken by chemicals through biological food chains. DDD insecticide was applied several times to Clear Lake to control gnats. The level of insecticide in the lake immediately following the last application was calculated to be 0.02 ppm or 1 part DDD to 50 million parts of water. "Residue levels of DDD in samples taken from the lake 13 months after treatment were 10 ppm in plankton, 903 ppm in fat of plankton-eating fish, 2690 ppm in fat of carnivorous fish, and 2134 ppm in fat of fish-eating birds. These residues represent a 500-fold increase in levels in plankton and a

100,000-fold increase in fish-eating birds over levels occurring in lake water after treatment."

Reinert (1967) suggested that perhaps pesticide accumulation in an aquatic community reflects a greater capacity of organisms at higher trophic levels to concentrate the chemical directly from the water than organisms at lower trophic levels, rather than reflecting biotransfer through the food chain.

Laboratory studies with ^{14}C-labeled DDT and Atlantic salmon in Canada have demonstrated that DDT can be taken from water quickly by fish and can be distributed to the major organs of the fish in about 5 min (Premdas and Anderson, 1963). High concentrations of DDT were detected in hearts, kidneys, gonads, liver, eggs, and swim bladder. Holden (1962) in England conducted a similar study using ^{14}C-labeled DDT and showed that DDT in a very dilute suspension in water is absorbed by fish mainly through the gills. The concentration in gill tissue was several hundred times higher than in water being passed over the gills, suggesting a rapid rate of accumulation from the water. Butler (1965) demonstrated that DDT may be stored by the oyster during a 40-day exposure at levels 70,000 times greater than the 0.1 ppb concentration in the surrounding waters. Chadwick and Brocksen (1969) made a careful study of the uptake of dieldrin into fish (*Cottus perplexus*), tubificid worms (*Tubifex* sp.), and midge larvae (Chironomidae). They showed that the accumulation of dieldrin by the fish was dependent on the concentration in the water and demonstrated that although fish could accumulate dieldrin either from their food or from the water directly the two routes were not additive. They further concluded that direct entry into the fish from the water was much more important than that through the food. In their experiments, a maximum of 16% of the dieldrin accumulated by the fish could have come from the contaminated food. Based on these results, they concluded that dieldrin in water is immediately and directly available for accumulation by fish, whereas accumulation via ingestion of food is much slower, possibly because of the limited amounts of food the fish can consume. They suggested that when pesticides are present in water for a very short time adsorption onto algae and food organisms might play more important contributory role in determining the residues found in fish. The amounts of organochlorine insecticides in fish tissues are remarkably constant, but there seems to be less residues in the tissue of marine fish than in freshwater fish; fish that migrate from fresh water to sea water tend to have a greater residue burden than other marine fish.

Pesticides in the aquatic environment constitute both direct and indirect hazards to man as well as to aquatic animals. The organisms at the end of the food chain may be exposed to a concentration of pesticide far higher than that originally found in the environment. This sort of a concentration mechanism is of much greater importance in aquatic than in terrestrial organisms and indeed may result in concentrations of pesticides in our own food supply.

4. Factors Involved in Insecticide Disappearance

The term *persistence* when applied to a pesticide means that the pesticide does not readily disappear from the environment. This does not mean, however, that the pesticide does not disappear at all. For example, DDT is considered to be persistent in the environment, yet it slowly disappears as a result of a number of microbiological and chemical processes. *Nonpersistence* is a term used to indicate that a pesticide disappears rapidly from the environment and that no toxic residues remain.

A nonpersistent pesticide may be unstable in the environment for a variety of reasons; it may be readily decomposed by soil microorganisms, it may be chemically unstable, or it may disappear as a result of the interplay of several processes. The insecticide parathion is considered to be relatively nonpersistent in the environment primarily because it is chemically unstable, especially in aqueous solutions. Microorganisms do not contribute substantially to parathion's nonpersistence even though it has been reported that some microorganisms can, under the appropriate conditions, degrade parathion to nontoxic substances. On the other hand, soil microorganisms may be almost entirely responsible for the disappearance from the environment of the substituted urea herbicides.

4.1. Biotransformations

Living organisms in the biosphere are capable of obtaining energy through the degradation of organic molecules. In fact, the basis of all life depends on the orderly degradation and synthesis of organic molecules for the production of energy and vitally needed cellular components. Microorganisms are important in this process because they are capable of degrading organic substances that accumulate in the biosphere. In doing so, they not only provide energy and cellular components for themselves but also release bound molecules for use by other organisms. According to Dagley (1972), "microbes are able to degrade chemical structures of many types, and it is probably true to say that every organic molecule that is synthesized by living matter may also be degraded, in turn, by microbes." It is not known if microorganisms can decompose all man-made organic molecules, but it seems unlikely that they would be able to do so without a long induction period. Dagley (1972) believes that a man-made compound will be biodegradable only when the relevant microbes are able to use the enzymatic apparatus acquired during evolution and designed to exploit the diverse sources of energy found in nature. This means that if man-made organic molecules are not similar to natural organic substances, then they would not be readily degraded in the environment. On the other hand, one could argue that because of the diversity of nature it seems unlikely that very many organic substances can be synthesized by man which are radically different from those found in nature.

It is apparent that the biotransformations of organic molecules in the environment are critical for sustaining life and that the accumulation of inert, man-made organic molecules is a justifiable cause for concern. Fortunately, most pesticides that have been studied do disappear in time from the environment, but it is not entirely clear whether microorganisms are always involved. Processes such as chemical decomposition and photodecomposition also enhance the disappearance of pesticides from the environment.

It is now clear that many organisms associated with the microflora can effect the degradation of pesticides and that soil microorganisms, particularly various fungi and bacteria, are probably the most important organisms in this group. Thus a bacterium classified as an *Arthrobacter* species isolated from soil can dehalogenate the herbicide dalapon, and the fungus *Aspergillus fumigatus* can degrade the herbicide simazine.

4.1.1. Microbiological Degradation

It is of considerable interest that the major types of biochemical reactions associated with the microbial degradation of pesticides are basically similar to those which have been described for higher animals in earlier sections of this book (Chapters 2–5). These include dehalogenation, oxidative reactions such as epoxidation, dealkylation, and hydroxylation, reduction, ester hydrolysis, and condensate or conjugate formation. The general nature of most of these reactions has been described in some detail, although most of the data have been derived from laboratory rather than from field experiments.

Most pesticide-degrading soil microorganisms have been isolated from soil by utilizing enrichment culture techniques, and these are commonly exposed to the pesticide molecules in an aqueous soil system. In such experiments, the investigators generally follow the disappearance of the pesticides by removing aliquots from the experimental mixture at various intervals for analysis. In many instances, investigators have attempted to simulate actual field conditions in their laboratory experiments.

a. Dehalogenation. Soil microorganisms are capable of degrading several chlorinated pesticides by dehalogenation. DDT is slowly converted to DDE by *Aerobacter aerogenes* (Wedemeyer, 1967), and under anaerobic conditions a large number of microorganisms appear capable of converting DDT to DDD (Menzie, 1969). Wedemeyer (1967) has reported the further degradation of DDT to 4,4′-dichlorobenzophenone. The dehydrochlorination of lindane to γ-pentachlorocyclohexene in moist soil (Yule *et al.*, 1967) has been attributed to soil microorganisms, and the bacteria *Clostridium sporogenes* and *Bacillus coli* produce trace amounts of benzene and monochlorobenzene from lindane (Allan, 1955). Apparently some soil microorganisms can further metabolize lindane, as MacRae *et al.* (1967) have reported the production of CO_2 from lindane in submerged soils.

b. Oxidation. Microorganisms are capable of carrying out a number of different oxidative reactions, and although these have not yet been shown to be catalyzed by the same basic mechanism (as they have in the case of the microsomal enzyme system of higher animals) epoxidation, dealkylation, and hydroxylation reactions will be discussed under this general subheading.

It has long been known that microorganisms have the ability to convert chlorinated hydrocarbon insecticides with isolated double bonds to their corresponding epoxides (e.g., the conversion of aldrin to dieldrin and heptachlor to heptachlor epoxide) (Bartha *et al.*, 1967). The process of β-oxidation of the alkane chain of a variety of phenoxyalkanoic acids is similarly well established (Gutenmann *et al.*, 1964) and probably involves the formation of an intermediate hydroxy derivative. The aromatic ring of the phenoxyacetic acids is also susceptible to hydroxylation. Thus *Aspergillus niger* is able to convert phenoxybutyric and phenoxyvaleric acids to 2-hydroxyphenoxyacetic and 3-(2-hydroxyphenoxy)propionic acids, respectively, and can also effect hydroxylations without further metabolism of the hydroxy acids (Byrde and Woodcock, 1957).

Microorganisms frequently dealkylate pesticides and can remove alkyl moieties attached to carbon, nitrogen, oxygen, or sulfur. Since the aromatic portion of the phenoxyalkanoic herbicides is often found as the phenol in soils (Kearney *et al.*, 1967; Loos, 1969), considerable effort has been devoted to studies on the metabolism and fate of the alkyl group of these compounds. The actual mechanism involved in microbial dealkylation is not completely understood, although Geissbuhler (1969) theorizes that it involves the formation of a *N*-hydroxymethyl intermediate; unfortunately, cell-free preparations from soil microorganisms capable of *N*-dealkylation have not yet been reported.

c. Reduction. Pesticides containing nitro substituents are subject to reduction to amines. Lichtenstein and Schulz (1964) reported that amino parathion was formed from parathion in soils, and according to these investigators soil yeasts were responsible for the conversion. Many organisms can also reduce PCNB to pentachloroaniline (Chacko *et al.*, 1966), a reaction which according to Ko and Farley (1969) is greatly enhanced under conditions when the soil is submerged in water.

d. Hydrolysis. Many pesticides, particularly the organophosphorus insecticides, are degraded by amide or ester hydrolysis mediated by soil microorganisms. Matsumura and Boush (1966) isolated several microorganisms capable of degrading malathion through hydrolysis of the carboxylester group and by processes involving demethylation.

e. Ring Cleavage. The cleavage of aromatic rings is a common reaction when soil microorganisms degrade pesticides and is one not usually encountered in higher animals. The sequence of steps involved in ring cleavage of the phenoxyalkanoic acids herbicides has been investigated in detail (Loos, 1969; Menzie, 1969), and it has been determined that the catechols produced from 2,4-D and 4-chlorophenoxyacetic acid are cleaved to form the corresponding chloromuconic acids. Apparently, these substances are to some extent dehalogenated prior to ring cleavage.

The methylcarbamate insecticide carbaryl is probably metabolized by a mechanism similar to that proposed for naphthalene (Fernley and Evans, 1958; Kaufman, 1970). This substance is hydrolyzed to α-naphthol prior to hydroxylation and ring cleavage, and in all probability these reactions yield compounds identical to those resulting from the metabolism of naphthalene.

f. Condensation or Conjugation Formation. Although it may not be considered by some as a metabolic pathway, certain microorganisms can condense or conjugate pesticides. Burchfield and Storrs (1957) reported that dyrene was metabolized to a conjugated product containing amino and sulfhydryl groups in a reaction involving the ring chlorine, and cell suspensions of *Saccharomyces cerevisiae* can metabolize sodium dimethyldithiocarbamate to γ-(dimethylthiocarbamoylthio)-α-aminobutyric acid (Sijpesteijn *et al.*, 1962).

4.1.2. Metabolism in Higher Plants and Animals

The absorption and metabolism of pesticides in higher plants and animals clearly play an important role in determining the fate and longevity of a large number of compounds in the environment (Casida and Lykken, 1969; Menn, 1972).

As far as animals are concerned, the processes of absorption and distribution and the enzymatic mechanisms through which they are able to degrade a large number of pesticides are reasonably well understood and have been covered in detail in earlier chapters of this book (see Chapters 1–5). To avoid unnecessary repetition, these will not be discussed here and the reader is referred to these chapters for further information.

As in animals, the absorption and distribution of pesticides in plants are largely functions of the polarity of the penetrating molecule. Most nonpolar lipophilic pesticides can penetrate the cuticle of leaves, fruits, stems, roots, and seeds to some extent (Finlayson and MacCarthy, 1965) and polar compounds can be absorbed by the roots in aqueous solution. Depending on their lipophil–hydrophil balance, absorbed pesticides may or may not be translocated to other parts of the plant (systemic action) in the phloem or xylem. Once inside the plant, the pesticide is subject to metabolism by a variety of enzymes. Although there is little information available on the nature of the plant degradative enzymes, it is of interest that the intermediate and terminal metabolites of many pesticides are similar or in many cases identical to those produced by animals. The absorption and metabolism of pesticides by plants therefore may be an important factor in the disappearance of pesticides from the environment.

Because of their extreme lipophilic character, chlorinated hydrocarbon pesticides have only a limited ability to penetrate plant tissues, and although the chlorinated herbicides have received some attention (Kearney and Kaufman, 1969) little is known of their metabolic fate (Casida and Lykken, 1969). Harrison *et al.* (1967) have shown that plant foliage dehydrochlorinates DDT to DDE within 1 week of application, and after several weeks some of the DDT present undergoes reductive dechlorination to DDD, which in turn is hydroxylated and further oxidized to dichlorobenzophenone. Apparently, not all plants will metabolize DDT, as Upshall and Goodwin (1964) have reported that DDT remains largely unmetabolized in barley plants.

Plants resistant to the herbicidal action of 2,4-D are often able to metabolize it quite rapidly, and indeed this constitutes one of the reasons for the selective action of the material. Degradation may occur through metabolic attack at the side chain or at other sites on the molecule.

Systematic insecticides, particularly the organophosphorus compounds and the methylcarbamates, frequently undergo metabolic transformations in a variety of plants. Thus the insecticide imidan is rapidly degraded in certain plant tissues (Menn and McBain, 1964), phthalic acid being a metabolite in the foliage of field-sprayed apple trees. Metcalf *et al.* (1968) studied the metabolism of the carbamate carbofuran in isolated cotton leaves and intact plants and showed that it underwent hydroxylation at the benzylic carbon. The 3-hydroxy metabolite was subsequently oxidized to 3-ketocarbofuran, which in turn was decarbamylated to the phenol. The carbamate carbaryl is also readily metabolized by bean plants, only about 6% remaining 28 days after application (Menn, 1972).

4.2. Photochemical Mechanisms

It has long been known that light in the ultraviolet (UV) region of the spectrum contains energy capable of inducing chemical transformations in organic molecules. Thus, in recent years, as the usage of pesticides has increased, there has been considerable interest in research dealing with the photodecomposition of pesticides.

Before a substance can undergo a photochemical reaction, it must have the ability to absorb energy from the appropriate portion of the spectrum. When energy is absorbed from UV light, electrons in the molecule are excited and the resulting disturbance may cause a breakage of existing chemical bonds or the formation of new ones; alternatively, the result may be fluorescence, or the absorbed energy may simply be lost as heat. Crosby and Li (1969) have reported that most common herbicides exhibit their principal absorption maxima in the region between 220 and 400 nm. Absorption of light within this range is sufficient to cause a chemical change.

Even though the emission spectrum of the sun is quite broad, the amount of solar energy reaching the earth below the wavelength of 295 nm is almost negligible. UV light is considered to include wavelengths between 40 and 4000 Å (4–400 nm). Light of wavelengths greater than 450 nm representing an energy of less than 65 kcal/mole cannot usually be expected to bring about chemical changes, even when the compounds are able to absorb sufficient energy in this region (Crosby and Li, 1969).

Although it has now been demonstrated that many pesticides can undergo photodecomposition, most of the studies have been conducted under laboratory conditions because it is extremely difficult to obtain reproducible data in field experiments. As a result, a variety of UV sources have been used and it is sometimes difficult to compare the data in different published reports and to assess their relevance to the photodecomposition likely to occur under natural conditions. On the other hand, it is now well accepted that at least some pesticides disappear from the environment as a result of their photodecomposition and ultimate degradation to nontoxic products.

4.2.1. Photoreactions

Most of the photochemical reactions of pesticides are photooxidations, but it is believed that some photoreductions may also occur. Since oxygen is electronegative, and a diradical, it interacts with other molecules in the presence of light by several mechanisms (Livingston, 1961), including direct free radical reactions, energy transfers to oxidizable substrates, and the excitation of oxygen itself. When free radicals are generated from light energy, they can react with molecular oxygen to form peroxy radicals, which in turn are able to induce further reactions such as the abstraction of hydrogen from organic substrates and the generation of radical chains.

For a photoreaction to occur, it is not always necessary for the reacting substance to directly absorb radiant energy of the appropriate exciting wavelength. In some instances, an intermediate substance, often referred to as a *sensitizer*, may be involved, which absorbs light energy and subsequently transfers it to the compound in question. The sensitizer therefore serves as an intermediate in the energy transfer required to elicit the reaction.

4.2.2. Photodecomposition of Pesticides

Many pesticides have now been reported to undergo photoreactions. Rosen *et al.* (1966) and Robinson *et al.* (1966) have demonstrated that the chlorinated hydrocarbon dieldrin can be transformed into a substance called photodieldrin as a result of a photoaddition reaction. Upon more intense illumination, Henderson and Crosby (1967) found that dieldrin could also undergo a reductive dechlorination to form a pentachloro derivative. Methoxychlor, another chlorinated hydrocarbon insecticide, is rapidly photooxidized to give primarily a mixture of 4,4′-dimethoxybenzophenone, *p*-methoxybenzoic acid, and *p*-methoxyphenol (Crosby and Leitis, 1969; Li and Bradley, 1969). It has also been reported (Harrison *et al.*, 1967) that DDT undergoes a similar reaction in the environment, but apparently the conversion is slow under these conditions. From these experiments, it is easy to understand why the instability to light has affected the ability of methoxychlor to assume many of the uses of DDT. Similarly, photooxidation of the pyrethrins (Chen and Casida, 1969) and rotenone (Jones and Haller, 1931; Cahn *et al.*, 1945) has restricted the use of these low-hazard compounds to applications where losses due to direct exposure to sunlight are relatively unimportant.

Although most photodecomposition products of pesticides appear to be less toxic than their parent compounds, some are more toxic. Photodieldrin is considerably more toxic than its parent substance, dieldrin, and the organophosphorus insecticide parathion is converted to the extremely toxic oxidation product paraoxon. According to Frawley *et al.* (1958), however, the overall toxicity of parathion is eventually reduced in the environment, possibly by photochemical isomerization as well as other reactions.

Based on their findings that water hyacinth plants were more susceptible to the butyl ester of 2,4-D in shade than in sunlight, Penfound and Minyard (1947) were the first to report that light intensity might affect the biological activity of phenoxy herbicides. Experiments with potted red kidney bean plants also indicated that the ultimate survival of the treated plants was more frequent at high light intensities. Although these results have been questioned because of the lack of appropriate controls, there now seems little doubt that 2,4-D and other phenoxy alkanoic acid herbicides decompose in aqueous solution under the influence of light. Bell (1956) demonstrated that in solution 2,4-D irradiated with short-wavelength UV light was decomposed to five degradation products. In subsequent studies, Crosby and Tutass (1966) positively identified the ether-soluble photolysis products of UV-irradiated 2,4-D. Since there is very little absorption by 2,4-D above 300 nm, it seems likely that a sensitizer is involved in the decomposition of this particular herbicide. There is some evidence to suggest that 2,4-D is decomposed in artificial ponds under the influence of UV light (Aly and Faust, 1964*a*), but there is no evidence to indicate that it is photodecomposed under natural conditions (Aly and Faust, 1964*b*).

Field experiments have sometimes also been useful in demonstrating the photodecomposition of herbicides. Slade (1965) have shown that on the leaves of several crop plants paraquat undergoes photolysis in sunlight to form 1-methylpyridinium-4-carboxylate, and Kuwahara *et al.* (1965) demonstrated that PCP decomposed to several major photolysis products in rice field water after several days of exposure to sunlight. The loss of herbicidal activity of organoborates under arid conditions has also been linked to sunlight (Rake, 1961).

The major difficulty in clearly establishing the extent to which pesticides are decomposed by sunlight is the fact that the photodecomposition products are in many instances identical to the metabolic products produced by plants and microorganisms.

It is clear, however, that such reactions can and do occur and that they contribute, at least to some extent, to the decomposition of pesticides in the environment. Until adequate experiments are designed to distinguish the roles of sensitizers, metabolism, volatilization, and numerous other environmental factors, it will be difficult to extrapolate laboratory data to what occurs under actual field conditions.

4.3. Physical Mechanisms

4.3.1. Volatilization

Volatilization is not a mechanism responsible for the disappearance of a pesticide from the environment, but it can play an important role in moving a pesticide from one place to another. This may involve transfer from a place of inactivity to one more conducive to degradation or *vice versa*. Whether or not a pesticide will volatilize depends of course on its physical characteristics. The boiling point of a liquid pesticide is a measure of its volatility, and liquid pesticides differ greatly among themselves in the magnitude of their vapor pressures at a given temperature. At ordinary temperatures, many pesticides have relatively high vapor pressures and consequently low boiling points and as a result evaporate rapidly under environmental conditions.

The vapor pressures of pesticides applied directly to soils may be greatly modified by their adsorption to soil particles. Spencer and Cliath (1969, 1970*a*,*b*) have concluded that the vapor densities of solid-phase dieldrin and lindane without soil were several times higher than those predicted from published vapor pressures.

Many investigators have also reported higher rates of volatilization of insecticides (Bowman *et al.*, 1965; Harris and Lichtenstein, 1961) and herbicides (Deming, 1963; Fang *et al.*, 1961; Gray and Weierich, 1965; Kearney *et al.*, 1964; Parochetti and Warren, 1966) from wet than from dry soils. Hartley (1969) describes this phenomenon as being due to the displacement of the pesticides from their adsorption sites by water, whereas others (Acree *et al.*, 1963; Bowman *et al.*, 1965) suggest that this loss is due to a codistillation phenomenon, which implies that the evaporation of water enhances volatilization from soils and other surfaces. From the work of Spencer *et al.* (1969) and Spencer and Cliath (1970*b*), it seems probable that the increased volatility in wet soils is due to displacement of the pesticides from the soil surfaces and a resulting increase in the vapor density or partial pressure of the insecticide. This interpretation is further strengthened by the work conducted by Igue (1969), who found that the soil water content affected volatilization losses of organochlorine insecticides simply through competition for adsorption sites. It was also concluded that water loss as such does not increase the rate of volatilization by codistillation but increases the pesticide concentration in the solid solution and soil air, and therefore makes the pesticide more available for volatilization into the atmosphere.

4.3.2. Adsorption to Particulate Matter

By *adsorption*, the physical chemist means the adherence of atoms, molecules, or ions of any kind to the surface of a solid or liquid. Since pesticides are organic substances, they can be readily adsorbed to various surfaces including those of plants and soil particles. Adsorption can play an important intermediate role in the decomposition of pesticides in the environment, since the process fixes the molecules to surfaces so that they may be readily avilable for chemical or metabolic decomposition.

Pesticides may come in contact with plant materials by direct application, by adsorption or fallout from the atmosphere, by root uptake and translocation, or by partitioning from soil solution (Grover and Hance, 1969; Lichtenstein *et al.*, 1968; Nash and Beall, 1970; Tamés and Hance, 1969; Ware *et al.*, 1968). Transfer of the pesticides to microflora may occur by parasitism, saprophytism, or partitioning (Chacko *et al.*, 1966; Ko and Lockwood, 1968).

The adsorption process is highly complex and may involve ion exchange, protonation, hemisalt formation, ion-dipole or coordination interactions, hydrogen bonding, van der Waals forces, and π bonding. Because of their physical and chemical structure, which provides a relatively large surface-to-volume ratio, soil minerals are mainly responsible for the adsorption of pesticides by clay soils. Once adsorbed to soil particles, many pesticides may be decomposed by purely chemical systems, as opposed to the biochemical reactions mediated by soil microorganisms. Kearney and Helling (1969) have reviewed some of the reactions involved in the decomposition and inactivation of pesticides by soils and cite several examples in which the rate of chemical hydrolysis of pesticides in the presence of soil particles is enhanced compared to that occurring in water. Castro and Belser (1966) have demonstrated that wet soil accelerates the hydrolysis of *cis*- and *trans*-1,3-dichloropropene, and Armstrong and Chesters (1968) have shown that atrazine is hydrolyzed in moist clay; these workers concluded that the catalytic activity of some soil components is responsible for the enhanced rate of hydrolysis.

Infrared studies (Mortland and Raman, 1967) have indicated that the catalytic hydrolysis of several organophosphorus insecticides occurs in copper-containing soils, and studies by Konrad *et al.* (1967) suggest that the hydrolysis of diazinon in Poygan silty clay is a function of the pH of the soil.

Although the precise role of pesticide adsorption to soil particles remains unknown, it is probable that it represents an important factor in their environmental decomposition.

4.3.3. Solubility in Aqueous Media

Despite the aqueous insolubility of many pesticides, water plays an important role in their disappearance from the environment by making them more readily available to both chemical and biological degradation. Even with the most insoluble materials, one mechanism by which this occurs is the displacement by water of pesticides adsorbed to soil particles and their consequently greater availability for microbial degradation.

Some pesticides, particularly herbicides, are quite soluble in water, and here water serves as a carrier medium so that the chemicals are more readily translocated in various plant species.

During recent years, concern has been expressed regarding the potential movement (leaching) of insecticidal residues in water from upper soil layers into lower uncontaminated soil strata where they could affect water sources. Similarly, lake and river waters can undoubtedly be contaminated with runoff water from adjacent agricultural land containing soil particles to which insecticidal residues are adsorbed. In attempts to establish the fate of such residues, Lichtenstein *et al.* (1966) found that the insecticides parathion, guthion,

Di-Syston, and DDT vary considerably in their persistence in lake and soil waters and that numerous factors including aqueous solubility, pH of the contaminated water, and microbiological activities all play a contributory role.

4.3.4. Soil Cultivation

Soil cultivation can also affect the ultimate fate of pesticides, especially those that are water insoluble and relatively volatile. In a study to establish the importance of this factor (Lichtenstein *et al.*, 1971*a,b*), the insecticides DDT and aldrin were applied in 1960 at a rate of 4 lb/acre and were rototilled into the soil at depths of 4–5 inches. Following the application, one-half of each insecticide-treated experimental plot was disced daily through the summer season of 1960 to determine whether frequent soil cultivation would enhance loss of DDT or aldrin residues. Neither soil was disced in the following years. When the soils were analyzed 10 years later, the data showed that the nondisced portions of the plots still contained 44% of the DDT and 11% of the aldrin (as dieldrin) which had been originally applied. In contrast, the disced portions of the soil contained 31% of the applied DDT and only 5% of the applied aldrin (as dieldrin).

In another series of long-term field tests, Lichtenstein *et al.* (1971*b*) studied the effects of a dense cover crop (alfalfa) and of repeated soil cultivation on the fate of aldrin and heptachlor residues in soils. The insecticides were initially applied to the upper 4–5 inches of soil. Eleven years after the applications were made, analyses indicated that, compared to the alfalfa-covered soil, soil cultivation resulted in a 76–82% reduction of residues derived from the original applied pesticides.

4.4. Chemical Mechanisms

Purely chemical transformations of pesticides in the environment are, at best, poorly understood, and because of the close similarities in the products generated by such reactions and those arising from biological systems it is often difficult to differentiate between the two. Since it is difficult to find optimal environmental conditions under which chemical reactions might proceed, the problem has been approached by subjecting pesticides to a variety of chemical reactions (alkylations, oxidations, reductions, dehydrohalogenations) in the laboratory. Although this is not entirely satisfactory, adequate information is now available to show that pesticides can be degraded in the environment by purely chemical means.

Whether or not a pesticide will undergo chemical transformation in the environment depends on a variety of factors—e.g. pH, the presence of water, and the presence of catalysts—in addition to its innate chemical reactivity. Because of its structural and physical characteristics, soil constitutes an effective medium for the conduct of such reactions, and therefore most studies on the chemical decomposition of pesticides in the environment involve soil. A major obstacle to distinguishing chemical from biochemical reactions in soils has been the difficulty in obtaining a sterile soil system that has not undergone extensive physical and/or chemical destruction due to the sterilization process. Nevertheless, the results clearly indicate that pesticides do undergo a variety of chemical reactions in soil and most probably in other segments of the environment.

4.4.1. Hydrolysis

Most organophosphorus insecticides are susceptible to hydrolysis, and the rate of hydrolysis is determined largely by the nature of the group attached to the phosphorus atom (O'Brien, 1969). The half-lives of some organophosphorus compounds at pH 8 and 25°C are TEPP, 73 hr, paraoxon, 22,200 hr, and parathion, 203,000 hr.

There is now sufficient evidence to show that certain organophosphorus insecticides are cleaved by chemical hydrolysis in solids (Konrad *et al.*, 1967; Gunner, 1967; Tiedje and Alexander, 1967). Gunner (1967) has demonstrated that the primary products of the chemical reactions of diazinon in soil are a substituted pyrimidine and diethylphosphorothioic acid and that subsequent degradation of these two substances is caused by soil microorganisms. Tiedje and Alexander (1967) have shown that the organophosphorus compound malathion is chemically degraded in soil to malathion monoacid. Other evidence suggests that malathion, demeton, mevinphos, and probably most other organophosphorus insecticides are primarily decomposed by chemical hydrolysis in soils.

According to Konrad *et al.* (1967), various components of the soil may catalyze the hydrolysis of organophosphorus insecticides. This group of investigators followed the hydrolysis of diazinon in acid and soil solutions and concluded that, in soils, hydrolysis is catalyzed by adsorption onto some soil component. This conclusion was based on the discovery that diazinon hydrolysis in a Poygan silty clay at pH 7.2 was rapid and amounted to about 11% per day. Under laboratory conditions, diazinon hydrolysis is rapid in acid solutions at pH 2.0, while at pH 6.0 the insecticide is quite stable.

Mortland and Raman (1967) observed catalytic hydrolysis of organophosphorus insecticides in both cupric chloride solutions and copper–montmorillonite clay suspensions and found that organic soils and clay minerals which bind metallic cations more strongly than montmorillonite permit little or no hydrolysis. In this regard, Wagner-Jauregg *et al.* (1955) demonstrated that the half-life of a phosphate ester was decreased by a factor of 100 in the presence of millimolar amounts of a number of amino acid chelates and by a factor of more than 300 by histidine chelate.

The fungicides mylone and vapam undergo rapid hydrolysis in soil (Drescher and Otto, 1968; Turner and Corden, 1963), the conversion of vapam being more than 87% complete within a few hours in sandy loam.

The *s*-triazine herbicides can also be hydrolyzed to their nonphytotoxic hydroxy analogues, and evidence is now available to suggest that in soils these reactions are chemical rather than biological. Armstrong *et al.* (1967) demonstrated that atrazine hydrolysis occurred in sterilized soil at a pH of 3.9 and found that at the same pH the hydrolysis rate was tenfold greater in the presence of the soil than in its absence. Skipper *et al.* (1967) found in greenhouse studies that 20% of ^{14}C-labeled atrazine was accounted for as hydroxyatrazine after 2–4 weeks of incubation in sterile soils, and Harris (1967) found that elevated temperature greatly enhances this conversion.

The work of Armstrong *et al.* (1967) showed that the rate of atrazine degradation was greatest in soils of high organic matter and low pH and that the clay content of the soil had only a slight effect on the reaction.

From studies on soils of either high clay or high organic matter content, Harris (1967) concluded that montmorillonite protects simazine from hydrolysis through adsorption and that organic matter increases the degradation rate by acting as a catalytic agent.

4.4.2. Dealkylation

Based to a large extent on the reactions of various materials with Fenton's reagent (H_2O_2, EDTA, Fe^{2+}) under laboratory conditions, several workers have suggested the possibility of dealkylation occurring in the environment. Fenton's reagent is a source of hydroxy radicals and is known to react with [5-^{14}C]amitrole to give $^{14}CO_2$, unlabeled urea, and cyanamide (Plimmer *et al.*, 1967). Plimmer *et al.* (1968) have suggested that dealkylation of one or both of the side chains of hydroxy atrazine occurs by a mechanism of this type, and since labeled carbon was trapped in an organic base when methyl-labeled diuron was treated with Fenton's reagent there is some evidence that dealkylation occurs with the *N,N*-dimethyl ureas. Plimmer *et al.* (1968) also made preliminary electron spin resonance (ESR) studies of the interactions of atrazine and hydroxy atrazine with silt loam. Changes in the ESR spectrum were noted in each case, and hydroxy atrazine incorporated into the silt loam gave an EST spectrum with a sharp signal characteristic of a free radical. Although it appears that free radicals may be formed in soil, the overall significance of these radicals with respect to pesticide degradation remains unclear.

4.4.3. Oxidation

There is not a great deal of information about the oxidation of pesticides in soils, but there are some reports which indicate that oxidations do, in fact, occur. Lichtenstein and Schultz (1960) found that there was a slow conversion of aldrin to dieldrin in organic fractions. Likewise, decomposition of 3-aminotriazole (Burchfield and Storrs, 1956) and the *s*-oxidation of phorate (Getzin and Chapman, 1960) have been reported.

4.4.4. Dehalogenation

Lord (1948) reported that DDT and lindane could be dehydrochlorinated by soil constituents, and reduced porphyrins and nitrogenous soil constituents have been shown to reductively dechlorinate chlorinated hydrocarbon insecticides such as DDT (Miskus *et al.*, 1965).

From the preceding discussion, it is evident that pesticides undergo a variety of chemical and biological reactions, in the environment, and that as a result they are frequently decomposed to harmless residues. On the basis of our present knowledge it is difficult to say whether or not these degradative reactions can proceed to completion. On the other hand, of the pesticides which have been or are being used, even the most persistent substances such as DDT eventually undergo some chemical changes. A few pesticides are converted to more toxic substances in the environment, but this is generally transitory and eventually these substances are degraded by natural processes. In conclusion, it appears that if pesticide usage is properly regulated and controlled the accumulation of harmful residues in the environment will be minimal.

5. *Occupational Exposure and Hazards*

Many pesticides are potentially toxic to higher animals and if used carelessly can be dangerous to farmers, farm workers, pesticide applicators, and others exposed to them. The increasing use of organophosphorus compounds as substitutes for DDT and other chlorinated hydrocarbon insecticides has resulted in a new dimension of occupational hazard in the agricultural industry, and it is not surprising that especially in the state of California, where approximately 20–22% of all pesticides sold annually in the country are used, organophosphorus poisonings are being reported with increasing frequency. These unfortunate incidents can be attributed, in part, to a lack of understanding of the fate of these pesticides after they are released into the environment.

Most of the organophosphorus pesticides are highly toxic to mammals and are readily absorbed through the skin as well as taken in by inhalation and ingestion. A 1970 California Department of Public Health report, *Occupational Disease in California Attributed to Pesticides and Other Agricultural Chemicals*, states:

> In the 15 years since 1956, these compounds [organophosphorus] were responsible for nearly 4000 reports of nonfatal illness, including over 3000 systemic poisonings. The 332 reports in 1970 was the largest number since 1959, when a very extensive epidemic of parathion residue poisoning affected numerous crews of pickers and 455 reports were received.

A study on pesticide-related illnesses completed by the California Department of Health and the California Department of Food and Agriculture has revealed that a total of 1474 cases were reported by physicians between January 1 and December 31, 1973; of these, 665 were classified as systemic illnesses (Table IV). This represents a large increase in this type of illness and might be due to a number of factors, among which is the increased use of

Table IV. Reports of Pesticide-Related Occupational Illness Received by California Department of Health, 1973

Occupation	Systemic	Skin	Eye/Skin	Eye	Totals
Totals	665	452	33	324	1474
Ground applicators	187	103	13	121	424
Mixer, loader	121	19	3	22	165
Field worker	45	94	0	18	157
Nursery, greenhouse	18	71	1	22	112
Formulating plant	41	15	2	5	63
Warehousing, loading trucks	33	8	1	9	51
Gardener	14	16	2	34	66
Firefighting	41	0	0	1	42
Fumigation	52	13	1	5	71
Creosote application	1	24	2	9	36
Drift	10	5	0	11	26
Structural pest control	11	5	0	8	24
Flaggers	16	3	0	1	20
Cleaning and machine repair	10	6	1	5	22
Aerial application	10	0	1	3	14
Other	55	70	6	50	181

organophosphates and carbamates and the increased use of particularly toxic combinations (e.g., lannate–phosdrin mixtures).

5.1. Field Worker Exposure

Although some information is available on residues of organophosphorus pesticides in plants and soils, relatively little is known concerning the levels at which such residues pose a health hazard for people exposed to them. In the past, it has been generally assumed that if a crop is safe for human consumption it is safe to harvest. The fallacy of this line of reasoning is that in the course of harvesting the farm worker is exposed over large parts of his body to parts of the plant which are not harvested, such as leaves and bark; he inhales air which contains hazardous vapors and dusts; he comes into contact with contaminated fruits, soil, and water. Organophosphorus compounds can exert their toxic effects through repeated exposure, and low-level daily exposures can cause cholinesterase activity to decrease at a rate faster than the enzyme can be regenerated. Under these conditions, a critical level of inhibition can occur and overt illness can result. Hazards to farm workers were reported in several studies (Sumerford *et al.*, 1953; Hayes *et al.*, 1957; Quinby *et al.*, 1958; Durham and Wolfe, 1962, 1963; West, 1964; Durham, 1965; Hartwell *et al.*, 1964; Wolfe *et al.*, 1967; El-Refai, 1971).

Quinby and Lemmon (1958) summarized 11 episodes of poisoning from contact with parathion residues involving a total of more than 70 workers who were engaged in harvesting, thinning, cultivating, and irrigating such crops as apples, grapes, citrus, and hops. Although six of the outbreaks occurred within 2 days of pesticide application, the remainder resulted from exposure to residues from 8 to 33 days old. Percutaneous absorption was considered to be the primary route by which parathion was taken up.

Following an outbreak of illness among peach harvesters, 186 peach orchard workers were studied in relation to pesticide application practice and fruit harvesting procedures representative of the orchards in which they worked (Milby *et al.*, 1964). The findings suggested that observed illnesses were "the result of residue accumulation related to total amount of parathion applied during the entire growing season."

Nemac *et al.* (1968) measured the amount of methyl parathion on the hands and arms of two entomologists who had entered a cotton field to make insect counts following application of the pesticide to the field. When the application rate was 2 lb/acre and the workers entered the field for a 5-min period 2 hr after application, the total methyl parathion washed off the hands and arms in acetone ranged from 2.024 to 10.192 mg/worker. When the application rate was 1.5 lb/acre and the time before entering the field was increased to 24 hr, the amounts of methyl parathion on the hands and arms of the two workers were 0.163 and 0.351 mg. Both workers were right-handed, and in each case the amount of pesticide on the right limb was significantly higher than that on the left. These measurements gave no indication of the actual amount of methyl parathion absorbed into the bodies of the workers, which would presumably not appear in the acetone washes. That some had indeed penetrated through the skin was clear from the substantial depression of blood cholinesterase which coincided with the workers' exposure. Aside from its brutal directness, this report clearly illustrates that the level of exposure is related to the concentration at which the insecticide is applied and/or the interval between application and entry into the field and that this kind of exposure to methyl parathion causes measurable physiological changes in human beings.

On September 17, 1970, 35 workers entered an orange grove in Kern County, California. Parathion had been applied to this grove on August 11–14, 1970, at the rate of 9 lb of active ingredients per acre. The application was carried out in a manner that met legal requirements and the entrance of the workers on September 17 was in accordance with current practices and regulations. However, 4 hr after they began picking oranges, the workers complained of feeling ill. Public officials investigating the incident reported:

> Saturday morning, September 19, 1970, Don Mengle, Research Specialist, Community Studies on Pesticides, Department of Public Health, Everett Gwinn, Tulare County Agricultural Inspector, and myself contacted Dr. Mathews at Lindsay District Hospital. He provided a list of the workers seen in their emergency room, and the results of cholinesterase tests received so far from their laboratory. Dr. Mathews indicated that some of the workers were dizzy and vomiting, but none displayed eye symptoms. The patients were treated with atropine.

> At noon we met with Seldon Morley in McFarland and interviewed Dr. A. W. Carlson, Jr., who had admitted twelve of the sickest workers to the Delano District Hospital between 11:30 a.m. and 2:00 p.m. on Thursday, September 17, 1970. He described their symptoms as cramps, vomiting, dizziness, and pinpoint pupils. All patients were fed intravenously and treated with atropine. The most severe cases were treated with 2-PAM. Blood samples for cholinesterase tests were run at the Mercy Hospital in Bakersfield. (California Department of Agriculture, Memorandum, September 21, 1970)

Although incidents of pesticide poisoning are all too common according to the 1970 California Department of Public Health Report, this particular incident was soon followed by hearings of the California Department of Agriculture, and the interval required between the time of application and the time that workers can legally enter treated fields was subsequently increased.

In the summer of 1970, the Agricultural Experiment Station of the University of California carried out a research project entitled "Post-application Effects of Certain Organophosphorus Pesticides on Grape Pickers" (Bailey, 1972). Fifty acres of Thompson seedless grapes in five replicated plots of 10 acres each at the Kearney Horticultural Field Station in Fresno County were treated with a combination of ethion and guthion at rates of 1.60 and 1.90 lb/acre, respectively. Harvesting was carried out by volunteer workers, with appropriate legal and medical safeguards, and blood samples were drawn by licensed medical personnel. A group of volunteer field workers had their cholinesterase levels determined at three intervals prior to harvest and three times during the 5-day harvest period. After the first two venipunctures, there was growing reluctance among the workers to continue giving blood samples, although 23 workers persevered for the series of six tests and only three dropped out. Significant cholinesterase depression was observed in some cases, ranging as high as 70% in a few instances (the California Department of Health considers a 25% depression to be significant).

Bailey (1972) described in detail two field studies conducted by the University of California regarding the effects of pesticide residues on farm laborers. He indicated that "to date, there is no workable system for fully protecting the California farm laborers working in fields treated with pesticides. They can be poisoned in the field, even if they are not actually applying pesticides at the time." Of 12 alternatives proposed to help avoid worker exposure, Bailey (1972) suggests that the most promising is the establishment of worker reentry periods.

Since the 1950s, the U.S. Department of Agriculture (USDA) and subsequently the Environmental Protection Agency (EPA), under the Federal Insecticide, Fungicide, and Rodenticide Act (FIFRA), have required certain label restrictions on pesticide products. These restrictions concern protective clothing, vacating of fields during spray operations, and precautions for reentry into fields which have been treated. This concern for the protection of all persons who might be exposed to pesticides during and after application has been an important aspect of the FIFRA registration process. The FIFRA, as amended by the Federal Environmental Pesticide Control Act of 1972, reemphasizes EPA's legislative mandate in this area.

Another organization concerned with safe and healthful working conditions for employees is the Occupational Safety and Health Administration (OSHA) of the Department of Labor, and this will hold the farm owner responsible for upholding farm worker protection standards. In contrast, standards promulgated by EPA will be designed to protect all individuals coming into contact with pesticides and pesticide residues, including the farm owner, the farm worker, and all other pesticide applicators.

In a rather surprising move, OSHA declared on April 30, 1973, that an emergency situation existed and subsequently issued an "Emergency Temporary Standard for Exposure to Organophosphorus Pesticides" in the *Federal Register* of May 1, 1973 (Vol. 38, No. 88, pp. 10715–17717). This document identified 21 organophosphorus pesticides and specified "field reentry safety intervals" for citrus (oranges, lemons, grapefruit), peaches, grapes, tobacco, and apples treated with the pesticides. This standard was to become effective June 18, 1973.

Since the promulgation of the emergency standard, the Florida Peach Growers Association and numerous other organizations have filed petitions for reconsideration and revocation of the standard. They charge that OSHA has no authority on which to pass emergency standards, that no grave danger from exposure exists, that the standards are unreasonable, and that the Department of Labor's jurisdiction in establishing reentry intervals is questionable under the Federal Environmental Pesticide Control Act (FEPCA) (which is administered by the Environmental Protection Agency).

In light of these circumstances, the application of the standard published on May 1, 1973, was suspended by OSHA pending the issuance of the new standard (*Federal Register*, Vol. 38, No. 115, p. 15729, June 15, 1973). Meanwhile, on June 15 the House Agriculture Committee voted 20 to 5 against the Labor Department's planned new regulations on farm worker exposure to pesticides.

On June 29, OSHA promulgated a new emergency temporary standard after determining that "(1) certain pesticides listed in the standard published on May 1, 1973, are not necessarily highly toxic, and (2) the standard published on May 1, 1973, is broader than necessary as it applies to such pesticides which are not considered highly toxic, and in the manner it protects workers from the pesticides which are considered highly toxic" (*Federal Register*, Vol. 38, No. 125, pp. 17214–17216, June 29, 1973). This document identified the major differences between the May 1 standard and the new emergency temporary standard, which included only 12 of the 21 organophosphorus pesticides listed earlier. These amendments were to become effective July 13, 1973.

On June 27, 1974, the occupational safety and health standards contained in 29 CFR Part 1910 were published in the *Federal Register* (Vol. 39, p. 23502). The republication included the emergency temporary standard for exposure to pesticides, which had been promulgated pursuant to Section 6(c) of the William-Steiger Occupational Safety and Health Act of 1970, on May 1, 1973, and revised on June 29, 1973. However, this emergency standard was challenged under Section 6(f) of the Act by petitions filed in ten United States Courts of Appeals. All of the court suits were consolidated in the Fifth Circuit, and on January 9, 1974, in the case of "Florida Peach Growers Association, Inc. *v.* United States Department of Labor," that court vacated the temporary standard on exposure to pesticides. Consequently, OSHA decided to delete from Part 1910 the standard on pesticides rendered inoperative by the court's decision and the reference to it in a related section (*Federal Register*, Vol. 39, p. 28878, August 12, 1974).

Both EPA and OSHA had held hearings to investigate the question of improved farm worker protection from pesticide hazards. After review of the records of the hearings together with written views, arguments, and data, and after consultation with the Department of Labor, USDA, and other interested federal agencies, the EPA issued final worker protection standards for agricultural pesticides on May 10, 1974 (*Federal Register*, Vol. 39, p. 16888). Under the existing regulation, restrictions against workers entering treated fields have been required for many pesticide products. These restrictions include label requirements specifying a permissible field reentry time, requirements for protective clothing, warnings against unnecessarily exposing workers to the risks posed by pesticides, and specific exposure precautions. The EPA standards (CFR 40, Sec. 170.3) state:

> (a) Application. No owner or lessee shall permit the application of a pesticide in such a manner as to directly or through drift expose workers or other persons except those knowingly involved in the application. The area being treated must be vacated by unprotected persons.
>
> (b) Reentry times. (1) No owner or lessee shall permit any worker not wearing protective clothing (under §170.2 (d)) to enter a field treated with pesticides until sprays have dried or dusts have settled, unless exempted from such requirements, or a longer reentry time has been assigned to that pesticide.
>
> (2) Pesticides containing the following active ingredients have a reentry time of at least the interval indicated:

		Hours
(i)	Ethyl parathion	48
(ii)	Methyl parathion	48
(iii)	Guthion	24
(iv)	Demeton	48
(v)	Azodrin	48
(vi)	Phosalone	24
(vii)	Carbophenothion	48
(ix)	EPN	24
(x)	Bidrin	48
(xi)	Endrin	48
(xii)	Ethion	24

> (3) The preceding requirements of this part notwithstanding, workers should not be permitted to enter treated fields if special circumstances exist which would lead a reasonable man to conclude that such entry would be unsafe.

The controversy over the promulgation of the "emergency temporary standards" clearly demonstrates the urgent need for systematic and realistic research findings regarding the farm worker reentry problems associated with the use of organophosphorus pesticides on various crops.

5.2. Applicator Exposure

Pest control workers are often subjected to relatively high levels of toxic pesticides when actively engaged in pest control operations. Much of the safety in relation to pesticides rests on the user or applicator of the compounds.

There are four major routes of pesticide entry into the human body: dermal, respiratory, oral, and through cuts or abrasions in the skin.

5.2.1. Dermal Route

The dermal route is considered to be the most important route of entry into the body during pest control operations regardless of the formulations being used. It has been undoubtedly responsible for many poisonings of workers, especially from the more toxic organophosphorus compounds.

Wolfe *et al.* (1972) studied the direct exposure of agricultural spraymen to pesticides. Values for potential dermal and respiratory exposure and for total exposure in terms of fraction of toxic dose were determined for 11 different pesticides during orchard spraying with air-blast application equipment in central Washington. They indicated that potential dermal exposure to each compound was much greater than potential respiratory exposure.

Durham *et al.* (1972) investigated the absorption and excretion of parathion by spraymen. Significant amounts of the metabolite *p*-nitrophenol were detected in the urine of spraymen as long as 10 days after the last exposure to the insecticide, and it was stated that "Tests considering only one route of exposure at a time indicated that the dermal route represents a potentially greater source of absorption than the respiratory route for orchard spraymen using liquid parathion formulations under the conditions of this study. However, with equivalent absorbed dosages the respiratory route is more hazardous."

The importance of protecting specific body areas has not been clearly defined in the past. The most useful and probably most accurate estimations or measurements on the precutaneous penetration of pesticides in man accomplished thus far have been made by Maibach *et al.* (1971). Using radiolabeled pesticides, they were able to determine the approximate fraction of an applied dose absorbed through the skin. The results indicated that the area of greatest absorption of parathion in man is the scrotum, where approximately 100% of an applied dose was absorbed. It also points out that the head and neck area should be given more attention. In this area, absorption of parathion was found to be from 32.2% to 36.3% of the applied dose, much more than at other areas of the body studied (with the exception of the armpit and scrotum). Feldmann and Maibach (1974) studied the skin absorption of 12 pesticides: five organophosphates, three chlorinated hydrocarbons, two carbamates, and two herbicides. ^{14}C-labeled pesticides and herbicides were applied to the forearms of human subjects, and the urinary excretion of ^{14}C was measured. All pesticides tested were absorbed, the greatest absorption occurring with carbaryl and the least with diquat.

5.2.2. Respiratory Route

Protection of the respiratory route is extremely important where toxic dusts and vapors or very small spray droplets are prevalent or where application is in confined spaces. The role of the respiratory route as a source of pesticide exposure has been studied in the past by both direct and indirect methods (Durham and Wolfe, 1962; Wolfe *et al.*, 1966, 1972; Jegier, 1964; Simpson, 1965; Batchelor and Walker, 1954). Little has been done to compare the exposures resulting from different pest control activities. Oudbier *et al.* (1974) studied the respiratory exposure to pesticides during several use activities to determine which activity presented the greatest hazard and found that the period of pesticide mixing and tank filling is of most importance as far as respiratory exposure is concerned. There appears to be more potential for dermal exposure when liquid formulations are used.

5.2.3. Oral Route

Little experimental work has been done to define the magnitude of oral exposure. The most serious exposures of this type may be brought about by the inadvertent splashing of liquid concentrate into the mouth while pouring and measuring pesticides. In preliminary studies to develop techniques for measuring oral exposure, Wolfe (1972) indicated that analysis of the saliva of exposed individuals appeared to give some indication of contamination.

5.2.4. Entry Through Cuts or Abrasions

Entry through cuts or abrasions has perhaps received the least amount of attention to date, but any break in the skin may allow a more direct entry of pesticide into the circulation.

6. References

Acree, F., Jr., Beroza, M., and Bowman, M. C., 1963, Codistillation of DDT with water, *J. Agr. Food Chem.* **11**:278–280.

Akesson, N. B., and Yates, W. E., 1964, Problems relating to application of agricultural chemicals and resulting drift residues, *Ann. Rev. Entomol.* **9**:285–318.

Akesson, N. B., Bayer, D. E., and Yates, W. E., 1970, Reducing contamination risks from aerial pesticide applications, *in: Proceedings of the 7th International Congress of Plant Protection*, pp. 800–801.

Alabama Water Improvement Commission, 1961, *A Report on Fish Kills Occurring on Choccolocco Creek and the Coosa River During May 1961.*

Allan, J., 1955, Loss of biological efficiency of cattle-dipping wash containing benzene hexachloride, *Nature (London)* **175:**1131–1132.

Aly, O. M., and Faist, S. D., 1964*a*, Study fate of 2,4-D in lakes, reservoirs [*sic*], *New Jersy Agr.* **46(1):**12.

Aly, O. M., and Faust, S. D., 1964*b*, Studies on the fate of 2,4-D and ester derivatives in natural surface waters, *J. Agr. Food Chem.* **12:**541–546.

Armstrong, D. E., and Chesters, G., 1968, Adsorption-catalyzed chemical hydrolysis of atrazine, *Environ. Sci. Technol.* **2:**683–689.

Armstrong, D. E., Chesters, G., and Harris, R. F., 1967, Atrazine hydrolysis in soil, *Soil Sci. Soc. Am. Proc.* **31:**61.

Bailey, J. B., 1972, The effects of pesticide residues on farm laborers, *Agrichem Age* **15(10):**6–8, 10.

Bartha, R., Lanzilotta, R. P., and Pramer, D., 1967, Stability and effects of some pesticides in soil, *Appl. Microbiol.* **15:**67–75.

Batchelor, G. S., and Walker, K. C., 1954, Health hazards in use of parathion in fruit orchards of north central Washington, *AMA Arch. Ind. Hyg.* **10:**522–528.

Bell, G. R., 1956, Photochemical degradation of 2,4-dichlorophenoxy acetic acid and structurally related compounds in the presence and absence of riboflavin, *Bot. Gaz.* **118:**133–136.

Bengtson, S. A., 1974, DDT and PCB residues in air-borne fallout and animals in Iceland, *Ambio* **3:**84-86.

Bowman, M. C., Schechter, M. S., and Carter, R. L., 1965, Behavior of chlorinated insecticides in a broad spectrum of soil types, *J. Agr. Food Chem.* **13:**360–365.

Burchfield, H. P., and Storrs, E. E., 1956, Chemical structures and dissociation constants of amino acids, peptides, and proteins in relation to their reaction rates with 2,4-dichloro-6-(*o*-chloranilino)-*s*-triazine, *Contrib. Boyce Thompson Inst.* **18:**395–418.

Burchfield, H. P., and Storrs, E. E., 1957, Effects of chlorine substitution and isomerism on the interactions of *s*-triazine derivatives with conidia of *Neurospora sitophilia, Contrib. Boyce Thompson Inst.* **18:**429–452.

Burnett, R., 1971, DDT residues: Distribution of concentrations in *Emerita analoga* (Stimpson) along coastal California, *Science* **174:**606–608.

Butler, P. A., 1965, *Commercial Fishery Investigations,* U.S. Wildlife Service Circular No. 226, p. 65.

Byrde, R. J. W., and Woodcock, D., 1957, Fungal detoxication. 2. The metabolism of some phenoxy-*n*-alkylcarboxylic acids by *Aspergillus niger, Biochem. J.* **65:**682–686.

Cahn, R. S., Phipers, R. F., and Brodaty, E., 1945, Stability of derris in insecticidal dusts: The solvent–powder effect, *J. Soc. Chem. Ind.* (*London*) **64:**33–40.

California Water Resources Control Board, 1971, *A Review of Pesticide Monitoring Programs in California,* prepared by Ad Hoc Working Group of the Pesticide Advisory Committee to the State Department of Agriculture, Sacramento, Calif.

Caro, J. H., and Taylor, A. W., 1971, Pathways of loss of dieldrin from soils under field conditions, *J. Agr. Food Chem.* **19:**379–384.

Casida, J. E., and Lykken, L., 1969, Metabolism of organic pesticide chemicals in higher plants, *Ann. Rev. Plant Physiol.* **20:**607–636.

Castro, C. E., and Belser, N. O., 1966, Hydrolysis of *cis*- and *trans*-1,3-dichloropropene in wet soil, *J. Agr. Food Chem.* **14:**69–70.

Chacko, C. I., Lockwood, J. L., and Zabik, M., 1966, Chlorinated hydrocarbon pesticides: Degradation by microbes, *Science,* **154:**893–895.

Chadwick, G. G., and Brocksen, R. W., 1969, Accumulation of dieldrin by fish and selected fish-food organisms, *J. Wildlife Manag.* **33:**693–700.

Chen, Y. L., and Casida, J. E., 1969, Photodecomposition of pyrethin I, allethrin, phalthrin, dimethrin: Modifications in the acid moiety, *J. Agr. Food Chem.* **17:**208–215.

Cohen, J. M., and Pinkerton, C., 1966, Widespread translocation of pesticides by air transport and rain-out, *in: Organic Pesticides in the Environment,* pp. 163–176, American Chemical Society, Advances in Chemistry Series 60, Washington, D.C.

Corneliussen, P. E., 1970, Pesticide residues in total diet samples (V), *Pestic. Monit. J.* **4:**89–105.

Corneliussen, P. E., 1972, Pesticide residues in total diet samples (VI), *Pestic. Monit. J.* **5:**313–330.

Cory, L., Fjeld, P., and Serat, W., 1970, Distribution patterns of DDT residues in the Sierra Nevada Mountains, *Pestic. Monit. J.* **3:**204–211.

Crosby, D. G., and Leitis, E., 1969, Photolysis of chlorophenylacetic acids, *J. Agr. Food Chem.* **17:**1036–1040.

Crosby, D. G., and Li, M. Y., 1969, Herbicide photodecomposition, *in: Degradation of Herbicides* (P. C. Kearney and D. D. Kaufman, eds.), Dekker, New York.

Crosby, D. G., and Tutass, H. O., 1966, Photodecomposition of 2,4-dichlorophenoxyacetic acid, *J. Agr. Food Chem.* **14:**596–599.

Dagley, S., 1972, Microbial degradation of stable chemical structures: General features of metabolic pathways, *in: Degradation of Synthetic Organic Molecules in the Biosphere*, pp. 1–16, National Academy of Sciences, Washington, D.C.

Deming, J. M., 1963, Determination of volatility losses of C^{14}-CDAA from soil surfaces, *Weeds* **11:**91–96.

Drescher, N., and Otto, S., 1968, Ueber den Abbau von Dazomet im Boden, *Residue Rev.* **23:**49–54.

Duggan, R. E., and Corneliussen, P. E., 1972, Dietary intake of pesticide chemicals in the United States (III), June 1968–April 1970, *Pestic. Monit. J.* **5:**331–341.

Durham, W. F., 1965, Pesticide exposure levels in man and animals, *Arch. Environ. Health* **10:**842–846.

Durham, W. F., and Wolfe, H. R., 1962, Measurement of the exposure of workers to pesticides, *Bull. WHO* **26:**75–91.

Durham, W. F., and Wolfe, H. R., 1963, An additional note regarding measurement of the exposure of workers to pesticides, *Bull. WHO* **29:**279–281.

Durham, W. F., Wolfe, H. R., and Elliott, J. W., 1972, Absorption and excretion of parathion by spraymen, *Arch. Environ. Health* **24:**381–387.

Edwards, C. A., 1973*a*, *Environmental Pollution by Pesticides*, Plenum Press, New York.

Edwards, C. A., 1973*b*, *Persistent Pesticides in the Environment*, 2nd. ed., CRC Press, Cleveland.

Edwards, C. A., and Thompson, A. R., 1973, Pesticides and the soil fauna, *Residue Rev.* **45:**1–79.

Edwards, C. A., Thompson, A. R., Benyon, K. I., and Edwards, M. J., 1970, Movement of dieldrin through soils. I. From arable soils into ponds, *Pestic. Sci.* **1:**169–173.

El-Refai, A. R., 1971, Hazards from aerial spraying in cotton culture area of the Nile River, *Arch. Environ. Health* **22:**328–333.

Fang, S. C., Thiesen, P., and Freed, V. H., 1961, Effects of water evaporation, temperature and rates of application on the retention of ethyl-*N,N*-di-*n*-propylthiolcarbamate in various soils, *Weeds* **9:**569–574.

Feldmann, R. J., and Maibach, H. I., 1974, Percutaneous penetration of some pesticides and herbicides in man, *Toxicol. Appl. Pharmacol.* **28:**126–132.

Fernley, H. N., and Evans, W. C., 1958, Oxidative metabolism of polycyclic hydrocarbons by soil pseudomonads, *Nature (London)* **182:**373–375.

Finlayson, D. G., and MacCarthy, H. R., 1965, The movement and persistence of insecticides in plant tissue, *Residue Rev.* **9:**114–152.

Foy, C. L., and Bingham, S. W., 1969, Some research approaches toward minimizing herbicidal residues in the environment, *Residue Rev.* **29:**105–135.

Frawley, J. P., Cook, J. W., Blake, J. R., and Fitzhugh, O. G., 1958, Effect of light on chemical and biological properties of parathion, *J. Agr. Food Chem.* **6:**28–31.

Geissbuhler, H., 1969, The substituted ureas, *in: Degradation of Herbicides* (P. C. Kearney and D. D. Kaufman, eds.), pp. 79–111, Dekker, Wahington, D.C.

Getzin, L. W., and Chapman, R. K., 1960, The fate of phorate in soils, *J. Econ. Entomol.* **53:**47–51.

Gray, R. A., and Weierich, A. J., 1965, Factors affecting the vapor loss of EPTC from soils, *Weeds* **13:**141–147.

Grover, R., and Hance, R. J., 1969, Adsorption of some herbicides by soil and roots, *Can. J. Plant Sci.* **49:**378–380.

Gunner, H., 1967, *The influence of Rhizosphere Microflora on the Transformation of Insecticides by Plants*, Massachusetts Agricultural Experiment Station, Annual Report of the Cooperative Regional Project NE-53.

Gutenmann, W. H., Loos, M. A., Alexander, M., and Lisk, D. J., 1964, Beta oxidation of phenoxyalkanoic acids in soil, *Soil Sci. Soc. Am. Proc.* **28:**205–207.

Harris, C. I., 1967, Fate of 2-chloro-*s*-triazine herdicides in soil, *J. Agr. Food Chem.* **15:**157–162.

Harris, C. R., and Lichtenstein, E. P., 1961, Factors affecting the volatilization of insecticides from soil, *J. Econ. Entomol.* **54:**1038–1045.

Harrison, H. L., Loucks, O. L., Mitchell, J. W., Parkhurst, D. F., Tracy, C. R., Watts, D. D., and Wannacone, V. J., Jr., 1970, Systems studies of DDT transport, *Science*, **170:**503–508.

Harrison, R. B., Holmes, D. C., Roburn, J., and Tatton, J. O'G., 1967, The fate of some organochlorine pesticides on leaves, *J. Sci. Food Agr.* **18:**10–15.

Hartley, G. S., 1969, Evaporation of pesticides, *in: Pesticidal Formulations Research, Physical and Colloidal Chemical Aspects.* (R. F. Gould, ed.), pp. 115–134, Advances in Chemistry Series 86, American Chemical Society, Washington, D.C.

Hartwell, W. V., Hayes, G. R., and Funckes, A. J., 1964, Respiratory exposure to volunteers to parathion, *Arch. Environ. Health* **8:**820–825.

Hayes, W. J., Jr., Dixon, E. M., Batchelor, G. S., and Upholt, W. M., 1957, Exposure to organic phosphorus sprays and occurrence of selected symptoms, *Pub. Health Rep.* **72:**787–784.

Henderson, G. L., and Crosby, D. G., 1967, Photodecomposition of dieldrin and aldrin, *J. Agr. Food Chem.* **15:**888–893.

Holden, A. V., 1962, A study of absorption of C^{14}-labeled DDT from water by fish, *Ann. Appl. Biol.* **50:**467–477.

Hsieh, D. P. H., Archer, T. E., Munnecke, D., and McGowan, E., 1972, Decontamination of noncombustible agricultural pesticide containers by removal of emulsifiable parathion, *Environ. Sci. Technol.* **6:**826–829.

Hunt, E. G., 1966, Biological magnification of pesticides, *in: Scientific Aspects of Pest Control*, pp. 251–262, National Academy of Sciences–National Research Council, NRC Publication No. 1402, Washington, D.C.

Hunt, E. G., and Bischoff, A. I., 1960, Inimical effects on wildlife of periodic DDD applications to Clear Lake. *Calif. Fish Game* **46:**91.

Igue, K., 1969, Volatility of organochlorine insecticides from soil, Ph.D. thesis, University of California, Riverside.

Jegier, Z., 1964, Exposure to guthion during spraying and formulation, *Arch. Environ. Health*, **8:**565–569.

Jones, H. A., and Haller, H. L., 1931, "Yellow compounds" resulting from the decomposition of rotenone in solution, *J. Am. Chem. Soc.* **53:**2320–2324.

Kallman, B. J., Cope, O. B., and Navarre, R. J., 1962, Distribution and detoxication of toxaphene in Clayton Lake, New Mexico, *Trans. Am. Fish. Soc.* **91:**14–22.

Kaufman, D. D., 1970, Pesticide metabolism, *in: Pesticides in the Soil: Ecology, Degradation and movement*, pp. 73–86, International Symposium, Michigan State University, East Lansing, Mich.

Kaufman, D. D., and Kearney, P. C., 1970, Microbial degradation of triazine herbicides, *Residue Rev.* **32:**235–265.

Kaufman, D. D., Kearney, P. C., and Sheets, T. J., 1965, Microbial degradation of simazine, *J. Agr. Food Chem.* **13:**238–242.

Kearney, P. C., and Helling, C., 1965, Reactions of pesticides in soils, *Residue, Rev.* **25:**25–44.

Kearney, P. C., and Kaufman, D. D., (eds.), 1969, *Degradation of Herbicides*, Dekker, New York.

Kearney, P. C., Sheets, T. J., and Smith, J. W., 1964, Volatility of seven *s*-triazines, *Weeds*, **12:**83–87.

Kearney, P. C., Kaufman, D. D., and Sheets, T. J., 1965, Metabolites of simazine by *Aspergillus fumigatus, J. Agr. Food Chem.* **13:**369–372.

Kearney, P. C., Kaufman, D. D., and Alexander, M., 1967, Biochemistry of herbicide decomposition in soils, *in: Soil Biochemistry* (G. H. Peterson, ed.), pp. 318–342, Dekker, New York.

Kennedy, M. V., Stojanovic, B. J., and Shuman, F. L., Jr., 1969, Chemical and thermal methods for disposal of pesticides, *Residue Rev.* **29:**89–104.

Kilgore, W. W., Marei, N., and Winterlin, W., 1972, Parathion in plant tissues: New considerations, *in: Degradation of Synthetic Organic Molecules in the Biosphere: Natural, Pesticidal and Various Other Man-Made Compounds,* pp. 291–312, Proceedings of a Conference, San Fransisco, June 12–13, 1971, National Academy of Sciences, Washington, D.C.

Ko, W. H., and Farley, J. D., 1969, Conversion of pentachloronitrobenzene to pentachloronitroaniline in soil and the effect of these compounds on soil microorganisms, *Phytopathology* **59:**64–67.

Ko, W. H., and Lockwood, J. L., 1968, Accumulation and concentration of chlorinated hydrocarbon pesticides by microorganisms in soil, *Can. J. Microbiol.* **14:**1075–1078.

Konrad, J. G., Armstrong, D. E., and Chesters, G., 1967, Soil degradation of diazinon, a phosphorothioate insecticide, *Agron. J.* **59:**591–594.

Kuhr, R. J., Davis, A. C., and Bourke, J. B., 1974, DDT residues in soil, water, and fauna from New York apple orchards, *Pestic. Monit. J.* **7:**200–204.

Kuwahara, M., Kato, N., and Munakata, K., 1965, The photochemical reaction products of pentachlorophenol, *Agr. Biol. Chem.* (*Tokyo*) **29:**880–882.

Lasher, C., and Applegate, H. G., 1966, Pesticides at Presidio. III. Soil and water, *Texas J. Sci.* **18:**386–395.

Li, C. F., and Bradley, R. L., 1969, Degradation of chlorinated hydrocarbon pesticides in milk and butteroil by ultraviolet energy, *J. Dairy Sci.* **52:**27–30.

Li, M. Y., and Fleck, R. A., 1972, *The Effects of Agricultural Pesticides in the Aquatic Environment, Irrigated Croplands, San Joaquin Valley,* Environmental Protection Agency, Pesticide Study Series 6, Washington, D.C.

Lichtenstein, E. P., and Schulz, K. R., 1960, Epoxidation of aldrin and heptachlor in soils as influenced by autoclaving, moisture, and soil types, *J. Econ. Entomol.* **53:**192–197.

Lichtenstein, E. P., and Schulz, K. R., 1964, The effects of moisture and microorganisms on the persistence and metabolism of some organophosphorus insecticides in soils, with special emphasis on parathion, *J. Econ. Entomol.* **57:**618–627.

Lichtenstein, E. P., Schulz, K. R., Skrentny, R. F., and Tsukano, Y., 1966, Toxicity and fate of insecticide residues in water, *Arch. Environ. Health* **12:**199–212.

Lichtenstein, E. P., Fuhremann, T. W., and Schulz, K. R., 1968, Effects of carbon on insecticide adsorption and toxicity in soills, *J. Agr. Food Chem.* **16:**348–355.

Lichtenstein, E. P., Fuhremann, T. W., and Schulz, K. R., 1971*a*, Persistence and vertical distribution of DDT, lindane, and aldrin residues, 10 and 15 years after a single soil application, *J. Agr. Food Chem.* **19:**718–721.

Lichtenstein, E. P., Schulz, K. R., and Fuhremann, T. W., 1971*b*, Effects of a cover crop and soil cultivation on the fate and vertical distribution of insecticide residues in soil, 7 to 11 years after soil treatment, *Pestic. Monit. J.* **2:**218–222.

Livingston, R., 1961, Photochemical autooxidation, *in: Autooxidation and Antioxidants,* Vol. I (W. O. Lundberg, ed.), pp. 249–298, Interscience, New York.

Loos, M. A., 1969, Phenoxyalkanoic acids, *in: Degradation of Herbicides.* (P. C. Kearney and D. D. Kaufman, eds.), pp. 1–49, Dekker, New York.

Lord, K. A., 1948, Decomposition of DDT by basic substances, *J. Chem. Soc.* (*London*), pp. 1657–1661.

Macek, K. J., 1970, Biological magnification of pesticide residues in food chains, *in: Biological Impact of Pesticides in the Environment* (J. W. Gillet, ed.), pp. 17–21, Oregon State University Press, Corvallis, Ore.

MacRae, I. C., Raghu, K., and Castro, T. F., 1967, Persistence and biodegradation of four common isomers of benzene hexachloride in submerged soils, *J. Agr. Food Chem.* **15:**911–914.

Maibach, H. I., Feldmann, R. J., Milby, T. G., and Serat, W. F., 1971, Regional variation in percutaneous penetration in man, *Arch. Environ. Health* **23:**208–211.

Matsumura, F., and Boush, G. M., 1966, Malathion degradation by *Trichoderma viride* and a *Pseudomonas* species, *Science,* **153:**1278–1280.

Menn, J. J., 1972, Absorption and metabolism of insecticide chemicals in plants, *in: Degradation of Synthetic Organic Molecules in the Biosphere*, pp. 206–243, National Academy of Sciences, Wahington, D.C.

Menn, J. J., and McBain, J. B., 1964, Metabolism of phthalimidomethyl-*O,O*-dimethylphosphorodithioate (imidan) in cotton plants, *J. Agr. Food Chem.* **12:**162–166.

Menzie, C. M., 1969, *Metabolism of Pesticides*, Bureau of Sport Fisheries and Wildlife, Special Scientific Report—Wildlife No. 127, Washington, D.C.

Metcalf, R. L., Fukuto, T. R., Collins, C., Borck, K., El-Aziz, S., Munoz, R., and Cassil, C. C., 1968, Metabolism of 2,2-dimethyl-2,3-dihydrobenzofuranyl-7-*N*-methylcarbamate (furadan) in plants, insects, and mammals, *J. Agr. Food Chem.* **16:**300–311.

Milby, T. H., Ottoboni, F., and Mitchell, H. W., 1964, Parathion residue poisoning among orchard workers, *J. Am. Med. Assoc.* **189:**351–356.

Miskus, R. P., Blair, D. P., and Casida, J. E., 1965, Insecticide metabolism: Conversion of DDT to DDD by bovine rumen fluid, lake water, and reduced porphyrins, *J. Agr. Food Chem.* **13:**481–483.

Mitchell, W. G., Parsons, D. A., Sand, P. F., Lynch, D. D., and Cook, W. S., Jr., 1970, *An Exploratory Study of Pesticide Migration from Waste Disposal Pits, 1967*, Department of Agriculture, Agricultural Research Service, Washington, D.C.

Moriarty, F., Pollutants and food chains. *New Scientist* **53(787):**594–596.

Mortland, M. M., and Raman, K. V., 1967, Catalytic hydrolysis of some organic phosphate pesticides by copper (II), *J. Agr. Food Chem.* **15:**163–167.

Nash, R. G., and Beall, M. L., Jr., 1970, Chlorinated hydrocarbon insecticides: Root uptake versus vapor contamination of soybean foliage, *Science* **168:**1109–1111.

National Academy of Sciences Committee on Persistent Pesticides, 1969, *Report*, National Research Council, Division of Biology and Agriculture, Washington, D.C.

Nemac, S. F., Adkisson, P. L., and Dorouth, H. W., 1968, Methyl parathion absorbed on the skin and blood cholinesterase levels of persons checking cotton treated with ultra-low-volume sprays, *J. Econ. Entomol.* **61:**1740–1742.

Nicholson, H. P., 1959, Insecticide pollution of water resources, *J. Am. Water Works Assoc.* **51:**981–986.

Nicholson, H. P., 1969, Occurrence and significance of pesticide residues in water, *J. Wash. Acad. Sci.* **59:**77–85.

Nicholson, H. P., 1970, The pesticide burden in water and its significance, *in: Agricultural Practices and Water Quality*, pp. 183–193, Iowa State University Press, Ames, Ia.

Nicholson, H. P., and Hill, D. W., 1970, Pesticide contaminants in water and their environmental impact, *in: Relationship of Agriculture to Soil and Water Pollution*, Cornell University Conference on Agricultural Waste Management, Cornell University, Ithaca, N.Y.

O'Brien, R. D., 1969, *Insecticides: Action and Metabolism*, Academic Press, New York.

Oudbier, A. J., Bloomer, A. W., Price, H. A., and Welch, R. L., 1974, Respiratory route of pesticide exposure as a potential health hazard, *Bull. Environ. Contam. Toxical.* **12:**1–9.

Parochetti, J. V., and Warren, W. J., 1966, Vapor losses of IPC and CIPC, *Weeds* **14:**281–285.

Penfound, W. T., and Minyard, V., 1947, Relations of light intensity to effect of 2,4-dichlorophenoxyacetic acid on water hyacinth and kidney bean plant, *Bot. Gaz.* **109:**231–234.

Peterle, T. J., 1969, DDT in Antarctic snow, *Nature (London)* **224:**620.

Plimmer, J. R., Kearney, P. C., Kaufman, D. D., and Guardia, F. S., 1967, Amitrole decomposition by free radical-generating systems and by soils, *J. Agr. Food Chem.* **15:**996–999.

Plimmer, J. R., Kearney, P. C., and Rowlands, J. R., 1968, Free-radical oxidation of *s*-triazines: Mechanism of *N*-dealkylation, *Abst. Amer. Chem. Soc.*, Atlantic City, N.J.

Premdas, F. H., and Anderson, J. M., 1963, The uptake and detoxification of C^{14}-labeled DDT in Atlantic salmon, *J. Fish. Res. Board Can.* **20:**827–837.

Quinby, G. E., and Lemmon, A. B., 1958, Parathion residues as a cause of poisoning in crop workers, *J. Am. Med. Assoc.* **166:**740–746.

Quinby, G. E., Walker, K. C., and Durham, W. F., 1958, Public health hazards involved in the use of organic phosphorus insecticides in cotton culture in the delta area of Mississippi, *J. Econ. Entomol.* **51:**831–838.

Rake, D. W., 1961, Some studies on photochemical and soil microorganism decomposition of granular organo-borate herbicides, *Weed Soc. Am. Abst.*, pp. 48–49.

Reese, C. D., 1972, *Pesticides in the Aquatic Environment*, Environmental Protection Agency, Office of Water Programs, Washington, D.C.

Reinert, R. E., 1967, The accumulation of dieldrin in an alga (*Scenedesmus obliquus*), daphnia (*Daphnia magna*), guppy (*Lebistes reticulatus*) food chain, Ph.D. dissertation, University of Michigan, *Diss. Abst.* **28B:**2210-B.

Risebrough, R. W., Huggett, R. J., Griffin, J. J., and Goldberg, E. D., 1968, Pesticides: Transatlantic movements in the Northeast Trades, *Science* **159:**1233–1236.

Robinson, J., Richardson, A., Bush, B., and Elgar, K. E., 1966, A photoisomerization product of dieldrin, *Bull. Environ. Contam. Toxicol.* **1:**127–132.

Rosen, J. D., Sutherland, D. J., and Lipton, G. R., 1966, The photochemical isomerization of dieldrin and endrin and effects on toxicity, *Bull. Environ. Contam. Toxicol.* **1:**133–140.

Sijpesteijn, A. K., Kaslander, J., and Von der Kech, G. J. M., 1962, On the conversion of sodium dimethyldithiocarbamate into its α-aminobutyric acid derivative by microorganisms, *Biochim. Biophys. Acta*, **62:**587–589.

Simpson, G. R., 1965, Exposure to orchard pesticides, *Arch Environ. Health*, **10:**884–885.

Skipper, H. D., Gilmour, C. M., and Furtick, W. R., 1967, Microbial versus chemical degradation of atrazine in soil, *Soil Sci. Soc. Am. Proc.* **31:**653.

Slade, P., 1965, Photochemical degradation of paraquat, *Nature* (*London*) **207:**515–516.

Spencer, W. F., and Cliath, M. M., 1969, Vapor density of dieldrin, *Environ. Sci. Technol.* **3:**670–674.

Spencer, W. F., and Cliath, M. M., 1970*a*, Vapor density and apparent vapor pressure of lindane, *J. Agr. Food Chem.* **18:**529–530.

Spencer, W. F., and Cliath, M. M., 1970*b*, Desorption of lindane from soil as related to vapor density, *Soil Sci. Soc. Am. Proc.* **34:**574–578.

Spencer, W. F., Cliath, M. M., and Farmer, W. J., 1969, Vapor density of soil-applied dieldrin as related to soil water content, temperature and dieldrin concentration, *Soil Sci. Soc. Am. Proc.* **33:**509–511.

Spencer, W. F., Farmer, W. J., and Cliath, M. M., 1973, Pesticide volatilization, *Residue Rev.* **49:**1–47.

Stevens, L. J., Colier, C. W., and Woodham, D. W., 1970, Monitoring pesticides in soils from areas of regular, limited, and no pesticide use, *Pestic. Monit. J.* **4:**45–166.

Stojanovic, B. J., Kennedy, M. V., and Shuman, F. L., Jr., 1972, Edaphic aspects of the disposal of unused pesticides, pesticide wastes, and pesticide containers, *J. Environ. Qual.* **1:**54–62.

Sumerford, W. T., Hayes, W. J., Jr., Johston, J. M., Walker, K. C., and Spillane, J. T., 1953, Cholinesterase response and symptomatology from exposure to organic phosphorous insecticides, *Arch. Ind. Hyg.* **7:**383.

Tamés, R. S., and Hance, R. J., 1969, The adsorption of herbicides by roots, *Plant and Soil* **30:**221–226.

Terriere, L. C., Kiigemagi, U., Gerlach, A. R., and Borovicka, R. L., 1966, The persistence of toxaphene in lake water and its uptake by aquatic plants and animals, *J. Agr. Food Chem.* **14:**66–69.

Thorton, G. F., and Walker, B. A., 1970, *Summary of Pesticide Use and Pesticide Container Disposition in Tennessee*, Tennessee Department of Agriculture Publication, Ellington Agriculture Center, Nashville, Tenn.

Tiedje, J. M., and Alexander, M., 1967, Microbial degradation of organophosphorus insecticides and alkyl phosphates, *Abst. Am. Soc. Agron. Ann. Meet.*, p. 94.

Turner, N. J., and Corden, M. E., 1963, Decomposition of sodium *N*-methyldithiocarbamate in soil, *Phytopathology* **53:**1388–1394.

Upshall, D. G., and Goodwin, T. W., 1964, Some biochemical investigations into the susceptibility of barley varieties to DDT, *J. Sci. Food. Agr.* **15:**846–855.

U.S. Department of Health, Education, and Welfare, 1969, *Reports of the Secretary's Commission on Pesticides and Their Relationship to Environmental Health,* Parts I and II, Washington, D.C.

Wagner-Jauregg, T., Hackley, B. E., Jr., Lieg, T. A., Owens, O. O., and Proper, R., 1955, Model reactions of phosphorus-containing enzyme inactivators. IV. The catalytic activity of certain metal salts and chelates in the hydrolysis of diisopropyl fluorophosphate, *J. Am. Chem. Soc.* **77:**922–929.

Ware, G. W., Esteson, B. J., and Cahill, W. P., 1968, An ecological study of DDT residues in Arizona soils and alfalfa, *Pestic. Monit. J.* **2:**129–132.

Wedemeyer, G., 1967, Dechlorination of 1,1,1-trichloro-2,2-bis(*p*-chlorophenyl)ethane by *Aerobacter aerogenes.* I. Metabolic products, *Appl. Microbiol.* **15:**569–574.

West, I., 1964, Occupational disease of farm workers, *Arch Environ. Health* **9:**92–98.

West I., 1966, Biological effects of pesticides in the environment, *in: Organic Pesticides in the Environment,* pp. 38–53, American Chemical Society, Advances in Chemistry Series 60, Washington, D.C.

Westlake, W. E., and Gunther, F. A., 1966, Occurrence and mode of introduction of pesticides in the environment, *in: Organic Pesticides in the Environment,* pp. 110–121, American Chemistry Society, Advances in Chemistry Series 60, Washington, D.C.

Wheatley, G. A., 1973, Pesticides in the atmosphere, *in: Environmental Pollution by Pesticides* (C. A. Edwards, ed.), pp. 365–408, Plenum Press, New York.

Wolfe, H. R., 1972, Protection of workers from exposure to pesticides, *Pest Control* **40(2):**17–18, 20, 38, 40, 42.

Wolfe, H. R., Durham, W. F., and Armstrong, J. F., 1967, Exposure of workers to pesticide, *Arch. Environ. Health* **14:**622–633.

Wolfe, H. R., Armstrong, J. F., Staiff, D. C., and Comer, S. W., 1972, Exposure of spraymen to pesticides, *Arch. Environ. Health* **25:**29–31.

Yule, W. W., Chiba, M., and Morley, H. V., 1967, Fate of insecticide residues: Decomposition of lindane in soil, *J. Agr. Food Chem.* **15:**1000–1004.

Chemical Index

Throughout the text of the book insecticides, drugs, and other chemicals are frequently referred to only by common name or manufacturers' designation. Since these provide little or no information concerning chemical class or structure, the following index has been compiled to afford the reader simple and rapid access to this information. Chemicals are listed alphabetically by the common names appearing in the text, and the chemical names are in general accordance with the principles of the Chemical Abstract Service of the American Chemical Society. The index was compiled mainly from the following:

E. E. Kenaga and W. A. Allison, 1969, Commercial and experimental organic insecticides, *Bull. Entomol. Soc. Am.* 15:85.

E. Y. Spencer, 1973, Guide to the chemicals used in crop protection, 6th ed., Research Branch, Canada Department of Agriculture, Ottawa, Canada.

Common Name or Designation	Chemical Name	Structure
Acethion	*O,O*-Diethyl-*S*-carboethoxymethyl phosphorodithioate	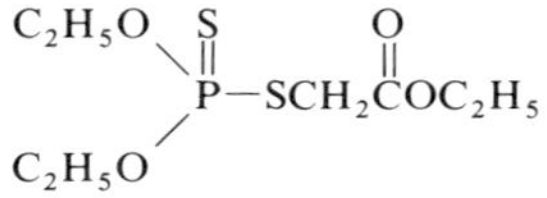
Abate	*O,O,O′,O′*-Tetramethyl-*O,O′*-thiodi-*p*-phenylene phosphorothioate	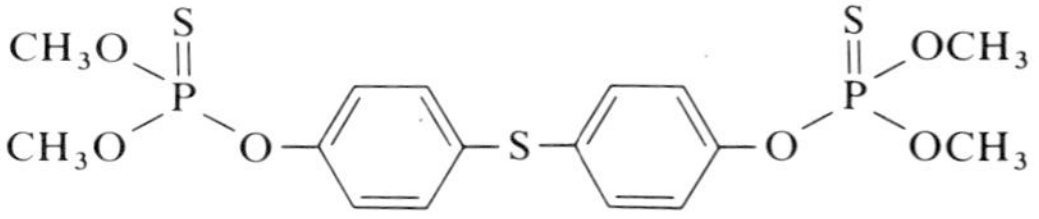
Aldicarb	2-Methyl-2-(methylthio)propionaldehyde-*O*-(methylcarbamoyl)oxime	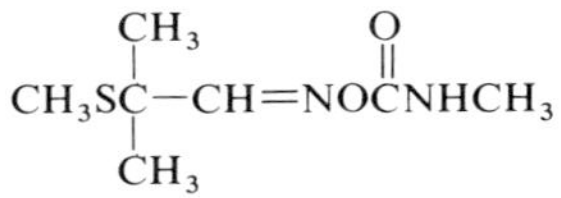
Aldrin	1,2,3,4,10,10-Hexachloro-1,4,4a,5,8,8a-hexahydro-1,4-*endo*, *exo*-5,8-dimethanonaphthalene	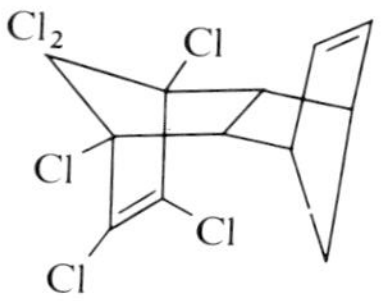
Allethrin	*dl*-2-Allyl-4-hydroxy-3-methyl-2-cyclopenten-1-one ester of *cis* and *trans-dl*-chrysanthemic acid	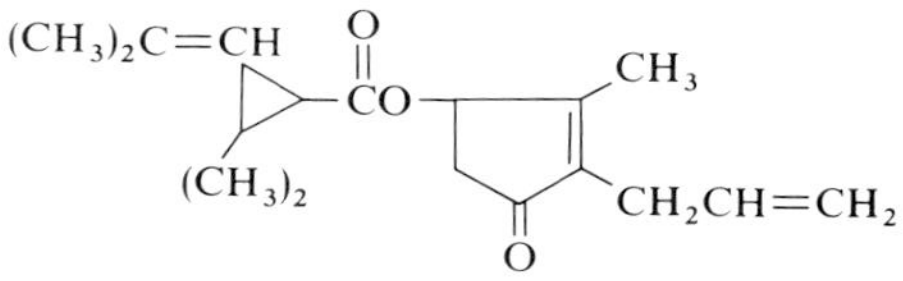

Alodan	1,2,3,4,7,7a-Hexachloro-5,6-bis(chloromethyl)-2-norbornene	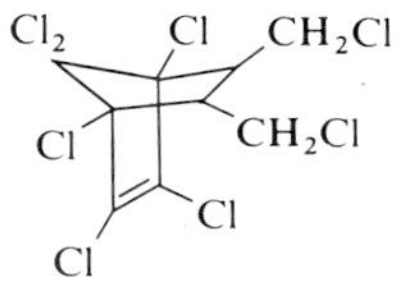
Aminocarb	4-Dimethylamino-*m*-tolyl methylcarbamate	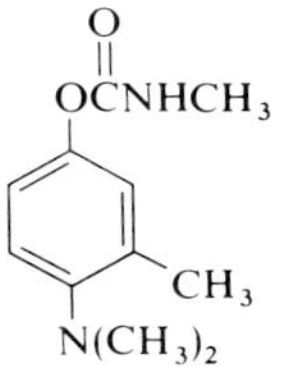
Aminopyrine	4-Dimethylamino-1,5-dimethyl-2-phenyl-3-pyrazolone	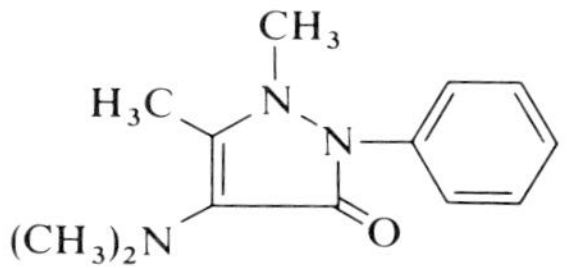
Amiton	*O*,*O*-Diethyl-*O*-2-diethylaminoethyl phosphorothioate	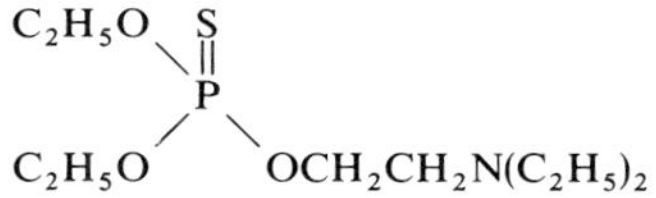
Amobarbital (sodium)	Sodium 5-ethyl-5-isoamylbarbiturate	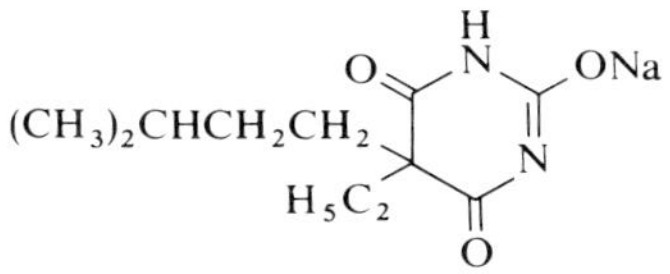

Common Name or Designation	Chemical Name	Structure
Amytal	(See Amobarbital, sodium)	
Anabasine	*l*-3-(2′-Piperidyl) pyridine	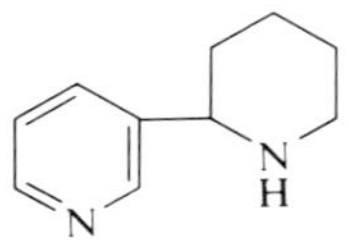
Antipyrine	1,5-Dimethyl-2-phenyl-3-pyrazolone	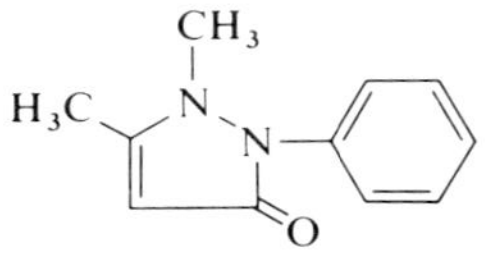
Apholate	2,2,4,4,6,6-Hexakis(1-aziridinyl)-2,2,4,4,6,6-hexahydro-1,3,5,2,4,6-triazatriphosphorine	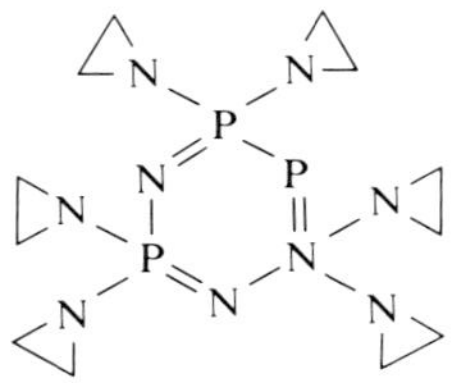
Arprocarb	(See Propoxur)	
Aspon	*O,O,O,O*-Tetrapropyl dithiopyrophosphate	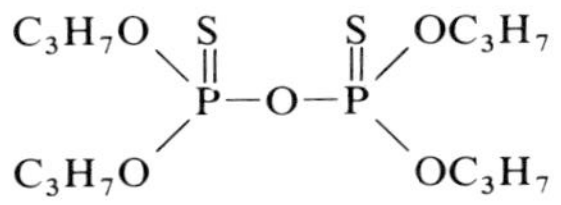

Atrazine	2-Chloro-4-ethylamino-6-isopropylamino-1,3,5-triazine	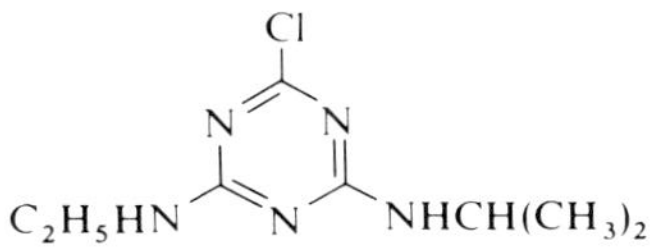
Atropine	*dl*-Tropyl tropate	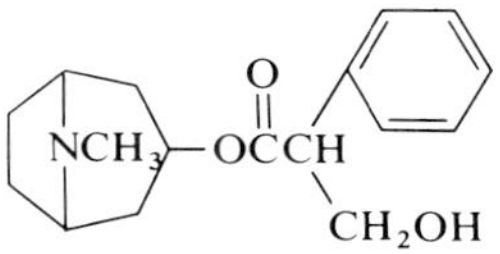
Azinphosethyl	*O,O*-Diethyl analog of azinphosmethyl	—
Azinphosmethyl	*O,O*-Dimethyl-*S*-[4-oxo-1,2,3-benzotriazin-3(4H)-ylmethyl] phosphorodithioate	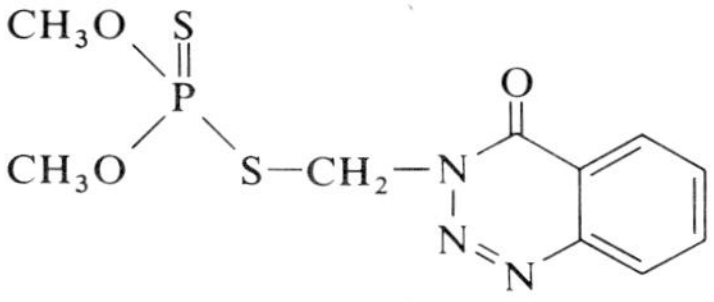
Azodrin	(See Monocrotophos)	—
Banol	(See Carbanolate)	—
Barbital (sodium)	Sodium 5,5-diethylbarbiturate	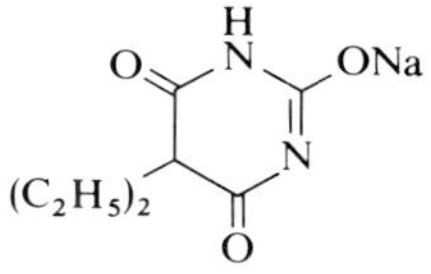

Common Name or Designation	Chemical Name	Structure
Barthrin	6-Chloropiperonyl (+)-*trans*-chrysanthemate	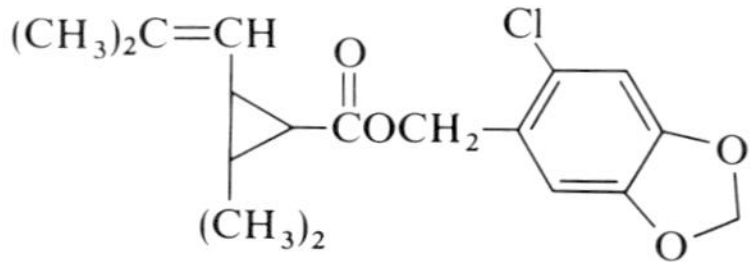
Baygon	(See Propoxur)	—
BHC	(See HCH isomers)	—
Bidrin	(See Dichrotophos)	—
Bioresmethrin	5-Benzyl-3-furylmethyl (+)-*trans*-chrysanthemate	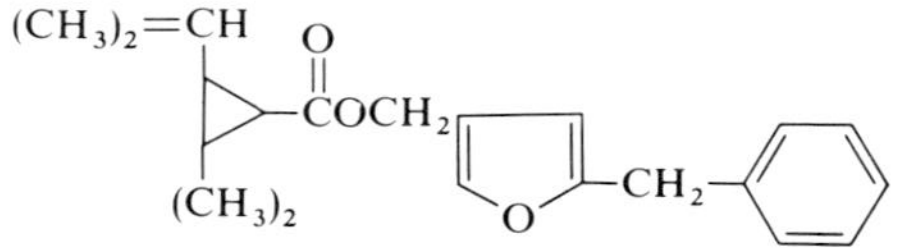
Bromophos	*O*-(4-Bromo-2,5-dichlorophenyl)-*O*,*O*-dimethyl phosphorothioate	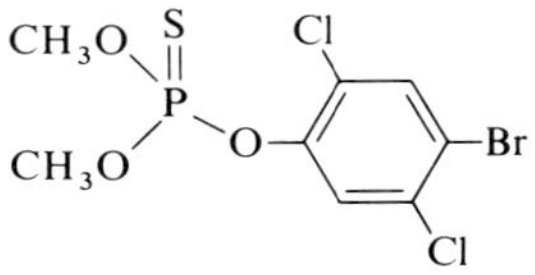
Bux	Commercial insecticide contains mixture of *m*-(1-methylbutyl)-phenyl methylcarbamate and *m*-(1-ethylpropyl)phenyl methylcarbamate in a 4:1 ratio	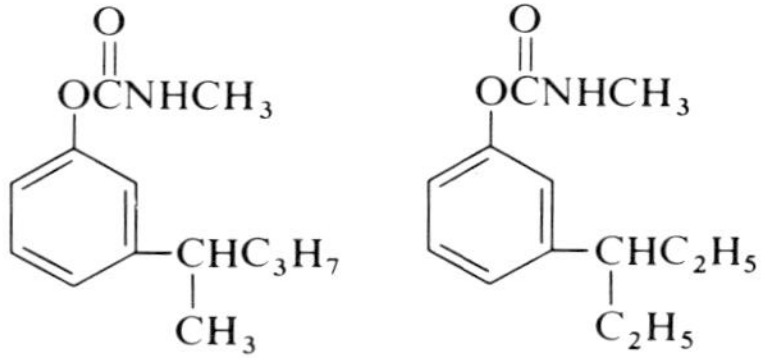

Butacarb	3,5-Di-*tert*-butylphenyl methylcarbamate	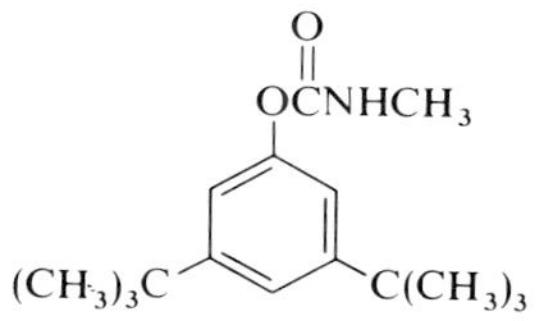
Carbanolate	6-Chloro-3,4-dimethylphenyl methylcarbamate	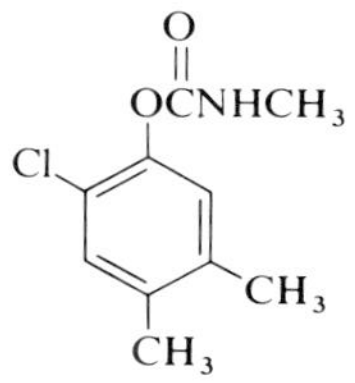
Carbaryl	1-Naphthyl methylcarbamate	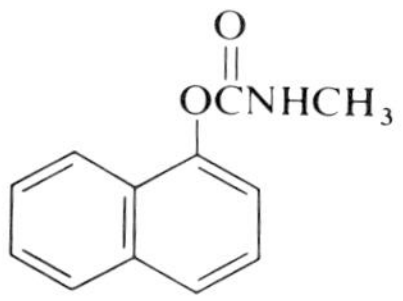
Carbophenothion	*S*-[(*p*-Chlorophenylthio)methyl]-*O*,*O*-diethyl phosphorodithioate	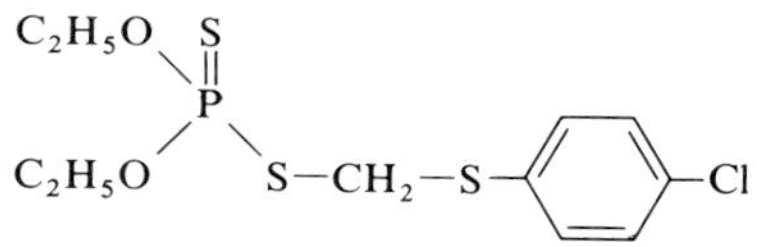
Carbofuran	2,3-Dihydro-2,2-dimethyl-7-benzofuranyl methylcarbamate	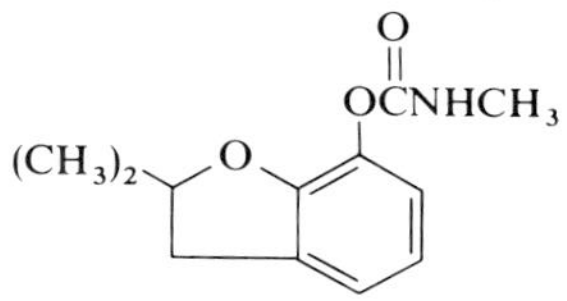

Common Name or Designation	Chemical Name	Structure
Cartap	Thiocarbamic acid *S,S*-[2-(dimethylamino)trimethylene]ester hydrochloride	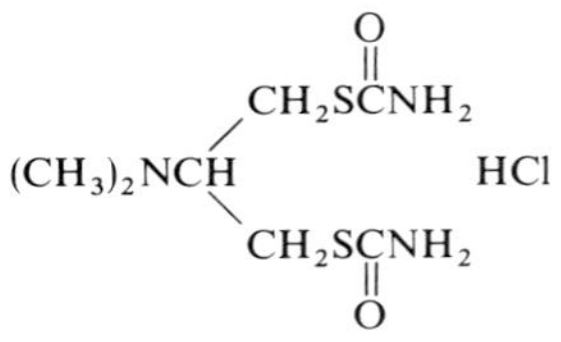
Chlorcyclizine (HCl)	1-(4-Chlorobenzhydryl)-4-methylpiperazine dihydrochloride	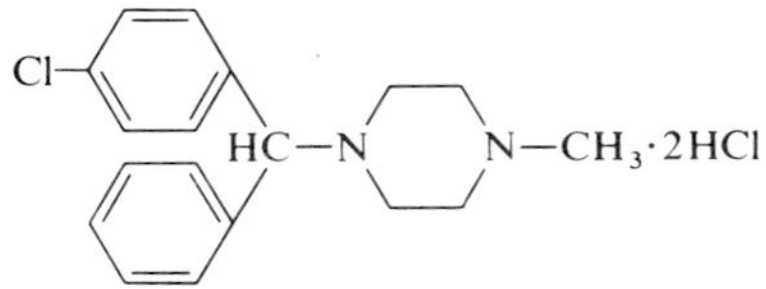
Chlordane	1,2,4,5,6,7,8,8-Octachloro-2,3,3a,4,7,7a-hexahydro-4,7-*endo*-methanoindene	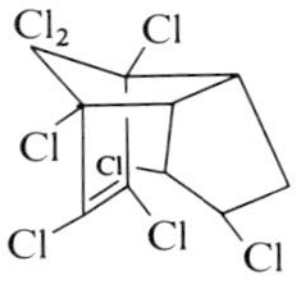
Chlordecone	(See Kepone)	—
Chlordene	4,5,6,7,8,8-Hexachloro-3a,4,7,7a-tetrahydro-4,7-*endo*-methanoindene	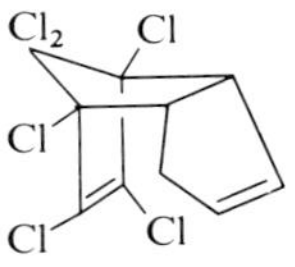

Chlordimeform	*N'*-(4-Chloro-*o*-tolyl)-*N*,*N*-dimethylformamidine	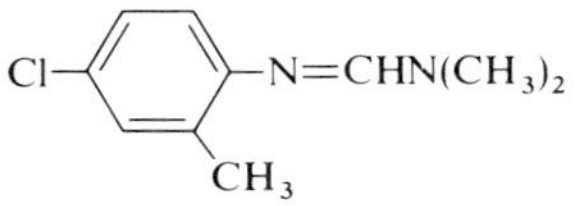
Chlorfenamidine	(See Chlordimeform)	—
Chlorfenvinphos	*O*-2-Chloro-1-(2,4-dichlorophenyl)vinyl-*O*,*O*-diethyl phosphate	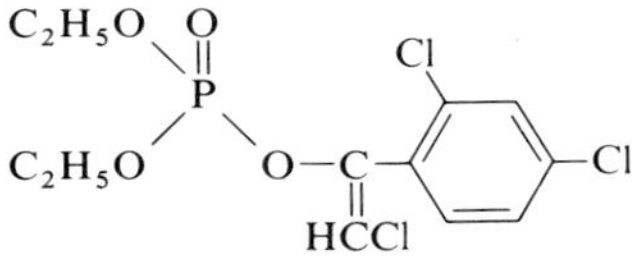
Chlorpromazine (HCl)	10-(3'-Dimethylaminopropyl)-3-chlorophenothiazine HCl	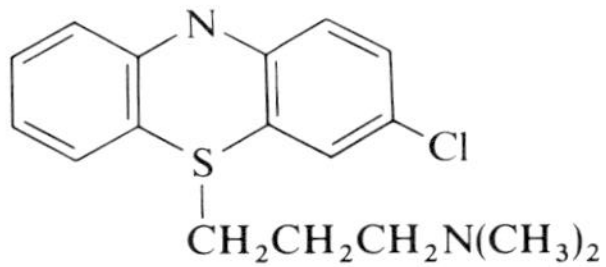
Chlorthion	*O*-3-Chloro-4-nitrophenyl-*O*,*O*-dimethyl phosphorothioate	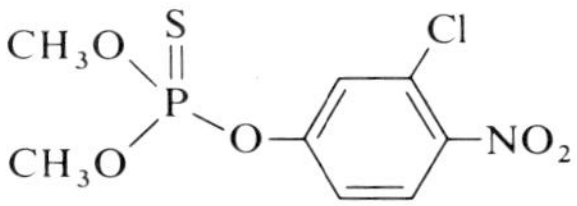
Co-Ral	(See Coumaphos)	—
Coroxon	Oxon analog of Coumaphos	—
Coumaphos	*O*,*O*-Diethyl-*O*-3-chloro-4-methyl-2-oxo-2H-1-benzopyran-7-yl phosphorothioate	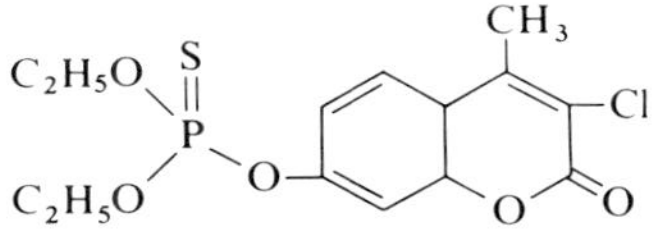

Common Name or Designation	Chemical Name	Structure
Cyclethrin	(±)-3-(Cyclopent-2-enyl)-4-keto-2-methylcyclopent-2-enyl (+)-*trans*-chrysanthemate	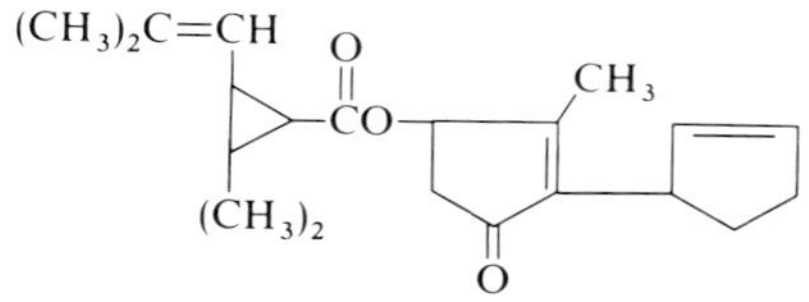
2,4-D	2,4-Dichlorophenoxyacetic acid	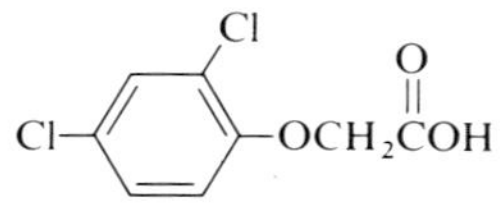
Dalapon	Sodium 2,2-dichloropropionate	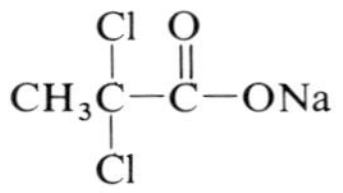
DDA	Bis-*p*-chlorophenyl acetic acid	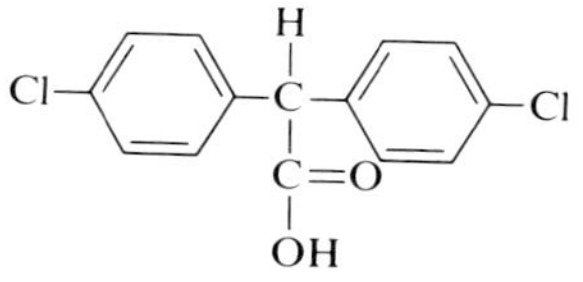
DDD	2,2-Bis(*p*-chlorophenyl)-1,1-dichloroethane	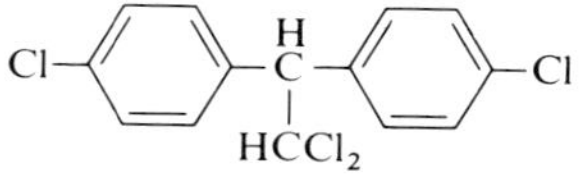

DDE	2,2-Bis(*p*-chlorophenyl)-1,1-dichloroethylene	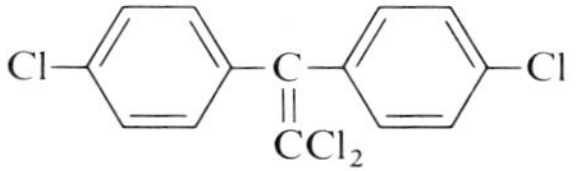
DDMU	2,2-Bis(*p*-chlorophenyl)-1-chloroethylene	
DDT	2,2-Bis(*p*-chlorophenyl)-1,1,1-trichloroethane	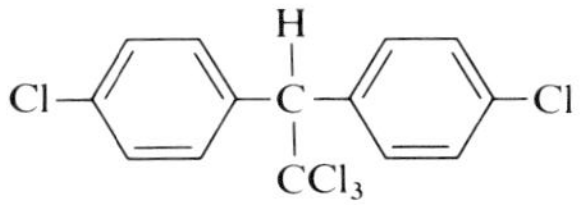
DDVP	(See Dichlorvos)	—
Decamethonium	Decomethylene-bis(trimethylammonium iodide)	$[(CH_3)_3N^+(CH_2)_{10}N^+(CH_3)_3]2I^-$
DEF	*S,S,S*-Tributyl phosphorotrithioate	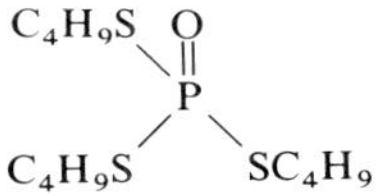
Delnav	(See Dioxathion)	—
Demeton	*O,O*-Diethyl-*S*- (and *O*-) 2-[(ethylmercapto)ethyl] phosphorothioates	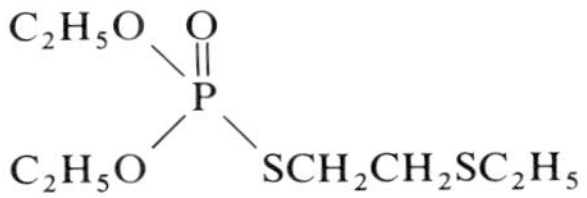 (thiolo isomer)

Common Name or Designation	Chemical Name	Structure
DFP	*O,O*-Diisopropyl phosphorofluoridate	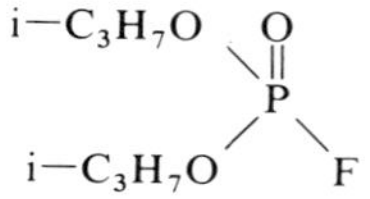
Diazinon	*O,O*-Diethyl-*O*-(2-isopropyl-4-methyl-6-pyrimidyl)-phosphorothioate	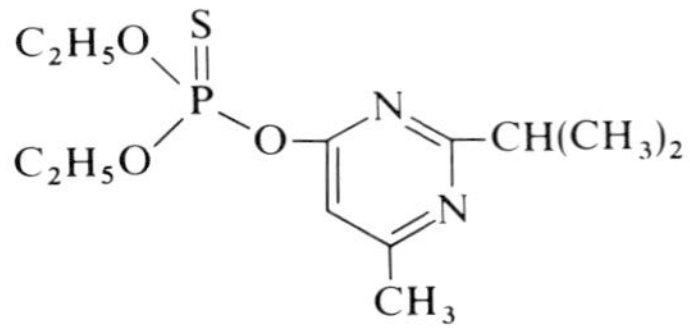
Diazoxon	Oxon analog of Diazinon	—
Dicofol	1,1-Bis(*p*-chlorophenyl)-2,2,2-trichloroethanol	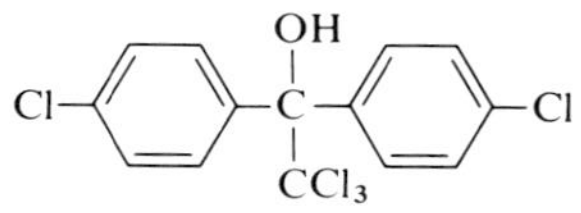
Dicrotophos	3-(Dimethoxyphosphinoxy)-*N,N*-dimethyl-*cis*-crotonamide	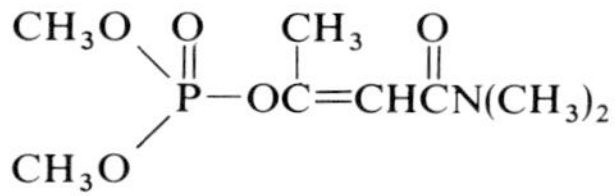
Dichlorvos	*O*-2,2-Dichlorovinyl-*O,O*-dimethyl phosphate	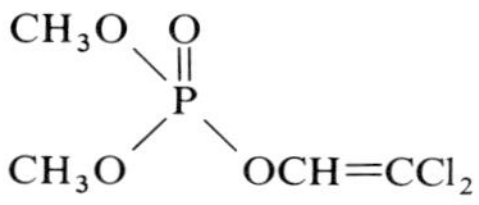

Dieldrin (HEOD)	1,2,3,4,10,10-Hexachloro-*exo*-6,7-epoxy-1,4,4a,5,6,7,8,8a-octahydro-1,4-*endo*, *exo*-5,8-dimethanonaphthalene	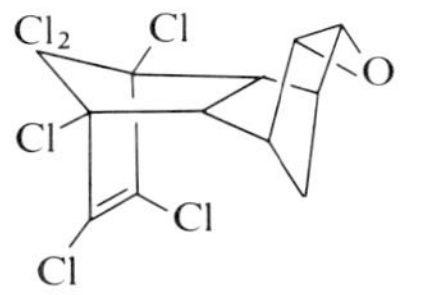
Dimethoate	*O*,*O*-dimethyl-*S*-(*N*-methylcarbamoylmethyl)-phosphorodithioate	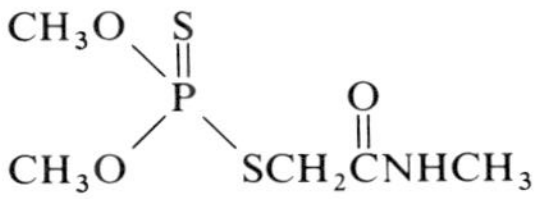
Dimethoxon	Oxon analog of Dimethoate	—
Dimethrin	2,4-Dimethylbenzyl(+)-*trans*-chrysanthemate	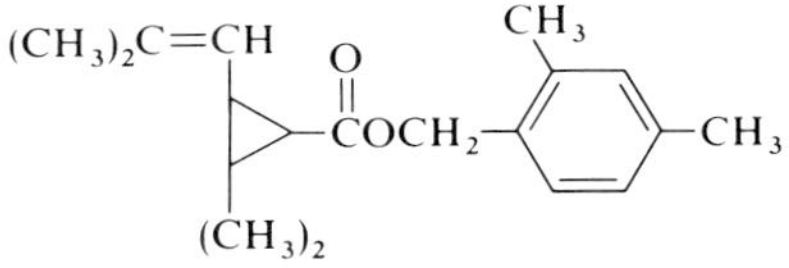
Dimetilan	1-(Dimethylcarbamoyl)-5-methyl-3-pyrazolyl dimethylcarbamate	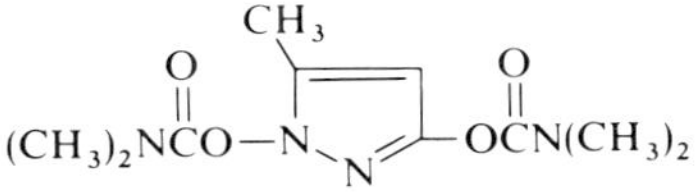
Dioxathion	2,3-*p*-Dioxane(dithiol)-*S*,*S*-bis(*O*,*O*-diethylphosphoro dithioate	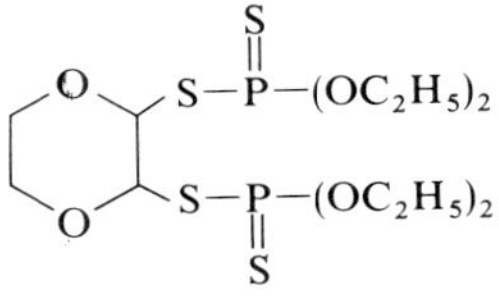

Common Name or Designation	Chemical Name	Structure
Diphenylhydantoin (sodium)	5,5-Diphenyl-2,4-imidazolidinedione sodium	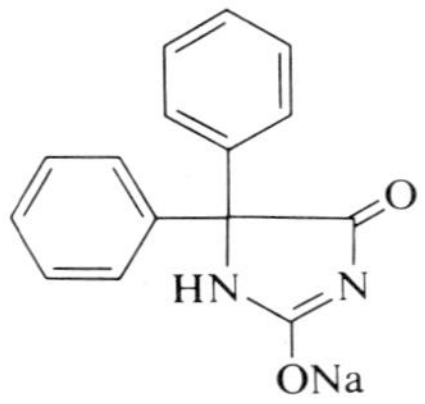
Dipterex	(See Trichlorfon)	—
Disulfoton	*O,O*-Diethyl-*S*-2-[(ethylthio)ethyl]phosphorothioate	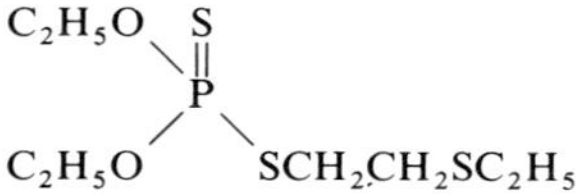
Disyston	(See Disulfoton)	—
Dithion	(See Sulfotep)	—
DMC	4,4′-Dichloro-α-methylbenzhydrol	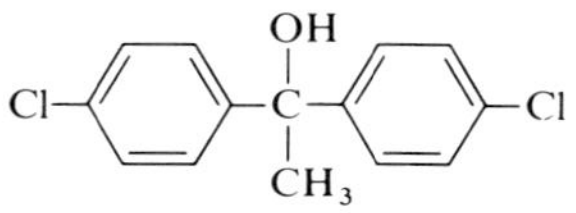
Dyfonate	*O*-Ethyl-*S*-phenyl ethylphosphonodithioate	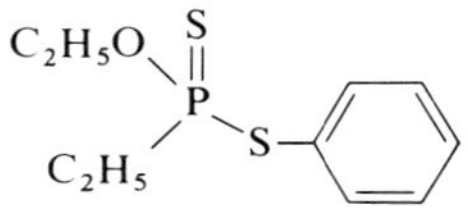

Endosulfan	6,7,8,9,10,10-Hexachloro-1,5,5a,6,9,9a-hexahydro-6,9-methano-2,4,3-benzodioxathiepin-3-oxide	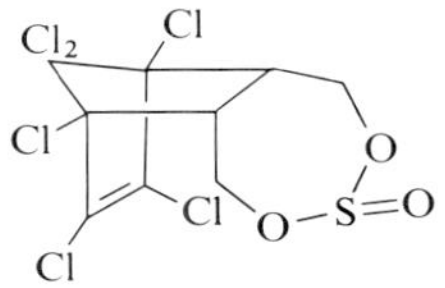
Endrin	1,2,3,4,10,10-Hexachloro-6,7-epoxy-1,4,4a,5,6,7,8,8a-octahydro-1,4-*endo,endo*-5,8-dimethanonaphthalene	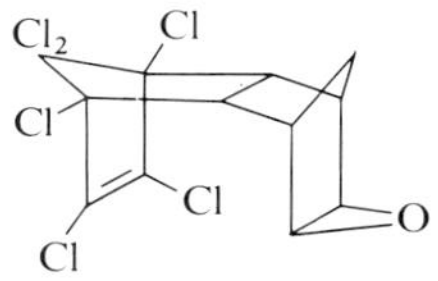
EPN	*O*-Ethyl-*O*-*p*-nitrophenyl phenylphosphorothioate	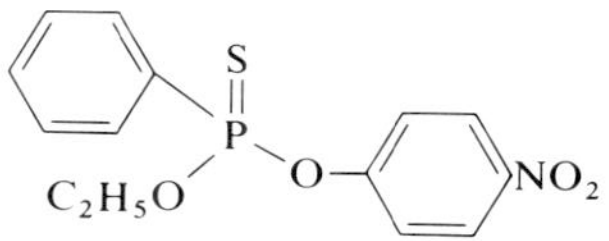
EPNO	Oxon analog of EPN	
Eserine	5-Methylcarbamoyloxy-1,3a,8-trimethyl-2,3,3a,8a-tetrahydro-pyrrolo-[2,3-b]-indol	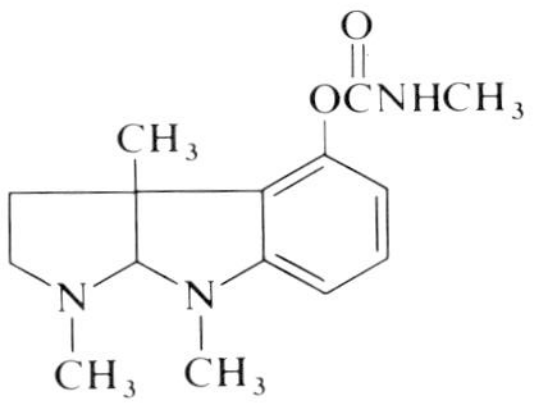
Ethion	*O,O,O′,O′*-Tetraethyl-*S,S′*-methylenebisphosphorodithioate	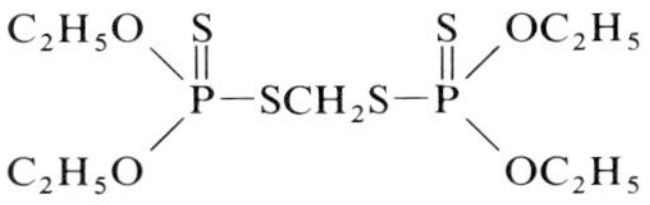

Common Name or Designation	Chemical name	Structure
Ethyl Chlorthion	*O,O*-Diethyl analog of Chlorthion	—
Ethyl Dichlorvos	*O,O*-Diethyl analog of Dichlorvos	—
Ethyl Guthion	(See Azinphosethyl)	—
Famphur	*O*-[4-(dimethylsulfamoyl)phenyl]-*O,O*-dimethyl phosphorothioate	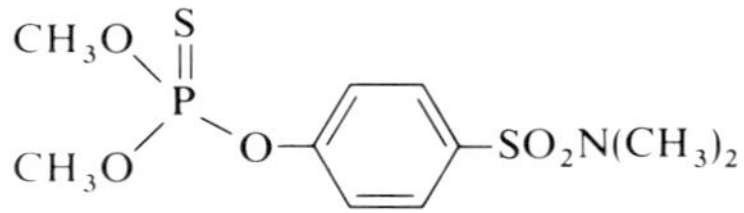
FDMC	4,4′-Dichloro-α-trifluoromethylbenzhydrol	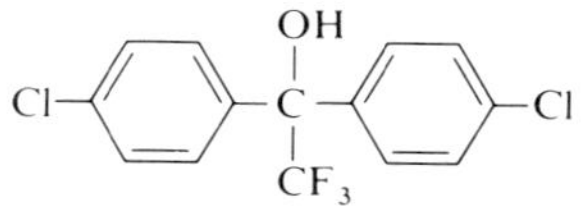
Fenethcarb	3,5-Diethylphenyl methylcarbamate	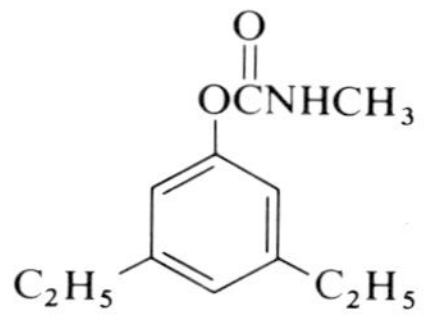
Fenitrothion	*O,O*-Dimethyl-*O*-(3-methyl-4-nitrophenyl)phosphorothioate	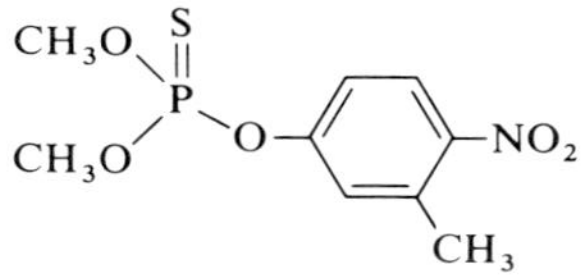

Fenthion	*O*,*O*-Dimethyl-*O*-[4-(methylthio)-*m*-tolyl]phosphorothioate	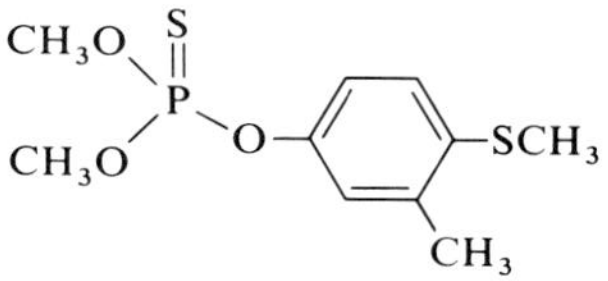
Folex	*S*,*S*,*S*-Tributyl phosphorotrithioite	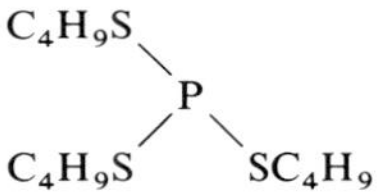
Furadan	(See Carbofuran)	—
Gardona	(See Tetrachlorvinphos)	—
Griseofulvin	7-Chloro-4,6-dimethoxycoumaran-3-one-2-spiro-1′-(2′-methoxy-6′-methylcyclohex-2′-en-4′-one)	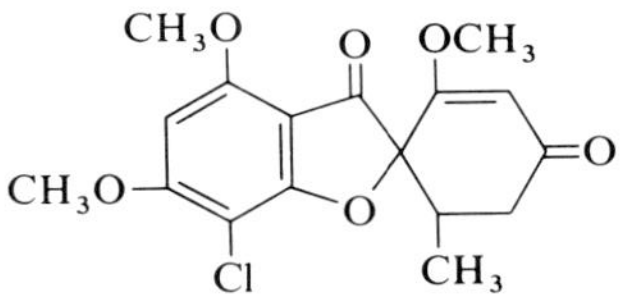
Guthion	(See Azinphosmethyl)	—
HCH	Isomers of 1,2,3,4,5,6-hexachlorocyclohexane	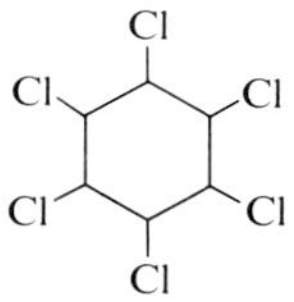
Hemel	2,4,6-Dimethylamino-s-triazine	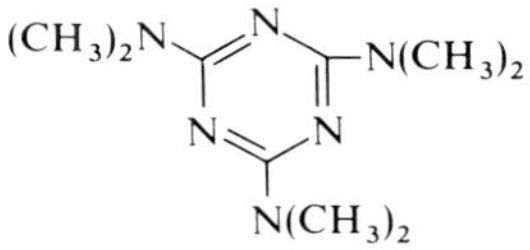

Common Name or Designation	Chemical Name	Structure
Hempa	Hexamethylphosphorictriamide	$(CH_3)_2N-\overset{\overset{O}{\parallel}}{\underset{\underset{N(CH_3)_2}{\mid}}{P}}-N(CH_3)_2$
Heptabarbital (sodium)	Sodium 5-(1-cyclohepten-1-yl)-5-ethylbarbiturate	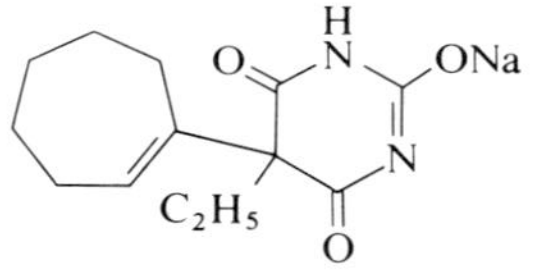
Heptachlor	1,4,5,6,7,8,8-Heptachloro-3a,4,7,7a-tetrahydro-4,7-*endo*-methanoindene	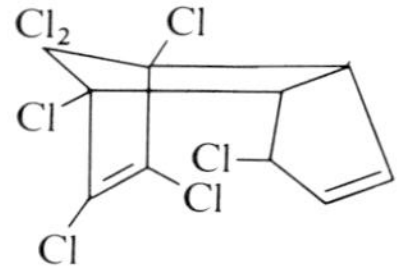
Hexamethonium	Hexamethylene-bis(trimethylammonium bromide)	$[(CH_3)_3N^+[CH_2]_6N^+(CH_3)_3]2Br^-$
Hexobarbital (sodium)	Sodium 5-(1-cyclohexen-1-yl)-1,5-dimethylbarbiturate	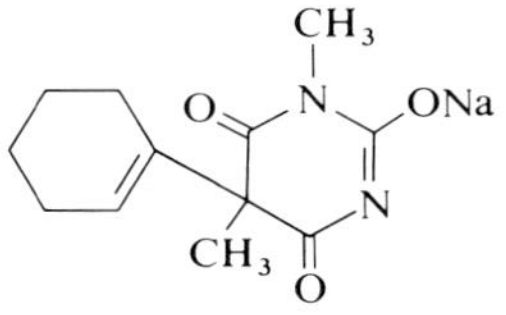
Hydroprene	Ethyl-3,7,11-trimethyldodeca-2,4-dienoate	

Imidan	(See Phosmet)	—
Isobenzan	*exo*-1-*exo*-3,4,5,6,7,8,8-Octachloro-1,3,3a,4,7,7a-hexahydro-4,7-methanoisobenzofuran	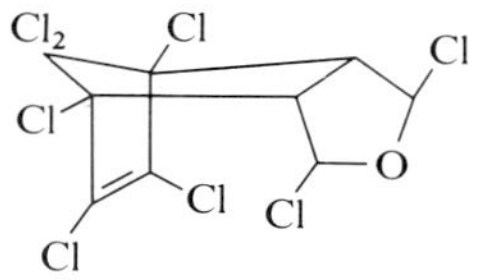
Isodrin	1,2,3,4,10,10-Hexachloro-1,4,4a,5,8,8a-hexahydro-1,4-*endo,endo*-5,8-dimethanonaphthalene	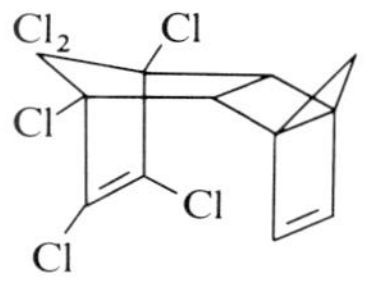
Isolan	1-Isopropyl-3-methyl-5-pyrazolyl dimethylcarbamate	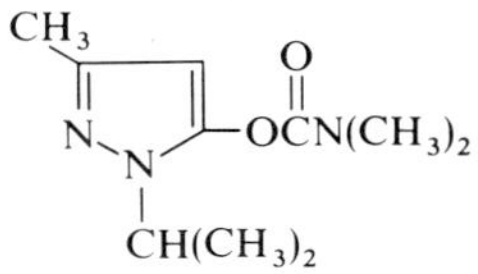
Isopropyl Diazinon	*O,O*-Diisopropyl analog of Diazinon	—
Isopropyl Paraoxon	*O,O*-Diisopropyl analog of Paraoxon	—
Kelthane	(See Dicofol)	—
Kepone	Decachlorooctahydro-1,3,4-metheno-2H-cyclobuta[c,d]pentalene-2-one	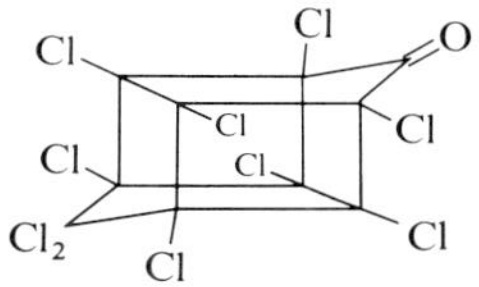

Common Name or Designation	Chemical Name	Structure
Landrin	Commercial insecticide contains SD 8786 (18%) and SD 8530 (75%)	
Lethane 60	2-Thiocyanoethyl esters of C_{10}–C_{18} aliphatic acids	
Lilly 18947	2-[(4,6-Dichloro-2-biphenylyl)]oxy triethylamine	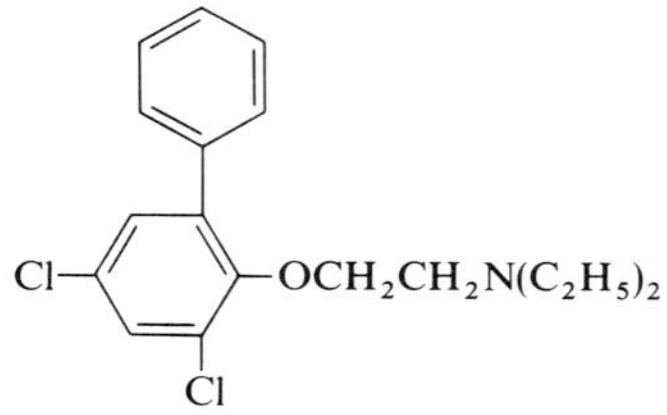
Lindane	(See HCH, γ-isomer)	—
Malaoxon	Oxon analog of Malathion	—
Malathion	*O,O*-Dimethyl-*S*-1,2-di(carboethoxy)ethyl phosphorodithioate	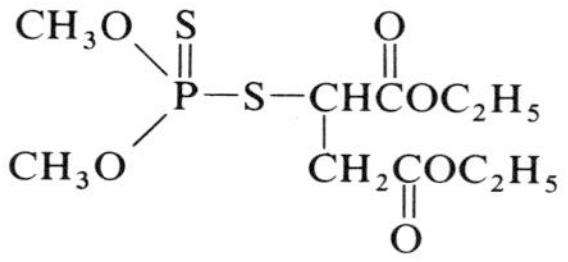
Menazon	*S*-[(4,6-Diamino-1,3,5-triazine-2-yl)methyl]-*O,O*-dimethyl phosphorodithioate	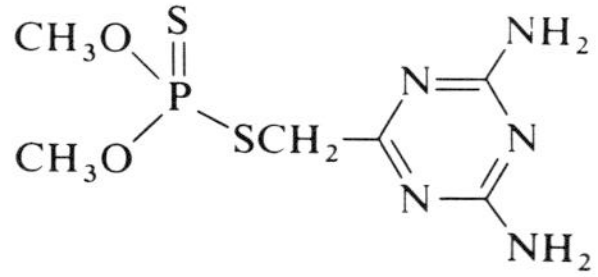

Meobal	3,4-Dimethylphenyl methylcarbamate	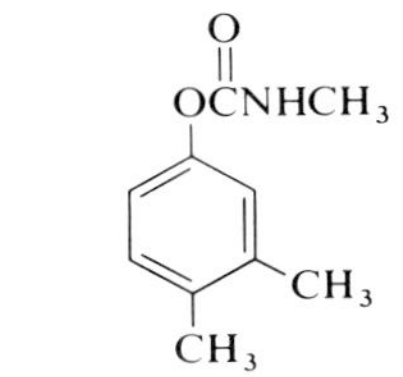
Mercaptodimethur	4-Methylthio-3,5-xylyl methylcarbamate	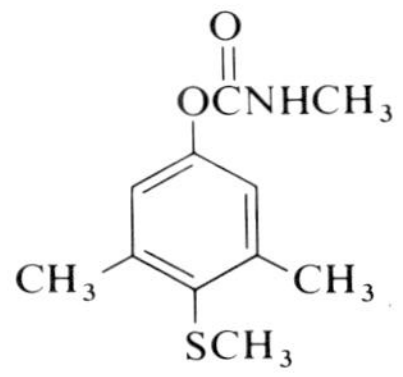
Mesurol	(See Mercaptodimethur)	—
Metaphoxide	Tris(2-methyl-1-aziridinyl)phosphine oxide	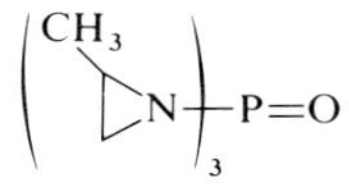
Methiocarb	(See Mercaptodimethur)	—
Methiochlor	2,2-Bis(*p*-methylthiophenyl)-1,1,1-trichloroethane	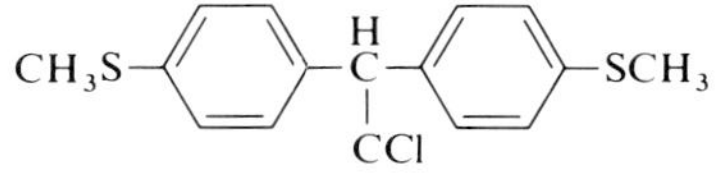
Methomyl	*S*-Methyl-*N*-[(methylcarbamoyl)oxy]thioacetimidate	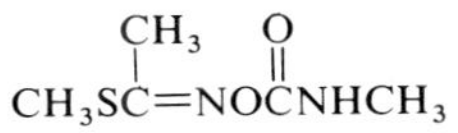

Common Name or Designation	Chemical Name	Structure
Methoprene	Isopropyl-11-methoxy-3,7,11-trimethyl dodeca-2,4-dienoate	
Methoxychlor	2,2-Bis(*p*-methoxyphenyl)-1,1,1-trichloroethane	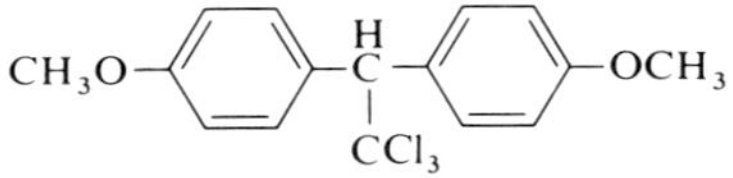
Methyl Acethion	*O*,*O*-Dimethyl analog of Acethion	—
Methyl Paraoxon	Oxon analog of Methyl Parathion	—
Methyl Parathion	*O*,*O*-Dimethyl-*O*-4-nitrophenyl phosphorothioate	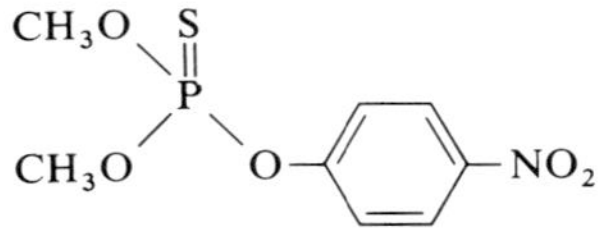
Mevinphos	*O*,*O*-Dimethyl-1-carbomethoxy-1-propen-2-yl phosphate	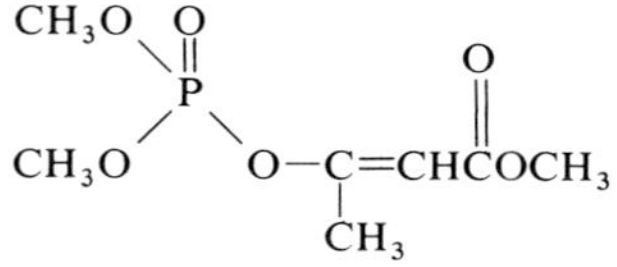
Mexacarbate	4-Dimethylamino-3,5-xylyl methylcarbamate	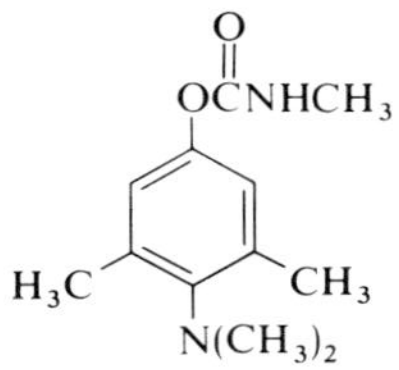

MGK 264	*N*-(2-ethylhexyl)-5-norbornene-2,3-dicarboximide	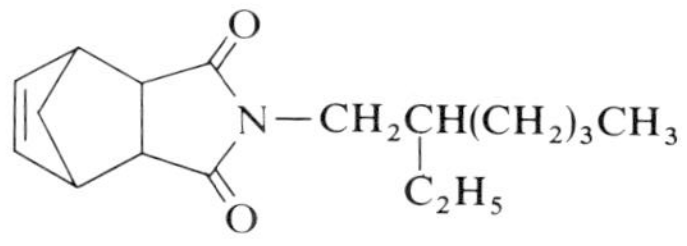
Mipafox	*N*,*N*-Diisopropylphosphorodiamidic fluoride	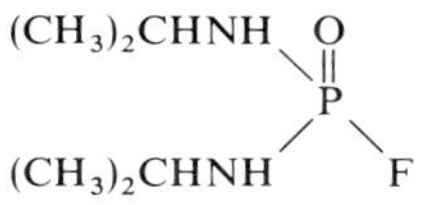
Mirex	Dodecachlorooctahydro-1,3,4-metheno-2H-cyclobuta-[c,d]pentalene	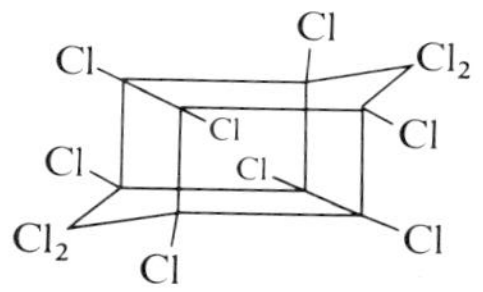
Mobam	Benzo[b]thien-4-yl methylcarbamate	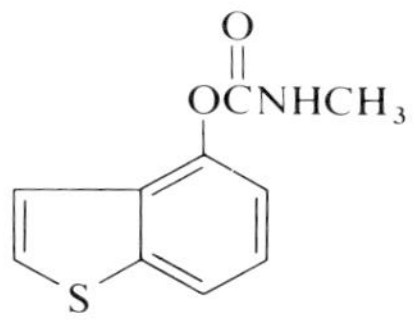
Monocrotophos	3-(Dimethoxyphosphinoxy)-*N*-methyl-*cis*-crotonamide	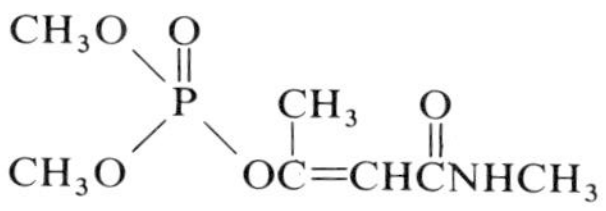

Common Name or Designation	Chemical Name	Structure
Mylone	3,5-Dimethyl-tetrahydro-1,3,5,2H-thiadiazine-2-thione	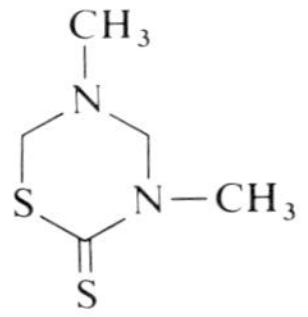
Neopynamin	(See Tetramethrin)	—
Nereistoxin	4-*N*,*N*-Dimethylamino-1,2-dithiolane	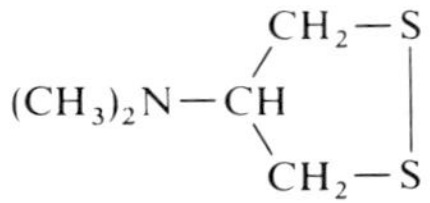
Nicotine	*l*-3(1-Methyl-2-pyrrolidyl)pyridine	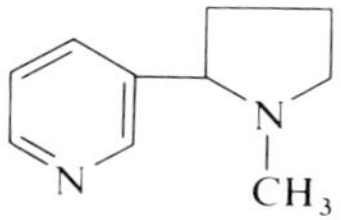
Nornicotine	*l*-3-(2-Pyrrolidyl)pyridine	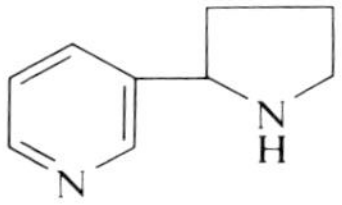
NRDC-108	5-Benzyl-3-furylmethyl-2,2,3,3-tetramethylcyclopropanecarboxylate	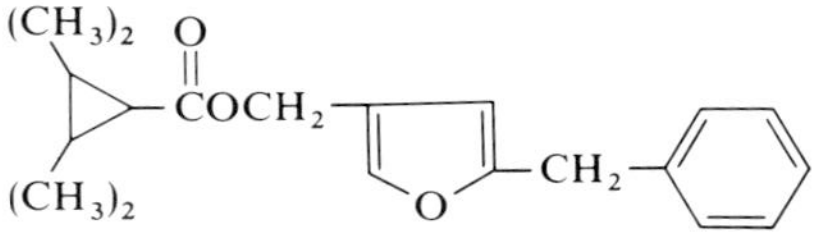

NRDC-119	5-Benzyl-3-furylmethyl(+)-*cis*-chrysanthemate	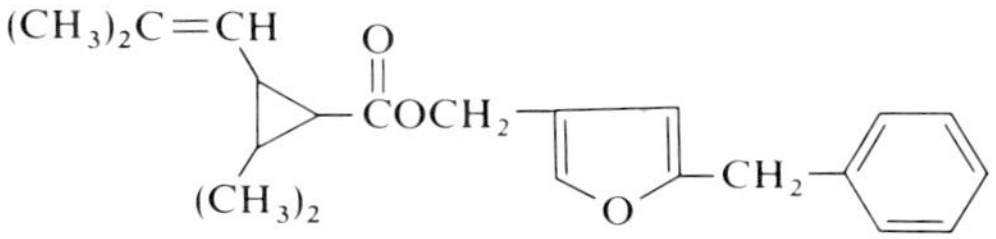
OMPA	(See Schradan)	—
Ovex	*p*-Chlorophenyl *p*-chlorobenzenesulphonate	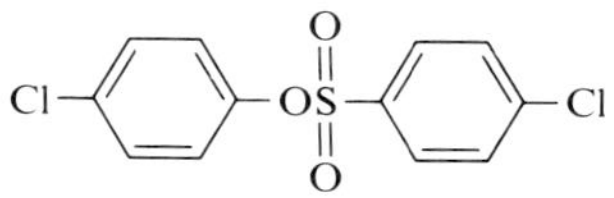
2-PAM	Pyridine-2-aldoxime methiodide	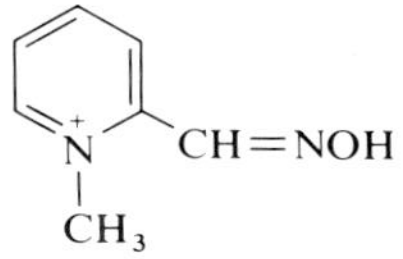
Paraquat	1,1′-Dimethyl-4,4′-bipyridylium ion	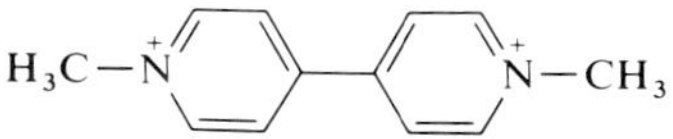
Paraoxon	Oxon analog of Parathion	—
Parathion	*O*,*O*-Diethyl-*O*-4-nitrophenyl phosphorothioate	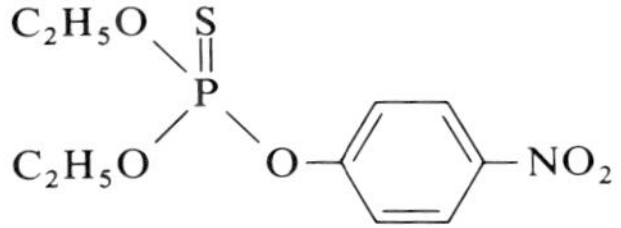

Common Name or Designation	Chemical name	Structure
PCP	Pentachlorophenol	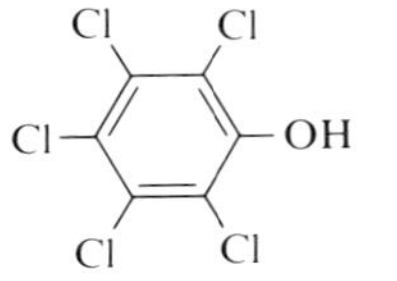
Pentobarbital (sodium)	Sodium 5-ethyl-5-(1-methylbutyl) barbiturate	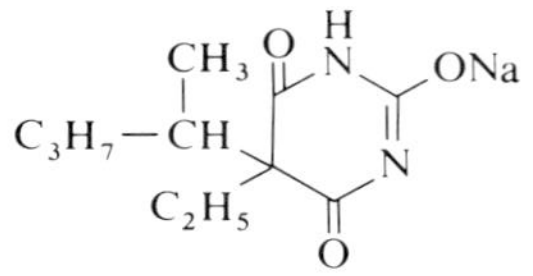
Permethrin	(3-Phenoxybenzyl)-3-(2,2-dichlorovinyl)-2,2-dimethylcyclopropanecarboxylate	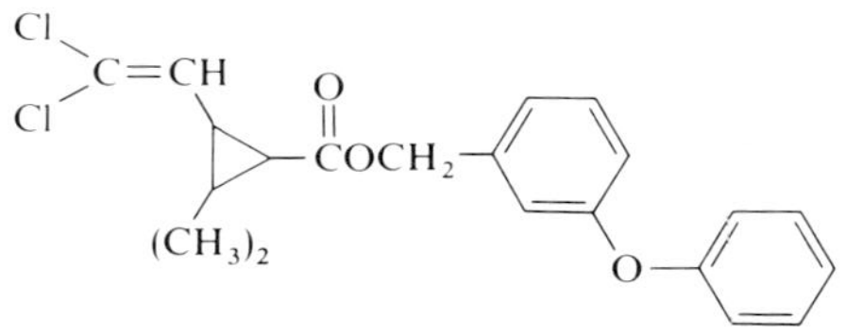
Phenobarbital (sodium)	Sodium 5-ethyl-5-phenyl barbiturate	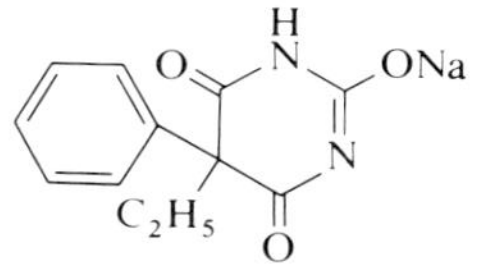

Phenylbutazone	1,2-Diphenyl-3,5-dioxo-4-*n*-butylpyrazolidin	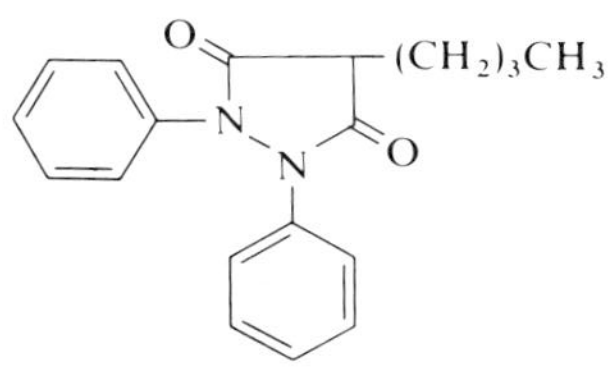
Phenytoin	(See Diphenylhydantoin, sodium)	—
Phorate	*O*,*O*-Diethyl-*S*-[(ethylthio)methyl]phosphorodithioate	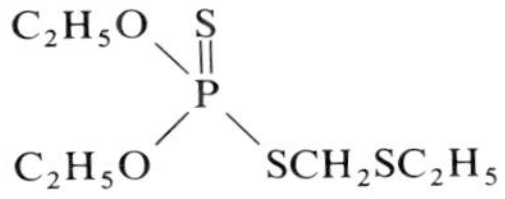
Phosalone	*O*,*O*-Diethyl-*S*-[(6-chlorobenzoxazalone-3-yl)methyl]-phosphorodithioate	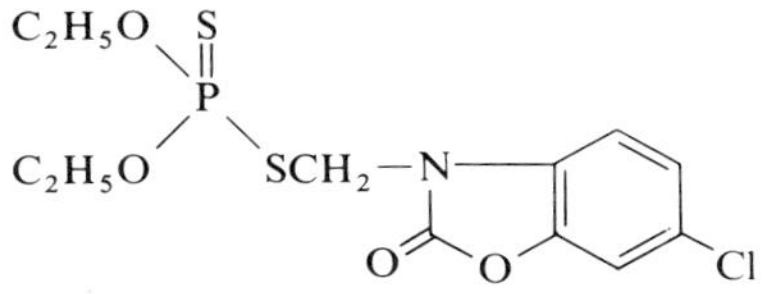
Phosdrin	(See Mevinphos)	—
Phosmet	*O*,*O*-Dimethyl-*S*-phthalimidomethyl phosphorodithioate	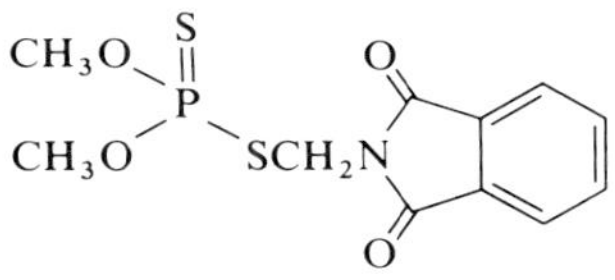
Phosphamidon	*O*-2-Chloro-2-diethylcarbamoyl-1-methylvinyl-*O*,*O*-dimethyl phosphate	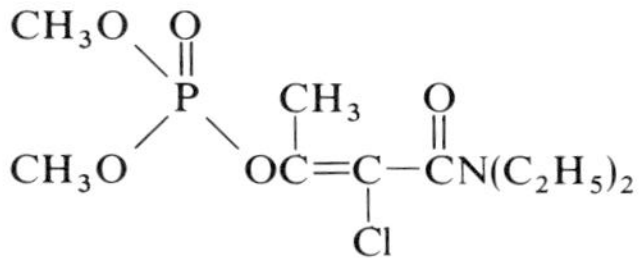

Common Name or Designation	Chemical Name	Structure
Phoxim	Phenylglyoxylonitrile oxime-*O*,*O*-diethyl phosphate	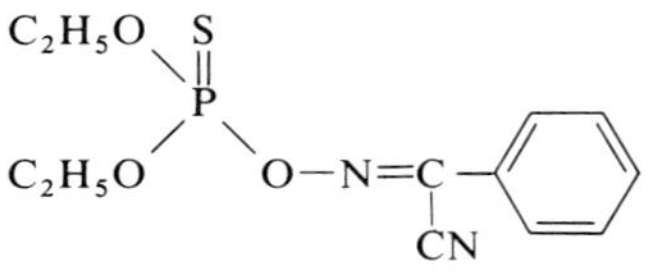
Phthalthrin	(See Tetramethrin)	—
Piperonyl butoxide	3,4-Methylenedioxy-6-propylbenzyl *n*-butyl diethyleneglycol ether	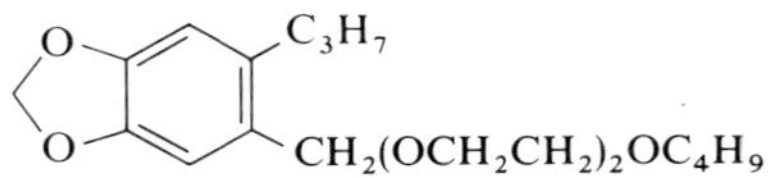
Procaine (HCl)	p-Amino-*N*-(2-diethylaminoethyl)benzamide hydrochloride	
Propoxur	2-Isopropoxyphenyl methylcarbamate	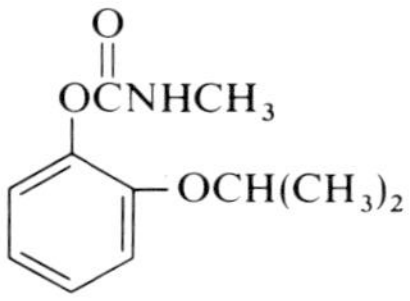
Propyl isome	Di-*n*-propyl-6,7-methylenedioxy-3-methyl-1,2,3,4-tetrahydronaphthalene-1,2-dicarboxylate	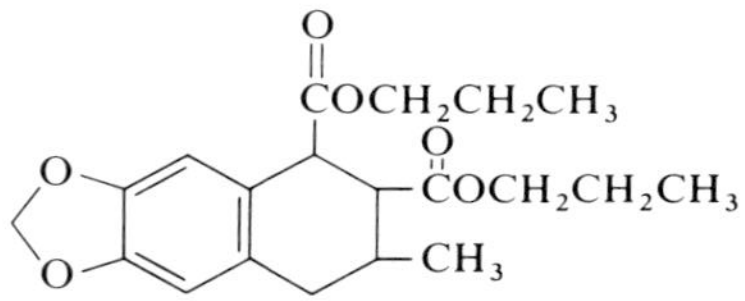

n-Propyl Paraoxon	*O,O*-Di-*n*-propyl analog of Paraoxon	—
Pyrethrin I	Pyrethrolone ester of chrysanthemic acid	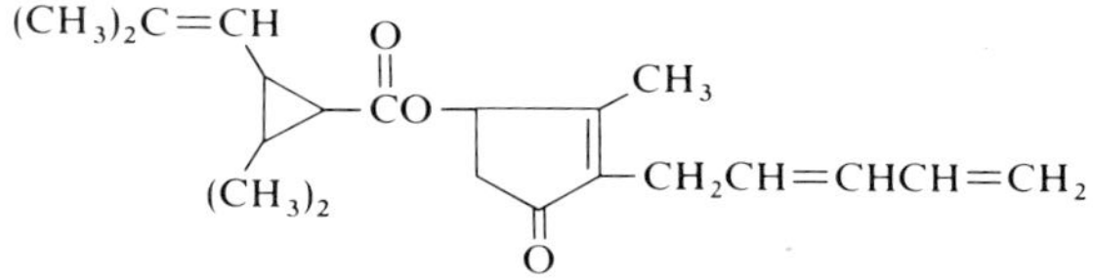
Pyrethrin II	Pyrethrolone ester of pyrethric acid	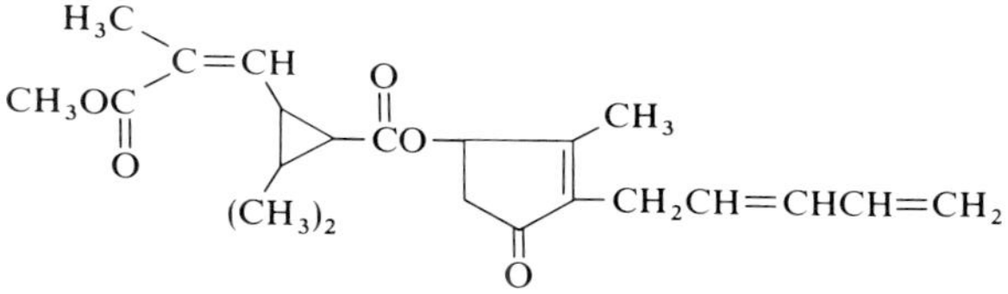
Pyrethrum	Dried flowers of *Chrysanthemum cinerariaefolium* containing several insecticidal components (See Pyrethrin I, II)	—
Rabon	(See Tetrachlorvinphos)	—
Ronnel	*O,O*-Dimethyl-*O*-2,4,5-trichlorophenyl phosphorothioate	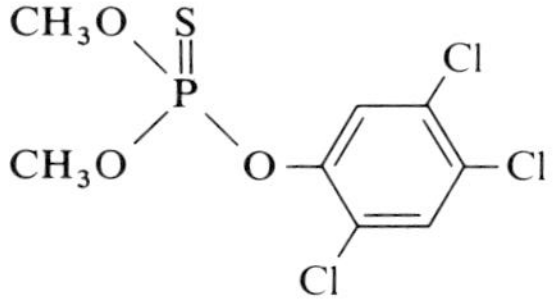
Ruelene	*O*-4-*tert*-Butyl-2-chlorophenyl-*O*-methyl methylphosphoramidate	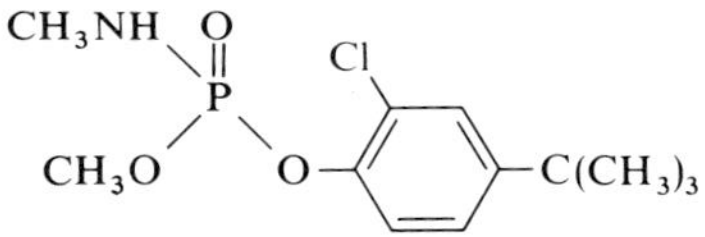

Common Name or Designation	Chemical Name	Structure
Ryania	Insecticidal material from ground stem wood of various *Ryania* species. Contains several alkaloids including ryanodine	—
Sabadilla	Insecticidal material from seeds of sabadilla (*Schoenocaulon officinale* Gray). Contains several veratrine and cevadine alkaloids.	—
Sarin	*O*-Isopropyl methylphosphorofluoridate	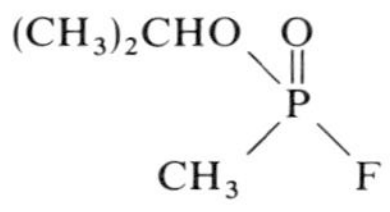
SBP-1382	Mixture of Bioresmethrin and NRDC-119	—
Schradan	Octamethylpyrophosphoramide	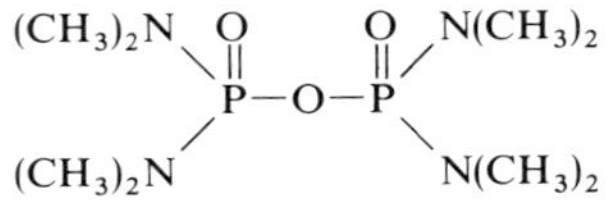
SD 8530	3,4,5-Trimethylphenyl methylcarbamate	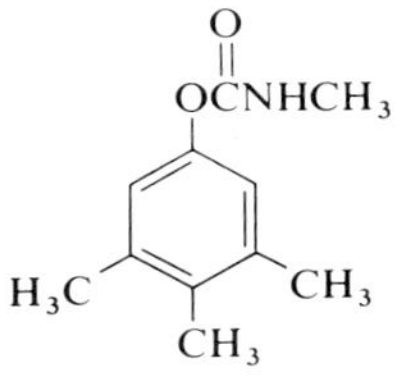

SD 8786	2,3,5-Trimethylphenyl methylcarbamate	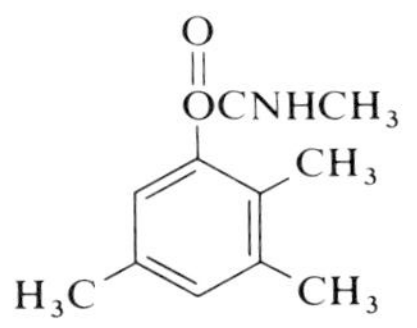
SD 9003	2,4,5-Trimethylphenyl methylcarbamate	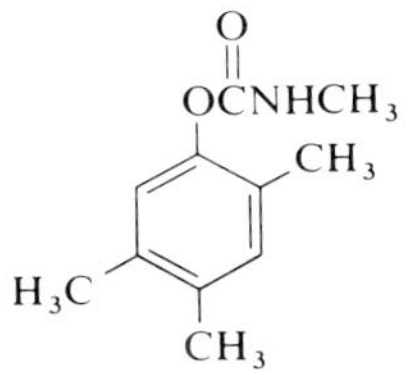
SD 11319	3-(Dimethoxyphosphinoxy)-*cis*-crotonamide	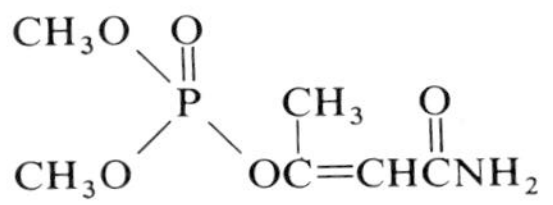
Seconal (sodium)	Sodium 5-allyl-5-(1-methylbutyl- barbiturate	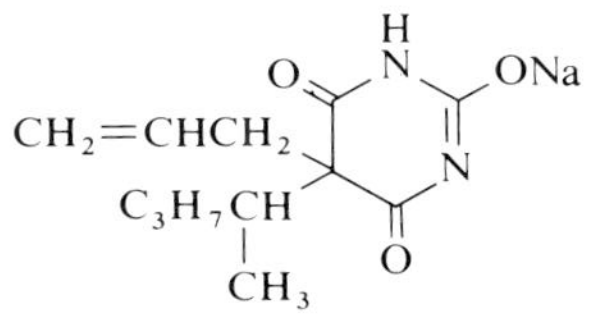
Sesamex	2-(3,4-Methylenedioxyphenoxy)-3,6,9-trioxaundecane	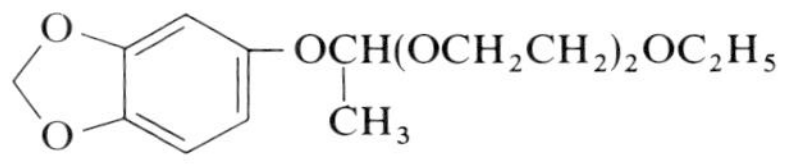

Common Name or Designation	Chemical Name	Structure
Simazine	2-Chloro-4,6-bisethylamino-1,3,5-triazine	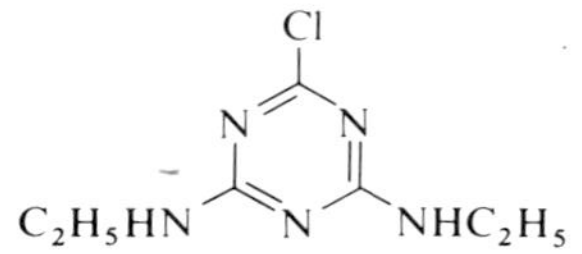
SKF-525A	2-(Diethylamino)ethyl-2,2-diphenylpentanoate	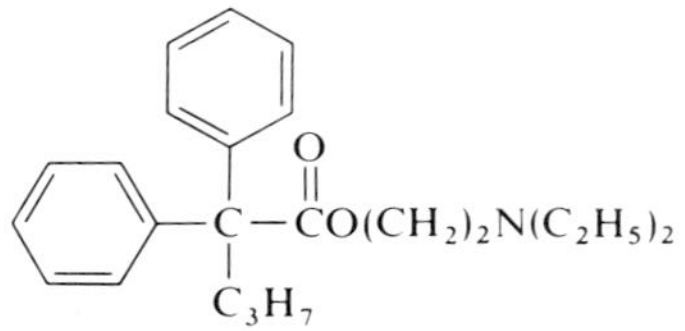
Soman	*O*-1,2,2-Trimethylpropyl methylphosphonofluoridate	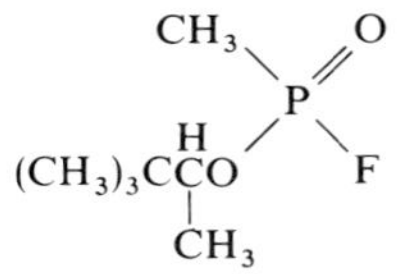
Stauffer R 20458	1-(4′-Ethylphenoxy)-6,7-epoxy-3,7-dimethyl-2-octene	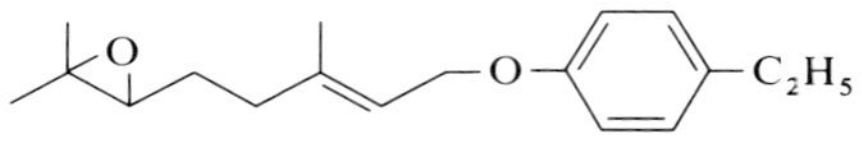
Stirofos	(See Tetrachlorvinphos)	—
Strobane	Mixture of polychlorinated terpenes	—

Sulfotep	Bis-*O*,*O*-diethylphosphorothionic anhydride	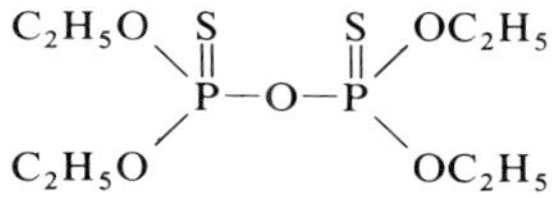
Sulfoxide	1,2-Methylenedioxy-4-[2-(octylsulfinyl)propyl]benzene	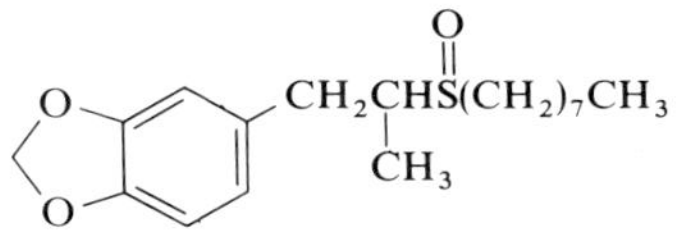
Sumioxon	Oxon analog of Fenitrothion	—
Sumithion	(See Fenitrothion)	—
SV_1	*O*,*O*-Diethyl-*O*-phenyl phosphorothioate	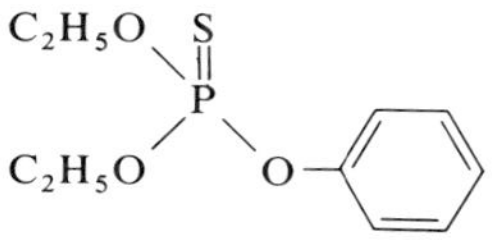
Systox	(See Demeton; commercial product contains 65% thiono and 35% thiolo isomers)	—
Tabun	*O*-Ethyl dimethylphosphoramidocyanidate	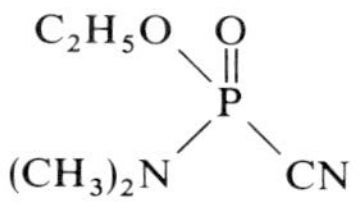
TBTP	(See Folex)	—
TDE	(See DDD)	—

Common Name or Designation	Chemical Name	Structure
Tem	2,4,6-Tris(1-aziridinyl)-s-triazine	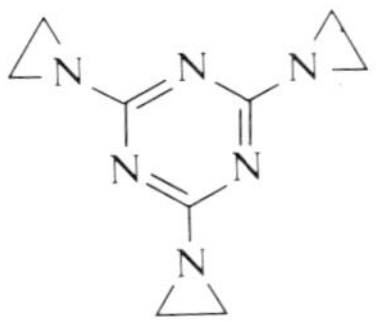
Temik	(See Aldicarb)	—
Tepa	Tris(1-aziridinyl)phosphine oxide	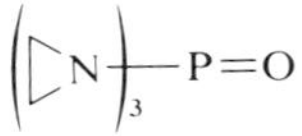
TEPP	Diethylphosphoric anhydride	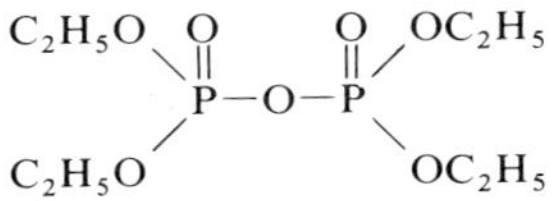
Tetrachlorvinphos	*O*-2-Chloro-1-(2,4,5-trichlorophenyl)vinyl-*O*,*O*-dimethyl phosphate	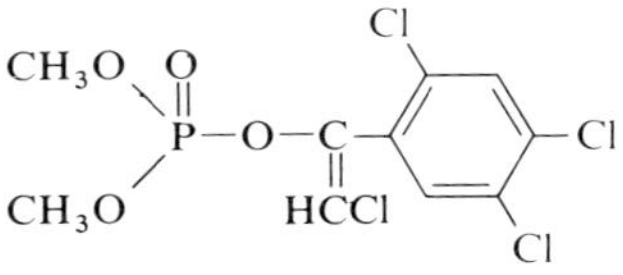
Tetram	Oxalate salt of Amiton	
Tetramethrin	*N*-Hydroxymethyltetrahydrophthalimide (+)-*trans*-chrysanthemate	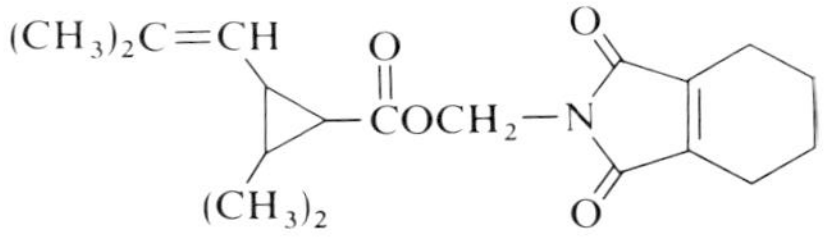

Thanite	Isobornyl thiocyanoacetate	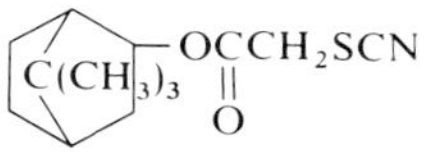
Thiotepa	Tris(1-aziridinyl)phosphine sulfide	
Tirpate	2,4-Dimethyl-1,3-dithiolane-2-carboxaldehyde-*O*-(methylcarbamoyl)oxime	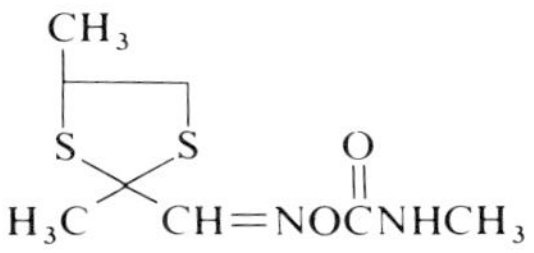
TOCP	(See TOTP)	—
Tolbutamide	*N*-*p*-tolylsulfonyl-*N*′-*n*-butylurea	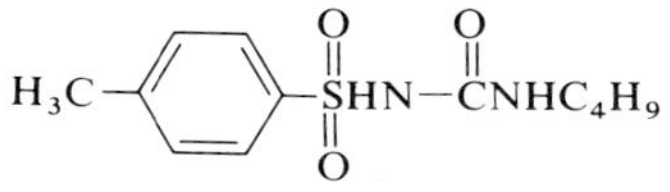
TOTP	Tri-*o*-tolyl phosphate	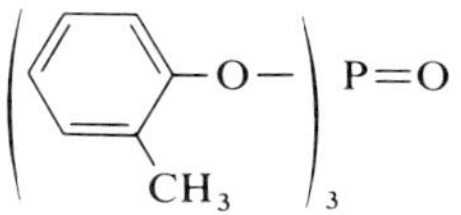
Toxaphene	Chlorinated camphene (67–69% chlorine). Insecticidal mixture consists of complex mixture of isomers	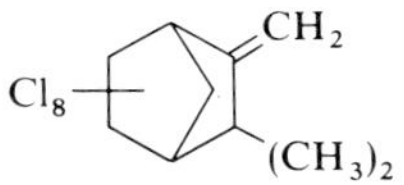
TPP	*O*,*O*,*O*-Triphenyl phosphate	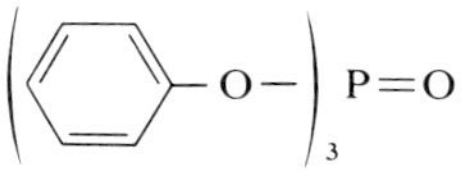

Common Name or Designation	Chemical name	Structure
Trichlorfon	*O,O*-Dimethyl-(1-hydroxy-2,2,2-trichloroethyl)phosphonate	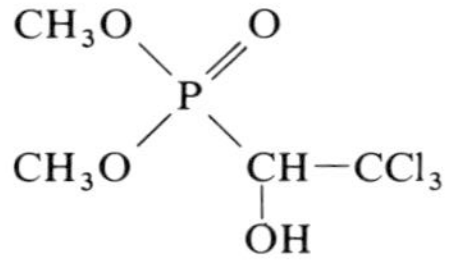
Trithion	(See Carbophenothion)	—
Tropital	Piperonal bis[2-(2-butoxyethoxy)ethyl]acetal	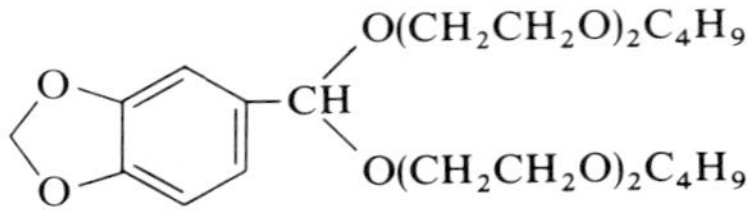
UC 10854	3-Isopropylphenyl methylcarbamate	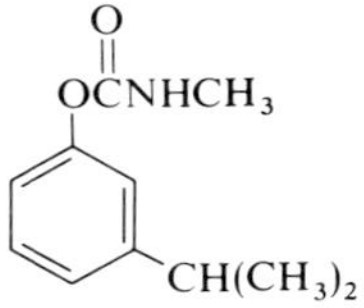
Vapam	Sodium methyldithiocarbamate	
Warf antiresistant	*N,N*-Di-*n*-butyl-*p*-chlorobenzenesulfonamide	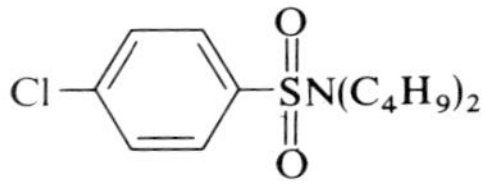
Zectran	(See Mexacarbate)	—

Zinophos	*O*,*O*-Diethyl-*O*-2-pyrazinyl phosphorothioate	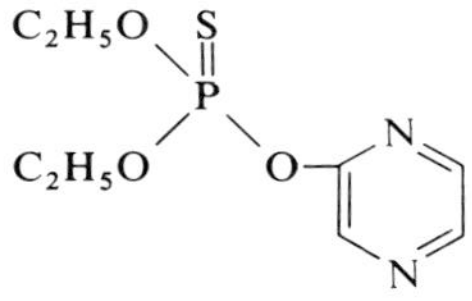
ZR-515	(See Methoprene)	—

Index